COSMIC RAY PHYSICS

INTERSCIENCE MONOGRAPHS AND TEXTS IN PHYSICS AND ASTRONOMY

Edited by **R. E. MARSHAK**

Volume I: E. R. Cohen, K. M. Crowe, and J. W. M. DuMond, **The Fundamental Constants of Physics**

Volume II: G. J. Dienes and G. H. Vineyard, **Radiation Effects in Solids**

Volume III: N. N. Bogalubov and D. V. Shirkov, **Introduction to the Theory of Quantized Fields**

Volume IV: J. B. Marion and J. L. Fowler, *Editors*, **Fast Neutron Physics** *In two parts*
Part I: Techniques **Part II: Experiments and Theory**

Volume V: D. M. Ritson, *Editor*, **Techniques of High Energy Physics**

Volume VI: R. N. Thomas and R. G. Athay, **Physics of the Solar Chromosphere**

Volume VII: Lawrence H. Aller, **The Abundance of the Elements**

Volume VIII: E. N. Parker, **Interplanetary Dynamical Processes**

Volume IX: Conrad L. Longmire, **Elementary Plasma Physics**

Volume X: R. Brout and P. Carruthers, **Lectures on the Many-Electron Problem**

Volume XI: A. I. Akhiezer and V. B. Berestetskii, **Quantum Electrodynamics**

Volume XII: John L. Lumley and Hans A. Panofsky, **The Structure of Atmospheric Turbulence**

Volume XIII: Robert D. Heidenreich, **Fundamentals of Transmission Electron Microscopy**

Volume XIV: G. Rickayzen, **Theory of Superconductivity**

Volume XV: Raymond J. Seeger and G. Temple, *Editors*, **Research Frontiers in Fluid Dynamics**

Volume XVI: C. S. Wu and S. A. Moszkowski, **Beta Decay**

Volume XVII: Klaus G. Steffen, **High Energy Beam Optics**

Volume XVIII: V. M. Agranovich and V. L. Ginzburg, **Spacial Dispersion in Crystal Optics and the Theory of Excitons**

Volume XIX: A. B. Migdal, **Theory of Finite Fermi Systems and Applications to Atomic Nuclei**

Volume XX: B. L. Moiseiwitsch, **Variational Principles**

Volume XXI: L. S. Shklovsky, **Supernovae**

Volume XXII: S. Hayakawa, **Cosmic Ray Physics**

Volume XXIII: J. H. Piddington, **Cosmic Electrodynamics**

Volume XXIV: R. E. Marshak, Riazuddin, and C. P. Ryan, **Theory of Weak Interactions in Particle Physics**

Volume XXV: N. Austern, **Direct Nuclear Reaction Theories**

INTERSCIENCE MONOGRAPHS AND TEXTS
IN PHYSICS AND ASTRONOMY

Edited by R. E. MARSHAK
University of Rochester, Rochester, New York

VOLUME XXII

COSMIC RAY PHYSICS
Nuclear and Astrophysical Aspects

SATIO HAYAKAWA
Department of Physics
Nagoya University, Nagoya, Japan

WILEY-INTERSCIENCE a division of
John Wiley & Sons New York · London · Sydney · Toronto

Library of Congress Catalog Card Number: 69-19930

SBN 471 36320 0

Printed in the United States of America

10 9 8 7 6 5 4 3 2 1

Preface

Cosmic rays have been studied from various viewpoints. Since the early days of cosmic ray research the nuclear-physical aspect has been emphasized, and high-energy interactions have been studied with the aid of cosmic rays. On this basis we can understand the morphological features of cosmic rays on the earth as well as in space. There are two important aspects of cosmic rays in space, the astrophysical and the geophysical; the geophysical includes problems of cosmic rays in interplanetary space. It is not possible to discuss all three aspects thoroughly in one volume; hence the last one, which is related to the earth's magnetosphere and interplanetary space, is omitted. This is partly because of the rapidity of development of research in this field. Astrophysical research is also rapidly developing but is included because without it the book would be only a skeleton. Moreover, an understanding of the astrophysical aspect of cosmic rays demands a knowledge of high-energy interactions. It thus makes sense to put the astrophysical and nuclear-physical aspects together in one book.

This book begins with a historical introduction (Chapter 1). However, it is not intended as pure history but to introduce the general topic of cosmic rays. The purpose of Chapter 2 is to present the essential quantitative information in a form that can be readily used. Although quantum theory and high-energy physics are needed for a thorough understanding, readers may use the formulas, tables, and graphs as in a handbook. Chapter 3 is a mixture of a review and a treatise on very-high-energy interactions. Most of the subjects discussed are still under investigation, and a personal bias can hardly be avoided. In Chapters 4 and 5 the morphology of cosmic rays on earth is described in textbook style, with the intention that the results given there will be used also for applications to other branches of science. Chapter 6 is a treatise on the astrophysical aspect of cosmic rays. Personal bias may be stronger than in Chapter 3, since many experimental results and theoretical analyses are not yet conclusive.

As explained above, each chapter has a different character. This seems inevitable if subjects in developing stages are to be included in the book.

Because of this character, each chapter is arranged to be as independently readable as possible. Tables, figures, and equations are numbered and references are given in each chapter. However, readers who wish to read a chapter independently may be required to have a reasonable standard of knowledge in one or two fields related to cosmic rays. If they follow the order of the chapters, the first-year graduate level may be sufficient.

References in Chapter 1 are extensive but not complete. In many cases only the earliest and the latest papers in a subject under discussion are mentioned. Review articles are often quoted in the hope that readers will find important references therein. The policy of selection is to help readers who wish to obtain further details. References that were available in September 1966 are included; this was the time when the book was completed. Later information was added only when a topic required radical revision.

This book is largely based on my lectures at Osaka City University, Kyoto University, Nagoya University, Brandeis University, and others—as well as on many review articles. A number of important results were promptly made available by many people. Their contributions have been indispensable to the preparation of this book. On this occasion I should like to express my heartfelt thanks to K. Aizu, Y. Fujimoto, H. Hasegawa, the late T. Hatanaka, K. Ito, Y. Kamiya, K. Kikuchi, T. Kitamura, M. Koshiba, M. Matsuoka, S. Miyake, J. Nishimura, H. Obayashi, M. Oda, H. Okuda, K. Suga, D. Sugimoto, M. Taketani, Y. Tanaka, Y. Terashima, Y. Yamamoto, and K. Yokoi. My thanks are due also to Professor R. E. Marshak for his valuable suggestions on the selection of topics.

Satio Hayakawa

Nagoya, Japan

Contents

COSMIC RAY PHYSICS

CHAPTER 1

Historical Survey

The most exciting part of cosmic-ray physics is the discovery of new phenomena. Indeed, most of the elementary particles were discovered in cosmic rays, and their properties were qualitatively known before accelerators were used to study them. Qualitative knowledge is of primary importance as an attitude in cosmic-ray studies, thus leaving a fruitful field of particle physics to studies by means of accelerators. This characteristic feature of cosmic-ray physics has caused changes in the subjects of main interest. In the 1930s and 1940s the central problem in cosmic rays was the nature of *elementary particles and their interactions at high energies*. In the 1950s the *geophysical and astrophysical aspects* have replaced the nuclear aspect as the center of cosmic-ray studies, though there still remain important problems concerning interactions at extremely high energies. This chapter describes how the central problems in cosmic-ray physics have changed with time. The description is not purely historical but is intended to give a bird's-eye view of cosmic-ray physics, particularly in Sections 1.7 through 1.10.

1.1 Discovery of Cosmic Rays*

Like many other new phenomena, the discovery of cosmic rays came by chance in the course of studying another subject. In their investigation of atmospheric electricity Elster (Elster 00) and Geitel (Geitel 00) noticed an unknown source of ions in the air. Independently, Wilson (Wilson 00, 01), through his study of the *ionization chamber*, suspected the existence of an ionizing agency that could penetrate a thick layer of

* Complete references are given, for example, in Y. Nishina, Y. Sekido, M. Takeuchi, and T. Ichimiya, *Cosmic Rays* (in Japanese), *Iwanami Shoten*, 1941; and in D. J. X. Montgomery, *Cosmic Ray Physics*, Princeton University Press, Princeton, N.J., 1949.

earth. Both of these works appeared in 1900, in the same year that quantum theory was originated by Planck. This is an accidental coincidence as far as the individuals are concerned, but it may be regarded as a necessary consequence in view of the history of science. Apparently, these investigations were closely connected with a series of studies on gaseous electronics conducted by J. J. Thomson at the Cavendish Laboratory, Cambridge, and were further stimulated by the discoveries of ionizing radiation toward the end of the nineteenth century.

Having noticed the residual ionization left after the removal of all possible causes, they were forced to conclude that there should be some ionizing agency that was capable of continuously ionizing a gas. As they observed ionization even when the ionization chamber was shielded, they suspected that the ionization could be caused by an as yet unknown radiation of great penetrability other than X-rays and radioactivity. Moreover, Wilson speculated that the radiation might come from extraterrestrial sources.

One might think that this statement of Wilson would have been accepted as the discovery of a new radiation later called cosmic radiation. However, it seems to have been felt that further verification was needed.

The widely accepted discovery in 1911 of cosmic rays was made by Hess (Hess 11, 12), who used a balloon-borne pressurized ionization chamber. He first observed a slight decrease of ionization as the balloon went up and then definitely found a rapid increase that persisted up to the highest altitude he reached, about 5 km. This result made it evident that the ionizing agency came from a high level; hence this was called *Höhenstrahlung*. It was not too speculative, after Hess' experiment, to call this radiation cosmic radiation, implying that its source was extraterrestrial.

This strange radiation of extremely high penetrability was further investigated by Kolhörster (Kolhörster 13) with more refined techniques. His experiment confirmed that the ionization increased up to an altitude of 9 km, and the absorption coefficient was much smaller than that of γ-rays. This may be regarded as the confirmation of the existence of cosmic rays.

Hess and Kolhörster continued their studies on cosmic rays even during World War I, but progress in this field was retarded until the middle 1920s. In the early 1920s some workers attempted to locate the source of cosmic rays by studying the sidereal-time variation of the intensity of cosmic rays, although no positive result was obtained. In the late 1920s, however, epoch-making progress was made due partly to the application of new techniques.

Skobeltzyn (Skobeltzyn 27) applied the *cloud chamber* to cosmic-ray studies and observed tracks of charged particles. Bothe and Kolhörster (Bothe 28, 29) used *Geiger-Müller counters* and identified a charged particle that penetrated two or more counters. These experiments were important not only because cosmic rays were found to involve a great number of charged particles, in contrast to the generally accepted belief that the high penetrability would imply the γ-ray nature of cosmic rays, but also because two powerful instruments, the cloud chamber and the Geiger-Müller counter, were added to the ionization chamber that had been used previously.

Another important achievement was made by Clay (Clay 27), who succeeded in observing the *latitude effect* in a voyage from Amsterdam to Java. Since the latitude effect indicated a lower intensity of cosmic rays near the equator, where the horizontal component of the geomagnetic field is stronger, Clay's result meant that *primary cosmic rays*, before entering into the terrestrial atmosphere, were charged particles that were affected by the geomagnetic field. The motion of a charged particle in the geomagnetic field was originally investigated theoretically by Størmer with the aim of explaining the aurora in relation to the beautiful model experiment of Birkeland. However, Størmer's theory is not directly applicable to the aurora, because the energy of an auroral particle is too low to penetrate into the auroral zone, located near the geomagnetic latitude of 65°. Then the theory found a suitable field of application in cosmic-ray physics. This was therefore elaborated by Størmer (Størmer 30) himself and by Lemâitre and Vallarta (Lemâitre 33, Vallarta 33), and has become an interesting branch of theoretical physics.

1.2 Development of Quantum Electrodynamics

Cosmic rays were believed, rather naively, to be γ-rays, because of their highly penetrating character. In those days the γ-ray was the most penetrating radiation known, but the mechanism of its interaction with matter was not well understood. The β-ray and the α-ray were known to be less penetrating, and their ability to penetrate was well explained by Bohr's theory (Bohr 13, 15) of the energy loss of a charged particle.

According to Bohr's theory, energy loss decreases rather rapidly with increasing energy and tends to level off at relativistic energies. In accordance with this theory the high penetration of cosmic rays was attributable to their high energy. Indeed, the energy loss in air of a relativistic particle with unit charge is about 2 MeV per g-cm^{-2}, and the thickness of the atmosphere is 1030 g-cm^{-2} in the normal state; a particle of energy greater than 2 GeV can penetrate the atmosphere vertically, provided that energy loss is due exclusively to the mechanism treated by Bohr; namely, *by ionization*.

In the relativistic energy region, however, the validity of any theory was not always accepted. Since the positon* introduced by Dirac's theory was not observed, this theory itself was suspected to be inadequate, even though the theory of *Compton scattering* developed by Klein and Nishina (Klein 29) was able to account for the angular distribution and energy dependence of the Compton cross section if negative-energy states were taken into account. Even with this success it was hard to convince some critical people, until the positon was observed in cosmic rays by means of a cloud chamber (Anderson 32).

Around 1930 quantum mechanics was little applied to problems of cosmic rays, partly because of the scepticism in applying quantum theory to relativistic phenomena and partly because quantum theoreticians were preoccupied with atomic problems. A quantum-mechanical treatment of energy loss by ionization by Bethe (Bethe 30), which has become one of the standard theoretical formulas in cosmic-ray physics, was developed almost independently of cosmic rays. This would explain why Bohr's formula was not seriously applied to cosmic-ray phenomena. Few people seemed to have thought that cosmic rays could be accounted for within the framework of existing theory.

The discovery of the positon alarmed both theoreticians and cosmic-ray experimenters. The former began to pay great attention to cosmic rays as a field open to the development of theory, and the latter recognized quantum mechanics as an important tool, closely related to their own experiments.

The investigation of cosmic rays along this direction was facilitated by two new experimental devices: (*a*) the fast *coincidence method* of multiple counter discharges developed by Rossi (Rossi 30a, b) and (*b*) the *counter-controlled expansion* of a cloud chamber by Blackett and Occhialini (Blackett 33). These two methods are particularly useful for investigating the interaction of charged particles with matter, because an impinging charged particle is selected by the discharges of aligned counters in coincidence.

Before the coincidence method became common in cosmic-ray experiments the absorption of cosmic rays had been measured by means of ionization chambers with absorbers of varying thicknesses above

* In this book we use the following terminology; the positive and negative electrons are called the positon and the negaton, respectively, and the electron stands for either of them, irrespective of charge.

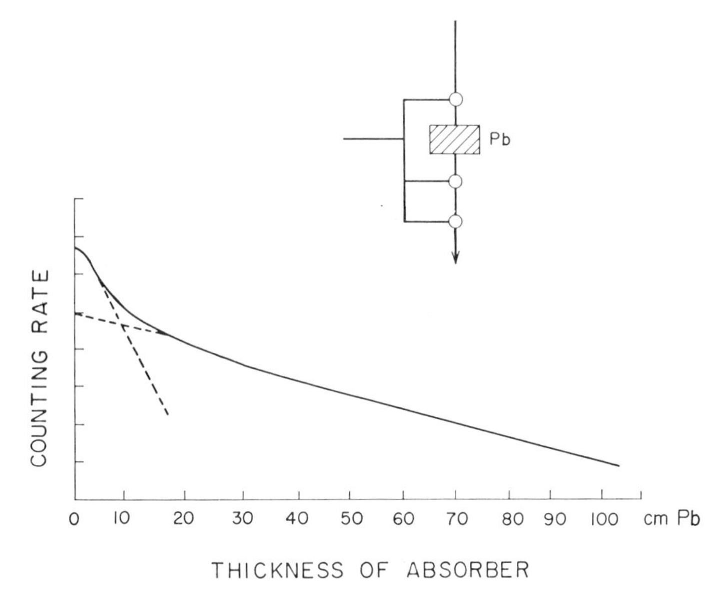

Fig. 1.1 Separation of the soft and the hard components by the triple coincidence method. The solid curve represents the counting rate in arbitrary units versus the thickness of a lead absorber inserted between two counters. The counting rate is decomposed into two parts as indicated by the dashed curves; the steeper one represents the counting rate due to the soft component, whereas the less steep one is due to the hard component.

them. An early experiment by Hoffmann (Hoffmann 26, 27) with this method, however, showed a complicated behavior. The intensity of cosmic rays does not always decrease monotonically with increasing absorber thickness but often exhibits a peak at a small thickness. Since such an increase was observed with cosmic rays going from a light material to a heavier one, this phenomenon was called the *transition effect*. Further investigations with ionization chambers showed a transition effect for cosmic rays going from a heavy material to a lighter one, in which the intensity decreases more slowly or even increases, approaching the intensity that is expected for the light absorber alone (Steinke 28, 30; Schindler 31).

More quantitative studies were made by Auger (Auger 35, 36) and Rossi (Rossi 35) with a counter telescope consisting of aligned Geiger-Müller counters in coincidence. Having put absorbers between counters, as shown schematically in Fig. 1.1, these authors measured the intensity with increasing thickness. The curve of intensity versus absorber thickness showed a rapid decrease in the first 10 cm of lead and a gradual decrease at greater thicknesses. Thus the absorption curve was interpreted as consisting of two parts, corresponding to rapidly and gradually absorbed components, called the *soft* and the *hard components*, respectively. The absorption coefficient of the hard component was found to be roughly proportional to the density, more exactly to the density times Z/A, where Z and A are the atomic and mass numbers of the absorber, respectively. On the other hand, the absorption of the soft component seemed anomalous—the mass-absorption coefficient was expressed approximately by a constant plus a term proportional to Z.

The Z/A absorption law for the hard component was well accounted for in terms of ionization loss, the theory of which was developed by Bethe (Bethe 30) and Bloch (Bloch 33) on the basis of quantum mechanics.

The usefulness of quantum electrodynamics was further revealed in treating the soft component. Heitler (Heitler 36) and Nishina et al. (Nishina 34) worked out various radiation processes, including the creation of electron pairs. These processes were extensively investigated by Bethe and Heitler (Bethe 34), who presented the cross sections for *bremsstrahlung* and *pair creation*, known as the Bethe-Heitler formulas. In their theory of quantum electrodynamics they adopted the perturbation approximation, in which the interaction between the electron and the radiation field was expressed by a power series of $\alpha = \frac{1}{137}$. If the energy of an electron or a photon is as high as $mc^2/\alpha = 137\ mc^2$, where mc^2 is the rest energy of the electron, the higher order terms would not always be negligibly small. In those days, therefore, it was suspected that there was a breakdown of quantum electrodynamics at energies higher than 137 mc^2, as was the case for energies higher than 2 mc^2 around 1930.

This scepticism was partly supported by cosmic-ray experiments. The absorption of the soft component was nicely interpreted in terms of bremsstrahlung of the electron. However, a startling phenomenon called *showers* was discovered by Rossi (Rossi 33) with a counter coincidence method and was also observed by Blackett and Occhialini (Blackett 33) with a counter-controlled cloud chamber.

Rossi further confirmed that a shower was a characteristic phenomenon associated with the soft component. If a shower were produced by a higher order process of electron-photon interaction, the probability of such an event occurring would be extremely small, on the order of a high power of α. Actually, however, almost all soft particles were observed to produce showers when they passed through a lead plate. Morevoer, most of the shower-producing particles seemed to have energies as great as 100 MeV or greater. These observed facts threw

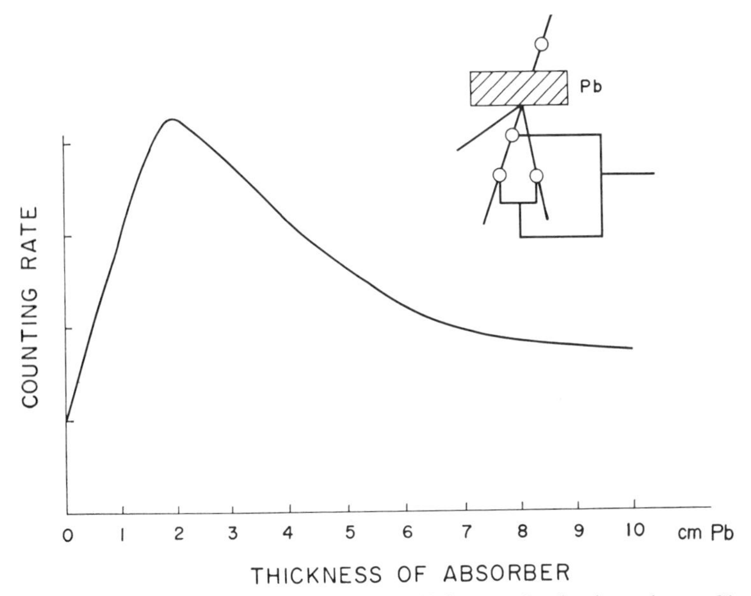

Fig. 1.2 Shower-counting rates versus the thickness of a lead producer. Showers are detected by the coincidences of three counters below the lead plate.

doubt on the validity of quantum electrodynamics, so that the Bethe-Heitler formula would not be applicable at high energies. It was also suspected that the small absorption coefficient for the hard component might also be due to the breakdown of quantum electrodynamics.

A criticism of this scepticism was raised by Weizsäcker (Weizsäcker 34) and Williams (Williams 35). They considered bremsstrahlung in the coordinate system in which an impinging electron was at rest. Then the emission of photons is the dipole radiation from an oscillating electron, as in the classical theory of Lorentz, electron motion being caused by the Coulomb force of a nucleus passing by. Transforming back to the laboratory system, one obtains bremsstrahlung that essentially agrees with the prediction of the Bethe-Heitler theory. In this way Weizsäcker and Williams made it clear that bremsstrahlung is essentially a low-energy phenomenon, so that the breakdown of the quantum-mechanical theory by Bethe and Heitler would be shared with that of classical electrodynamics.

No matter whether shower phenomena implied a breakdown or not, it was recognized as giving a consistent picture of many cosmic-ray phenomena. By means of a shower-detecting arrangement of Geiger-Müller counters shown in Fig. 1.2, Rossi (Rossi 33) observed the frequency of the showers versus the thickness of a shower-producing

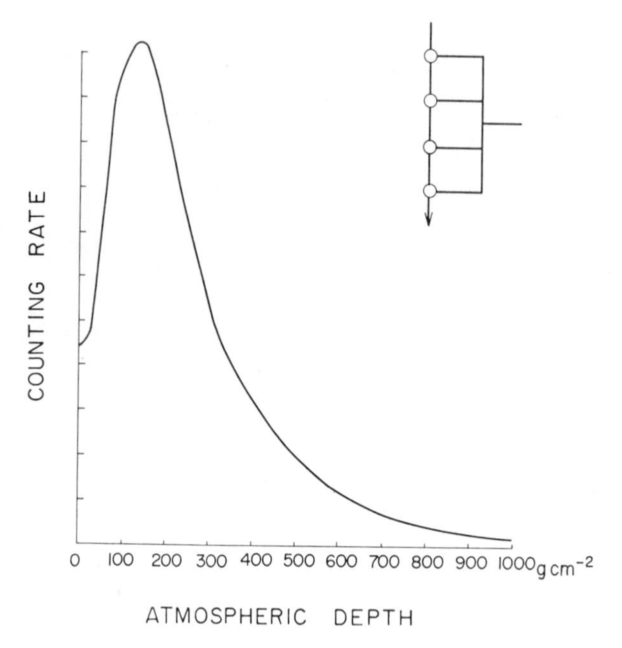

Fig. 1.3 Pfotzer curve, the vertical intensity of cosmic rays in the atmosphere versus atmospheric thickness, measured by the fourfold coincidence method.

material. The frequency-versus-thickness curve thus obtained, called the *Rossi curve* (Fig. 1.2), was so similar to the *transition curve* that the *transition effect* was interpreted as due to the increase in the number of particles resulting from shower production. Moreover, particles produced in a shower were found to belong to the soft component and consequently to be shower producing if their energies were high enough. It could therefore be concluded that the shower was a phenomenon involving electrons and possibly photons; consequently, if this assumption turned out to be correct, the shower should be treated by quantum electrodynamics.

The *transition effect*, which could be understood in terms of the multiplication of particles due to shower production, was also seen in the atmosphere. Regener (Regener 33) sent a Geiger-Müller counter with a balloon as high up as 25 km and observed the intensity leveling off. Pfotzer (Pfotzer 36) extended this observation to higher altitudes by means of a balloon-borne counter telescope. He then found a maximum in intensity at about 15 km, beyond which the intensity decreased rapidly. A curve for the *vertical intensity* versus the atmospheric depth is therefore called the *Pfotzer curve* and is shown in Fig. 1.3.

A similar curve was obtained also by Millikan and his collaborators (Bowen 34, 37, 38*a*) with an ionization chamber that measured the *omnidirectional intensity*. They made a number of observations at different latitudes and found that transition curves varied with latitude.

The increasing experimental evidence for showers stimulated their theoretical interpretation, whether it indicated breakdown or an ordinary process. Finally, the latter point of view dominated when the *cascade theory* was developed by Bhabha and Heitler (Bhabha 37) and by Carlson and Oppenheimer (Carlson 37) independently by slightly different mathematical methods. According to the cascade theory, an impinging electron emits a photon by bremsstrahlung, and subsequently the photon creates a pair of electrons. Since the energy of a parent particle is approximately equally shared with the two secondary particles in the bremsstrahlung and the pair-creation processes, the energies of individual particles decrease rather slowly in comparison with the increase in the number of particles. This explains the initial rise of the transition curve. When the energies of individual particles become so low that they can no longer create new particles in competition with absorption processes, the number of particles decreases, as seen in the decline of the Rossi curve. In this way the cascade theory achieved a remarkable success in explaining the Rossi curve not only qualitatively but also quantitatively; the theory nicely explained the shape of the Rossi curve (except at great thicknesses), and also its Z and energy dependences. This also explained the general trend of the Pfotzer curve at high altitudes.

Thanks to the cascade theory, the behavior of the soft component was essentially understood. After that theoretical efforts took two directions (*a*) the elaboration of the cascade theory and (*b*) the interpretation of cosmic-ray phenomena. As an example of the latter, careful analyses of the Pfotzer curve made by Heitler (Heitler 37) and Nordheim (Nordheim 38) may be worth mentioning. They assumed a power energy spectrum for incident electrons, and the power index of the spectrum was determined by comparison between the Pfotzer and cascade curves. They, however, failed to explain the Pfotzer curve at low altitudes, the predicted intensity being too low and the observed one being due mainly to the hard component.

Then the existence of the hard component became mysterious because of its high penetration and its inability to radiate photons. Both Heitler and Nordheim suspected that the hard component should consist of particles much heavier than the electron, because the bremsstrahlung cross section was known to be inversely proportional to the square of mass. Before this suggestion Williams (Williams 34) thought that protons might constitute the hard component, but he abandoned this idea because negative particles were observed to be almost equally abundant as positive ones. Shortly after this Yukawa (Yukawa 35) proposed the renowned theory of the meson on an entirely different basis, but unfortunately his theory was not known for a long time to most cosmic-ray physicists, so that they had to expend much effort in vain.

1.3 Geophysical Aspects of Cosmic Rays

In deriving the energy spectrum of cosmic rays the Pfotzer curve gave only an indirect clue. A more direct method was found by using the geomagnetic field as a momentum spectrometer.

Immediately after the discovery of the latitude effect by Clay, as mentioned in Section 1.1, Størmer (Størmer 30) developed his theory, once intended for application to the aurora, so that it became applicable to the geomagnetic effect on cosmic rays. Størmer's theory is based on two integrals of motion, corresponding to the conservation of energy and of canonical angular momentum, which determine the allowed region of motion. This simply gives the *cutoff energy*, the minimum energy for a particle of a given angle of incidence at a given latitude.

A more advanced analysis of orbits by Lemâitre and Vallarta (Lemâitre 33, Vallarta 33) showed that not all of the orbits in the allowed region were accessible to the earth but that some traversed the earth, so that there appeared shadow regions. This served to determine more precise values of the cutoff energies. Since the density of particles is not affected by a static magnetic field, according to the Liouville theorem, the energy spectrum can be determined by observing the intensities of primary rays at a number of latitudes. In practice, however, the intensities were observed only at low altitudes. Compton (Compton 33) obtained about 10-percent difference of the ionization rates at sea level between a latitude of 50° and the equator. The difference was found to increase with increasing altitude.

Above 50° the intensity increase with latitude was not observed. This was interpreted by Janossy (Janossy 37) as a low-energy cutoff of the primary energy spectrum due to a solar magnetic field. The high magnetic-field strength of the sun adopted in those days was consistent with the cutoff rigidity of 3 GV.

Although the theory of the geomagnetic effect was based on the assumption of a dipole field, the actual geomagnetic field was already known to deviate from the field produced by a magnetic moment located at the center of the earth. Millikan and Neher (Millikan 36) constructed the intensity contour on the basis of their worldwide survey by means of a standard ionization chamber, called the Neher type. This resembled the distribution of the horizontal component of the geomagnetic field so well that practically all primary cosmic rays could be regarded as consisting of charged particles.

They further extended ionization measurements to very high altitudes by a balloon-borne ionization chamber at several latitudes (Bowen 37, 38b). The intensity-altitude curve they obtained at a high latitude is essentially the same as the Pfotzer curve, with a correction taking account of the fact that they measured the omnidirectional intensity. The curve becomes less steep as the latitude decreases, thus exhibiting the increase of the latitude effect with increasing altitude. This not only confirmed what was anticipated from the result of Compton but also gave a primary energy spectrum that was in essential agreement with that deduced from the Pfotzer curve with the aid of the cascade theory. It was therefore believed that primary cosmic radiation consisted of electrons, though it remained to determine whether they were negatons or positons.

The charge sign of the primary particles was determined by observing the *east-west asymmetry* (Johnson 33, Rossi 34). Having analyzed the east-west asymmetries observed at various altitudes and latitudes, Johnson (Johnson 35) concluded that practically all of the primary cosmic rays were positively charged, because the intensity from the west was found to be stronger than that from the east. The asymmetry was observed to increase with altitude particularly for the hard component. This led Johnson (Johnson 39) to suggest that the primary particles of positive charge could be protons.

We have described thus far such static properties of cosmic rays as can be interpreted by taking the effects of the terrestrial magnetic field into account. The magnetic properties of the space surrounding the earth have been studied by means of cosmic rays, since the continuous observations of cosmic rays began at various stations in the middle 1930s; among them the network organized by the Department of Terrestrial Magnetism, Carnegie Institution, was most powerful.

The most striking result observed was a worldwide decrease in cosmic-ray intensity associated with a *magnetic storm* in April 1937 (Forbush 37). Such a phenomenon was found to occur quite often, but not always, and is now called a *Forbush decrease*. This was explained as being due to the formation of a ring current, which was, in Chapman's theory of the geomagnetic storm, responsible for the decrease in the horizontal component of the geomagnetic field.

Though not so striking, various types of *time variations* in the cosmic-ray intensity were observed. They were found to have periods of half a day and a day (Schonland 37), and also a recurrence tendency with an interval of 27 days (Monk 39), approximately equal to the rotation period of the sun. A variation with sidereal time was not confirmed, although Compton and Getting (Compton 35) inferred its existence by taking the rotation of the Galaxy into account.

The observed periodic variations could be attributed to extra-terrestrial origins, but an atmospheric origin could not be discarded because the atmospheric state changes with the same periods. In fact, atmospheric pressure was known to anticorrelate with cosmic-ray intensity (Myssowsky 26), and the annual variation showed an anti-correlation with temperature (Schonland 37). Even the latitude effect was considered as due partly to the *temperature effect* (Compton 37). This was not understood until Blackett (Blackett 38) pointed out the decay effect of mesons, whereas the *barometer effect* was well accounted for in terms of the absorption of cosmic rays due to an increase in the amount of air.

1.4 Discovery of the Meson

The temperature effect is only a minor part in the story of the *meson*. The meson was, as is well known, introduced by Yukawa (Yukawa 35) as the heavy-quantum counterpart of the light quantum. The latter is simply the photon, which results from the quantization of the electro-magnetic field. The heavy quantum is analogous to the light quantum in the sense that the former results from the quantization of a field as an intermediary of the nuclear force. However, it differs from the light quantum in mass. Thus the introduction of the meson seems to have been motivated mainly by the quantization of the field and by the nuclear force; it had little to do with cosmic rays, although Yukawa mentioned toward the end of his paper that the heavy quantum could be produced only in cosmic rays because of its high threshold energy.

More directly, Yukawa may have been motivated by the theory of β-decay. He incorporated Fermi's theory of β-decay with the nuclear force by introducing the meson, following Heisenberg's idea of the nuclear force arising from the exchange of electrons but amending its great disagreement with the actual strength of the nuclear force. He was then able to predict the essential properties of the meson—its mass m and its coupling constant g on the basis of the range and the strength of the nuclear force, respectively. He predicted its lifetime τ from the lifetime of RaE. The values of these quantities were

$$m \simeq 200\, m_e, \qquad \frac{g^2}{\hbar c} \simeq \tfrac{1}{10}, \qquad \tau \simeq 10^{-6} \text{ sec},$$

where m_e is the electron mass and $g^2/\hbar c$ is the analog of $\alpha = e^2/\hbar c$.

A meson of such properties was recognized to exist when cosmic-ray experiments forced the assumption of the existence of the heavy electron (Yukawa, 37, Oppenheimer 37). Two groups were observing the energy loss of cosmic-ray particles when they passed through lead and iron plates in cloud chambers. They measured the curvatures of tracks before entering, and after coming out of, a plate, so that they were able to obtain the energy loss as a function of the momentum of a particle. Both groups confirmed an increasing energy loss with increasing energy, as expected from the radiation loss of the electron, but they were surprised by the smallness of the energy loss above 10^8 eV and by the inability to produce showers. These results were essentially the same for both, but their interpretations were different.

Blackett and Wilson (Blackett 37) suggested that this could be an indication of the breakdown of quantum electrodynamics at energies above 137 mc^2, a mysterious number widely discussed before the cascade theory. On the other hand, Neddermeyer and Anderson (Neddermeyer 37) concluded that there existed a new particle with a mass about 100 times the electron mass, by comparison with the momentum-versus-energy-loss curve, as in Fig. 1.4. Meanwhile, Street and Stevenson (Street 37) and Nishina, Takeuchi, and Ichimiya (Nishina 37) independently observed tracks that stopped in their cloud chambers. This enabled them to obtain a value for the meson mass of about 175 times the electron mass. Subsequently a number of meson tracks were observed at various laboratories, one of which was found in a photograph published in 1933. During 1938 about 10 more tracks suitable for mass measurements were obtained, and most of them gave mass values of about 200 times the electron mass.

The instability of the meson was exhibited not only in the temperature effect but also in the absorption anomaly; the absorption for inclined particles was found to be too large compared with that for vertical particles if the absorption were caused only by the mass absorption by the air. This was interpreted by Kulenkampff (Kulenkampff 38) as due to the instability of the meson, because the probability of meson decay should be larger for inclined particles, which pass over a longer path. This was checked by comparing the intensities of cosmic rays absorbed in the atmosphere and by a dense absorber of equivalent mass. These

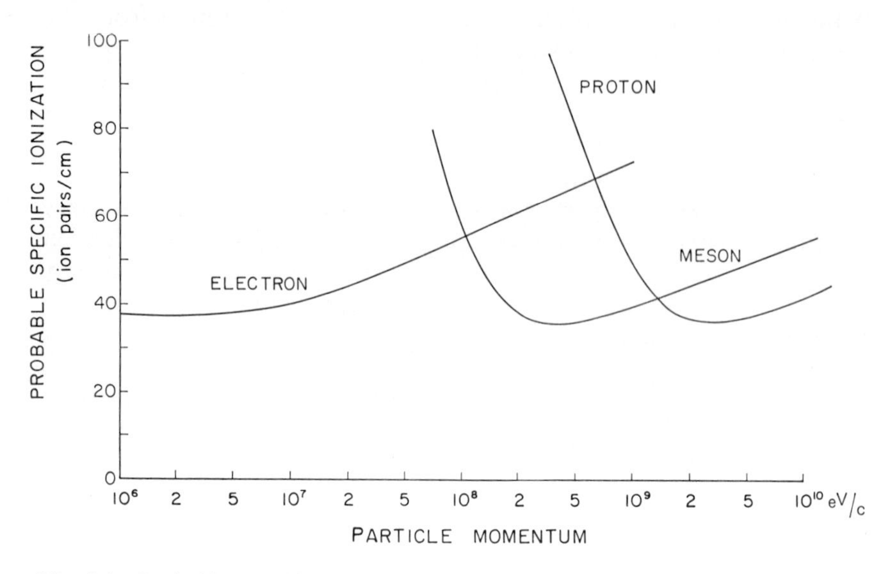

Fig. 1.4 Probable specific ionization in air versus particle momentum. Three curves represent the relations for the electron, the meson, and the proton.

experiments gave the mean life of the meson as about 10^{-6} sec (Rossi 39).

These two properties of the meson, the mass and the lifetime, were quantitatively in agreement with Yukawa's prediction. This stimulated theoreticians to develop meson theory in more detail. Theoretical results were encouraging, because the anomalous magnetic moment of the nucleon, as well as the spin dependence of nuclear forces, could be derived qualitatively at least (Fröhlich 38; Kemmer 38a,b; Yukawa 38a,b). The introduction of the neutral meson was made to explain the charge independence of nuclear forces (Yukawa 38a,b).

The success of the meson hypothesis, as well as the achievement of the cascade theory, were fully taken into account by Euler and Heisenberg (Euler 38) in devising a consistent picture of cosmic rays. According to them, the primary particles are positons with a power energy spectrum, $E^{-2.87}\,dE$; and a positon produces a cascade shower, in which many photons are produced at an atmospheric depth of about 100 g-cm^{-2}. These photons produce mesons by their interactions with air nuclei. Electrons produced by the primary positons as well as by low-energy mesons through their decay form the soft component at high altitudes, thus explaining the Pfotzer curve. High-energy mesons penetrate down to low altitudes and form the hard component, which explains the tail of the Pfotzer curve as well as its energy spectrum at sea level. The soft component at low altitudes arises mainly from electrons produced by meson decay as well as by the *knock-on process* analyzed by Bhabha (Bhabha 36). These two sources of electrons quantitatively explain the intensity of the soft component relative to that of the hard component.

It seems quite strange that little attention was paid to the penetration of mesons deep underground, because they would have to be absorbed by strong nuclear interactions if the meson were the one predicted by Yukawa. This would mean that the third important property of Yukawa's meson, the strong nuclear interaction, was not manifested in cosmic-ray phenomena. This became a central point in the further development of meson theory. Another difficulty emphasized by Euler and Heisenberg was the *burst phenomenon*, which was thought by them to indicate the *multiple production* of mesons.

1.5 Difficulties in Meson Theory

The burst phenomenon was first noticed by Hoffmann (Hoffmann 27) as a sudden increase of the ionization current in an ionization chamber and was sometimes called in German "Hoffmannsche Stoss." Most of them were interpreted as due to cascade showers, as described in Section 1.2. According to the careful analysis by Euler and Heisenberg (Euler 38), however, bursts occurred even beneath a thick absorber, where few electrons and photons should survive to produce showers. They intended to take this as evidence for the multiple production that could indicate the breakdown of meson theory (Heisenberg 36). The puzzling nature of the burst was emphasized by the observation that bursts occurred even at zero thickness of an absorber (Montgomery 38, 39a).

A thorough analysis by Euler (Euler 40a) showed that bursts were complicated phenomena, attributable to several causes, some of which were of little theoretical interest. Small bursts were interpreted in terms of nuclear disintegrations taking place in the wall and the gas of an ionization chamber; they could be identified with the stars observed on photographic plates. Bursts observed with a thin-wall ionization chamber were explained as due to cascade showers developed in the atmosphere.

The star was first observed by Blau and Wambacher (Blau 37) in photographic plates exposed at mountain altitudes. A star consists of several heavily ionizing tracks emerging from a point. These tracks were considered to be protons and α-particles produced by a nuclear disintegration. The nuclear disintegration was interpreted by Bagge (Bagge 41) in terms of the evaporation of a nucleus excited by a cosmic-ray particle which he believed to be a photon because of the similarity in the altitude dependence of star frequencies and the intensity of the soft component.

A second cause is the shower in the atmosphere, which was first recognized by Auger et al. (Auger 38) and by Kolhörster et al. (Kolhörster 38) through the coincidence of Geiger-Müller counters widely separated horizontally and was thus called the *extensive air shower* (EAS). Its lateral extension was interpreted by Euler and Wergeland (Euler 40b) in terms of the multiple scattering of electrons in the atmosphere.* An air shower of high particle density could be originated by an electron of extremely high energy and was observed as a burst.

A third cause is now known to be the *penetrating shower*, which was first observed by Fussel in his cloud-chamber photograph, as quoted in the article by Euler and Heisenberg (Euler 38). The penetrating shower differs from the electronic shower in the high penetration of secondary particles and their small angular divergence when they emerge from a lead plate in a cloud chamber. The same phenomenon was also recognized by means of a set of Geiger-Müller counter trays separated by thick absorbers (Janossy 40, 41; Wataghin 40). Although the relation between the penetrating shower and the burst was not clear in those days, they were interpreted as indicating the multiple production of mesons.

A fourth cause was distinguished from the others in that bursts of large sizes were observed even at great absorber thickness. Euler and Heisenberg (Euler 38) intended to attribute this to the multiple production of mesons, which would be a phenomenon outside the framework of quantum field theory. Heisenberg proposed two possible cases to which the existing theory would not be applicable: (*a*) if the interaction energy became larger than the energy of free fields because of a coupling constant with the dimensions of length (Heisenberg 36) and/or (*b*) if the absolute value of the transferred four-momentum exceeded a certain value because of the presence of the universal length (Heisenberg 38, Wataghin 34). The first possibility was seriously considered by Oppenheimer and his collaborators (Oppenheimer 40) in connection with bursts and cosmic-ray phenomena underground.

As mentioned above, the intensity of cosmic rays underground was found to decrease rather slowly, approximately as a power of depth, $d^{-1.8}$ (Ehmert 37; Wilson 38, 39*a*; Clay 39). This was explained by Euler and Heisenberg (Euler 38) in terms of the ionization loss of energy

* The first suggestion made by Clay and Blackett is quoted in N. Arley, *Proc. Roy. Soc.*, **A168**, 519 (1938).

of the hard component with an integral energy spectrum, $E^{-1.8}$. This means that the mean free path for the hard component to suffer a violent collision is greater than 3×10^4 g-cm^{-2}, the corresponding cross section per nucleon being less than 10^{-28} cm^2, and no extra energy-loss process is important for energy up to 10^{11} eV. However, an anomaly in the absorption curve could be noticed at $d \simeq 5 \times 10^4$ g-cm^{-2}, beyond which the intensity-depth curve appeared to be steeper. This could be regarded as an indication of an excess energy loss setting in at about 10^{11} eV.

Oppenheimer et al. (Oppenheimer 40) related this anomaly to the generation of bursts, which could be associated with a large loss of energy. Further they attributed this to a characteristic feature of the vector meson, whose bremsstrahlung cross section was known to increase linearly with energy. However, a more careful analysis by Christy and Kusaka (Christy 41a,b) showed that this should not be the case but that the frequency of bursts with a thick absorber (Schein 39) could be accounted for in terms of knock-on electrons for mesons of any spin as well as the bremsstrahlung of mesons of spin 0 or $\frac{1}{2}$. The burst frequency would be too large, by a factor of more than 10, if the meson had a spin of unity or larger. No explanation was found for the anomaly in the intensity-depth curve, as far as the energy spectrum of mesons were of the same power shape above 10^{11} eV, even taking into account the energy loss due to pair creation*.

Although some attention was paid to the absorption anomaly, the penetration itself was not yet considered too seriously, and there was more interest in the direct observation of the nuclear scattering of mesons. Such an experiment was performed by Wilson (Wilson 39b), who observed the deflection of mesons by a lead plate in a cloud chamber. The angular distribution of all deflected mesons but one was well accounted for in terms of multiple Coulomb scattering. One anomalous event showed a large-angle scatter associated with a heavily ionizing track that could be identified as a recoil proton—evidence of a nuclear interaction. This resulted in a nuclear-interaction cross section of the meson on the order of 10^{-28} cm^2.

On the other hand, an unsophisticated application of meson theory yielded a meson-nucleon scattering cross section of several times 10^{-27} cm^2. Dozens of theoretical papers were written with the aim of reducing the cross section. These works may be grouped into two classes, both attempting the elimination of the self-energy inherent in quantum

* S. Tomonaga, quoted in S. Hayakawa and S. Tomonaga, *Prog. Theor. Phys.* **4**, 287 (1949).

field theory. One group, which included Heisenberg (Heisenberg 38), emphasized the importance of inertia caused by the self-field of a nucleon. Another group, exemplified by the clear presentation of Heitler (Heitler 41) and Wilson (Wilson 41), paid much attention to damping. The former was developed by Tomonaga (Tomonaga 41, 46, 47) into the strong- and intermediate-coupling theory, which also implied a damping effect. This theory provided a basis for the renormalization theory developed after World War II.

Apart from the involved theoretical work, direct evidence could be obtained through analysis of experimental results. If mesons were created by photons, as was believed in the late 1930s, the production cross section should be quite large, not much smaller than the cross section for creating an electron pair. The large cross section leads to a strong electromagnetic interaction of the meson and consequently to a strong absorption, in contradiction to observed facts (Nordheim 39b).

Meanwhile, Schein et al. (Schein 41) obtained rather direct evidence with their balloon-borne experiments that primary cosmic rays consisted mainly of protons, as was anticipated by Johnson (Johnson 39) through his analysis of the east-west effect. Before Schein's experiment the intensity of the hard component had been observed to pass through a maximum (Dymond 39) or flatten (Ehmert 40) at great heights, so that the primary particles had been believed to be electronics, which should result in a maximum of the intensity. This belief had been strengthened by a quantitative analysis by Serber (Serber 38) with his improved theory of cascade showers, although disagreement was later found with the intensity-altitude relation observed near the magnetic equator (Neher 42). With the arrangement shown in Fig. 1.5, Schein et al. (Schein 41) observed hard particles by means of a coincidence (1, 2, 3, 4, 5) while they checked the association of showers with coincidences (1, 2, 4, 6) and (2, 4, 5, 6). The intensity of the hard component thus obtained was much lower than the total intensity at high altitudes, as compared with the Pfotzer curve in Fig. 1.5, and was found to increase monotonically to the highest altitude. They further found that only a few percent of hard particles are associated with showers*. These two observations led them to conclude that the *primary cosmic rays were protons* and that the soft component was secondary, originated by the primary protons, probably through the production of mesons that subsequently decayed into electrons (Carlson 41).

According to this result, mesons are produced by protons with a

* S. N. Vernov, *J. Exp. Theor. Phys.*, **19**, 621 (1949), has found an association of showers with penetrating particles by means of a similar experimental method.

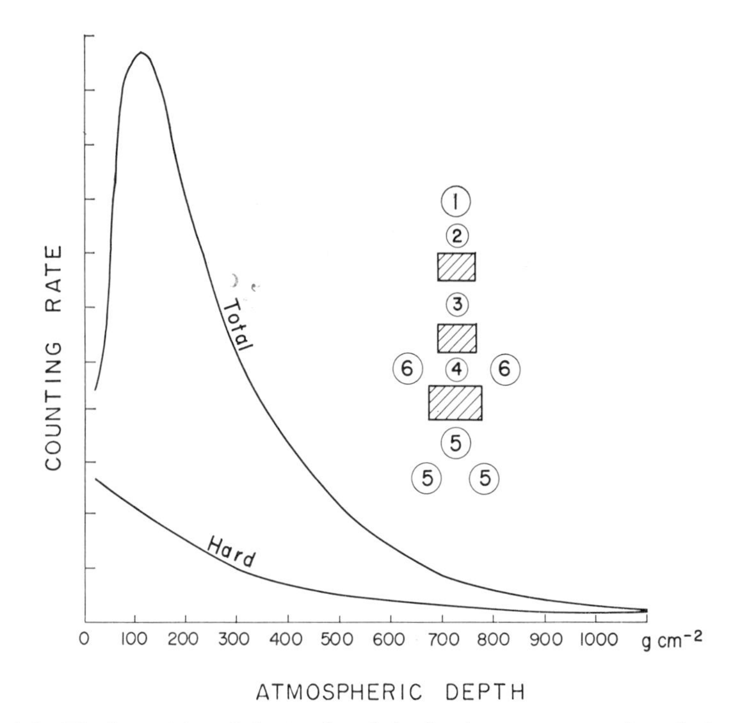

Fig. 1.5 The intensities of the total and the hard components of vertical cosmic rays in the atmosphere. The curve for the total component is the Pfotzer curve shown in Fig. 1.3, whereas that for the hard component is obtained by the coincidence of counters 1, 2, 3, and 4 in anticoincidence with counters 5 and 6, whereby particles producing showers in the two top lead plates and those penetrating the bottom lead plate are rejected, thus giving the counting rate of non-shower-producing particles of various ranges.

rather short mean free path, about 100 g-cm^{-2} in air. This is consistent with Yukawa's hypothesis that the meson strongly interacts with the nucleon but does not remove the difficulty that, once produced, the meson interacts so weakly with matter that it has a small scattering cross section as well as a high penetration. There was a suggestion that the meson-nucleon intreaction could be energy dependent, so that the interaction would be strong for meson production caused by high-energy protons, whereas it would become weak for mesons whose energies were low as a result of multiple production. This idea was, however, not favored, because the interaction should be strong for the nuclear binding that is operative at low energies.

In addition to the above difficulty concerning the weak interaction of the meson with matter, another difficulty arose in connection with the lifetime of the meson, which had been regarded as a success of meson theory. The quantitative investigation of nuclear β-decay revealed that lifetimes were strongly influenced by nuclear matrix elements. If this is properly taken into account, the lifetime for the spontaneous decay of the meson would have to be about 10^{-8} sec rather than 10^{-6} sec (Nordheim 39a; Yukawa 39a,b). This discrepancy was seriously considered and various attempts for removing it were proposed. Some of them will be described below.

Indirect Coupling of the Meson with the Electron and the Neutrino. The first is a theoretical approach that has turned out to be correct if the meson is replaced by the π-meson. In contrast to the original scheme proposed by Yukawa, in which the meson was assumed to couple directly with an electron and a neutrino,

$$\text{nucleon} \rightleftarrows \text{nucleon} + \text{meson} \qquad \text{meson} \rightarrow \text{electron} + \text{neutrino},$$

Sakata (Sakata 40b, 41) suggested a slightly different coupling scheme; namely,

$$\text{meson} \rightleftarrows \text{nucleon} + \text{antinucleon} \rightarrow \text{electron} + \text{antineutrino}.$$

In the latter β-decay takes place as in the original scheme of Fermi, and the spontaneous-decay life of the meson may be modified by adjusting the types and the strengths of coupling. For example, Sakata found for the pseudoscalar meson with pseudovector coupling the mass dependence that is now well known; with the aid of this feature, he pointed out that the discrepancy could be removed, although he doubted its responsibility for nuclear forces.

This partial success was not taken too seriously, however, because it did not help to reduce the meson-nucleon cross section. On the other hand, Sakata's scheme was applied by Sakata and Tanikawa (Sakata 40a) to the decay of the neutral meson, in which a nucleon antinucleon pair decays into photons. The number of photons then emitted was pointed out to be subject to invariance under charge conjugation, called the Furry theorem. This seems to have been the first explicit statement on the selection rules concerning mesons.

The introduction of the neutral meson was, as mentioned in Section 1.4, motivated by the charge independence of nuclear forces, but its observability in cosmic rays was expected to be prohibitively difficult because of its very short lifetime. On the basis of a detailed analysis of the Pfotzer and Schein curves, Tamaki (Tamaki 42) showed, against Carlson and Schein (Carlson 41), that the decay electrons of mesons produced by primary protons were able to account for only a small fraction of the observed soft component. Then Taketani (Taketani 43,

48) pointed out that photons from neutral mesons were indeed responsible for an additional contribution to the soft component. He also showed that the introduction of neutral mesons was favorable for the explanation of the neutrino loss that Tamaki (Tamaki 43) analyzed by comparing incident energy and energy dissipated in the atmosphere and in the earth.

Density Effect. An entirely different approach was a critical examination of experiments on meson lifetime, which was obtained by comparing the intensities of mesons passing through a dense medium and a tenuous medium with the same amount of matter. It was suspected by Fermi (Fermi 40) and by Halpern and Hall (Halpern 40) that the difference might be accounted for by the density dependence of loss by ionization. They pointed out that loss by ionization should be reduced in a dense medium due to polarization caused by an impinging particle; but this effect, called the *density effect*, was found to contribute too little to account for the observed difference. However, the theory provided a quantitative method of measuring high-energy particles.

Decay and Capture of the Meson. Having been stimulated by the question of lifetime, a more direct measurement of the lifetime was attempted by the delayed-coincidence method, in which the time interval between the stopping of a meson and the emission of its decay electron was measured by delaying the signal of the meson stopping. The first successful experiment was performed by Rasetti (Rasetti 41), who obtained a mean lifetime of 1.5 ± 0.3 μsec, after an earlier but unsuccessful attempt by Montgomery et al. (Montgomery 39b). An ingenious method of delayed coincidences invented by Rossi and Nereson (Rossi 42, 43) resulted in a very accurate value of the lifetime, 2.15 ± 0.07 μsec, which within statistical error agrees with the value now adopted.

Before the direct measurement of the lifetime, a question was raised by Yukawa and Okayama (Yukawa 39c) as to whether the absorption of mesons in matter could modify the decay of mesons. They estimated the time for a meson to stop in a medium and pointed out that a meson would be captured before its decay in heavy media such as lead. They further interpreted the failure to observe decay electrons by Montgomery et al. (Montgomery 39b) in this way.

Although their conclusion is no longer accepted as it stood, their work stimulated Tomonaga and Araki (Tomonaga 40) to refine the theory of *meson capture*. They took the effect of the nuclear Coulomb field into account and showed that positive mesons should decay into electrons, whereas negative ones were likely to be captured even in a tenuous medium. For this argument they referred to cloud-chamber pictures, which showed only decay positons (Kunze 33; Ehrenfest 38; Neddermeyer 38; Williams 40a,b) but no negatons (Nishina 37, Maier-Leibnitz 39). They referred to the experiment by Montgomery et al. and further suggested that the absence of decay electrons implied the existence of a meson that was different from the ordinary cosmic-ray meson.

The cloud-chamber photographs of meson decay referred to above provided direct evidence for the decay scheme, which was believed to be meson → electron + neutrino. The two-body decay was not well confirmed, but an event obtained by Shutt et al. (Shutt 42) with a pressurized cloud chamber supported this, because the energy of a decay electron was estimated to be greater than 30 MeV.

1.6 Two-Meson Hypothesis

The two-meson hypothesis was proposed at this stage by Sakata, Sakata and Inoue, and Tanikawa (Sakata 43, 46; Tanikawa 43, 47) and later independently by Marshak and Bethe (Marshak 47).

In 1942 the difficulties generally recognized were (*a*) a small interaction cross section of mesons compared with the large production cross section and (*b*) a long lifetime of mesons compared with the short lifetime expected from nuclear β-decay. A solution to these difficulties was sought in three directions. The first was to regard the phenomena concerned as being outside the applicability of quantum field theory, as had been done before when difficulties were encountered. The second was to modify the formalism or the method of calculation, as in damping theory and strong-coupling theory. The third was to introduce a new model, as has often been found successful in nuclear physics. The third method succeeded on this occasion.

The essential point of the new model was to assume two kinds of mesons, one being identified with Yukawa's proposal and the other being the hard component observed in cosmic rays. A difference between Sakata's and Tanikawa's was found in the types of cosmic-ray mesons —fermions and bosons, respectively. Although these alternatives did not seem to make a significant difference in those days, we shall describe the former one, which is now recognized as correct.

Let the nuclear and the cosmic-ray mesons be designated by Y and m, respectively. With these conventional abbreviations, Yukawa's theory of the meson is characterized by the following interactions:

$$P \rightleftarrows N + Y^+, \qquad N \rightleftarrows P + Y^-, \qquad P(N) \rightleftarrows P(N) + Y^0, \tag{g}$$

$$Y^- \to e^- + \bar{\nu}, \qquad Y^+ \to e^+ + \nu, \tag{g'}$$

where g and g' are the respective coupling constants and induce nuclear forces and nuclear β-decays, respectively. The cosmic-ray meson is a decay product of the nuclear meson, as

$$Y^{\pm} \to m^{\pm} + n, \tag{γ}$$

where n is a neutral meson whose mass is tentatively assumed to be zero. The coupling constant γ is determined so as to give a scattering cross section as large as 10^{-28} cm^2. Therefore $\gamma^2/\hbar c \simeq 10^{-2}$ is far larger than the value adopted at present. The decay of the cosmic-ray meson takes place through the interactions g' and γ as

$$m^{\pm} \to e^{\pm} + \nu(\bar{\nu}) + n.$$

The value of $\gamma^2/\hbar c \simeq 10^{-2}$ is responsible for the well-recognized discrepancy of a factor of 10^2 in the scattering cross section, but it gives too small a lifetime, about 10^{-21} sec for the decay of $Y^{\pm}$.

The above was the main part of the two-meson hypothesis. In addition, Sakata and Inoue argued for this hypothesis on the following grounds:

1. The three-body decay is in favor of the slow-decay electrons observed by some authors (Maier-Leibnitz 39); this is now found to be true.
2. The $\frac{1}{2}$ spin of the cosmic-ray meson is consistent with the conclusion drawn by Christy and Kusaka (Christy 41b) through their analysis of bursts.
3. The large mass of the nuclear meson seems to be in favor of the nuclear-force range expected from nucleon-nucleon scattering experiments.
4. The spontaneous decay of the nuclear meson results mainly in $m^+ + m^-$ and $n + n$, so that the absence of cascade showers stated by Schein et al. (Schein 41) can be well understood, because otherwise the decay into photons could give rise to showers. This is not acceptable nowadays, as described above.

Unfortunately, this work was unknown in other countries because of World War II. Quite independently, the third difficulty in Yukawa's theory of the meson was made clear by Italian physicists.

The work of Tomonaga and Araki (Tomonaga 40) described in the preceding section stimulated experiments, and attempts were made to observe the fraction of mesons that stop without producing decay electrons. In the experiments by Rasetti, and by Nereson and Rossi, half of the mesons that stopped were found to give electrons, thus confirming the theoretical prediction of Tomonaga and Araki. Later, however, Conversi, Pancini, and Piccioni (Conversi 45, 47) observed meson decay by separating positive and negative mesons and found that even negative mesons that stopped in carbon decayed into electrons, whereas they did not when stopping in lead or in iron. This was a surprising disagreement with the prediction of Tomonaga and Araki, according to which negative mesons should be captured even in the lightest material. Further experiments showed an increasing probability of capture with increasing atomic number of the medium, and the capture probability was found to be much smaller than the prediction (Sigurgeirson 47, Valley 47a). A more refined theory of capture made this discrepancy much clearer (Fermi 47a). The slowing-down time of a meson in a medium was found to be too short to account for this observation (Fermi 47b).

It was at this time that Marshak and Bethe (Marshak 47) proposed their two-meson hypothesis without knowing of the Japanese work. Since they knew the third difficulty concerning the capture of mesons, details were, of course, different from those of Sakata et al. For the nuclear meson they adopted the meson-pair theory for a conventional reason; consequently, the nuclear meson was assumed as a fermion, and a counterparticle was neutral and interacted strongly with nucleons. The nuclear meson was then assumed to decay into the cosmic-ray meson and the neutral counterparticle. Since the cosmic-ray meson was a boson in this case, its capture by a nucleon should take place without the emission of any particle. Since they knew the capture probability of the cosmic-ray meson, its interaction with the nucleon was very weak, and the coupling constant for this process was much smaller than the value of γ in Sakata's theory, in essential agreement with present knowledge. The lifetime for the nuclear meson decaying into the cosmic-ray meson was estimated to be longer than 10^{-8} sec, because they noticed no anomaly in the intensity-depth curve of underground cosmic rays.

The last point was criticized by Greisen (Greisen 48) and Hayakawa (Hayakawa 48b), who considered the fact that the intensity-depth curve changed its slope at several hundred meters of water equivalent and attributed this to the decay of nuclear mesons. Nuclear mesons produced in the upper atmosphere decay into cosmic-ray mesons only if their lifetime is longer than the mean free time for nuclear collisions. Since the lifetime increases with energy due to the relativistic effect, the decrease in decay probability is inversely proportional to energy, so that the energy spectrum of cosmic-ray mesons changes its slope, thus resulting in the changed slope of the intensity-depth curve. In this way they were able to estimate the lifetime of the nuclear meson to be about 10^{-8} sec.

Meanwhile, direct evidence for the two-meson hypothesis was obtained by Lattes, Occhialini, and Powell (Lattes 47) by means of newly developed, sensitive photographic plates. In this photographic emulsion they were able to observe meson tracks and discovered the interesting events that showed a stopped meson emitting another meson. They named the former the *π-meson* and the latter the *μ-meson*. It then became evident that the π-meson was what had been proposed by Yukawa and the μ-meson was what was observed in cosmic rays as the hard component. In fact, most mesons that stopped were associated with no secondary track, since they were μ-mesons; and some mesons were found to produce stars, since they were negative π-mesons captured by nuclei.

There followed the elaboration of the two-meson theory. Firstly, the definite range of μ-mesons produced by π-mesons at rest assured the two-body decay of the π-meson. Secondly, the absence of secondary particles associated with stopped μ-mesons, as remarked by Piccioni (Piccioni 48a) led to the assumption that most of the rest energy of a μ-meson should be taken away by a neutral particle, leaving little energy for nuclear excitation. Thirdly, the *positive temperature effect* observed far underground (Forro 47) was interpreted in terms of the π-μ decay, because the decay probability should increase with increasing path length for the nuclear collision of π-mesons and consequently with increasing temperature (Miyazima 48). In concluding this section we mention the discovery of mesons artificially produced with a synchrocyclotron (Gardner 48).

1.7 Meson Production and Meson-Nucleon Interactions

1.7.1 Penetrating Showers

A central problem of meson production was the question whether or not it was multiple. Some of the earliest attempts at this question were mentioned in Section 1.5 (Janossy 40, 41; Wataghin 40). A typical example of experimental studies of meson production is shown in Fig. 1.6 (Janossy 43a; Broadbent 47a,b).

Fig. 1.6*a* shows an experimental arrangement by which penetrating showers produced in absorber II give coincidences of three or more counters in *B*, two or more in *C*, and two or more in *D*. Whether the initiating particle is charged or neutral was determined by examining the discharge of a counter in tray *A* in the absence of absorber I. It was thus found that about one-third were neutral. By varying the thickness

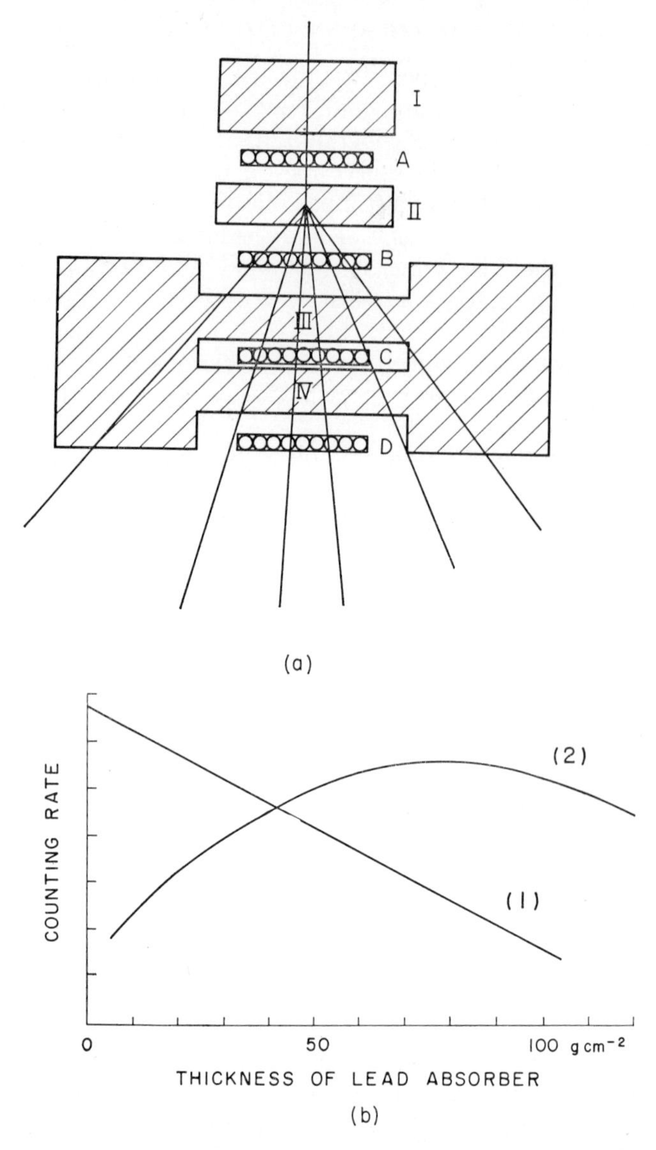

Fig. 1.6 Absorption and transition curves of hard, shower-producing particles. A typical counter hodoscope for the study of penetrating showers (Broadbent 47a) is shown in (*a*). Curve 1 in (*b*) was obtained from the shower-counting rate versus the thickness of the lead layer I, whereas curve 2 was obtained without I and by varying the thickness of absorber II. The latter is based on E. P. George and A. C. Jason, *Proc. Phys. Soc.* (London), **A63**, 1081 (1950).

of absorber I and selecting particles that did not produce secondary particles in it, curve 1 in Fig. 1.6*b* was obtained, its slope giving the *interaction mean free path* of shower-producing particles. By varying the thickness of absorber II, for fixed absorber I, the transition curve 2 was obtained. The maximum in the transition curve was observed at about 150 g-cm^{-2} of lead.

The presence of the transition effect was, as in the case of the electronic shower, interpreted as evidence for the multiple production of mesons. This does not necessarily mean multiple production by collision with a single nucleon but may be explained in terms of multiple production by collision with a nucleus as the consequence of successive single-production processes from individual nucleons in a nucleus. This mechanism of meson production was suggested by Janossy (Janossy 43b) and is called the *plural-production* process.

Janossy's theory was born out of the *damping theory* of Heitler (Heitler 41), according to which the meson-production cross section was predicted to decrease at a rate inversely proportional to energy and to forbid multiple production because of the strong damping effect. According to this theory the cross section was so large at moderate energies that an impinging nucleon would interact simultaneously with several nucleons in a target nucleus. Thus the target nucleus could become opaque, so that the cross section should be proportional to $A^{2/3}$, where A is the mass number of the nucleus. In this way Janossy was able to explain two important features—production from a nucleus and the $A^{2/3}$ dependence of the cross section—in terms of his plural-production theory.

However, these two features may also be predicted in terms of genuine multiple production, as frequently emphasized by Heisenberg. Moreover, the development of renormalization theory and the detailed examination of meson theory made it difficult to adopt the result of the damping theory as it stood.

After the discovery of π- and μ-mesons, interest in meson production as indication of the strong interaction of the meson declined. On the other hand, there arose new problems of the nuclear interactions of π- and μ-mesons. Two historical events—the nuclear scattering of a meson observed by Wilson (Wilson 39b) and a pair of penetrating particles produced in a cloud chamber by Braddick and Hensby (Braddick 39)—can now be regarded as the nuclear scattering of a proton and π-meson production by a μ-meson, respectively. The nuclear interactions of these two kinds of mesons were best revealed by stars taking place in their capture by nuclei.

1.7.2 *Interaction of μ-Mesons*

Quantitative studies of the capture process were made immediately after the existence of π- and μ-mesons became known. The slowing down of a meson to a velocity as small as the velocity of the outer electrons of an atom had been discussed by Yukawa and Okayama (Yukawa 39c), as described in Section 1.5. As the velocity decreases the impact time becomes so long compared with the revolution period of atomic electrons that these electrons cannot be excited. When the velocity of a meson becomes smaller than that of the outermost electrons, the energy loss no longer increases with decreasing energy but begins to decrease (Bohr 48). Then atomic electrons may be regarded as a Fermi gas, and only electrons near the surface of the Fermi sea can be excited (Fermi 47b). The fraction of electrons excited decreases proportionally to the velocity of the meson, and the energy transferred is also proportional to it. Hence the energy loss is proportional to the kinetic energy, so that the characteristic time for slowing down is given by $m_\mu \hbar^3/m_e^2 e^4 \simeq 5 \times 10^{-15}$ sec, where m_μ is the mass of the μ-meson. When the meson becomes thermal, it goes around atoms and is finally captured by an atomic electric field. The trapping is caused by radiative and Auger transitions, and *Auger electrons* are observed with a certain probability (Cosyns 49). When the meson comes down to a low-lying level of a *mesic atom*, the radiative transition becomes predominant, as was observed in a cloud chamber.

The transition processes between low-lying levels were studied by Wheeler (Wheeler 49), who emphasized the importance of investigating the mesic atom from various aspects. As the meson comes down to the K-orbit, there occurs the nuclear capture of the μ-meson. Since the radius of the K-orbit is inversely proportional to Z and the number of capturing protons is Z, the capture probability is proportional to Z^4, provided that the K-orbit lies outside a nucleus. For a heavy nucleus, however, the K-orbit lies inside a nucleus, so that Z must be replaced by Z_{eff}, which tends to 37 for large Z (Wheeler 47, Tiomno 49).

The absolute value of the capture probability was obtained by comparing the lifetimes of positive and negative mesons, and using the fact that the observed decay rate of negative mesons should be the sum of the spontaneous-decay rate and the capture rate (Valley 47b, Ticho 48). The capture rate thus obtained for carbon was found to be about equal to the spontaneous-decay rate. The Z_{eff}^4 law was also confirmed by the same experiments. The magnitude of the coupling constant involved in the nuclear capture of the μ-meson was noticed to be approximately equal to that for the nuclear β-decay.

The rest energy of a negative μ-meson is mostly taken away by a neutrino, and only a small fraction is left in a nucleus. This is not at all negligible, leaving about 15 to 20 MeV for nuclear excitation, according to a simplified theoretical analysis based on the Fermi-gas model of the nucleus (Rosenbluth 49). The excitation energy should be released by evaporating nucleons. In fact, Sard et al. (Sard 48) observed neutrons from the capture of negative μ-mesons in lead. In photographic emulsions exposed underground, George and Evans (George 51) found protons associated with the track ends of μ-mesons.

These observations confirmed the capture process analogous to K-capture of electrons; neither high-energy electrons nor high-energy photons were found in association with the capture (Piccioni 48a,b; Hincks 48). The similarity between the μ-meson–nucleon and the electron-nucleon interactions, both being operative among four fermions, led to the assumption of an analogous interaction between the μ-meson and the electron, although the experimental evidence available in those days was not in favor of the three-body decay of the μ-meson expected for the four-fermion interaction (Thompson 48). Shortly, however, the continuous-energy spectrum of decay electrons was observed both with a cloud-chamber experiment (Leighton 49) and with a delayed-coincidence method (Steinberger 49*a*), thus presenting evidence for the decay of a μ-meson into an electron and two massless particles presumed to be neutrinos.

This was considered as evidence for symmetry in the interactions of four fermions, in which at least two of them are either two of the electron, the neutrino, and the μ-meson (Klein 48, Yukawa 49). Hence these particles are called *leptons*. The interactions of leptons with the other particles, except for the electromagnetic interaction, are much weaker than the nuclear interaction and are of about the same strength irrespective of the particles taking part. Details of these interactions, called the *universal Fermi interaction*, were extensively analyzed by paying particular attention to *selection rules* in decay processes (Nakamura 50, Ruderman 49, Steinberger 49*b*). Further development was made by various investigations, mainly with artificially produced mesons, finally leading to the discovery of the nonconservation of parity, the *V-A* coupling of the universal Fermi interaction and the distinction between two kinds of neutrinos—one associated with the electron and the other with the μ-meson.

In spite of the weak interactions of the lepton, George et al. (George 49, 50) observed star production by energetic μ-mesons in photographic emulsions exposed underground. This was, however, interpreted by them in terms of photonuclear reactions induced by the Coulomb field of a μ-meson. *Neutrons* observed by Cocconi (Cocconi 49d, 51b) were also found to have the same origin.

Much effort was expended in investigating *anomalous interactions* of μ-mesons, differing from the ordinary ones described above. A number of authors claimed to have observed the production of an associated pair of penetrating particles by a μ-meson and others on anomalous scattering, which could not be accounted for in terms of Coulomb scattering (Fowler 58). However, these results are now regarded as spurious at least up to several GeV (Fukui 59). At very high energies the interactions of μ-mesons were found to be due essentially to electromagnetic interactions; namely, ionization, bremsstrahlung, and pair-creation processes. This was determined by analysis of underground cosmic rays (Hayakawa 49, Bollinger 50, Barret 52).

1.7.3 N-*Component*

In the early days of cosmic-ray research the nucleon was considered to be unimportant, except for secondary particles produced by nuclear disintegrations. As one of the secondary products the *neutron* was extensively investigated by Korff and his collaborators (Korff 39a,b,c; 41) with boron proportional counters. Owing to the $1/v$-law of the neutron cross section of ^{10}B, they measured the density of thermal and epithermal neutrons up to balloon altitudes. The result was analyzed by Bethe, Korff, and Placzek (Bethe 40) in terms of the slowing down of neutrons arising from nuclear evaporation and subsequent diffusion in the atmosphere.

Concerning nucleons of moderately high energies, Alikhanian and Alikhanov (Alikhanian 45) noticed copious heavily ionizing particles interpreted as protons at mountain altitudes, although they attributed them to new particles of varying masses, called *varitrons*. In the same energy range neutrons were recognized from stars produced by non-ionizing radiations in cloud chambers (Hazen 44, Powell 46). Further evidence was obtained by observing bursts in fast ionization chambers, by means of which a burst due to heavily ionizing particles was distinguished from that due to a shower (Bridge 47, 48; Rossi 47; Tatel 48; Hulsizer 49). The bursts thus observed at various altitudes from ground level to rocket altitudes made it possible to obtain the altitude dependence of the burst frequency supposedly produced by nuclear active particles.

Direct observations of nuclear events were made by means of photographic emulsions (Brown 49b, Camerini 49), in which stars were observed and identified with the nuclear disintegrations observed in ionization chambers. Since the number and energies of produced particles in a star can be observed in emulsions, a more quantitative analysis became possible. Among the secondary particles were *black prongs* consisting of α-particles and slow protons due to the nuclear evaporation process and *gray prongs*, regarded as due to fast protons knocked out by energetic nuclear active particles. *Thin prongs* were observed in electron-sensitive emulsions and were identified as single-charge relativistic particles. Theoretical studies of nuclear excitation (Goldberger 48) and the subsequent evaporation (Fujimoto 49a, LeCouteur 50) were made on the basis of the Fermi-gas model of the nucleus, and the results were successfully applied to analyses of cosmic-ray stars and of stars induced by the capture of negative π-mesons (Fujimoto 49b; Perkins 49).

With the aid of evaporation theory, most of the stars and ionization bursts were found to be produced by nucleons with energies of 0.1 to 1 GeV. Summarizing the burst frequencies at various altitudes, Rossi (Rossi 48) deduced the attenuation of these nucleons in the atmosphere below 10 km as being about 140 g-cm^{-2}.

Since most of these nucleons have energies below the geomagnetic cutoff, they must be produced by nuclear interactions of still higher energies. The thin tracks associated with stars in emulsions and penetrating showers provide the evidence for such nuclear interactions. A counter experiment with penetrating showers at balloon altitudes exhibited an exponential decrease of their frequencies with atmospheric depth with an attenuation length of about 120 g-cm^{-2} (Tinlot 48). Individual events of penetrating showers were obtained with cloud chambers (Shutt 46, Rochester 47a, Fretter 48).

There arose a problem about the nature of secondary particles, although most of them were believed to be protons and π-mesons. The existence of neutral π-mesons in *penetrating showers* was disclosed by the *mixed shower*, which consists of a mixture of penetrating and electronic cascade showers originating from a common point (Birger 49; Fretter 49). Conclusive evidence for the production of the neutral π-meson and its decay into two photons was obtained by a photographic observation of correlated photons (Carlson 50) in the stratosphere as well as by accelerator experiments. The relative abundances among relativistic charged particles were examined with photographic emulsions and about two-thirds of them were found to be π-mesons (Camerini 51).

Most of the π-mesons produced at high altitudes decay into μ-mesons; the rest of them, together with nucleons, form the nuclear-active N-component. As these particles multiply by successive collisions with air nuclei, a cascade process like the cascade shower of electrons and photons takes place in the atmosphere; this is called the N-cascade, since it consists of N-particles. A mathematical treatment of the N-cascade was developed by Heitler and Janossy (Heitler 49) in analogy to the theory of the electronic cascade.

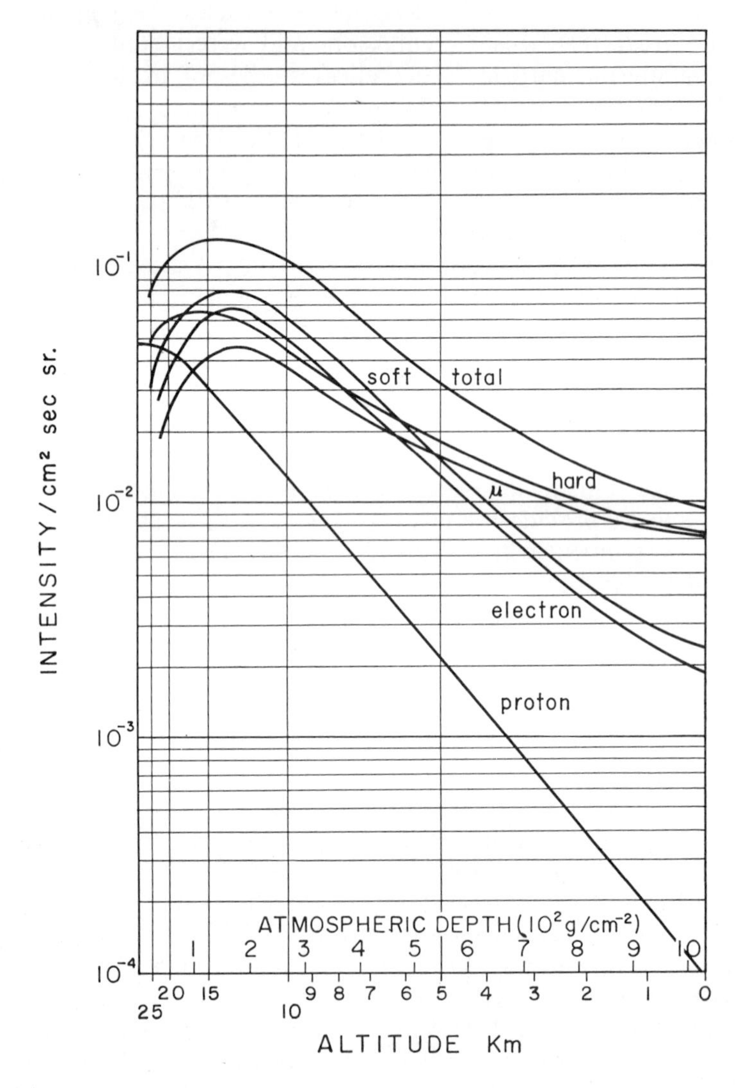

Fig. 1.7 The intensities of the various components of vertical cosmic rays in the atmosphere. The curves were obtained for the geomagnetic latitude of 28°N by S. Hayakawa, *Kagaku*, **22**, 278 (1952).

Because of the primary importance of the N-component, the genetic relation of various components in the atmosphere was analyzed by Rossi (Rossi 48). The primary cosmic rays produce the N-cascade, in which π-mesons take part. A part of the π-mesons in the N-cascade decay into μ-mesons, forming the main part of the hard component. The soft component consists chiefly of electrons from the decay of neutral π- and μ-mesons and also from the knock-on processes involving mesons at low altitudes. Thus the intensity versus atmospheric depth was analyzed as in Fig. 1.7, and the behavior of cosmic rays in the atmosphere was essentially understood. However, their behavior at very high energies was left for further studies.

Around 1950 much effort was made in investigating nuclear interactions in the GeV-energy region by means of cosmic rays. In addition to the achievement described above, a number of important results were obtained concerning the multiple production of mesons, the mass-number dependence of the nucleon-nucleus cross section, the interaction cross section of π-mesons, etc. Most of them were qualitative, and more accurate results obtained with newly constructed accelerators were more favored for investigating the properties of elementary particles.

1.7.4 Extensive Air Showers

The *extensive air shower* (EAS) is characteristic of cosmic rays. The energy of a primary particle may reach as high as 1 joule. Since primary cosmic radiation consists mainly of protons, EAS should be interpreted in terms of a shower initiated by a primary proton. Early attempts to explain the EAS by proton primaries were successful in accounting for its gross features, in which the electronic component was assumed to be produced by nuclear collisions of energetic nucleons either through the charge-exchange process or neutral π-meson production (Hayakawa 48a, Zatsepin 49); the latter was found to play a main role in the development of EAS. Quantitative knowledge was obtained by observing various phenomena that exhibit the structure of EAS, together with their theoretical analyses.

The most characteristic feature of the EAS is its lateral structure. The extent of the shower can be found by measuring the *decoherence curve*, the coincidence rate of two shower detectors as a function of the distance between them. This was observed at various altitudes by counter trays (Skobeltzyn 47a,b; Kraybill 49; Biehl 51; Campfell 52) and by ionization chambers (Williams 48, Blatt 49) and related to the *lateral distribution* of electrons with the aid of the structure function quantitatively derived by Molière (Molière 43). A more direct method is to measure the electron density as a function of the distance from the core, located by a core selector, whereby a core is characterized by the high density of electrons and the high concentration of penetrating particles. With such a core selector the structure function was observed at mountain altitudes by Cocconi et al. (Cocconi 49c), and the lateral distribution thus obtained was found to be in good agreement with the more refined theory of Nishimura and Kamata (Nishimura 50, 51, 52).

The most generally measured quantity was the density distribution, both with sets of counter trays (Cocconi 46, 49ab; Zatsepin 47; Ise 49; Zaharova 49; Broadbent 50) or ionization chambers (Lapp 43, Lewis 45). The integral density spectrum of $\Delta^{-1.4}$ to $\Delta^{-1.6}$, where Δ is the number of particles hitting unit area, can be converted to the integral size spectrum. The total number of electrons in an air shower is called the size and is evaluated by reference to the density Δ measured and the structure function. The size spectrum is further related to the energy spectrum of primary cosmic rays. Commonly observed showers with sizes of 10^4 to 10^7 correspond to primary energies of 10^{14} to 10^{17} eV.

The structure function is mainly affected by multiple scattering of electrons near the observing point, and the relation between size and primary energy is due essentially to the conservation of energy, so that these two tell us little about high-energy interactions. On the other hand, altitude dependence is rather directly affected by nuclear interactions at high energies. Equivalent information can be obtained also from the zenith-angle distribution. Such observations were made by several authors (Hilberry 41, Kraybill 49, Biehl 51).

In an early experiment by Auger et al. (Auger 38), the penetrating particles in EAS were already noticed. This fact was made clear by a cloud-chamber observation (Daudin 45) and by a set of counter trays (Rogoginsky 44; Cocconi 46; Broadbent 47a,b, 48). The fraction of penetrating particles among all the charged particles was found to be a few percent near the core (Cocconi 49c) and to increase to about 50 percent at several hundred meters from the core (Eidus 52), thus implying a concentration of about 10 percent in the entire shower. The relative number of penetrating particles was found to decrease slowly with increasing altitude.

The penetrating particles consist of nuclear active particles and μ-mesons. Their abundances were found to be nearly equal near the core (Greisen 50, McCusker 50, Sitte 50), but the relative abundance of μ-mesons was found to increase with the distance from the core (Fujioka 55).

μ-Mesons associated with EAS can be observed underground. Their lateral spread at 60 meters water equivalent was measured as about 60 meters (George 53), whereas that at 1600 meters water equivalent as only 13 meters (Barret 52). The μ-mesons that penetrate far underground are always associated with EAS and are produced at high altitudes. Having taken this fact into account, Barret et al. (Barret 52) estimated the primary energy responsible for EAS and μ-mesons in coincidence and derived the primary energy spectrum of up to 10^{16} eV.

The high-energy μ-mesons observed far underground are contained in the core of EAS and exhibit the structure of the core. Direct observations of cores were made by means of large ionization chambers, in which a great number of electrons were found (Hazen 52, Danis 54, Heineman 54). The lateral distribution of such electrons was observed to have a peak with a spread of about 1 meter, and the presence of multiple cores was uncertain.

The experimental information described above allowed the construction of a model of EAS and relation of the model to the mechanism of nuclear collisions at extremely high energies. Among various such attempts, one of the most comprehensive analyses was made by Dobrotin et al. (Dobrotin 53) on the basis of an extensive summary of experimental results. According to their study, EAS are in principle identical with the bulk of cosmic rays in the sense that both are initiated by primary nuclear particles that produce both the N- and the electronic components simultaneously by collisions with air nuclei. The difference between them lies in quantity; for EAS the primary energy is so great that many secondary particles are observed in coincidence, whereas for the majority the probability of coincidence of two or more particles from a single primary particle is not appreciable. To be specific, an EAS is a gigantic electronuclear shower that is originated by a primary nuclear particle of extremely high energy. Its collision with an air nucleus creates a number of nuclear active particles, which develop themselves in an N-cascade. In each nuclear collision neutral π-mesons are also produced, but they immediately decay into photons, which result in a huge number of electrons and photons; although they are so numerous, the amount of energy carried by them in an EAS is about the same as that carried by N-particles. Charged π-mesons of relatively low energies in the N-cascade decay into μ-mesons, which form the penetrating particles with a small nuclear interaction.

1.7.5 Evidence of Multiple Production

Many features of EAS were considered to be in favor of multiple production of mesons. A number of jetlike events found in photographic emulsions also seemed to support multiple production

(Leprince-Ringuet 49a,b; Bradt 49, 50a; Lord 50). However, most of the high-energy collisions take place with heavy nuclei. Collision with a hydrogen nucleus or a nucleon at the periphery of a nucleus is identified when only a few black and grey prongs are associated with the event, because an interaction with a nucleus would have to result in the emission of a considerable number of knocked-on and evaporated nuclear fragments. At very high incident energy, however, the secondary particles produced by the collision with a nucleon in a nucleus have rather high energies and are emitted within a narrow angle, so that they produce tertiary particles again within a narrow angle. In this way few particles are knocked out with low energies in a wide angular region, and a cylindrical hole of small radius is left in the target nucleus. Then the energy available for exciting the nucleus left after the reaction is due chiefly to the increase of the surface area of the tunnel. The excitation energy thus acquired is so small that the number of evaporated particles is expected to be small. Heitler and Terreaux (Heitler 51, 53) thus argued that multiple production associated with a few slow particles should not always be interpreted as due to a nucleon-nucleon collision but could be accounted for in terms of a nucleon-nucleus collision.

At energies below 10^{11} eV, as described at the beginning of this section, multiple production could be explained as due to the nucleon-meson cascade process in a nucleus (Messel 52). It was therefore considered to be difficult to present unambiguous proof of multiple production by nucleon-nucleon collisions, although it was regarded as probable on account of the following experimental facts.

In cloud-chamber studies of penetrating showers, graphite (C) and paraffin (CH_2) were used as shower producers. The difference between multiple events in these two producers could be attributed to multiple production from hydrogen. Since the result was based on a small difference between two big numbers, evidence for production was not necessarily convincing.

In order to avoid the drawback inherent in the subtraction method, a hydrogen target was employed. Liquid hydrogen in a copper vessel was flown at a very high altitude, and multiple events were detected by means of a counter hodoscope (Vidale 51a,b). However, they did not necessarily indicate that multiple production was caused by proton-proton collisions; it could have been caused by primary α-particles. A multiple event was observed by Kusumoto et al. (Kusumoto 53) in a cloud chamber filled with high-pressure hydrogen and operated at mountain altitude; it was regarded as evidence for genuine multiple production, although the cause might have been alcohol mixed with the hydrogen gas.

Although the above facts were regarded as sufficient to establish the multiple-production process, clearcut evidence was only finally obtained by an accelerator experiment, in which double pion production was observed in a hydrogen diffusion chamber (Fowler 53a, 54).

1.7.6 Theory of Multiple Meson Production

Most of the important ideas on the theory of multiple production appeared earlier than, or as early as, 1950, before experimental evidence was established. As described in Section 1.5, Heisenberg (Heisenberg 36, 39) emphasized a nonlinear interaction resulting in a multiple process. In this theory the field quantity is regarded as an analogue of a field quantity in hydrodynamics such as the velocity potential; thus the nonlinearity in his meson field theory would give rise to the characteristic features of hydrodynamics. As the interactions between mesons are strong, the meson fluid may be equivalent to a fluid of a large Reynolds number, thus resulting in turbulent motion. By reference to the turbulence spectrum he had derived for an actual fluid (Heisenberg 48), Heisenberg (Heisenberg 49) inferred a power energy spectrum for mesons excited by the violent collision. He solved the nonlinear field equation in the one-dimensional approximation and obtained the spectrum as inferred from the isotropic turbulence of a real fluid.

According to this spectrum the average energy of mesons is as low as their rest energy, and consequently the multiplicity increases as the square root of the incident energy. Actually, however, the multiplicity was found to increase rather slowly with energy. In order to explain this feature Heisenberg (Heisenberg 52) considered the peripheral collision of small momentum exchange, according to which he obtained small multiplicities for large impact parameters.

Since Heisenberg's theory started from an unfamiliar nonlinear Lagrangian, the relation to current meson theory was not clear. In analogy to multiple photon production, a conventional method was developed by evaluating the difference between the mesic self-fields of a nucleon before and after the collision. The self-field was conveniently treated by means of the intermediate-coupling theory of Tomonaga (Miyazima 42). On the basis of this theory the multiplicity of mesons shaken off by sudden changes in momentum, spin, and isotopic spin of a nucleon was calculated. In this theory only the fundamental method of approach was presented, but the details adopted were not realistic enough.

A similar approach based on the weak-coupling theory was attempted by Lewis, Oppenheimer, and Wouthuysen (LOW theory) (Lewis 48).

According to the LOW theory many mesons are independently emitted from two scattered nucleons, with the average multiplicity increasing with energy as $E^{1/3}$ and the angular distribution being isotropic in the center-of-mass system.

A quite different approach was made by Fermi (Fermi 50), who emphasized the importance of the statistical nature of high-energy collisions. Since complicated processes occur in high-energy collisions, the detailed behavior relevant to field theory may be smeared out, and the matrix elements of the reactions under consideration are considered constant. The transition probability is then entirely determined by the final-state density or the phase-space volume. Applying thermodynamics to the collision complex, Fermi obtained the $E^{1/4}$ dependence of the multiplicity. This weak energy dependence was supported by comparison with experiment.

As a modification of Fermi's theory, Takagi (Takagi 52, Kraushaar 54) proposed a model of two fireballs due to the excitation of two nucleons after a collision. Since mesons are emitted isotropically in the respective rest systems of these two nucleons, the angular distribution in the laboratory system is considerably peaked; this results in too large a core of EAS at very high energy.

A natural development of Fermi's theory was made by Landau (Landau 53). In the thermodynamical theory of Fermi the concept of particle density was employed, but this could not be permitted until the phases of excited meson waves became random, because the particle number and the phase are subject to the uncertainty principle. In place of particle density Landau used entropy, which increased due to shock waves caused by the violent impact of two nucleons. After the shock waves reached the edges of the interacting volume, there began the expansion of a meson fluid that followed behind the shock waves. In the course of the expansion entropy is practically constant, because the viscosity is very small due to the small mean free path of mesons. When the free path becomes as large as the dimension of the system, as the energy density decreases the meson fluid begins to spread out. At this moment we can speak about the number of mesons as being proportional to the entropy produced by the initial shock. This naturally results in the multiplicity law derived by Fermi. However, the angular distribution is different in Landau's theory, because this is characterized by the nearly longitudinal expansion. The transverse momentum arises from the lateral expansion due to the fluid pressure and the thermal motion at the final moment. A detailed analysis of the expansion results in a peaked angular distribution, in which the angle of emission decreases with increase in the energy of an emitted particle. Since the final temperature is as low as the rest energy of the meson, the contribution of thermal motion to transverse momentum is quite small. In contrast to Fermi's prediction, the low final temperature leads to a small antinucleon-to-meson ratio, because pairs of nucleons and antinucleons are nearly frozen at such a low temperature.

Because of its theoretical foundation and its fair agreement with experiment, Landau's theory of multiple production became popular around 1955. The discovery of the constancy of transverse momentum by Nishimura (Nishimura 56) was partly motivated by Landau's theory.

1.8 Strange Particles

In the course of studying the mass of the meson a variety of mass values were obtained, mainly by means of cloud chambers with magnetic fields. In the measurement of magnetic rigidity, however, the curvature of a track is subject to multiple-Coulomb scattering, so that the mass value can be obtained only statistically. As was pointed out by Bethe (Bethe 46), all particles observed as having strange masses could be identified with known particles. Nevertheless, a particle observed by Leprince-Ringuet and L'Heritier (Leprince-Ringuet 44) was shown to have the most probable mass value of 990 m_e, due to the observation of a knock-on electron, and the probability that the particle was a proton was estimated to be only about 10 percent.

On the basis of a series of experiments by various means, Alikhanian et al. (Alikhanian 48) claimed to have found a number of new particles of various masses, which they called *varitrons*. An extensive experiment was made by means of a hodoscope with magnetic fields and absorbers, with which both the magnetic rigidity and the range of each particle were measured. However, a critical examination of this method by replacing a part of the hodoscope array with a cloud chamber revealed that most of the varitrons could result from spurious tracks due mainly to electrons accidentally passing through counters (Azimov 51; Vernov 51, 52).

1.8.1 Discovery of V-Particles

In comparison with the mass measurements mentioned above, decay events seem to be more favorable because of the possibility of using the energy-momentum balance. The first such example was obtained by Daudin (Daudin 44) in the course of studying penetrating showers with a cloud chamber. He observed a track with a kink and interpreted it as due to the decay of a charged particle of about 1000 m_e into a lighter charged particle and a neutral particle. This was before the discovery of π- and μ-mesons, and little attention was paid to this novel event.

Immediately after the discovery of π- and μ-mesons, Rochester and Butler (Rochester 47b) found two strange events associated with penetrating showers, with tracks resembling forks, as sketched in Fig. 1.8. The event shown in Fig. 1.8*a* was interpreted as due to the decay of a neutral particle into two charged ones, whereas that in Fig. 1.8*b* was thought to be due to the decay of a positive particle into a lighter positive one and a neutral one. The masses of the parent particles, neutral and charged, were estimated to be about 1000 m_e. It was very lucky that these two events were found among 50 photographs. In spite of great efforts to find analogous events in cloud-chamber photographs, nothing was found in the subsequent two years.

On the other hand, particles of strange masses were found in photographic emulsions. The first of them was discovered by Leprince-Ringuet (Leprince-Ringuet 48) as a particle producing a large nuclear star at its end point. The mass of the primary particle was estimated to be greater than 700 m_e by reference to the energy transferred to six star particles, including one π^- meson. Although a few analogous events were later observed these were not necessarily regarded as conclusive, because the direction of flight of a short track identified as incoming could not be determined unambiguously.

Unambiguous evidence was obtained by the Bristol group (Brown 49a) as the three-meson decay of a charged particle of about 1000 m_e. A particle with track length of 3 mm came to rest and decayed into two relativistic particles of mesic mass and one π^--meson producing a two-prong star, as shown in Fig. 1.9. The relativistic particles were detected

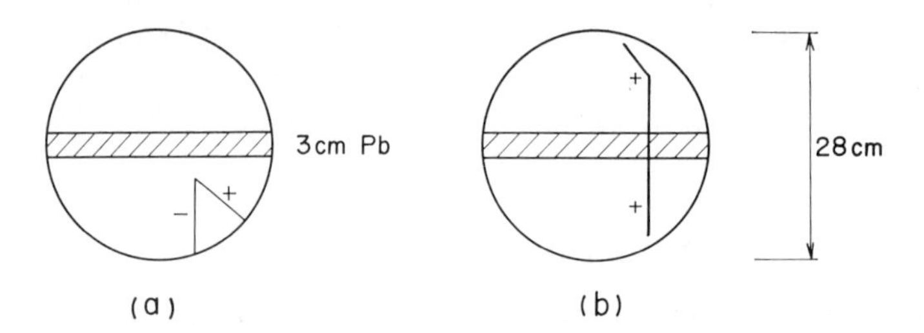

Fig. 1.8 Cloud-chamber photographs of *V*-particles: a neutral *V*-particle decays into two charged particles, (*a*); a charged *V*-particle decays into a charged particle and a neutral particle, (*b*); reprinted from (Rochester 47b).

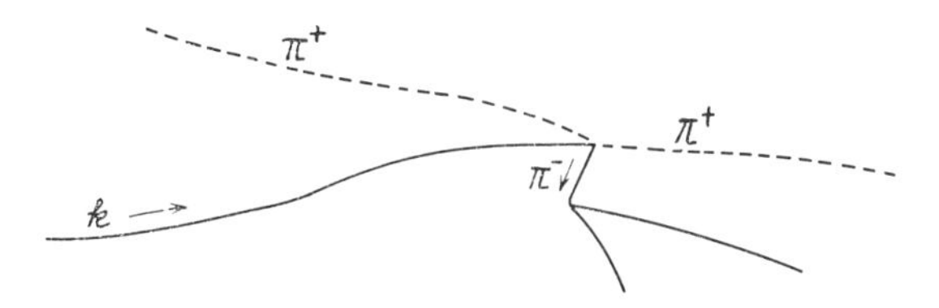

Fig. 1.9 A τ-decay event observed by the Bristol group (Brown 49a). Track k represents a τ-meson that stops and decays into two π^+-mesons (indicated by the dashed lines) and a π^--meson, which produces a two-prong star.

because electron-sensitive emulsions had been used, whereas Bradt and Peters seemed to have failed in the detection of daughter particles of a similar parent particle at rest because electron-sensitive emulsions were not available to them. However, the name "τ-meson," proposed by the latter authors, has come into common use—rather than "k-particle," which was used by the Bristol group.

1.8.2 Variety of V-particles

When the new particles were still very rare, a cloud-chamber experiment at mountain altitude increased their number. Seriff et al. (Seriff 50) operated a cloud chamber controlled by penetrating showers and obtained 28 forked tracks in 8000 photographs, whereas they had obtained only 6 forks in 3000 photographs at sea level. Of 34 cases altogether, 30 were neutral and 4 were charged. Analysis of these events led to the conclusion that the lifetime of these particles was on the order of 10^{-10} sec and that most of the decays were two-body and radiationless. Then any doubt disappeared about the production of such new particles by nuclear interactions. The new particles were named V-particles after the shape of their tracks.

Stimulated by the above observation, the Manchester group (Armenteros 51a,b) also made a successful experiment at mountain altitude and found two groups of neutral V-particles, $V_1{}^0 \to p + \pi^-$ and $V_2{}^0 \to \pi^+ + \pi^-$, their masses being determined as 2200 and 800 m_e, respectively. There was an event that could be interpreted as a cascade decay, $V^- \to \pi^- + V^0$ and $V^0 \to p + \pi^-$. A more convincing example of the cascade decay was found later (Cowan 54).

As a modification of the V-event a decay particle was observed to come from the point where a charged particle stopped (Annis 52). This was called the S-particle, a kind of charged V-particle.

In competition with cloud-chamber studies, the use of photographic plates also achieved great success. In the course of studying the μ-e decay, a μ-meson was found to come from the end of a heavy track, and this was concluded to be due to a new particle, which was called the *κ-particle*, which decayed into a μ-meson and two neutral particles (O'Ceallaigh 51). The decay mode was based on two examples that gave μ-mesons of different energies. Charged *V*-particles decaying into $\pi^+ + \pi^0$ (Menon 54) and $n(p) + \pi^+(\pi^0)$ (Bonetti 53) were observed also with emulsions.

As a result of this flood of new particles, a nomenclature was proposed at the Bagnères de Bigorre International Conference on Cosmic Rays in 1953. All the strange particles were divided into two groups according to their masses; those heavier than the nucleon were called *hyperons*, or *H*-particles; whereas those heavier than the π-meson but lighter than the nucleon were called *heavy mesons*, or *K*-particles. They are collectively designated by *Y* and *K*, respectively. The hyperons consist of a neutral species Λ^0, a charged species $\Sigma^{\pm}$, and the species that gives rise to the cascade decay, Ξ. The heavy mesons are distinguished by their decay products, which are written as subscripts of *K*. The first suffix, π or μ, stands for a charged secondary, and the second suffix stands for the number of secondary particles. The neutral *K*-particle with two secondaries, one of which is the μ-meson or the charged π-meson, is written with the subscript $\mu 2$ or $\pi 2$. The conventional nomenclature employing Greek letters is also used; for example,

$$K^0_{\pi 2}(\theta^0) \rightarrow \pi^+ + \pi^-, \qquad K_{\pi 3}(\tau) \rightarrow \pi^+ + \pi^- + \pi^+.$$

In addition, $K_{\mu 2}$ and K_{e3} were confirmed after the conference.

As may be seen from the agenda of the 1953 International Conference on Cosmic Rays, the central problem in the early 1950s, was the new unstable particles. In great contrast, this problem was not all discussed in the 1955 conference, which was devoted to cosmological and geophysical problems. This change was foreseen when multi-GeV accelerators came into use.

1.8.3 Theories of Strange Particles

Evidence obtained only from cosmic rays was so qualitative that the various models proposed for understanding the new unstable particles were inevitably speculative. The most difficult point was recognized to be their long lifetimes in comparison with their strong nuclear interaction. The nuclear interaction was found to be only slightly weaker than that of π-mesons, as seen from the formation of the *hyperfragment* (Crussard 53, Danysz 53, Tidman 53, Bonetti 54), a nucleus formed with nucleons and a hyperon.

The long lifetime was possible only by forbidding the inverse processes through a strong interaction. Selection rules based on space reflection, charge conjugation, and so forth were not always responsible for forbidding rapid decays. A suggestion was made that V-particles were produced in pairs, so that a strong decay process was forbidden on energetic grounds (Nambu 51, Oneda 51). However, cosmic-ray experiments seemed to be against pair production.

As the beam energy of accelerators increased, the pair production of a Λ^0 and a θ^0 was observed in a hydrogen-filled cloud chamber bombarded by π^--mesons with energies of 1.5 GeV, and the production cross section of about 1 mb and more precise Q-values were measured (Fowler 53b).

The pair production and other properties of V-particles were incorporated into a theoretical scheme by Gell-Mann (Gell-Mann 53) and by Nakano and Nishijima (Nakano 53). There are several conserved quantities in the V-particle problem. Both the nucleon number N and charge Q are strictly conserved, whereas the isotopic spin I is conserved only for the strong interaction. The electromagnetic interaction violates the isotopic-spin conservation but conserves its z-component I_z. Even the latter is violated in weak interactions, thus forming the hierarchy of interactions. In the absence of V-particles the charge is related to I_z and to the baryon number N as

$$Q = I_z + \frac{N}{2}.$$

If integer and half-odd isotopic spins are attributed to the hyperon and the heavy meson, respectively, the above relation does not hold in the presence of the V-particle. Then a new quantum number S was introduced, as follows:

$$Q = I_z + \frac{N}{2} + \frac{S}{2}.$$

This was later called *strangeness* because of its relationship to the strange particles. One can now assign $S=0$ to the nucleon and the π-meson, and $S=\pm 1$ to the strange particles. The choice of ± 1 can be made by reference to the experimental fact that most of the charged heavy mesons observed in cosmic rays are positive. This leads to $I_z = \frac{1}{2}$ and $S=1$ for the heavy meson, and $I_z = 0$ or 1 and $S=-1$ for the hyperon, except for the cascade particle. Since the cascade particle decays slowly into a lighter hyperon, it must have a different value of S, say $S=-2$. Then $I_z = \frac{1}{2}$; consequently the existence of the negative and the neutral cascade particles was predicted. With such assignment of isotopic spins, the branching ratios of various production processes were also predicted (Nishijima 54, 55).

Stimulated by the success of Gell-Mann and Nishijima, various people looked for the underlying basis of this rather phenomenological theory. Here we mention only that the compound model of elementary particles (Markov 55, Sakata 56) and the nonconservation of parity (Lee 56) stemmed from studies of the strange particles.

Both Sakata's theory of the compound model and parity nonconservation were motivated by a serious puzzle concerning the nature of the heavy meson. As far as the decay processes were concerned there seemed to be many kinds of heavy mesons; but their other properties, such as mass and lifetime, appeared to be similar to each other. This fact could also be interpreted as due to the existence of many different decay modes of a single kind of heavy meson. In the latter interpretation, however, $K_{\pi 2}(\theta)$ and $K_{\pi 3}(\tau)$ can hardly be attributed to two different decay modes of a single particle, because the possible spin and parity of $K_{\pi 2}$ are $0^+, 1^-, 2^+, \ldots$, according to the conservation of angular momentum and parity, whereas $K_{\pi 3}$ is most likely 0^-. Accordingly, $K_{\pi 2}$ and $K_{\pi 3}$ seemed to be different insofar as their spins and parities were concerned.

In order to know whether the heavy meson is of a single kind with a variety of decay modes or whether it consists of different particles it is necessary to (*a*) measure the masses of heavy mesons of various modes, (*b*) measure their lifetimes, and (*c*) observe the relative frequencies of occurrence of different decay modes for various production processes and energies. A systematic study for this purpose was organized among European cosmic-ray physicists in 1954—1955. This organization was called the *G*-stack collaboration, because a giant emulsion stack was used in order to follow decay products to the end of their ranges. The results thus obtained were presented at the Pisa conference in June 1955, in which 36 authors were involved (Davies 55).

The *G*-stack collaboration was primarily concerned with (*a*) mentioned above and part of (*c*). The collaborators were able to collect about 80 *K*-mesons suitable for identifying decay modes. The results thus obtained for *K*-mesons with single-charge daughters are summarized in Table 1.1.

The results were unable to give definite answers to the above questions, but the mass value obtained was not significantly different from the τ mass, 966 m_e having been accepted in those days.

In early 1955 the Bevatron began to produce *K*-mesons with momentum of 360 MeV/*c*. From a focused *K*-beam with emulsions 300 *K*-decays were quickly obtained (Birge 55). After the Pisa conference the

Table 1.1 Result of the *G*-stack Collaboration

Decay Mode	Number of Events	Frequency (%)	Mass (in electron mass)
$K_{\mu 2} \to \mu + \nu$	38	50 to 70	977 ± 6
$K_{\pi 2} \to \pi^+ + \pi^0$	24	15 to 30	966 ± 3
$K_{e3} \to e + \pi^0 + \nu$	5	~9	—
$K_{\mu 3} \to \mu + \pi^0 + \nu$	5	~3	—
$K_{\pi 3} \to \pi^+ + \pi^0 + \pi^0$	6	~1	—

strange particle was no longer a central problem in cosmic rays and was studied mainly with accelerators. Quantitative knowledge thus obtained supported the single kind of *K*-mesons in all respects but the τ-θ puzzle. The puzzle was finally resolved by introducing the nonconservation of parity in weak interactions.

1.9 Origin of Cosmic Rays

The origin of cosmic rays is mysterious in that a single particle possesses high energy, the highest reaching as high as 1 joule. Theoretical studies of this problem were developed with increasing pieces of information on primary cosmic radiation.

1.9.1 Early Speculations

The most fantastic suggestion on the origin of cosmic rays was made by Millikan et al. (Millikan 42) in their interpretation of the energy spectrum of primary cosmic rays derived from the latitude effect observed by means of their balloon-flight experiments. The energy spectrum they obtained appeared to consist of several knees at energies respectively corresponding to the rest energies of light nuclei. Hence they thought that cosmic rays might be due to the spontaneous annihilation of nuclei, such as carbon, nitrogen, and oxygen. Neither the observation nor the interpretation is convincing, and this mass-annihilation hypothesis is of historical interest only.

However, Klein (Klein 44) considered this hypothesis rather seriously and looked for the possibility that matter and antimatter forming galaxies and antigalaxies, respectively, could annihilate to release an enormous amount of energy. Although this process cannot be responsible for particles with energies above 10^{12} eV, Klein's idea has been revived occasionally as a possible mechanism of energy release.

A different possibility for converting potential energy into kinetic energy was suggested by Hoyle (Hoyle 47), who assumed the ejection of superheavy nuclei in supernova outbursts. Although the lifetime of such a superheavy nucleus is a question of nuclear physics, the fission of a superheavy nucleus could give considerable energy to a fragment. However, this, too, cannot explain the presence of energetic particles. Nevertheless, Hoyle's suggestion that the energy released by supernova outbursts may account for the energy density of cosmic rays remains valid.

1.9.2 Heavy Nuclei in Primary Cosmic Rays

The astrophysical significance of cosmic rays became clear when the Minnesota and Rochester groups (Bradt 48, Freier 48a,b) discovered heavy nuclei in primary cosmic rays by means of balloon-borne emulsions and cloud chambers. The very existence of heavy nuclei implies that there should not be any violent process to destroy nuclei in the course of acceleration; the mass-annihilation process is therefore quite unlikely. The gross features of the relative abundances of these nuclei were similar to those of galactic elements (Bradt 50b).

This may indicate that the source of cosmic rays is not an extraordinary part of the Galaxy but rather an ordinary part. If, however, the fragmentation of heavy cosmic-ray nuclei by collisions with interstellar matter is taken into account, the similarity would result from the fragmentation of the heaviest nuclei, say the iron group, which alone would be accelerated at the sources. In the latter case the mean path length traversed by cosmic rays would have to be as long as, or longer than, the collision mean free path of protons.

1.9.3 Magnetic Trapping and Electromagnetic Acceleration

In order for cosmic rays to traverse a great thickness of interstellar matter, the path of a particle cannot be straight but must be curled in a complicated way. This is possible if there exist magnetic fields that trap cosmic rays. The trapping is also responsible for the essential isotropy of cosmic rays.

The existence and the importance of celestial magnetic fields were emphasized particularly by Alfvèn (Alfvèn 50b) as early as the 1930s. The acceleration of cosmic rays by a changing magnetic field, now called the betatron acceleration, was proposed by Swann (Swann 33) in 1933. However, serious concern with celestial magnetic fields began with investigations of the origin of cosmic rays.

According to Alfvèn (Alfvèn 49) a reasonable estimate of the magnetic-field strength may be obtained by assuming the equipartition of energy between hydrodynamic turbulent motion and magnetic fields, which couple to form hydromagnetic waves. In interstellar space the magnetic-field strength thus derived may be as high as 10^{-5} gauss. This is strong enough to trap extremely-high-energy particles in the Galaxy. If, however, one notices the fact that the energy density of cosmic rays is as large as that of turbulent motion, the efficiency of converting turbulent energy into cosmic-ray energy could hardly be explained by any known mechanism.

Thus Richtymer and Teller (Richtymer 49) proposed a *solar origin* of cosmic rays, according to which particles are trapped in the solar system for a period longer than thousands of years. Then the energy given to cosmic rays is only a very small fraction of the radiation energy, which is not at all unreasonable. As for the trapping field, Alfvèn (Alfvèn 49, 50a) suggested that the solar magnetic field is frozen into a beam of plasma ejected from the sun and fills up the solar system. Since the magnetized plasma travels outward, there arises an electric field that causes the polarization of charge in the conductive plasma beam, while the potential difference is maintained through a circuit going around the sun. This electric-potential difference could be as high as 10^9 volts, so that acceleration would be possible. This model was more concerned with magnetic storms and their effect on variations in cosmic-ray intensity.

An alternative theory was the Galactic origin of cosmic rays proposed by Fermi (Fermi 49), who pointed out the *statistical acceleration* resulting from collisions with interstellar clouds. A cosmic-ray particle spiraling along a magnetic line of force is either reflected at a position where the field strength increases so as to make a magnetic mirror or the field line is curved to turn the direction of motion along the curved line of force. If such a turning point moves with velocity v, a particle gains or loses energy by a fraction $\pm(v/c)$, depending on whether the collision with the moving turning point is head on or overtaking. Since the frequency of head-on collisions is greater than that of overtaking ones by $2(v/c)$, the statistical average of fractional energy changes results in a resultant energy gain of $2(v/c)^2$ per collision. Hence the total energy of a particle increases exponentially with the number of collisions or with the age of the particle. The energy gain continues until the particle escapes from the accelerating region. The distribution of energy arises from the age distribution; a Poisson distribution of age gives a power energy spectrum.

This was a great success in a number of respects. Firstly, the Galactic magnetic fields anticipated theoretically by Alfvèn were incorporated with the origin of cosmic rays. Secondly, cosmic radiation was found to have an important bearing in astrophysics as a tool for exploring Galatic magnetic fields. Thirdly, the observed isotropy and power energy spectrum were well explained by the turbulent motion of interstellar magnetic fields.

The first point served to develop a new field of astrophysics, called cosmic electrodynamics, or cosmic aerodynamics, and also to stimulate plasma physics. The second point was verified by more direct means; namely, the polarization of starlight caused by its scattering by paramagnetic grains aligned in a magnetic field (Hall 49, Hiltner 49, Spitzer 49). This then led to the observation that magnetic fields exist along spiral arms of the Galaxy as well as in wandering interstellar clouds.

As for the origin of cosmic rays, Fermi's theory was generally taken for granted, particularly according to the following observations. The energy densities (averaged over the Galaxy) of starlight, turbulent motion, magnetic fields, and cosmic rays are of the same order of magnitude, 1 eV-cm^{-3}. Since the stellar energy from nuclear fusion reactions is regarded as the most powerful energy source and this energy is emitted into the interstellar space mainly as starlight, one might think it quite strange that the energy density of starlight is nearly equal to that of the others. However, if one considers the lifetimes of these energy carriers in the Galaxy, only starlight has a short lifetime compared with the others, provided that cosmic rays are stored for a long period due to magnetic trapping. Thus only a very small fraction of stellar energy is found to be converted into other modes—turbulent motion, magnetic fields, and cosmic rays. The equality of energy densities of the first two may be regarded as evidence for their equipartition, and the equality of cosmic-ray to energy that of the other two supports Fermi's mechanism of acceleration.

1.9.4 Difficulties in Fermi's Theory

After the general features of the origin of cosmic rays were understood, there came the investigation of details. Then it was found that not all parts of Fermi's theory were correct but required revision in many respects.

Fermi himself pointed out a difficulty with the injection of heavy nuclei. Since the Fermi acceleration is rather slow, the injection energy should be high enough for the rate of energy gain to be greater than the rate of loss, which is due mainly to the ionization process. This condition is difficult to satisfy for heavy nuclei, which are subject to a large loss by ionization. Another difficulty may arise with high-energy heavy nuclei if their energy spectrum has a shape similar to that of protons. If the mean lifetime of cosmic rays is determined by nuclear collisions with interstellar matter, it would have to decrease with increasing nuclear mass, because the cross section increases with mass number. Therefore the energy spectrum would be expected to be steeper for heavier nuclei (Fan 51). However, experiments appeared to favor the same slope irrespective of mass number (Kaplon 52). Definite evidence concerning the thickness of interstellar matter traversed by cosmic rays could be obtained by examining the abundance of the light nuclei—lithium, beryllium, and boron—which are very rare in the Galaxy. If heavy nuclei traversed a great thickness of matter, they should disintegrate by nuclear collisions into smaller fragments, including the light nuclei. However, the light nuclei observed by Bradt and Peters (Bradt 50b) with balloon-borne nuclear emulsions were found to be produced in air by heavier primary nuclei; Bradt and Peters also claimed that light nuclei were practically absent from primary cosmic rays. From this they concluded that the thickness traversed would be less than 1 g-cm^{-2}. This is rather thin, but it is thick enough to strip all orbital electrons from heavy nuclei by atomic collisions. Later investigations gave a finite abundance of the light nuclei, corresponding to a matter thickness of about 3 g-cm^{-2}, but this was less than the mean free path for nuclear collisions.

Another difficulty would arise if nuclear collisions occurred as frequently as in Fermi's theory. One may raise the question of why electrons are not accelerated in the same way as ions. Feenberg and Primakoff (Feenberg 47) investigated this problem prior to the above theories and pointed out a mechanism for eliminating electrons by their Compton collisions with starlight. Earlier still Pomeranchuk (Pomeranchuk 40) noted another elimination mechanism—the magnetic bremsstrahlung of electrons with ultrahigh energies in the earth's magnetic field. The search for electrons and photons in primary cosmic rays was made by observing bursts in an ionization chamber covered by a lead plate (Hulsizer 49) and also by observing cascade showers in a cloud chamber with lead plates (Critchfield 50). In the latter experiment the upper limit to the intensity of electrons and photons with energies higher than 1 GeV was found to be about 1 percent of the proton intensity. The reduction of electrons to such a small fraction was considered to be possible if cosmic rays propagate in the Galaxy for a long enough period or if they are confined to the solar system, where the photon intensity is sufficiently large. Even if the electrons accelerated by the Fermi mechanism could be eliminated in this way, electrons produced through π-μ-e decays from π-mesons resulting from nuclear collisions with interstellar matter give a considerable contribution (Hayakawa 52, 53). Compton collisions with stellar photons and magnetic bremsstrahlung in Galactic magnetic fields were found to be not effective enough in reducing the electron intensity below several percents of the proton intensity if the mean lifetime of cosmic rays was determined by nuclear collisions. It was further remarked by Hayakawa (Hayakawa 52) that the same mechanism gives γ-ray through the production of neutral π-mesons.

All the above difficulties can be overcome if the thickness of matter traversed is small, so that the mean lifetime of cosmic rays is not determined by nuclear collisions but by escape from the Galaxy. If the latter is the case, the mean lifetime may not be long enough to accelerate particles to high energy. However, an argument against the Fermi acceleration was presented even if the mean lifetime is sufficiently long. According to observations of interstellar clouds, collisions with which are responsible for acceleration, the average cloud velocity is about 10 km-sec^{-1} which is far smaller than the value required for the Fermi acceleration. Thus Unsöld (Unsöld 50) was led to suggest a local acceleration possibly in conjunction with nonthermal radio emission, such as was observed in intensive solar flares associated with the generation of cosmic rays (Forbush 46).

1.9.5 Solar Cosmic Rays

Cosmic rays associated with solar flares provide definite evidence for the celestial generation of cosmic rays, but stars like the sun are rather poor cosmic-ray producers and can contribute only a very small fraction of the Galactic cosmic rays. If solar cosmic rays were confined within the solar system, they would account for the observed intensity, at least a part of it only at low energies.

Solar cosmic rays were first observed as an unusual increase in cosmic-ray intensity that was associated with an intensive solar flare on February 28, 1942. The sunspot that produced the flare seemed to make another outburst on March 7, as was inferred from a sudden ionospheric disturbance (SID). This again produced cosmic rays. Similar events were found on July 25, 1946, and November 19, 1949. On all these occasions stations at low magnetic latitudes failed to observe the unusual increases. This indicates that the energies of solar cosmic rays are rather low, few particles having energies greater than 10 GeV. The number of relativistic particles produced in such a flare is estimated to be as high as 10^{32}.

If solar cosmic rays are trapped in the solar system, they are gradually lost by collisions with planets, with a lifetime of about 10^4 years (Kane

49, Treiman 54). This is long enough to account for cosmic-ray density in terms of solar production and magnetic trapping in the solar system. This does not hold for high-energy particles. The sidereal-time variation of extensive air showers was found to be negligible (Daudin 49). Having analyzed the time variation of energetic μ-mesons observed at a great depth, Cocconi (Cocconi 51a) pointed out that the essentially isotropic distribution could not be accounted for in terms of solar origin because the radius of curvature of such high-energy particles in the interplanetary magnetic field should be larger than the radius of the solar system.

1.9.6 Radio Astronomy and Cosmic Rays

Since solar cosmic rays were found to be responsible for a minor part of the total cosmic rays, a model of Galactic origin had to be found. Having taken into account the disklike shape of the Galaxy, Morrison et al. (Morrison 54) pointed out that escape through the disk surfaces should determine the mean lifetime of Galactic cosmic rays. The mean lifetime of about 10^6 years thus determined is too short to allow the Fermi acceleration.

The difficulty seemed to disappear, at least in part, by reference to the radio astronomical observation of general Galactic radio emission. Having analyzed the intensity distribution of nonthermal radio emission, Pikelner (Pikelner 53) concluded that there was spherical distribution of radio sources; radio emission is strongest in the direction of the Galactic center, but it is considerable even at high Galactic latitudes. As the cause of radio emission magnetic bresmsstrahlung of energetic electrons was suggested earlier by Kiepenheuer (Kiepenheuer 50) and Ginzburg (Ginzburg 51). On this basis the spherical distribution of radio emission indicates that magnetic fields extend outside the flat Galaxy observed optically and relativistic electrons fill up this spherical system (Ginzburg 56). If the halo filled with relativistic electrons exists, it is quite reasonable to assume that nuclear particles also exist in the spherical halo. They spend most of the time in the halo, where the matter density is quite small, and the thickness of matter traversed by cosmic rays can be small even if their age is considerable.

However, a difficulty was pointed out in the source of energy for acceleration in the halo. The halo contains tenuous gases and globular clusters of extreme age, and any celestial activity is believed to have practically ceased. If, therefore, cosmic rays took up the energy of turbulent motion during their life, no supply of energy would be left in the halo. Some energy sources are still required in the Galactic disk.

In order to revive magnetic acceleration Fermi (Fermi 54) proposed that a more efficient mechanism is operative along spiral arms. Owing to radio astronomical observations, especially of the 21-cm line of the hydrogen atom, neutral hydrogen was known to be distributed mainly in the spiral arms, along which magnetic fields were shown to exist. The stability of the arm as well as the polarization of starlight indicated a rather regular magnetic field with a strength of about 10^{-5} gauss along the arms. Slight irregularities should produce magnetic mirrors between which cosmic rays could be trapped. If two mirrors approach one another, a particle gains energy at every reflection by a mirror. This efficient mechanism is called the Fermi II acceleration, whereas the stochastic acceleration emphasized in his original paper was called the Fermi I acceleration. When particles eventually escape out of the approaching mirror trap, however, they may get into a receding mirror trap, so that their energies are lost. Therefore the resultant effect may not be much different from the result of the stochastic Fermi I acceleration.

In view of the difficulty with slow acceleration in the whole Galaxy, some particular sources of cosmic rays were looked for. Among various possibilities the supernova was noted to be the most likely because of the enormous amount of energy ejection (Haar 50). The *supernova origin* of cosmic rays was considered as a realistic one, when the strong radio emission of a supernova remnant, the Crab Nebula, was interpreted by Shklovsky (Shklovsky 53) and Ginzburg (Ginzburg 51a,b) in terms of the magnetic bremsstrahlung of relativistic electrons; on this basis they suggested the acceleration of nuclear particles also in supernova remnants. The Crab Nebula is a planetary nebula with two central stars, but they are not luminous enough to account for the luminosity of the nebula. The visible light of the Crab may also be attributed to magnetic bremsstrahlung, and consequently the light must be polarized. The polarization was indeed observed by Oort and Walraven (Oort 56), and relativistic electrons with energies as high as 10^{11} eV were shown to be responsible for the visible light.

The supernova origin was further supported by the overabundances of heavy nuclei in cosmic rays in comparison with their Galactic abundances (Hayakawa 56). A supernova is regarded as the last stage in the evolution of a star, in which the relative abundances of heavy elements increase. In a supernova outburst the matter forming the star is ejected into interstellar space, and part of it is accelerated to cosmic-ray energy. The former is diluted by preexisting matter, whereas the latter is not. The difference between the relative abundances of elements in cosmic rays and in general.

Another radio source that shows optical polarization is Virgo A identified with the extragalactic nebula M87 (Baade 56). It is evident that active extragalactic nebulae may be efficient cosmic-ray sources. There are a considerable number of strong radio sources such as Cygnus A, and their acceleration efficiency per volume seems to be as great as that of the Crab Nebula (Burbidge 56). These astronomical data provided a basis for the extragalactic origin of cosmic rays that could be responsible for their high-energy part. Extensive air showers of great size suggest the necessity of the extragalactic origin, because primary particles producing such large air showers have energies that are too high to be confined within the Galaxy.

Through studies of their origin in the last 10 years, cosmic rays have been recognized as an important part of astrophysics, particularly in the respect that a considerable part of the energy in the universe appears to be nonthermal, so that cosmic rays play a significant role in the evolution of the universe.

REFERENCES

Alfvèn 49	Alfvèn, H., *Phys. Rev.*, **75**, 1732 (1949).
Alfvèn 50a	Alfvèn, H., *Phys. Rev.*, **77**, 375 (1950).
Alfvèn 50b	Alfvèn, H., Cosmical Electrodynamics (1950).
Alikhanian 45	Alikhanian, A. I., and Alikhanov, A. I., *Jour. Exp. Theor. Phys. USSR*, **15**, 145 (1945).
Alikhanian 48	Alikhanian, A. I., Alikhanov, A. I., and Weissenberg, A., *Jour. Exp. Theor. Phys. USSR*, **18**, 301 (1948).
Anderson 32	Anderson, C. D., *Phys. Rev.*, **41**, 405 (1932).
Annis 52	Annis, M., et al., *Nuovo Cim.*, **9**, 624, (1952),
Armenteros 51a	Armenteros, R., et al., *Nature*, **167**, 501 (1951).
Armenteros 51b	Armenteros, R., et al., *Phil. Mag.*, **42**, 1113 (1951).
Auger 35	Auger, P., *Compt. rend.*, **200**, 739 (1935).
Auger 36	Auger, P., Leprince-Ringuet, L., and Ehrenfest, P., *J. phys. radium*, **7**, 58 (1936).
Auger 38	Auger, P., Maze, R., and Grivet-Meyer, T., *Compt. rend.*, **206**, 1721 (1938).
Azimov 51	Azimov, S., et al., *Dokl. Akad. Nauk USSR*, **78**, 447 (1951).
Baade 56	Baade, W., *Bull. Astr. Inst. Netherland*, **12**, 312 (1956).
Bagge 41	Bagge, B., *Ann. Physik*, **39**, 512 (1941).
Barret 52	Barret, P. H., et al., *Rev. Mod. Phys.*, **24**, 133 (1952).
Bethe 30	Bethe, H. A., *Ann. Physik*, **5**, 325 (1930).
Bethe 34	Bethe, H. A., and Heitler, W., *Proc. Roy. Soc.* (London), **A146**, 83 (1934).
Bethe 40	Bethe, H. A., Korff, S. A., and Placzek, G., *Phys. Rev.*, **57**, 573 (1940).
Bethe 46	Bethe, H. A., *Phys. Rev.*, **70**, 821 (1946).
Bhabha 36	Bhabha, H. J., *Proc. Roy. Soc.* (London), **A154**, 195 (1936).

Bhabha 37	Bhabha, H. J., *Proc. Roy. Soc.* (London), **A159**, 432 (1937).
Biehl 51	Biehl, A. T., and Neher, H. V., *Phys. Rev.*, **83**, 1169 (1951).
Birge 55	Birge, R. W., et al., *Phys. Rev.*, **99**, 329 (1955).
Birger 49	Birger, N. G., et al., *Jour. Exp. Theor. Phys. USSR*, **19**, 826 (1949).
Blackett 33	Blackett, P. M. S., and Occhialini, G. P. S., *Proc. Roy. Soc.* (London), **A139**, 699 (1933).
Blackett 37	Blackett, P. M. S., and Wilson, J. G., *Proc. Roy. Soc.* (London), **A160**, 304 (1937).
Blackett 38	Blackett, P. M. S., *Phys. Rev.*, **54**, 973 (1938).
Blatt 49	Blatt, J. M., *Phys. Rev.*, **75**, 1584 (1949).
Blau 37	Blau, M., and Wambacher, H., *Nature*, **140**, 585 (1937).
Bloch 33	Bloch, F., *Z. Physik*, **81**, 363 (1933).
Bohr 13	Bohr, N., *Phil. Mag.*, **25**, 10 (1913).
Bohr 15	Bohr, N., *Phil Mag.*, **30**, 581 (1915).
Bohr 48	Bohr, N., *Kgl. Dansk. Vid. Selsk. USSR*, **18**, No. 9 (1948).
Bollinger 50	Bollinger, L. M., *Phys. Rev.*, **79**, 207 (1950).
Bonetti 53	Bonetti, A., et al., *Nuovo Cim.*, **10**, 345, 736 (1953).
Bonetti 54	Bonetti, A., et al., *Nuovo Cim.*, **11**, 210, 330 (1954).
Bothe 28	Bothe, W., and Kolhörster, W., *Naturwiss.*, **16**, 1044, 1045, (1928).
Bothe 29	Bothe, W., and Kolhörster, W., *Phys. Zeits.*, **30**, 516 (1929).
Bowen 34	Bowen, I. S., Millikan, R. A., and Neher, H. V., *Phys. Rev.*, **46**, 641 (1934).
Bowen 37	Bowen, I. S., Millikan, R. A., and Neher, H. V., *Phys. Rev.*, **52**, 80 (1937).
Bowen 38a	Bowen, I. S., Millikan, R. A., and Neher, H. V., *Phys. Rev.*, **53**, 217 (1938).
Bowen 38b	Bowen, I. S., Millikan, R. A., and Neher, H. V., *Phys. Rev.*, **53**, 855 (1938).
Braddick 39	Braddick, J. J., and Hensby, G. S., *Nature*, **144**, 1012 (1939).
Bradt 48	Bradt, H. L., and Peters, B., *Phys. Rev.*, **74**, 1828 (1948).
Bradt 49	Bradt, H. L., Kaplon, M. F., and Peters, B., *Phys. Rev.*, **76**, 1735 (1949).
Bradt 50a	Bradt, H. L., Kaplon, M. F., and Peters, B., *Helv. Phys. Acta*, **23**, 24 (1950).
Bradt 50b	Bradt, H. L., and Peters, B., *Phys. Rev.*, **80**, 943 (1950).
Bridge 47	Bridge, H., and Rossi, B., *Phys. Rev.*, **71**, 379 (1947).
Bridge 48	Bridge, H., et. al., *Phys. Rev.*, **73**, 1252 (1948).
Broadbent 47a	Broadbent, D., and Janossy, L., *Proc. Roy. Soc.* (London), **A190**, 497 (1947).
Broadbent 47b	Broadbent, D., and Janossy, L., *Proc. Roy. Soc.* (London), **A191**, 517 (1947).
Broadbent 48	Broadbent, D., and Janossy, L., *Proc. Roy. Soc.* (London), **A192**, 364 (1948).
Broadbent 50	Broadbent, D., Kellerman, E. W., and Hakkem, M., *Proc. Phys. Soc.* (London), **63A**, 864 (1950).
Brown 49a	Brown, R., et al., *Nature*, **163**, 82 (1949a).
Brown 49b	Brown, R., et al., *Phil. Mag.*, **40**, 862 (1949b).
Burbidge 56	Burbidge, G. R., *Astrophys. J.*, **124**, 416 (1956).

Camerini 49 Camerini, U., et al., *Phil. Mag.*, **40**, 1073 (1949).

Camerini 51 Camerini, U., et al., *Phil. Mag.*, **42**, 126, 1241 (1951).

Campfell 52 Campfell, I. D., and Prescott, J. R., *Proc. Phys. Soc.* (London), **65A**, 258 (1952).

Carlson 50 Carlson, A. G., Hooper, J. E., and King, D. R., *Phil Mag.*, **41**, 701 (1950).

Carlson 37 Carlson, J. F., and Oppenheimer, J. R., *Phys. Rev.* **51**, 220 (1937).

Carlson 41 Carlson, J. F., and Schein, M., *Phys. Rev.*, **59**, 840 (1941).

Christy 41a Christy, R. F., and Kusaka, S., *Phys. Rev.*, **59**, 405 (1941).

Christy 41b Christy, R. F., and Kusaka, S., *Phys. Rev.*, **59**, 414 (1941).

Clay 27 Clay, J., *Proc. Amsterdam*, **30**, 1115 (1927).

Clay 39 Clay, J., *Rev. Mod. Phys.*, **11**, 128 (1939).

Cocconi 46 Cocconi, G., Loverdo, A., and Tongiorgi, V., *Phys. Rev.*, **70**, 846 (1946).

Cocconi 49a Cocconi, G., and Cocconi, V. T., *Phys. Rev.*, **75**, 1058 (1949).

Cocconi 49b Cocconi, G., Cocconi, V. T., and Greisen, K., *Phys. Rev.*, **75**, 1063 (1949).

Cocconi 49c Cocconi, G., Cocconi, V. T., and Greisen, K., *Phys. Rev.*, **76**, 1020 (1949).

Cocconi 49d Cocconi, V. T., *Phys. Rev.*, **76**, 517 (1949d).

Cocconi 51a Cocconi, G., *Phys. Rev.*, **83**, 135 (1951).

Cocconi 51b Cocconi, G., and Cocconi, V. T., *Phys. Rev.*, **84**, 29 (1951).

Compton 33 Compton, A. H., *Phys. Rev.*, **43**, 387 (1933).

Compton 35 Compton, A. H., and Getting, I. A., *Phys. Rev.*, **47**, 817 (1935).

Compton 37 Compton, A. H., and Turner, R. N., *Phys. Rev.*, **52**, 799 (1937).

Conversi 45 Conversi, M., Pancini, E., and Piccioni, O., *Phys. Rev.*, **68**, 232 (1945).

Conversi 47 Conversi, M., Pancini, E., and Piccioni, O., *Phys. Rev.*, **71**, 209 (1947).

Cosyns 49 Cosyns, M. G. E., et al., *Proc. Phys. Soc.* (London), **62A**, 801, (1949).

Cowan 54 Cowan, E. W., *Phys. Rev.*, **94**, 161 (1954).

Critchfield 50 Critchfield, C. L., Ney, E. P., and Oleska, S., *Phys. Rev.*, **79**, 402 (1950).

Crussard 53 Crussard, J., and Morellet, D., *Compt. rend.*, **236**, 64 (1953).

Danis 54 Danis, W. P., Hazen, W. E., and Heineman, R. E., *Nuovo Cim.*, **12**, 233 (1954).

Danysz 53 Danysz, M., and Pnieuski, J., *Phil. Mag.*, **44**, 348 (1953).

Daudin 49 Daudin, A. and Daudin, J., *J. phys. radium*, **10**, 394 (1949).

Daudin 44 Daudin, J., *Ann. Phys.*, **19**, 110 (1944).

Daudin 45 Daudin, J., *Ann. Phys.*, **20**, 563 (1945).

Davies 55 Davies, J. H., et al., *Nuovo Cim.*, **2**, 1063 (1955).

Dobrotin 53 Dobrotin, N. A., et al., *Usp. Fiz. Nauk USSR*, **49**, 185 (1953).

Dymond 39 Dymond, F. G., *Nature*, **144**, 782 (1939).

Ehmert 37 Ehmert, A., *Z. Physik*, **106**, 751 (1937).

Ehmert 40 Ehmert, A., *Z. Physik*, **115**, 326 (1940).

Ehrenfest 38 Ehrenfest, P., *Compt. rend.*, **206**, 428 (1938).

Eidus 52 Eidus, L. H., et al., *Jour. Exp. Theor. Phys., USSR*, **22**, 440 (1952).

Elster 00	Elster, J., *Phys. Zeits.*, **2**, 560 (1900).
Euler 38	Euler, H., and Heisenberg, W., *Erg. Exakt. Naturwiss.*, **17**, 1 (1938).
Euler 40a	Euler, H., *Z. Physik*, **116**, 73 (1940).
Euler 40b	Euler, H., and Wergeland, H., *Astrophysica Norvegica*, **3**, 165 (1940).
Fan 51	Fan, C. Y., *Phys. Rev.*, **82**, 211 (1951).
Feenberg 47	Feenberg, E., and Primakoff, H., *Phys. Rev.*, **73**, 449 (1947).
Fermi 40	Fermi, E., *Phys. Rev.*, **57**, 485 (1940).
Fermi 47a	Fermi, E., Teller, E., and Weisskopf, V. F., *Phys. Rev.*, **71**, 314 (1947).
Fermi 47b	Fermi, E., and Teller, E., *Phys. Rev.*, **72**, 399 (1947).
Fermi 49	Fermi, E., *Phys. Rev.*, **75**, 1169 (1949).
Fermi 50	Fermi, E., *Prog. Theor. Phys.*, **5**, 570 (1950).
Fermi 54	Fermi, E., *Astrophys. J.*, **119**, 1 (1954).
Forbush 37	Forbush, S. E., *Phys. Rev.*, **51**, 1108 (1937).
Forbush 46	Forbush, S. E., *Phys. Rev.*, **70**, 771 (1946).
Forro 47	Forro, M., *Phys. Rev.*, **72**, 868 (1947).
Fowler 58	Fowler, G. N., and Wolfendale, A. W., *Prog. Elem. Part. and Cosmic Ray Phys.*, **III**, 105 (1958).
Fowler 53a	Fowler, W. B., et al., *Phys. Rev.*, **90**, 758 (1953).
Fowler 53b	Fowler, W. B., et al., *Phys. Rev.*, **93**, 861 (1953).
Fowler 54	Fowler, W. B., et al., *Phys. Rev.*, **95**, 1026 (1954).
Freier 48a	Freier, P., et al., *Phys. Rev.*, **74**, 213 (1948).
Freier 48b	Freier, P., et al., *Phys. Rev.*, **74**, 1818 (1948).
Fretter 48	Fretter, W. B., *Phys. Rev.*, **73**, 41 (1948).
Fretter 49	Fretter, W. B., *Phys. Rev.*, **76**, 511 (1949).
Fröhlich 38	Fröhlich, H., Heitler, W., and Kemmer, N., *Proc. Roy. Soc.* (London), **A166**, 154 (1938).
Fujimoto 49a	Fujimoto, Y., and Yamaguchi, Y., *Prog. Theor. Phys.*, **4**, 468 (1949).
Fujimoto 49b	Fujimoto, Y., Hayakawa, S., and Yamaguchi, Y., *Prog. Theor. Phys.*, **4**, 575, 576 (1949).
Fujioka 55	Fujioka, G., *J. Phys. Soc. Japan*, **10**, 245 (1955).
Fukui 59	Fukui, S., Kitamura, T., and Watase, Y., *Phys. Rev.*, **113**, 315 (1959).
Gardner 48	Gardner, E., and Lattes, C. M. G., *Science*, **109**, 270 (1948).
Geitel 00	Geitel, H., *Phys. Zeits.*, **2**, 116 (1900).
Gell-Mann 53	Gell-Mann, M., *Phys. Rev.*, **92**, 833 (1953).
George 49	George, E. P., and Trent, P., *Nature*, **164**, 838 (1949).
George 50	George, E. P., and Evans, J., *Proc. Phys. Soc.* (London), **63A**, 1248 (1950).
George 51	George, E. P., and Evans, J., *Proc. Phys. Soc.* (London), **64A**, 193 (1951).
George 53	George, E. P., MacAnuff, J. W., and Sturgess, J. W., *Proc. Phys. Soc.* (London), **66A**, 346 (1953).
Ginzburg 51	Ginzburg, V. L., *Dokl. Akad. Nauk USSR*, **76**, 377 (1951).
Ginzburg 53a	Ginzburg, V. L., *Usp. Fiz. Nauk USSR*, **51**, 343 (1953)
Ginzburg 53b	Ginzburg, V. L., *Dokl. Akad. Nauk USSR*, **92**, 1133 (1953).
Ginzburg 56	Ginzburg, V. L., *Izv. Akad. Nauk USSR*, **20**, 5 (1956).

Goldberger 48 Goldberger, M. L., *Phys. Rev.*, **74**, 1269 (1948).

Greisen 48 Greisen, K. I., *Phys. Rev.*, **73**, 521 (1948).

Greisen 50 Greisen, K. I., Walker, W. D., and Walker, S. P., *Phys. Rev.*, **80**, 535 (1950).

Haar 50 Haar, D. ter, *Rev. Mod. Phys.*, **22**, 119 (1950).

Hall 49 Hall, J. S., *Science*, **109**, 166 (1949).

Halpern 40 Halpern, O., and Hall, H., *Phys. Rev.*, **57**, 459 (1940).

Hayakawa 48a Hayakawa, S., and Tomonaga, S., *J. Sci. Res. Inst. Japan*, **43**, 67 (1948).

Hayakawa 48b Hayakawa, S., *Prog. Theor. Phys.*, **3**, 199 (1948).

Hayakawa 49 Hayakawa, S., and Tomonaga, S., *Prog. Theor. Phys.*, **4**, 287, 496 (1949).

Hayakawa 52 Hayakawa, S., *Prog. Theor. Phys.*, **8**, 571 (1952).

Hayakawa 53 Hayakawa, S., and Kobayashi, S., *J. Geomag. Geol.*, **5**, 83 (1953).

Hayakawa 56 Hayakawa, S., *Prog. Theor. Phys.*, **15**, 111 (1956).

Hazen 44 Hazen, W. E., *Phys. Rev.*, **65**, 67 (1944).

Hazen 52 Hazen, W. E., *Phys. Rev.*, **85**, 455 (1952).

Heineman 54 Heineman, R. E., *Phys. Rev.*, **96**, 161 (1954).

Heisenberg 36 Heisenberg, W., *Z. Physik*, **101**, 533 (1936).

Heisenberg 38 Heisenberg, W., *Z. Physik*, **110**, 251 (1938).

Heisenberg 39 Heisenberg, W., *Z. Physik*, **113**, 61 (1939).

Heisenberg 48 Heisenberg, W., *Z. Physik*, **124**, 624 (1948).

Heisenberg 49 Heisenberg, W., *Z. Physik*, **129**, 569 (1949).

Heisenberg 52 Heisenberg, W., *Z. Physik*, **133**, 65 (1952).

Heitler 36 Heitler, W., *Quantum Theory of Radiation*, 1st ed., Oxford University Press, London, 1936.

Heitler 37 Heitler, W., *Proc. Roy. Soc.* (London), **A161**, 261 (1937).

Heitler 41 Heitler, W., *Proc. Camb. Phil. Soc.*, **37**, 291 (1941).

Heitler 49 Heitler, W., and Janossy, L., *Proc. Phys. Soc.* (London), **62A**, 374 (1949).

Heitler 51 Heitler, W., and Terreaux, C., *Helv. Phys. Acta*, **24**, 551 (1951).

Heitler 53 Heitler, W., and Terreaux, C., *Proc. Phys. Soc.* (London), **66A**, 929 (1953).

Hess 11 Hess, V. F., *Phys. Zeits.*, **12**, 998 (1911).

Hess 12 Hess, V. F., *Phys. Zeits.*, **13**, 1084 (1912).

Hilberry 41 Hilberry, N., *Phys. Rev.*, **60**, 1 (1941).

Hiltner 49 Hiltner, W. A., *Science*, **109**, 165 (1949).

Hincks 48 Hincks, E. P., and Pontecorvo, B., *Phys. Rev.*, **73**, 257 (1948).

Hoffmann 26 Hoffmann, G., *Ann. der Phys.*, **80**, 779 (1926).

Hoffmann 27 Hoffmann, G., *Ann. der Phys.*, **82**, 413 (1927).

Hoyle 47 Hoyle, F., *Monthly Notices Roy. Astron. Soc.*, **106**, 384 (1947).

Hulsizer 49 Hulsizer, R., *Phys. Rev.*, **76**, 164 (1949).

Ise 49 Ise, J., and Fretter, W. B., *Phys. Rev.*, **76**, 933 (1949).

Janossy 37 Janossy, L., *Z. Physik*, **104**, 430 (1937).

Janossy 40 Janossy, L., and Ingleby, P., *Nature*, **145**, 511 (1940).

Janossy 41 Janossy, L., *Proc. Roy. Soc.* (London), **A179**, 361 (1941).

Janossy 43a Janossy, L., and Rochester, G. D., *Proc. Roy. Soc.* (London), **A182**, 180 (1943); **183**, 181 (1944).

Janossy 43b Janossy, L., *Phys. Rev.*, **64**, 345 (1943).

Johnson 33	Johnson, T. H., *Phys. Rev.*, **43**, 307, 381 (1933).
Johnson 35	Johnson, T. H., *Phys. Rev.*, **48**, 287 (1935).
Johnson 39	Johnson T. H., *Rev. Mod. Phys.*, **11**, 208 (1939).
Kane 49	Kane, E. O., Shanley, T. J. B., and Wheeler, J. A., *Rev. Mod. Phys.*, **21**, 51 (1949).
Kaplon 52	Kaplon, M. F., et al., *Phys. Rev.*, **85**, 295 (1952).
Kemmer 38a	Kemmer, N., *Proc. Roy. Soc.* (London), **A166**, 127 (1938).
Kemmer 38b	Kemmer, N., *Proc. Camb. Phil. Soc.*, **34**, 354 (1938).
Kiepenheuer 50	Kiepenheuer, K. O., *Phys. Rev.*, **79**, 738 (1950).
Klein 29	Klein, O., and Nishina, Y., *Z. Physik*, **52**, 853 (1929).
Klein 44	Klein, O., *Arkiv Mat. Astronom. Fys.*, **31A**, No. 14 (1944).
Klein 48	Klein, O., *Nature*, **161**, 897 (1948).
Kolhörster 13	Kolhörster, W., *Phys. Zeits.*, **14**, 1153 (1913).
Kolhörster 38	Kolhörster, W., Matthes, J., and Weber, E., *Naturwiss*, **26**, 576 (1938).
Korff 39a	Korff, S. A., and Danforth, W. E., *Phys. Rev.*, **55**, 980 (1939).
Korff 39b	Korff, S. A., *Phys. Rev.*, **56**, 210, 1241 (1939).
Korff 39c	Korff, S. A., *Rev. Mod. Phys.*, **11**, 211 (1939).
Korff 41	Korff, S. A., *Phys. Rev.*, **59**, 949 (1941).
Kraushaar 54	Kraushaar, W. L., and Mark, L. J., *Phys. Rev.*, **93**, 326 (1954).
Kraybill 49	Kraybill, H. L., *Phys. Rev.*, **76**, 1092 (1949).
Kulenkampff 38	Kulenkampff, H., *Verhand. deutsch. phys. Gesell.*, **3**, 19, 92 (1938).
Kunze 33	Kunze, P., *Z. Physik*, **83**, 1 (1933).
Kusumoto 53	Kusumoto, O., et al., *Phys. Rev.*, **90**, 998 (1953).
Landau 53	Landau, L. D., *Izv. An. USSR*, **17**, 51 (1953).
Lapp 43	Lapp, L. G., *Phys. Rev.*, **64**, 129 (1943).
Lattes 47	Lattes, C. M. G., Occhialini, G. P. S., and Powell, C. F., *Nature*, **160**, 453, 486 (1949).
LeCouteur 50	LeCouteur, K. J., *Proc. Phys. Soc.* (London), **63A**, 259 (1950).
Lee 56	Lee, T. D., and Yang, C. N., *Phys. Rev.*, **104**, 254 (1956).
Leighton 49	Leighton, R. B., Anderson, C. D., and Seriff, A. J., *Phys. Rev.*, **75**, 1432 (1949).
Lemâitre 33	Lemâitre, G., and Vallarta, M. S., *Phys. Rev.*, **43**, 87 (1933).
Leprince-Ringuet 44	Leprince-Ringuet, L., and L'Heritier,, M., *Compt. rend.*, **219**, 618 (1944).
Leprince-Ringuet 48	Leprince-Ringuet, L., et al., *Compt. rend.*, **226**, 1897 (1948).
Leprince-Ringuet	Leprince-Ringuet, L., et al., *Phys. Rev.*, **76**, 1273 (1949).
Leprince-Ringuet	Leprince-Ringuet, L., et al., *Compt. rend.*, **229**, 163 (1949).
Lewis 48	Lewis, H. W., Oppenheimer, J. R., and Wouthuysen, S. A., *Phys. Rev.*, **73**, 127 (1948).
Lewis 45	Lewis, L. G., *Phys. Rev.*, **67**, 228 (1945).
Lord 50	Lord, J. J., Fainberg, J., and Schein, M., *Phys. Rev.*, **80**, 970 (1950).
Maier-Leibnitz	Maier-Leibnitz, H., *Z. Physik*, **112**, 569 (1939).
Markov 55	Markov, A. M., *Dokl. Akad. Nauk USSR*, **101**, 449 (1955).
Marshak 47	Marshak, R. E., and Bethe, H. A., *Phys. Rev.*, **72**, 506 (1947).
McCusker 50	McCusker, C. B. A., *Proc. Phys. Soc.* (London), **63A**, 1240 (1950).

Menon 54	Menon, M. K. G., and O'Ceallaigh, C., *Proc. Roy. Soc.* (London), **A221**, 292 (1954).
Messel 52	Messel, H., Potts, R. B., and McCusker, C. R., *Phil Mag.*, **43**, 889 (1952).
Millikan 36	Millikan, R. A., and Neher, V. H., *Phys. Rev.*, **50**, 15 (1936).
Millikan 42	Millikan, R. A., Neher, V. H., and Pickering, W. H., *Phys. Rev.*, **61**, 397 (1942).
Miyazima 42	Miyazima, T., and Tomonaga, S., *Sci. Pap. Inst. Phys. Chem. Res.*, **40**, 21 (1942).
Miyazima 48	Miyazima, T., *Prog. Theor. Phys.*, 3, 99 (1948).
Molière 43	Molière, G., *Kosmische Strahlung*, ed. by W. Heisenberg, 1943.
Monk 39	Monk, A. T., and Compton, A. H., *Rev. Mod. Phys.*, **11**, 173 (1939).
Montgomery 38	Montgomery, C. G., and Montgomery, D. D., *J. Franklin Inst.*, **226**, 623 (1938).
Montgomery 39a	Montgomery, C. G., and Montgomery, D. D., *Rev. Mod. Phys.*, **11**, 255 (1939).
Montgomery 39b	Montgomery, C. G., et al., *Phys. Rev.*, **55**, 1117, 56, 635 (1939).
Morrison 54	Morrison, P., Olbert, S., and Rossi, B., *Phys. Rev.*, **94**, 440 (1954).
Myssowsky 26	Myssowsky, L., and Tuwin, L., *Z. Physik*, **39**, 146 (1926).
Nakamura 50	Nakamura, S., et al., *Prog. Theor. Phys.*, **5**, 740 (1950).
Nakano 53	Nakano, T., and Nishijima, K., *Prog. Theor. Phys.*, **10**, 581 (1953).
Nambu 51	Nambu, Y., Nishijima, K., and Yamaguchi, Y., *Prog. Theor. Phys.*, **6**, 619 (1951).
Neddermeyer 37	Neddermeyer, S. H., and Anderson, C. D., *Phys. Rev.*, **51**, 884 (1937).
Neddermeyer 38	Neddermeyer, S. H., and Anderson, C. D., *Phys. Rev.*, **54**, 88 (1938).
Neher 42	Neher, H. V., and Pickering, W. H., *Phys. Rev.*, **61**, 407 (1942).
Nishijima 54	Nishijima, K., *Prog. Theor. Phys.*, **12**, 107 (1954).
Nishijima 55	Nishijima, K., *Prog. Theor. Phys.*, **13**, 285 (1955).
Nishimura 50	Nishimura, J., and Kamata, K., *Prog. Theor. Phys.*, **5**, 899 (1950).
Nishimura 51	Nishimura, J., and Kamata, K., *Prog. Theor. Phys.*, **6**, 262, 628 (1951).
Nishimura 52	Nishimura, J., and Kamata, K., *Prog. Theor. Phys.*, **7**, 175 (1952).
Nishimura 56	Nishimura, J., *Soryushiron Kenkyu*, Japan **12**, 24 (1956).
Nishina 34	Nishina, Y., Tomonaga S., and Sakata, S., *Sci. Pap. Inst. Phys. Chem. Res.* Japan, **24**, No. 17 (1934)
Nishina 37	Nishina, Y., Takeuchi, M., and Ichimiya, T., *Phys. Rev.*, **52**, 1198 (1937).
Nordheim 38	Nordheim, L. W., *Phys. Rev.*, **53**, 694 (1938).
Nordheim 39a	Nordheim, L. W., *Phys. Rev.*, **55**, 506 (1939).
Nordheim 39b	Nordheim, L. W., and Hebb, M. H., *Phys. Rev.*, **56**, 494 (1939).
O'Ceallaigh 51	O'Ceallaigh, C., *Phil. Mag.*, **42**, 1032 (1951).
Oneda 51	Oneda, S., *Prog. Theor. Phys.*, **6**, 633 (1951).
Oort, 56	Oort, J. H., and Walraven, Th., *Bull. Astr. Inst. Netherland*, **12**, 285 (1956).
Oppenheimer 37	Oppenheimer, R. J., and Serber, R., *Phys. Rev.*, **51**, 1113 (1937).

Oppenheimer 40	Oppenheimer, R. J., Snyder, H. S., and Serber, R., *Phys. Rev.*, **57**, 75 (1940).
Perkins 49	Perkins, D. H., *Phil. Mag.*, **40**, 601 (1949).
Pfotzer 36	Pfotzer, G., *Z. Physik*, **102**, 23 (1936).
Piccioni 48a	Piccioni, O., *Phys. Rev.*, **73**, 411 (1948).
Piccioni 48b	Piccioni, O., *Phys. Rev.*, **74**, 1754 (1948).
Pikelner 53	Pikelner, S. B., *Dokl. Akad. Nauk USSR*, **88**, 229 (1953).
Pomeranchuk 40	Pomeranchuk, I., *J. Phys. USSR*, **2**, 65 (1940).
Powell 46	Powell, W. M., *Phys. Rev.*, **69**, 385 (1946).
Rasetti 41	Rasetti, F., *Phys. Rev.*, **60**, 198 (1941).
Regener 33	Regener, E., *Phys. Zeits*, **34**, 306, 820 (1933).
Richtmyer 49	Richtmyer, R. D., and Teller, E., *Phys. Rev.*, **75**, 1729 (1949).
Rochester 47a	Rochester, G. D., Butler, C. C., and Runcorn, S. K., *Nature*, **159**, 227 (1947).
Rochester 47b	Rochester, G. D., and Butler, C. C., *Nature*, **160**, 855 (1947).
Rogoginsky 44	Rogoginsky, A., *Phys. Rev.*, **65**, 291 (1944).
Rosenbluth 49	Rosenbluth, M. N., *Phys. Rev.*, **75**, 532 (1949).
Rossi 30a	Rossi, B., *Nature*, **125**, 636 (1930).
Rossi 30b	Rossi, B., *Lincei Rend.*, **11**, 831 (1930).
Rossi 33	Rossi, B., *Z. Physik*, **82**, 151 (1933).
Rossi 34	Rossi, B., *Ric. Sci.*, **5**, 569 (1934).
Rossi 35	Rossi, B., *Proc. Int. Conf. Phys.*, Vol. I, 1935, p. 238.
Rossi 39	Rossi, B., *Rev. Mod. Phys.*, **11**, 296 (1939).
Rossi 42	Rossi, B., and Nereson, N., *Phys. Rev.*, **62**, 417 (1942).
Rossi 43	Rossi, B., and Nereson, N., *Phys. Rev.*, **64**, 199 (1943).
Rossi 47	Rossi, B., and Williams, R. W., *Phys. Rev.*, **72**, 172 (1947).
Rossi 48	Rossi, B., *Rev. Mod. Phys.*, **20**, 537 (1948).
Ruderman 49	Ruderman, M., and Finkelstein, R., *Phys. Rev.*, **76**, 1453 (1949).
Sakata 40a	Sakata, S., and Tanikawa, Y., *Phys. Rev.*, **57**, 548 (1940).
Sakata 40b	Sakata, S., *Phys. Rev.*, **58**, 576 (1940).
Sakata 41	Sakata S., *Proc. Phys. Math. Soc. Japan*, **23**, 281, 291 (1941).
Sakata 43	Sakata, S., *Report of the Symposium on Meson Theory*, 1943.
Sakata 46	Sakata, S., and Inoue, T., *Prog. Theor. Phys.*, **1**, 143 (1946).
Sakata 56	Sakata, S., *Prog. Theor. Phys.*, **16**, 686 (1956).
Sard 48	Sard, R. D., et al., *Phys. Rev.*, **74**, 97 (1948).
Schein 39	Schein, M., and Gill, P. S., *Rev. Mod. Phys.*, **11**, 267 (1939).
Schein 41	Schein, M., Jesse, W. P., and Wollan, E. O., *Phys. Rev.*, **59**, 615 (1941).
Schindler 31	Schindler, H., *Z. Physik*, **72**, 625 (1931).
Schonland 37	Schonland, B. F. J., Delatizky, B. and Gaskell, J. *Terr. Mag.*, **42**, 137 (1937).
Serber 38	Serber, R., *Phys. Rev.*, **54**, 217 (1938).
Seriff 50	Seriff, A. J., et al., *Phys. Rev.*, **78**, 290 (1950).
Shklovsky 53	Shklovsky, I. S., *Dokl. Akad. Nauk USSR*, **90**, 983 (1953).
Shutt 42	Shutt, R. P., De Benedetti, S., and Johnson, T. H., *Phys. Rev.*, **62**, 552 (1942).
Shutt 46	Shutt, R. P., *Phys. Rev.*, **69**, 128, 261 (1946).
Sigurgeirson 47	Sigurgeirson, T., and Yamakawa, A., *Phys. Rev.*, **71**, 319 (1947).
Sitte 50	Sitte, K., *Phys. Rev.*, **78**, 721 (1950).
Skobeltzyn 27	Skobeltzyn, D. V., *Z. Physik*, **43**, 354 (1927).

Skobeltzyn 47a	Skobeltzyn, D. V., Zatsepin, G. T., and Miller, V. V., *Phys. Rev.*, **71**, 315 (1947a).
Skobeltzyn 47b	Skobeltzyn, D. V., Zatsepin, G. T., and Miller, V. V., *Jour. Exp. Theor. Phys. USSR*, **17**, 939 (1947b).
Spitzer 49	Spitzer, L., and Tukey, J. W., *Science*, **109**, 461 (1949).
Steinberger 49a	Steinberger, J., *Phys. Rev.*, **75**, 1136 (1949).
Steinberger 49b	Steinberger, J., *Phys. Rev.*, **76**, 1186 (1949).
Steinke 28	Steinke, E., *Z. Physik*, **48**, 647 (1928).
Steinke 30	Steinke, E., *Phys. Zeits.*, **31**, 1019 (1930).
Størmer 30	Størmer, C., *Z. Astrophys.*, **1**, 237 (1930).
Street 37	Street, J. C., and Stevenson, E. C., *Phys. Rev.*, **52**, 1003 (1937).
Swann 33	Swann, W. F. G., *Phys. Rev.*, **43**, 217 (1933).
Takagi 52	Takagi, S., *Prog. Theor. Phys.*, **7**, 123 (1952).
Taketani 43	Taketani, M., *Report of the Symposium on Meson Theory* (1943).
Taketani 48	Taketani, M., *Prog. Theor. Phys.*, **3**, 349 (1948).
Tamaki 42	Tamaki, H., *Riken Iho*, **21**, 891 (1942).
Tamaki 43	Tamaki, H., *Report of the Symposium on Meson Theory* (1943).
Tanikawa 43	Tanikawa, Y., *Report of the Symposium on Meson Theory* (1943).
Tanikawa 47	Tanikawa, Y., *Prog. Theor. Phys.*, **2**, 220 (1947).
Tatel 48	Tatel, H. E., and Van Allen, J. A., *Phys. Rev.*, **73**, 87 (1948).
Thompson 48	Thompson, R. W., *Phys. Rev.*, **74**, 490 (1948).
Ticho 48	Ticho, H. K., *Phys. Rev.*, **74**, 1337 (1948).
Tidman 53	Tidman, D. A., et al., *Phil. Mag.*, **44**, 350 (1953).
Tinlot 48	Tinlot, J. H., *Phys. Rev.*, **73**, 1476; **74**, 1197 (1948).
Tiomno 49	Tiomno, J., and Wheeler, J. A., *Rev. Mod. Phys.*, **21**, 153 (1949).
Tomonaga 40	Tomonaga, S., and Araki, G., *Phys. Rev.*, **58**, 90 (1940).
Tomonaga 41	Tomonaga, S., *Sci. Pap. Inst. Phys. Chem. Res. Japan*, **39**, 247 (1941).
Tomonaga 46	Tomonaga, S., *Prog. Theor. Phys.*, **1**, 83 (1946).
Tomonaga 47	Tomonaga, S., *Prog. Theor. Phys.*, **2**, 6, 63 (1947).
Treiman 54	Treiman, S. B., *Phys. Rev.*, **93**, 544 (1954).
Unsöld 50	Unsöld, A., *Phys. Rev.*, **82**, 857 (1950).
Vallarta 33	Vallarta, M. S., *Phys. Rev.*, **44**, 1 (1933).
Valley 47a	Valley, G. E., *Phys. Rev.*, **71**, 772 (1947a).
Valley 47b	Valley, G. E. *Phys. Rev.*, **72**, 772 (1947b).
Vernov 51	Vernov, S. N., Dobrotin, N. A., and Zatsepin, G. T., *Jour. Exp. Theor. Phys. USSR*, **21**, 1045 (1951).
Vernov 52	Vernov, S. N., Dobrotin, N. A., and Zatsepin, G. T., *Jour. Exp. Theor. Phys. USSR*, **22**, 499 (1952).
Vidale 51a	Vidale, M., and Schein, M., *Nuovo Cim.*, **8**, 774 (1951a).
Vidale 51b	Vidale, M., and Schein, M., *Phys. Rev.*, **84**, 593 (1951b).
Wataghin 34	Wataghin, G., *Z. Physik*, **88**, 92 (1934).
Wataghin 40	Wataghin, G., de Souza Santos, M., and Pompeia, P. A., *Phys. Rev.*, **57**, 61, 339 (1940).
Weizsäcker 34	Weizsäcker, C. F., von, *Z. Physik*, **88**, 612 (1934).
Wheeler 47	Wheeler, J. A., *Phys. Rev.*, **71**, 320, 462 (1947).
Wheeler 49	Wheeler, J. A., *Rev. Mod. Phys.*, **21**, 133 (1949).
Williams 34	Williams, E. J., *Phys. Rev.*, **45**, 729 (1934).
Williams 35	Williams, E. J., *Kgl. Dansk. Vid. Selsk. USSR*, **13**, No. 4 (1935).
Williams 40a	Williams, E. J., and Roberts, G. E., *Nature*, **145**, 102 (1940).

Williams 40b	Williams, E. J., and Evans, G. R., *Nature*, **145**, 818 (1940).
Williams 48	Williams, R. W., *Phys. Rev.*, **74**, 1689 (1948).
Wilson 41	Wilson, A. H., *Proc. Camb. Phil. Soc.*, **37**, 301 (1941).
Wilson 00	Wilson, C. T. R., *Proc. Camb. Phil. Soc.*, **11**, 52 (1900).
Wilson 01	Wilson, C. T. R., *Proc. Roy. Soc.* (London), **A68**, 151; **A69**, 277 (1901).
Wilson 39b	Wilson, J. G., *Proc. Roy. Soc.* (London), **A172**, 517 (1939).
Wilson 38	Wilson, V. C., *Phys. Rev.*, **53**, 337 (1938).
Wilson 39a	Wilson, V. C., *Rev. Mod. Phys.*, **11**, 231 (1939).
Yukawa 35	Yukawa, H., *Proc. Phys. Math. Soc. Japan*, **17**, 48 (1935).
Yukawa 37	Yukawa, H., *Proc. Phys. Math. Soc. Japan*, **19**, 912 (1937).
Yukawa 38a	Yukawa, H., Sakata, S., and Taketani, M., *Proc. Phys. Math. Soc. Japan*, **20**, 319 (1938).
Yukawa 38b	Yukawa, H., et al., *Proc. Phys. Math. Soc. Japan*, **20**, 720 (1938).
Yukawa 39a	Yukawa, H., and Sakata, S., *Proc. Phys. Math. Soc. Japan*, **21**, 138 (1939).
Yukawa 39b	Yukawa, H., and Sakata, S., *Nature*, **143**, 761 (1939).
Yukawa 39c	Yukawa, H., and Okayama, T., *Sci. Pap. Inst. Phys. Chem. Res. Japan*, **36**, 385 (1939c).
Yukawa 49	Yukawa, H., *Rev. Mod. Phys.*, **21**, 474 (1949).
Zaharova 49	Zaharova, V. P., and Eidus, L. H., *Dokl. Akad. Nauk USSR*, **65**, 477 (1949).
Zatsepin 47	Zatsepin, G. T., et al., *Jour. Exp. Theor. Phys. USSR*, **17**, 1125 (1947).
Zatsepin 49	Zatsepin, G. T., *Dokl. Akad. Nauk USSR*, **67**, 993 (1949).

CHAPTER 2

Interactions of High-Energy Particles with Matter

The interactions of high-energy particles with matter are the very subject of high-energy physics. It is not our aim to discuss this subject thoroughly, and we confine ourselves to describing important theoretical and experimental results that are useful for cosmic-ray physics.

2.1 Coulomb Scattering

2.1.1 Angular Distribution

Rutherford scattering can be described by the classical orbit picture. The deflection angle θ^* in the center-of-mass system (CMS) is given as a function of the impact parameter b by†

$$\tan\frac{\theta^*}{2}=\frac{ZZ'e^2}{Mv^{*2}b}, \tag{2.1.1}$$

where Ze and $Z'e$ are the respective charges of colliding particles, and M and v^* are the reduced mass and the relative velocity, respectively. The differential cross section is given by

$$\sigma_R(\theta)^*\, d\Omega^*=2\pi\, b\, db=\frac{Z^2Z'^2e^4}{4(Mv^{*2})^2\sin^4(\theta^*/2)}\, d\Omega^*. \tag{2.1.2}$$

The relativistic extension of the Rutherford formula above is called the Mott formula:

$$\sigma_M(\theta^*)=\frac{Z^2Z'^2e^4}{4p^{*2}v^{*2}}\frac{1-\beta^{*2}\sin^2(\theta^*/2)}{\sin^4(\theta^*/2)}, \tag{2.1.3}$$

where $\beta^*=v^*/c$, c being the velocity of light.

† A quantity in the center-of-mass system is distinguished by an asterisk or the subscript c.

Further correction is needed for large Z and Z' because a strong Coulomb force results in the distortion of the wave, so that the Born approximation, equivalent to the classical approximation, becomes too crude. The scattering cross section thus modified, on account of the Coulomb distortion (McKinley 48, Feshbach 52, Dalitz 51),

$$\sigma_p(\theta^*) = \sigma_R(\theta^*)\left\{1 - \beta^{*2}\sin^2\left(\frac{\theta^*}{2}\right) + \pi\frac{ZZ'e^2}{\hbar c}\beta^*\sin\left(\frac{\theta^*}{2}\right)\left[1 - \sin\left(\frac{\theta^*}{2}\right)\right]\right\}, \tag{2.1.4}$$

is considered as adequate for the Coulomb scattering of two point charges with any energy.

In the scattering of a relativistic electron by a heavy nucleus we can make the following approximation: We can first neglect the mass of the electron, so that the relations in Appendix B (Section B8) for reactions induced by a massless particle are applicable. A correction that is needed for transformation into the laboratory system is the change in solid angle elements. As in (B.17) or (B.49*b*), the laboratory cross section is given by†

$$\sigma(\theta) = \sigma(\theta^*)\frac{p}{p^*} = \sigma(\theta^*)\left[1 + \frac{2E_0}{M}\sin^2\left(\frac{\theta}{2}\right)\right]^{-1}, \tag{2.1.5}$$

where E_0 is the incident energy, and the last relation is obtained by reference to (B.56). By introducing $\sigma_p(\theta^*)$ with $\beta^* = 1$ in (2.1.4) into $\sigma(\theta^*)$ in (2.1.5) we obtain the laboratory cross section for a point nucleus.

Actually the nucleus is of finite size; hence the nuclear-form factor appears in the cross section, which is called the Rosenbluth formula (Rosenbluth 50), as

$$\sigma_F(\theta) = \sigma_p(\theta)\left\{F_1{}^2 + \frac{q^2}{4M^2}\left[2(F_1 + \mu F_2)^2\tan^2\left(\frac{\theta}{2}\right) + \mu^2 F_2{}^2\right]\right\}, \tag{2.1.6}$$

where q is momentum transfer represented for incident momentum p_0 as

$$q = 2p^*\sin\left(\frac{\theta^*}{2}\right) = \frac{2p_0\sin(\theta/2)}{[1 + (2E_0/M)\sin^2(\theta/2)]^{1/2}}; \tag{2.1.7}$$

F_2 is the magnetic-form factor, which represents the distribution of magnetic moment with anomalous part μ (this is important only for the lightest nuclei, such as hydrogen), and F_1 is the charge-form factor and is connected with the charge distribution $\rho(r)$ as

† We often put $c = \hbar = 1$ if no confusion arises.

$$F_1(q) = \frac{4\pi}{q} \int_0^\infty \rho(r) \sin(qr) r\, dr. \tag{2.1.8}$$

This can be expressed in terms of moments as

$$F_1(q) = 1 - \tfrac{1}{6} q^2 \langle r^2 \rangle + \cdots. \tag{2.1.8$'$}$$

This implies that in the case of small momentum transfer we can determine only the rms radius, but not the shape of the charge distribution.

The nuclear-size effect is important at large angles. At small angles the cross section diverges as θ^{-2}, exhibiting a characteristic feature of the Coulomb field. Actually, however, the nuclear Coulomb field is screened by atomic electrons, so that the effective field is represented by

$$V_{sc}(r) = \frac{ZZ'e^2}{r} e^{-r/a}, \tag{2.1.9}$$

where the screening radius a is given by

$$a = \frac{\hbar^2}{me^2} Z^{-1/3} = 0.529 \times 10^{-8} Z^{-1/3} \text{ cm}. \tag{2.1.10}$$

Because of the screening effect, the scattering cross section is obtained as

$$\sigma(\theta) = \sigma_F(\theta)\left(1 + \frac{1}{qa}\right)^{-2}. \tag{2.1.11}$$

The contributions from various effects are compared according to the magnitude of momentum transfer q in the following way:

Screening: $q < \dfrac{1}{a}$ or $\theta < \dfrac{1}{pa} \sim 10^4$ eV-pc^{-1},

Nuclear size: $q \gtrsim \dfrac{1}{\langle r^2 \rangle^{1/2}}$ or $\theta \gtrsim \dfrac{1}{p\langle r^2 \rangle^{1/2}} \sim 10^8$ eV-pc^{-1}

Magnetic force: $q \gtrsim M$ or $\theta \gtrsim \dfrac{M}{p} \sim A\, 10^9$ eV-pc^{-1}.

The screening effect becomes more complicated if the Rutherford radius, $b_R = ZZ'e^2/2p^*v^*$, is as large as the screening radius. Since this is the case only at low energies, we do not discuss its details (Bohr 48), but remark only that the cross section approaches the atomic one.

The nuclear size can be determined by a great number of methods.†

† Papers from the International Congress on Nuclear Sizes and Density Distributions, *Rev. Mod. Phys.*, **30**, 412 (1958).

The results thus obtained seem to be in essential agreement with each other. The nuclear charge, including the charge distribution, is now best measured by electron scattering. If the rms radius of the nuclear charge is accounted for in terms of the uniform distribution, the nuclear charge radius is expressed by

$$R = r_0 A^{1/3}, \qquad r_0 \simeq (1.2 \sim 1.3) \times 10^{-13} \text{ cm}. \tag{2.1.12}$$

The value of r_0 above is valid except for light nuclei, such as lithium and beryllium. The distribution of transferred momenta in electron scattering favors the Fermi or the Woods–Saxon-type distribution, at least for heavy nuclei:

$$\rho(r) = \rho_0[1 + e^{(r-R)/t}]^{-1}, \tag{2.1.13}$$

where R is the radius at which the charge density falls to one-half of the central value and t is the thickness of the tapered part. The parameters are thus obtained as

$$R = r_1 A^{1/3}, \qquad r_1 = (1.08 \pm 0.02) \times 10^{-13} \text{ cm}, \qquad t = (2.2 \sim 2.8) \times 10^{-13} \text{ cm}, \tag{2.1.14}$$

varying slightly from one nucleus to another.

For the proton and the neutron detailed analyses have been made of both the charge and the magnetic-moment distributions, with emphasis on understanding the nucleon structure and its bearing on elementary particle theory. All of these distributions except the charge distribution of the neutron are nearly the same and may be expressed by an exponential function:

$$\rho_p(r) = \rho_0 \exp\left(-\sqrt{12}\,\frac{r}{\langle r^2\rangle^{1/2}}\right), \qquad \langle r^2\rangle^{1/2} = (0.85 \pm 0.06) \times 10^{-13} \text{ cm}. \tag{2.1.15}$$

The charge density of the neutron is nearly zero everywhere.

The magnetic force arising from the normal magnetic moment of the Dirac particle is taken into account in the Mott formula. The Rosenbluth formula given in (2.1.6) takes into account the anomalous magnetic moment, which is μ times the nuclear Bohr magneton. It is well known for the nucleon that

$$\mu_p = 1.79, \qquad \mu_n = -1.91. \tag{2.1.16}$$

The magnetic moments of nuclei are of the same order of magnitude, depending sensitively on nuclear structure. Since the mass of nuclei is very large, their contribution is negligible in almost all practical cases.

We have thus far assumed that the two particles participating in scattering are different. If they are like fermions, the scattering cross section is expressed by

$$\frac{d\sigma(\theta^*)}{d\Omega^*} = \frac{Z^2e^4}{p^{*2}v^{*2}}\frac{1}{4}(1+\beta^{*2})^2\left[\frac{4}{\sin^4\theta^*} - \frac{3}{\sin^2\theta^*} + \frac{p^{*4}/m^4}{(1+\beta^{*2})^2}\left(1+\frac{4}{\sin^2\theta^*}\right)\right] \tag{2.1.17}$$

Since $\sin^2\theta^* = \sin^2(\pi - \theta^*)$, (2.1.17) exhibits a symmetry with respect to two outgoing particles. The above formula applies to electron-electron scattering, provided that the charge distribution of the electron is neglected. A correction due to the finite-size effect is similar to that in (2.1.6). At nonrelativistic energies the scattering of two bosons is of practical interest, such as in α-α scattering. The nonrelativistic formula is expressed by

$$\frac{d\sigma^*}{d\Omega^*} = \frac{Z^2e^4}{16p^{*2}v^{*2}}\left[\left(\sin^4\frac{\theta^*}{2}\right)^{-1} + \left(\cos^4\frac{\theta^*}{2}\right)^{-1} + \left(\sin^2\frac{\theta^*}{2}\cos^2\frac{\theta^*}{2}\right)^{-1}\right]. \tag{2.1.18}$$

For fermions we need to change only the sign of the last term in the square brackets, since this term arises from the interference effect.

2.1.1 Energy Distribution of Knock-on Electrons

Instead of observations of the angular distribution, the energy distribution of recoil particles is a quantity that is often measured. The recoil kinetic energy is uniquely related by (B.13) and (B.54) to the center-of-mass-system angle and the laboratory-system angle, respectively, by†

$$W_r = \gamma_c^2\beta_c^2(1+\cos\theta^*)m = \frac{2\beta_c^2\cos^2\theta}{1-\beta_c^2\cos^2\theta}m, \tag{2.1.19}$$

where m is the mass of a recoil particle initially at rest. This applies to knock-on electrons. The energy and the emission angle of a knock-on electron serve to determine the mass and the energy of an incident particle. The maximum energy transfer is given by

$$W_m = 2\beta_c^2\gamma_c^2 m = \frac{2mp^2}{m^2+M^2+2mE}. \tag{2.1.20}$$

† The total and kinetic energies of a particle are denoted by E and W, respectively.

The following two limiting cases are of practical interest:

$$W_m \simeq p \quad \text{for} \quad m \ll M \quad \text{and} \quad E \gg \frac{M^2}{m}, \tag{2.1.20a}$$

$$W_m \simeq 2m\beta^2\gamma^2 = 2m\left(\frac{p}{M}\right)^2 \quad \text{for} \quad m \ll M \quad \text{and} \quad p \ll \frac{M^2}{m}. \tag{2.1.20b}$$

The cross sections for an incident particle of total energy E_0 and velocity β_0 to produce a recoil electron of kinetic energy W_r are given, depending on the spin of the incident particle, by

$$\sigma(E_0, W_r)\, dW_r = 2\pi Z'^2\left(\frac{e^2}{mc^2}\right)^2 \frac{mc^2}{\beta_0^{\,2}} \frac{dW_r}{W_r^{\,2}}$$

$$\times \begin{cases} \left(1 - \beta_0^{\,2}\dfrac{W_r}{W_m}\right) & \text{for spin 0,} \quad (2.1.21a) \\[2ex] \left[1 - \beta_0^{\,2}\dfrac{W_r}{W_m} + \dfrac{1}{2}\left(\dfrac{W_r}{E_0}\right)^2\right] & \text{for spin } \tfrac{1}{2}, \quad (2.1.21b) \\[2ex] \left[\left(1 - \beta_0^{\,2}\dfrac{W_r}{W_m}\right)\left(1 + \dfrac{1}{3}\dfrac{mW_r}{M^2}\right) + \dfrac{1}{3}\left(\dfrac{W_r}{E_0}\right)^2\left(1 + \dfrac{1}{2}\dfrac{mW_r}{M^2}\right)\right] & \\ & \text{for spin 1.} \quad (2.1.21c) \end{cases}$$

Here an important quantity appears in the expression for the cross section:

$$r_e = \frac{e^2}{mc^2} = 2.82 \times 10^{-13}\ \text{cm}, \tag{2.1.22}$$

which is called the classical electron radius. The frequency of knock-on per g-cm^{-2} is represented by the factor

$$2\pi r_e^{\,2}\frac{NZ}{A}\frac{mc^2}{\beta^2 W_r} = 0.153\,\frac{Z}{A}\frac{mc^2}{\beta^2 W_r}\ \text{g-cm}^{-2}, \tag{2.1.23}$$

where $N = 6.03 \times 10^{23}$ g^{-1} is the Avogadro number.

Since knock-on probability increases rapidly with decreasing recoil energy, the spin-dependent factors in (2.1.21) are not too important in practical use. Furthermore, at high recoil energies the effects of the charge distribution and the anomalous magnetic moment become as important as the spin-dependent effect.

A slight modification of (2.1.21*b*) is necessary for the collision of the positon with the negaton.† In the relativistic limit the cross section is expressed by

$$Q_{e^-e^+}(E, E_r)\, dE_r = 2\pi r_e^2 \frac{mc^2}{E_r^2}\, dE_r \left[1 - \frac{E_r}{E} + \left(\frac{E_r}{E}\right)^2\right]^2, \qquad (2.1.24)$$

in which $E_r \simeq W_r$ in this approximation.

In electron-electron (practically negaton-negaton) scattering the exchange effect is exhibited by the following formula:

$$Q_{e-e}(E, E_r)\, dE_r = 2\pi r_e^2 \frac{mc^2 E^2\, dE_r}{(E - E_r)^2 E_r^2} \left[1 - \frac{E_r}{E} + \left(\frac{E_r}{E}\right)^2\right]^2$$

$$= 2\pi r_e^2 mc^2\, dE_r \left[\frac{E}{E_r(E - E_r)} - \frac{1}{E}\right]^2. \qquad (2.1.25)$$

The expressions (2.1.25) as well as (2.1.24) are, of course, subject to the size effect or the validity of quantum electrodynamics at small distances. Such an effect could be observed with an extremely small cross section.

In most cases recoil electrons of small energy are emitted nearly perpendicular to the incident direction, as can be seen from (2.1.19). Such electrons are called δ-rays. The number of δ-rays per path length is proportional to the square of the charge of the incident particle and to the inverse square of the incident velocity. These features are useful in the identification of an incident particle.

2.1.3 *Multiple Scattering*

A small value of energy transfer corresponds to the small scattering angle. In this case the Mott formula can be approximated by

$$\sigma_M(\theta)\, d(\cos\theta) \simeq \frac{8\pi Z'^2 Z^2 e^4}{p^2 v^2 \theta^3}\, d\theta. \qquad (2.1.26)$$

The probability that a particle is deflected with angles greater than θ after traversing a thickness of x g-cm^{-2} is given by

$$P_s(>\theta, x) = 4\pi \frac{N}{A} \left(\frac{Z'Ze^2}{pv\theta}\right)^2 x. \qquad (2.1.27)$$

The deflection may also be caused by the statistical accumulation of small-angle scatters. The mean-square angle of deflection after traversing

† The ordinary electron (e) and the positron ($\bar{e}$ or e^+) are called the negaton and the positon, respectively. They are collectively called the electron.

x g-cm^{-2} is obtained as

$$\langle\theta^2\rangle = 8\pi \frac{N}{A}\left(\frac{ZZ'e^2}{pv}\right)^2 \ln\left(\frac{\theta_2}{\theta_1}\right)x, \tag{2.1.28}$$

where θ_1 and θ_2 are the lower and upper limits of scattered angles discussed later. If we compare the thicknesses giving rise to a given squared deflection angle due to single- and multiple-scattering processes, the latter is found to be larger by a factor of $2\ln(\theta_2/\theta_1)$. Moreover, the rms angle of deflection due to multiple scattering increases proportional to $x^{1/2}$. Hence the observed deflection is mainly accounted for in terms of *multiple scattering* rather than *single scattering*.

The number of scatterings that result in multiple scattering may be estimated as follows. Single scattering occurs most frequently near the minimum angle θ_1, which is connected to the effective maximum-impact parameter equal to the shielding radius a as

$$\theta_1 = \frac{\hbar}{ap} = \frac{e^2}{\hbar c} Z^{1/3} \frac{mc}{p}. \tag{2.1.29}$$

Hence $(\theta_2/\theta_1)^2$ approximately gives the number of scatterings. By substituting θ_1 into (2.1.27) we obtain the thickness in which a particle is scattered at angles greater than θ_1; $P_s(>\theta_1, x) = 1$ gives us

$$x \simeq \frac{A}{Z^{4/3}}\left(\frac{v}{c}\right)^2 \frac{1}{4\pi r_e^2 Z'^2 N}\left(\frac{e^2}{\hbar c}\right)^2 \simeq \left(\frac{v}{c}\right)^2 10^{-5}\text{ g-cm}^{-2}.$$

This is so small that multiple scattering is important in most cases except for large deflection angles.

The characteristic angle and energy, X_s and E_s, for multiple scattering are given by expressing (2.1.28) as

$$\langle\theta^2\rangle = \left(\frac{E_s}{vp}\right)^2 \frac{x}{X_s}, \tag{2.1.30}$$

where

$$E_s = \left(\frac{4\pi\hbar c}{e^2}\right)^{1/2} mc^2 Z'^2 = 21 Z'^2 \quad \text{Mev},$$

$$\frac{1}{X_s} = 4\frac{e^2}{\hbar c}\frac{NZ^2}{A} r_e^2 \ln\left(\frac{\theta_2}{\theta_1}\right)^{1/2}. \tag{2.1.31}$$

Here the maximum-deflection angle θ_2 corresponds to the effective minimum-impact parameter equal to the nuclear radius, say (2.1.12):

$$\theta_2 = \frac{\hbar}{pr_0 A^{1/3}} = \frac{\hbar}{mcr_0 A^{1/3}}\frac{mc}{p}. \tag{2.1.32}$$

Thus we obtain

$$\left(\frac{\theta_2}{\theta_1}\right)^{1/2}=\left[\frac{\hbar c}{e^2}\left(\frac{1}{ZA}\right)^{1/3}\frac{\hbar}{mcr_0}\right]^{1/2}\simeq 2.1\times 10^2 Z^{-1/3}\left(\frac{Z}{A}\right)^{1/6}. \tag{2.1.33}$$

After this is substituted into (2.1.31), X_s is essentially equal to the radiation length given later if Z^2 is replaced by $Z(Z+1)$.

Adopting the radiation length in place of X_s,

$$\frac{1}{X}=4\frac{e^2}{\hbar c}\frac{NZ(Z+1)}{A}r_e^2\ln(191Z^{-1/3}), \tag{2.1.34}$$

we can roughly express the probability of a particle's undergoing single scattering in a layer dx as

$$[f(\theta)\,d\theta]_{dx}=\frac{dx}{2X}\frac{E_s^2}{v^2p^2}\frac{1}{\ln(191Z^{-1/3})}\frac{d\theta}{\theta^3}. \tag{2.1.35}$$

Multiple scattering caused by successive single scattering is represented by a two-dimensional random-walk process. The probability of finding a particle between θ and $\theta+d\theta$ is therefore expressed by

$$P(\theta)\,d\theta=\frac{2}{\langle\theta^2\rangle}e^{-\theta^2/\langle\theta^2\rangle}\,\theta\,d\theta. \tag{2.1.36}$$

In many cases we observe scattered angles projected onto a plane. The mean-square projected angle $\langle\varphi^2\rangle$ is related to $\langle\theta^2\rangle$ as

$$\langle\varphi^2\rangle=\frac{\langle\theta^2\rangle}{2}. \tag{2.1.37}$$

The distribution of the projected angles is given by

$$P(\varphi)\,d\varphi=\frac{1}{\sqrt{2\pi\langle\varphi^2\rangle}}e^{-\varphi^2/2\langle\varphi^2\rangle}\,d\varphi. \tag{2.1.38}$$

The *Williams formula* (Williams 39, 40) given by (2.1.36) or (2.1.38) is useful for semiquantitative purposes. More accurately, however, the shielding effect (Molière 47, 48) and the nuclear-size effect (Cooper 55) have to be taken into account. The single-scattering law for the projected angle is thus obtained as

$$\begin{aligned} f(\varphi)\,d\varphi &=\frac{x}{8X}\left(\frac{E_s}{vp}\right)^2\frac{1}{\ln(191Z^{-1/3})}\frac{F(\varphi/\varphi_2)}{(\varphi^2+\varphi_1{}^2)^{3/2}}\,d\varphi \\ &=\frac{1}{2}Q\frac{F(\varphi/\varphi_2)}{(\varphi^2+\varphi_1{}^2)^{3/2}}\,d\varphi, \end{aligned} \tag{2.1.39}$$

$$Q=\frac{x}{4X}\left(\frac{E_s}{vp}\right)^2\frac{1}{\ln(191Z^{-1/3})},$$

where

$$\varphi_1 = 1.14 \frac{e^2}{\hbar c} \frac{mc}{p} Z^{1/3} \left[1.13 + 3.76 \left(\frac{Ze^2}{\hbar v}\right)^2\right]^{1/2} \simeq \theta_1, \qquad (2.1.40)$$

$$\varphi_2 = \frac{\hbar}{pr_0} A^{1/3} = \theta_2 .$$

The term $F(\varphi/\varphi_2)$ is the nuclear-form factor and consists of the coherent and the inelastic parts, the latter being evaluated by the closure theorem. Thus the nuclear-form factor is expressed by the coherent part alone as

$$F\left(\frac{\varphi}{\varphi_2}\right) = F_c\left(\frac{\varphi}{\varphi_2}\right) + Z^{-1}\left[1 - F_c\left(\frac{\varphi}{\varphi_2}\right)\right]. \qquad (2.1.41)$$

The coherent form factor $F_c(\varphi/\varphi_2)$ is a rapidly decreasing function for large φ/φ_2. For the uniform charge distribution this is expressed as

$$F_c(y) = \left[\frac{3}{y^3}(\sin y - y \cos y)\right]^2, \qquad y = \frac{\varphi}{\varphi_2}. \qquad (2.1.42)$$

However, this is not realistic, as described above, especially because this gives two sharp diffraction minima. Because of a more realistic charge distribution, the coherent form factor may be given numerically, as in Table 2.1.

Table 2.1 Numerical Values of the Coherent Nuclear-Form Factor

$y = \varphi/\varphi_2$	0	1	2	3	$\gtrsim 4$
$F_c(y)$	1.00	0.82	0.50	0.15	$12/y^4$

The introduction of the form factor makes the derivation of the multiple-scattering formula extremely complicated. The formula is expressed as a function of η, which is proportional to φ:

$$\eta = (2GQ)^{-1/2}\varphi, \qquad (2.1.43)$$

where G is the solution of

$$G = -\frac{1}{2} \ln \left(\frac{\gamma^2 \varphi_1{}^2}{2G\varphi_e}\right), \qquad \gamma = \text{Euler constant}$$

and is given numerically by

$$G \simeq 5.66 + 1.24 \log_{10} \left[\frac{Z^{4/3} A^{-1} x(\text{g/cm}^2)}{1.13\beta^2 + 3.76(Z/137)^2}\right]. \qquad (2.1.44)$$

Since $GQ \simeq \langle\varphi^2\rangle$, η is closely connected with the scattered angle in units of $\langle\varphi^2\rangle^{1/2}$. The multiple-scattering formula is thus expressed as

$$M(\eta)\,d\eta = \frac{1}{\sqrt{\pi}} e^{-\eta^2}\left[1 + \frac{q(\eta_0, \eta)}{4G}\right] + \frac{1}{\sqrt{\pi}}\frac{1}{4G}\int_{\eta_0}^{\infty} \zeta^{-3} F\left(\frac{\zeta}{\eta_2}\right) T(\eta, \zeta)\,d\zeta, \tag{2.1.45}$$

where $\eta_0 \simeq \frac{1}{4}$ is a number corresponding to an angle. The first term on the right-hand side of (2.1.45) is nothing but the Williams formula with

$$\langle\varphi^2\rangle = GQ = \frac{x}{X}\left(\frac{E_s}{vp}\right)^2 \frac{G}{4\ln(191Z^{-1/3})}. \tag{2.1.46}$$

The second term indicates a correction due mainly to the screening effect, in which

$$q(\eta_0, \eta) = 2(2\eta^2 - 1)\left[\ln\left(\frac{\eta_0}{1.26}\right) + \int_0^{2\eta_0\eta} \frac{\cosh\xi - 1}{\xi}\,d\xi\right]$$

$$+ 6\eta^2 - \frac{1}{\eta_0{}^2}(\cosh 2\eta_0\eta) - 1 - \frac{2\eta}{\eta_0}\sinh 2\eta_0\eta, \tag{2.1.47}$$

with

$$\int_0^{2\eta_0\eta} \frac{\cosh t - 1}{t}\,dt \simeq \frac{(2\eta_0\eta)^2}{2\cdot 2} + \frac{(2\eta_0\eta)^4}{4\cdot 4!} + \cdots.$$

The last term in (2.1.45) represents the nuclear-size effect and has to be evaluated numerically by taking account of

$$T(\eta, \zeta) = e^{-(\eta+\zeta)^2} + e^{-(\eta-\zeta)^2} - 2e^{-\eta^2}. \tag{2.1.48}$$

This term is important at large angles and represents the effect of single scattering.

Table 2.2 Values of $q(\eta_0 = \frac{1}{4}, \eta)$

η	$q(\eta_0 = \frac{1}{4}, \eta)$
0	3.238
0.5	1.595
1.0	−3.218
1.5	−11.339
2.0	−22.07
2.5	−34.65
3.0	−47.40
3.5	−57.48
4.0	−60.35

Table 2.3 Values of $T(\eta, \zeta)\zeta^{-3} = [e^{-(\eta+\zeta)^2} + e^{-(\eta-\zeta)^2} - 2e^{-\eta^2}]\zeta^{-3}$

ζ	$\eta=0$	$\eta=1$	$\eta=2$	$\eta=3$	$\eta=4$	$\eta=5$	$\eta=6$	$\eta=7$
			$\times 10^{-1}$	$\times 10^{-2}$	$\times 10^{-5}$	$\times 10^{-8}$	$\times 10^{-12}$	
0.25	−7.755	2.793	10.540	1.9112	3.650	0.8463	0.2504	
0.50	−3.539	1.1875	5.656	1.3507	3.649	1.2620	0.5793	
0.75	−2.040	0.5936	4.113	1.4421	6.079	3.385	2.537	
1.00	−1.2642	0.2826	3.314	1.8069	12.319	11.251	13.887	
	$\times 10^{-1}$	$\times 10^{-2}$	$\times 10^{-1}$	$\times 10^{-2}$	$\times 10^{-4}$	$\times 10^{-6}$	$\times 10^{-9}$	$\times 10^{-13}$
1.25	−8.094	10.751	2.730	2.382	2.659	0.3999	0.08138	0.02241
2.50	−5.301	1.3329	2.199	3.116	5.719	1.4178	0.4756	0.2159
1.75	−3.557	−3.087	1.6845	3.907	11.810	4.827	2.670	1.9982
2.00	−2.454	−4.597	1.2042	4.595	22.89	15.426	14.067	17.360
	$\times 10^{-2}$	$\times 10^{-2}$	$\times 10^{-2}$	$\times 10^{-2}$	$\times 10^{-3}$	$\times 10^{-4}$	$\times 10^{-7}$	$\times 10^{-10}$
2.25	−17.447	−4.619	7.926	5.000	4.106	0.4561	0.6858	0.13953
2.50	−12.775	−4.034	4.7499	4.983	6.746	1.2355	3.062	1.0273
2.75	−9.612	−3.313	2.564	4.516	10.079	3.044	12.438	6.879
3.00	−7.406	−2.657	1.2268	3.703	13.625	6.784	45.71	41.68
	$\times 10^{-2}$	$\times 10^{-2}$	$\times 10^{-3}$	$\times 10^{-3}$	$\times 10^{-2}$	$\times 10^{-3}$	$\times 10^{-5}$	$\times 10^{-7}$
3.25	−5.826	−2.125	5.039	27.36	1.6598	1.3625	1.5136	17.2276
3.50	−4.665	−1.7116	1.6040	18.159	1.8165	2.458	4.503	1.1161
3.75	−3.793	−1.3942	0.19227	10.800	1.7814	3.975	12.003	4.905
4.00	−3.125	−1.1494	−0.28618	5.744	1.5625	5.748	28.62	19.283
	$\times 10^{-2}$	$\times 10^{-3}$	$\times 10^{-4}$	$\times 10^{-4}$	$\times 10^{-3}$	$\times 10^{-3}$	$\times 10^{-4}$	$\times 10^{-5}$
4.25	−2.605	−9.584	−3.947	27.27	12.237	7.423	6.093	0.6769
4.50	−2.195	−8.074	−3.808	11.539	8.547	8.547	11.567	2.118
4.75	−1.8662	−6.865	−3.370	4.341	5.317	8.765	19.558	5.906
5.00	−1.6000	−5.886	−2.921	1.4455	2.943	8.000	29.43	14.65
	$\times 10^{-3}$	$\times 10^{-3}$	$\times 10^{-4}$	$\times 10^{-6}$	$\times 10^{-4}$	$\times 10^{-3}$	$\times 10^{-3}$	$\times 10^{-4}$
5.25	−13.821	−5.085	−2.530	42.04	14.486	6.492	3.924	3.232
5.50	−12.021	−4.422	−2.201	10.120	6.335	4.681	4.681	6.335
5.75	−10.520	−3.870	−1.9268	1.4347	2.460	2.997	4.941	11.026
6.00	−9.259	−3.406	−1.6959	−0.5713	0.8479	1.7031	4.630	17.031
	$\times 10^{-3}$	$\times 10^{-3}$	$\times 10^{-4}$	$\times 10^{-7}$	$\times 10^{-6}$	$\times 10^{-4}$	$\times 10^{-3}$	$\times 10^{-3}$
6.25	−8.192	−3.014	−1.5004	−9.050	25.93	8.586	3.848	2.326
6.50	−7.283	−2.679	−1.3339	−8.813	7.029	3.838	2.836	2.836
6.75	−6.503	−2.392	−1.1911	−8.000	1.689	1.5208	1.8465	3.055
7.00	−5.831	−2.145	−1.0680	−7.193	0.3591	0.5340	1.0726	2.916
	$\times 10^{-3}$	$\times 10^{-3}$	$\times 10^{-5}$		$\times 10^{-8}$	$\times 10^{-6}$	$\times 10^{-4}$	$\times 10^{-3}$
7.25	−5.248	−1.9307	−9.613		6.729	16.610	5.500	2.465
7.50	−4.741	−1.7440	−8.683		1.0809	4.576	2.498	1.8461
7.75	−4.297	−1.5806	−7.869		0.1195	1.1162	1.0048	1.2200
8.00	−3.906	−1.4370	−7.155			0.24103	0.3577	0.7185
	$\times 10^{-3}$	$\times 10^{-3}$	$\times 10^{-5}$			$\times 10^{-9}$	$\times 10^{-6}$	$\times 10^{-5}$
8.25	−3.562	−1.3103	−6.524			46.07	11.273	37.33
8.50	−3.257	−1.1980	−5.965			7.792	3.143	17.162
8.75	−2.985	−1.0983	−5.468			1.1660	0.7756	6.981
9.00	−2.743	−1.0092	−5.025			0.1543	0.1093	2.512

The numerical values of $q(\eta_0, \eta)$ given by Cooper and Rainwater (Cooper 55) are reproduced in Table 2.2. The values of the integral containing $T(\eta, \zeta)\zeta^{-3}$ are also given in Table 2.3 for a number of values of Z, with use of the numerical values of $T(\eta, \zeta)$ given by them.

2.2 Energy Loss by Ionization

2.2.1 *Average Energy Loss*

A charged particle passing through matter loses its energy as a result of the excitation and ionization of atoms. Summing up the contributions of individual atoms, we obtain the *Bethe-Bloch formula* for the rate of energy loss (Bethe 30, 32; Bloch 33):

$$-\frac{dE}{dx}\bigg|_{\text{B-B}} = \frac{NZ}{A}\frac{2\pi Z'^2 e^4}{mv^2}\left[\ln\frac{2mv^2 W\gamma^2}{I^2(Z)} - \beta^2\right]. \tag{2.2.1}$$

This gives the energy loss per g-cm^{-2} in a medium consisting of atoms of atomic number Z, mass number A, and average ionization potential $I(Z)$. It must be remarked that quantities concerning the incident particle are charge $Z'e$ and velocity v, the latter appearing also through $\beta = v/c$ and $\gamma = (1-\beta^2)^{-1/2}$, as well as the upper limit of energy transfer W. The value of W depends on which part of the energy-loss rate one is concerned with; for example, for primary ionization the value of W is the maximum energy transferred to δ-rays that stop within a given distance from the trajectory of an incident particle.

The energy loss due to energy transfer greater than W is obtained from (2.1.21) by integrating energy transferred to knock-on electrons as

$$-\frac{dE}{dx}\bigg|_{>W} = \frac{NZ}{A}\int_W^{W_m} Q(E, W_r)W_r\,dW_r \simeq \frac{NZ}{A}\frac{2\pi Z'^2 e^4}{mv^2}\left(\ln\frac{W_m}{W} - \beta^2\right) \tag{2.2.2}$$

in which small terms are neglected since $W \ll W_m$. By adding (2.2.1) and (2.2.2) we obtain the average energy loss,

$$-\frac{dE}{dx} = \frac{NZ}{A}\frac{2\pi Z'^2 e^4}{mv^2}\left[\ln\frac{2mv^2\gamma^2 W_m}{I^2(z)} - 2\beta^2\right]. \tag{2.2.3}$$

For the incident electron we have to use (2.1.25) for $Q(E, W_r)$ instead of (2.1.21), so that (2.2.2) is modified to

$$-\frac{dE}{dx}\bigg|_{e,>W} = \frac{NZ}{A}\frac{2\pi e^4}{mv^2}\left(\ln\frac{E}{W} + \frac{9}{8} - 2\ln 2\right) \tag{2.2.2'}$$

for $\beta^2 \simeq 1$. Because $W_m = E/2$, we obtain in place of (2.2.3)

$$-\left.\frac{dE}{dx}\right|_e = \frac{NZ}{A}\frac{2\pi e^4}{mv^2}\left[\ln\frac{mv^2\gamma^2 W_m}{I^2(z)} + \frac{9}{8} - \beta^2\right]. \tag{2.2.3'}$$

In the above formula the energy-loss rate is proportional to

$$L \equiv \frac{NZ}{A} 2\pi \left(\frac{e^2}{mc^2}\right)^2 mc^2 = 0.0765\left(\frac{2Z}{A}\right) \text{ MeV/g-cm}^{-2}, \tag{2.2.4}$$

where $2Z/A \simeq 1$ should be noticed. The dependence on incident energy is implied in L/β^2, $\ln(\beta\gamma)^2$, $\ln W_m$, and $-2\beta^2$ or $-\beta^2$. A quantity depending on the medium is represented by

$$B \equiv \ln\frac{mc^2}{I^2}. \tag{2.2.5}$$

We have thus far neglected the fact that the electric field of an incident particle is modified in a dielectric medium. Actually properties of a dielectric medium are represented by the *density effect*. (Fermi 39, 40; Halpern 40, 48). This is taken into account by introducing a term δ in the square brackets in (2.2.3):

$$-\frac{dE}{dx} = L\frac{Z'^2}{\beta^2}\left(B + 0.69 + 2\ln\frac{p}{Mc} + \ln W_m - 2\beta^2 - \delta\right), \tag{2.2.6}$$

where M is the mass of the incident particle, arising from $\beta\gamma = p/Mc$. In (2.2.5), (2.2.6), and hereafter quantities concerning energy, such as mc^2, W_m, and I, are expressed in units of MeV. The corresponding expression for the electron is given by

$$-\left.\frac{dE}{dx}\right|_e = \frac{L}{\beta^2}\left(B + 0.43 + 2\ln\frac{p}{mc} + \ln E - \beta^2 - \delta\right). \tag{2.2.6'}$$

The quantity δ that represents the density effect may be expressed by (Sternheimer 52, 53, 54, 56)

$$\begin{aligned} \delta &= 4.606y + C + a(y_1 - y)^b \quad \text{for} \quad y_0 < y < y_1. \\ \delta &= 4.606y + C \quad \text{for} \quad y > y_1, \end{aligned} \tag{2.2.7}$$

where $y = \log_{10}(p/Mc)$. The first term is nothing but $2\ln(p/Mc)$, which cancels the relativistic increase $2\ln(p/Mc)$ in the Bethe-Bloch formula.

The numerical values of L, B, C, I, a, b, y_0, and y_1 are given by

Sternheimer (Sternheimer 52, 53, 54, 56) for a number of media of practical importance, and are reproduced in Table 2.4. The graphical representations of (2.2.6) and (2.2.6′) are shown in Fig. 2.1 through 2.4.

Table 2.4 Values of Parameters in the Ionization-Energy-Loss Equations (2.2.6) and (2.2.7)

Material	I/I_H	L	B	$-C$	a	b	y_1	y_0
Lithium	2.87	0.0664	19.63	3.07	0.374	3.05	2	−0.05
Beryllium	4.71	0.0681	18.64	2.83	0.413	2.82	2	−0.10
Graphite	5.74	0.0768	18.25	3.22	0.531	2.63	2	−0.05
Magnesium	11.5	0.0758	16.86	4.54	0.0938	3.56	3	0.10
Aluminum	12.0	0.0740	16.77	4.21	0.0906	3.51	3	0.05
Iron	24.8	0.0715	15.32	4.62	0.127	3.29	3	0.10
Copper	27.7	0.0701	15.09	4.74	0.119	3.38	3	0.20
Silver	48.5	0.0669	13.98	5.75	0.251	2.88	3	0.20
Tin	52.1	0.0647	13.83	6.28	0.404	2.52	3	0.20
Tungsten	72.9	0.0618	13.16	6.03	0.0283	3.91	4	0.30
Gold	83.5	0.0615	12.89	6.31	0.0436	3.62	4	0.30
Lead	86.8	0.0608	12.81	6.93	0.0652	3.41	4	0.40
Uranium	97.4	0.0594	12.58	6.69	0.0652	3.37	4	0.30
Anthracene	4.94	0.0810	18.55	3.11	0.420	2.86	2	0.11
Stilbene	4.81	0.0818	18.60	3.12	0.423	2.86	2	0.12
Polystyrene	4.69	0.0826	18.65	3.15	0.429	2.85	2	0.13
Polyethylene	4.04	0.0876	18.95	2.94	0.393	2.86	2	0.12
Lucite	5.08	0.0829	18.49	3.21	0.456	2.78	2	0.14
Toluene	4.58	0.0834	18.70	3.30	0.454	2.83	2	0.17
Xylene	4.50	0.0839	18.73	3.25	0.444	28.4	2	0.16
Water	5.45	0.0853	18.35	3.47	0.519	2.69	2	0.23
Silver chloride	36.1	0.0686	14.57	5.77	0.0177	4.21	4	0.33
Silver bromide	42.2	0.0671	14.25	5.95	0.0235	4.03	4	0.30
Emulsion	27.4	0.0698	15.12	5.55	0.0220	4.01	4	0.23
Lithium iodide	46.8	0.0643	14.05	6.66	0.525	2.32	3	0.08
Sodium iodide	41.3	0.0656	14.30	6.49	0.452	2.44	3	0.18
Hydrogen	1.40	0.1524	21.07	9.50	0.505	4.72	3	1.85
Helium	3.24	0.0767	19.39	11.18	2.13	3.22	3	2.21
Nitrogen	6.69	0.0768	17.94	10.68	0.125	3.72	4	1.86
Oxygen	7.65	0.0768	17.67	10,80	0.130	3.72	4	1.90
Neon	9.56	0.0761	17.23	11.72	0.258	38.18	4	2.14
Argon	16.8	0.0692	16.00	12.27	0.0255	4.36	5	2.02
Krypton	36.3	0.0661	14.56	13.12	0.0771	3.57	5	2.12
Xenon	55.7	0.0632	13.70	13.57	0.150	3.07	5	1.90
Methane	3.27	0.0958	19.37	9.56	0.0552	4.22	4	1.55
Ethylene	4.04	0.0876	18.95	9.52	0.0700	3.94	4	1.54
Acetylene	4.69	0.0826	18.65	9.95	0.0841	3.91	4	1.61
Carbon dioxide	7.08	0.0768	17.82	10.32	0.0865	4.03	4	1.72

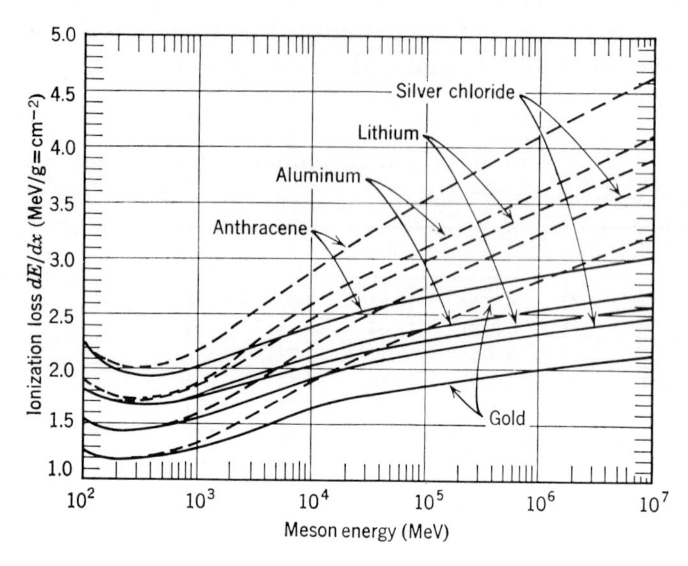

Fig. 2.1 Rates of energy loss by ionization for the muon in solid materials. The broken curves represent the ionization loss without the density effect (Sternheimer 52).

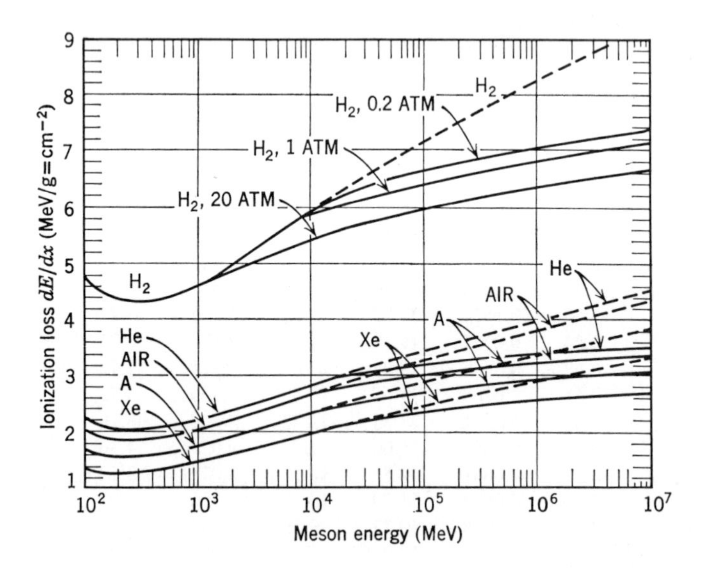

Fig. 2.2 Rates of energy loss by ionization for the muon in gases. The broken curves represent the ionization loss without the density effect (Sternheimer 52).

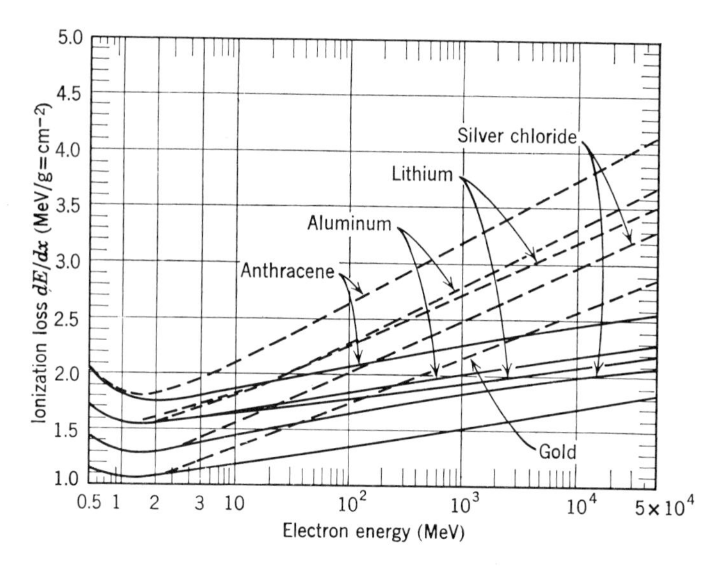

Fig. 2.3 Rates of energy loss by ionization for the electron in solid materials. The broken curves represent the ionization loss without the density effect (Sternheimer 52).

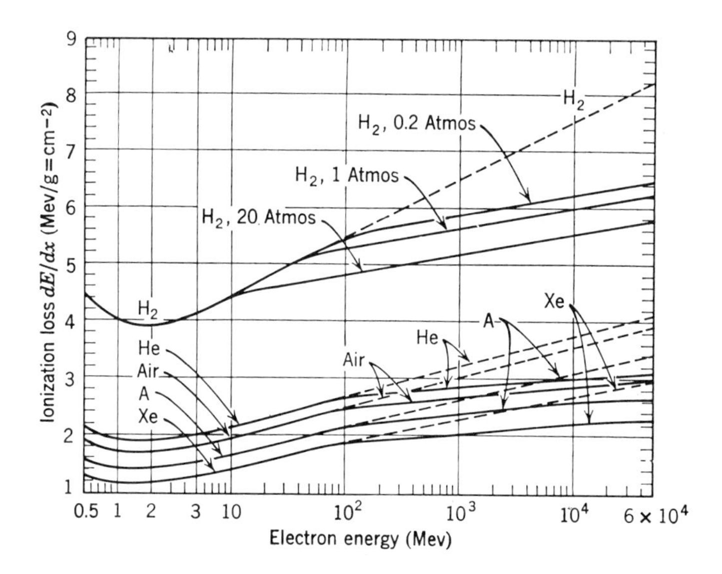

Fig. 2.4 Rates of energy loss by ionization for the electron in gases. The broken curves represent the ionization loss without the density effect (Sternheimer 52).

Equation (2.2.6) includes the energy loss due to *Čerenkov radiation.* Its contribution may be obtained taking account of that the optical transition to the first excited state is most effective. Using the values of the oscillator strength f, the excitation energy E_x, and the half-width of the optical spectrum w for the first excited state, the Čerenkov energy deposited at distances larger than b from a passing particle is approximately expressed as

$$-\left.\frac{dE}{dx}\right|_{>b,\,\check{C}} \simeq \frac{2fL}{3\beta^2} \ln \frac{cE_x}{2\omega_p f^{1/2} wb}, \tag{2.2.8}$$

where $\omega_p = (4\pi e^2 n/m)^{1/2}$ is the plasma frequency for the electron density n. Since oscillator strength decreases with increasing Z, the contribution of Čerenkov loss is negligible in heavy elements, but in hydrogen $(-dE/dx)\,|_{>b,\,\check{C}} \simeq 0.13$ MeV/g-cm^{-2} for $b = 0.1$ cm.

2.2.2 *Fluctuations of Energy Loss*

The energy loss given by (2.2.6) is the so-called average energy loss. Actually, however, the energy loss after passage through a thin absorber fluctuates due to the smallness of the cross section for large energy transfer (Landau 44). The probability that a particle of incident energy E_0 has an energy between E and $E + dE$ after traversing a thickness of x g-cm^{-2}, $w(E_0, E, x)$, obeys the integrodifferential equation

$$\frac{\partial w(E_0, E, x)}{\partial x} = \int_0^\infty [w(E_0, E + E', x) Q(E + E', E') - w(E_0, E, x) Q(E, E')]\, dE'. \tag{2.2.9}$$

The average energy loss is obtained by expanding $w(E_0, E + E', x)$ in a power series of E', neglecting the dependence of Q on the incident energy and taking the first term only. The second term in the power series indicates the spread of energy loss due to fluctuations. Taking the first two terms of the expansion, we can reduce (2.2.9) to a diffusion equation

$$\frac{\partial w(E_0, E, x)}{\partial x} = \frac{\varepsilon_a}{x} \frac{\partial w(E_0, E, x)}{\partial E} + \frac{1}{2} \Delta^2 \frac{\partial^2 w(E_0, E, x)}{\partial E^2}, \tag{2.2.10}$$

where ε_a/x is the average energy loss given by (2.2.6) and

$$\Delta^2 = \int_0^\infty E'^2 Q(E, E')\, dE'. \tag{2.2.11}$$

The solution of (2.2.10) is

$$w(E_0, E, x) = \frac{1}{(2\pi\, \Delta^2 x)^{1/2}} \exp\left[-\frac{(E - E_a)^2}{2\, \Delta^2 x}\right], \tag{2.2.12}$$

where

$$E_a = E_0 - \varepsilon_a .$$

Since this was first obtained by Landau, the fluctuations of energy loss are called the Landau fluctuations.

For a thin absorber the distribution of energy lost is no longer symmetric because the approximation leading to (2.2.10) is not valid (Symon 48, Rossi 52).

Asymmetry of the fluctuations arises if $Lx/\beta^2 W_m$ is small. In this case the most probable energy loss ε_p is different from the average loss ε_a; the former is given by

$$\varepsilon_p = \frac{Lx}{\beta^2}\left(B + 1.06 + 2 \ln \frac{p}{Mc} + \ln (Lx/\beta^2) - \beta^2 - \delta\right). \tag{2.2.13}$$

The distribution of energy lost is expressed in terms of two parameters ν and λ, which are given in Figs. 2.5 and 2.6 as functions of $Lx/\beta^2 W_m$, These parameters are connected with

$$\Delta_0 = \frac{Lx}{\beta^2}\, \nu$$

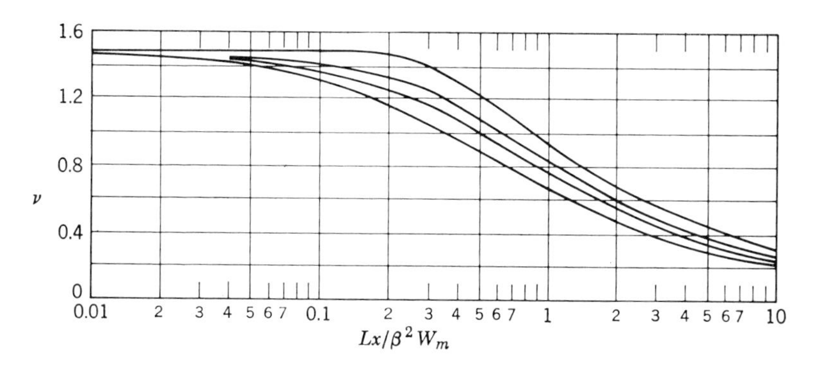

Fig. 2.5 Values of the parameter ν against $Lx/\beta^2 W_m$ in the Landau fluctuations. The four curves, from top to bottom, refer to $\beta^2 = 0$, 0.4, 0.7, and 1 (Rossi 52, Fig. 2.7.1).

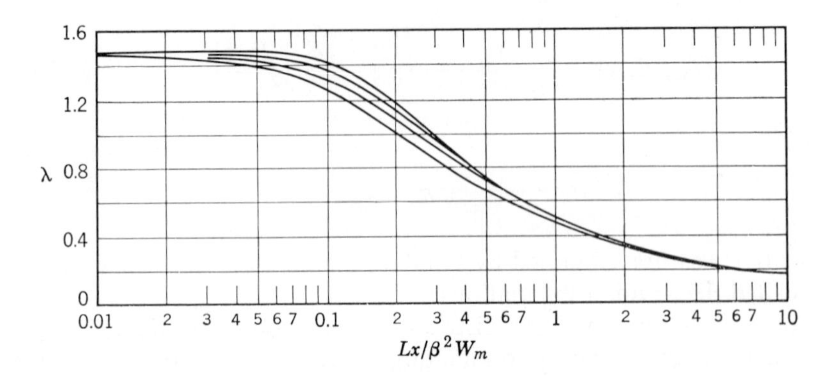

Fig. 2.6 Values of parameter λ against $Lx/\beta^2 W_m$ in the Landau fluctuations. The four curves, from top to bottom, refer to $\beta^2 = 0, 0.4, 0.7$, and 1 (Rossi 52, Fig. 2.7.1).

and F, respectively, the latter being shown in Fig. 2.7. With the aid of these quantities the differential and integral probability distributions of energy lost, ε, are shown in Figs. 2.8*a* and *b*, respectively.

Figure 2.8 indicates that the distribution approaches a Gaussian function for $\lambda \simeq 0$, corresponding to a thick absorber. Landau's solution corresponds to the one with $\lambda \simeq 1.48$, the most asymmetric limit. However, the above result applies neither to a very thin nor to a very thick absorber.

2.2.3 *Energy Loss at Low Energies*

The formulas discussed above apply to particles of high velocities. As the velocity decreases various complicated effects come in the energy-loss mechanism (Bethe 53).

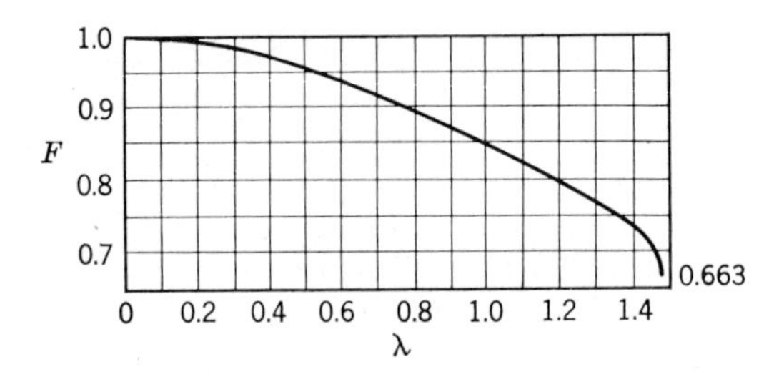

Fig. 2.7 Relation between F and λ in the Landau fluctuations (Rossi 52, Fig. 2.7.2).

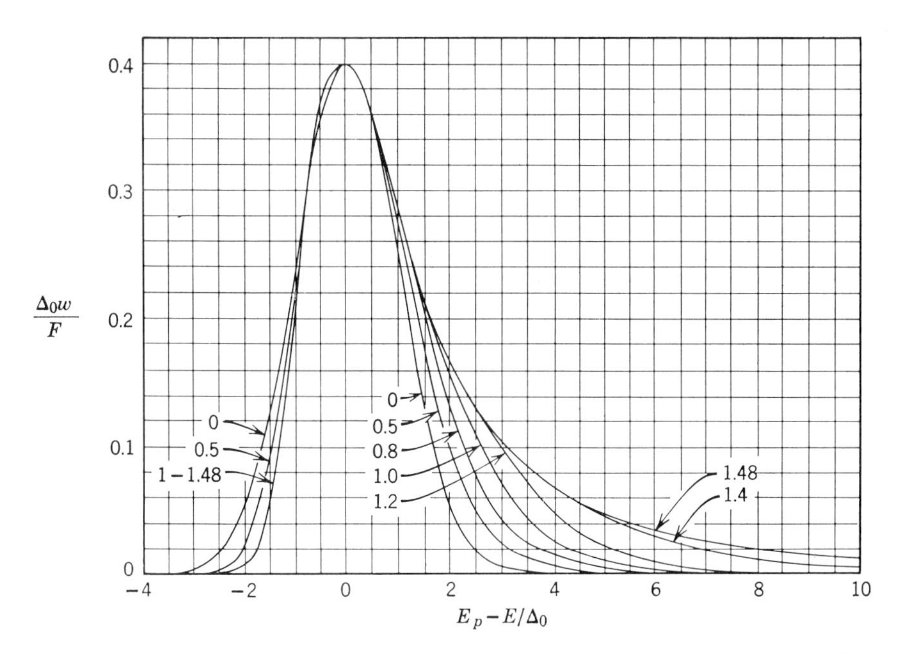

Fig. 2.8*a* Distribution of $\Delta_0 w/F$ as a function of $(E_p - E)/\Delta_0$ in the Landau fluctuations, w. The figures attached to the curves represent the values of λ (Rossi 52, Fig. 2.7.2).

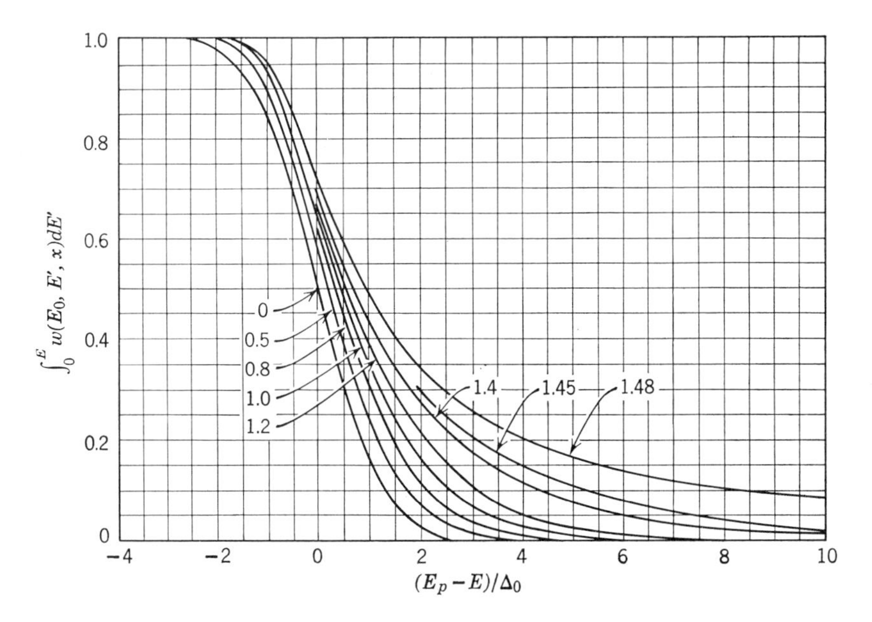

Fig. 2.8*b* Integral distribution of energy loss as a function of $(E_p - E)/\Delta_0$. The figures attached to the curves represent the values of λ (Rossi 52, Fig. 2.7.2).

The Born approximation is no longer valid if $Z'e^2/\hbar v \gtrsim 1$. Then terms in (2.2.3) are replaced by

$$-\ln\gamma - \beta^2 \to \psi(1) - \Re\psi\left(1 + i\frac{Z'e^2}{\hbar v}\right), \tag{2.2.14}$$

where ψ is the logarithmic derivative of the gamma function. For $Z'e^2/\hbar v \ll 1$, (2.2.14) vanishes, whereas for $Z'e^2/\hbar v \gg 1$, because of $\psi(1) = -\ln C = -0.577$ and $\Re\psi(1 + iZ'e^2/\hbar v) \simeq \ln(Z'e^2/\hbar v)$, the energy-loss formula is reduced to

$$-\frac{dE}{dx} = \frac{4\pi Z'^2 e^4}{mv^2}\frac{NZ}{A}\ln\frac{Cmv^2\hbar}{Z'e^2 I}, \tag{2.2.15}$$

which was originally derived by Bohr with a classical method.

When the incident velocity becomes as low as the velocity of a K-electron, the energy transfer to K-electrons becomes difficult. Thus a correction for the ineffectiveness of K-ionization has to be made as

$$Z\ln\frac{2mv^2}{I} \to Z\ln\frac{2mv^2}{I} - C_K(\eta, \eta_K). \tag{2.2.16}$$

The correction term C_K depends on incident energy through

$$\eta = \frac{mv^2}{2Z_{\text{eff}}^2 I_{\text{H}}}, \tag{2.2.17}$$

and on the ionization potential of the K-shell, E_K, through

$$\eta_K = \frac{E_K}{Z_{\text{eff}}^2 I_{\text{H}}}, \tag{2.2.18}$$

where I_{H} is the ionization potential of the hydrogen atom, and $Z_{\text{eff}} \simeq Z - 0.3$ is the effective nuclear charge in the K-shell. The numerical value of the correction term is shown in Fig. 2.9 and is expressed asymptotically as

$$C(\eta, \eta_K) \simeq \frac{C(\eta_K)}{\eta} + \frac{D}{\eta^2} \tag{2.2.19}$$

for large η.

Since K-ionization is separately treated, the average excitation potential, I in (2.2.16), is expressed as

$$\ln I = \left(1 - \frac{1 + f(\eta_K)/2}{Z}\right)\ln I' + \frac{1 + f(\eta_K)/2}{Z}\ln I_K, \tag{2.2.20}$$

where $f(\eta_K)$ is the oscillator strength of all transitions from two K-states

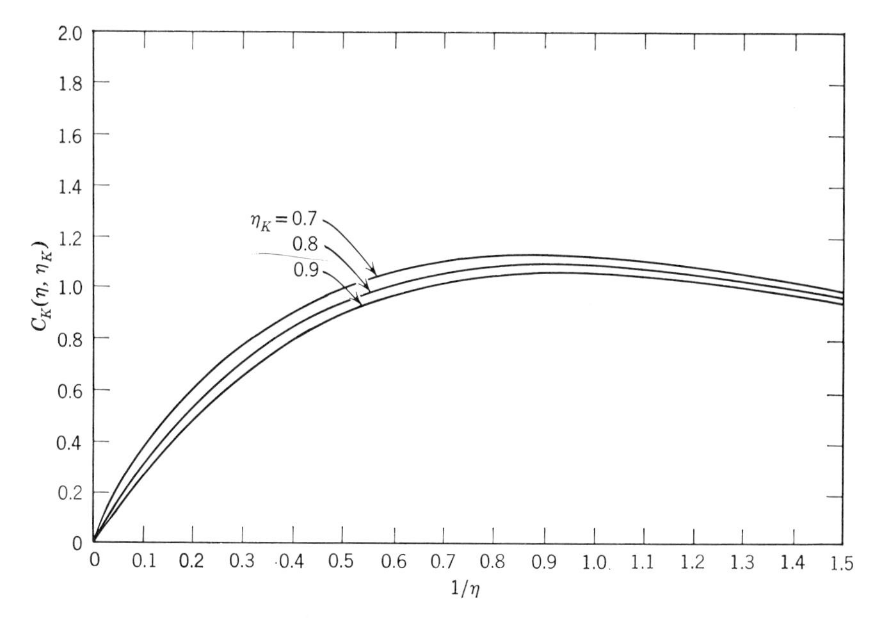

Fig. 2.9 Correction term due to K-ionization, $C_K(\eta, \eta_K)$ defined in (2.2.16) as a function of $1/\eta$, where η is proportional to energy as given in (2.3.17). Three curves refer to $\eta_K = 0.7$, 0.8, and 0.9, respectively (Bethe 53, Fig. 1*b*).

into the continuous state and I_K is the average excitation potential of the K-shell:

$$I_K = \lambda(\eta_K) Z_{\mathrm{eff}}{}^2 I_{\mathrm{H}}, \tag{2.2.21}$$

where $\lambda(\eta_K)$ is a numerical factor of about unity.

If the contribution from the K-electrons is extracted from the logarithmic factor, this is expressed as

$$[1 + \tfrac{1}{2} f(\eta_K)] \ln \frac{2mv^2}{I_K} - C(\eta, \eta_K) \simeq A(\eta_K) \ln \eta + B(\eta_K) - \frac{C(\eta_K)}{\eta} - \frac{D}{\eta^2}, \tag{2.2.22}$$

in which $D \simeq 2$. The numerical values of η_K, f, and λ for a number of media and the values of A, B, C, etc., as functions of η_K are given in Tables 2.5 and 2.6, respectively.

At still lower energies the binding effects of outer shells have to be taken into consideration. The logarithmic factor may be expressed as a sum of contributions from respective shells n,

$$\ln \frac{2mv^2}{I} \rightarrow \sum_n \tfrac{1}{2}(Z_n + f_n) \ln \frac{2mv^2}{I_n}.$$

Table 2.5 Values of the K-ionization Potential, Oscillator Strength, and Average K-Excitation Potential

	Air	Silicon	Calcium	Zinc	Molyb-denum	Neo-dynium	Tungsten	Uranium
Z	7.22	14	20	30	42	60	74	92
η_K	0.665	0.724	0.761	0.797	0.828	0.858	0.869	0.888
f	1.79	1.54	1.41	1.30	1.22	1.14	1.12	1.09
λ	0.967	0.944	0.937	0.929	0.926	0.923	0.922	0.921

Table 2.6 Constants in the K-Stopping-Power Formulas(2.3.19) through (2.3.22)

η_K	A	B	C	$\eta_K{}^2 f$	λ
H atom	2	2.579	1.000	—	1.102
0.7	1.813	2.598	1.067	0.799	0.953
0.75	1.722	2.495	1.100	0.812	0.939
0.8	1.646	2.402	1.120	0.826	0.929
0.9	1.525	2.240	1.131	0.850	0.920
1.0	1.435	2.110	1.119	0.870	0.918

The logarithm above becomes negative for atomic electrons with velocities smaller than $\sqrt{I/2m}$. This means that electrons effective for energy loss are those with velocities smaller than $v = \sqrt{I/2m}$. The number of such electrons is evaluated by the Thomas-Fermi model as $2n_{\text{T-F}}(2v)$, which replaces the logarithmic factor

$$\ln \frac{2mv^2}{I} \rightarrow 2n_{\text{T-F}}(2v) = \frac{4Z^{1/3}\hbar v}{e^2}. \tag{2.2.23}$$

Hence the energy-loss formula is expressed as

$$-\frac{dE}{dx} = \frac{16\eta Z'^2 e^2 \hbar}{mv} \frac{NZ^{1/3}}{A}. \tag{2.2.24}$$

If the incident velocity becomes smaller than the maximum velocity in the Fermi gas of electrons, v_F, only a small fraction of electrons can acquire energy. Electrons responsible for energy loss are only those that are near the Fermi surface, their density being about $n \simeq m^3 v_F{}^2 v$, and the energy transfer is about $\Delta E \simeq m v_F v$. The cross section may be equal to that of Rutherford scattering, $\sigma \simeq Z'^2(e^2/mv_F{}^2)^2$. Hence the energy-loss rate is roughly given by

$$-\frac{dE}{dt} \simeq n\sigma v_{\mathrm{F}}\,\Delta E \simeq Z'^2 \frac{me^4}{2\hbar^2}\frac{mv^2}{\hbar}. \tag{2.2.25}$$

A more elaborate expression given by Fermi and Teller (Fermi 47) is

$$-\frac{dE}{dx} = -\frac{1}{\rho v}\frac{dE}{dt} \simeq \frac{4Z'^2}{3\pi} I_{\mathrm{H}} \frac{mv}{\hbar\rho} \ln \frac{\hbar v_{\mathrm{F}}}{e^2}, \tag{2.2.26}$$

where ρ is the mass density of the medium.

The transition from (2.2.24) to (2.2.26) determines an energy at which energy loss reaches the maximum value. For the mass of an incident particle, M, this occurs roughly at

$$W_{\mathrm{max}} \simeq \frac{1}{2} M v_{\mathrm{F}}^2 \simeq \frac{1}{2} Mc^2 \left(\frac{e^2}{\hbar c}\right)^2 Z^{2/3}, \tag{2.2.27}$$

thus increasing with atomic number. The maximum energy loss may be estimated by putting $v = v_{\mathrm{F}}$ either in (2.2.24) or (2.2.26). As an element becomes heavier, therefore, the maximum energy loss decreases, and the energy at which it occurs shifts toward higher values.

At low energies the charge-transfer process is more important than the ionization process thus far discussed. An approximate cross section for the capture of atomic electrons is given for $v > v_0$ by Bohr (Bohr 48) as

$$\sigma_c = 4\pi a_0^2 Z'^5 Z^{1/3} \left(\frac{v_0}{v}\right)^6, \tag{2.2.28}$$

where $a_0 = \hbar^2/me^2 = 0.53 \times 10^{-8}$ cm is the Bohr radius and $v_0 = e^2/\hbar$.

The atom or ion formed by capturing an electron may lose the electron. For light materials the cross section for electron loss is expressed as

$$\sigma_l = 4\pi a_0^2 Z(Z+1) Z'^{-2} \left(\frac{v_0}{v}\right)^2, \tag{2.2.28$'$}$$

whereas for intermediate Z materials

$$\sigma_l = \pi a_0^2 Z^{2/3} Z'^{-1} \frac{v_0}{v} \tag{2.2.29}$$

because of the screening effect.

Experimentally the capture and loss cross sections are obtained from the equilibrium ratio of fast moving ions and atoms (Allison 53). The cross sections thus obtained for protons in hydrogen and in air are shown in Fig. 2.10. The experimental cross sections are in rough agreement with the theoretical ones.

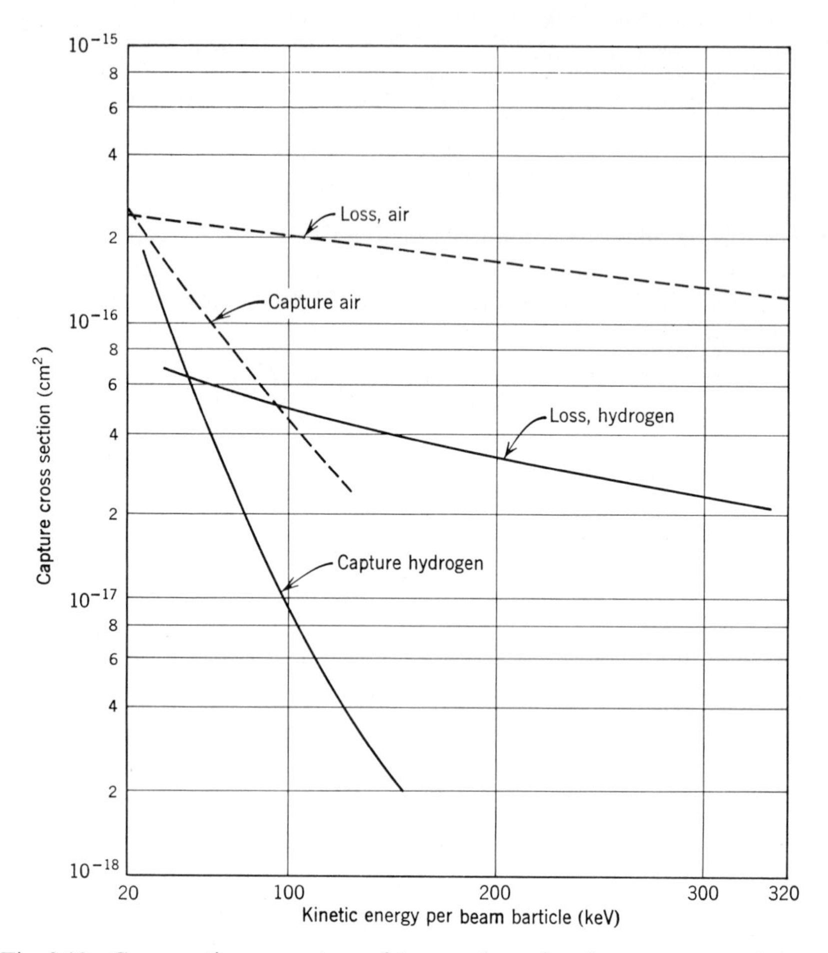

Fig. 2.10 Cross sections per atom of traversed gas for electron capture (σ_c) and loss (σ_l) of electrons by hydrogen ions in motion (Allison 53).

A further correction is found for chemical binding, though it may be as small as 1 percent or smaller. Practically, molecular stopping power may be equated to a simple sum of the stopping powers of the constituent atoms. No appreciable difference is found for stopping powers in different phases of the same material.

The above discussions based mainly on theory may help to clarify the qualitative features of energy loss at low energies. In contrast to the theory of energy loss at high energies, however, theories for low energies are not quantitative enough for accurate use. For quantitative purposes it is advisable to refer to experimental results; for example, those summarized by Allison and Warshaw (Allison 53). The stopping powers of protons with energies between several tens of keV and a few MeV are shown in Fig. 2.11 for a number of solid materials and in Fig. 2.12 for gases. For the sake of convenience, the former is represented in units of keV-cm^2-mg^{-1}, whereas the latter is in units of eV-cm^2-atom^{-1}.

2.2.4 Range-Energy Relation

Quantities concerning the incident particle appearing in the energy-loss formula discussed above are only its charge and velocity, although at extremely high energy its mass comes in the maximum energy transfer. Practically therefore the *average range* of a particle of mass M and charge Z' for a decrease of its energy from E to E' is expressed by

$$\begin{aligned} R(E \to E') &= \int_{E'}^{E} \frac{dE}{-dE/dx} = \frac{M}{Z'^2} \int_{v'}^{v} f(v)\, dv \\ &= \frac{M}{Z'^2} [F(v) - F(v')], \end{aligned} \tag{2.2.30}$$

where v and v' are velocities corresponding to E and E', respectively, and the energy-loss rate is expressed as $-dE/dx = (Z'^2/M)/f(v)$.

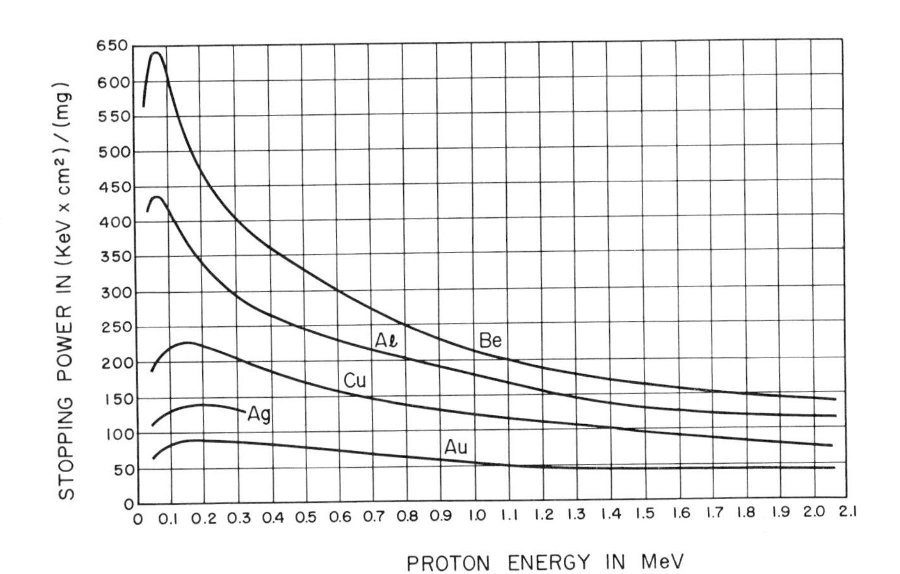

Fig. 2.11 Stopping powers of metals for protons (Allison 53). Ordinate: stopping power in keV-cm^2-mg^{-1}; abscissa: proton energy in MeV.

If the range of a standard particle of mass M_0 and charge Z_0' is known, the range of any particle can be obtained by

$$R_{M,Z'}(E) = \frac{MZ_0'^2}{M_0 Z'^2} R_{M_0,Z'_0}\left(\frac{M_0}{M} E\right). \tag{2.2.31}$$

An approximate range-energy relation can be obtained from the energy-loss formula expressed as

$$-\frac{dE}{dx} = Z'^2 a\left(\frac{c}{v}\right)^2 b(v), \tag{2.2.32}$$

in which $b(v)$ is weakly dependent on velocity v.

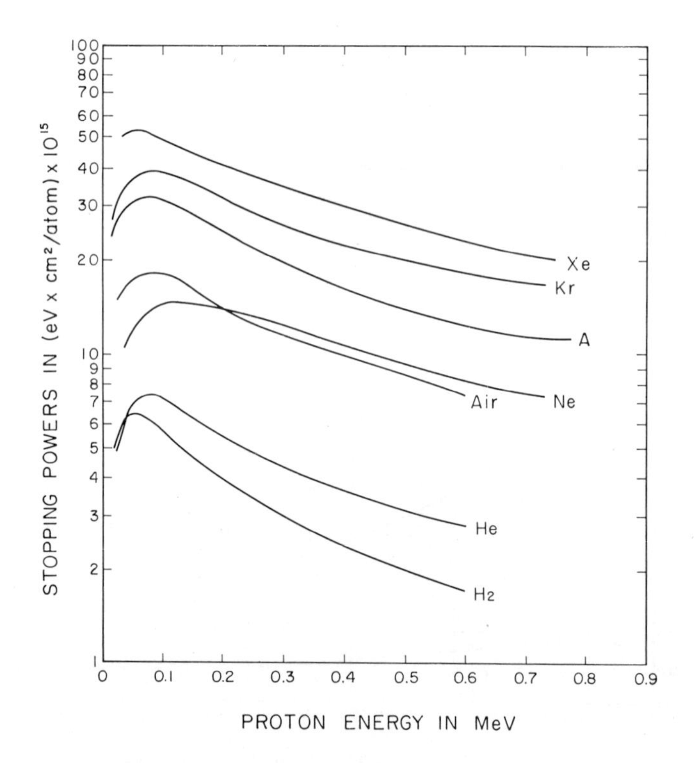

Fig. 2.12 Stopping powers of gases for protons. Ordinate: stopping powers in eV-cm²-atom⁻¹ × 10¹⁵; abscissa: proton energy in MeV.

At extremely relativistic energy the range spent at nonrelativistic energy is negligibly small. By substituting (2.2.32) into (2.2.30) therefore, we obtain

$$R(E) = \frac{1}{Z'^2 a} \int_0^E \frac{dE}{b(v)} \simeq \frac{E}{Z'^2 a}, \tag{2.2.33}$$

where $(c/v)^2 b(v) \simeq 1$ is assumed as constant in deriving the last relation. This is approximately the case for high-energy muons and β-rays, as long as the radiation energy loss is negligible.

At nonrelativistic energy we have

$$R(W) \simeq \frac{W^2}{Z'^2 a M c^2}, \tag{2.2.34}$$

neglecting the velocity dependence of b. However, the energy region in which (2.2.34) is valid is so narrow that $R(W)$ is proportional to a smaller power of energy; for example, the range of a proton between a few MeV and a few hundred MeV is proportional to $W^{1.8}$ and that at lower energy to $W^{1.5}$. The latter energy dependence reveals the screening effect, due to which $b(v) \propto v$, as is shown in (2.2.24). At still lower energy the correction for the capture and loss of electrons is appreciable. Since this effect is different for different incident particles, a small constant term must be added in (2.2.31).

The actual range fluctuates about the average range, as is well known from the Bragg curve. Because the mean-square deviation from the average range, Δx^2, is related to the mean-square deviation from the average energy loss ΔE^2 through

$$\Delta E^2 = \left(\frac{dE}{dx}\right)^2 \Delta x^2,$$

we have

$$\Delta x^2 = \int \frac{d}{dx}(\Delta E^2) \left(\frac{dE}{dx}\right)^{-2} dx = \int \frac{d}{dx}(\Delta E^2) \left(\frac{dE}{dx}\right)^{-3} dE. \tag{2.2.34}$$

The mean-square deviation of energy loss is obtained from (1.2.20*b*) and (1.2.21*a*) as

$$\frac{d}{dx}\Delta E^2 = \frac{NZ}{A} \int_I^{W_m} Q(E, W_r) W_r^2 \, dW_r = \frac{NZ}{A} 4\pi e^2 Z'^2, \tag{2.2.35}$$

which should be substituted into (2.2.34). Neglecting the energy dependence of $b(v)$ in (2.2.32), we have the rms range fluctuations for light stopping materials.

$$\frac{\sqrt{\Delta R^2}}{R} = \frac{4m}{M} \frac{1}{\ln(2mv^2/I)}. \tag{2.2.36a}$$

For heavy stopping materials we have

$$\frac{\sqrt{\Delta R^2}}{R} = \frac{3m}{4M}. \tag{2.2.36b}$$

At high energy a weak energy dependence of $b(v)$ is taken into account, thus giving

$$\frac{\sqrt{\Delta R^2}}{R} = 0.24\left(\frac{Mc^2}{E}\right)^{0.1} \frac{2m}{M^{1/2}} \tag{2.2.36c}$$

However, an experimental result is about twice the above one.

The *range straggling* discussed above implies that actual ranges are distributed about the average range according to the Gaussian law. If we observe a beam of N particles therefore, the number of particles stopping within the average range R is $N/2$. The curve for the tangent of a particle number versus range, drawn at R and $N/2$, intersects the range axis at $R+S$. The quantity

$$S = \sqrt{\frac{\pi\, \Delta R^2}{2}} \tag{2.2.37}$$

is called "straggling."

Near the end of the range the multiple Coulomb scattering becomes so effective that the range observed by varying the thickness of absorbers is different from that along the track. Hence the residual range, r_s, at which the mean-square scattering angle is as large as

$$\langle\theta^2\rangle = \left(\frac{E_s}{2W}\right)^2 \frac{r_s(W)}{X_s} = \frac{\pi^2}{4},$$

may be subtracted from the range R to obtain the observed range R_{obs},

$$R_{\text{obs}} = R - r_s. \tag{2.2.38}$$

Correspondingly, the mean-square deviation from the observed range may be defined as

$$\Delta R_{\text{obs}}^2 = \Delta R_i^2 + r_s^2, \tag{2.2.39}$$

where ΔR_i^2 is the mean-square deviation due to the ionization process given in (2.2.34) or (2.2.36).

At high energy other processes such as radiation and nuclear collisions are significant and most of them are subject to large fluctuations. These processes are important for identifying an incident particle as well as for determining its energy. For electrons the radiation loss process is so important that the ionization loss process alone is of little practical use.

Taking this into account, we show the range-energy relation of particles heavier than the electron in Fig. 2.13. For practical purposes the nomograph reproduced in Fig. 2.14 is very useful.

2.2.5 *Čerenkov and Transition Radiations*

As briefly mentioned in Section 2.2.1, the ionization process may be associated with Čerenkov radiation. This takes place in frequency regions in which there holds the relation

$$\varepsilon_1(\omega)\beta^2 > 1, \tag{2.2.40}$$

where $\varepsilon_1(\omega)$ is the real part of the dielectric constant. If the imaginary part thereof is neglected, $\sqrt{\varepsilon_1(\omega)}$ is the refractive index of a medim, and the wave number of a photon is given by $k = (\omega/c)\sqrt{\varepsilon_1(\omega)}$. When the phase velocity of the photon, $\omega/k = c/\sqrt{\varepsilon_1}$ is smaller than the particle velocity $c\beta$, condition (2.2.40) is satisfied and a free photon is emitted. The energy-momentum conservation results in the photons being emitted at angle θ, which is given for a particle of momentum p by

$$\cos\theta = \frac{1}{\beta\sqrt{\varepsilon_1}} + \frac{\hbar\omega\sqrt{\varepsilon_1}}{2cp(1-\varepsilon_1^{-1})}; \tag{2.2.41}$$

the last term represents a quantum effect and is negligible in most practical cases.

The number of photons emitted per unit length in a frequency range $d\omega$ is expressed by

$$\frac{d^2n(\omega)}{dL\,d\omega} = \frac{(Z'e)^2}{\hbar v^2}\frac{\varepsilon_1(\omega)}{|\varepsilon(\omega)|^2}[\varepsilon_1(\omega)\beta^2 - 1] = \frac{(Z'e)^2}{\hbar c^2}\sin^2\theta. \tag{2.2.42}$$

The Čerenkov light emitted in dielectrics usually lies in an optical region, whereas that emitted in magnetized plasma lies below the electron cyclotron frequency.

In a heterogeneous medium the radiation of photons is possible even if (2.2.40) does not hold. The simplest case is the transition radiation predicted by Ginzburg and Frank (Ginzburg 46), which arises when a

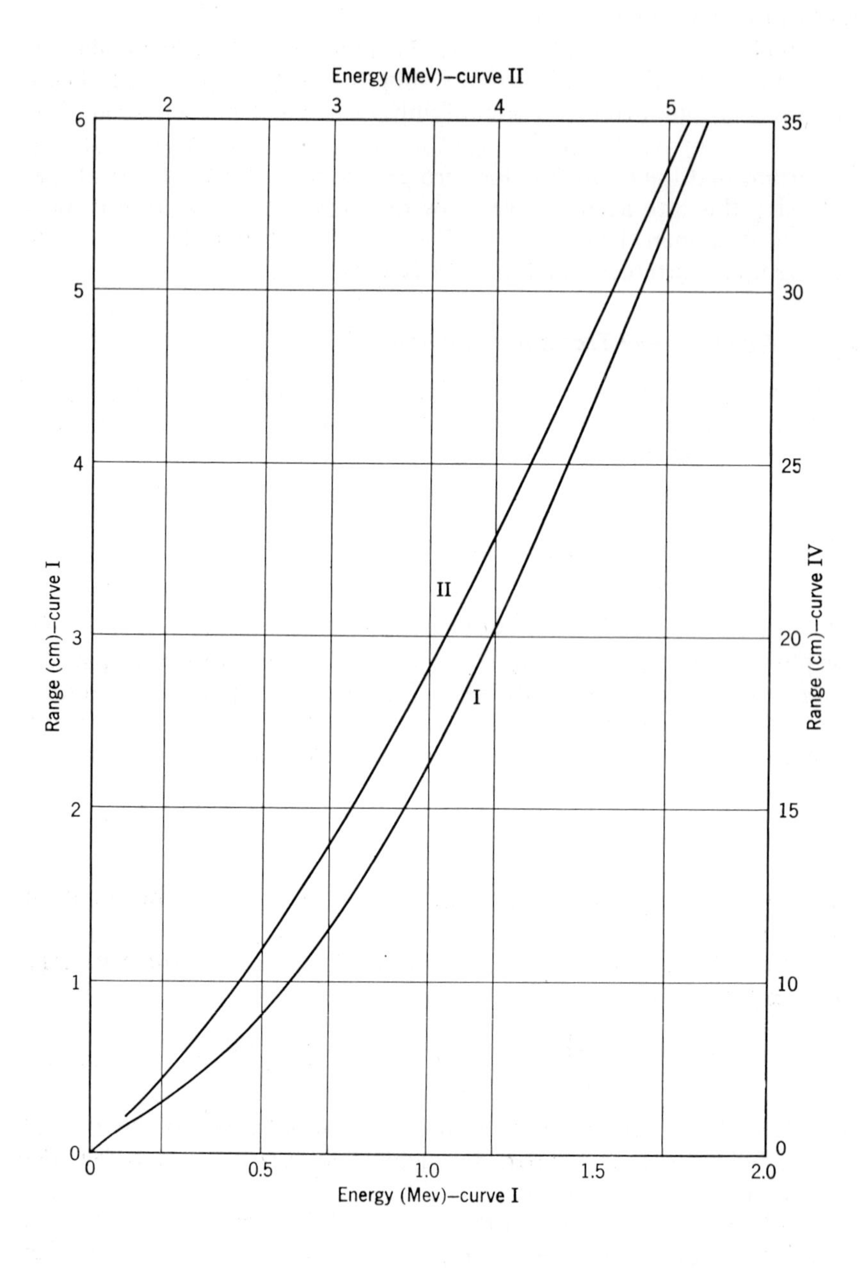
Energy (MeV)—curve II
2
3
4
5
6
5
4
3
2
1
0
35
30
25
20
15
10
0
Range (cm)—curve I
Range (cm)—curve IV
II
I
0
0.5
1.0
1.5
2.0
Energy (Mev)—curve I

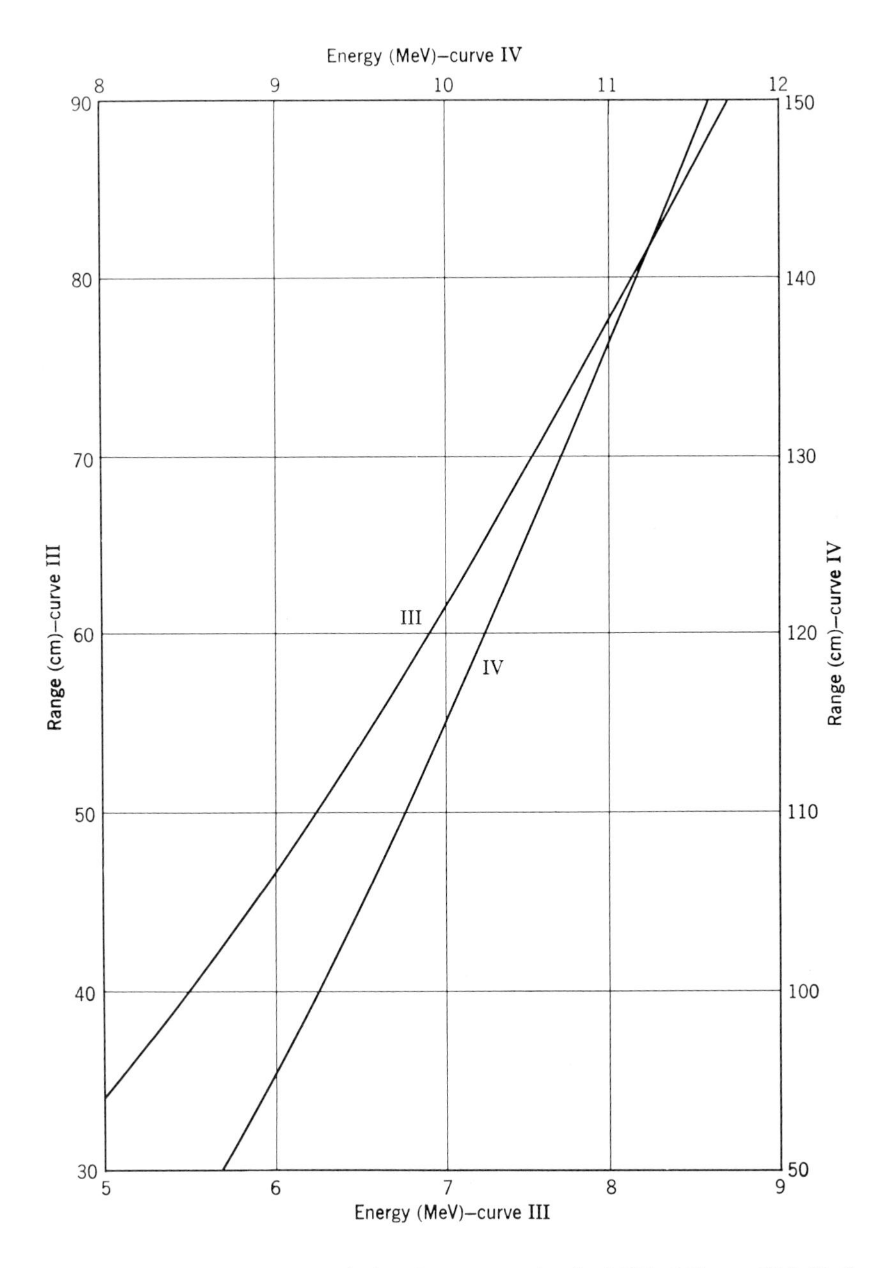

Fig. 2.13*a*–1 Range-energy relation for protons in air (15°C, 760 mm Hg) (Bethe 53, Fig. 3*a*, *b*).

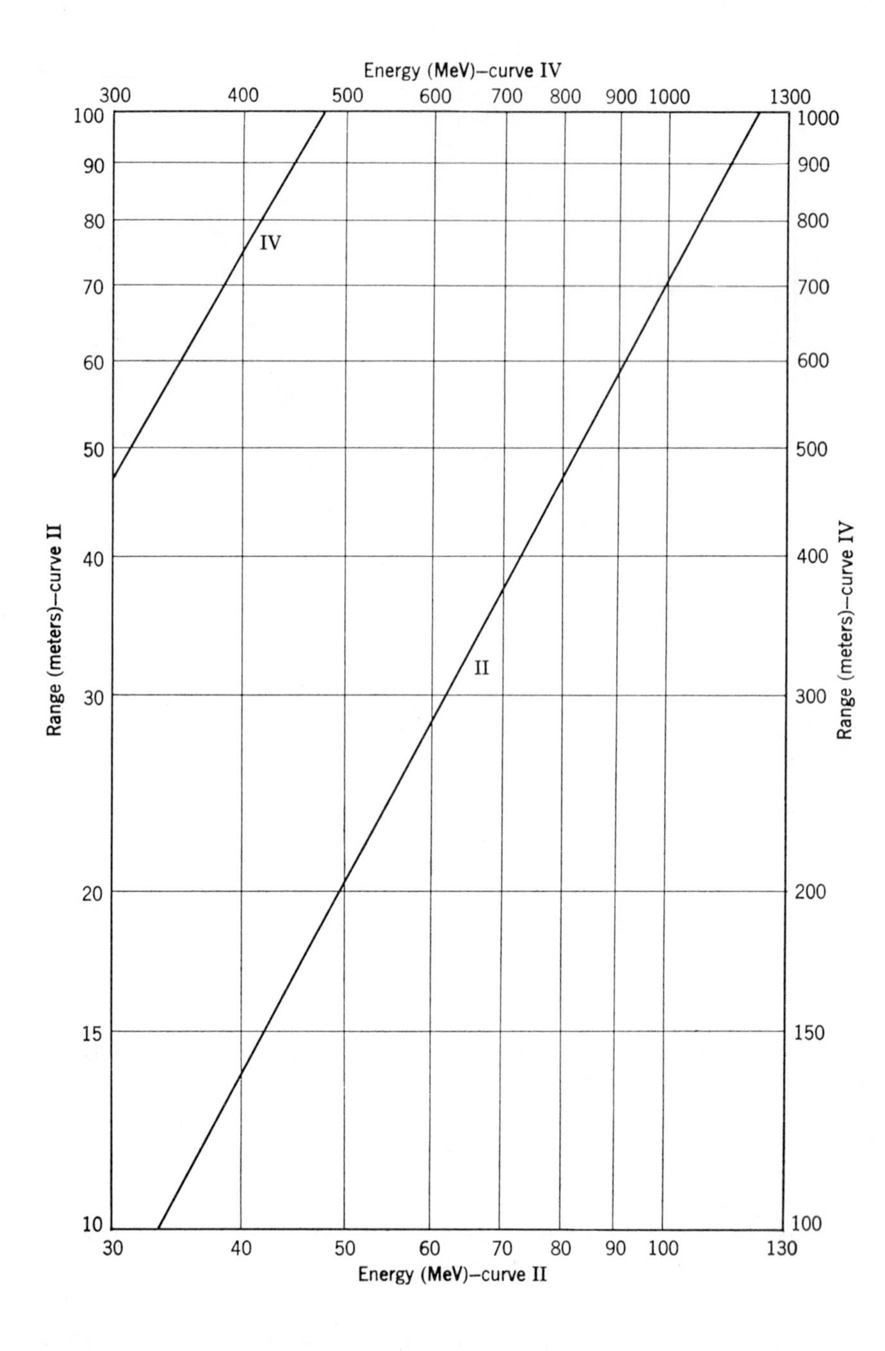

Energy (MeV)–curve IV
300
400
500
600
700
800
900
1000
1300
100
90
80
70
60
50
40
30
20
15
10
1000
900
800
700
600
500
400
300
200
150
100
IV
II
Range (meters)–curve II
Range (meters)–curve IV
30
40
50
60
70
80
90
100
130
Energy (MeV)–curve II

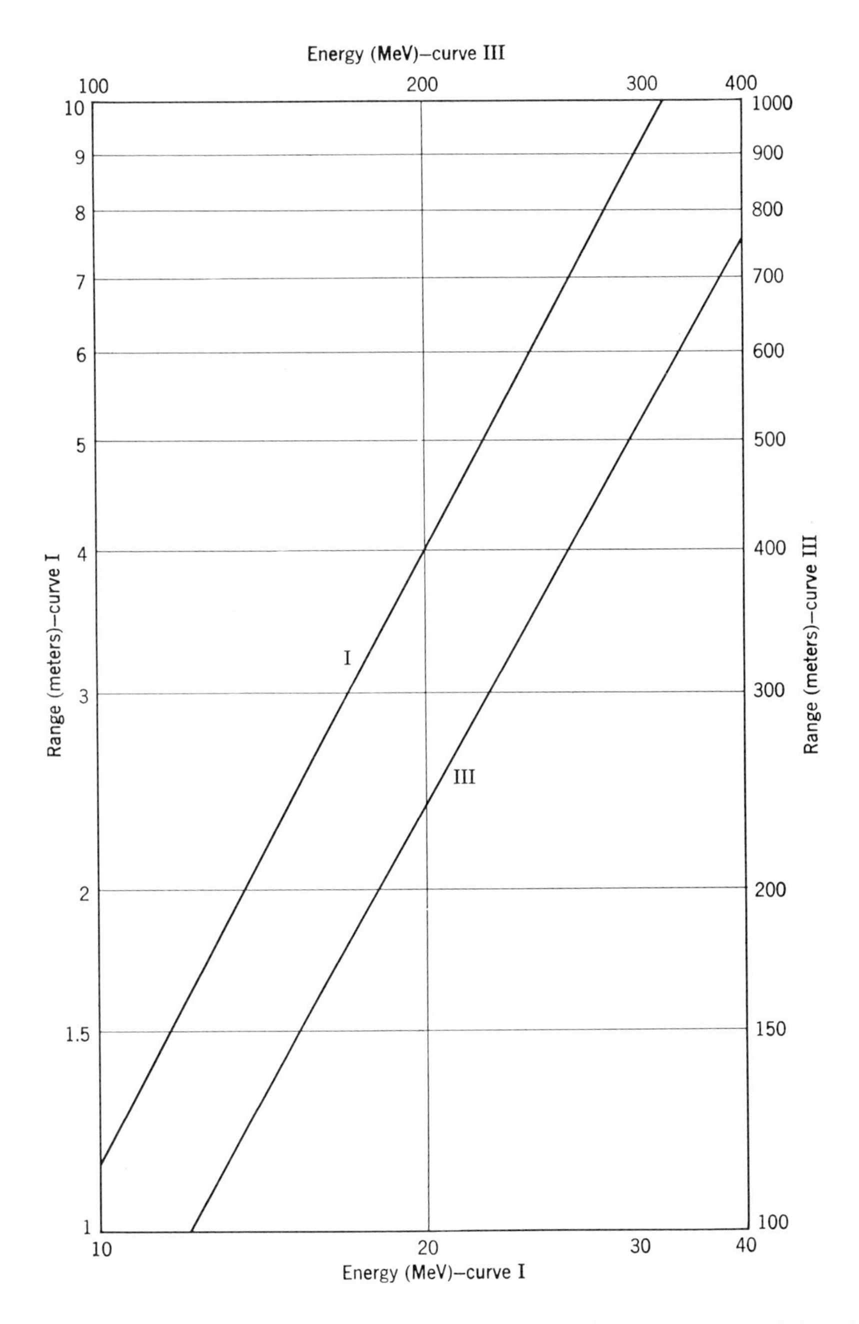

Fig. 2.13*a*–2 Range-energy relation for protons in air (15°C, 760 mm Hg) (Beeth 53, Fig. 3*c*, *d*).

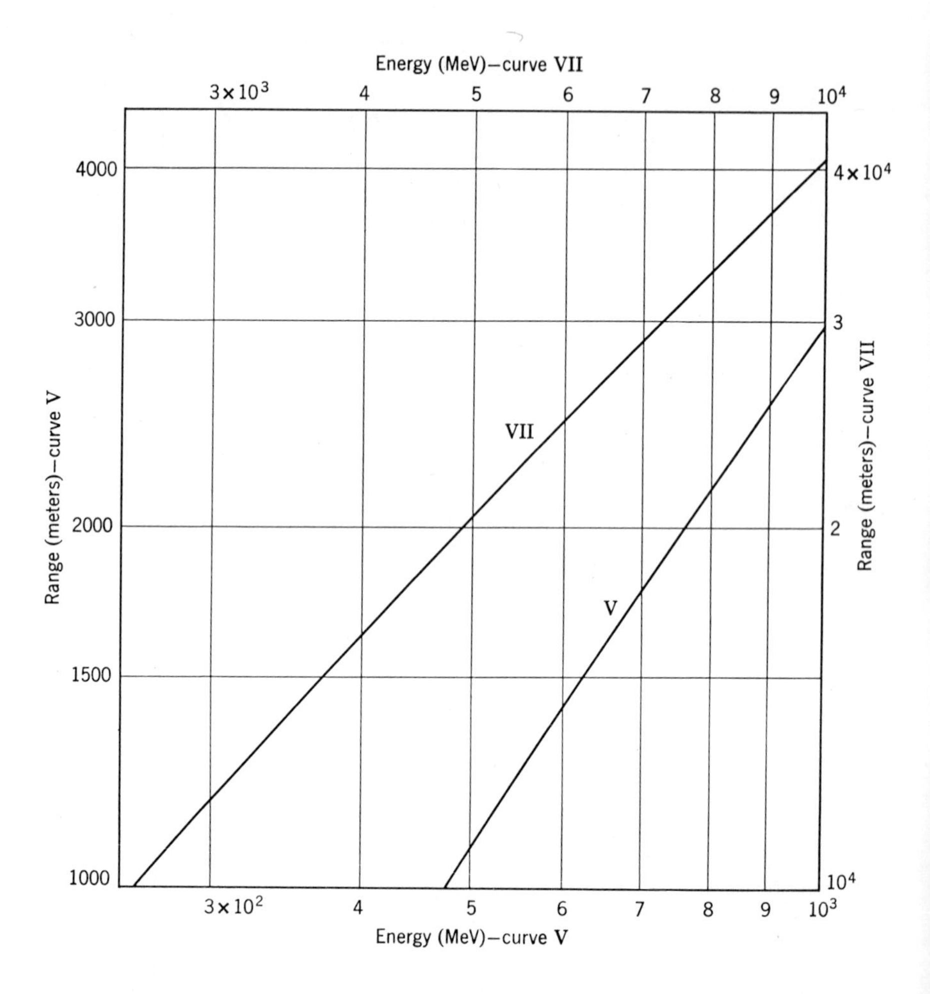

Energy (MeV)—curve VII
3×10³
4
5
6
7
8
9
10⁴
4000
3000
2000
1500
1000
4×10⁴
3
2
10⁴
Range (meters)—curve V
Range (meters)—curve VII
VII
V
3×10²
4
5
6
7
8
9
10³
Energy (MeV)—curve V

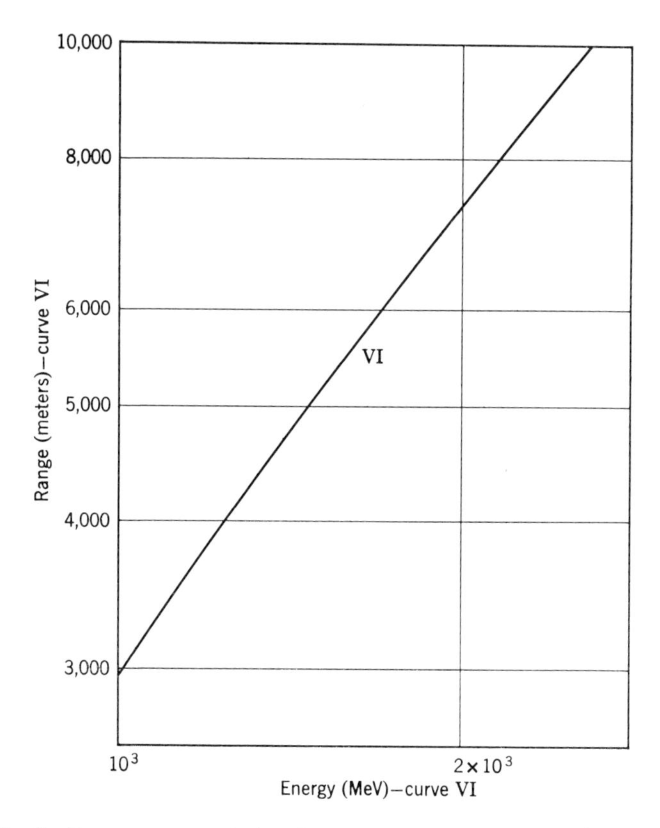

Fig. 2.13*a*–3 Range-energy relation for protons in air (15°C, 760 mm Hg) (Bethe 53, Fig. 3*e*, *f*).

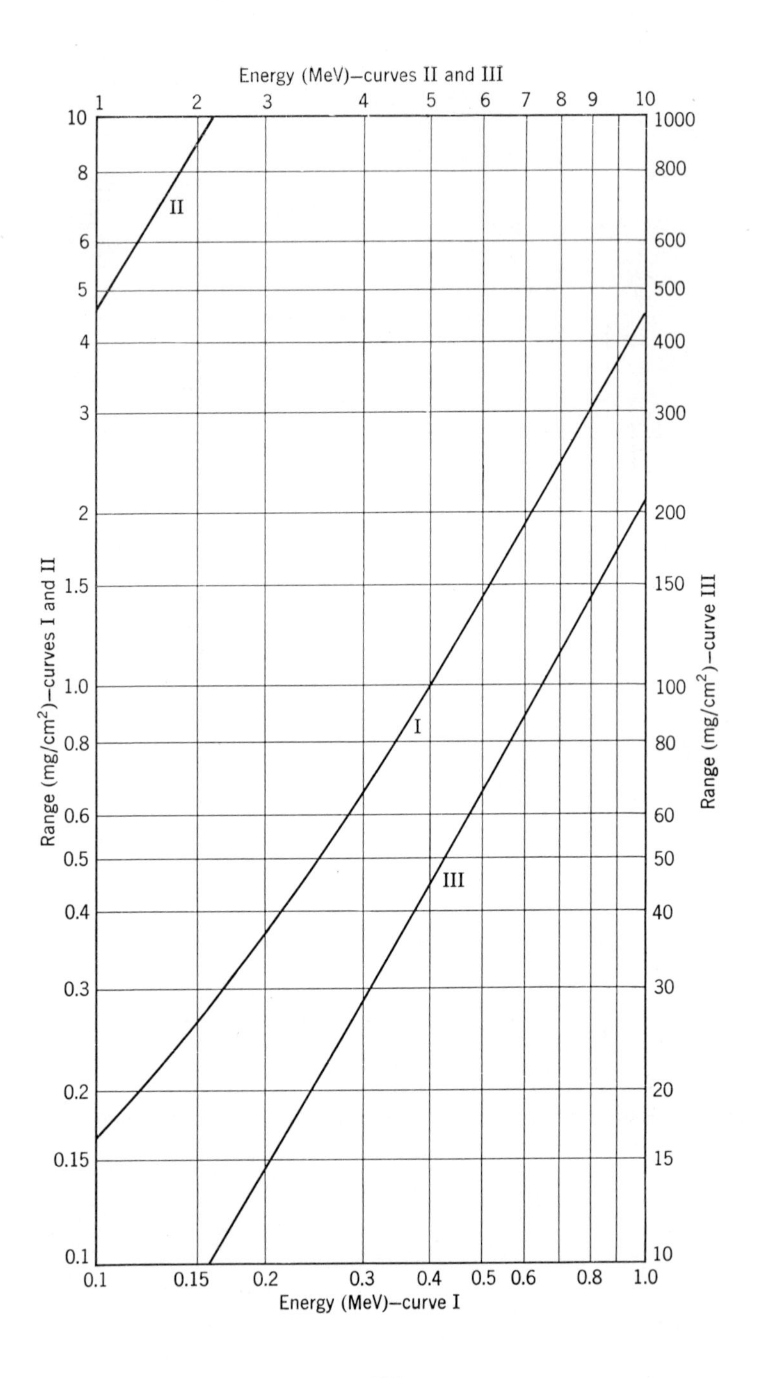

Energy (MeV)—curves II and III
1
2
3
4
5
6
7
8
9
10
10
8
6
5
4
3
2
1.5
1.0
0.8
0.6
0.5
0.4
0.3
0.2
0.15
0.1
1000
800
600
500
400
300
200
150
100
80
60
50
40
30
20
15
10
II
I
III
Range (mg/cm²)—curves I and II
Range (mg/cm²)—curve III
0.1
0.15
0.2
0.3
0.4
0.5
0.6
0.8
1.0
Energy (MeV)—curve I

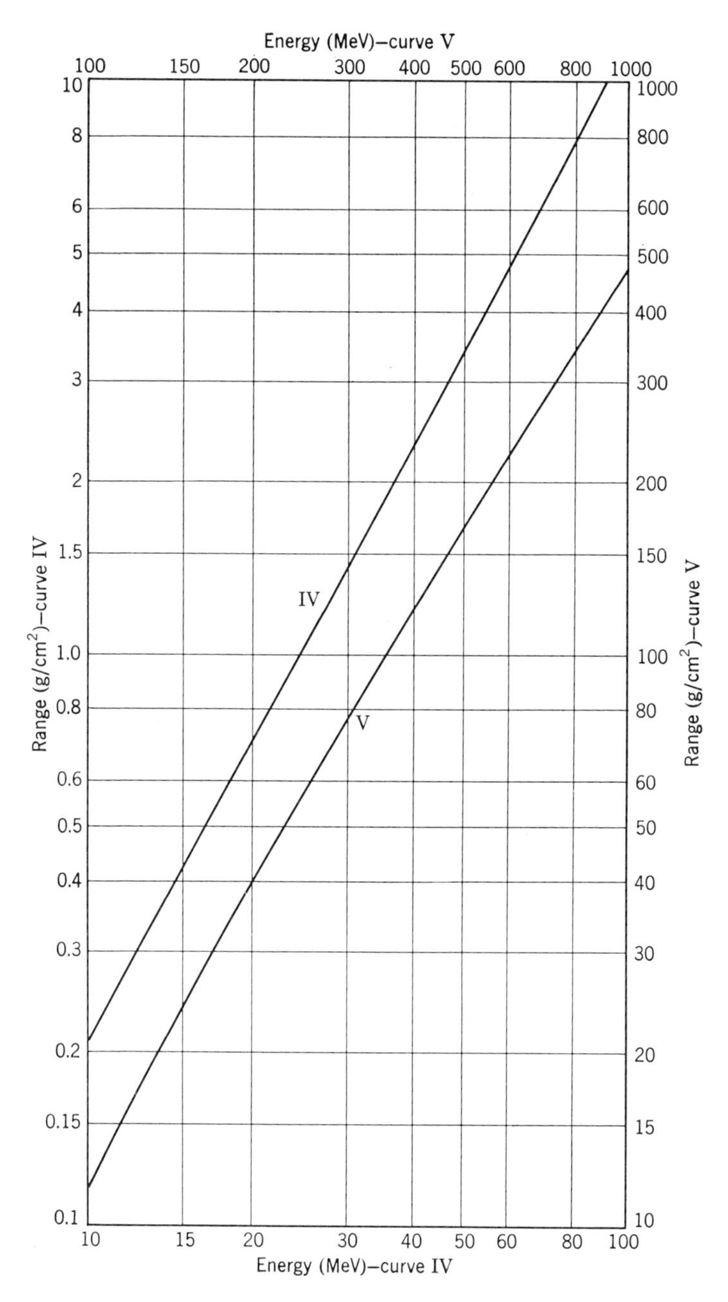

Fig. 2.13*b* Range-energy relation for protons in copper (Bethe 53, Fig. 8*a*).

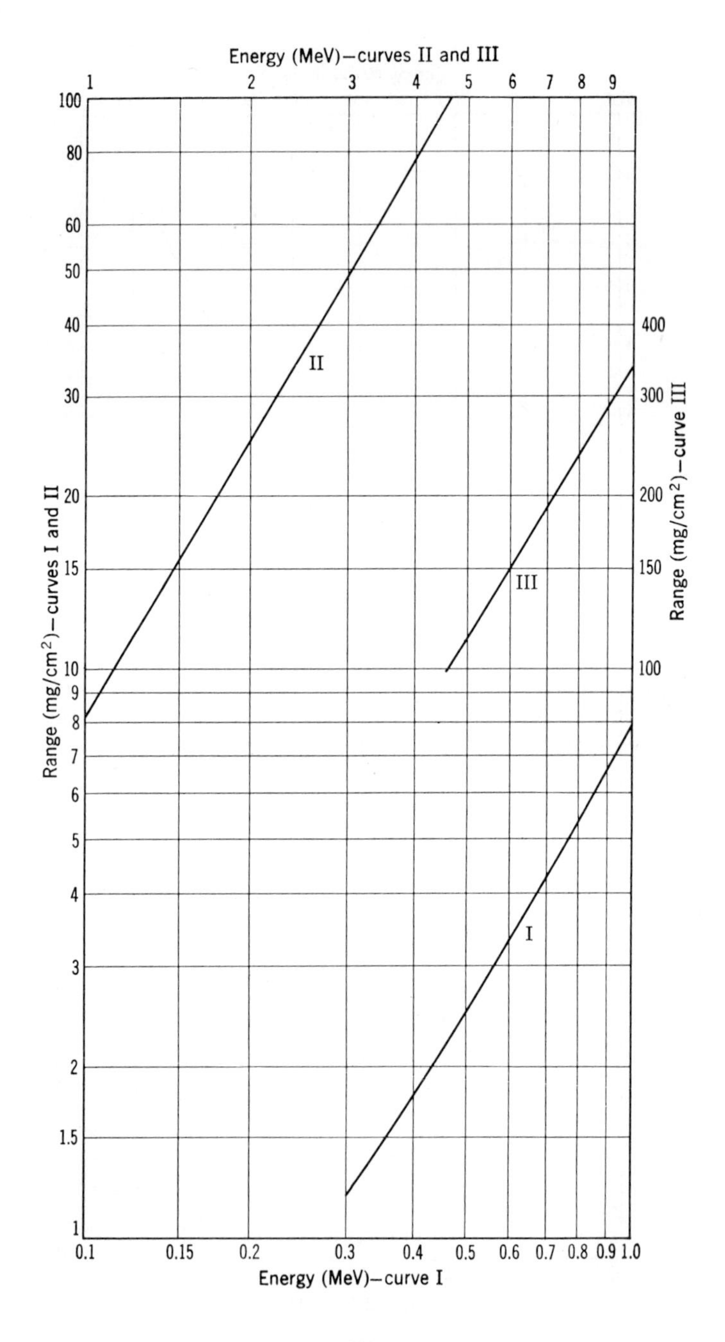
Energy (MeV)—curves II and III
1
2
3
4
5
6
7
8
9
100
80
60
50
40
30
20
15
10
9
8
7
6
5
4
3
2
1.5
1
Range (mg/cm²)—curves I and II
400
300
200
150
100
Range (mg/cm²)—curve III
II
III
I
0.1
0.15
0.2
0.3
0.4
0.5
0.6
0.7
0.8
0.9
1.0
Energy (MeV)—curve I

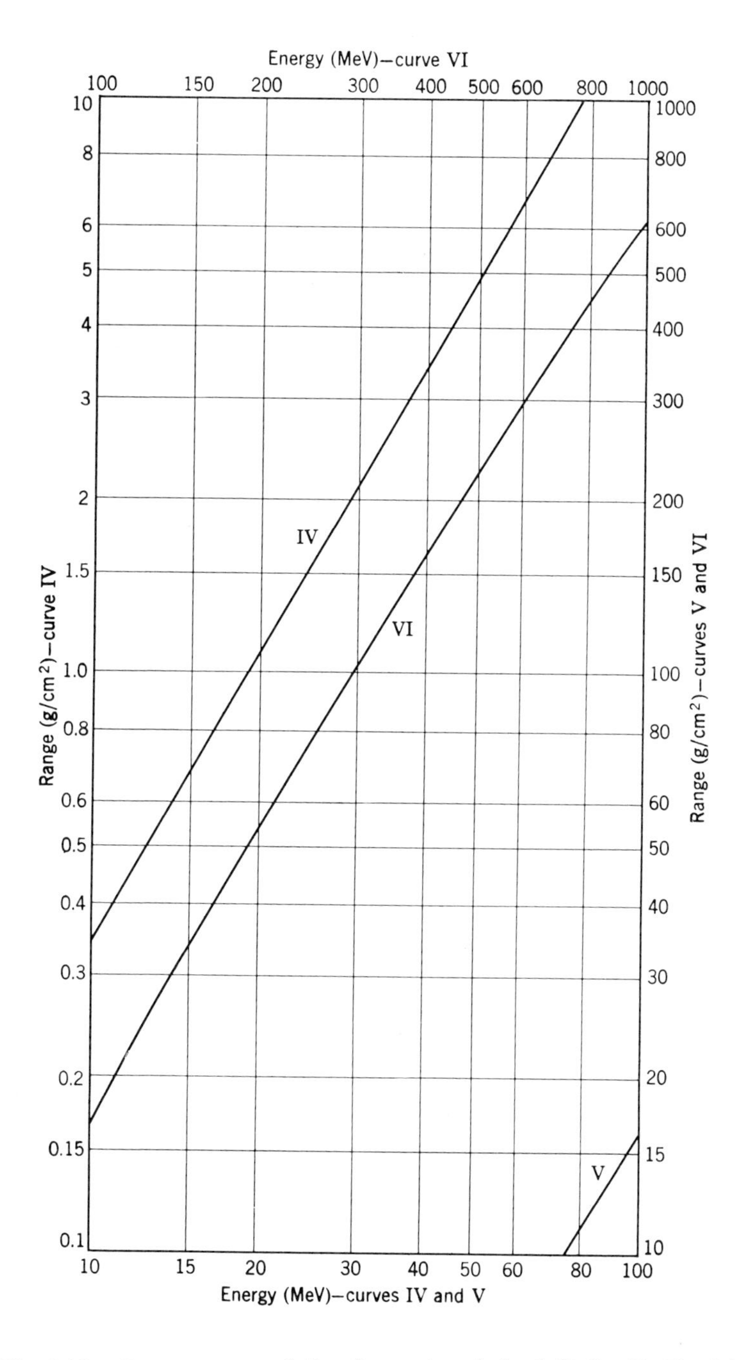

Fig. 2.13*c* Range-energy relation for protons in lead (Bethe 53, Fig. 8*c*).

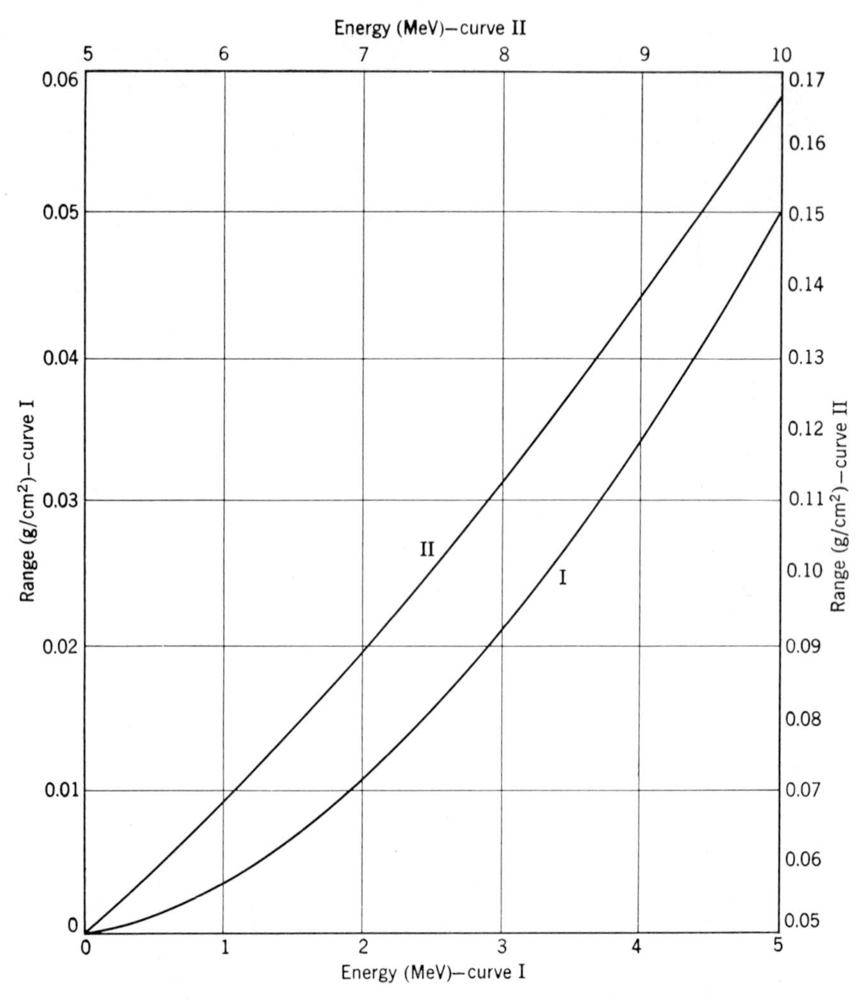

Fig. 2.13*d*–1 Range-energy relations for protons in aluminum (Bethe 53, Fig. 9*a*).

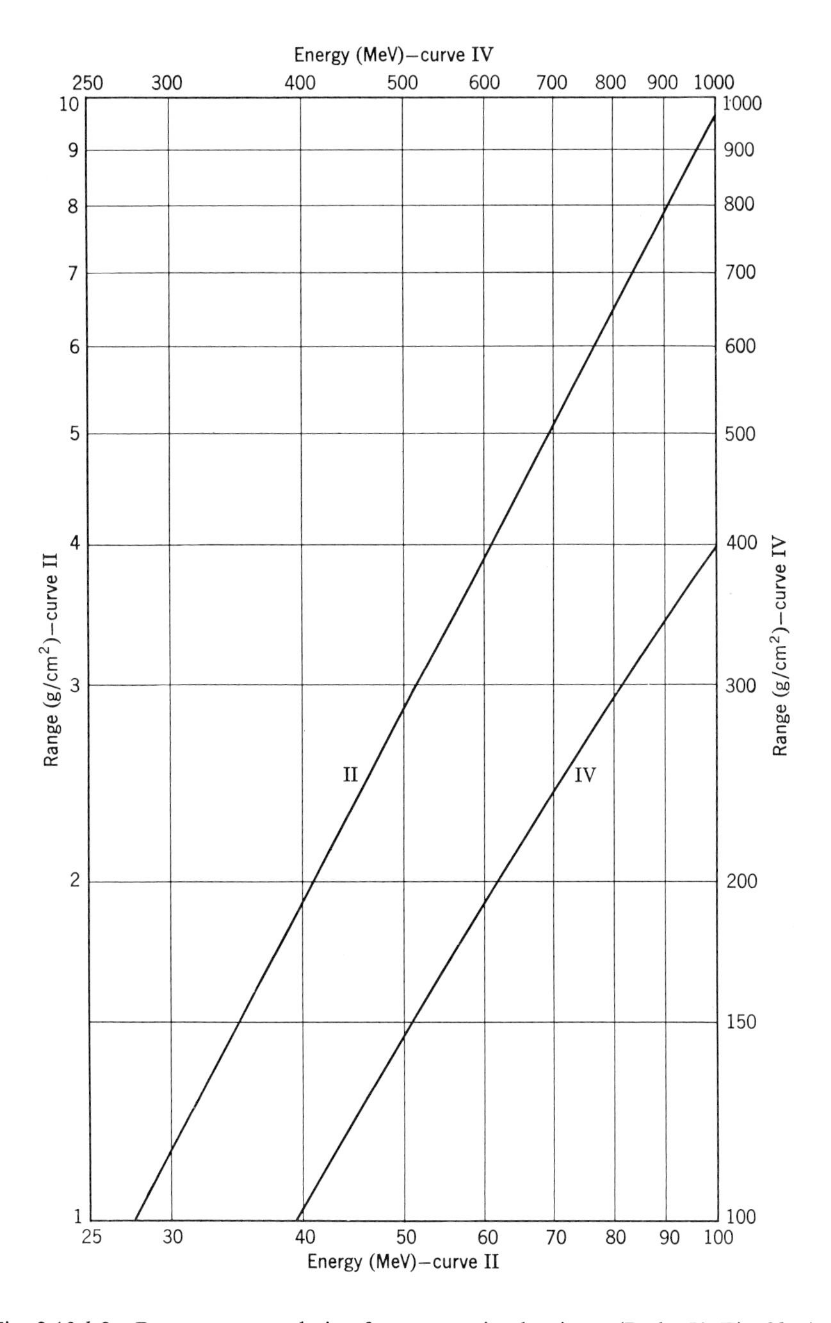

Fig. 2.13*d*–2 Range-energy relation for protons in aluminum (Bethe 53, Fig. 9*b*, *c*).

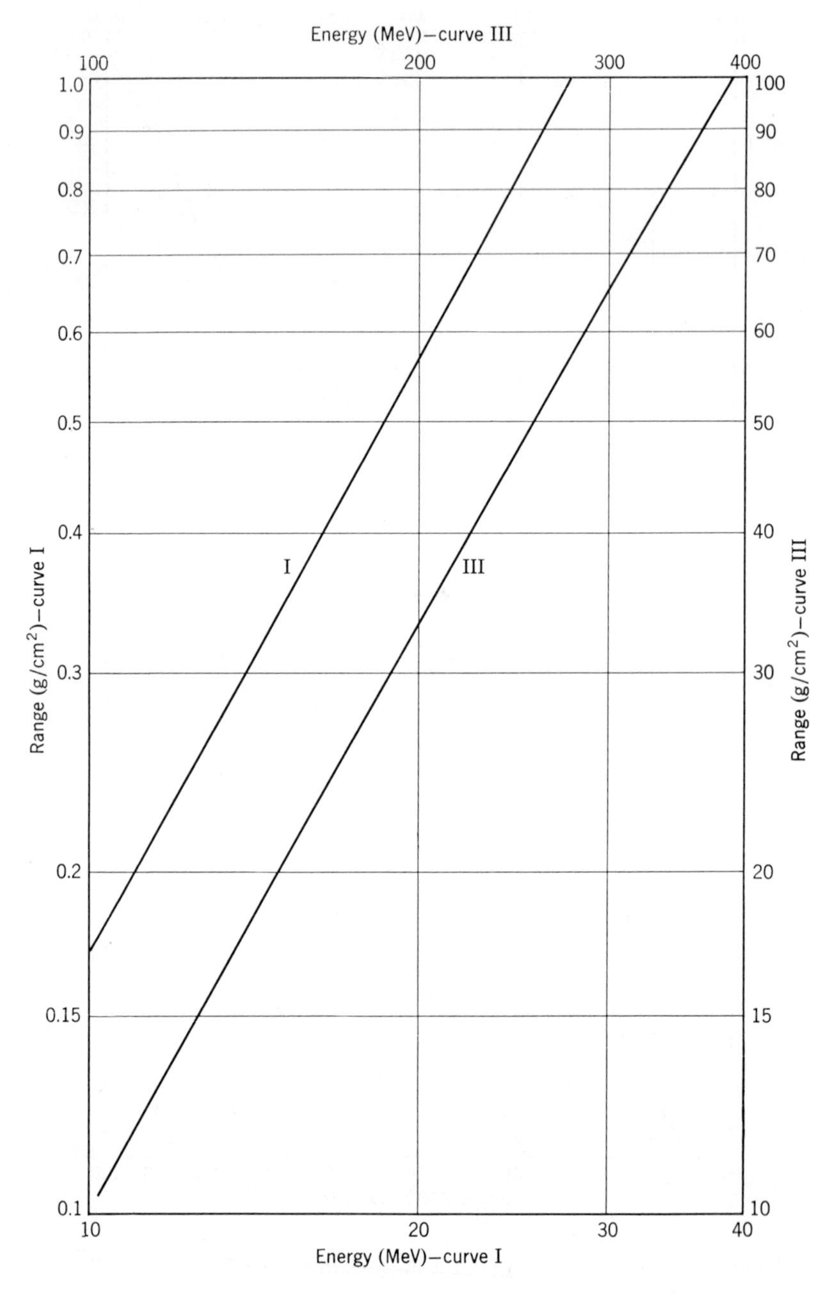

Fig. 2.13*d*–2 (*continued*)

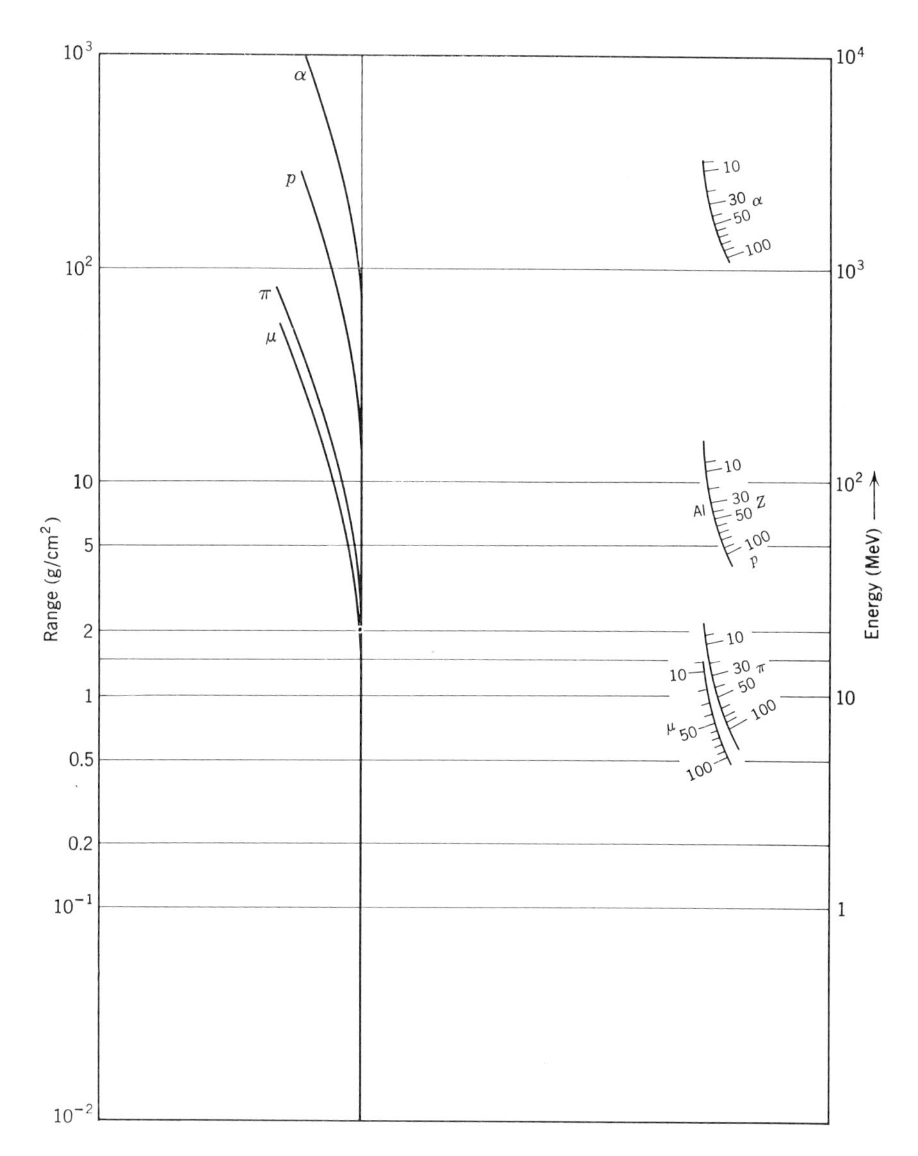

Fig. 2.14 Nomograph for approximate determination of energy from range. Left scale: range in g-cm^{-2}; middle scale: kinetic energy in MeV; right-hand scales: atomic number Z of stopping material and mass of particle. To use, connect range, energy, and Z by a straight line (Bethe 53, Fig. 10).

charged particle crosses the border of two media with different dielectric constants. The number of photons emitted per particle in a frequency range $d\omega$ and in solid angle element $d\Omega$ is given by

$$\frac{d^2n(\omega,\theta)}{d\omega\,d\Omega}=\frac{(Z'e)^2v^2}{\pi^2\hbar c^3\omega}\sin^2\theta\cos^2\theta\frac{|\varepsilon-1|^2}{(1-\beta^2\cos^2\theta)^2}$$
$$\times\left|\frac{1-\beta^3(\varepsilon-\sin^2\theta)^{1/2}-\beta^2(\varepsilon-\cos^2\theta)}{(\varepsilon\cos\theta)+(\varepsilon-\sin^2\theta)^{1/2}(1-\varepsilon\beta^2\sin^2\theta)}\right|^2 \qquad (2.2.43)$$

If a medium is a thin foil of thickness a, the last factor in (2.2.43) is replaced by a complicated quantity on $a\omega/v$. The photon yield for $a \ll v/\omega$ is approximately greater than (2.2.43) by a factor of $(a\omega/2v)^2/\varepsilon_2$.

In a medium in which the dielectric constant changes periodically with cell length a in the direction of particle motion, the periodic structure contributes to the momentum balance by a term $2\pi r/a$, where r is an arbitrary integer. Thus the emission angle given in (2.2.41) is modified for $\hbar\omega \ll cp$ to

$$\cos\theta=\frac{1}{\beta\sqrt{\varepsilon_1}}-\frac{2\pi rc}{a\omega\sqrt{\varepsilon_1}}. \qquad (2.2.44)$$

If the medium consists of alternating layers of dielectric constants $1-\omega_p^2/\omega^2$ and $1-\omega_p^2/\omega^2-\omega_p'^2/\omega^2$ with $\omega_p'^2 \ll \omega_p^2$, where

$$\omega_p=(4\pi n_e e^2/m_e)^{1/2}$$

is the plasma frequency, the frequencies of photons to be emitted are restricted, according to (2.2.44), to

$$\frac{\omega_p^2}{\omega_0}\left\{1-\left[1-(1-\beta)\frac{2\omega_p^2}{\omega_0^2}\right]^{1/2}\right\}^{-1}\geq\omega\geq\frac{\omega_p^2}{\omega_0}\left\{1+\left[1-(1-\beta)\frac{2\omega_p^2}{\omega_0^2}\right]^{1/2}\right\}^{-1} \qquad (2.2.45)$$

where $\omega_0=2\pi cr/a$. In the extremely relativistic case this relation is simplified to

$$2\omega_0\gamma^2\gtrsim\omega\gtrsim\frac{\omega_p^2}{2\omega_0},\qquad \gamma\equiv\frac{E}{m_e c^2}. \qquad (2.2.45')$$

For $a\simeq 1\mu$ the energy of the radiation is distributed in the optical as well as in the X-ray regions. The threshold energy of a particle for the transition radiation is obtained from (2.2.45′) as

$$\gamma\geq\frac{\omega_p}{2\omega_0}=\frac{a}{2rc}\sqrt{n_e e^2/\pi m_e}. \qquad (2.2.46)$$

The photon yield per unit length is approximately given by (Ter-Mikaelyan 61)

$$\frac{d^2n(\omega)}{dL\, d\omega} = \frac{2}{\pi} \frac{Z'^2 e^2}{\hbar c a \omega} \left[2 \ln \left(\frac{\omega_p}{\omega} \gamma \right) - 1 \right]. \tag{2.2.47}$$

The number of photons emitted by a particle as transition radiation may exceed unity if several hundred thin foils are used as radiators.

2.3 Photoelectric and Compton Effects

The interactions of photons with matter differ from the ionization process of charged particles in the sense that at each photon interaction the incident photon is absorbed or scattered at large angles, whereas in the latter only a small amount of energy is lost without appreciable deflection. Therefore the attenuation of a photon beam can be represented by an exponential law

$$I = I_0 \exp[-\mu x]. \tag{2.3.1}$$

The *absorption coefficient* μ consists of a number of interaction cross sections,

$$\mu = \frac{N}{A} \sum_i \sigma_i, \tag{2.3.2}$$

where σ_i is the cross section per atom for process i. Which of these processes is most significant depends on the energy of the photons. At low energy the most important process is Rayleigh scattering, on which is superimposed resonance scattering. Beyond the ionization potential the *photoelectric effect* becomes most important.

2.3.1 Photoelectric Effect

The photoelectric cross section of photons with energy k for the K-shell is expressed as (Heitler 54)

$$\sigma_K = \sigma_{\mathrm{Th}} \frac{64}{Z^2} \left(\frac{\hbar c}{e^2} \right)^3 \left(\frac{E_K}{k} \right)^{7/2} g(\eta), \tag{2.3.3}$$

where

$$\sigma_{\mathrm{Th}} = \frac{8\pi}{3} r_e^2 = 6.65 \times 10^{-25}\ \mathrm{cm}^2 \tag{2.3.4}$$

is the cross section for *Thomson scattering* and E_K is the ionization potential for the K-shell, which is occupied by two electrons; $g(\eta)$ represents the deviation from the Born approximation,

$$g(\eta) = 2\pi\sqrt{E_K/k}\,\frac{\exp(-4\eta\cot^{-1}\eta)}{1-\exp[-2\pi\eta]},$$
$$\eta = \sqrt{E_K/(k-E_K)} = \frac{Z_{eff}e^2}{\hbar v}, \tag{2.3.5}$$

where $k - E_K$ is the kinetic energy of an ionized electron and v is its velocity. This factor is important near the absorption edge; at the edge $(\eta \to \infty)$, $g(\infty) = 2\pi\exp(-4) = 0.12$, and even at $k = 50\,E_K$, $g(\eta) \simeq \frac{2}{3}$.

Below the K absorption edge the ionization from $L, M, \ldots$ shells determines the photoelectric cross section. The cross section for other shells may be approximately evaluated by changing the ionization potential and Z_{eff}. If the cross sections are compared at the same value of E_i/k, where E_i is the ionization energy of the shell concerned, the cross section increases with decreasing energy because of a decrease of effective Z. At energies higher than the K absorption edge the contribution from other shells is so small that the total photoelectric cross section may be practically obtained as

$$\sigma_{photo} \simeq \tfrac{5}{4}\sigma_K. \tag{2.3.6}$$

At relativistic energies the cross section is given in the Born approximation by

$$\sigma_K = \sigma_{Th}\tfrac{3}{2}Z^5\left(\frac{e^2}{\hbar c}\right)^4\left(\frac{mc^2}{k}\right)^5(\gamma^2-1)^{3/2}$$
$$\times\left\{\frac{4}{3}+\frac{\gamma(\gamma-2)}{\gamma+1}\left[1-\frac{1}{2\gamma\sqrt{\gamma^2-1}}\ln\left(\frac{\gamma+\sqrt{\gamma^2-1}}{\gamma-\sqrt{\gamma^2-1}}\right)\right]\right\}, \tag{2.3.7}$$

where

$$\gamma = (1-\beta^2)^{-1/2} = \frac{k+mc^2}{mc^2}.$$

For $k \gg mc^2$ this is reduced to

$$\sigma_K = \sigma_{Th}\,\tfrac{3}{2}Z^5\left(\frac{e^2}{\hbar c}\right)^4\frac{mc^2}{k}. \tag{2.3.7'}$$

For heavy elements the Born approximation is no longer valid. A correction factor multiplied by (2.3.7′) in the case of $k \gg mc^2$ is

$$g_{rel}(\eta) = \exp[-\pi\eta + 2\eta^2(1-\ln\eta)].$$

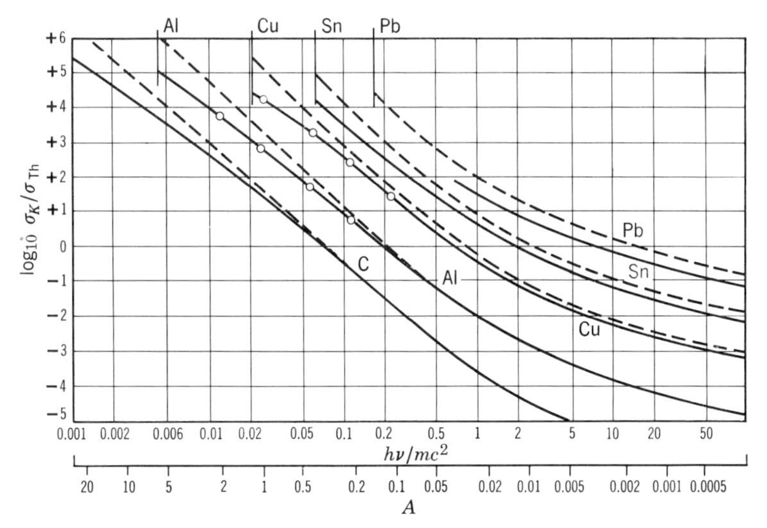

Fig. 2.15 Photoelectric cross section as a function of γ-ray energy for various atoms. The lower abscissa scale gives the wavelength in Angstrom units. Solid curves, exact calculation; dotted curves, Born approximation (Heitler 54, Fig. 8).

The cross sections thus calculated for several elements are shown in Fig. 2.15.

2.3.2 *Compton Effect*

As the energy of a photon increases, the Compton effect becomes more important than the photoelectric one. The cross section for an incident photon of energy k_0 to be scattered at angle θ into the solid angle $d\Omega$ is given by

$$d\sigma_c = \tfrac{1}{2} r_e^{\,2} \frac{k^2}{k_0^{\,2}} \left(\frac{k_0}{k} + \frac{k}{k_0} - \sin^2 \theta \right) d\Omega, \tag{2.3.8}$$

where k is the energy of a scattered photon and is uniquely determined by the scattered angle through the well-known relation

$$k = \frac{k_0}{1 + q(1 - \cos\theta)}, \qquad q \equiv \frac{k_0}{mc^2}. \tag{2.3.9}$$

This indicates that the scattered energy falls off rapidly with increasing θ for $\theta > \sqrt{mc^2/k_0}$.

The angular distribution is obtained by substituting k in (2.3.8) as

$$d\sigma_c = \tfrac{1}{2}r_e^2 \frac{1+\cos^2\theta}{[1+q(1-\cos\theta)]^2}\left\{1+\frac{q^2(1-\cos\theta)^2}{(1-\cos^2\theta)[1+q(1-\cos\theta)]}\right\} d\Omega \tag{2.3.10}$$

For small q the angular distribution follows the $(1+\cos^2\theta)$ law, as in the case of Thomson scattering. For large q the cross section becomes small for $\theta > \sqrt{mc^2/k_0}$.

The energy distribution of scattered photons is obtained by eliminating θ from (2.3.8) with the aid of (2.3.9) as follows:

$$d\sigma_c = 2\pi r_e^2 \frac{dk}{k}\frac{1}{q}\left[1+\left(\frac{k}{k_0}\right)^2 - \frac{2(q+1)}{q^2} + \frac{1+2q}{q^2}\frac{k}{k_0} + \frac{1}{q^2}\frac{k_0}{k}\right]. \tag{2.3.11}$$

For small q, k is limited to near k_0, but for large q the distribution becomes flatter. The average energy of scattered photons decreases gradually from k_0 to

$$\langle k \rangle = \frac{\frac{4}{3} - \dfrac{3}{2q}}{\ln(2q+1)+\frac{1}{2}} k_0 \quad \text{for} \quad q \gg 1 \tag{2.3.12}$$

as q increases.

The total cross section is

$$\sigma_c = r_e^2 \frac{1}{q}\left\{\left[1-\frac{2(q+1)}{q^2}\right]\ln(2q+1) + \frac{1}{2} + \frac{4}{q} - \frac{1}{2(2q+1)^2}\right\}. \tag{2.3.13}$$

At two extreme limits this is reduced to

$$\sigma_c \simeq \sigma_{\mathrm{Th}}\left(1-2q+\frac{26}{5}q^2+\cdots\right) \quad \text{for} \quad q \ll 1, \tag{2.3.13a}$$

$$\sigma_c \simeq \sigma_{\mathrm{Th}}\frac{3}{8}\frac{1}{q}(\ln 2q + \tfrac{1}{2}) \quad \text{for} \quad q \gg 1, \tag{2.3.13b}$$

respectively. Thus the Compton cross section decreases from the Thomson limit to (2.3.13b) as incident energy increases.

The Compton cross section is compared with the photoelectric cross sections for a number of elements in Fig. 2.16. This shows the relative importance of these cross sections.

The absorption coefficient is obtained by taking account of the fact that Z electrons per atom are responsible for the Compton scattering.

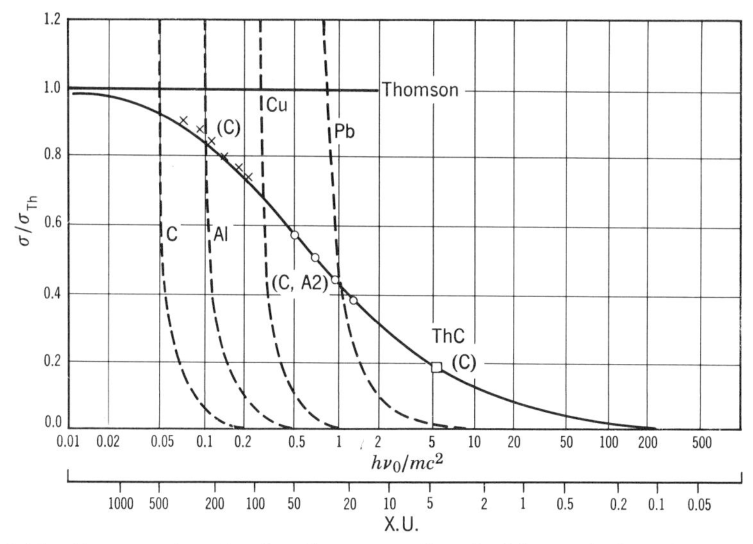

Fig. 2.16 Cross section for the Compton effect (solid curve), in units of $\sigma_{\mathrm{Th}} = (8\pi/3)r_e^2$, as a function of the incident γ-ray energy in units of mc^2 (lower scale gives wavelength). The dotted curves give the photoelectric cross section for various elements in the same units. Crosses, circles, and square represent experimental points (Heitler 54, Fig. 13).

It may be convenient to give the numerical value for Thomson scattering:

$$\mu_{\mathrm{Th}} = \frac{NZ}{A}\sigma_{\mathrm{Th}} = 0.2003\,\frac{2Z}{A}\ \mathrm{cm}^2/\mathrm{g}. \tag{2.3.14}$$

It may be worthwhile to remark that the Compton effect does not cause net absorption; it is a kind of energy loss of a photon. In measuring the transmission of photons therefore, we observe secondary photons together with primary ones. With the secondary photons taken into account, the transmission coefficient is expressed as

$$T = B\exp(-\mu x), \tag{2.3.15}$$

in which B represents the *build-up factor*; it lies between 1 and $(\mu x)^2$.

2.4 Bremsstrahlung and Pair Creation

The Coulomb scattering of an electron is associated with its acceleration, which results in the emission of electromagnetic radiation. This is called *bremsstrahlung*.

2.4.1 Bremsstrahlung Cross Section under the Born Approximation

A quantum-mechanical calculation with the Born approximation gives us the Bethe-Heitler formula for the bremsstrahlung cross section (Bethe 34):

$$\sigma_r(E, k)\, dk = 4Z^2\alpha r_e^{\,2} \frac{dk}{k} F(E, u), \qquad \alpha \equiv \frac{e^2}{\hbar c}, \tag{2.4.1}$$

where $u \equiv k/E$, the energy of an emitted photon divided by that of an incident electron. The function $F(E, u)$ depends on a parameter,

$$\xi \equiv 100 \frac{mc^2}{E} \frac{u}{1-u} Z^{-1/3}, \tag{2.4.2}$$

that is inversely proportional to the incident energy as well as proportional to the ratio of the radiated photon energy to the outgoing electron energy. Depending on the value of ξ, $F(E, u)$ is given by (Rossi 52)

$$F(E, u) = [1 + (1-u)^2 - \tfrac{2}{3}(1-u)] \left[\ln \left(\frac{2E}{mc^2} \frac{1-u}{u}\right) - \frac{1}{2}\right]$$

$$\text{for } \xi \gg 1 \text{ (no screening)}, \tag{2.4.3a}$$

$$F(E, u) = [1 + (1-u)^2 - \tfrac{2}{3}(1-u)] \left[\ln \left(\frac{2E}{mc^2} \frac{1-u}{u}\right) - \frac{1}{2} - c(\xi)^{-1/2}\right]$$

$$\text{for } 2 < \xi < 15, \tag{2.4.3b}$$

$$F(E, u) = [1 + (1-u)^2]\left[\frac{f_1(\xi)}{4} - \tfrac{1}{3} \ln Z\right] - \tfrac{2}{3}(1-u) \left[\frac{f_2(\xi)}{4} - \tfrac{1}{3} \ln Z\right]$$

$$\text{for } \xi < 2, \tag{2.4.3c}$$

$$F(E, u) = [1 + (1-u)^2 - \tfrac{2}{3}(1-u)] \ln (191Z^{-1/3}) + \tfrac{1}{9}(1-u)$$

$$\text{for } \xi \simeq 0 \text{ (complete screening)}. \tag{2.4.3d}$$

In the intermediate cases, (2.4.3*b*) and (2.4.3*c*), three complicated functions of ξ appear. Their numerical values are shown in Fig. 2.17.

The expressions (2.4.3) show that the energy spectrum of radiated photons is rather flat. This is seen from Fig. 2.18, in which $F(E, u)/\ln (191Z^{-1/3})$ is plotted against u. At high incident energy, as seen in Fig. 2.19, the u-dependence in $F(E, u)$ may be neglected for the purpose of rough approximation.

At nonrelativistic energy the cross section is expressed as a function

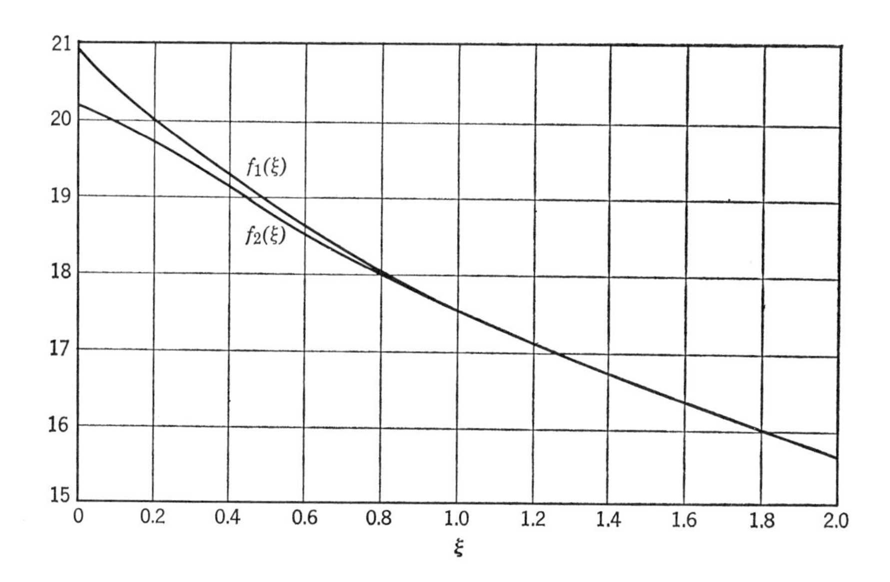

Fig. 2.17*a* The functions $f_1(\xi)$ and $f_2(\xi)$ in (2.4.3*c*) (Bethe 34).

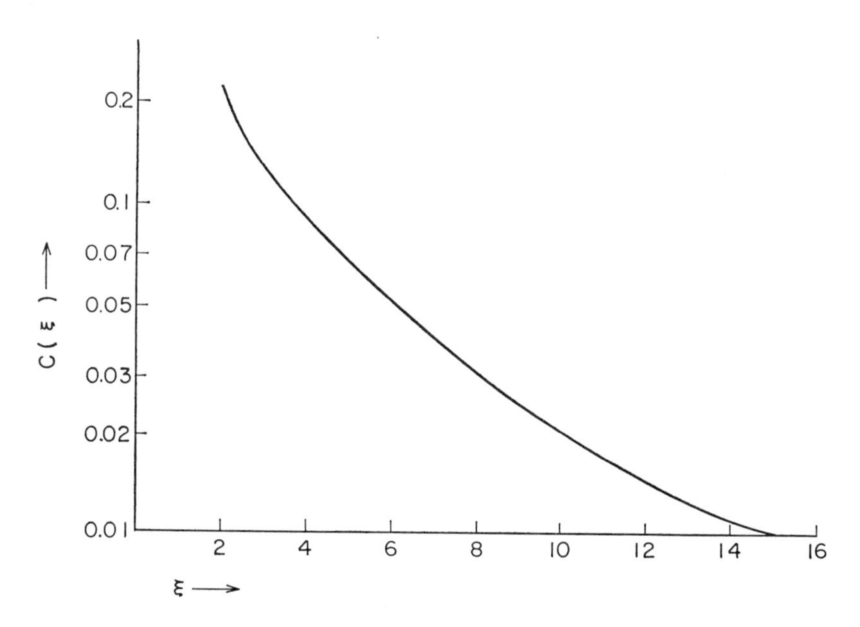

Fig. 2.17*b* The function $c(\xi)$ in (2.4.3*b*) (Bethe 34).

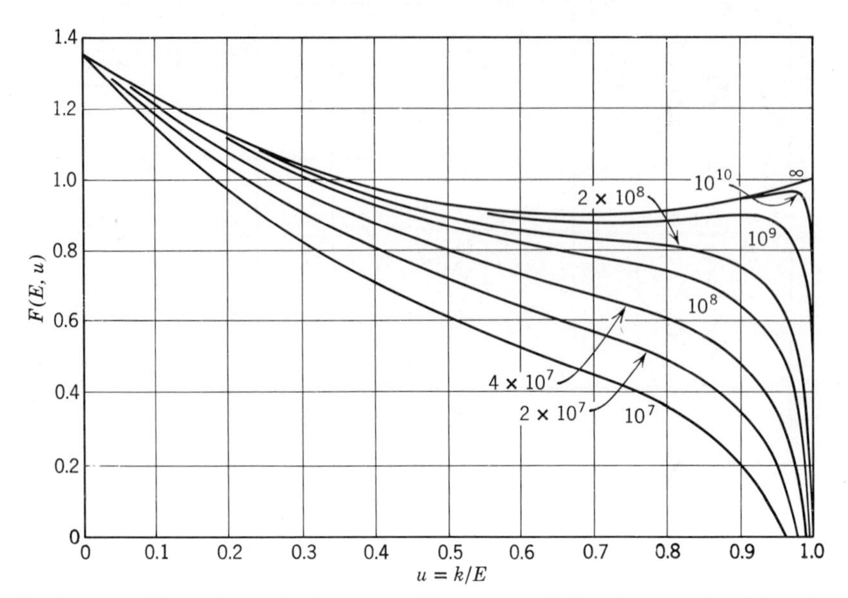

Fig. 2.18*a* Differential radiation probability per radiation length of air for electrons of various energies. The numbers attached to the curves indicate the total energy E of the primary electron (Rossi 41).

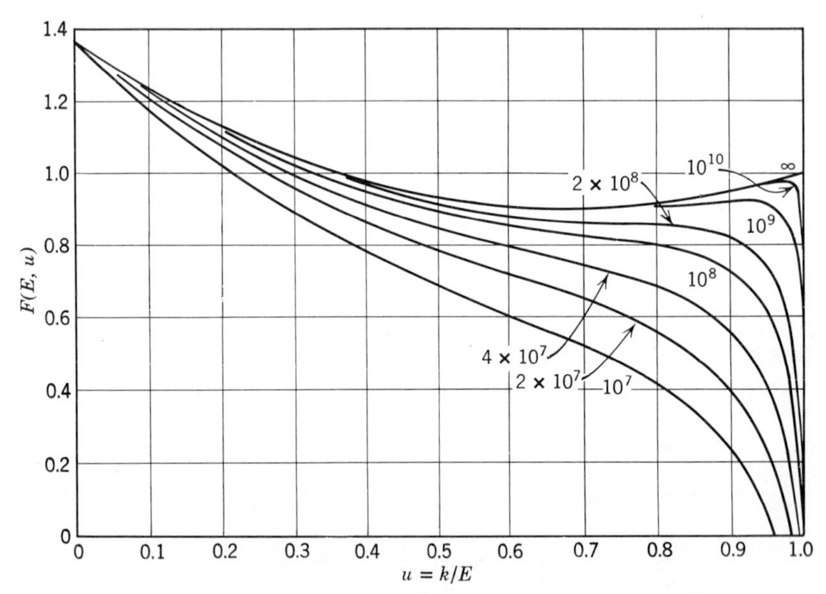

Fig. 2.18*b* Differential radiation probability per radiation length of lead for electrons of various energies. The numbers attached to the curves indicate the total energy E of the primary electron (Rossi 41).

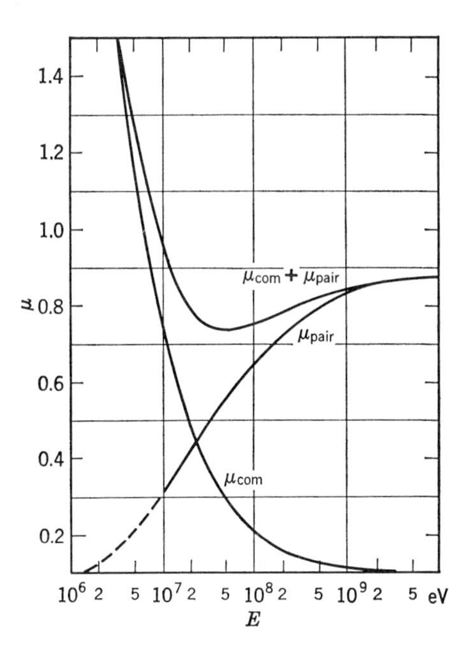

Fig. 2.19*a* The total probability per radiation length of air for Compton scattering (μ_{Com}), for pair production (μ_{pair}), and for either effect ($\mu_{Com} + \mu_{pair}$) (Rossi 41).

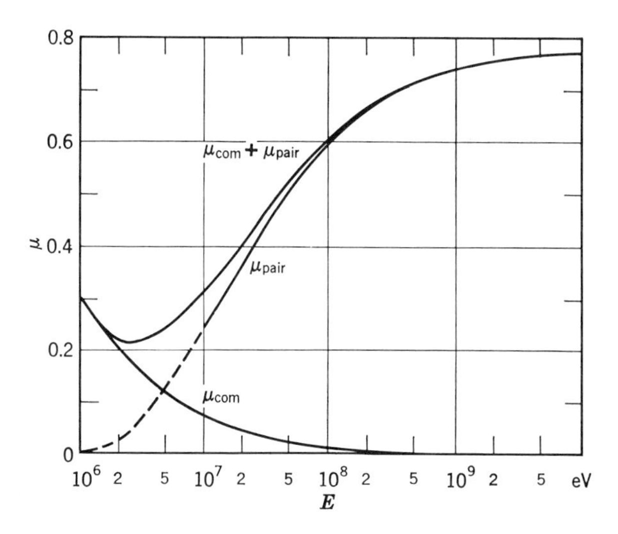

Fig. 2.19*b* The total probability per radiation length of lead for Compton scattering (μ_{Com}), for pair production (μ_{pair}), and for either effect ($\mu_{Com} + \mu_{pair}$) (Rossi 41).

of the kinetic energy, $W = E - mc^2$, rather than the total energy; thus $F(W, k)$ is given by

$$F(W, k) = \frac{2}{3} \frac{mc^2}{W} \ln \frac{(\sqrt{W} + \sqrt{W - k})^2}{k} \tag{2.4.3'}$$

2.4.2 *Energy Loss by Radiation*

The average energy loss per g-cm^{-2} due to bremsstrahlung is

$$-\left.\frac{dE}{dx}\right|_{\text{rad}} = \int_0^{E - mc^2} \frac{N}{A} \sigma_r(E, k) k \, dk. \tag{2.4.4}$$

At low energy ($\xi \gg 1$) the cross section without screening, given by (2.4.3*a*), may be used, whereas at high energy ($\xi \ll 1$) the cross section with complete screening given by (2.4.3*d*), may be of practical use. Thus we obtain

$$-\left.\frac{dE}{dx}\right|_{\text{rad}} = 4 \frac{NZ^2}{A} \alpha r_e^2 E \left[\ln\left(\frac{2E}{mc^2}\right) - \tfrac{1}{3}\right] \quad \text{for} \quad mc^2 \ll E \ll 137 mc^2 Z^{-1/3}, \tag{2.4.5a}$$

$$-\left.\frac{dE}{dx}\right|_{\text{rad}} = 4 \frac{NZ^2}{A} \alpha r_e^2 E \left[\ln(191 Z^{-1/3}) + \tfrac{1}{18}\right] \quad \text{for} \quad 137 mc^2 Z^{-1/3} \ll E. \tag{2.4.5b}$$

At about $E = 137 mc^2 Z^{-1/3}$ the integral in (2.4.4) has to be evaluated numerically.

Bremsstrahlung takes place also in collisions with atomic electrons (Wheeler 39). The cross section for this process is essentially the same as that for bremsstrahlung in the nuclear Coulomb field but for the absence of the Z^2 dependence and a small correction term. The correction term in the square brackets in (2.4.5*a*, *b*) may be as small as $-\ln 2$. Hence the effect of atomic electrons may be taken into account by adding a term proportional to Z in place of Z^2 in (2.4.5); consequently the radiation loss is proportional to $Z(Z+1)$ in place of Z^2. It is thus suggested to introduce the *radiation length* X_0 through

$$\frac{1}{X_0} = 4 \frac{NZ(Z+1)}{A} \alpha r_e^2 \ln(191 Z^{-1/3}), \tag{2.4.6}$$

which gives the mean length for the radiation loss of an electron. Numerical values of the radiation length in various substances are given in Table 2.7. The radiation length of a mixture consisting of substances having radiation lengths X_i and of fractional weights w_i is given by

$$\frac{1}{X_0} = \sum_i \frac{w_i}{X_i}. \tag{2.4.7}$$

By the use of the radiation length, the distribution of the fractional photon energy is expressed as

$$\psi_{\text{rad}}(u) \equiv \frac{F(E, u)}{u \ln (191 Z^{-1/3})} = \frac{1}{u}\left[1 + (1-u)^2 - (1-u)\left(\tfrac{2}{3} - 2b\right)\right], \tag{2.4.8}$$

Table 2.7 Radiation Length and Critical Energy for Various Materials[a]

Material	Z	A	Radiation Length (g-cm^{-2})	(cm)	Critical Energy[b,c] (MeV)
Hydrogen	1	1.008	62.8	[d]7500	350
Helium	2	4.003	93.1	[d]5600	250
Lithium	3	6.940	83.3	15.6	138
Carbon (graphite)	6	12.010	43.3	16.9	79
Nitrogen	7	14.008	38.6	[d]331	85
Oxygen	8	16.000	34.6	[d]258	75
Aluminium	13	26.980	24.3	9.10	40
Silicon	14	28.090	22.2	9.52	37.5
Iron	26	55.85	13.9	1.77	20.7
Copper	29	63.54	13.0	1.46	18.8
Bromine	35	79.916	11.5	3.71	15.7
Silver	47	10.988	9.0	0.86	11.9
Iodine	53	126.910	8.5	1.74	10.7
Tungsten	74	183.92	6.8	0.35	8.08
Lead	82	207.21	6.4	0.57	7.40
Compound Material					
Air:					
Nitrogen, 75.52%			37.1	[d]308	81
Oxygen, 23.14%					
Argon, 1.3%					
Silicon dioxide			27.4	10.3	47.3
Water			36.4	36.4	73.0
Lithium hydride			80.0	113	157
Nuclear emulsion G5 (58% RH)[e]			11.4	2.98	16.4

[a] From J. Nishimura 67.

[b] Density effect is included except for gaseous materials; all gases are assumed to be at 20°C and 1-atmosphere pressure.

[c] The critical energy is approximately 500 MeV/Z.

[d] In meters.

[e] Recalculated.

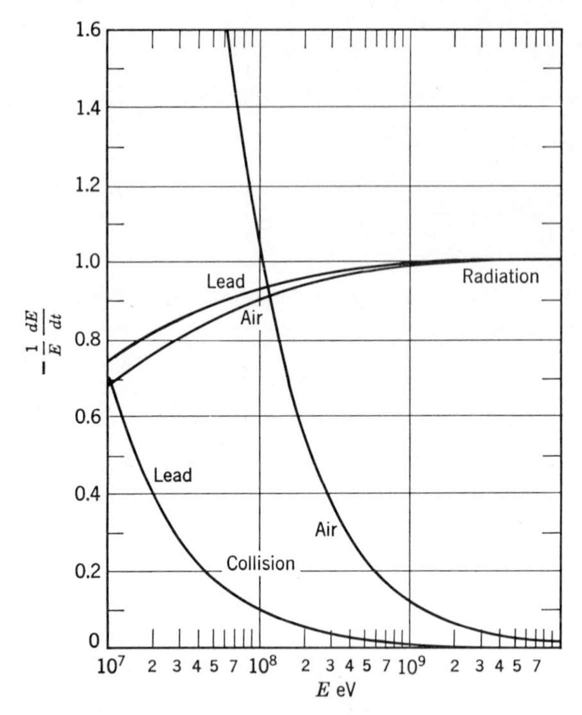

Fig. 2.20 Fractional energy loss by ionizing collision and that by radiation for electrons, per radiation length of air or lead (Rossi 41).

with

$$b \equiv \tfrac{1}{18} \ln (191 Z^{-1/3}).$$

The second expression of (2.4.8) is valid for the case of complete screening. The average fractional energy-loss rate is correspondingly expressed as

$$-\frac{X_0}{E}\frac{dE}{dx} = 1 + b. \tag{2.4.9}$$

The fractional energy loss by radiation is compared with that by ionization in Fig. 2.20. Loss by radiation is greater than loss by ionization above a certain energy, which is defined as the *critical energy* ε_0:

$$-\left.\frac{dE}{dx}\right|_{\text{rad}}(\varepsilon_0) = -\left.\frac{dE}{dx}\right|_{\text{ion}}(\varepsilon_0). \tag{2.4.10}$$

The critical energy is equal to loss by ionization in the radiation length and decreases with increasing Z; its numerical values are given in Table 2.7.

It may be of practical use to give the approximate Z-dependences of the radiation length and critical energy:

$$X_0 \simeq \left[\frac{A}{Z(Z+1)(7 \sim 5)}\right] \times 10^3 \text{ g-cm}^{-2},$$
$$\varepsilon_0 \simeq (6 \text{ to } 8) \times 10^2/(Z+1)\text{MeV}. \tag{2.4.11}$$

The energy loss by a radiative collision is so large that fluctuations in energy loss are considerable. It is not practical to define the range due to radiation loss. Instead we use the probability that an electron of energy E_0 has an energy between E and $E+dE$ after traversing t radiation lengths. The probability is expressed as

$$w(E_0, E, t)\, dE = \frac{dE}{E_0} \frac{[\ln (E_0/E)]^{(t/\ln 2)-1}}{\Gamma(t/\ln 2)}, \tag{2.4.12}$$

where Γ is the gamma function. In this formula the length may be measured better in units of $X_0 \ln 2$. This is the length in which an electron loses half its energy on the average.

2.4.3 *Bremmstrahlung of the Muon*

It is well known that bremsstrahlung is effective only for light particles such as the electron, because its cross section is inversely proportional to the square of mass. However, bremsstrahlung becomes important for muons of high energies, because other interactions are of minor effect. Since the spin of the muon is $\frac{1}{2}$, the formula for the electron applies as it is if the electron mass is replaced by the muon mass. In addition the minimum-impact parameter has to be chosen as the nuclear radius, because the Compton wavelength of the muon is smaller than the nuclear radius except for hydrogen.

Corresponding to (2.4.3*a*) therefore, the radiation cross section is given by

$$\sigma_r(E, k)\, dk = 4Z^2\alpha\left(\frac{m}{\mu}\right)^2 r_e^{\,2} \frac{dk}{k} [1 + (1-u)^2 - \tfrac{2}{3}(1-u)] \left[\ln \left(\frac{2E}{\mu c^2} \frac{\hbar}{\mu c R} \frac{1-u}{u}\right) - \tfrac{1}{2}\right], \tag{2.4.13}$$

where μ is the muon mass and R the nuclear radius.

2.4.4 *Pair Creation*

The production of a negaton-positon pair is the inverse process of bremsstrahlung. If a positon produced in the former is replaced by an incoming negaton and then the whole process is reversed, we get the latter process. According to the detailed balancing, these cross sections are connected by

$$\sigma_{\text{pair}}(k, E) = \sigma_r(E, k)\left(\frac{E^2}{k^2}\right). \tag{2.4.14}$$

By putting $v \equiv E/k$, where E is the energy of one of the produced electrons, we obtain from (2.4.14) the pair-creation cross section as

$$\sigma_{\text{pair}}(k, E) = 4\alpha Z^2 r_e^{\,2} \frac{dE}{k} G(k, v). \tag{2.4.15}$$

Analogous to (2.4.3), the function $G(k, v)$ is expressed as

$$G(k, v) = [v^2 + (1-v)^2 + \tfrac{2}{3}v(1-v)]\left[\ln\left(\frac{2k}{mc_2}\right) v(1-v) - \tfrac{1}{2}\right] \tag{2.4.16a}$$

for $\xi \gg 1$ (no screening),

$$G(k, v) = [v^2 + (1-v^2) + \tfrac{2}{3}v(1-v)]\left[\ln\left(\frac{2k}{mc^2}\right) v(1-v) - \tfrac{1}{2} - c(\xi)\right] \tag{2.4.16b}$$

for $2 < \xi < 15$,

$$G(k, v) = [v^2 + (1-v)^2]\left[\frac{f_1(\xi)}{4} - \tfrac{1}{3}\ln Z\right] + \tfrac{2}{3}v(1-v)\left[\frac{f_2(\xi)}{4} - \tfrac{1}{3}\ln Z\right] \tag{2.4.16c}$$

for $\xi < 2$, and

$$G(k, v) = [v^2 + (1-v)^2 + \tfrac{2}{3}v(1-v)]\ln(191Z^{-1/3}) - \tfrac{1}{9}v(1-v) \tag{2.4.16d}$$

for $\xi \simeq 0$ (complete screening).

Since the infrared divergence is absent, the total cross section can now be given by

$$\sigma_{\text{pair}}(k) = \int_{mc^2}^{k-mc^2} \sigma_{\text{pair}}(k, E)\, dE$$

$$= 4\alpha Z^2 r_e^{\,2} \times \begin{cases} \left[\frac{7}{9}\ln\frac{2k}{mc^2} - \frac{109}{54}\right] & \text{for } mc^2 \ll k \ll 137mc^2Z^{-1/3}, \quad (2.4.17a) \\ \left[\frac{7}{9}\ln(191Z^{-1/3}) - \frac{1}{54}\right] & \text{for } k \gg 137mc^2Z^{-1/3}. \quad (2.4.17b) \end{cases}$$

The absorption cross section (2.4.17) gives the absorption probability per radiation length, $\mu_{\text{pair}}(k) = (N/A)\sigma_{\text{pair}}(k)/X$. This is compared with that for Compton scattering in Fig. 2.20.

The absorption coefficient of energetic photons consists essentially of these two processes. This has a minimum at a certain energy that depends on the absorber, as seen in Fig. 2.20.

The effect of atomic electrons can be introduced in a way that is analogous to the case of bremsstrahlung. At high energy the contribution from atomic electrons is considered to be identical with that in bremsstrahlung; the cross section is proportional to Z instead of Z^2, and a correction term in the square brackets in (2.4.17*a*, *b*) is approximately $-\ln 2$. For practical purposes therefore, the pair-creation cross section per atom is given by replacing Z^2 in (2.4.17) by $Z(Z+1)$. At low energy, however, we have to be cautious about the difference in the threshold energy. It is not $2mc^2$, but $4mc^2$. Hence the relative contribution from atomic electrons decreases with decreasing energy. Detailed discussions on this problem are given in the papers by Suh and Hart (Suh 59, Hart 59), and papers cited there.

The energy spectrum of electrons produced is nearly flat. This is expressed in the case of complete screening by

$$\psi_{\text{pair}}(v) = \frac{G(k, v)}{\ln(191Z^{-1/3})} = v^2 + (1-v)^2 + (\tfrac{2}{3} - 2b)v(1-v). \quad (2.4.18)$$

This is of course symmetric with respect to the interchange of v and $1-v$. Numerical values of $\psi_{\text{pair}}(v)$ are shown in Fig. 2.21.

The angle between two produced electrons is sometimes useful for determining the energy of an incident photon. Since the pair-creation process is effective at small impact parameters on the order of $\hbar/mc$, the transverse momentum of a produced electron is about mc. Consequently the angle of emission is on the order of mc^2/k. More exactly, the rms angle of emission is given by

$$\langle\theta^2\rangle^{1/2} = g(k, E, Z)\left(\frac{mc^2}{k}\right)\ln\left(\frac{k}{mc^2}\right), \quad (2.4.19)$$

where $g(k, E, Z)$ is a function on the order of unity. This depends primarily on $v = E/k$ and weakly on Z, as shown in Fig. 2.22. The rms angle of divergence of two electrons can be obtained easily with the aid of (2.4.19) and Fig. 2.22. The rms angle of emission of a photon in bremsstrahlung is given by an expression analogous to (2.4.19), but with a slightly different functional form of $f(E, k, Z)$ in place of g.

There are a number of corrections to be made if the cross sections for bremsstrahlung and pair creation are to be obtained with high

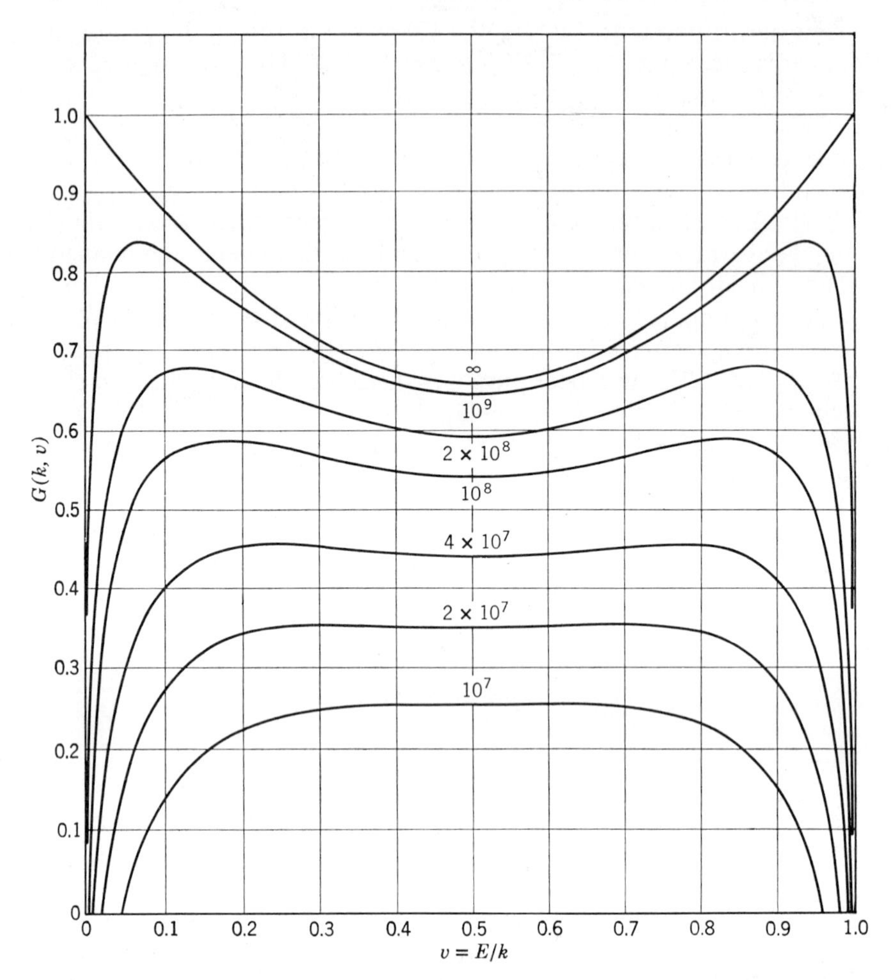

Fig. 2.21*a* Differential probability of pair production per radiation length of air for photons of various energies. The numbers attached to the curves indicate the energy k of the primary photon (Rossi 41).

accuracy. Some of these corrections are due to the screening effect and the contribution from atomic electrons, both of which have already been discussed. In addition, the corrections given below may also be of practical importance.

2.4.5 Coulomb Effect

The cross sections given above are derived by using the Born approximation. If the velocity of an electron is so small that $Ze^2/\hbar v$ is consider-

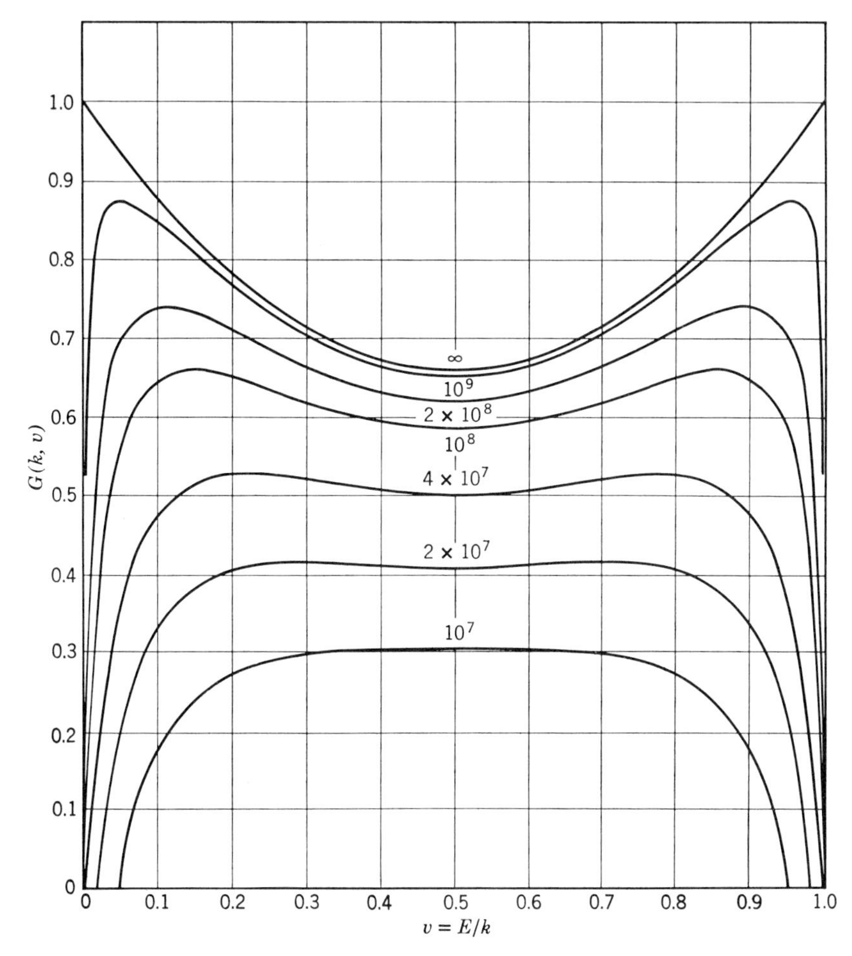

Fig. 2.21*b* Differential probability of pair production per radiation length of lead for photons of various energies. The numbers attached to the curves indicate the energy k of the primary photon (Rossi 41).

able, the distortion of an electron wave has to be taken into account. This effect is more important for pair creation than for bremsstrahlung, because there is a chance that one of two created electrons has a low energy.

Near the threshold energy the Coulomb effect can be expressed by the Coulomb penetration factor (Nishina 34):

$$C(\eta_+, \eta_-) = \frac{\eta_+ \eta_-}{[\exp(\eta_+) - 1][1 - \exp(-\eta_-)]}, \qquad (2.4.20)$$

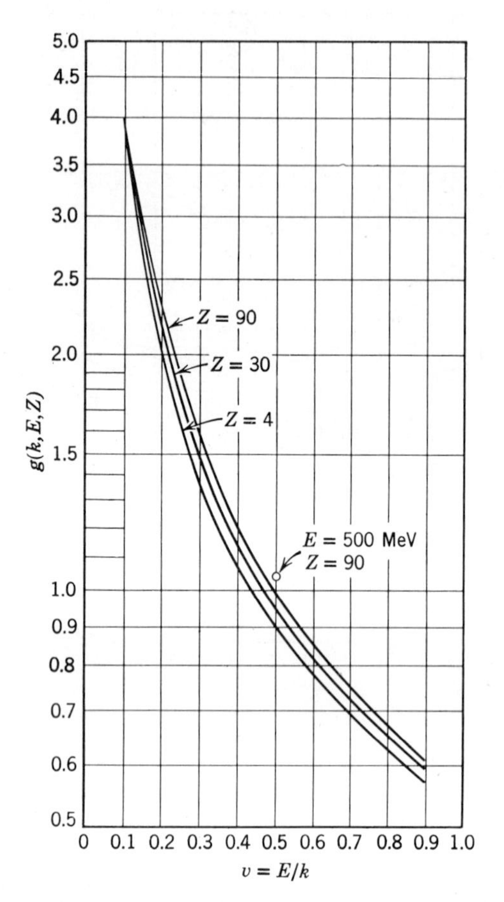

Fig. 2.22 The quantity $g(k, E, Z)$ in (2.4.19) plotted as a function of $v = E/k$. The three curves refer to elements with atomic numbers 4, 30, and 90, respectively, and are valid for 50 MeV $< E <$ 300 MeV. The circle represents the value of g for $Z = 90$ and $E = 500$ MeV. From M. Stearns, *Phys. Rev.*, **76**, 836 (1949)

multiplied by the Born-approximation cross section. Here

$$\eta = \frac{2\pi Ze^2}{\hbar v}, \tag{2.4.21}$$

and the plus and minus signs refer to a positon and a negaton, respectively. This indicates that the energy distribution is greatly distorted from the symmetric one in such a way that the maximum of the energy distribution of negatons is shifted toward a low energy.

The same effect in bremsstrahlung applies both for incident and outgoing electrons. The correction factor is expressed by

$$C(\eta_0, \eta) = \frac{\eta}{\eta_0} \frac{1 - \exp(-\eta_0)}{1 - \exp(-\eta)}, \tag{2.4.22}$$

where the subscript 0 refers to the incident electron. For bremsstrahlung of a positon the signs of η_0 and η must be changed.

The total pair-creation cross section in lead is semiempirically given above 5 MeV by (Bethe 54, Davies 54)

$$\sigma_{\text{pair}} = \sigma_{\text{B-H}} - 4.0 + \frac{46}{k}, \tag{2.4.23}$$

where $\sigma_{\text{B-H}}$ is the Born approximation or the Bethe-Heitler cross section in barns, given in (2.4.17), and k is measured in MeV. The second term, -4.0 barns, on the right-hand side in (2.4.23), is a correction roughly proportional to $(Z\alpha)^2$. In the relativistic case the logarithmic term in $G(k, v)$ in (2.4.16) must be corrected for a term $-C(Z)$,

$$C(Z) = (Z\alpha)^2 \sum_{n=1}^{\infty} \frac{1}{n[n^2 + (Z\alpha)^2]}. \tag{2.4.24}$$

For small $(Z\alpha)^2$ this is evaluated as

$$C(Z) = 1.20(Z\alpha)^2 = 0.63 \times 10^{-4} Z^2, \tag{2.4.24$'$}$$

thus being proportional to $(Z\alpha)^2$; but for large $(Z\alpha)^2$ the correction term is smaller than that given by (2.4.24′); $C(Z) = 0.925(Z\alpha)^2$ for lead. The total cross section (2.4.17) is therefore corrected as

$$\sigma_{\text{pair}}(k) = 4\alpha Z^2 r_e^{\,2} \left\{ \tfrac{7}{9} \left[\ln \frac{2k}{mc^2} - C(Z) \right] - \tfrac{109}{54} \right\} \quad \text{(no screening)}, \tag{2.4.25a}$$

$$= 4\alpha Z^2 r_e^{\,2} \{ \tfrac{7}{9} [\ln (191 Z^{-1/3}) - C(Z)] - \tfrac{1}{54} \} \quad \text{(complete screening)}. \tag{2.4.25b}$$

In bremsstrahlung the Coulomb correction is negligible in the case of complete screening, whereas it is analogous to that in pair creation, like (2.4.25*a*), in the case of no screening. In the case of partial screening the Coulomb effect is less in bremsstrahlung than in pair creation; in the latter the Coulomb correction is independent of the screening effect.

An experiment on the absorption of photons with 1 GeV gives absorption coefficients in various substances in good agreement with the theoretical prediction (Malamud 59). Contributions from various processes are theoretically calculated in Table 2.8.

Table 2.8 Photon Cross Sections from Various Processes at 1 GeV

Substance	Z	Compton (mb)	Pair Creation (Born) (barn)	Atomic (mb)	Coulomb (mb)	Total (barn)	Experiment/ Theory
Hydrogen	1	1.12	0.00870	9.51	1×10^{-4}	0.01933	1.004 ± 0.031
Lithium	3	3.36	0.07201	28.38	0.90×10^{-2}	0.10375	0.994 ± 0.023
Beryllium	4	4.52	0.12597	37.15	2.8×10^{-2}	0.16764	1.000 ± 0.11
Carbon	6	6.72	0.27674	54.47	0.14	0.33795	1.002 ± 0.009
Aluminum	13	14.7	1.2383	111.4	2.9	1.3515	1.001 ± 0.013
Titanium	22	25	3.421	180	25	3.601	1.023 ± 0.014
Copper	29	33	5.832	230	78	6.017	0.986 ± 0.014
Molybdenum	42	47	11.905	321	349	11.924	1.021 ± 0.013
Tin	50	67	16.626	375	667	16.401	0.997 ± 0.008
Tantalum	73	80	34.38	520	2690	32.29	0.992 ± 0.019
Lead	82	90	43.04	580	4100	39.61	0.993 ± 0.010
Uranium	92	100	53.54	640	6140	48.14	0.985 ± 0.011

Because the accuracy of experiment is as good as 1 percent, the following corrections to the Bethe-Heitler cross section are considered important. The contribution of atomic electrons is found to be in good agreement with the prediction by the Wheeler-Lamb theory (Wheeler 39). The Coulomb correction is important for $Z > 20$, as is expected from $C(Z) \simeq (Z\alpha)^2$. Contributions from nuclear photodisintegrations as well as photopion production are fractions of 1 percent. Corrections of less than 1 percent may arise also from the effects of chemical binding, atomic shielding, higher order radiative processes, etc.

A possible deviation from the theoretical cross section (2.4.25) seems to be found at energy levels below 100 MeV. The experimental cross section appears to increase with decreasing energy. This may be accounted for by (2.4.23) or (2.4.20), but no detailed theory has been worked out yet.

2.4.6 Landau-Pomeranchuk Effect

At very high energy the cross sections for bremsstrahlung and pair creation are suppressed in dense media. This is a kind of density effect, called the Landau-Pomeranchuk effect (Landau 53). It is due to the fact that the radiation process terminates when a radiating electron is scattered by nearby atoms.

In the rest system of an incident energy with energy E_0, the time needed for radiating a photon of frequency ω is $1/\omega$. In the laboratory system this increases by a Lorentz factor E_0/mc^2. After a photon of energy k is emitted, the electron energy decreases to $E_0 - k$. The photon energy k is related to ω as $k = \hbar\omega(E_0 - k)/mc^2$. Hence the radiation time is given by $\hbar E_0(E_0 - k)/m^2c^4k$. The length for a photon to become free is expressed as

$$l = \frac{E_0(E_0 - k)}{kmc^2} \frac{\hbar}{mc}. \tag{2.4.26}$$

While an electron passes through distance l, it is scattered by Coulomb fields of nearby atoms. The rms scattering angle is given, according to (2.1.28), by

$$\langle\theta^2\rangle_l^{1/2} = \left[8\pi \frac{N}{A} \frac{Z^2e^4}{p^2v^2} \ln\left(\frac{\theta_2}{\theta_1}\right)\rho l\right]^{1/2}, \tag{2.4.27}$$

where $pv = E$ may be taken as equal to the mean value of the electron energies before and after radiation, $E = (2E_0 - k)/2$. This is compared with the angle of deflection of the electron by radiation, $\simeq mc/p$. The ratio of these two angles is represented by a parameter:

$$s \equiv \frac{mc/4p}{\langle\theta^2\rangle_l^{1/2}}. \tag{2.4.28}$$

If the value of $4s$ is smaller than unity, the radiation probability is reduced.

On the basis of the above qualitative considerations the result of a more detailed theory (Migdal 56) is presented below. By introducing a parameter s representing the scattering probability, $F(E, u)$ in (2.4.3d) is now modified to

$$F(E, u) = \tfrac{1}{3}\{u^2H(s) + 2[1 + (1 - u)^2]\phi(s)\}\xi(s). \tag{2.4.29}$$

The functions $H(s)$ and $\phi(s)$ are given by

$$H(s) = 48s^2\left[\frac{\pi}{4} - \frac{1}{2}\int_0^\infty e^{-st} \frac{\sin(st)}{\sinh(t/2)}\, dt\right] \tag{2.4.30a}$$

and

$$\phi(s) = 12s^2 \int_0^\infty \coth\left(\frac{t}{2}\right) e^{-st} \sin(st)\, dt - 6\pi s^2, \tag{2.4.30b}$$

respectively. Their asymptotic values are expressed by

$$H(s \to 0) = 12\pi s^2, \qquad H(s \to \infty) = 1 - \frac{0.022}{s^4}, \tag{2.4.30a'}$$

$$\phi(s \to 0) = 6s, \qquad \phi(s \to \infty) = 1 - \frac{0.012}{s^4}, \tag{2.4.30b'}$$

respectively. Numerical values of $H(s)$ and $\phi(s)$ are tabulated in Table 2.9 and are plotted in Fig. 2.23.

Table 2.9 Values of $H(s)$ and $\phi(s)$

s	$\phi(s)$	$H(s)$
0	0	0
0.05	0.258	0.094
0.1	0.446	0.206
0.2	0.686	0.475
0.3	0.805	0.695
0.4	0.880	0.800
0.5	0.931	0.875
0.6	0.954	0.917
0.7	0.965	0.945
0.8	0.975	0.963
0.9	0.985	0.975
1.0	0.990	0.985
1.5	0.998	0.994
2.0	0.999	0.998

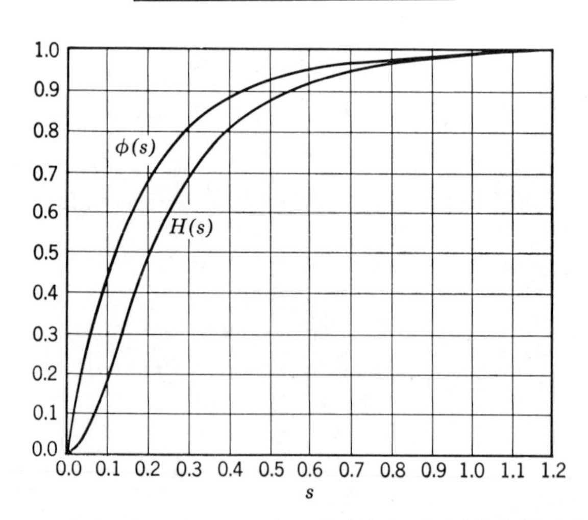

Fig. 2.23 Values of the functions $H(s)$ and $\phi(s)$ in (2.4.29) (Migdal 56).

Function $\xi(s)$ in (2.4.29) represents the logarithmic factor in (2.4.13). Depending on the magnitude of s, it is given as

$$\xi(s) = \begin{cases} \ln(191Z^{-1/3}) & \text{for} \quad s > 1, \\ \ln(191Z^{-1/3}) - \frac{1}{2}n(s) & \text{for} \quad 1 \geq s \geq \left(\dfrac{Z^{1/3}}{191}\right)^2, \\ 2 & \text{for} \quad s \leq \left(\dfrac{Z^{1/3}}{191}\right)^2. \end{cases} \tag{2.4.31}$$

The modification of the pair-creation cross section can be obtained by changing $E_0 - k$ into $k - E_0$. A remark may be necessary for E; this is now equal to $k/2$. Therefore we have

$$G(k, v) = \tfrac{1}{3}[H(s) + 2\{v^2 + (1-v)^2\}\phi(s)]\xi(s). \tag{2.4.32}$$

Table 2.9 and Fig, 2.23 show that the Landau-Pomeranchuk effect becomes appreciable for $4s < 1$. When $k = E_0/2$ and $\xi \simeq 1$, the energy above which the effect becomes important is given by

$$\frac{E_0}{mc} \gtrsim \frac{X_0}{2\pi\rho}\frac{Z}{Z+1}\frac{me^2}{\hbar^2} \simeq 3.0 \times 10^7 \frac{X_0}{\xi}\frac{Z}{Z+1}. \tag{2.4.33}$$

For lead ($X_0/\rho \simeq 0.5$ cm) and for air ($X_0/\rho \simeq 3 \times 10^4$ cm), the Landau-Pomeranchuk effect is considerable above 10^{13} and 5×10^{17} eV, respectively. For the emission of low-energy photons this effect becomes appreciable at lower energies.

2.4.7 *Interference Effects Due to the Lattice Structure of a Medium*

If a medium is a crystal, there arises an interference effect that is essentially identical with that in the diffraction of X-rays, electrons, or neutrons. Both in bremsstrahlung and in pair creation the recoil momentum of a nucleus may be so small as to be comparable to the reciprocal lattice constant. Then the outgoing amplitudes due to a great number of atoms are subject to interference, so that a Laue-Bragg interference takes place. The recoil momentum along the incident direction is the smallest, as small as

$$q_{\parallel} \simeq \frac{k(mc)^2 c}{E(E-k)} \quad \text{(bremsstrahlung)}, \tag{2.4.34a}$$

$$\simeq \frac{k(mc)^2 c}{E_+ E_-} \quad \text{(pair)}. \tag{2.4.34b}$$

The interference becomes appreciable for

$$\frac{2\pi\hbar}{a} \gtrsim q_{\parallel} \simeq \frac{(mc)^2 c}{E} \text{ (bremsstrahlung) for } k = \frac{E}{2}, \qquad (2.4.35a)$$

$$\simeq \frac{4(mc)^2 c}{k} \text{ (pair) for } E_+ = E_- = \frac{k}{2}, \qquad (2.4.35b)$$

where a is the lattice constant. Because the lattice constant a is about 3×10^{-8} cm, the interference effect is expected to occur at energies above 75 MeV for bremsstrahlung and above 300 MeV for pair creation. For the emission of a low-energy photon we expect a greater effect.

A detailed theory of the interference effect was worked out by Überall (Überall 56, 57), after whom this is usually called the Überall effect. According to his result the interference effect is considerable at emission angles around $q_{\parallel}(\hbar^2/me^2)Z^{-1/3}$; the cross section is larger than the Bethe-Heitler one at this angle and falls off in both sides. In bremsstrahlung the energy spectrum of emitted photons goes up toward low energy and shows a zigzag shape near the upper end. The energy spectrum of pair electrons has sharp zigzag peaks near both energy limits and deviates in the middle from the Bethe-Heitler one, depending on angle.

The Überall effect gives rise also to the polarization of a radiated photon. The degree of polarization may be as high as 30 percent at low photon energies, depending, of course, on the direction relative to crystal axes and emitted angles.

There are a number of effects that are analogous to conventional diffraction phenomena; one of these is the lattice vibration effect, which must be taken into account in quantitative theory. Experiments thus far carried out have not checked these details yet, but they have proved the essential correctness of Überall's theory.

2.4.8 Reduction of Ionization by Pair Electrons of Small Separation

Another effect of a medium is the nonadditivity of the energy loss by ionization of pair electrons emitted with an extremely small angle, though this is not directly related to the pair-creation process. If the distance between two charged particles is as small as, or smaller than, the shielding radius in a medium through which they propagate, the electromagnetic fields due to respective particles are subject to interference. Therefore the ionization loss is not a simple sum of (2.2.6) and (2.2.6′) but is reduced by a term that depends on the distance d and the shielding radius $D = (4\pi e^2 n/mc^2)^{-1/2}$, where n is the density of electrons (Chudakov 55). Thus the ionization loss of a pair of electrons with the same energy E is given by (Mito 57, Yekutieli 57)

$$-\frac{dE}{dx}\bigg|_{\text{pair}} = 2\frac{L}{\beta^2}\left[B + 3\ln\left(\frac{E}{mc^2}\right) - 0.13 - \tfrac{3}{2}\beta^2 - \delta - K_0\left(\frac{d}{D}\right)\right], \quad (2.4.36)$$

where K_0 is the modified Bessel function of the second kind. Because $D \simeq 10^{-6}$ cm in a dense material, the reduction of ionization is expected in the first 100 microns from the origin of a pair produced by a photon with energy of about 10^{10} eV. This reduction effect is sometimes called the *Chudakov effect* and has been verified experimentally. This effect may be of use in measuring photon energy.

2.5 Other Electromagnetic Processes

2.5.1 Direct Pair Creation by a Charged Particle

Virtual photons associated with an energetic charged particle are responsible for the creation of pairs of electrons. Since this is a higher order process than bremsstrahlung, its cross section for the electron is smaller, by a factor on the order of α, than the radiation cross section. For a heavy particle, however, the direct pair-creation cross section may be greater than the radiation cross section because the latter is inversely proportional to the square of mass, whereas the former is essentially independent of the mass of an incident particle. This may be explained by the fact that pair creation by a photon has a cross section on the order of $Z^2\alpha r_e^2$, whereas the density of virtual photons is on the order of α, with a cross section about $Z^2\alpha^2 r_e^2$. In comparison with this the bremsstrahlung cross section for a charged particle of mass M is on the order of $Z^2\alpha r_e^2(m/M)^2$. For $M^2 > m^2/\alpha$ therefore, direct pair creation has a larger cross section than bremsstrahlung.

Although the cross section is larger, the energy loss due to direct pair creation is not as large as we might at first expect. Since the effective impact parameter for direct pair creation is about $\hbar/mc$, which is compared with that for bremsstrahlung, $\hbar/Mc$, the energy transferred to the pair at the effective impact parameter is on the order of $(m/M)E$, where E is the incident energy, in contrast to the large fractional energy transfer in bremsstrahlung. Hence the average energy loss times the cross section in direct pair creation is on the order of $Z^2\alpha^2 r_e^2(m/M)E$, in comparison with $Z^2\alpha r_e^2(m/M)^2 E$ in bremsstrahlung. Thus the former is larger than tge latter if $M > m/\alpha$.

The quantitative calculation of the pair-creation cross section is rather difficult, mainly because four particles emerge in the final state. Most existing theoretical cross sections (Bhabha 35, Nishina 35) may be in error by a factor of 2 or so. The formula given below (Murota 56) may be recommended as most reliable, its inaccuracy being claimed to be at most 20 percent.

The differential cross section for the production of a pair of electrons with energies ε_+ and ε_- is expressed by a function of

$$u = \frac{\varepsilon_+ + \varepsilon_-}{E}, \qquad v = \frac{\varepsilon_+ - \varepsilon_-}{\varepsilon_+ + \varepsilon_-}, \tag{2.5.1}$$

where E is the energy of an incident particle of mass M, as

$$\frac{d\sigma}{du\,dv} = \frac{2}{3\pi}(Z\alpha)^2 r_e^{\,2} L\left(\frac{1+(1-u)^2}{2}\{[(2+v^2)+x(3+v^2)]\ln\frac{1+x}{x} - (3+v^2)\} + \frac{1-u}{1+x}(1-v^2) + \tfrac{1}{2}u^2 \times\left[\frac{x}{1+x} + (2+v^2) - x(3+v^2)\ln\frac{1+x}{x}\right]\right), \tag{2.5.2}$$

where

$$x = \tfrac{1}{4}\left(\frac{M}{m}\right)^2 \frac{u^2(1-v^2)}{1-u}.$$

L contains a logarithmic term representing an effect of the long tail of the Coulomb field; namely,

$$L = \ln\left[a\frac{Eu(1-v^2)}{2mc^2\sqrt{1+x}}\right] - 1 \quad \text{for} \quad \frac{2}{u(1-v^2)}\frac{mc^2}{E}(1+x) \gg \alpha Z^{1/3}$$

$$\text{(no screening)}, \tag{2.5.3a}$$

where a is a constant on the order of unity, and

$$L = \ln(191Z^{-1/3}\sqrt{1+x}) \quad \text{for} \quad \frac{2}{u(1-v^2)}\frac{mc^2}{E}(1+x) \ll \alpha Z^{1/3}$$

$$\text{(complete screening)}. \tag{2.5.3b}$$

The integration of (2.5.2) over v is carried out numerically for the muon passing through a medium of $Z = 12.9$ (a representative value of earth), and the result is shown in Fig. 2.24. This is the differential cross section for energy transfer to a pair of electrons and is a rapidly decreasing function of u; approximately u^{-3} for $u > 10^{-2}$ and approximately u^{-2} for $u < 10^{-2}$. The theoretical cross section was found to be

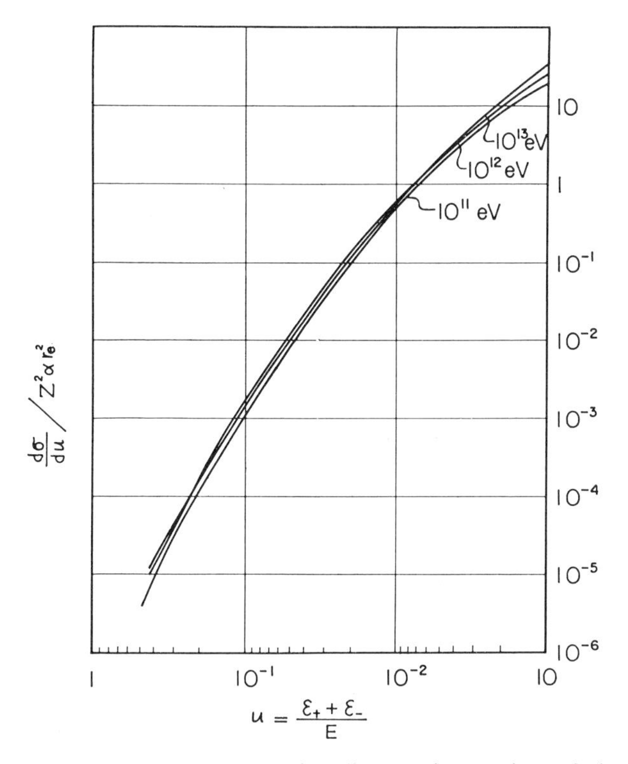

Fig. 2.24 Differential cross section for direct pair creation of the muon in a medium of $Z = 12.9$. The numbers attached to the curve indicate the energy of the muon. From A. Ueda, *Uchūsen Kenkyū*, **8**, No. 4, 595 (1963).

in good agreement with the experimental result obtained with a multiplate cloud chamber exposed to underground muons (Chaudhuri 63).

The total cross section in the case of no screening is expressed, depending on the energy transferred to a pair of electrons, u, by

$$\sigma_a^{(ns)} = \frac{4}{3\pi} Z^2 \alpha^2 r_e^{\,2} \left[\tfrac{7}{9} \ln^3 \left(\frac{E}{Mc^2} \right) - C_1 \ln^2 \left(\frac{E}{Mc^2} \right) + C_2 \ln \left(\frac{E}{Mc^2} \right) \right] \tag{2.5.4a}$$

for $u \lesssim (m/M)$, and

$$\sigma_b^{(ns)} = \frac{4}{3\pi k^2} Z^2 \alpha^2 r_e^{\,2} \left(\tfrac{5}{2} \ln 2 + \tfrac{3}{2} \right) \ln \left(k' \frac{E}{Mc^2} \right) \tag{2.5.4b}$$

for $u \gtrsim (m/M)$,
where k and k' are constants on the order of unity; C_1 and C_2 are

given by

$$C_1 = 0.85 - \tfrac{4}{3}\ln k'$$

and

$$C_2 = 20.81 - \tfrac{14}{3}\ln^2 k_1 - \tfrac{62}{9}\ln k' + \tfrac{28}{3}\ln\tfrac{3}{4}\ln k_2,$$

where k_1 and k_2 are again on the order of unity. In the above formulas uncertainty is implied in k, k', k_1, and k_2, all of which appear in logarithms. In the case of complete screening the cross section is expressed as

$$\sigma_a^{(a)} = \frac{4}{3\pi} Z^2\alpha^2 r_e^2 \ln(137Z^{-1/3})$$

$$\left[\tfrac{7}{3}\ln\left(k'\frac{E}{Mc^2}\right)\ln\left(\frac{E}{191Z^{-1/3}Mc^2}\right) + \tfrac{7}{9}\ln^2(191Z^{-1/3})\right] \tag{2.5.5a}$$

for $u \lesssim (m/M)$, and

$$\sigma_b^{(s)} = \frac{4}{3\pi k^2} Z^2\alpha^2 r_e^2 (\tfrac{5}{2}\ln 2 + \tfrac{3}{2})\ln(k'191Z^{-1/3}) \tag{2.5.5b}$$

for $u \gtrsim (m/M)$.

If the incident particle is an electron, it is necessary to take the exchange effect into account. However, the effective impact parameter

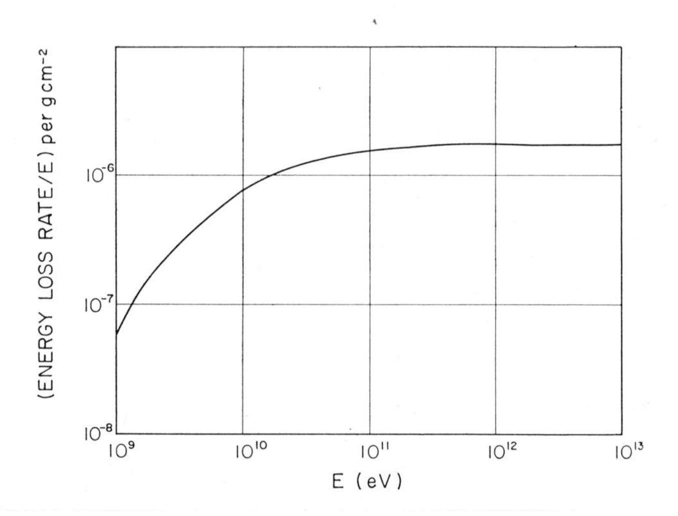

Fig. 2.25 The energy-loss rate by direct pair creation of the muon in a medium of $Z = 12.9$ and $A = 26.3$. From A. Ueda, *Uchūsen Kenkyū*, **8**, No. 4, 596 (1963).

is so large that this is practically negligible. Since the region corresponding to (2.5.4*b*) or (2.5.5*b*) does not exist, the cross section is simply given by (2.5.4*a*) or (2.5.5*a*), according to whether the screening is not effective or complete.

The energy loss due to direct pair creation will be obtained by reference to the differential cross section. However, this is so complicated that a simple analytic expression of the energy-loss formula cannot be given; only its numerical values for the muon are shown in Fig. 2.25. As remarked above, this is as large as the radiation energy loss at high energies ($1.76 \times 10^{-6}E$ eV/g-cm^{-2} at $E = 10^{12}$ eV), so that we must take this into account for muons of energies above 10^{12} eV.

As in the case of bremsstrahlung, it may be necessary to consider various correction factors that arise, for example, from the Coulomb effect and the Landau-Pomeranchuk effect. However, no serious attempt to evaluate such correction factors has yet been made.

2.5.2 *Annihilation of Positons*

The behavior of the positon at high energies is nearly the same as that of the negaton. With decreasing energy, however, the excess absorption of the positon becomes appreciable because the probability of annihilation with an electron in matter increases.

The annihilation cross section of a positon of energy γmc^2 by collision with a free electron at rest is given by (Dirac 30)

$$\sigma_{\text{an}} = \pi r_e^2 \frac{1}{\gamma + 1} \left[\frac{\gamma^2 + 4\gamma + 1}{\gamma^2 - 1} \ln (\gamma + \sqrt{\gamma^2 - 1}) - \frac{\gamma + 3}{\sqrt{\gamma^2 - 1}} \right]. \quad (2.5.6)$$

As a result of annihilation two photons are emitted, preferentially in the forward and backward directions, in the center-of-mass system; in the laboratory system one photon possesses a high energy, whereas the other has an energy on the order of mc^2. At nonrelativistic energies two photons are emitted in opposite directions, and their polarizations are mutually perpendicular.

At extreme relativistic energies the expression (2.5.6) can be approximated as

$$\sigma_{\text{an}} \simeq \pi r_e^2 \frac{1}{\gamma} [\ln (2\gamma) - 1], \quad (2.5.7a)$$

whereas at nonrelativistic energies

$$\sigma_{\text{an}} \simeq \pi r_e^2 \frac{1}{\beta}. \quad (2.5.7b)$$

At low energies, therefore, the annihilation rate is equal to (N/A) $Z\pi r_e^2 c$, independent of velocity. The lifetime of a positon might thus be expected to be about 0.5×10^{-10} sec.

Actually, however, the lifetime was found to be longer than expected (Deutsch 51). This is due to the formation of positronium, a bound system consisting of a positon and a negaton. Its lifetime is estimated to be 1.2×10^{-10} sec for the ^1S-state, from which two photons are emitted, and 1.4×10^{-7} sec for the ^3S-state, from which three photons are emitted, both more accurately depending on the electronic structure of the absorbing material.

In correspondence to the photoelectric effect the annihilation may take place with a bound electron by emitting one photon. The one-photon annihilation cross section is given for a K-electron by

$$\sigma_{\mathrm{an}}(K) = 4\pi Z^5 \alpha^4 r_e^2 \frac{1}{\beta\gamma(\gamma+1)^2}\left[\gamma^2 + \tfrac{2}{3}\gamma + \tfrac{4}{3} - \frac{\gamma+2}{\beta\gamma}\ln(1+\beta)\gamma\right] \quad (2.5.8)$$

At extremely relativistic and nonrelativistic energies this is reduced respectively to

$$\sigma_{\mathrm{an}}(K) \simeq \frac{4\pi Z^5 \alpha^4 r_e^2}{\gamma} \qquad (\gamma \gg 1), \quad (2.5.9a)$$

$$\sigma_{\mathrm{an}}(K) \simeq \frac{4\pi}{3} Z^5 \alpha^4 r_e^2 \beta \qquad (\beta \ll 1). \quad (2.5.9b)$$

Since $Z^5\alpha^4 \ll 1$, this is smaller than the cross section of two-photon annihilation.

2.5.3 *Collision Between Two Photons*

Pair creation by the collision of two photons is a process that is the inverse of two-photon annihilation. In the head-on collision with a photon of energy ε its threshold energy is $(mc^2)^2/\varepsilon$. The cross section is given by

$$\sigma = \frac{\pi}{4} r_e^2 \left(\frac{mc^2}{k^*}\right)^2 \beta^* \left[2(\beta^{*2} - 2) + \frac{3-\beta^{*4}}{\beta^*}\ln\left(\frac{1+\beta^*}{1-\beta^*}\right)\right], \quad (2.5.10)$$

where k^* is the energy of a colliding photon and $c\beta^*$ is the velocity of an outgoing electron in the center-of-momentum system.

If a pair of electrons are produced only virtually, the collision between two photons results in photon-photon scattering. Its cross section is roughly α^2 times (2.5.10), but it decreases below the pair-creation threshold with decreasing energy as $(k^*/mc^2)^6$. Since the scattering cross section does not exceed 10^{-29} cm^2, this process is of little practical interest.

The scattering of photons by a Coulomb field—namely, Delbrück scattering—is an analogous process. The cross section may be further multiplied by $Z^4\alpha^2$, but a more accurate estimate gives us about 10^{-28} cm^2 for lead at about $k = mc^2$.

2.6 Cascade Shower

Since the energy distributions of both photons and surviving electrons in bremsstrahlung and those of electrons in pair creation are rather flat, the secondary electrons and photons from these processes have energies high enough to produce tertiary particles by the same processes with rather high probabilities. If a primary particle has a high energy, these processes continue to take place successively, and the number of particles produced thereby increases. Such a sequence of multiplicative processes begins to decline as some of the particles receive energies that are too low to produce further particles effectively. Then the number of the particles reaches a maximum and decreases as the thickness of matter traversed increases still further. This phenomenon is called the *cascade shower* and is characteristic of the interactions of high-energy electrons and photons.

Conventionally the number of electrons or photons in a cascade shower is expressed as a function of the primary energy of an electron or a photon, E_0, and of the thickness of matter traversed, t, as $N(E_0, t)$. If we are interested in electrons or photons with energies greater than E (as in most practical cases), we put E in the argument as $N(E_0, E, t)$. It is convenient to express E_0 and E in units of the critical energy and t in units of radiation length, because the functional form then becomes nearly independent of the nature of the matter. When we are specifically concerned with the number of electrons or photons, we may distinguish N by subscripts, such as N_β or N_γ. Similarly, we may distinguish between the energies of an electron and a photon by E and W, respectively.

The number of particles, N, is not completely determined by giving E_0, E, and t. Actually the value of N fluctuates to a considerable extent. Thus we define N as the average value. The probability of finding n particles is denoted as $P(n, E_0, E, t)$. The average number N is therefore given by

$$N(E_0, E, t) = \sum_{n=0}^{\infty} nP(n, E_0, E, t). \tag{2.6.1}$$

The magnitude of fluctuations is represented by the variance

$$\sigma^2 = \sum_{n=0}^{\infty} n^2 P(n) - N^2. \tag{2.6.2}$$

In the above we have been concerned only with the one-dimensional cascade shower. In practice particles in a shower spread laterally, largely because of the Coulomb scattering of electrons. Often there is interest in the number of particles whose angles with respect to the shower axis or to the incident direction are smaller than θ and in the number of particles whose lateral distances from the axis are smaller than r. Their average numbers are denoted as $N(E_0, E, \theta, t)$ and $N(E_0, E, r, t)$, respectively. By definition,

$$N(E_0, E, \theta = \infty, t) = N(E_0, E, r = \infty, t) = N(E_0, E, t). \tag{2.6.3}$$

The differential angular and lateral distributions are often used:

$$f(\theta) = \frac{1}{N(\theta = \infty) \sin\theta} \frac{dN(\theta)}{d\theta}, \qquad f(r) = \frac{1}{N(r = \infty) 2\pi r} \frac{dN(r)}{dr}. \tag{2.6.4}$$

The material dependence of the lateral distribution is almost eliminated if r is measured in Molière units, which are defined by

$$r_M \equiv X_0 \frac{\varepsilon_0}{E_s}. \tag{2.6.5}$$

This gives roughly the rms spread of electrons of energy equal to the critical energy while they travel through one radiation length. Fluctuations are as important in the three-dimensional cascade as in the one-dimensional case, and analogous expressions hold for the fluctuation problem in the former case, too.

2.6.1 *Characteristic Features of Analytic Solutions*

The cascade process provides an interesting example of a stochastic process and has been treated by many authors. Analytic methods of solving the cascade problem are believed to give essentially correct results in light media. In heavy media the energy dependence of the cross sections and the back scattering of electrons play important roles, so that the analytic treatment of the problem is prohibitively difficult, except in the case of particles with energies considerably higher than the critical energy. Therefore numerical methods are needed to obtain results of practical use. For the fluctuation problem the numerical method is found in all cases to be superior to the analytic method.

The analytic method can be worked out with some approximations. Rossi and Greisen (Rossi 41, 52) introduced two ranks of approximation, by which solutions are obtainable with relative ease. Approximation A means that all cross sections taking part in the cascade process are expressed in fractional form; in other words, the loss of energy by ionization and the Compton scattering are neglected, and the asymptotic forms of complete screening are used for the radiation and the pair-creation cross sections. By the use of this approximation the average behavior of a shower is expressed by a function of E_0/E, or

$$N_A(E_0, E, t) = N\left(\frac{E_0}{E}, t\right). \qquad (2.6.6)$$

As E approaches the critical energy the ionization loss becomes very important. This is included in approximation B—but as a constant energy-loss term. Then the average behavior is expressed as

$$N_B(E_0, E, t) = N\left(\frac{E_0}{\varepsilon_0}, \frac{E}{\varepsilon_0}, t\right). \qquad (2.6.7)$$

With approximation B it is not easy to obtain an explicit solution for finite E. The one-dimensional theory for these approximations was worked out by Rossi and Greisen (Rossi 41, 52), and by Belenkij (Belenkij 48).

In order to solve the three-dimensional cascade we further make the Landau approximation, which assumes a small angular divergence. Even with this approximation the mathematical procedure of obtaining the structure function, $N(E_0, E, \theta \text{ or } r, t)$, is so involved that we usually obtain only its moments (Belenkij 48):

$$\langle\theta^n\rangle = \int \theta^n f(\theta) \sin\theta \, d\theta, \qquad \langle r^n\rangle = \int r^n f(r) \, 2\pi r \, dr. \qquad (2.6.8)$$

The structure function has been obtained by Kamata and Nishimura (Kamata 58) (hence it is called the N-K function) in some limited cases; for example, for $E_0 = \infty$ in general and for finite E_0 near the axis. The Landau approximation can be avoided, and the structure function at the shower maximum is also obtained (Kamata 58).

Because these approximations are permissible when we deal with particles of rather high energies even in heavy media, the analytic method is most useful in determining the lateral distribution of energy flow and the number of electrons found within a given distance from the axis (Kamata 58). In these problems theoretical results are important in estimating the energy of a primary particle.

The mathematical detail of the analytic method is quite involved and is not given in this book. Instead we recommend that readers refer to a comprehensive treatment presented by Nishimura (Nishimura 67). Important numerical results are summarized in Appendix C.

An important parameter in the analytic expression is the age parameter, denoted by s. This represents the age of a shower and increases as it develops; $s=0$ at the starting point, $s=1$ at the maximum, and $s>1$ in the declining stage of a shower. The shower curve as a function of s exhibits similarity independent of energy; for example, the slope of a shower curve is expressed approximately by a function of s alone as (Greisen 56)

$$\frac{\partial N}{\partial t} \simeq \lambda_1(s) \simeq \tfrac{1}{2}(s-1-3\ln s), \tag{2.6.9}$$

except at very small values of s.

The value of s is determined as the saddle point in the inverse Mellin transformation. By introducing parameters representing energy and lateral spread respectively as

$$y=\ln\frac{E_0}{E} \quad \text{or} \quad y_0=\ln\frac{E_0}{\varepsilon_0}, \qquad x=\ln\frac{r}{r_{\rm M}}, \tag{2.6.10}$$

where y and y_0 are used in approximations A and B, respectively, we obtain the value of s from

$$\lambda_1'(s)t+y \simeq \lambda_1'(s)t+y_0+x \simeq 0. \tag{2.6.11}$$

In deriving (2.6.11) we have taken account of

$$\frac{E}{\varepsilon_0}=\frac{r_{\rm M}}{r}; \tag{2.6.12}$$

in other words, the structure function is a function of $Er/\varepsilon_0 r_{\rm M}$. The approximate expression of $\lambda_1(s)$ in (2.6.9) yields

$$s \simeq \frac{3t}{t+2y}=\frac{3t}{t+2y_0+2x}. \tag{2.6.13}$$

In the one-dimensional cascade in approximation B, y in the first expression of (2.6.11) and the second expression in (2.6.13) is replaced by y_0.

The shower curve is expressed as

$$N(E_0, E, t) \simeq \frac{0.135}{y^{1/2}} \exp\,[\lambda_1(s)t+sy] \simeq \frac{0.135}{y^{1/2}} \exp\,[t(1-\tfrac{3}{2}\ln s] \tag{2.6.14a}$$

and

$$N(E_0, 0, t) \simeq \frac{0.31}{y_0^{1/2}} \exp\,[\lambda_1(s)t + sy_0] \simeq \frac{0.31}{y_0^{1/2}} \exp\,[t(1 - \tfrac{3}{2} \ln s)]. \tag{2.6.14b}$$

The shower maximum is reached at

$$\lambda_1(s) = 0, \qquad (s=1;\ t_m \simeq y \text{ or } y_0). \tag{2.6.15}$$

At the shower maximum the integral energy spectrum of electrons can be given more precisely as

$$N_{\max}(E_0, E) = \frac{0.31(E_0/\varepsilon_0)}{y_0^{1/2}} \phi(\varepsilon), \tag{2.6.16}$$

where

$$\phi(\varepsilon) = 1 - \varepsilon e^{\varepsilon}[-Ei(-\varepsilon)], \qquad -Ei(-\varepsilon) = \int_{\varepsilon}^{\infty} e^{-x} \frac{dx}{x}, \qquad \varepsilon = \frac{2.3E}{\varepsilon_0}.$$

The asymptotic expressions of the energy spectrum are

$$\begin{aligned} \phi(\varepsilon) &\simeq 1 - \varepsilon e^{\varepsilon} \left[\ln\left(\frac{1}{\gamma\varepsilon}\right) + \varepsilon - \frac{\varepsilon^2}{2.2!} + \cdots \right] \quad \text{for} \quad \varepsilon < 1, \\ \phi(\varepsilon) &\simeq \frac{1}{\varepsilon}\left(1 - \frac{2}{\varepsilon} + \cdots\right) \quad \text{for} \quad \varepsilon > 1, \end{aligned} \tag{2.6.17}$$

where $\gamma = 1.7811$ and $\ln \gamma = 0.5772$.

The energy spectrum in the general case is roughly expressed as

$$N(E_0, E, t) \propto E^{-s}. \tag{2.6.18}$$

Noticing that the structure function depends on rE, we obtain a rough lateral distribution

$$f(r) \propto r^{s-2}. \tag{2.6.19}$$

Both (2.6.18) and (2.6.19) show the usefulness of introducing the age parameter. At s greater than 2 the lateral distribution becomes flat. The flat region may be defined as the core, and its radius is given for $s = 2$ and $\lambda_1'(s = 2) = \frac{1}{4}$ as

$$r_c = r_{\mathrm{M}} \left(\frac{\varepsilon_0}{E_0}\right) \exp \frac{t}{4}. \tag{2.6.20}$$

More precisely, the lateral distribution is expressed as

$$f\left(\frac{r}{r_M}\right) = C(s)\left(\frac{r}{r_M}\right)^{s-2}\left(\frac{r}{r_M}+1\right)^{s-4.5}, \tag{2.6.21}$$

where $C(s)$ is the normalization factor whose numerical values are given in Table 2.10 and are approximately represented by

$$C(s) \simeq 0.443 s^2 (1.90 - s) \quad \text{for} \quad s < 1.6$$
$$C(s) \simeq 0.366 s^2 (2.07 - s)^{5/4} \quad \text{for} \quad s < 1.8.$$

Table 2.10 Values of $C(s)$

s	0.50	0.75	1.00	1.25	1.50
$C(s)$	$16/10\pi^2$	$231/256\pi$	$5/4\pi$	$45/32\pi$	$4/\pi^2$

The expression (2.6.21) is called the N-K-G function (after Nishimura-Kamata-Greisen) and is found to reproduce well the exact one for $0.6 < s < 1.0$, whereas at larger s it is too small at small distances but still good for $r > r_M$.

2.6.2 *Monte Carlo Method*

The analytic method fails to give correct answers for shower particles of low energies in heavy materials. This is because the cross section depends considerably on energy, and the scattering of low-energy electrons is quite large. The energy dependence of the cross section is such that the cross section for photon absorption has a minimum at an energy at which the cross section for pair creation is comparable to that for the Compton effect. Therefore the attenuation of a shower at great depths is determined essentially by the absorption length of photon of this energy; most electrons at such great depths are produced by such photons. The scattering of low-energy electrons is so large that electrons with energies as low as the critical energy or lower are scattered nearly isotropically. It is therefore expected that few electrons with such energies come out in the direction of the shower, but many of them go backward. Such complicated features of showers can hardly be treated by the analytic method.

The analytic method is also inappropriate for the fluctuation problem, although considerable effort has been made for mathematical development (Arley 48). Analytic solutions were obtained only with simplified models. A number of numerical calculations have therefore been aimed at solving the fluctuation problem.

The numerical method consists in the construction of model showers that are supposed to be randomly chosen from a population of actual showers. The random choice is made by selecting one of many possible events, each of which is expected to occur with equal probability. This method is thus called the Monte Carlo method.

The first attempt at the Monte Carlo calculation of cascade showers was made by using a wheel of chance, according to which the rotation angle of a wheel is controlled by a signal of a cosmic-ray particle (Wilson 52). Unfortunately some of the results were not properly presented in the paper (Wilson 52), but they were corrected in a recent paper (Thom 64).

A series of comprehensive calculations have been achieved by Messel and his collaborators (Butcher 58, 60; Crawford 62; Messel 62). The results are so extensive that only a few representative ones are reproduced here. The results available in these papers may be seen from Table 2.11, in which only references are given.

These numerical calculations are made for showers developing in air, aluminum, nuclear emulsion, and lead. These media are assumed to be homogeneous, and the numbers of electrons and photons expected to

Table 2.11 Numerical Results on Cascade Showers

	Butcher 58	Butcher 60	Messel 62	Adachi 64	Crawford 62
Medium	Air	Air	Lead, emulsion	Air	Lead
Primary energies[a] (MeV)	5×10^n ($n = 1, 2, 3$)	50; 1, 2, 5×10^n ($n = 2, 3, 4$)	β : 1000	10, 32, 100 $300 \times E$	50, 100, 200 500, 1000
Secondary energies[b] (MeV)	5, 50	5; 1, 2, 5×10^n ($n = 1, 2, 3, 4$)	1, 2, 5×10^n ($n = 1, 2$)	$\beta + \gamma$	>10 <10
t	≤ 10	≤ 10	≤ 10	≤ 8	≤ 10
Tables and graphs	$N(E_0, E, t)$ $P(n, E_0, E, t)$ $P(\geq n, E_0, E, t)$ $P(\leq n, E_0, E, t)$ (c)	$N(E_0, E, t)$ $P(n, E_0, E, t)$ $P(\geq n, E_0, E, t)$ $P(\leq n, E_0, E, t)$ $\Delta N(E_0, E, t)$ (c)	$N(E_0, E, t)$ $P(n, E_0, E, t)$ $N(E_0, E, r, t)$ $\Delta N(E_0, E, r, t)$ $N(E_0, E, \theta, t)$ $\Delta N(E_0, E, \theta, t)$ $P(n, E_0, E, r, t)$ $P(n, E_0, E, \theta, t)$	$N(E_0, E, t)$ $P(n, E_0, E, t)$ $N(E_0/E, r, t)$ $\Sigma E(E_0/E, t)$ ΣEr; ΣEr^2, $\Sigma E(E_0/E, r, t)$ Σr (c)	$N(E_0, E, t)$ $\Sigma E(t)$ T

[a] Of both electrons (β) and photons (γ) if not specified.
[b] Of electrons alone if not specified.
[c] Results of approximation A are given.

be present at given thicknesses of the media are calculated. The angular and radial distributions are also calculated. The results are checked by experiments with monoenergetic beams of electrons that produce showers in multi-lead-plate spark chambers (Kajikawa 63), cloud chambers (Becklin 64, Thom 64), and X-ray films (Murata 65), as well as with scintillation probes in various materials (Kantz 54). With the first two methods the tracks of individual electrons can be counted for each shower, and consequently the results are suitable for checking fluctuations. With the last two methods only the average behavior can be studied—but with better precision (especially for the lateral distribution), because these detectors are essentially homogeneous. We shall also present some illustrative results of these experiments

The numerical as well as experimental methods have not only proved the correctness of cascade theory but also provided a basis for the application of cascade theory, for example, to the measurement of γ-ray energy and to the design of electron and γ-ray detectors. Since the available results are limited to the energy region below 1 GeV, they are mainly used for laboratory experiments. For cosmic-ray use we need to know the features of cascade showers at higher energies. This is possible by combining the analytic and numerical solutions, because the former are believed to be correct as long as low-energy particles are cut off. However, the fluctuation problem cannot be solved by this means, and the Monte Carlo calculation is still needed, even in approximation A. This has been achieved by aiming at the analysis of γ-ray events observed in emulsion chambers, by which it has been possible to detect both γ-rays that are produced directly by neutral pion decays and those that have been affected by cascade processes in air above the detector (Adachi 64). The results of this calculation are compared with those of another one (Butcher 60) as well as with the analytic solutions (Nishimura 67) when comparison is possible.

2.6.3 Results of Numerical Calculations

The necessity of the numerical calculation may be revealed by the comparison of shower curves with given E_0/E. In approximation A the average number of shower particles versus depth is a function of E_0/E but does not depend separately on E_0 and E. For showers in light media this approximation is expected to be valid if E is greater than the critical energy. Actually, however, the result depends considerably on the value of E_0 and E, as seen in Fig. 2.26, where the cascade curves are drawn for $E_0/E = 100$ but for $E_0 = 1$, 10, and 50 GeV. Agreement with the result of approximation A is obtained only when $E \gtrsim \varepsilon_0$ for

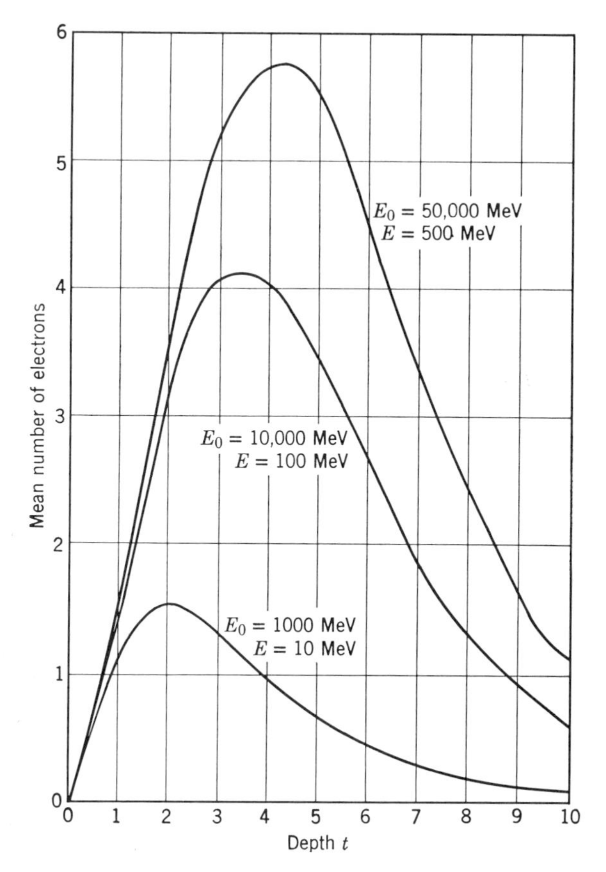

Fig. 2.26 The average number of electrons with energies greater than $E = 10$, 100, and 500 MeV is plotted against the depth t in radiation lengths for three typical cases of photon-initiated showers with primary energies $E_0 = 1000$, 10,000, and 50,000 MeV in air. In the three cases $E_0/E = 100$ (Butcher 60).

$E_0/E \leq 100$; for a large value of E_0/E, such as $E_0/E = 1000$, agreement is never reached.

Fluctuations in the number of shower particles are shown in Figs. 2.27 and 2.28. Figure 2.27*a* shows the probability of finding n electrons with energies greater than 10 MeV in a shower initiated by a photon of 1 GeV, whereas Fig. 2.27*b* gives the probability of finding n electrons with energies greater than 500 MeV in a shower initiated by an electron of 50 GeV. It should be remarked that the probability of not finding any particle is considerable. The distributions of the electron number at various thicknesses of nuclear emulsion are shown in Fig. 2.28. It is found to be approximately a Poisson distribution for $t \geq 3$, whereas the Furry distribution gives a good approximation at smaller values of t.

In heavy media such as lead even approximation B is not valid. The inadequacy of the analytic solution is most remarkably seen in the

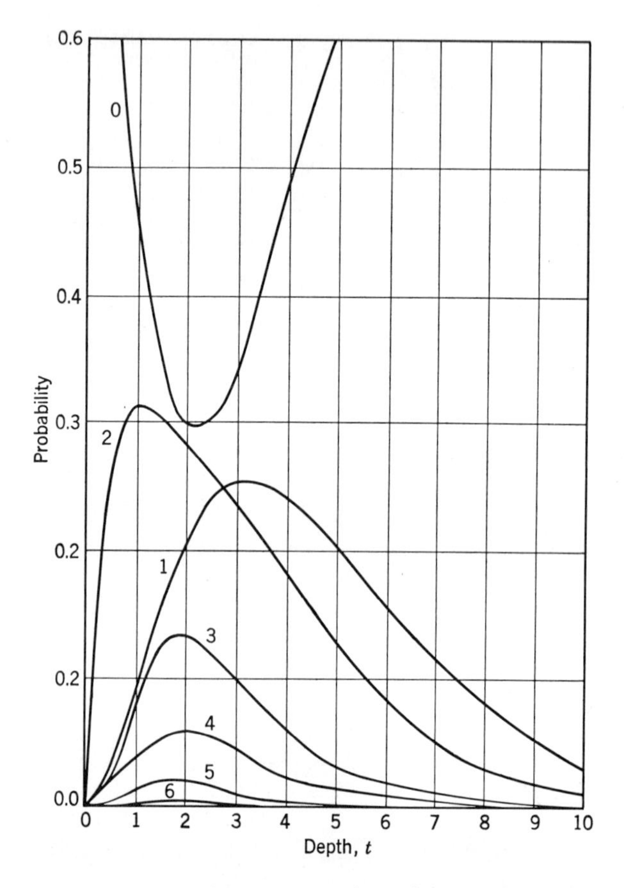

Fig. 2.27*a* The probability of finding exactly n electrons for $n = 0, 1, 2, 3, 4, 5$, and 6 is plotted against the depth t in radiation lengths in an air absorber. The shower is initiated by a photon with primary energy $E_0 = 1000$ MeV, and each of the n electrons must have an energy greater than 10 MeV (Butcher 60).

presence of backward particles, as indicated in Table 2.12. In this table particles with $E > 10$ MeV may be understood to be those that appear underneath the absorber, whereas those with $E < 10$ MeV are those that disappear in the absorber. It must be noticed that about 15 percent of the incident energy is still left in the shower even after

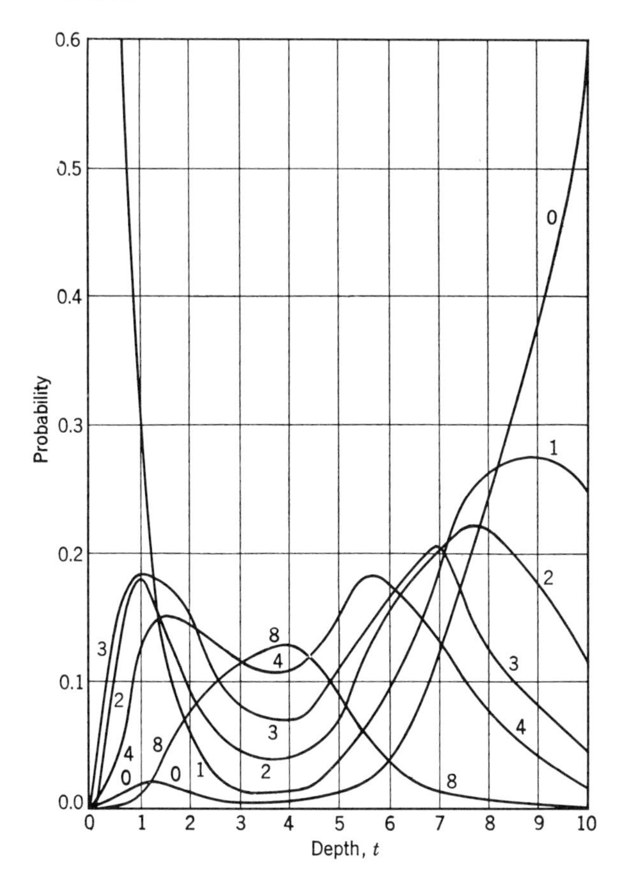

Fig. 2.27*b* The probability of finding exactly n electrons for $n = 0, 1, 2, 3, 4,$ and 8 is plotted against the depth t in radiation lengths in an aluminum absorber. The shower is initiated by an electron with primary energy $E_0 = 50{,}000$ MeV, and each of the n electrons must have an energy greater than 500 MeV (Butcher 60).

$t = 10$. In the tail part of the shower curve the attenuation of energy contained in a shower is expressed by

$$E = E_0 a e^{-bt} \tag{2.6.22}$$

for $t \geq 5$. This is dictated by the minimum cross section for the absorption of photons. With the values of a and b listed in Table 2.13, the penetration depth of a shower is given by

$$t_p = 5.0 \ln \left(\frac{E_0}{30.5}\right) \quad \text{for primary electrons} \tag{2.6.23}$$

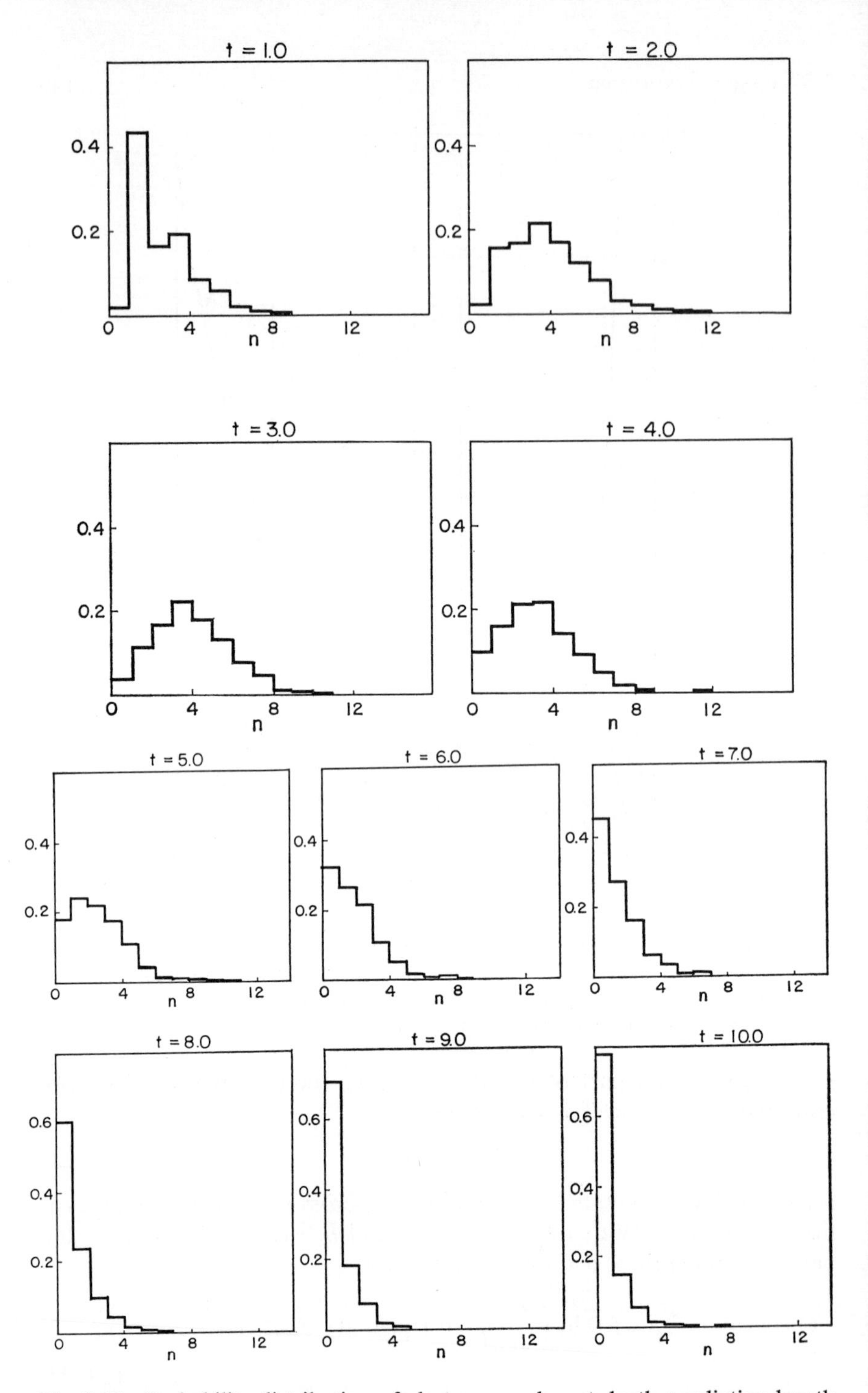

Fig. 2.28 Probability distribution of electron number at depth t radiation length in nuclear emulsion in a shower initiated by an electron of 1000 MeV. Abscissa: number of electrons with energies greater than 10 MeV; ordinate: probability of finding n electrons with energies greater than 10 MeV.

Table 2.12 Monte Carlo Results of Cascade Showers for a Lead Absorber with Primary 1000-MeV Photons

Depth	Energy loss[a] to				Ioniza-tion	Total	Electron Track Length	Energy Left in Cascade			
	Electrons $E < 10$	Photons $E < 10$	Backward Electrons	Backward Photons				Electrons $E > 10$	Photons $E > 10$	Total $E > 10$	Fraction Left
0.5	0.026	1.702	0.002	0.000				0.614	1.218	1.832	
	0.145	1.983	0.044	0.000	1.453	3.625	0.169	223.876	772.497	996.373	0.99637
1.0	0.234	5.234	0.010	0.000				1.286	2.146	3.432	
	1.524	5.697	0.152	0.000	4.099	11.472	0.489	319.237	665.666	984.903	0.98490
1.5	0.548	9.052	0.024	0.000				1.908	3.306	5.214	
	3.864	9.631	0.379	0.000	6.765	20.638	0.822	347.267	617.012	964.279	0.96428
2.0	1.058	13.142	0.050	0.002				2.774	4.542	7.316	
	7.271	14.025	0.775	0.024	10.039	32.134	1.238	358.111	574.063	932.173	0.93217
2.5	1.590	18.372	0.086	0.000				3.450	5.714	9.164	
	11.175	17.416	1.251	0.000	13.502	43.344	1.686	354.458	534.368	888.826	0.88883
3.0	2.232	21.668	0.110	0.000				3.790	6.926	10.716	
	15.803	20.309	1.735	0.000	15.965	53.813	2.009	323.545	511.400	834.945	0.83495
4.0	5.878	48.108	0.298	0.000				3.866	8.312	12.178	
	41.338	44.450	4.150	0.000	34.409	124.346	4.362	247.199	463.330	710.528	0.71053
5.0	6.498	47.728	0.378	0.002				3.588	3.702	12.290	
	45.392	43.248	5.321	0.031	34.075	128.067	4.347	210.473	371.818	582.291	0.58229
6.0	6.398	42.920	0.376	0.002				3.178	8.480	11.658	
	44.817	38.266	5.069	0.029	30.923	119.103	3.961	146.162	316.904	463.067	0.46307
7.0	6.156	37.052	0.386	0.006				2.470	7.708	10.178	
	43.147	32.602	5.378	0.065	26.183	107.374	3.367	103.700	251.891	355.592	0.35559
8.0	5.252	28.330	0.352	0.004				1.840	6.824	8.664	
	36.462	24.737	4.847	0.052	20.022	86.121	2.583	68.793	200.619	269.412	0.26941
9.0	4.330	21.080	0.240	0.004				1.378	5.744	7.122	
	29.943	18.188	3.397	0.072	14.681	66.281	1.900	43.114	159.926	203.040	0.20304
10.0	3.790	15.646	0.188	0.000				0.990	4.626	5.616	
	26.282	13.353	2.574	0.000	11.002	53.211	1.428	31.143	118.619	149.763	0.14976
Totals	43.992	310.204	2.500	0.020							
	307.165	283.906	35.070	0.273	223.117	849.531	28.362				

[a] The upper figure at each depth indicates the average number of particles; the lower figure, the average energy lost thereby.

and

$$t_p = 5.2 \ln\left(\frac{E_0}{22.7}\right) \quad \text{for primary photons.}$$

It may be seen from Table 2.13 that a considerable fraction of the energy dissipates into unseen parts. Only a quarter of the incident energy may be observable as electrons going forward. This results in the observable track length being smaller by a factor of 4 than the total track length given by the energy conservation. The observable track length given in Table 2.13 is expressed as

$$T = 0.032E_0 \qquad (E_0 \text{ in MeV}). \tag{2.6.24}$$

The remaining energy is lost as low-energy electrons and photons with average energies of 6.98 and 0.91 MeV, respectively. Back-scattered particles may or may not be observable, depending on the method of detection. Such electrons and photons have average energies of 14.0 and 13.1 MeV, respectively. Electrons going forward with energies greater than 10 MeV are usually observable. The numbers of electrons in showers are shown in Figure 2.29*a*, *b*. By comparing this with experimental results (Kajikawa 63, Thom 64), we see that the cutoff energy of 10 MeV is a little too high. In fact, the experimental results show a

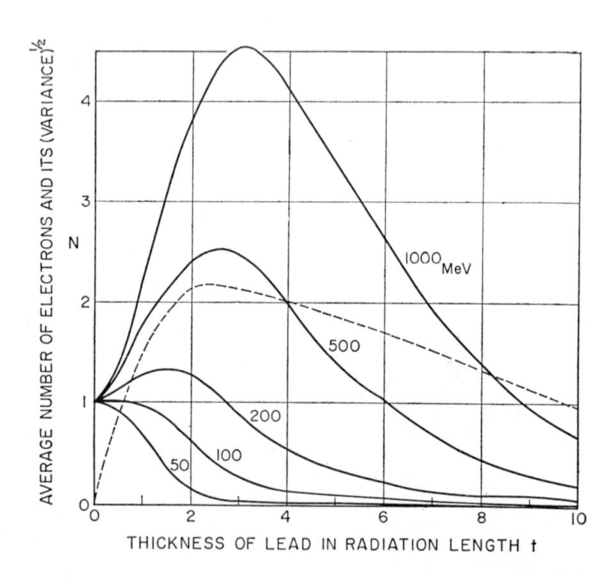

Fig. 2.29*a* Average number of cascade electrons with energies greater than 10 MeV. Primary electron: 50, 100, 200, 500, and 1000 MeV; absorber: lead. Dashed curve shows the square root of variance for the primary energy of 1000 MeV.

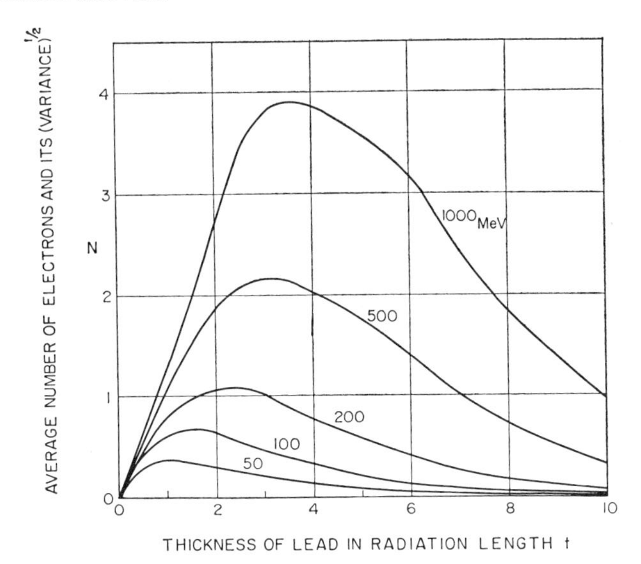

Fig. 2.29*b* Average number of electrons with energies greater than 10 MeV. Primary photon: 50, 100, 200, 500, and 1000 MeV; absorber: lead.

higher value of the track length as well as the deviation from the linear dependence as

$$T = 0.073 E_0^{0.92} \quad \text{for} \quad E_0 \leq 500 \text{ MeV}. \tag{2.6.25}$$

The deviation may be due partly to the energy still left in a shower after all absorbers have been traversed.

The numerical calculation of the lateral distribution may be of greater value because the analytic solution, called the Nishimura-Kamata (N-K) function, has been obtained only approximately for finite values of the incident energy. The comparison is made in Fig. 2.30 at the shower maximum, and the analytic solution is found to reproduce essential features of the lateral distribution. This can also be seen in the comparison of the mean-square spreads of electrons, as shown in Fig. 2.31 (Misaki 64). This, however, demonstrates that the convergence to the solution of $E_0/E \to \infty$ is very slow.

The comparison is also made with the Nishimura-Kamata-Greisen (N-K-G) function in Table 2.14. It is interesting to note that the values of s and α in Table 2.14 do not always vary smoothly with t. Moreover, the value of α almost always exceeds the upper limit $4.5 - s$ that is expected in the N-K-G function. This indicates that the N-G-K function is good only for $x < 1$ and s around unity.

Table 2.13 Average Behavior of Showers in Lead

Primary Energy (MeV)	(Track Length)/E_0	Percentage Energy Loss to				Ionization	Energy Decay		Penetration
		Electrons $E < 10$ MeV	Photons $E < 10$ MeV	Backward Electrons	Backward Photons		a	b	
Primary Electron									
50	0.03359	36.28	32.80	4.89	0.04	25.99	0.43 ± 0.02	0.331 ± 0.007	2.31
100	0.03237	38.21	31.93	4.61	0.05	25.20	0.87 ± 0.03	0.363 ± 0.004	5.96
200	0.03199	38.67	31.71	4.64	0.04	24.94	1.26 ± 0.02	0.339 ± 0.003	9.52
500	0.03188	39.19	31.09	4.89	0.05	24.78	1.93 ± 0.03	0.332 ± 0.002	13.76
1000	0.03155	40.37	30.44	4.81	0.02	24.37	2.35 ± 0.03	0.315 ± 0.002	17.33
Primary Photon									
50	0.03184	40.71	29.64	5.09	0.12	24.45	1.21 ± 0.02	0.351 ± 0.003	5.13
100	0.03261	38.16	31.29	5.23	0.05	25.27	1.59 ± 0.03	0.357 ± 0.003	7.75
200	0.03229	38.11	31.82	4.90	0.05	25.13	1.78 ± 0.02	0.320 ± 0.002	11.16
500	0.03191	40.23	30.07	4.94	0.02	24.74	2.10 ± 0.05	0.292 ± 0.004	15.94
1000	0.03144	41.28	29.78	4.72	0.03	24.19	2.54 ± 0.06	0.282 ± 0.003	19.64

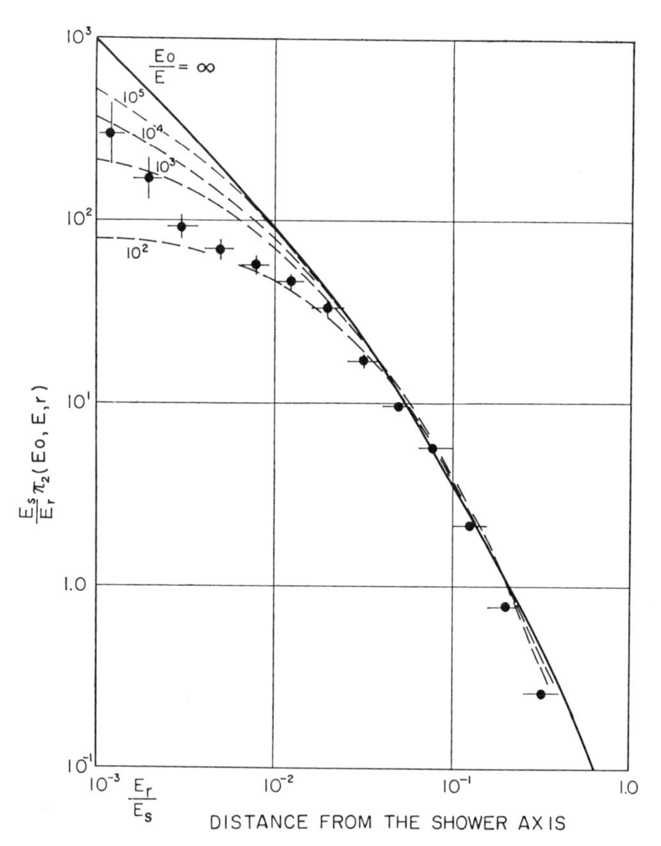

Fig. 2.30 Lateral distribution of shower electrons for finite primary energies at $s = 1.0$ in the approximation A. ⊕ —results of Monte Carlo calculation for $E_0/E = 10^2$ (Kamata 58).

Table 2.14 Values of Parameters in the Nishimura-Kamata-Greisen Function

$$f(x) = Cx^{s-2}(1 + x)^{-\alpha}, \quad x = r/r_M.$$

E_0	t	1	2	3	4	5	6	8
	s	1.0	1.2	1.1	1.2	1.5	1.1	1.7
10	α	21	10	6.2	5.5	5.8	3.6	4.1
	C	3.3	2.2	0.91	0.93	1.7	0.43	0.90
	s	0.43	0.85	1.1	1.3	1.4	1.2	1.7
32	α	24	11	9.0	7.6	6.3	4.8	5.3
	C	0.34	0.97	1.7	2.3	1.7	0.74	1.9
	s	1.8	0.51	0.80	1.0	1.2	1.2	1.6
100	α	43	12	8.5	6.9	6.5	6.1	6.0
	C	0.15	0.31	0.68	0.94	1.3	1.4	2.4
	s	−0.69	0.07	0.51	0.73	0.94	1.1	1.5
300	α	6.2	10	9.0	7.0	6.5	5.9	6.2
	C	0.001	0.05	0.28	0.48	0.81	0.94	2.2

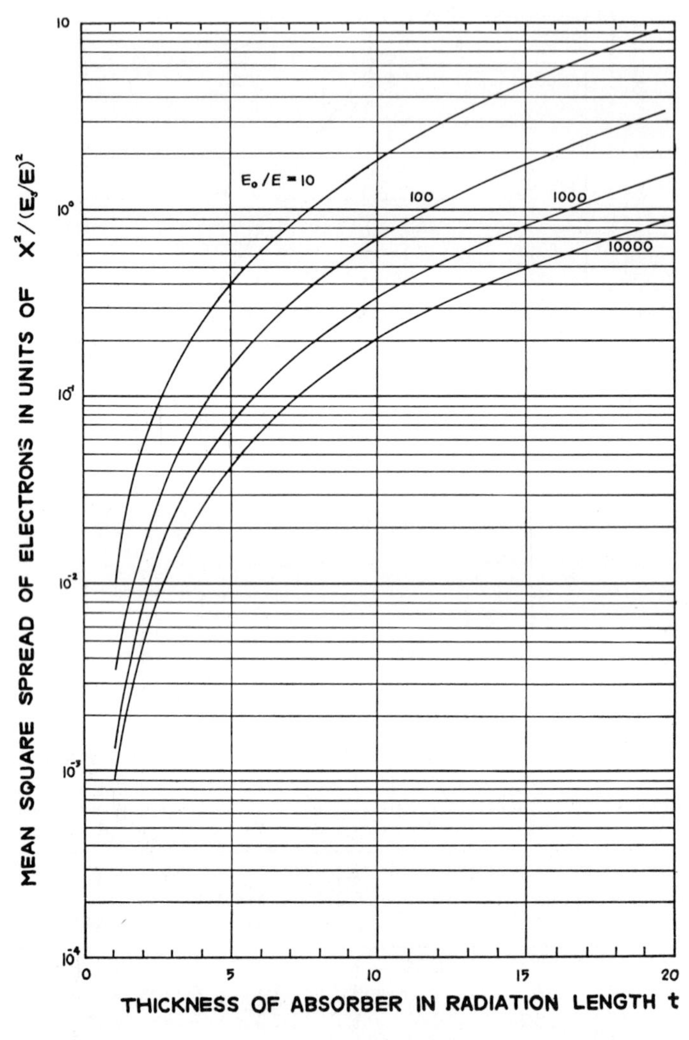

Fig. 2.31*a* Mean-square lateral spread of electrons with energies greater than E in a cascade shower initiated by photon of energy E_0 (Misaki 64).

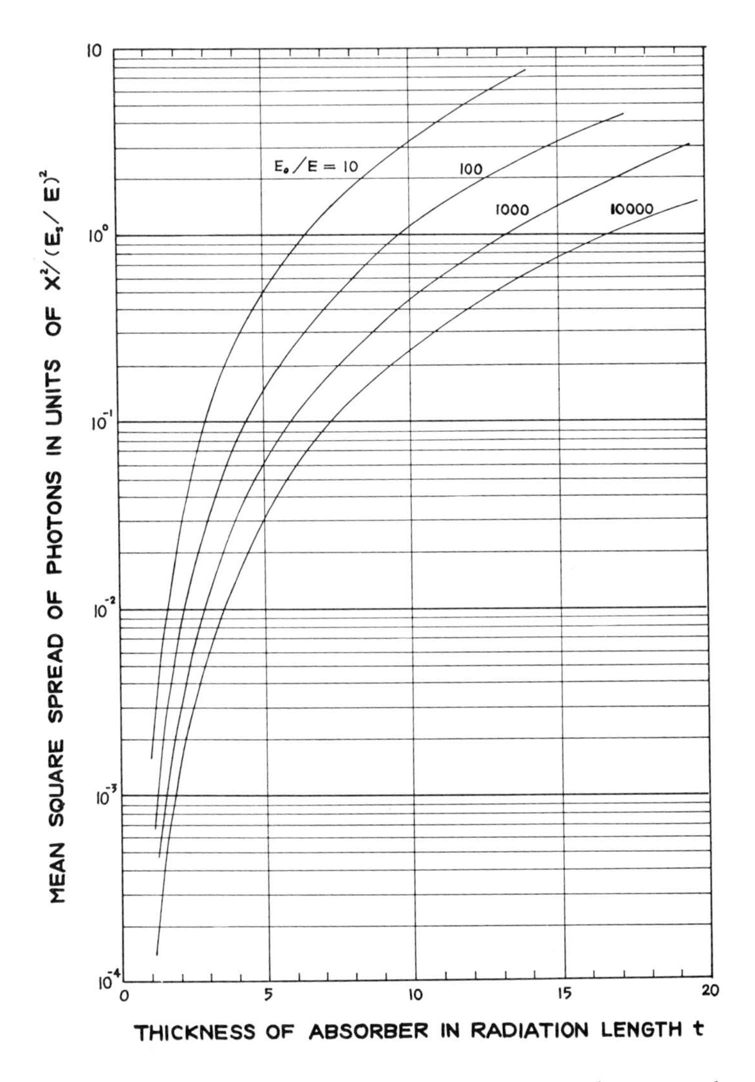

Fig. 2.31*b* Mean-square lateral spread of photons with energies greater than E in a cascade shower initiated by photon of energy E_0 (Misaki 64).

In heavy media the analytic solution is not expected to give a satisfactory result for the structure function. The fraction of energy contained within a given radius of lead is presented in Fig. 2.32 as a function of thickness (Murata 65). This demonstrates how much energy is lost out of the side wall of a cylinder.

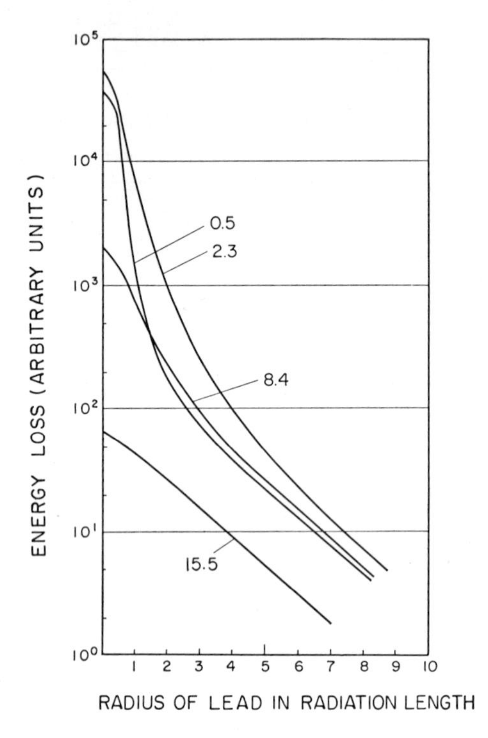

Fig. 2.32 Energy of a cascade shower contained in a given radius of lead. The thickness of lead in radiation lengths is shown for each curve. Showers are initiated by 200-MeV electrons, and the energy lost is measured with X-ray films (Murata 65).

2.7 Properties of Elementary Particles

It is not our aim to discuss the definition of the elementary particle, but to describe the known properties of elementary particles in the conventional sense. By this we mean particles that have mean lifetimes considerably longer than 10^{-23} sec. Therefore we do not include recently discovered excited states, because they have very short lifetimes. By this definition the 34 particles listed in Table 2.15 may be called elementary particles.

These 34 particles consist of 18 particles and their antiparticles, the one being transformed into the other by the charge or particle-antiparticle conjugation; two of them, the photon and the neutral pion, are self-conjugate. In other cases a particle-antiparticle pair can be distinguished from one another by quantum numbers.

The quantum numbers that are common to all the elementary particles are charge and spin. In addition, statistics (Bose or Fermi) is a well-defined concept, so that a particle may be called a boson or a fermion, according to whether it obeys Bose or Fermi statistics. Although the direct determination of statistics has been possible only in a few cases, the assignment is made on the basis of the general rule that the boson has an integral spin, whereas the fermion has a half-odd spin. The spin is an internal degree of freedom, which is $2s + 1$ for spin s, except for a massless particle. Although the spin of the photon is unity, its degree of freedom is 2, because it is massless; in other words, the longitudinal polarization of the photon is absent. The neutrino is believed to have a definite helicity and accordingly has only 1 degree of freedom.

Family. All these elementary particles interact with each other; as a result, one particle may be transformed into another one. In particular, the number of bosons may change indefinitely in the course of a transformation. On the other hand, the number of fermions has to change by an even number because of the conservation of statistics. There exists a rule that the difference between the numbers of particles and antiparticles is conserved among the ones belonging to a family. There are two such families, one being called the *baryon family*, to which belong nucleons and heavier fermions named hyperons, whereas the other is the *lepton family*, which consists of the neutrino, the electron, the muon, and their antiparticles. Taking the special situation of the photon into consideration, we introduce four families; namely, *photon*, *lepton*, *meson*, and *baryon*.

Baryon and Lepton Numbers. The concept of the family implies the conservation of particle number; for the lepton and the baryon families we define the numbers of leptons and baryons respectively as

$$\begin{aligned} L &= (\text{number of leptons}) - (\text{number of antileptons}), \\ B &= (\text{number of baryons}) - (\text{number of antibaryons}). \end{aligned} \tag{2.7.1}$$

The designation "particle" or "antiparticle" is rather arbitrary; however, the particle is defined as the one that is copiously found in nature, and the other particles can be defined through the conservation laws

Table 2.15 Properties of Elementary Particles

Symbol Charge, Antiparticle	Isospin Strangeness	Spin Parity	Mass (MeV)	Magnetic Moment ($e\hbar/2mc$)	Mean Life (sec)	Main Decay Modes	Branching Ratios (%)
				Photon Family			
γ^0	—	1^-	0	0	∞		
				Lepton Family			
$\nu_e{}^0\bar{\nu}_e$	—	$-\frac{1}{2}$ $+\frac{1}{2}$	$<2.5 \times 10^{-4}$	0	∞		
$\nu_\mu{}^0\bar{\nu}_\mu$	—	$-\frac{1}{2}$ $+\frac{1}{2}$	<2.5	0	∞		
e^-e^+	—	$\frac{1}{2}^+$	0.511006 ± 0.000002	$(1.0011609 \pm 0.0000024)$	∞		
$\mu^-\mu^+$	—	$\frac{1}{2}^+$	105.659 ± 0.002	(1.00116 ± 0.000005)	$2.2001 \pm 0.0008 \times 10^{-6}$	$e^-\bar{\nu}_e\nu_\mu$	100
$\pi^+\pi^-$	1, 0	0^-	139.580 ± 0.015	0	$2.551 \pm 0.026 \times 10^{-8}$	$\mu^+\nu_\mu$	100
π^0			134.974 ± 0.015	0	$1.78 \pm 0.26 \times 10^{-16}$	2γ γe^+e^-	98.8 1.2
				Meson Family			
K^+	$\frac{1}{2}$, 1		493.78 ± 0.17	0	$1.229 \pm 0.008 \times 10^{-8}$	$\mu^+\nu_\mu(\mu 2)$ $\pi^+\pi^0(\pi 2)$ $\mu^+\pi^0\nu_\mu(\mu 3)$ $e^+\pi^0\nu_2(e3)$ $\pi^+\pi^+\pi^-(\tau^+)$ $\pi^+\pi^0\pi^0(\tau^{+1})$	63.2 ± 0.4 21.3 ± 0.4 3.4 ± 0.2 4.9 ± 0.2 5.52 ± 0.08 1.68 ± 0.05
K^-							
K^0		0^-	497.7 ± 0.30	<0.04	$K_1{}^0$: $(0.881 \pm 0.010) \times 10^{-10}$	$\pi^+\pi^-$ $\pi^0\pi^0$	68.5 ± 1.0 31.5 ± 1.0

K^0					$K_2{}^0$: $(5.77 \pm 0.59) \times 10^{-8}$	$\pi^+\pi^-\pi^0$	13.6 ± 1.0
						$3\pi^0$	24.8 ± 3.0
						$\pi^\pm e^\mp \bar{\nu}_e(\nu_e)$	35.4 ± 2.7
						$\pi^\pm \mu^\mp \bar{\nu}_\mu(\nu_\mu)$	26.2 ± 2.6
						$\pi^+\pi^-$	$(2.1 \pm 0.3) \times 10^{-3}$
Nucleon Family							
p^+ $\bar{p}$	$\frac{1}{2}, 0$	$\frac{1}{2}^+$	938.256 ± 0.005	2.792816 ± 0.000034	∞		
n^0 $\bar{n}$			939.550 ± 0.005	-1.913184 ± 0.000066	1013 ± 26	$pe^-\bar{\nu}_e$	100
Λ^0 $\overline{\Lambda}^0$	0, 1	$\frac{1}{2}^+$	1115.44 ± 0.12	-0.73 ± 0.17	$(2.61 \pm 0.02) \times 10^{-10}$	$p\pi^-$	66.3 ± 1.0
						$n\pi^0$	33.6 ± 1.0
Σ^+ $\overline{\Sigma}^-$			1189.39 ± 0.14	4.3 ± 1.5	$(0.794 \pm 0.026) \times 10^{-10}$	$p\pi^0$	51.0 ± 2.4
						$n\pi^+$	49.0 ± 2.4
Σ^0 $\overline{\Sigma}^0$	1, 1	$\frac{1}{2}^+$	1192.3 ± 0.2		$10^{-14} >> 10^{-22}$	$\Lambda\gamma$	100
Σ^- $\overline{\Sigma}^+$			1197.20 ± 0.14		$(1.58 \pm 0.05) \times 10^{-10}$	$n\pi^-$	100
Hyperon Family							
Ξ^0 $\bar{\Xi}^0$	$\frac{1}{2}, 2$	$\frac{1}{2}^+$	1314.3 ± 1.0		$(3.05 \pm 0.38) \times 10^{-10}$	$\Lambda\pi^0$	100
Ξ^- $\bar{\Xi}^+$			1320.8 ± 0.2		$(1.75 \pm 0.05) \times 10^{-10}$	$\Lambda\pi^-$	100
Ω^- $\overline{\Omega}^+$	0, 3	$\frac{3}{2}^+$	1675 ± 3		$(1.3 \pm 0.7) \times 10^{-10}$	$\pi\Xi$	
						$\bar{K}\Lambda$	

given in (2.7.1). Since the proton, the neutron, and the negaton are copious, they are defined as particles. A charged particle can be distinguished from its antiparticle by charge, whereas the neutron can be distinguished from the antineutron by their opposite senses of magnetic moments. The neutrino is massless, so that its longitudinal polarization, or helicity, is a Lorentz-invariant quantity. Since a neutrino is found to be completely polarized, the neutrino can be distinguished from the antineutrino by helicity. In the β-decay caused by a fundamental process, $n \to p + e^- + \bar{\nu}_e$, the helicity of $\bar{\nu}_e$ is found to be $+1$; hence the helicity of the neutrino is -1. Observations of longitudinal polarizations of leptons produced in decay processes have led to the introduction of the concept of the lepton number, as given in Table 2.15. Another characteristic feature of a particle-antiparticle pair is their annihilation, which results in the transformation into bosons.

Strangeness. The families are useful also in classifying the interactions between particles. The photon is described by a vector field and is the source of the electromagnetic interaction. Baryons and mesons interact with each other more strongly, for some combinations of particles. The rule of combinations can be formulated by introducing the concept of strangeness characterized by integral numbers S. If the strangeness of a baryon or a meson is assigned as in Table 2.15, that of its antiparticle being of opposite sign, an interaction is strong for reactions conserving the strangeness. This results, for example, in the production of K and Λ or Σ in pair by the collision between zero-strangeness particles. The decay of a hyperon (Λ or Σ) into a nucleon and a pion is associated with the change of strangeness, $\Delta S = 1$; hence the interaction causing the decay process is weak, so that the lifetime is considerable. In the case of a lepton any interaction except the electromagnetic one is weak.

The strangeness quantum numbers of elementary particles are not restricted to the smallest possible values; $|S| = 3$ for the Ω hyperon. This is in sharp contrast to the charge quantum number, which is restricted to 0 or ± 1. It is therefore proposed to introduce an auxiliary quantum number Y, called the *hypercharge*, in place of strangeness through

$$Y = S + B. \tag{2.7.2}$$

The values of Y are restricted to 0 and ± 1 except for Ω.

The above considerations lead to the introduction of at least three classes of interactions; namely, the strong interaction characterized by $\Delta S = 0$, the electromagnetic one for any charged particle, and the weak interaction for $\Delta S = 1$ and for leptonic processes. The dimensionless squared coupling constants have values of 1, 10^{-2}, and 10^{-12} for the respective classes of interactions. We may further take into account weaker interactions for $\Delta S = 2$ and for gravitation.

Isospin. In strong interactions a set of particles that seem to be in charge multiplet states behave alike. In order to emphasize this fact particles belonging to one and the same set are regarded as different states of a particle, the states being characterized by components of isospin I. The third component of isospin is related to the charge number through

$$Q = I_3 + \tfrac{1}{2}B + \tfrac{1}{2}S = I_3 + \tfrac{1}{2}Y. \tag{2.7.3}$$

Isospin has the same mathematical structure as angular momentum and is conserved in any strong interaction. The isospins of mesons and baryons are assigned as in Table 2.15.

The electromagnetic interaction destroys the conservation of isospin but still conserves its third component. The weak interaction further destroys the conservation of I_3. In many decay processes the violation of isospin conservation is dictated by $|\Delta I| = \frac{1}{2}$.

Spin and Parity. In the strong and electromagnetic interactions parity as well as angular momentum are conserved. The spin and the parity of a particle are determined by reference to various reactions. In atomic physics the electron and the nucleon are known to have spin $\frac{1}{2}$. Their parities are taken as standard; namely, as even. The radiative processes give rise to the change of parity by $(-1)^L$ and $-(-1)^L$ for EL and ML radiations, respectively. The multipolarities of radiation can be determined from the angular distributions and correlations characteristic to respective multipolarities. A similar method applies for nonradiative processes. The spins and parities given in Table 2.15 are assigned in this way.

Charge Conjugation. The weak interaction is invariant neither under space inversion nor under charge conjugation. The charge conjugation transforms the particle to the antiparticle and vice versa. In association with this operation we can introduce the particle-conjugation parity C. This quantum number distinguishes between two states of a system consisting of a particle-antiparticle pair.

Analogous to that a system consisting of two particles with the same intrinsic parity has parity

$$P = (-1)^L, \tag{2.7.4}$$

as determined by the orbital angular momentum of this system, L, the particle-conjugation parity of a pair system is given by

$$C = (-1)^{L+S}, \tag{2.7.5}$$

where S is the total spin of the system. We can also introduce the particle conjugation parity of an n-photon system, which is given by

$$C = (-1)^{n_\gamma}. \tag{2.7.6}$$

The conservation of particle-conjugation parity restricts the number of photons emitted in the radiative decay of a pair system to either even or odd; for example, the positronium in ^1S decays into two photons, whereas that in ^3S decays into three photons. The selection rule in the radiative decay of the neutral pion is based on this conservation law.

Mass. Unlike the quantum numbers described above, the values of masses, lifetimes, and branching ratios for decays depend on experimental accuracy, which is being improved from time to time. The masses of elementary particles are determined by reference to the masses of the electron and the proton, which are measured macroscopically:

$$m_e c^2 = 0.511006 \pm 0.000002 \text{ MeV}, \tag{2.7.7a}$$

$$m_p c^2 = 938.256 \pm 0.005 \text{ MeV} = (1836.12 \pm 0.02) m_e c^2. \tag{2.7.7b}$$

The mass of the neutron is determined very accurately from $m_n - m_p$, and that of the muon from the energy level of muonic atoms as well as the observed magnetic moment. The masses of other particles are determined mainly with the aid of kinematics in various reactions. The mass values given in Table 2.15 were compiled in the middle of 1965.

Lifetimes and Decay Modes. Most elementary particles decay with finite lifetimes. If a charged particle is emitted in a decay, a photon may accompany it due to internal bremsstrahlung. The probability of emitting a photon with energy k is approximately given by

$$N(k)\, dk = \frac{2}{\pi} \frac{e^2}{\hbar c} \left[\frac{1}{\beta} \ln \left(\frac{1+\beta}{1-\beta} \right) - 2 \right] \frac{dk}{k}, \tag{2.7.8}$$

where $c\beta$ is the velocity of the particle. Such radiative decay modes are so common that they are not explicitly shown. Quite analogously, any secondary photon may be replaced by a negaton-positon pair produced by internal conversion. An example of such modes is the decay mode $\pi^0 \rightarrow e^- + e^+ + \gamma$, called the Dalitz pair. The branching ratio of the Dalitz pair in the neutral-pion decay is calculated to be 1.24 percent; the probabilities of internal conversion in other cases are not much different.

It is well known that the universality of four-fermion interactions seems to hold in leptonic processes; the magnitude of the coupling constant for the weak four-fermion interaction is often expressed as $g_F = 1.4 \times 10^{-49}$ erg-cm^3, but a dimensionless expression

$$\left(\frac{g_F}{\hbar c}\right)^2\left(\frac{m_\pi c}{\hbar}\right)^4 \simeq 5 \times 10^{-14}, \tag{2.7.9}$$

where m_π is the pion mass, seems to be more convenient.

The type of the weak interaction is investigated by reference to energy spectra, angular correlations, polarizations, and so on. The energy spectrum and the angular distribution of electrons from μ-decays provide a typical example. The momentum spectrum of electrons in μ-decays is expressed as

$$f(p) \propto p^2[(p_m - p) + \tfrac{2}{9}\rho(4p - 3p_m)], \tag{2.7.10}$$

where $p_m = m_\mu c/2$ is the maximum momentum of an electron. The symbol ρ is called the *Michel parameter*, and its value may vary from 0 to 1 depending on the type of interaction. The present experiments indicate $\rho \simeq \frac{3}{4}$, after radiative corrections.

The decay of the pion is thought to take place through the virtual creation of a pair of nucleons and their subsequent conversion into two leptons due to the weak four-fermion interaction. This qualitatively accounts for the mean lifetime as well as the branching ratio

$$\frac{\text{rate } (\pi \to e + \nu_e)}{\text{rate } (\pi \to \mu + \nu_\mu)} = (1.24 \pm 0.03) \times 10^{-4}. \tag{2.7.11}$$

The decay modes of such small branching ratios are not shown in Table 2.15.

A remark may be necessary about the neutral K-meson. Both K^0 and $\bar{K}^0$ are eigenstates of the strong interaction for production. In their decays they are not necessarily eigenstates, because the particle-conjugation parity is no longer a good quantum number in the weak interaction. In the decay process we have to refer to the eigenstate of the CP operation. It is therefore appropriate to introduce such eigenstates:

$$|K_1{}^0\rangle = \frac{1}{\sqrt{2}}(|K^0\rangle + |\bar{K}^0\rangle), \qquad |K_2{}^0\rangle = \frac{1}{\sqrt{2}}(|K^0\rangle - |\bar{K}^0\rangle); \tag{2.7.12}$$

they have even and odd CP-parities, respectively. Thus the decay processes of the neutral kaon are shown for $K_1{}^0$ and $K_2{}^0$ in Table 2.15. The result is that the beam of K^0 produced consists of a mixture of $K_1{}^0$ and $K_2{}^0$, the former decaying more quickly than the latter.

However, these two states are not strictly independent; they couple with one another by a very weak interaction. The coupling produces a small difference between their masses; namely,

$$|m_{K_1{}^0} - m_{K_2{}^0}|c^2 = (5.5 \pm 1.7) \times 10^{-6} \text{ eV}. \tag{2.7.13}$$

Table 2.16 Properties of Meson Resonances

Symbol and Charge	Isospin Strangeness	Spin Parity	Mass (MeV)	Full Width (Mev)	Main Decay Modes	Branching Ratios (%)
η^0	0, 0	0^-	548.9 ± 0.5	≤ 10	$\pi^+\pi^-\pi^0$	25.0 ± 1.6
					$3\pi^0 + \pi^0 2\gamma$	30.8 ± 2.3
					$\pi^+\pi^-\gamma$	5.5 ± 1.2
					2γ	38.6 ± 2.7
κ^0	$\frac{1}{2}$, 1	0^+	725 ± 2	≤ 12	$K\pi$	100
$\rho^\pm$					2π	100
					4π	<5
ρ^0	1, 0	1^-	765 ± 3	124 ± 4	$\pi\gamma$	<2
					e^+e^-	6.5×10^{-3}
ω^0	0, 0	1^-	782.8 ± 0.5	12.0 ± 1.7	$\pi^+\pi^-\pi^0$	88
					$\pi^+\pi^-\gamma$	<3.2
					Neutrals ($\pi^0\gamma$)	12 ± 2
					$\mu^+\mu^-$	<0.10
					e^+e^-	~ 0.01
K^*					$K\pi$	100
$\bar{K}^{*-}$	$\frac{1}{2}$, 1	1^-	891.4 ± 0.8	49 ± 2	$K\pi\pi$	<0.2
K^{*0}					$\kappa\pi$	<0.2
$\bar{K}^{*0}$						
X^0	0, 0	0^-	958.6 ± 1.6	<4	$\eta 2\pi$	76 ± 4
					$\pi^+\pi^-\gamma$	24 ± 4
ϕ^0	0, 0	1^-	1019.5 ± 0.3	3.3 ± 0.6	$K_1{}^0 K_2{}^0$	38 ± 3
					K^+K^-	30 ± 3
					$\pi\rho + 3\pi$	32 ± 8
$A_1{}^{\pm,0}$	1, 0	1^+	1072 ± 8	125	$\rho\pi$	~ 100
					$K\bar{K}$	<5
$B^{\pm,0}$	1, 0	$\geq$1s	1220	125 ± 17	$\omega\pi$	~ 100
					$\pi\pi$	<30
					$K\bar{K}$	<10
					4π	<50
f	0, 0	2^+	1253 ± 20	118 ± 16	$\pi\pi$	~ 100
					4π	<4
					$K\bar{K}$	<4
D	0, 0	?	1286 ± 6	40 ± 10	$K\bar{K}\pi$	100
$A_2{}^{\pm,0}$	1, 0	2^+	1324 ± 9	90 ± 10	$\rho\pi$	91
					$K\bar{K}$	5.5 ± 1.5
					$\eta\pi$	3.6 ± 3.0
K^*	$\frac{1}{2}$, 1	2^+	1405 ± 8	95 ± 11	$K\pi$	
f'	0, 0	2^+	1500	80	K_1K_1	
					$KK^*(890)$	

Table 2.17 Properties of Baryon Resonances

Symbol	Isospin Strange-ness	Spin Parity	Mass (MeV)	Full Width (MeV)	Main Decay Modes	Branching Ratios (%)
$N^*_{3/2}$	$\frac{3}{2}, 0$	$\frac{3}{2}^+$	1236.0 ± 0.4	120.0 ± 1.5	πN	100
$N^*_{1/2}$	$\frac{1}{2}, 0$	$\frac{3}{2}^-$	1518 ± 10	120	πN	75
					$\pi\pi N$	
					$\pi N^*_{3/2}$ (1236)	
$N^*_{5/2}$	$\frac{5}{2}, 0$		1560 ± 20	220 ± 20	$\pi\pi N$	
$N^*_{1/2}$	$\frac{1}{2}, 0$	$\frac{5}{2}^+$	1688	100	πN	85
					$\pi\pi N$	
					KA	
					ηN	<2
$N^*_{3/2}$	$\frac{3}{2}, 0$	$\frac{7}{2}^+$	1924	170	πN	<67
					ΣK	
					$\pi N^*_{3/2}$ (1236)	
$N^*_{1/2}$	$\frac{1}{2}, 0$	$\frac{7}{2}^-$	2190	~ 200	πN	~ 40
					$K\Lambda$	
$N^*_{3/2}$	$\frac{3}{2}, 0$	$\frac{9}{2}^-$	2360	~ 200	πN	~ 15
					Others	
$N^*_{1/2}$	$\frac{1}{2}, 0$	$\frac{9}{2}^+$	2645 ± 10	~ 200	πN	
					ηN	
$N^*_{3/2}$	$\frac{3}{2}, 0$	$\frac{11}{2}^+$	2825	260	πN	
Y_1^*	$1, -1$	$\frac{3}{2}^+$	1382.7 ± 0.5	44 ± 2	$\pi\Lambda$	90 ± 2
					$\pi\Sigma$	$10 + 2$
Y_0^*	$0, -1$	$\frac{1}{2}^-$	1405	35 ± 5	$\pi\Sigma$	~ 100
					$\pi\pi\Lambda$	<1
Y_0^*	$0, -1$	$\frac{3}{2}^-$	1518.9 ± 1.5	16 ± 2	$\pi\Sigma$	55 ± 7
					KN	29 ± 4
					$\pi\pi\Lambda$	16 ± 2
Ξ^*	$\frac{1}{2}, -2$	$\frac{3}{2}^+$	1529.7 ± 0.9	7.5 ± 1.0	$\Xi\pi$	100
Y_1^*	$1, -1$	$\geq\frac{3}{2}$	1660 ± 10	44 ± 5	$\bar{K}N$	15
					$\pi\Sigma$	30
					$\pi\Lambda$	5
					$\pi\pi\Sigma$	30
					$\pi\pi\Lambda$	20
Y_1^*	$1, -1$	$\frac{5}{2}^-$	1762 ± 17	75 ± 7	$\bar{K}N$	60
					$\pi\Lambda$	16
					$\pi\Sigma$	≤ 3
					$\pi Y_1^*(1385)$	10
					$\pi Y_0^*(1520)$	10
Y_0^*	$0, -1$	$\frac{5}{2}^+$	1815	70	$\bar{K}N$	80
					$\pi\Sigma$	<10
					$\pi\pi\Lambda$	<15
					$\eta\Lambda$	?
Ξ^*	$\frac{1}{2}, -2$	$\frac{3}{2}^-$	1816 ± 3	16 ± 4	$\pi\Xi^*$	25
					$\bar{K}\Lambda$	65
					$\pi\Xi$	5
					$\pi\pi\Xi$	5
Ξ^*	$\frac{1}{2}, -2$	$\frac{5}{2}^+$	1933 ± 16	140 ± 35	$\pi\Xi$	

The mass difference implies that the states of $K_1{}^0$ and $K_2{}^0$ alternate with a frequency of $|m_{K_1{}^0} - m_{K_2{}^0}|\, c^2/\hbar$. Thus $K_1{}^0$ is regenerated from the $K_2{}^0$ beam that is left after the essential elimination of $K_1{}^0$ by its fast decay. Recent experiments indicate, however, that the distinction between $K_1{}^0$ and $K_2{}^0$ by the presence and the absence of the two-pion decay is not appropriate, because the long-lifetime neutral kaon also decays into two pions with a small probability; this implies the violation of the *CP* conservation in K^0 decay.

Resonance States. A great number of resonance states have been observed through the energy dependences of cross sections as well as energy and angular correlations of secondary particles produced by strong interactions. It is an open question whether or not they may be called elementary particles. Quite a number of theories have been proposed to incorporate the resonance states into a series of elementary particles by introducing some symmetry principles. Whatever the arguments, a deep-lying reason for the inclusion of resonance states, is that they indicate that strong interactions can readily bring particles in resonances at several energies. Resonance states thus far found are as numerous as those of a light nucleus. Expecting a further increase in the number of resonance states in the near future and the revision of existing data, we list them in Tables 2.16 and 2.17 on the basis of firmly established results as of August 1965.*

The names of the resonance states are given rather conventionally. No attempt is made to systematize these resonance states, except that baryon resonances are grouped in Table 2.17 according to strangeness and isospin. Meson resonances are listed in Table 2.16 according to the order of their masses because classification into groups does not seem to be established as yet.

2.8 Collisions of Elementary Particles

The recent development of accelerators has made it possible to obtain various cross sections for elementary-particle reactions rather accurately. Here we collect data that are necessary for cosmic-ray physics.

2.8.1 Nucleon-Nucleon Collisions

Experimental results on the total cross sections, the angular distributions, and the polarizations of *p-p* and *n-p* collisions up to 6 GeV were

* A. H. Rosenfeld, A. Barbaro-Galtieri, W. H. Bostien, J. Kirz, and M. Roos, UCRL–8030–Part I.

summarized by Hess (Hess 58). Data at higher energies are supplemented on the basis of the latest reports. Here we are concerned mainly with the total cross sections and briefly describe qualitative features of the angular distributions and other properties of *p-p* and *n-p* collisions.

The total cross section consists of elastic and inelastic parts, the latter starting at the pion threshold of 290 MeV. The *p-p* scattering is subject to the Coulomb effect at small angles. This is subtracted, so that the cross section due only to nuclear forces is presented. According to the charge symmetry the *n-n* cross section is believed to be equal to the *p-p* one. In fact, the former, derived from the *n-d* cross section by taking the eclipsing correction into account, is found to agree with the latter. The same holds for the *p-n* cross section, which is essentially equal to the *n-p* cross section, provided that the eclipsing effect brings about a constant correction term of 6 ± 3 mb above 100 MeV. The total and elastic cross sections for *p-p* and *n-p* collisions are shown in Figs. 2.33 and 2.34, respectively. For *p-p* collisions the elastic and the inelastic cross sections are separately available at a number of energies, as shown in Fig. 2.33.

The energy dependence of the *p-p* cross section can be seen from

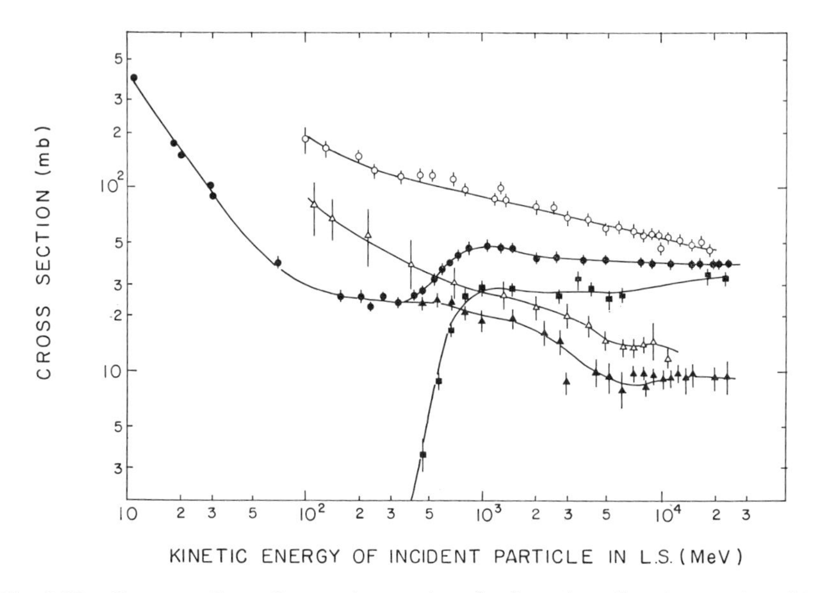

Fig. 2.33 Cross sections for proton-proton (*p-p*) and antiproton-proton ($\bar{p}$-*p*) collisions versus the kinetic energy of an incident particle in the laboratory system.

p-p: ●—total; ▲—elastic; ■—inelastic

$\bar{p}$-*p*: ○—total; △—elastic.

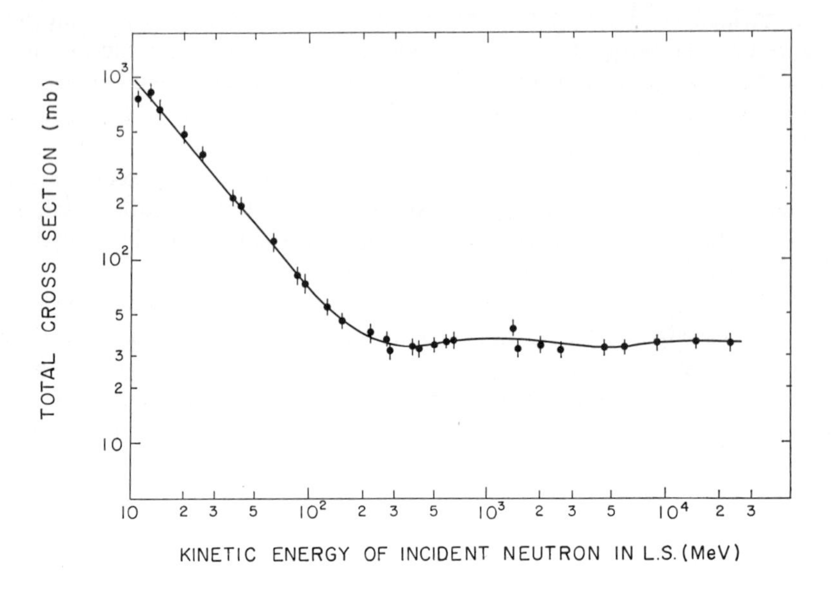

Fig. 2.34 Total cross section for a neutron-proton collision versus the kinetic energy of an incident neutron in the laboratory system.

Fig. 2.33. Below 100 MeV the $1/E$ law holds. Between 170 and 500 MeV the elastic cross section is essentially constant, whereas the total cross section begins to rise at 290 MeV due to the inelastic, pion-producing process. The inelastic cross section increases rapidly with energy and reaches a maximum at about 1 GeV. This causes a maximum in the total cross section, too. The elastic cross section begins to decrease at about 500 MeV and becomes smaller than the inelastic one at 800 MeV. Above that elastic scattering is due mainly to the shadow effect of inelastic processes, and the genuine elastic ones contribute but a little to the cross section, at most 1 mb. Above 10 GeV the total cross section appears to stay constant, at 39.5 ± 1.0 mb.

Experimental data on *n-p* collisions are scanty and less reliable for obvious reasons. We would find some similarity between the *n-p* and *p-p* cross sections because the isiopin triplet part of the former is equal to a half of the latter, according to the charge independence. Below 100 MeV the *n-p* total cross section follows also the $1/E$ law, but its absolute value is higher than that of the *p-p* cross section, mainly because the ^3S-state, which is absent in the *p-p* system, is effective in this energy region. The cross section reaches a minimum at about 300 MeV and then rises more slowly than the *p-p* cross section since the fraction of the inelastic cross section is about a half of that in the *p-p* case. The elastic scattering has not been well separated from the inelastic one at high energies. Above 10 GeV the total cross section again stays constant—but at a little lower value, about 35 mb, than the *p-p* cross section.

The antiproton-proton cross section is compared with the *p-p* cross section in Fig. 2.33. The former is much higher than the latter and consists of the annihilation and elastic parts, as will be discussed later. The total $\bar{p}$-p cross section decreases with increasing energy and approaches the *p-p* cross section. At the highest antiproton energy available at present, however, there still exists a gap of about 20 mb between these two cross sections. Data on $\bar{n}$-n and $\bar{p}$-n collisions are fewer but show the approximate equality between $\bar{p}$-p and $\bar{p}$-n or $\bar{n}$-p cross sections.

The angular distribution of *p-p* scattering is nearly isotropic up to 430 MeV except for the Coulomb part at small angles. Between 170 and 430 MeV the differential cross section is practically constant at 3.7 mb-sr^{-1}*. Above 500 MeV the forward peak becomes more and more pronounced with increasing energy. Above 10 GeV the forward peak is represented by a simple formula (Cocconi 61):

$$\left(\frac{d\sigma}{d\Omega}\right)_{p\text{-}p,\ \text{elastic}} = \left(\frac{\sigma_{\text{tot}}}{4\pi}\right)^2 k^2 \exp(-7.5q^2) \qquad \text{for} \quad q^2 < 0.5(\text{GeV}/c)^2$$

$$q^2 = 4(\hbar k)^2 \sin^2\left(\frac{\theta}{2}\right) \quad \text{in } (\text{GeV}/c)^2, \tag{2.8.1}$$

where k is the wave number and θ the scattered angle in the center-of-mass system.

For large momentum transfer ($q > 1$ GeV/c) the elastic cross section decreases more slowly as a function of q. A simpler expression of the differential cross section is obtained if this is expressed as a function of transverse momentum,

$$p_T = \hbar k \sin\theta. \tag{2.8.2}$$

Both q and p_T are Lorentz invariant, and the invariant expression of the differential cross section has an exponential form (Orear 64, Cocconi 64):

$$\frac{d\sigma}{d\Omega} = 34 \times e^{-p_T/0.151}\text{mb-sr}^{-1}, \tag{2.8.3}$$

where p_T is in GeV/c and

$$\langle p_T \rangle = 0.302 \text{ GeV}/c. \tag{2.8.4}$$

* $4\pi \times 3.7$ mb-sr$^{-1} = 2\sigma_{\text{tot}}$.

As a result the cross section at large angles decreases rather rapidly as energy increases.

For the *n-p* scattering an anisotropic part becomes appreciable at 20 MeV but nearly symmetric with respect to 90° up to 90 MeV. The minimum of the angular distribution shifts toward a smaller angle with increasing energy but turns back at about 200 MeV.

2.8.2 *Pion-Nucleon Scattering*

Beams of positive and negative pions are now available up to energies as high as 20 GeV. The total cross sections for pion-proton collisions have been measured rather accurately, as shown in Figs. 2.35 and 2.36.

The scattering cross sections show a number of peaks and valleys as energy increases. The first peaks appearing in three processes at about 190 MeV are very high and are due to a resonance in a state of $I = \frac{3}{2}$ and $J = \frac{3}{2}(P_{3/2})$, where J is the total angular momentum. Thus this is called the 3–3 resonance, one of the baryon resonances $N^*_{3/2}$ given in Table 2.17. In this way many peaks are identified with baryon resonances, as shown in Table 2.17.

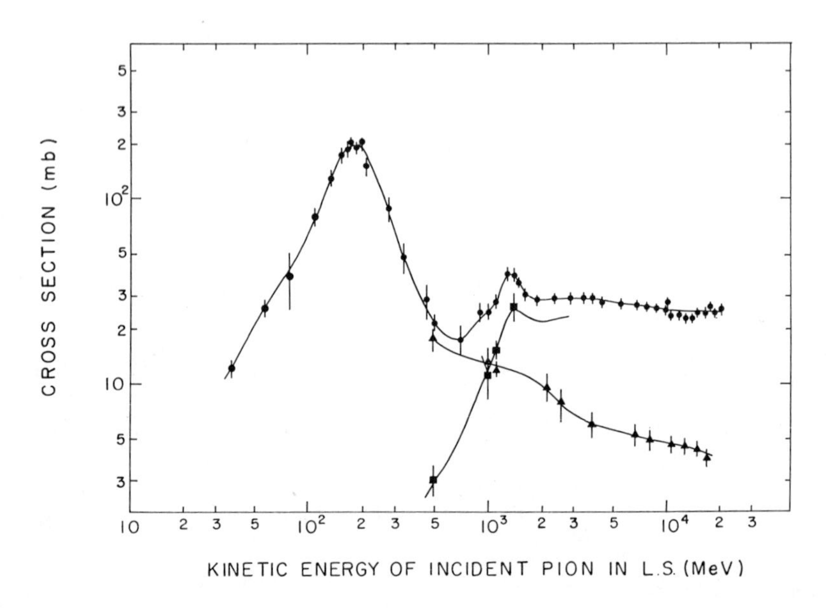

Fig. 2.35 Cross sections for a positive pion-proton collision versus the kinetic energy of an incident pion in the laboratory system.

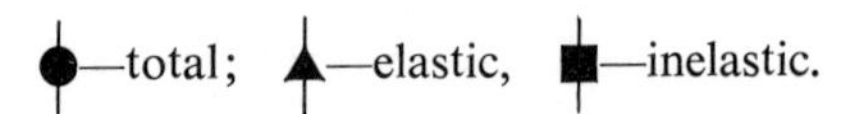

At high energies the π-p cross sections are found to be smooth and decrease monotonically with increasing energy as

$$\sigma(\pi\text{-}p) = a + \frac{b}{p}, \tag{2.8.5}$$

where p is the momentum in GeV/c of an incident pion. The values of a and b are obtained for the total π^+-p and π^--p cross sections respectively as

$$\begin{aligned} a^+ &= 22.26 \pm 0.33 \text{ mb}, \qquad b^+ = 25.10 \pm 2.83 \text{ mb-GeV}/c, \\ a^- &= 24.37 \pm 0.29 \text{ mb} \qquad b^- = 24.94 \pm 2.65 \text{ mb-GeV}/c, \end{aligned} \tag{2.8.6}$$

on the basis of experimental results for $p \simeq 4$ to 20 GeV/c.

In this energy region inelastic processes are predominant, whereas elastic processes are due mainly to diffraction scattering. The elastic cross section at large angles is very small, smaller than 1 mb-sr^{-1}, but shows an increase in the backward direction.

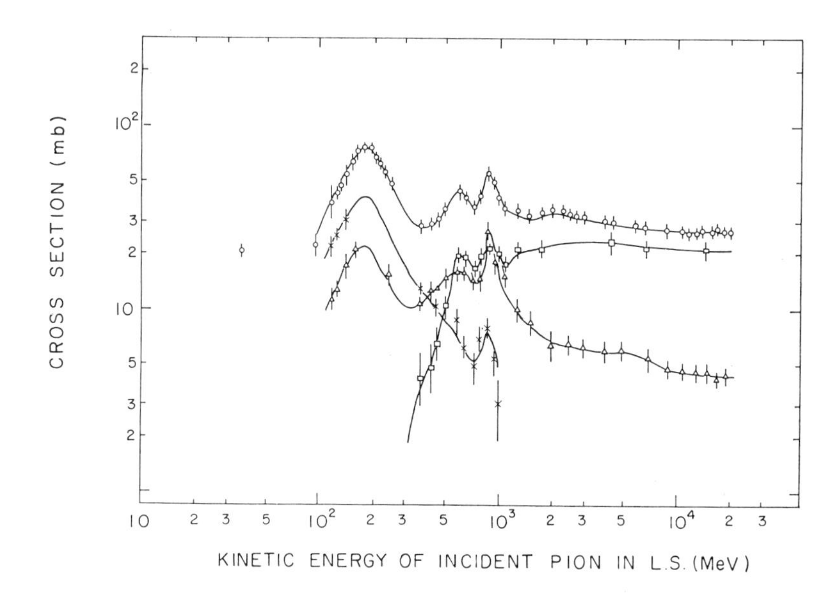

Fig. 2.36 Cross sections for a negative pion-proton collision versus the kinetic energy of an incident pion in the laboratory system.

○—total; △—elastic; ×—charge exchange; □—inelastic.

2.8.3 Photon-Nucleon Collision

Elastic scattering at low energies can be described in terms of Thomson scattering in the same way as photon-electron scattering. As energy increases pion production begins at about 150 MeV, so that elastic scattering is due mainly to the shadow effect of pion production. Since the elastic cross section is much smaller than the inelastic one above the pion threshold, we shall not pay much attention to the elastic process in considering cosmic-ray phenomena.

The pion-producing collision consists of the following processes:

$$\gamma + p(n) \to [n(p) + \pi^+(\pi^-)], \tag{2.8.7a}$$

$$\gamma + p(n) \to [p(n) + \pi^0], \tag{2.8.7b}$$

$$\gamma + p(n) \to (p \text{ or } n + \text{many } \pi). \tag{2.8.7c}$$

The photon-pion cross sections are in close relation to the pion scattering cross section. A peak in these cross sections at about 300 MeV is due to the 3–3 resonance in pion-nucleon scattering.

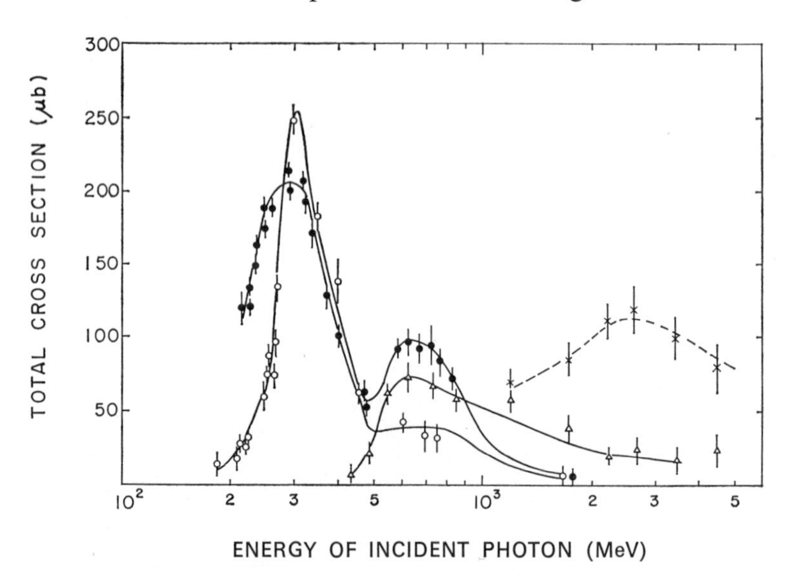

Fig. 2.37 Cross sections for a photon-proton collision versus the energy of an incident photon in the laboratory system.

● — $\gamma + p \to n + \pi^+$; ○ — $\gamma + p \to p + \pi^0$

△ — $\gamma + p \to p + \pi^+ + \pi^-$

× — $\gamma + p \to p + \pi^+ + \pi^- + n\pi^0 + n\pi^\pm$ ($n = 0, 1, \ldots$) plus strange particles.

After the maximum the cross sections decrease rapidly with energy, and show peaks and valleys as in pion-nucleon scattering. The cross sections for various processes are shown in in Fig. 2.37. The peaks correspond to the resonance states found by pion-nucleon scattering and are identified with baryon resonances in Table 2.17.

2.8.4 *Pion-pion Interactions*

In multiple pion production the pion-pion interaction reveals itself in the correlation of pions. Some of the resonances can be accounted for in terms of a resonant interaction between an incident pion and a virtual one in the pion cloud surrounding a nucleon. The energy and angular correlations of pions as the decay products of the $K_{\pi 3}$ and those produced by the $\bar{p}$-p annihilation also indicate the pion-pion interaction. On the basis of these pieces of information, the pion-pion cross section is shown to be

$$\sigma_{\pi\text{-}\pi} \simeq 35 \text{ mb} \tag{2.8.8}$$

at low pion energies. Many of the meson resonances given in Table 2.16 are found in this way.

2.8.5 *Kaon-Nucleon Collisions*

The cross sections for K^+ and K^- against the nucleon are considerably different from one another, mainly because the K^--$n(p)$ collision involves particle-emission channels (e.g., $K^- + p \rightarrow \Lambda + \pi^0$) even at zero energy.

The K^+-p cross section starts at a low value, about 12 mb, and rises slowly until it goes over a broad peak centered around 1.5 GeV/c. Above 10 GeV/c a gradual decrease is indicated, analogous to the pion-nucleon cross section. The separation between the elastic and inelastic cross sections has been only poorly made, but the latter is as small as 1 mb at about 0.8 GeV/c and increases with increasing incident momentum. The K^+-n cross section is found to be rather low at low momenta, but it becomes nearly equal to the K^+-p one above 0.5 GeV/c.

The K^--p cross section behaves in a quite different way. At low momenta it is represented by the characteristic $1/v$ law, but above 0.5 GeV/c it rises rapidly to about 50 mb at about 1 GeV/c. After subtracting the nonresonant background, we find a resonance at a center-of-mass kinetic energy of 380 MeV and with a half-width of 60 MeV. This is identified with $Y_0{}^*$ listed in Table 2.17. Above the resonance both the K^--p and the K^--n cross sections appear to be rather smooth, decreasing gradually with a weakly oscillating energy dependence.

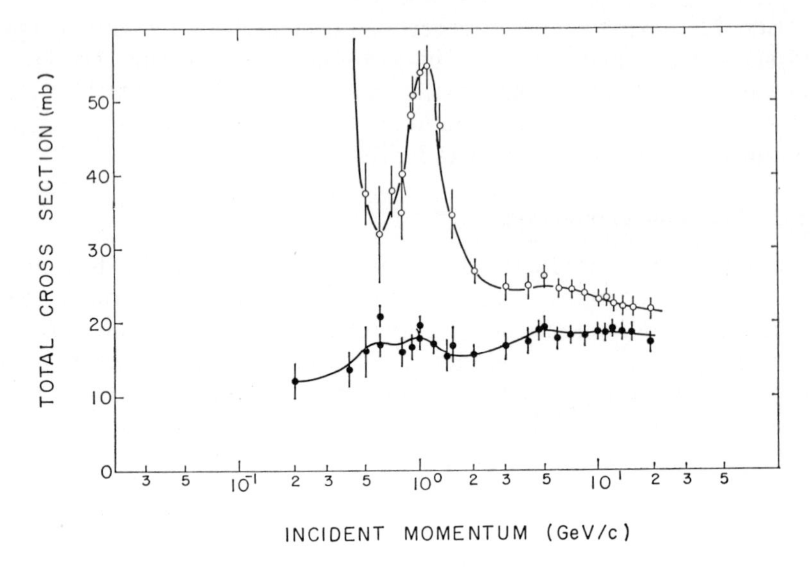

Fig. 2.38 Total cross sections for kaon-proton collisions versus the incident momentum of a kaon. ○—$K^- - p$, ●—$K^+ - p$.

These two seem to approach one another but stay higher than the K^+-p cross section even at 15 GeV/c. These cross sections against incident momentum are shown in Fig. 2.38.

2.8.6 *Hyperon-Nucleon Interactions*

Since hyperons have very short lifetimes, few of them can interact with matter before their decay; for example, on the basis of 14 Λ-p collisions in a bubble chamber the Λ-p cross section at several hundred MeV is shown to be (Alexander 61)

$$\sigma_{\Lambda\text{-}p} \simeq 20 \text{ mb}. \tag{2.8.9}$$

2.9 Collisions with Nuclei

Nuclear reactions at high energies do not depend very much on the level structure of a particular nucleus. The collision may be regarded as being initiated by the collision of an impinging particle with one of the nucleons in the nucleus, followed by successive collisions of a scattered particle, recoil particles, and, if any, produced particles. The whole process is therefore described in terms of a nuclear cascade inside a nucleus. Particles of low energies generated in the course of the cascade process cannot participate in the further development of the cascade but may be regarded as contributing to the nuclear excitation that is followed by the evaporation of low-energy particles. These processes can be described analytically at least for qualitative purposes (Messel 54, Kikuchi 60) and also by means of the Monte Carlo method more quantitatively (Dostrovsky 58, 59; Metropolis 58; Roos 61).

2.9.1 Nuclear Cascade Processes

The Monte Carlo method makes it possible to take into account realistic cross sections and geometrical details. With this method nuclear cascades induced by nucleons, pions, photons, and subsequent evaporation processes have been investigated with the aid of fast computers. An advantage of this method is sensitivity of the number of cascade nucleons emitted and the excitation energy to the nuclear radius; if the nuclear radius is increased from $1.3 \times 10^{-13}A^{1/3}$ cm to $1.4 \times 10^{-13}A^{1/3}$ cm, the average number of cascade nucleons from a heavy nucleus decreases by about 50 per cent, whereas the excitation energy left after the cascade process increases by 30 to 40 percent.

In the Monte Carlo calculations referred to here nuclei are assumed to be spherical and to have radii

$$R = r_0 A^{1/3}, \qquad r_0 = 1.3 \times 10^{-13} \text{ cm}. \tag{2.9.1}$$

Nucleons are assumed to be distributed uniformly inside such a sphere, and the Fermi energies for protons and neutrons are therefore obtained for given Z and $A - Z$, respectively. With these Fermi energies and the binding energies of the loosest nucleons, the nuclear potentials for protons and neutrons are obtained, and the kinetic energy of a nucleon inside a nucleus is given by taking account of such uniform potentials. Nucleons are regarded as participating in a cascade process until their energies decline to the cut-off energy E_c, which is equal to the average potential energy of protons and neutrons plus the Coulomb barrier for a proton. The value of E_c increases from about 40 MeV for light nuclei to about 50 MeV for very heavy nuclei as the Coulomb barrier increases.

The elementary cross sections that are responsible for the cascade process have been approximated as

$$\sigma_{pp} = \sigma_{nn} = \frac{10.63}{\beta_n^2} - \frac{29.92}{\beta_n} + 42.9 \text{ mb}, \tag{2.9.2a}$$

$$\sigma_{pn} = \frac{34.10}{\beta_n^2} - \frac{82.2}{\beta_n} + 82.2 \text{ mb}, \tag{2.9.2b}$$

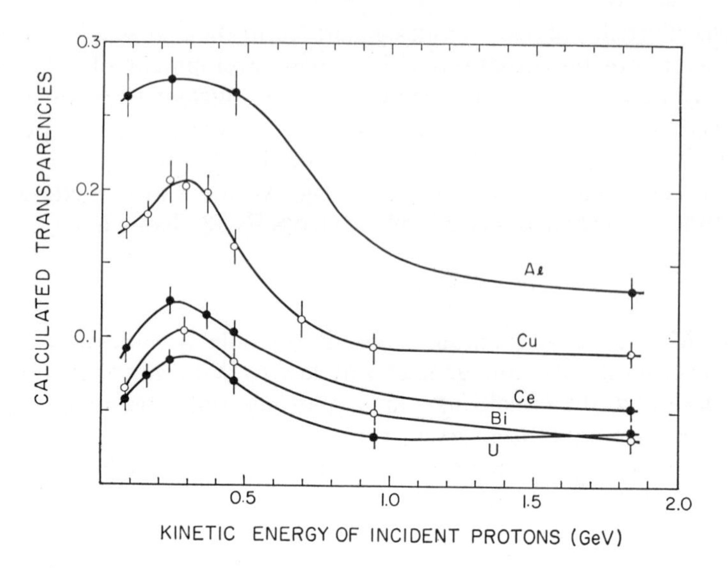

Fig. 2.39 Transparencies of nuclei for protons (Metropolis 58).

$$\sigma_{\pi^+p} = \sigma_{\pi^-n} = 3.7 + 286(\gamma_\pi - 1)^3 \text{ mb} \quad \text{for} \quad E_\pi \leq 51 \text{ MeV} \tag{2.9.2c}$$

$$\sigma_{\pi^-p} = \sigma_{\pi^+n} = 6.5 + 23.9(\gamma_\pi - 1) \text{ mb} \quad \text{for} \quad E_\pi \leq 51 \text{ MeV} \tag{2.9.2d}$$

$$\sigma_{\pi^0p} = \sigma_{\pi^0n} = \tfrac{1}{2}(\sigma_{\pi^+p} + \sigma_{\pi^-n}) \text{ mb}, \tag{2.9.2e}$$

$$\sigma_{\text{abs}} = \frac{16.4(0.14 + \eta_\pi^2)}{\eta_\pi} \text{ mb} \quad \text{for} \quad E_\pi \leq 51 \text{ MeV}, \tag{2.9.2f}$$

where $\beta_n c$ is the velocity of a nucleon, $\gamma_\pi m_\pi c^2$ is the total energy of a pion, and $\eta_\pi m_\pi c$ is the momentum thereof; σ_{abs} refers to the cross section for the absorption of charged pions by deuterons; the absorption cross section for neutral pions is one-half of σ_{abs} defined above. The nucleon cross sections given in (2.9.2) are valid only for elastic scattering —and so are pion cross sections below 51 MeV. At high energies the experimental cross sections shown in Section 2.8 are employed by approximating them by chains of segments.

The angular distributions in elastic scattering are represented in the center-of-mass system by

$$\frac{d\sigma}{d\Omega} \propto A \cos^4 \theta + B \cos^3 \theta + 1. \tag{2.9.3}$$

The values of A and B are taken from experiment.

With the elementary cross sections given above the particles taking part in a cascade process are followed by the Monte Carlo method. The results thus obtained are summarized below.

Transparency. The transparencies of a number of representative nuclei for protons are shown in Fig. 2.39. From their energy dependences we can see that transparencies anticorrelate with the elementary cross sections; for example, the transparencies for neutrons are slightly smaller than those for protons between 100 and 500 MeV. Even for uranium, transparencies are finite, and consequently the interaction cross section is always smaller than the geometrical one at least by 3 percent.

Numbers of Cascade Particles. The average number of emitted cascade nucleons per inelastic collision increases with increasing incident energy, as shown in Fig. 2.40 for incident protons. At low energies it decreases with increasing nuclear size, whereas at high energies the nuclear-size dependence behaves in the opposite way.

The ratio of the average numbers of neutrons to protons increases with nuclear size. For incident neutrons the *n-p* ratio is about twice as great.

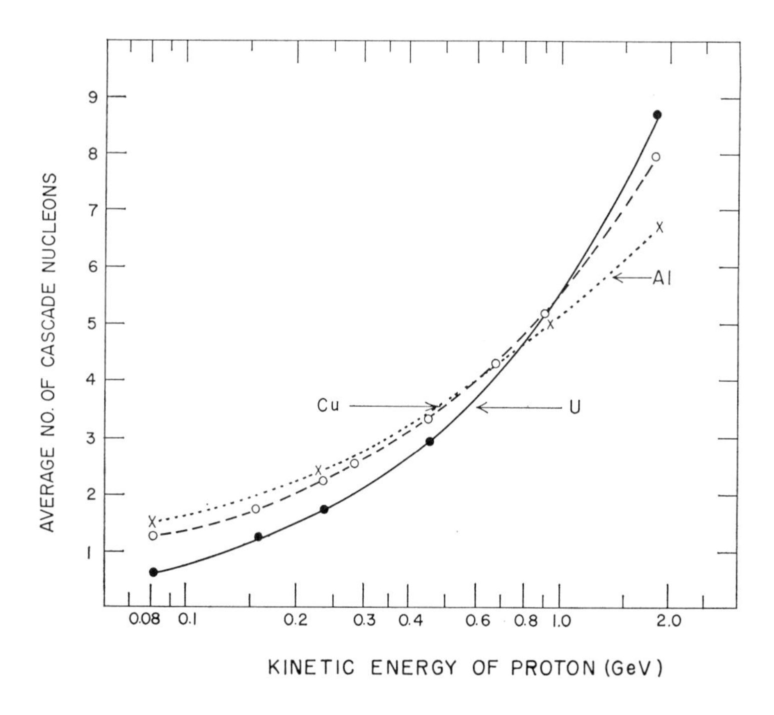

Fig. 2.40 Average number of cascade nucleons emitted from copper and uranium nuclei bombarded by protons (Metropolis 58).

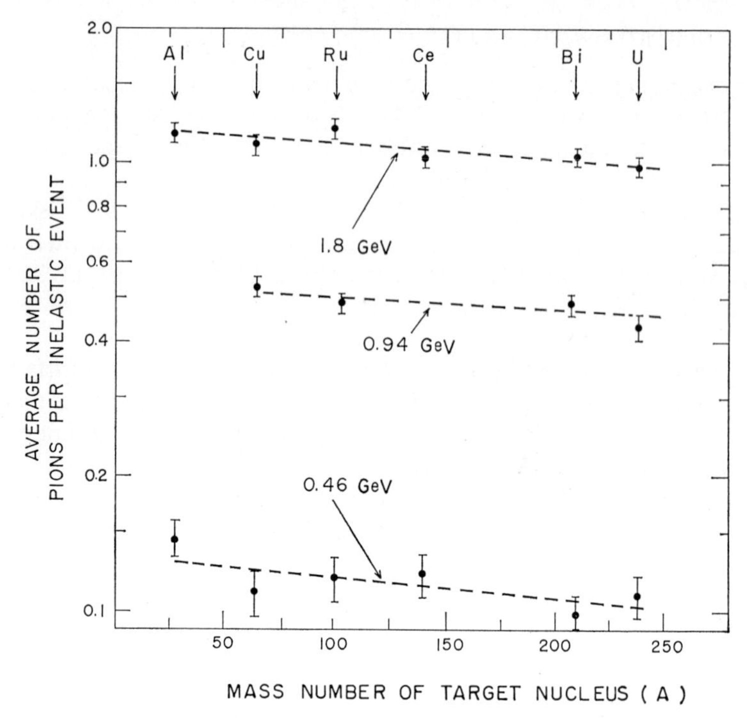

Fig. 2.41 Average number of pions emitted from various nuclei bombarded by protons with energies of 0.46, 0.94, and 1.8 GeV (Metropolis 58).

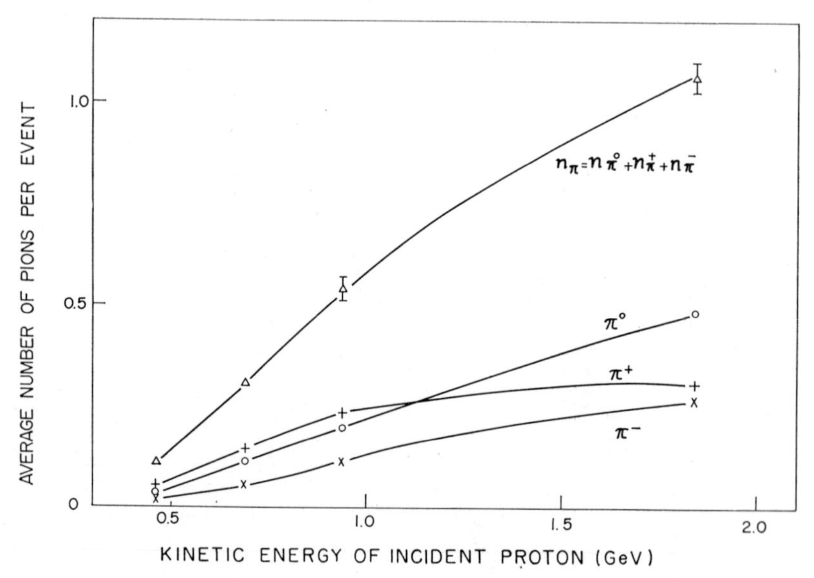

Fig. 2.42 Average numbers of π^+, π^-, and π^0 emitted from a ^{64}Cu target bombarded by protons.

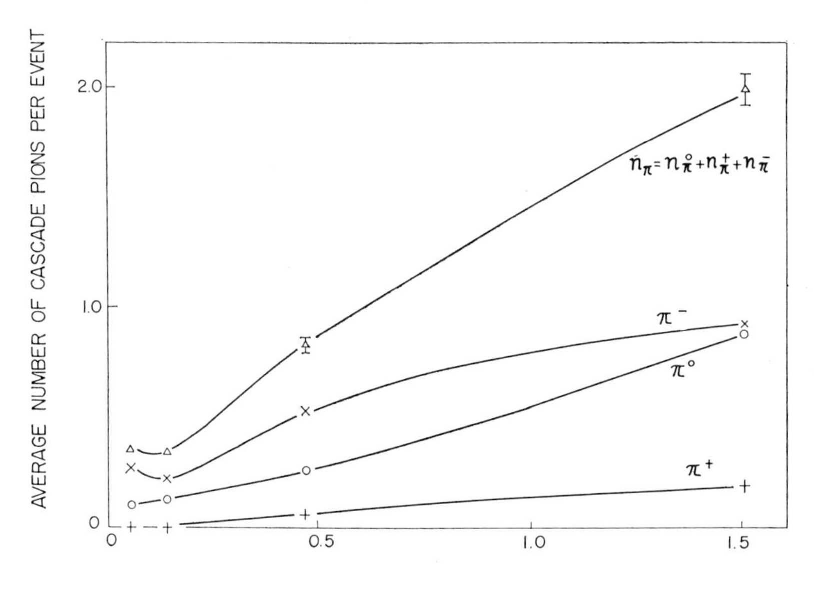
AVERAGE NUMBER OF CASCADE PIONS PER EVENT
2.0
1.0
0
0
0.5
1.0
1.5
$n_\pi = n_\pi^0 + n_\pi^+ + n_\pi^-$
π^-
π^0
π^+
KINETIC ENERGY OF INCIDENT π^- (GeV)

(a)

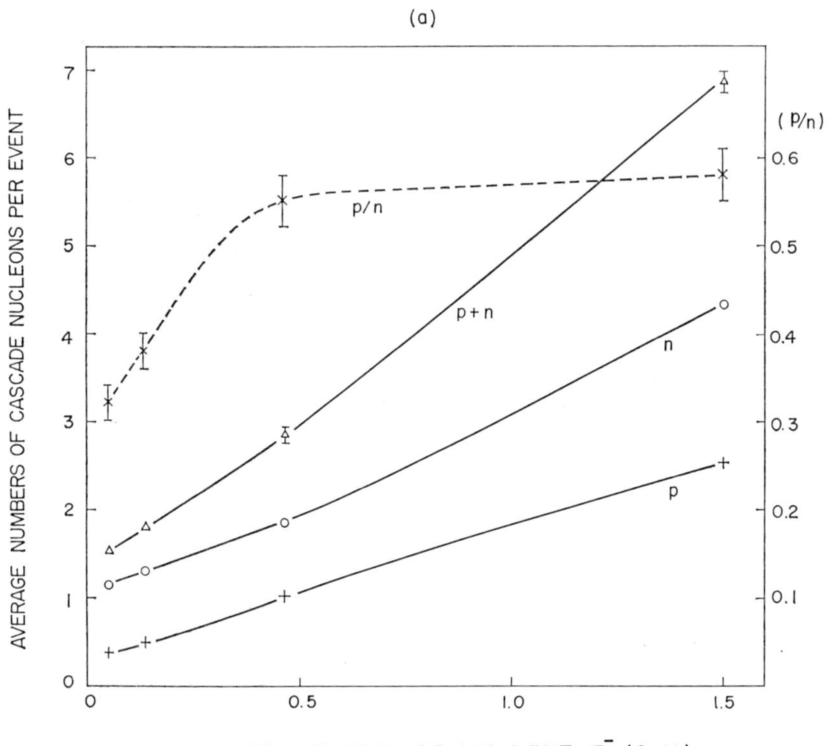
AVERAGE NUMBERS OF CASCADE NUCLEONS PER EVENT
7
6
5
4
3
2
1
0
(p/n)
0.6
0.5
0.4
0.3
0.2
0.1
0
0.5
1.0
1.5
p/n
p+n
n
p
KINETIC ENERGY OF INCIDENT π^- (GeV)

(b)

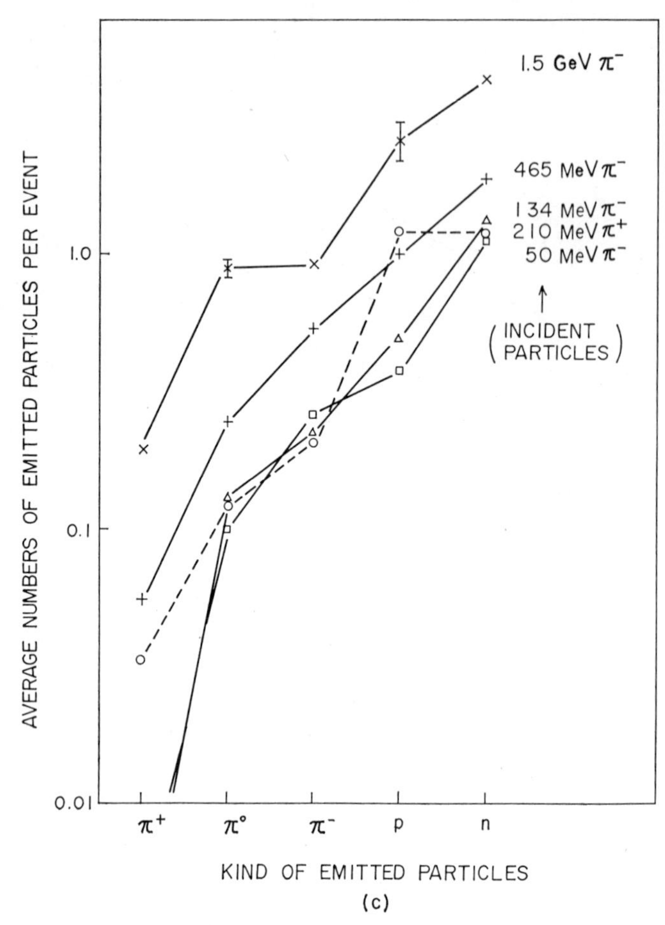

Fig. 2.43 Average numbers of cascade particles emitted from ^{100}Ru: π^+, π^-, and π^0 for incident π^-, (*a*); *p* and *n* for incident π^-, (*b*); comparison of yields of π^+, π^0, π^-, *p*, and *n* for incident π^+ and π^-, (*c*).

The average number of pions per proton-induced reaction is shown in Fig. 2.41. Among emitted pions, positive ones are most abundant below 1 GeV, whereas neutral ones are most abundant above that for incident protons, as indicated in Fig. 2.42 for copper.

For negative-pion-induced reactions on ruthenium the average numbers of respective particles emitted are shown in Fig. 2.43*a*. Positive pions are much fewer than others even for incident positive pions, mainly because of Coulomb effects. Nucleons are more than three times more abundant than pions, and the abundances of the former increase more rapidly than those of the latter, as shown in Fig. 2.43*b*. The yields of various kinds of particles are compared in Fig. 2.43*c*. The smallness of the pion yield is considered to be due to the reabsorption of pions in the same nucleus.

The actual number of emitted particles fluctuates about the average value to a considerable extent. The fluctuations are greater than those expected from the Poisson distribution and also depend on the energy available for the emission of nucleons. The fluctuations in cascade processes are represented by the yields of residual nuclei. From Fig. 2.44 we can see that the yields of nuclei with smaller mass numbers become appreciable as the incident energy increases, and they are nearly energy independent far above the threshold energies.

Angular Distribution. The angular distribution of emitted particles depends on their energy rather than the energy of an incident particle. It is more peaked in the forward direction as the energies of emitted particles increase. It also depends slightly on the size of a target nucleus, being flatter for heavier nuclei. The above properties are shown in Fig. 2.45, where the angular distributions of protons with energies between 30 and 90 MeV and above 90 MeV are shown for aluminum and uranium targets, respectively. It may be worthwhile to notice that

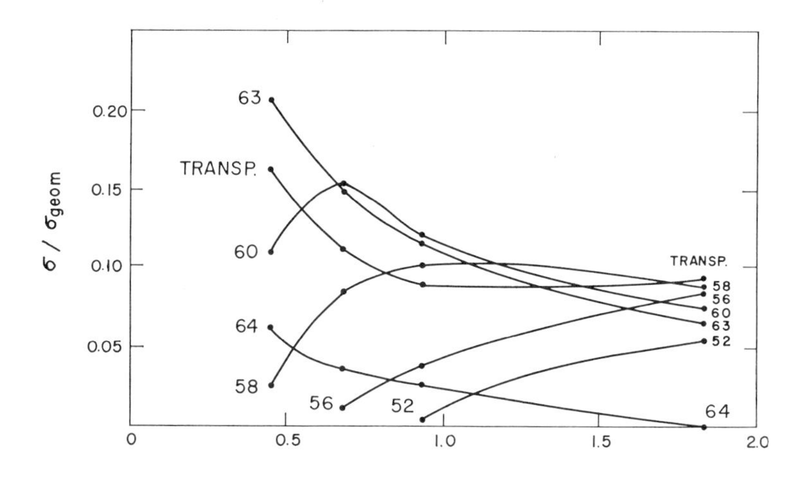

Fig. 2.44 Cross sections (relative to the geometrical ones) for the formation of residual nuclei of various mass numbers in the interactions of protons with ^{64}Cu. The transparencies are shown by the curve labeled "Transp"; the numbers attached to the curves represent the mass numbers of the residual nuclei (Metropolis 58).

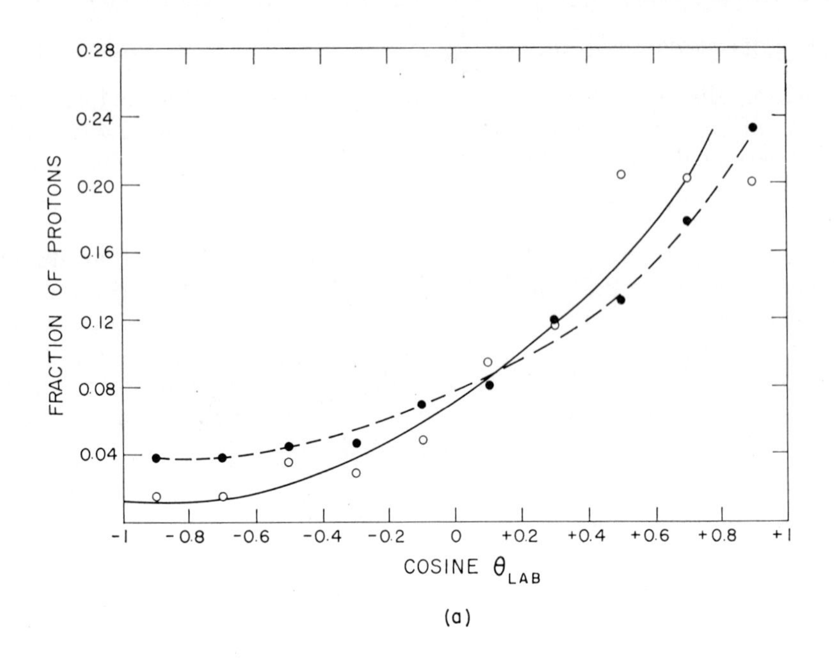

0.28
0.24
0.20
0.16
0.12
0.08
0.04
FRACTION OF PROTONS
-1
-0.8
-0.6
-0.4
-0.2
0
+0.2
+0.4
+0.6
+0.8
+1
COSINE θ_{LAB}

(a)

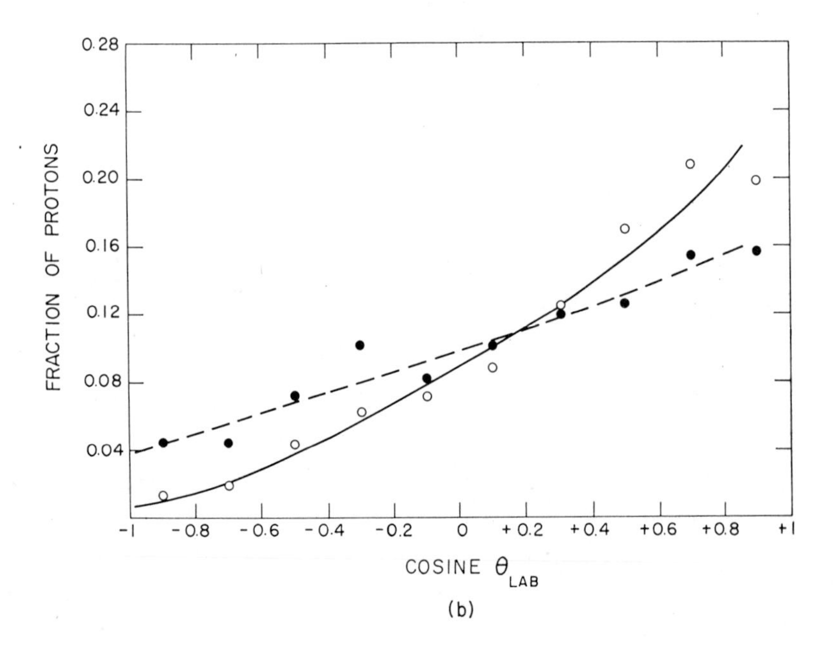

0.28
0.24
0.20
0.16
0.12
0.08
0.04
FRACTION OF PROTONS
-1
-0.8
-0.6
-0.4
-0.2
0
+0.2
+0.4
+0.6
+0.8
+1
COSINE θ_{LAB}

(b)

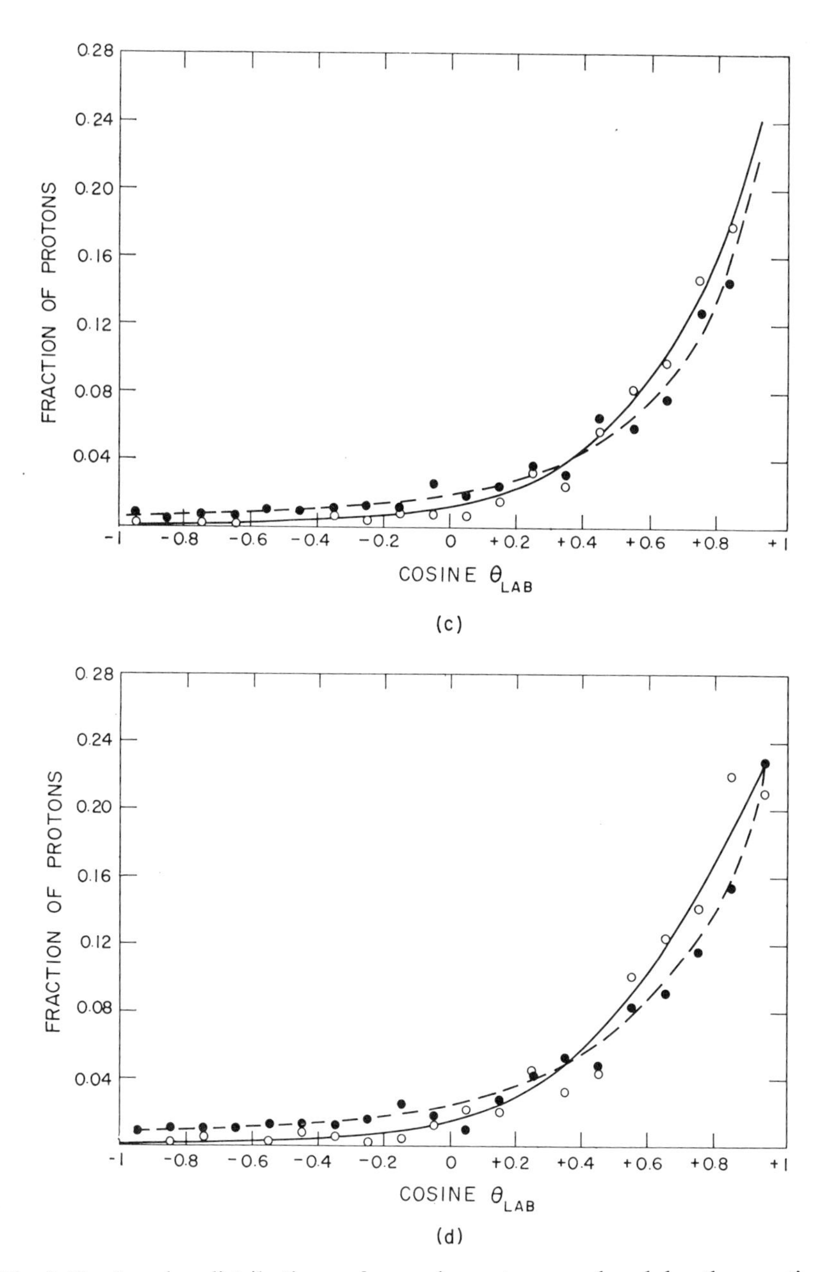

Fig. 2.45 Angular distributions of cascade protons produced by the reactions induced by protons of kinetic energies 460 MeV (—○—) and 1840 MeV (- - -●- - -). Abscissa: cosine of emission angle in the laboratory system (θ_{Lab}); ordinate: Fraction of protons per 0.2 interval of cos θ (Metropolis 58). Kinetic energies of emitted protons and the targets: 30 to 90 MeV, aluminum, (*a*); 30 to 90 MeV, uranium, (*b*); >90 MeV, aluminum, (*c*); >90 MeV, uranium, (*d*).

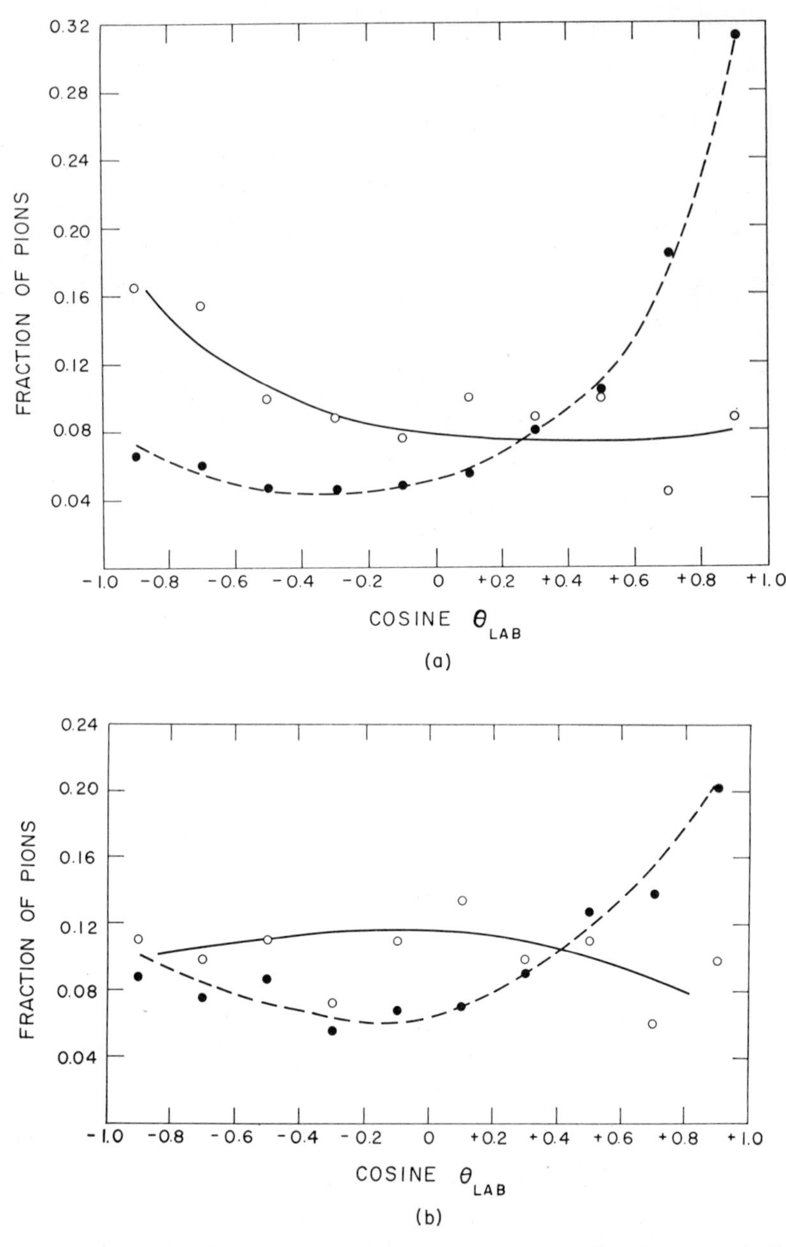

Fig. 2.46 Angular distribution of pions (all charges and all energies) emitted in the interactions induced by protons with kinetic energies of 460 MeV (○) and 1840 MeV (- - -●- - -). Abscissa: cosine of emission angle in the laboratory system (θ_{Lab}); ordinate: fraction of pions per 0.2 interval in cos θ (Metropolis 58). Aluminum target, (*a*); uranium target, (*b*).

the angular distribution of the low-energy protons is steeper for a lower incident energy. This is because the number of nuclear collisions necessary to produce secondary protons in this energy is smaller for a lower incident energy.

The emitted pions are more isotropic than nucleons. Since the multiplicity of collisions increases as a target nucleus becomes heavier, the angular distribution becomes more isotropic with increasing target mass. The forward peak for high incident energies is due to the collimated production of pions at the first collision. These features can be seen from Fig. 2.46.

Energy Distribution. The energy spectrum of emitted nucleons is a little steeper than dE/E between 30 MeV and a half of the incident energy, depending slightly on the incident energy and the target nucleus. A bump near one-half the incident energy indicates the contribution of single collisions. An increase of the incident energy results in the extension of the spectrum toward high energy. Obviously the spectrum is steeper as the target nucleus becomes heavier. The proton spectra for incident protons are shown in Fig. 2.47.

The pion spectra have not been calculated as accurately as the proton spectra. The integral spectra of pions for protons incident on copper and uranium are shown in Fig. 2.48. They are nearly flat below 100 MeV and decrease gradually toward high energy. This feature is also seen for incident pions.

Excitation Energy. The energy left after the emission of cascade particles is regarded as the excitation energy in a residual nucleus. The energy and mass-number dependences of the excitation energy are seen in Fig. 2.49. Obviously it increases with increasing target size. Its energy dependence is very weak up to 350 MeV because the increase in the incident energy is partly compensated for by the decrease in the nucleon-nucleon cross section. Above 350 MeV the excitation energy increases rapidly with increasing incident energy because of the onset of pion production, by which the incident energy is shared with a recoil nucleon and a pion that may be reabsorbed inside the same nucleus. A further increase above 1 GeV is due to the double-pion production that becomes important in this energy region.

The average excitation energy also depends on the number of emitted particles, as shown in Fig. 2.50 for a copper target. In general it increases with the number of cascade nucleons emitted, but the relation is reversed at low incident energies. If the number of cascade nucleons is fixed, the excitation energy increases with incident energy below a certain energy, reaches a broad maximum, and then decreases slowly. The regular

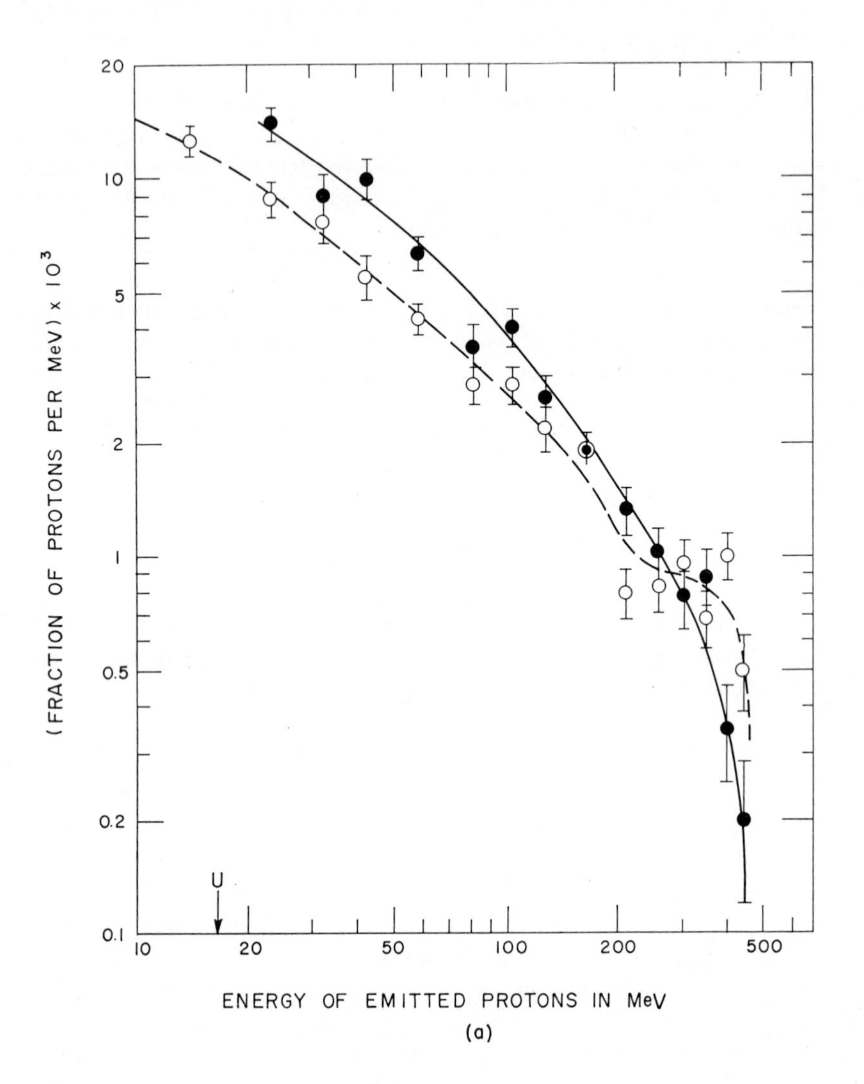
(FRACTION OF PROTONS PER MeV) x 10^3
20
10
5
2
1
0.5
0.2
0.1
U
10
20
50
100
200
500
ENERGY OF EMITTED PROTONS IN MeV
(a)

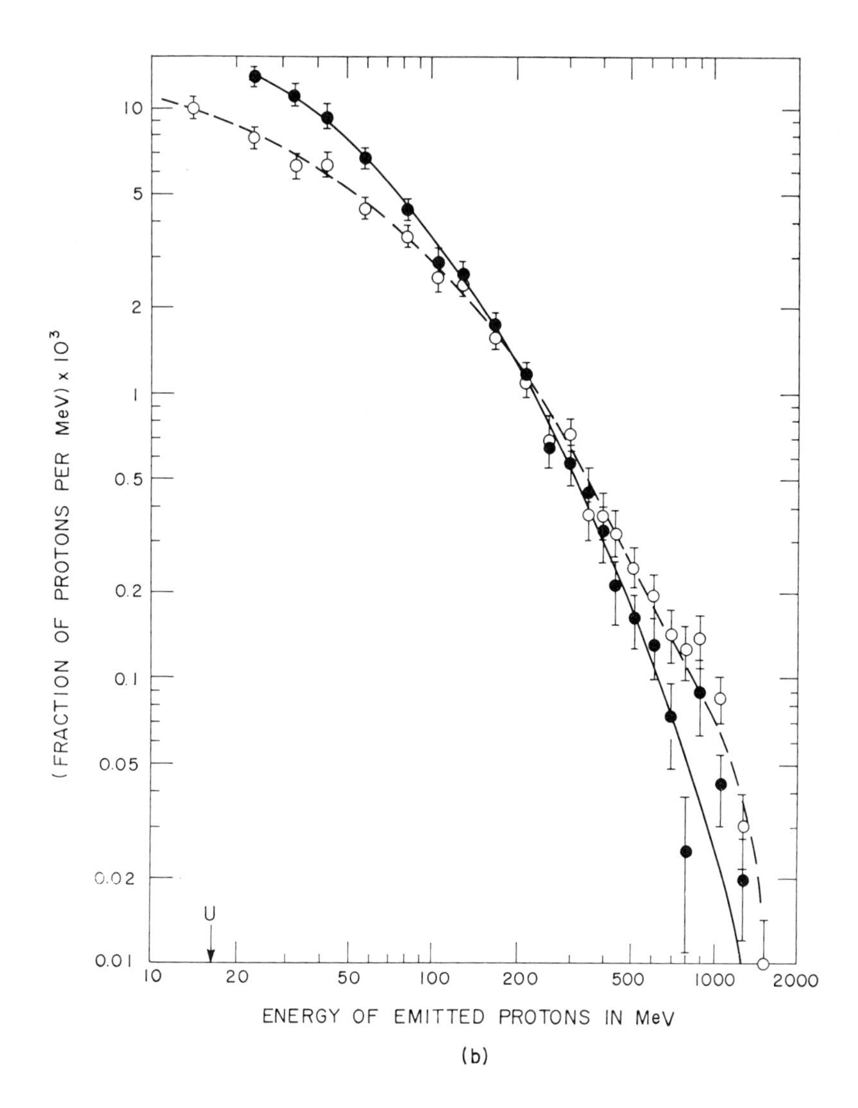

Fig. 2.47 Energy spectra of cascade protons emitted from aluminum (- -○- -) and uranium (—●—). The arrow marked U indicates the lowest kinetic energy of protons that can emerge from a uranium nucleus (Metropolis 58). Kinetic energies of incident protons: 460 MeV, (*a*); 1840 MeV, (*b*).

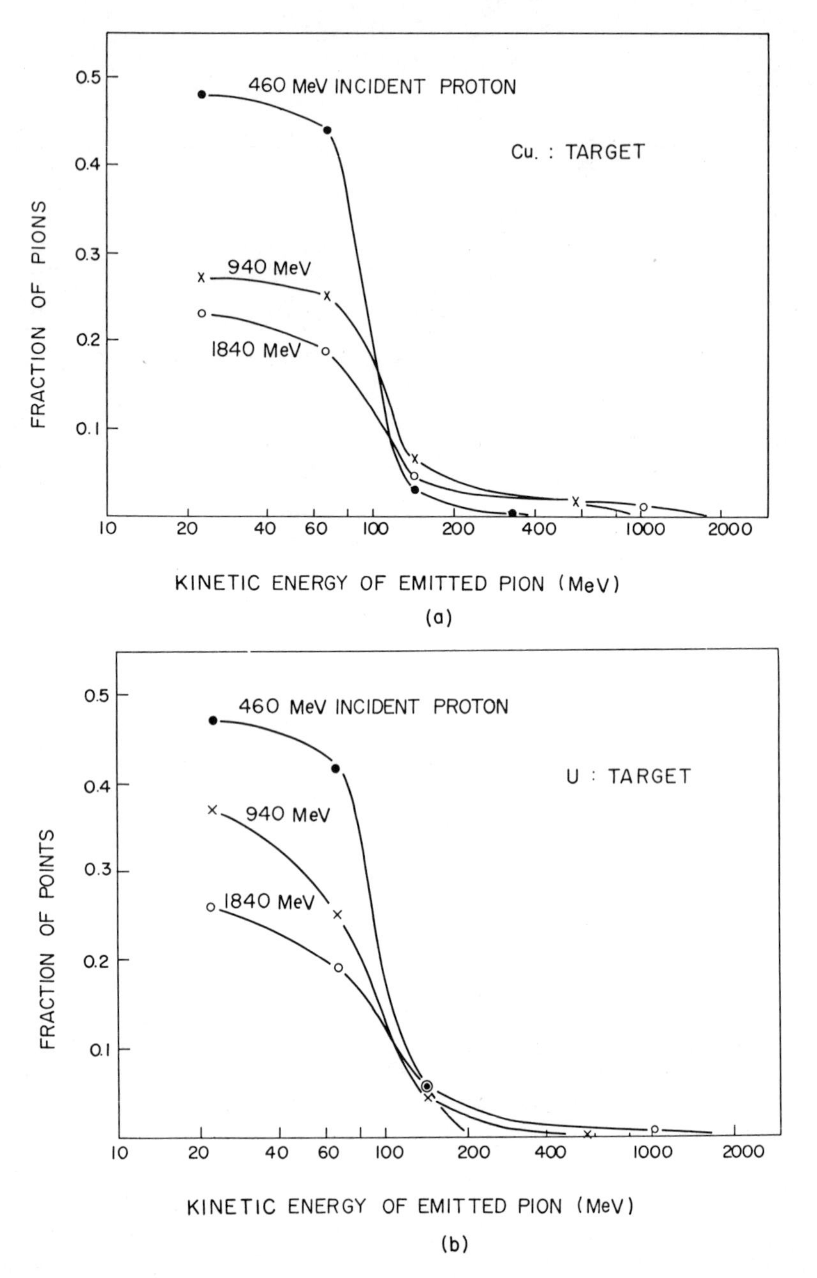

Fig. 2.48 Energy spectra of pions emitted in the interactions induced by protons with kinetic energies of 460 MeV (—●—), 940 MeV (—×—), and 1840 MeV (—○—). Abscissa: kinetic energy of pion emitted (MeV); ordinate: fraction of pions emitted in each of four energy intervals: 0 to 45, 45 to 90, 90 to 200, and >200 MeV. Copper target, (*a*); uranium target, (*b*).

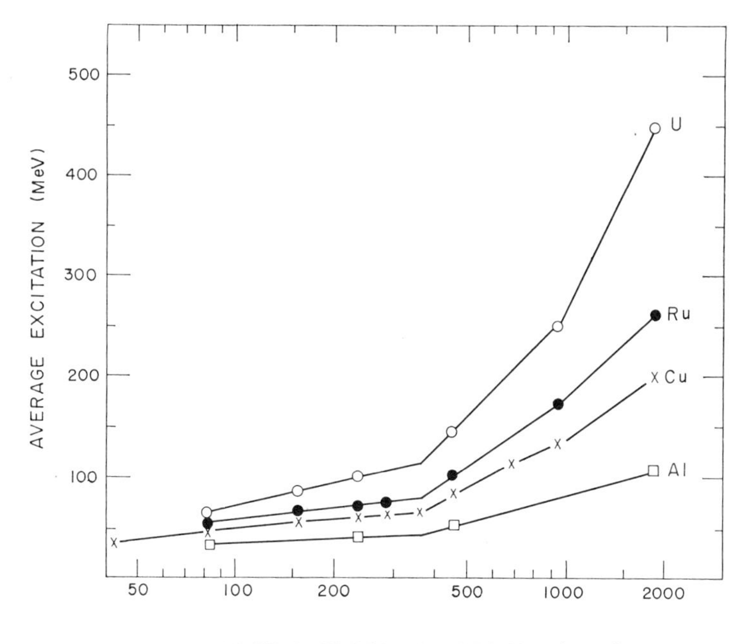

Fig. 2.49 Average excitation energies in residual nuclei in the interactions induced by protons (Metropolis 58).

dependence on the number of cascade nucleons appears at and above the maximum, and the incident energy at which the maximum appears increases with the number of cascade particles. These features can be seen in Fig. 2.50 and are easily understandable in terms of the cascade process.

The excitation energy is distributed in a wide range about the average value. The rms spread is found to be comparable to the mean value. The distribution becomes wider as the number of cascade nucleons increases. These features are shown for a copper target in Figs. 2.51 and 2.52.

2.9.2 Evaporation Process

The excitation energy left after the cascade process is shared with many nucleons, so that a nucleus can be described as a thermodynamic system. The probability per unit time for the emission of particle j with a kinetic energy between E and $E + dE$ is given by

$$P(E)\,dE = g_j \frac{m_j}{\pi^2 \hbar^3} \sigma E \frac{\rho_f}{\rho_i}\,dE, \tag{2.9.4}$$

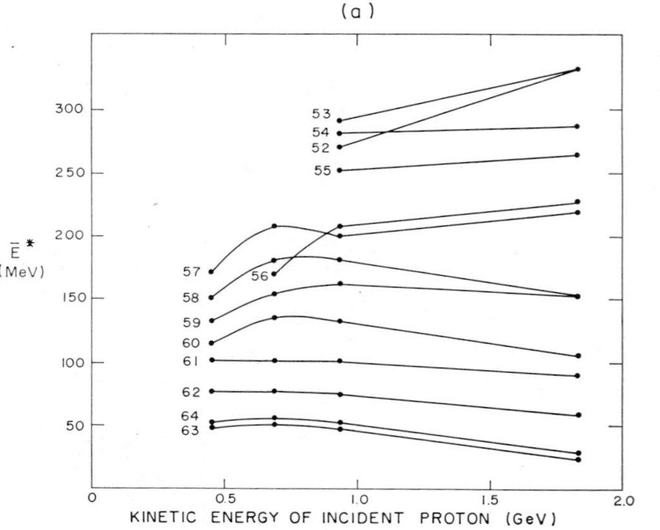

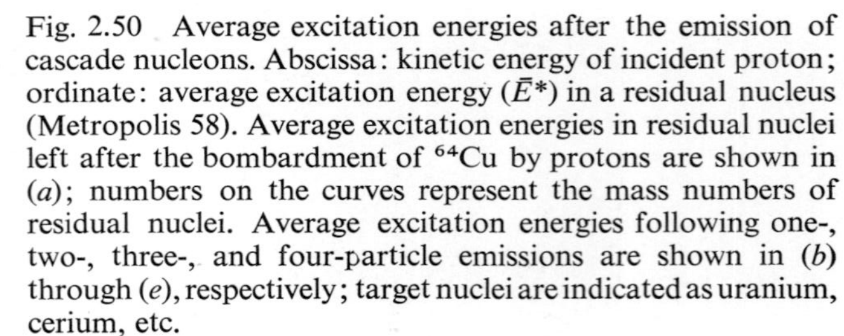

Fig. 2.50 Average excitation energies after the emission of cascade nucleons. Abscissa: kinetic energy of incident proton; ordinate: average excitation energy ($\bar{E}^*$) in a residual nucleus (Metropolis 58). Average excitation energies in residual nuclei left after the bombardment of ^{64}Cu by protons are shown in (*a*); numbers on the curves represent the mass numbers of residual nuclei. Average excitation energies following one-, two-, three-, and four-particle emissions are shown in (*b*) through (*e*), respectively; target nuclei are indicated as uranium, cerium, etc.

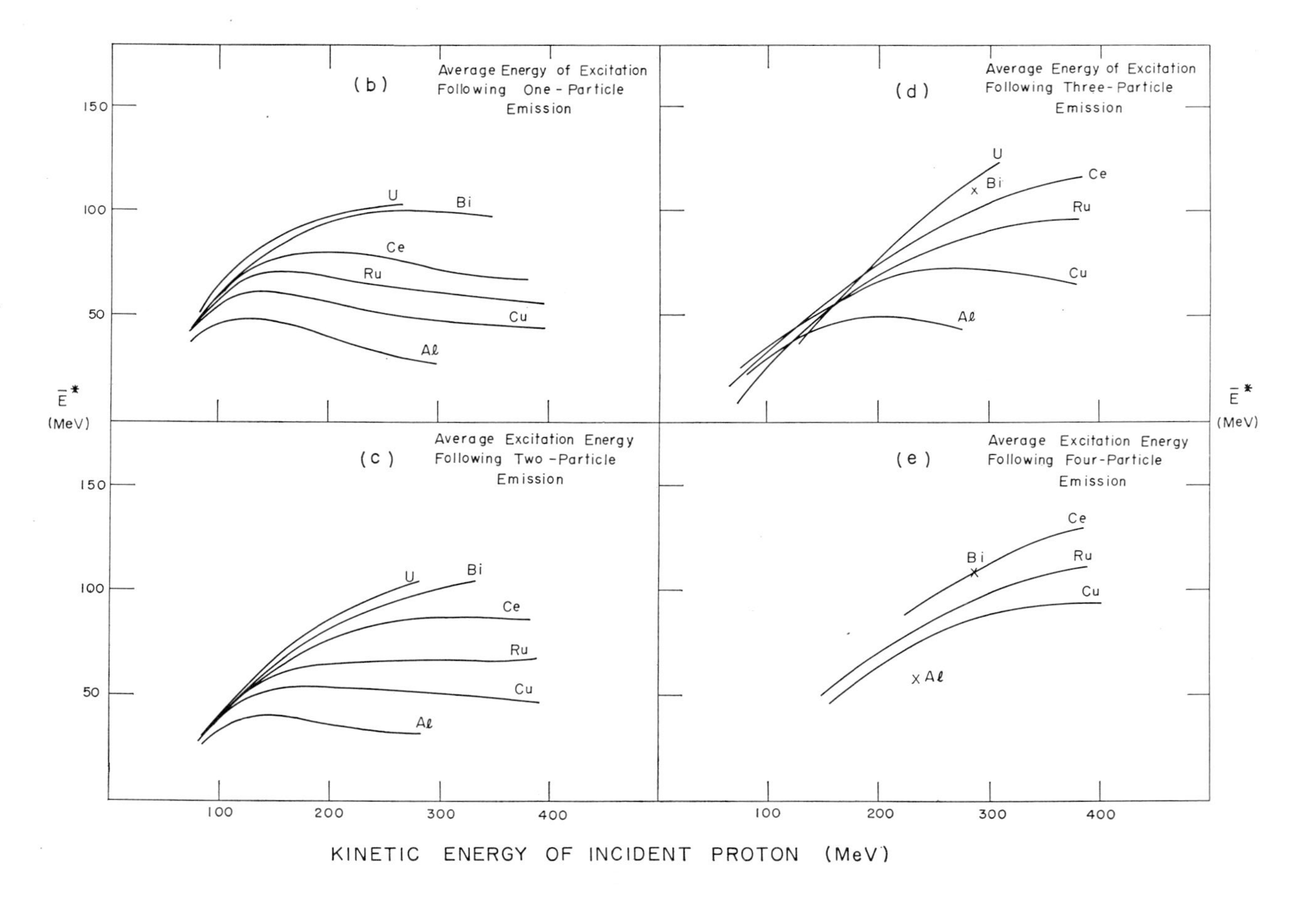

(b)
Average Energy of Excitation
Following One-Particle
Emission
U
Bi
Ce
Ru
Cu
Aℓ
(d)
Average Energy of Excitation
Following Three-Particle
Emission
U
Bi
Ce
Ru
Cu
Aℓ
(c)
Average Excitation Energy
Following Two-Particle
Emission
U
Bi
Ce
Ru
Cu
Aℓ
(e)
Average Excitation Energy
Following Four-Particle
Emission
Ce
Bi
Ru
Cu
Aℓ
150
100
50
Ē*
(MeV)
100
200
300
400
KINETIC ENERGY OF INCIDENT PROTON (MeV)

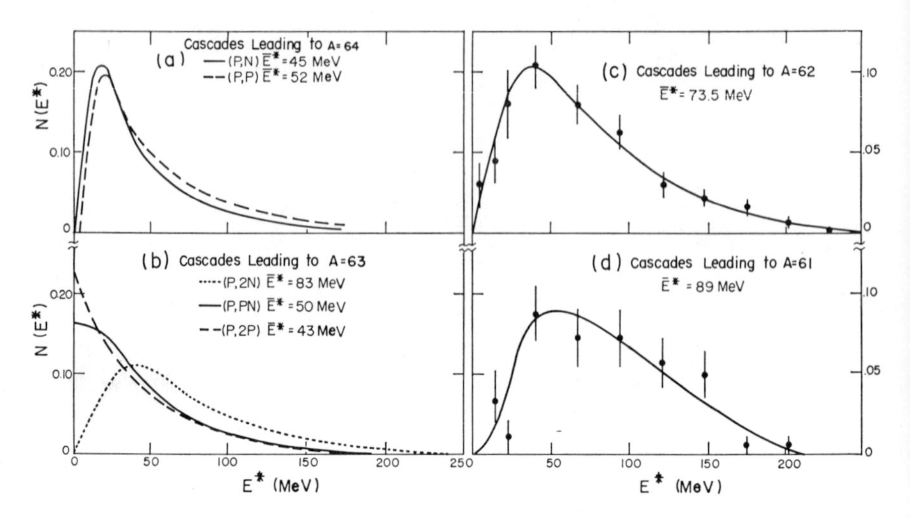

Fig. 2.51 Distribution of excitation energies associated with particular cascade processes in ^{64}Cu initiated by 286-MeV protons. Abscissa: excitation energy (E^*) in MeV; ordinate: probability of having excitation energy E^* per 10-MeV interval (Metropolis 58).

where g_j and m_j are the spin weight and the mass of particle j, respectively. The derivation of (2.9.4) is based on the detailed balancing between the emission and capture reactions of particle j:

$$(i) \rightleftharpoons (f) + j.$$

In (2.9.4) σ is the capture cross section; this is proportional to the level density of the initial nucleus, ρ_i, whereas the emission rate is proportional to the level density of the final nucleus, ρ_f.

Since we are interested only in the relative probability of emission, the final result is not much affected by the capture cross section but depends mainly on the final state density. The level density is represented by

$$\rho(E_x) = C \exp [2(aE_x)^{1/2}], \tag{2.9.5}$$

where E_x is the excitation energy of the nucleus concerned. The excitation energy decreases after the emission of a particle with energy E_j by

$$E_{x,\,i} - E_{x,\,f} = E_j + Q_j,$$

where Q_j is the separation energy of particle j. The maximum energy of a particle emitted is $E_{x,\,i} - Q_j$ for a neutron and $E_{x,\,i} - Q_j - V_j$ for a charged particle that is subjected to the Coulomb barrier V_j. The effective Coulomb barrier is lower than the classical value because of the finite penetrability. The level-density parameter a is subject to future investigation but may be chosen approximately as

$$a = \frac{A}{10}\,\text{MeV}^{-1} \quad \text{or} \quad \frac{A}{20}\,\text{MeV}^{-1}. \tag{2.9.6}$$

The nuclear temperature is given by $T = (E_x/a)^{1/2}$.

On the basis of the above model the evaporation process is followed by the Monte Carlo method. The main results thus obtained are summarized below.

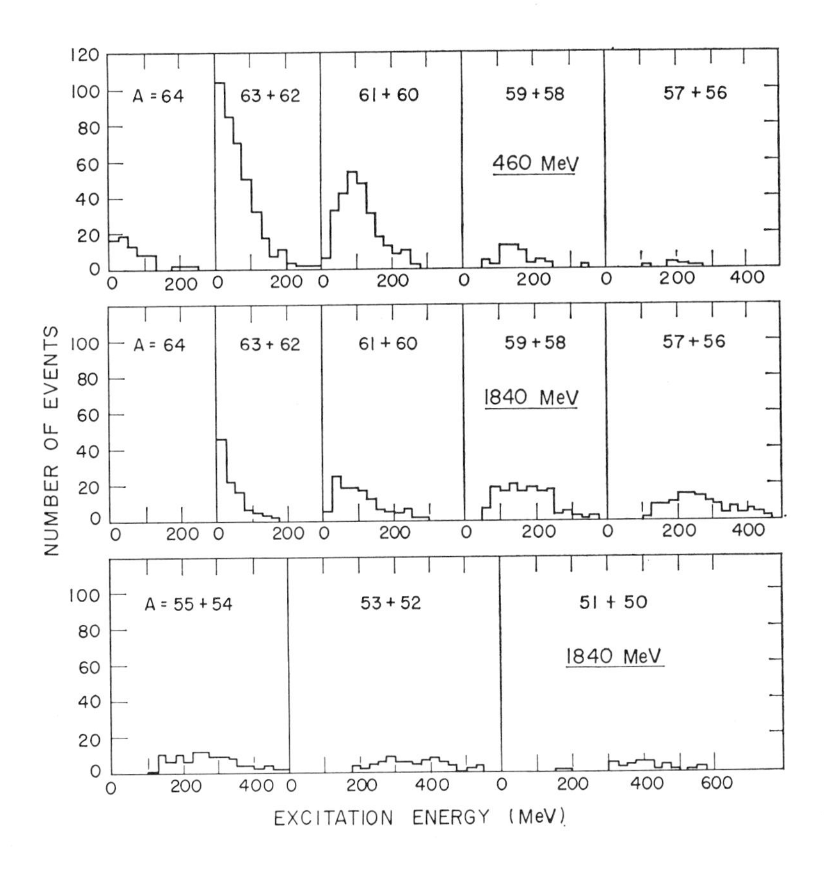

Fig. 2.52 Excitation-energy spectra for different residual nuclei in the interactions of 460- and 1840-MeV protons with ^{64}Cu (Metropolis 58).

Average Number of Evaporated Particles. The average numbers of neutrons and protons evaporated against A are shown for an initial temperature of $T_0 = 4$ MeV in Figs. 2.53*a* and *b*, respectively. The shell effect is revealed by the zigzag behavior. The number of particles evaporated depends also on the charge of the initial nucleus; the number of neutrons increases with neutron excess, whereas that of protons, with proton excess. This is found to be proportional to the deviation from the stable case as

$$n_j = A_j + B_j(Z - Z_A), \tag{2.9.7}$$

where Z_A is the charge of a nucleus of mass A on the stable line. The values of these coefficients are given in Table 2.18.

Table 2.18 Values of A_j and B_j; $n_j = A_j + B_j(Z - Z_A)$

A	A_n	B_n	A_p	B_p	A_α	B_α
64	4.13 ± 0.04	-0.89 ± 0.02	1.60 ± 0.06	0.57 ± 0.03	0.35 ± 0.03	0.09 ± 0.01
109	8.13 ± 0.09	-0.95 ± 0.04	2.00 ± 0.04	0.51 ± 0.02	0.57 ± 0.03	0.08 ± 0.01
181	16.0 ± 0.02	-0.99 ± 0.01	2.30 ± 0.04	0.39 ± 0.02	0.69 ± 0.04	0.08 ± 0.02
219	19.7 ± 0.04	-1.20 ± 0.02	2.50 ± 0.05	0.43 ± 0.02	1.10 ± 0.01	0.17 ± 0.01

The loss of charge, $\langle \Delta Z \rangle_{av}$, relative to the average mass loss $\langle \Delta A \rangle_{av}$ due to evaporation is seen from Fig. 2.54. The neutron deficiency increases with increasing mass of the starting nucleus. The loss of mass, $\langle \Delta A \rangle$, is nearly proportional to the mass number of the initial nucleus for the given initial temperature, and the shell effect essentially disappears in this relation. This relation is expressed by

$$\frac{\langle \Delta A \rangle}{A} = \left(\frac{a}{A}\right) \frac{T_0^2}{(8.9 + 0.97T_0)} \tag{2.9.8}$$

where T_0 is the initial temperature in MeV. Since $T_0^2 = E_0/a$, where E_0 is the initial excitation energy, the average de-excitation energy per evaporated nucleon is given by

$$\varepsilon \equiv \frac{E_0}{\langle \Delta A \rangle} = 8.9(\pm 0.15) + 0.97(\pm 0.025)T_0 \text{ MeV} \tag{2.9.9}$$

The average numbers of various types of particles emitted have been computed for ^{64}Cu, ^{109}Ag, and ^{181}Ta. They are plotted against the initial excitation energy in Fig. 2.55. More than half the particles emitted are found to be neutrons and about half the charged particles are protons. Among the rest, deuterons are nearly as abundant as heavier

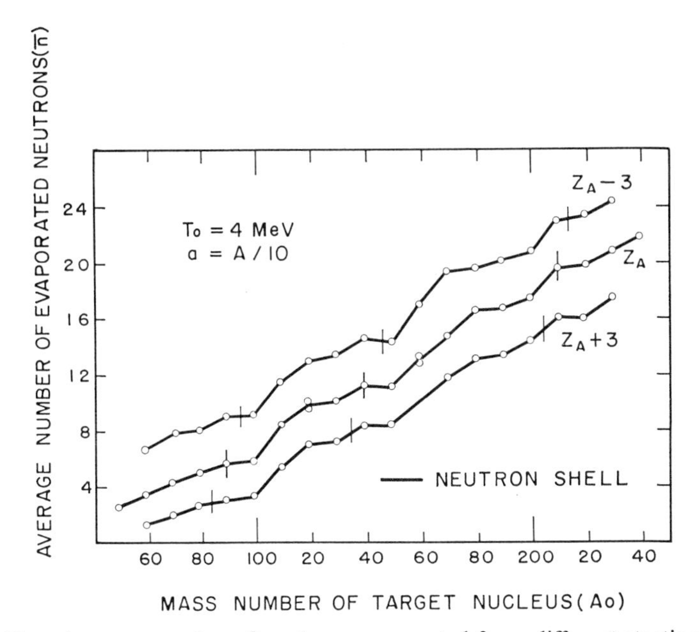

Fig. 2.53*a* Average number of neutrons evaporated from different starting nuclei. Initial temperature $T_0 = 4$ MeV, $a = A/10$. Vertical bars indicate neutron shells, and Z_A is the nuclear charge corresponding to the target mass number A_0 on the stability line (Dostrovsky 58).

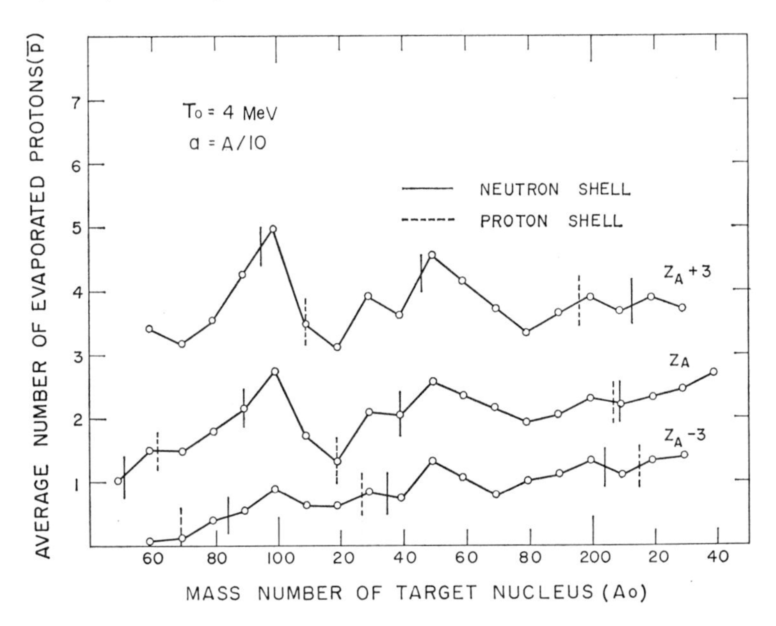

Fig. 2.53*b* Average number of protons evaporated from different starting nuclei. Initial temperature $T_0 = 4$ MeV, $a = A/10$. Vertical solid and dotted bars indicate neutron and proton shells, respectively, and Z_A is the nuclear charge corresponding to the target mass number A_0 on the stability line (Dostrovsky 58).

particles. This means that about one-quarter of the particles emitted are composite nuclei of average mass number 3, and consequently the average mass number of all particles is about 1.5. The relative abundance of composite nuclei emitted decreases with increasing target mass, because the excitation energy is preferentially taken away by neutrons that are in excess in a target nucleus, and because the emission of charged particles is suppressed by high Coulomb barriers. This also decreases with decreasing excitation energy, again because of the preferential emission of neutrons.

The actual number of evaporated particles fluctuates about their average number. The fluctuations are found to be smaller than those for a Poisson distribution, because most of the de-excitation energy is spent for the binding energy, as indicated by the temperature-independent term in (2.9.9). The rms spread in the number of emitted particles is about 10 percent for an excitation energy as high as 300 MeV and slowly decreases with decreasing excitation energy.

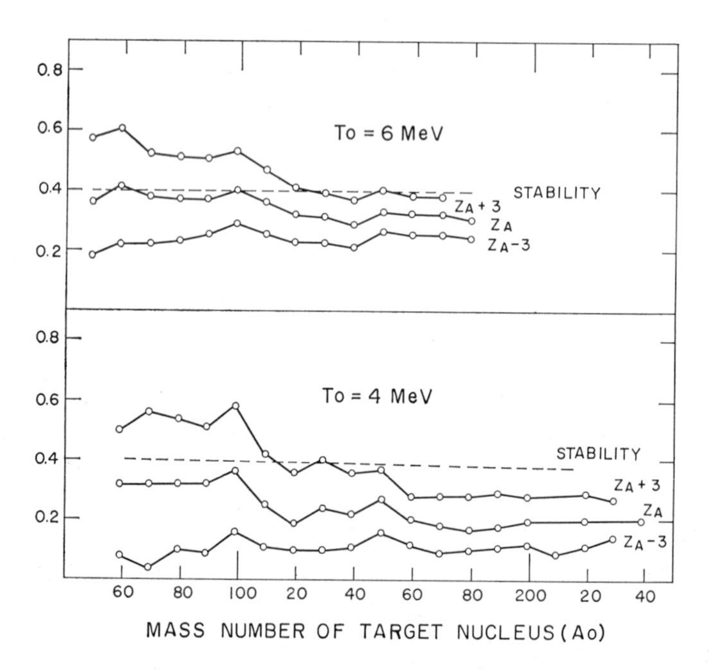

Fig. 2.54 The average charge loss $\langle \Delta Z \rangle_{av}$ relative to the average mass loss $\langle \Delta A \rangle_{av}$ as a function of the mass number of the target nucleus; $a = A/10$ and $T_0 = 6$ and 4 MeV. Dashed lines represent the stability lines of residual nuclei. Note that residual nuclei are neutron deficient (Dostrovsky 58).

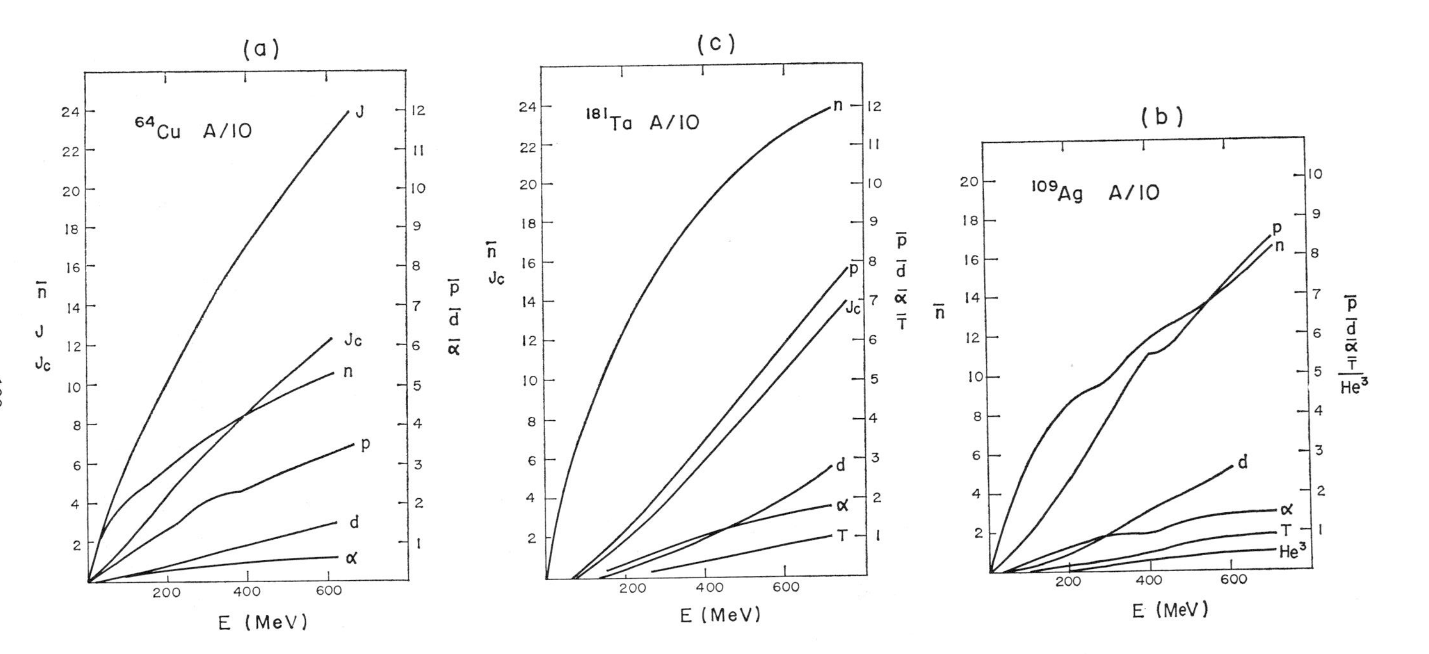

Fig. 2.55 The average numbers of various types of evaporated particles. The types of particles are indicated on the curves; J_c is the number of charged particles, J is the number of all particles; $a = A/10$. Abscissa: excitation energy in MeV; ordinate: number of evaporated particles (Dostrovsky 58). Target nuclei: ^{64}Cu, (*a*); ^{109}Ag, (*b*); ^{181}Ta, (*c*).

Energy Spectra of Emitted Particles. The average de-excitation energy per nucleon given in (2.9.9) may be written as

$$\varepsilon = \bar{Q} + \bar{V} + 2\bar{T}, \tag{2.9.9'}$$

in which $2\bar{T}$ represents the average kinetic energy of emitted nucleons for the average temperature $\bar{T}$. The energy spectrum is essentially

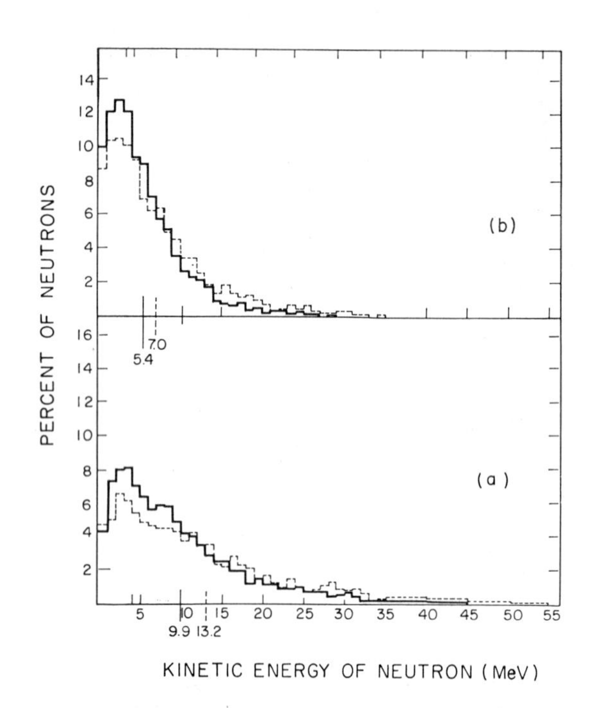

Fig. 2.56 Energy spectra of neutrons evaporated from ^{109}Ag: broken line, $a = A/20$; solid line, $a = A/10$. Vertical bars drawn on the abscissa represent the average values of the kinetic energy (Dostrovsky 58). Initial excitation: 700 MeV, (*a*); 200 MeV, (*b*).

governed by the energy-dependent factor in (2.9.4), $E \exp(-E/T)\, dE$, which has a maximum at T. Since the temperature decreases in the course of evaporation, the maximum is expected to appear near $\bar{T}$, which is a few MeV. This can be seen in the neutron spectra shown in Fig. 2.56. The dependence on the initial excitation energy reveals itself mainly in the high-energy tail.

For charged particles the low-energy part is cut by the Coulomb barrier. As its consequence, the high-energy tail is more pronounced than that in the neutron spectrum, as seen in Figs. 2.57 and 2.58, but few particles are emitted with energies greater than 30 MeV. Since the number of cascade particles emitted is considerable down to 20 MeV, the evaporation process takes main part in particle emission only below 20 MeV. Even in the case of low incident energies, the contribution of direct reactions is found to be considerable.

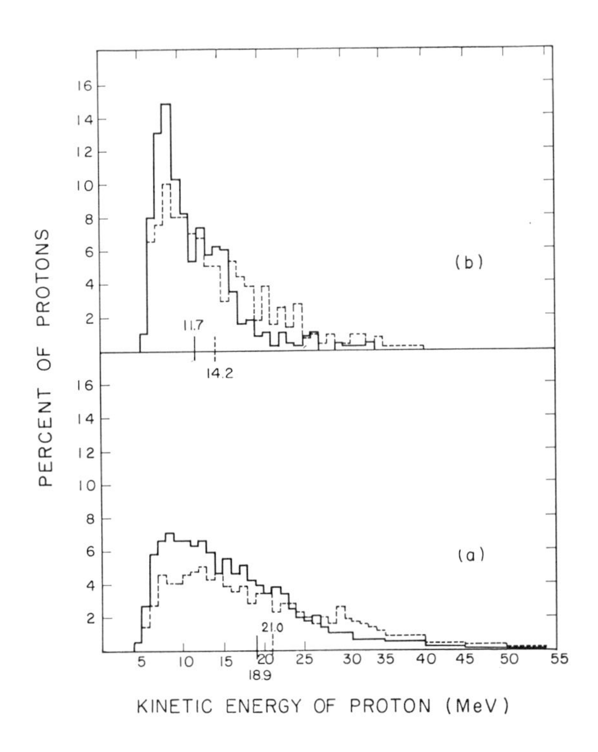

Fig. 2.57 Energy spectra of protons evaporated from ^{109}Ag: broken line, $a = A/20$; solid line, $a = A/10$. Vertical bars drawn on the abscissa represent the average values of the kinetic energy (Dostrovsky 58). Initial excitation: 700 MeV, (*a*); 200 MeV, (*b*).

The comparison of the Monte Carlo results with experimental facts indicates that agreement is good for nuclear cascade processes but seems to be rather poor for evaporation processes. The latter may be due mainly to the difficulty in separating the direct processes from evaporation. As far as the average behavior is concerned, however, the calculated results seem to be good enough to predict the essential features of nuclear reactions.

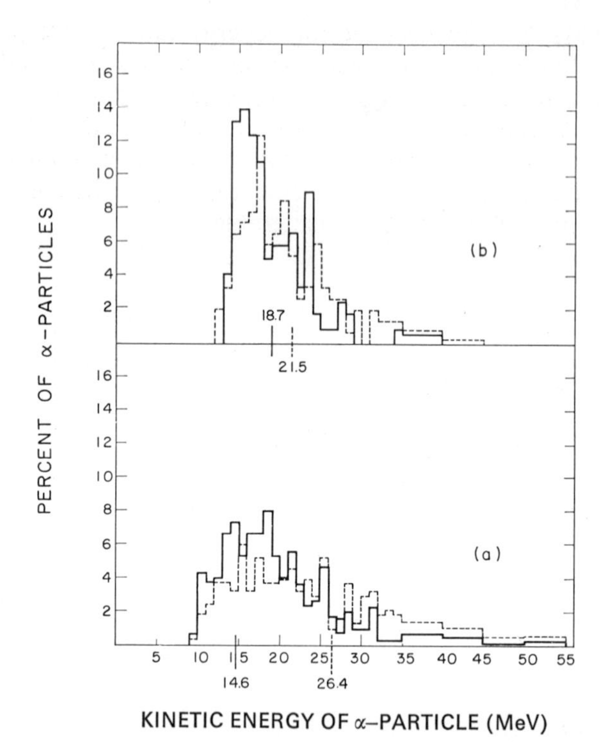

Fig. 2.58 Energy spectra of α-particles evaporated from ^{109}Ag: broken line, $a = A/20$; solid line, $a = A/10$. Vertical bars drawn on the abscissa represent the average values of the kinetic energy (Dostrovsky 58). Initial excitation: 700 MeV, (*a*); 200 MeV, (*b*).

2.9.3 *Empirical Method*

Although the Monte Carlo calculations of cascade and evaporation processes are extensive, the results thus far available are not convenient enough to predict the final products on high-energy nuclear reactions, because the connection of these two processes has not been fully investigated yet. It is therefore convenient to have an empirical relation.

Among the various empirical relations thus far proposed the so-called Rudstam formula is found to reproduce the observed yields of various nuclides. The cross section of yielding a nuclide of Z and A from copper is expressed as (Rudstam 62)

$$\sigma(Z, A) = \exp(PA - Q - R|Z - SA + TA^2|^{3/2}), \qquad (2.9.10)$$

where P is a parameter that decreases with increasing incident energy until it levels off above 1 GeV, whereas R and S are roughly energy independent; T is a small parameter that increases with energy at nonrelativistic incident energies; Q is a parameter for normalization. The values of these parameters are given in Table 2.19.

Table 2.19 Values of Parameters P, R, S, and T in the Rudstam Formula

Irradiation Energy (MeV)	P	R	S	T	ε
50	0.90 ± 0.15	1.4 ± 0.5	0.43 ± 0.08	-0.0007 ± 0.0013	1.96
90	0.46 ± 0.15	1.8 ± 1.0	0.46 ± 0.07	-0.0001 ± 0.0011	2.14
190	0.22 ± 0.10	2.2 ± 0.5	0.47 ± 0.04	0.0000 ± 0.0007	1.73
340	0.25 ± 0.02	1.9 ± 0.2	0.48 ± 0.01	0.0002 ± 0.0001	1.73
680	0.16 ± 0.02	1.7 ± 0.3	0.50 ± 0.01	0.0006 ± 0.0002	2.44
980	0.053 ± 0.035	2.0 ± 0.3	0.49 ± 0.02	0.0004 ± 0.0003	1.72
2200	0.066 ± 0.014	1.8 ± 0.2	0.49 ± 0.01	0.0005 ± 0.0001	1.66
5700	0.052 ± 0.006	2.0 ± 0.1	0.49 ± 0.01	0.0004 ± 0.0001	1.43
24,000	0.065 ± 0.008	2.0 ± 0.1	0.49 ± 0.01	0.0004 ± 0.0001	1.57

The Rudstam formula is based on the yields of radioactive nuclides, and the predicted yields are found to deviate from experimental ones within a factor of 2, as seen from ε in Table 2.19, where ε is defined by

$$\varepsilon = \langle (\ln \sigma_{\text{exp}} - \ln \sigma_{\text{calc}})^2 \rangle^{1/2}, \tag{2.9.11}$$

the rms deviation of the logarithm of the cross section.

The yields of nuclides with mass numbers from A_1 to A_2 are approximately obtained as

$$\sum_{A_1}^{A_2} \sigma(Z, A) \simeq \frac{1.79}{R^{2/3}} \frac{e^{-Q}}{P} (e^{PA_2} - e^{PA_1}). \tag{2.9.12}$$

2.9.4 Photon-Induced Reactions

The elementary processes of photoreactions at high energies consist of the absorption by a whole nucleus, photodisintegration of a binucleon system, pion production, and at still higher energies, kaon production. The total absorption cross section below the pion threshold k_{th} is given independently of details of the processes concerned by

$$\begin{aligned} \int_0^{k_{\text{th}}} \sigma_{\text{abs}}\, dk &= 2\pi^2 \frac{e^2\hbar}{Mc} \frac{NZ}{A} \left(1 + 0.1 \frac{A^2}{NZ}\right) \\ &\simeq 60 \frac{NZ}{A} \left(1 + 0.1 \frac{A^2}{NZ}\right) \quad \text{mb} - \text{MeV}, \end{aligned} \tag{2.9.13}$$

where $N = A - Z$ and M is the nucleon mass; $k_{\rm th}$ may be taken as being about 155 MeV.

The largest contribution to this cross section comes from the electric-dipole absorption that is called the giant resonance. A simple oscillator model gives a cross section

$$\int \sigma_{\rm abs}({\rm El})\, dk = 2\pi^2 \frac{e^2\hbar}{Mc} \frac{NZ}{A}. \tag{2.9.14}$$

This is found to agree rather well with experimental cross sections for $40 \lesssim A \lesssim 210$, whereas it is too small for $A > 210$ because of the contribution from photofission. The resonance energy is expressed approximately by

$$E_g \simeq 82 A^{-1/3} \quad \text{MeV}, \tag{2.9.15}$$

and the resonance width is about 8 MeV but fluctuates according to effects of the nuclear shell structure. It is often practical to give the absorption cross section averaged over the bremsstrahlung spectrum

$$\int \sigma_{\rm abs} \frac{dk}{k} \simeq 0.36 A^{4/3} \text{ mb}. \tag{2.9.16}$$

Above the giant resonance, at which the wavelength of an incident photon is comparable to the internucleon distance, the photon is mainly absorbed by a quasi-deuteron inside a nucleus. Therefore the cross section is proportional to the photodisintegration cross section of the deuteron, σ_d, as

$$\sigma_{\rm abs} \simeq 6.4 \frac{NZ}{A} \sigma_d. \tag{2.9.17}$$

The photodisintegration cross section of the deuteron rapidly decreases with increasing energy, reaches a minimum at about 120 MeV, then increases gradually, and passes through a maximum at about 230 MeV.

Above 170 MeV, however, pion production dominates over photodisintegration. The photopion process may be considered to be due to the sum of collisions of an incident photon with free nucleons inside a target nucleus. The motion of the target nucleons results in the smoothing out of the energy dependence of the photopion cross section.

The photon-nucleon collision results in the production of pions of mean energy $\langle E_\pi \rangle$ and of a recoil nucleon of mean energy $\langle E_n \rangle$; as representatives of these energy values the pion and nucleon energies at 90° in the center-of-mass system are plotted in Fig. 2.59. These pions and nucleons suffer absorption inside the target nucleus. The respective

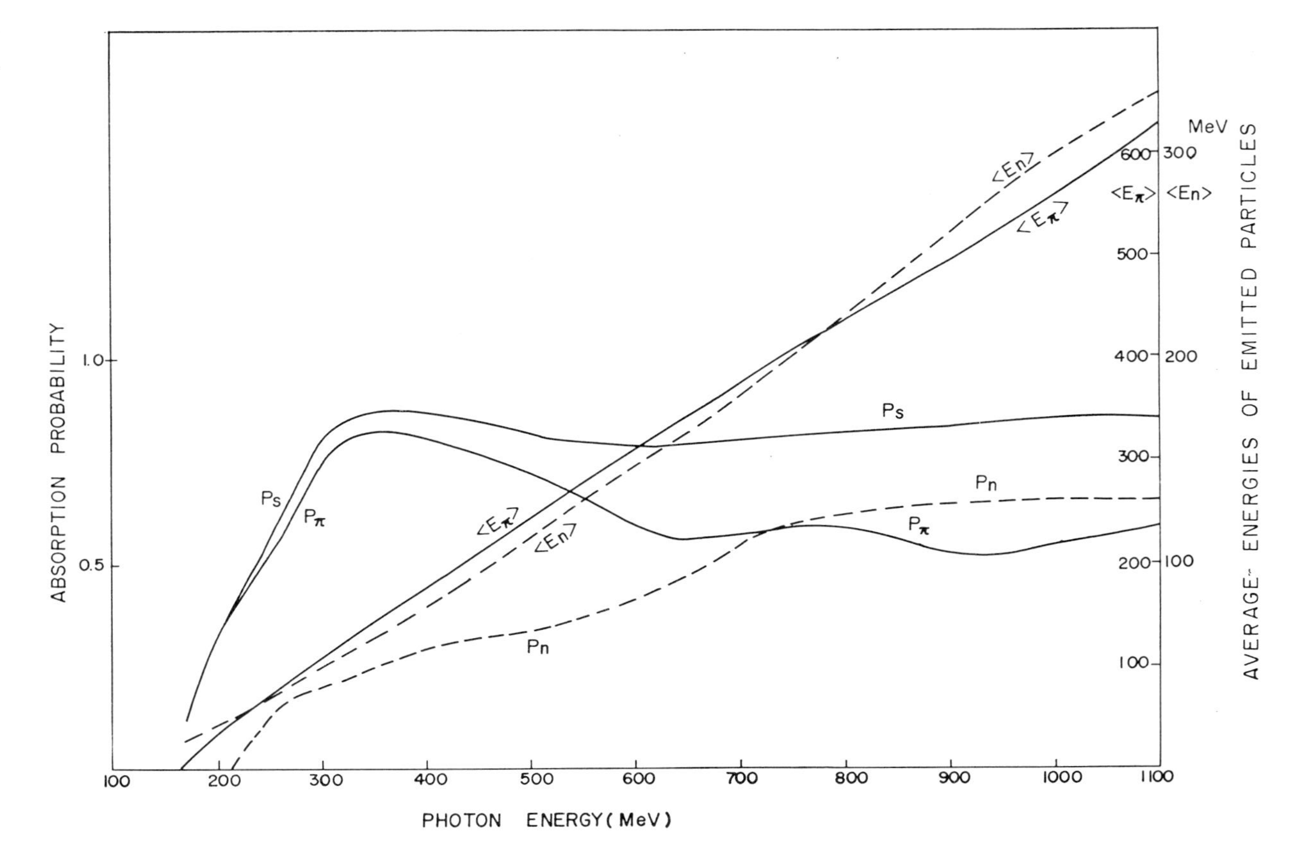

Fig. 2.59 Mean energies of pions and nucleons, $\langle E_\pi \rangle$ and $\langle E_n \rangle$, respectively, emitted from a nucleus of $A = 100$ by photon-induced reactions. The probabilities of the absorption of pions and nucleons inside a nucleus of $A = 100$, P_π, and P_n, respectively. The star-formation probability P_s is defined in (2.9.18) (Roos 61).

absorption propabilities, P_π and P_n, of a pion and a nucleon with $\langle E_\pi \rangle$ and $\langle E_n \rangle$ in a nucleus of $A = 100$, representative of heavy nuclei in nuclear emulsions, are also shown in Fig. 2.59, by reference to the Monte Carlo calculations. If either the pion or the nucleon is absorbed inside the nucleus, the photoreaction is observed as a star with the probability

$$P_s = P_\pi + P_n - P_\pi P_n. \tag{2.9.18}$$

The energy dependence of P_s is also plotted in Fig. 2.59.

The star-formation cross section is thus obtained as

$$\sigma_s = \sigma_f A P_s, \tag{2.9.19}$$

where σ_f is the free nucleon cross section. By integrating this over the bremsstrahlung spectrum we obtain the star-formation cross section for the irradiation by bremsstrahlung, since the definition of P_s includes the probability that either a pion or a nucleon produces at least one charged particle. This is compared with the experimental cross section for stars with two or more prongs, in Fig. 2.60.

Figure 2.60 also shows the cross section for single-prong events, which are due mainly to the (γ, p) reactions of the giant resonance and

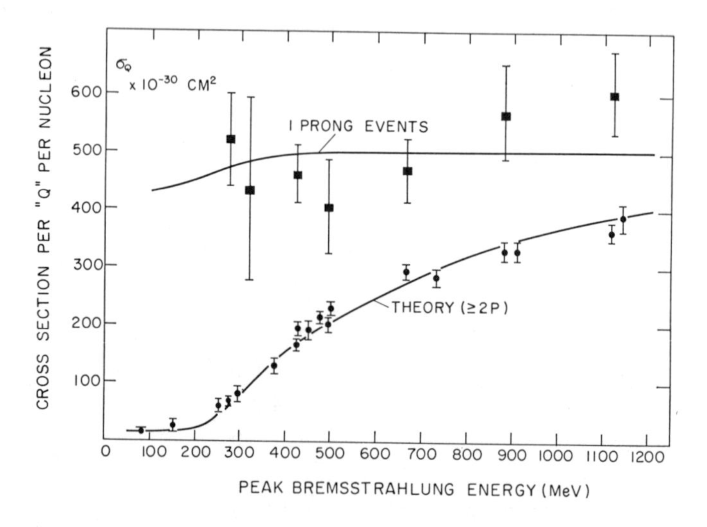

Fig. 2.60 Cross sections per Q per nucleon in emulsion by bremsstrahlung for single-prong events and for two or more prong stars. Most one-prong stars are produced by low-energy photons (Roos 61).

■—one-prong event; ●—≥ two-prong event.

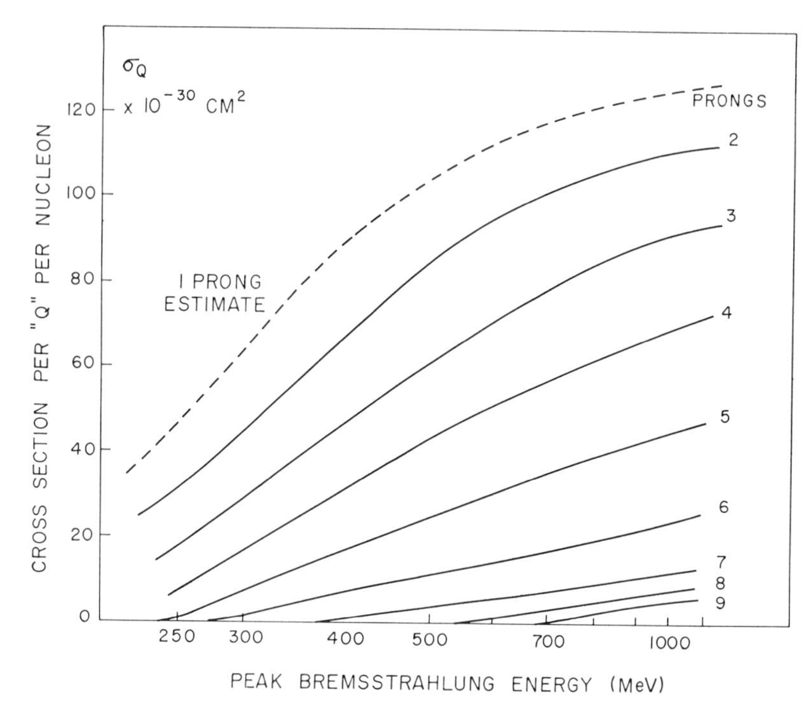

Fig. 2.61 Star-formation cross sections per Q per nucleon in emulsion by bremsstrahlung photons, σ_s/A defined in (2.9.19), with the numbers of charged particles emitted shown on the curves. The dashed curve represents the cross section estimated for one-prong stars produced by high-energy photons, and therefore it is much smaller than the one-prong star-formation cross section in Fig. 2.60 (Roos 61).

to the quasi-deuteron disintegrations. Their contribution is illustrated by the dashed curve for one-prong events in Fig. 2.61. A slow rise with increasing energy indicates the contributions of escaped charged pions or protons. The cross sections for various prong numbers are shown in Fig. 2.61. The fraction of zero-prong events is expected to be about 5 percent, because of pion- and proton-induced stars.

The above result allows us to calculate the energy dependence of the average prong number, as shown in Fig. 2.62. It is interesting to see that the average prong number does not depend on the kind of exciting particle and is nearly twice the number of cascade particles for pion- and proton-induced events. This would indicate that the mechanism of star production is the same for any incident particle; this consists of the nuclear cascade process and the subsequent evaporation process.

The importance of the cascade process can be seen also from the angular distribution of emitted protons. The forward-backward ratio

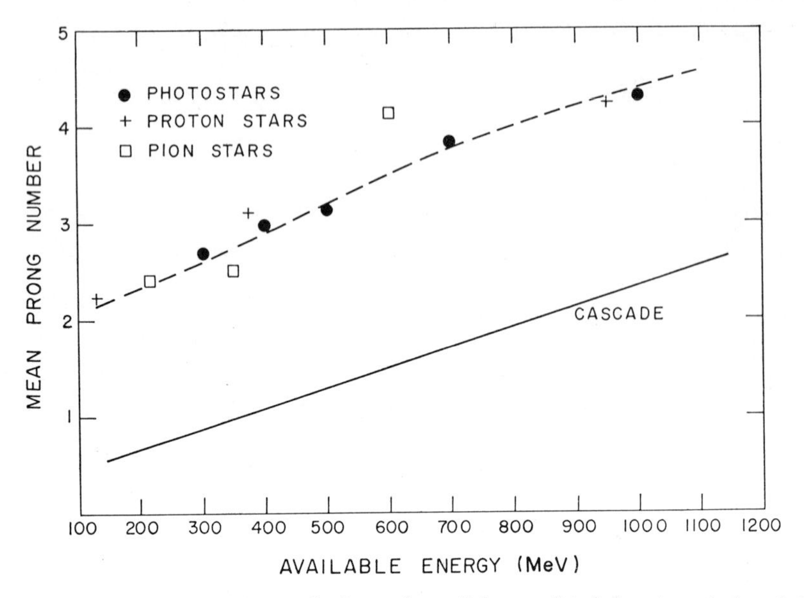

Fig. 2.62 Average numbers of charged particles emitted in stars induced by photons, protons, and pions. The solid curve represents the average number of cascade particles for pion- and proton-induced reactions (Roos 61).

of protons from photostars produced by 1150-MeV bremsstrahlung is observed to be 2.5 for protons with energies above 116 MeV; this decreases to 2.0 for energies between 30 and 116 MeV; however, it is greater than unity, 1.5, even for energies below 30 MeV. The energy spectrum of emitted protons appears to be evaporation-like; that is, like (2.9.4) as shown in Fig. 2.63, but a significant part of protons below 30 MeV are products of the cascade process. The high-energy tail seen therein is due, of course, to the cascade process. The rather small contribution of the evaporation process is consistent with the Monte Carlo calculation of the cascade process induced by high-energy particles, by which the excitation energy left after the nuclear cascade is found to be rather small—around 100 MeV. The relative contributions of these two processes can be seen in Fig. 2.62.

2.9.5 *Emission of Heavy Fragments*

In high-energy nuclear reactions fragments heavier than alpha particles are sometimes emitted. The phenomenon was first observed in cosmic-ray-induced stars in photographic emulsions (Bonetti 49, Hodgson 49), and its essential features were explored by further investigation of cosmic-ray stars (Perkins 50, Sörenson 51). According to these experiments neither the probability of emitting heavy fragments nor their energies are always consistent with those expected from evaporation theory. Moreover, they are preferentially emitted in the forward direction.

If these features were accounted for in terms of nuclear evaporation theory, we would have to assume that the fragments were evaporated from an excited nucleus left after a nuclear cascade and moving with velocity v (Skjeggestat 59). The energy spectrum of emitted fragments in such a case is obtained by modifying Weisskopf's spectrum (2.9.4) as

$$P(E) = \frac{1}{2(2mE)^{1/2}v}\left\{1 - e^{-[E-V+(2mE)^{1/2}v]/T}\,\frac{E - V + (2mE)^{1/2}v + T}{T}\right\} \tag{2.9.20a}$$

for

$$V - (2mV)^{1/2}v < E < V + (2mV)^{1/2}v$$

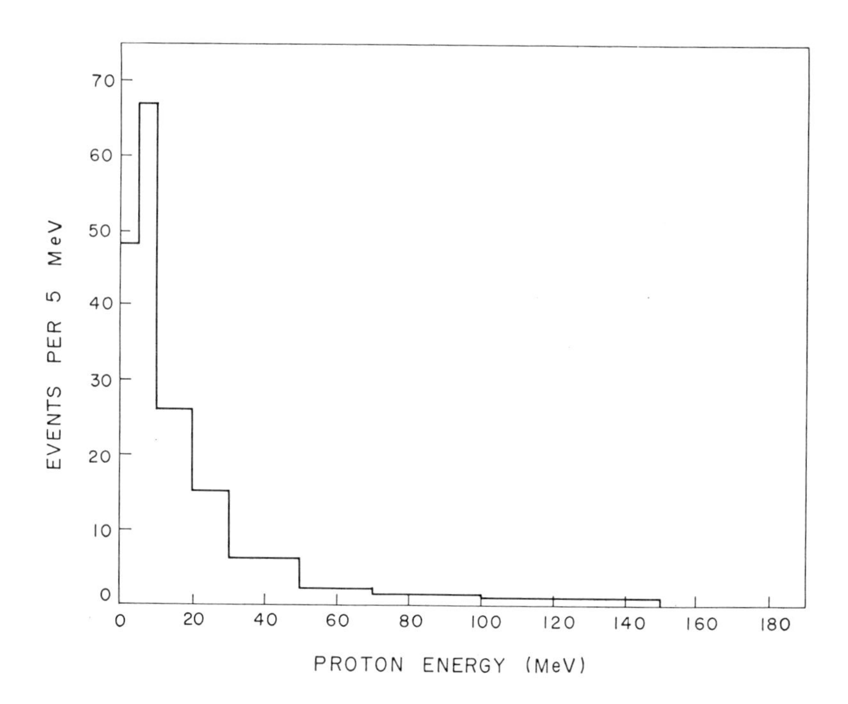

Fig. 2.63 Energy spectrum of protons produced in emulsion by 1150-MeV bremsstrahlung (Roos 61).

and

$$P(E)=\frac{1}{2(2mE)^{1/2}v}\left\{e^{-[E-V-(2mE)^{1/2}v]/T}\,\frac{E-V-(2mE)^{1/2}v+T}{T}\right.$$
$$\left.-e^{-[E-V+(2mE)^{1/2}v]/T}\,\frac{E-V+(2mE)^{1/2}v+T}{T}\right\} \tag{2.9.20b}$$

for

$$E>V+(2mV)^{1/2}v,$$

where m is the mass of the emitted fragment, E is its kinetic energy, and V is the effective Coulomb barrier. Under the same assumption the forward-backward ratio is given approximately by

$$\frac{F}{B}=\frac{\exp\,[(E-V)/T]-[(E-V+T)/T]}{(E-V+T)/T-\{[E-V+(2mE)^{1/2}v+T]/T\}\exp\,[-(2mE)^{1/2}v/T]} \tag{2.9.21a}$$

for $V<E<V+(2mV)^{1/2}v$

$$\frac{F}{B}=$$
$$\frac{\exp\,[(2mE)^{1/2}v/T]\{[E-V-(2mE)^{1/2}v+T]/T\}[(E-V+T)/T]}{[(E-V+T)/T]-\exp\,[-(2mE)^{1/2}v/T]\{[E-V+(2mE)^{1/2}v+T]/T\}} \tag{2.9.21b}$$

for $E>V+(2mV)^{1/2}v$.

In comparison with these formulas the energy spectra of fragments, helium, lithium, beryllium, and boron, and the forward-backward ratio for lithium are shown in Figs. 2.64 and 2.65, respectively. Since the low-energy part of the spectra is very difficult to determine for normal fragments, ^{8}Li hammer tracks are used for this purpose because they can be recognized from their hammerlike shape caused by two alpha-particles arising from $^8\text{Li}\rightarrow{}^8\text{Be}+e+\bar{\nu}_e$, $^8\text{Be}\rightarrow 2\alpha$ at their range ends. The energy spectrum of ^{8}Li fragments thus obtained is shown in Fig. 2.66.

The solid curves in Figs. 2.65 and 2.66 are obtained from (2.9.21) and (2.9.20), respectively, with

$$T=11.5\text{ MeV},\qquad V=6\text{ MeV},\qquad v=0.016\,c. \tag{2.9.22}$$

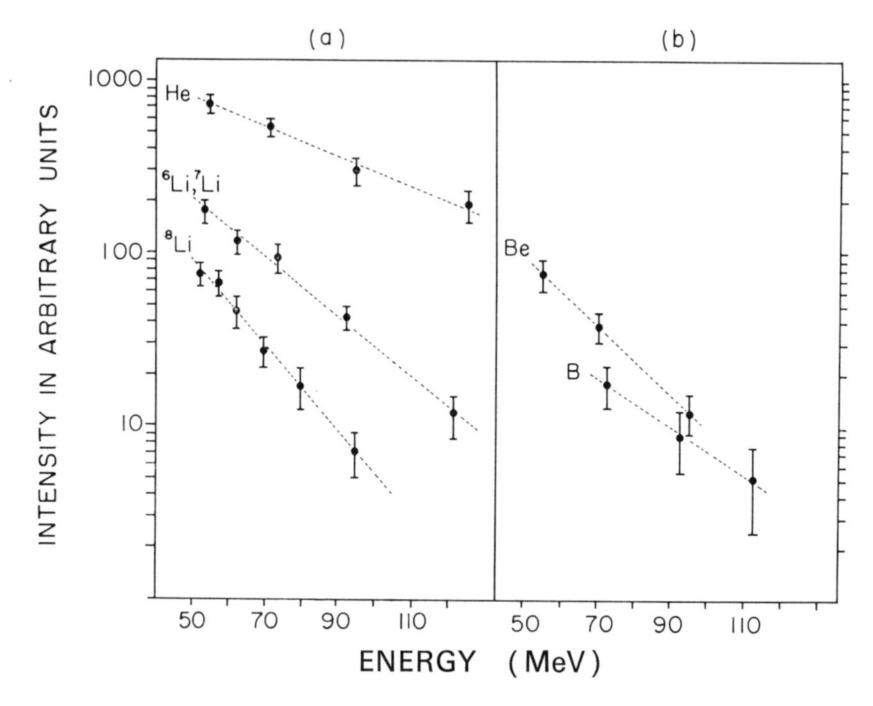

Fig. 2.64 Energy spectra of nuclear fragments produced by cosmic rays in nuclear emulsion (Skjeggestad 59).

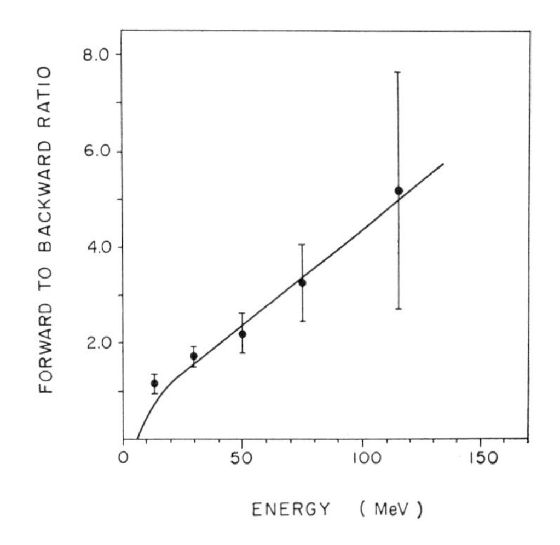

Fig. 2.65 Forward-backward ratio of lithium fragments. The solid curve represents the theoretical F/B for ^{8}Li given by (2.9.21) with $T = 11.5$ MeV, $V = 6$ MeV, and $v = 0.016c$ (Skjeggestad 59).

The agreement seen in these figures might be regarded as evidence for evaporation theory. However, the values of T and V given in (2.9.22) are not reasonable, although the value of v is acceptable.

A value of 11.5 MeV for T seems to be too large, because the maximum possible excitation energy for evaporation is the total excitation energy, which is about 800 MeV for a silver or bromine nucleus, corresponding to $T \simeq 10$ MeV. A value of 6 MeV for V is too small compared with the classical Coulomb barrier of about 16.5 MeV. This difference is too large to be accounted for in terms of an increase of the effective nuclear radius.

The yields of various fragments can also be compared with the consequence of evaporation theory. The relative yields of $(^6\text{Li} + {}^7\text{Li})/(^7\text{Be} + {}^9\text{Be} + {}^{10}\text{Be})$ and of $(^7\text{Be} + {}^9\text{Be} + {}^{10}\text{Be})/(^{10}\text{B} + {}^{11}\text{B})$ are calculated as functions of temperature in Fig. 2.67. The experimental value of the lithium-berryllium ratio, about 15, can be explained with a reasonable value of temperature, whereas that of the beryllium-boron ratio, about 3, seems to be too low compared with the theoretical value even at the highest possible temperature. In fact, a more elaborate calculation by the Monte Carlo method (Dostrovsky 60) has given qualitative agreement with radiochemical results for the yields of ^{6}He, ^{7}Be, and Li isotopes. At higher incident energies, 9 GeV, and for the production of heavier nuclei, $4 \leq Z < 9$, the yield seems to decrease

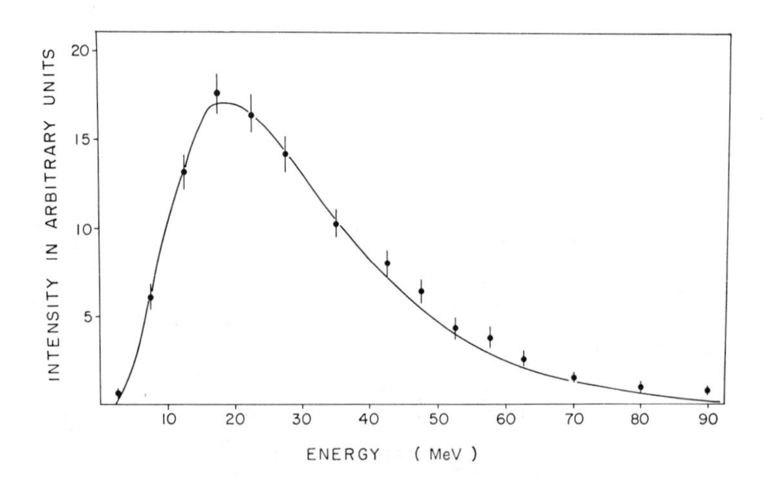

Fig. 2.66 Energy spectrum of ^{8}Li hammer tracks. The curve represents the theoretical spectrum given by (2.9.20) with $T = 11.5$ MeV, $V = 6$ MeV, and $v = 0.016c$ (Skjeggestad 59).

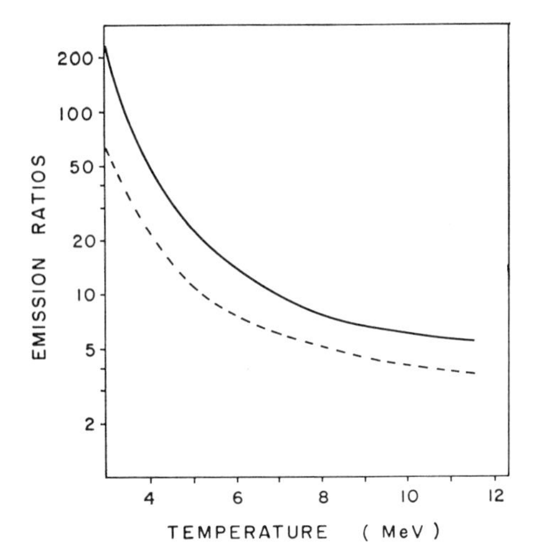

Fig. 2.67 Relative yields of nuclear fragments as functions of temperature. Solid curve: $(^7Be + ^9Be + ^{10}Be)/(^{10}B + ^{11}B)$; dashed curve: $(^6Li + ^7Li)/(^7Be + ^9Be + ^{10}Be)$ (Skjeggestad 59).

more slowly with increasing charge of the fragments than is expected from evaporation theory (Goritschev 62).

In addition to heavy fragments of high energies, considerable amounts of energetic light fragments such as deuterons are produced by bombardment with 25-GeV protons (Cocconi 60). Among the secondary particles produced at 15.9° from aluminum and platinum targets, the deuteron-proton ratios are observed as 2 and 3 percent, respectively, independent of their momenta between 2 and 6 GeV/c. The relative abundance of deuterons to protons is known from cosmic-ray jets to increase with decreasing deuteron energy, about 15 percent at 800 MeV and about 30 percent below 50 MeV.

REFERENCES

Adachi 64	Adachi, A., et al., *Progr. Theor. Phys. Suppl.* No. 32, 154 (1964).
Alexander 61	Alexander, G., et al., *Phys. Rev. Letters*, **7**, 348 (1961).
Allison 53	Allison, S. K., and Warshaw, S. D., *Rev. Mod. Phys.*, **25**, 779 (1953).
Arley 48	Arley, N., *Stochastic Processes and Cosmic Radiation*, New York, 1948.
Becklin 64	Becklin, E. E., and Earl, J. A., *Phys. Rev.*, **136**, B237 (1964).
Belenkij 48	Belenkij, S. Z., *Shower Processes in Cosmic Rays*, Moscow, Leningrad, 1948.
Bethe 30	Bethe, H., Ann. *Physik*, **5**, 325 (1930).

Bethe 32 — Bethe, H., *Z. Physik*, **76**, 293 (1932).

Bethe 34 — Bethe, H. A., and Heitler, W., *Proc. Roy. Soc.* (London), **A146**, 83 (1934).

Bethe 53 — Bethe, H. A., and Ashkin, J., *Experimental Nuclear Physics*, Vol. I, ed. by E. Segre, Wiley, New York, 1953, p. 166.

Bethe 54 — Bethe, H. A., and Maximon, L. C., *Phys. Rev.* **93**, 768 (1954).

Bhabha 35 — Bhabha, H. J., *Proc. Roy. Soc.* (London), **A152**, 559 (1935).

Bloch 33 — Bloch, F., *Z. Physik*. **81**, 363 (1933).

Bohr 48 — Bohr, N., *Kgl., Dansk. Vid. Selsk*. **18**, 8 (1948).

Bonetti 49 — Bonetti, A., and Dilworth, C., *Phil. Mag.*, **40**, 585 (1949).

Butcher 58 — Butcher, J. C., and Messel, H., *Phys. Rev.*, **112**, 2096 (1958).

Butcher 60 — Butcher, J. C., and Messel, H., *Nuclear Physics*, **20**, 15 (1960).

Chaudhuri 63 — Chaudhuri, N., and Sinha, M. S., *Proc. Int. Conf. on Cosmic Rays at Jaipur*, **6**, 101 (1963).

Chudakov 55 — Chudakov, A. E., *Izv. Akad. Nauk USSR*, **19**, 650 (1955).

Cocconi 61 — Cocconi, G., et al., *Phys. Rev. Letters*, **7**, 450 (1961).

Cocconi 64 — Cocconi, G., *Nuovo Cim.*, **33**, 643 (1964).

Cocconi 60 — Cocconi, V. T., et al., *Phys. Rev. Letters*, **5**, 19 (1960).

Cooper 55 — Cooper, L. N., and Rainwater, J., *Phys. Rev.*, **97**, 492 (1955).

Crawford 62 — Crawford, D. F., and Messel, H., *Phys. Rev.*, **128**, 2352 (1962).

Dalitz 51 — Dalitz, R. H., *Proc. Roy. Soc.* (London), **A206**, 509 (1951).

Davies 54 — Davies, H., Bethe, H. A., and Maximon, L. C., *Phys. Rev.*, **93**, 788 (1954).

Deutsch 51 — Deutsch, M., *Phys. Rev.*, **82**, 866 (1951).

Dirac 30 — Dirac, P. A. M., *Proc. Camb. Phil. Soc.*, **26**, 361 (1930).

Dostrovsky 58 — Dostrovsky, I., Rabinowitz, P., and Bivins, R., *Phys. Rev.*, **111**, 1659 (1958).

Dostrovsky 59 — Dostrovsky, I., Fraenkel, Z., and Friedlander, G. *Phys. Rev.*, **116**, 683 (1959).

Dostrovsky 60 — Dostrovsky, I., Fraenkel, Z., and Rabinowitz, P., *Phys. Rev.*, **118**, 791 (1960).

Fermi 39 — Fermi, E., *Phys. Rev.*, **56**, 1242 (1939).

Fermi 40 — Fermi, E., *Phys. Rev.*, **57**, 485 (1940).

Fermi 47 — Fermi, E., and Teller, E., *Phys. Rev.*, **72**, 399 (1947).

Feshbach 52 — Feshbach, H., *Phys. Rev.*, **88**, 295 (1952).

Ginzburg 46 — Ginzburg, V. L., and Frank, I., *J. Exp. Theor. Phys. USSR*, **16**, 15 (1946).

Goritschev 62 — Goritschev, P. A., et al., *Phys. Rev.*, **126**, 2196 (1962).

Greisen 56 — Greisen, K., *Progress in Cosmic Ray Physics*, Vol. III, North-Holland, Amsterdam, 1956.

Halpern 40 — Halpern, O., and Hall, H., *Phys. Rev.*, **57**, 459 (1940).

Halpern 48 — Halpern, O., and Hall, H., *Phys. Rev.*, **73**, 477 (1948).

Hart 59 — Hart, E. L., et al., *Phys. Rev.*, **115**, 672 (1959).

Heitler 54 — Heitler, W., *The Quantum Theory of Radiation*, Oxford at the Clarendon Press, 1954.

Hess 58 — Hess, W. N., *Rev. Mod. Phys.*, **30**, 368 (1958).

Hodgson 49 — Hodgson, P. and Perkins, D. H., *Nature*, **163**, 439 (1949).

Kajikawa 63 — Kajikawa, R., *J. Phys. Soc. Japan*, **18**, 1365 (1963).

Kamata 58 — Kamata, K., and Nishimura, J., *Prog. Theor. Phys. Suppl.* No. 6, 93 (1958).

Kantz 54 — Kantz, A., and Hofstadter, R., *Nucleonics*, **12**, 36 (1954).

Kikuchi 60 — Kikuchi, K., *Nuclear Physics*, **20**, 590, 601 (1960).

Landau 44	Landau, L., *J. Phys. USSR*, **8**, 201 (1944).
Landau 53	Landau, L., and Pomeranchuk, I., *Dokl. Akad. Nauk USSR*, **92**, 535, 735 (1953).
Malamud 59	Malamud, E., *Phys. Rev.*, **115**, 687 (1959).
Mckinley 48	McKinley, W. A., and Feshbach, H., *Phys. Rev.*, **74**, 1759 (1948).
Messel 54	Messel, H., *Prorgess in Cosmic Ray Physics*, Vol. II, North-Holland, Amsterdam, 1954, p. 133.
Messel 62	Messel, H., et al., *Nuclear Physics*, **39**, 1 (1962).
Metropolis 58	Metropolis, N., et al., *Phys. Rev.*, **110**, 185, 204 (1958).
Migdal 56	Migdal, A. B., *Phys. Rev.*, **103**, 1811 (1956).
Misaki 64	Misaki, A., *Prog. Theor. Phys.*, **31**, 717 (1964).
Mito 57	Mito, I., and Ezawa, H., *Prog. Theor. Phys.*, **18**, 437 (1957).
Molière 47	Molière, G., *Z. Naturforsch*, **2a**, 133 (1947).
Molière 48	Molière, G., *Z. Naturforsch*, **3a**, 78 (1948).
Murata 65	*Murata, Y., J. Phys. Soc. Japan*, **20**, 209 (1965).
Murota 56	Murota, T., Ueda, A., and Tanaka, H., *Prog. Theor. Phys.*, **16**, 482 (1956); Murota, T., and Ueda, A., ibid., 497.
Nishimura 67	Nishimura, J., *Handbuch der Physik*, Vol. XLVI/2, 1, (Springer-Verlag, Berlin, 1967).
Nishina 34	Nishina, Y., Tomonaga, S., and Sakata, S., *Sci. Pap. Inst. Phys. Chem. Res. Japan*, **24**, No. 17 (1934).
Nishina 35	Nishina, Y., Tomonaga, S., and Kobayashi, M., *Sci. Pap. Inst. Phys. Chem. Res. Japan*, **27**, 137 (1935).
Orear 64	Orear, J., *Phys. Rev. Letters*, **12**, 112 (1964).
Perkins 50	Perkins, D. H., *Proc. Roy. Soc.* (London), **A203**, 399 (1950).
Roos 61	Roos, C. E., and Peterson, V. Z., *Phys. Rev.*, **124**, 1610 (1961).
Rosenbluth 50	Rosenbluth, M. N., *Phys. Rev.*, **79**, 615 (1950).
Rossi 41	Rossi, B., and Greisen, K., *Rev. Mod. Phys.*, **13**, 240 (1941).
Rossi 52	Rossi, B., *High Energy Particles*, Prentice-Hall, Englewood Cliffs, N.J., 1952.
Rudstam 62	Rudstam, G., Brumins, E., and Papas, A. C., *Phys. Rev.*, **126**, 1852 (1962).
Skjeggestad 59	Skjeggestad, O., and Sörensen, S. O., *Phys. Rev.*, **113**, 1115 (1959).
Sörensen 51	Sörensen, S. O., *Phil. Mag.*, **42**, 325 (1951).
Sternheimer 52	Sternheimer, R. M., *Phys. Rev.* **88**, 851 (1952).
Sternheimer 53	Sternheimer, R. M., *Phys. Rev.*, **89**, 1148 (1953).
Sternheimer 54	Sternheimer, R. M., *Phys. Rev.*, **93**, 351 (1954).
Sternheimer 56	Sternheimer, R. M., *Phys. Rev.*, **103**, 511 (1956).
Suh 59	Suh, K. S., and Bethe, H. A., *Phys. Rev.*, **115**, 672 (1959).
Symon 48	Symon, K. R., Ph.D. Thesis, Harvard University, 1948.
Ter-Mikaelyan 61	Ter-Mikaelyan, M. L., *Nucl. Phys.*, **24**, 43 (1961).
Thom 64	Thom H., *Phys., Rev.*, **136**, B447 (1964).
Überall 56	Überall, H., *Phys. Rev.*, **103**, 1055 (1956).
Überall 57	Überall, H., *Phys. Rev.*, **107**, 223 (1957).
Wheeler 39	Wheeler, J. A., and Lamb, W. E., *Phys. Rev.*, **55**, 858 (1939).
Williams 39	Williams, E. J., *Proc. Roy. Soc.* (London), **A169**, 531 (1939).
Williams 40	Williams, E. J., *Phys. Rev.*, **58**, 292 (1940).
Wilson 52	Wilson, R. R., *Phys., Rev.*, **86**, 261 (1952).
Yekutieli 57	Yekutieli, G., *Nuovo Cim.*, **5**, 1381 (1957).

CHAPTER 3

Very-High-Energy Interactions

In this chapter we shall discuss high-energy interactions induced by nuclear active particles, photons, muons, and neutrinos. We shall restrict ourselves here to information obtained in rather direct ways. Indirect information obtained through involved analyses of complicated phenomena will be discussed in Chapters 4 and 5. Most of this chapter is devoted to discussion of nuclear interactions.*

The mechanism of nuclear collisions at extremely high energies is not well understood, in spite of its great importance in investigating the very nature of elementary particles. This is because experimental knowledge is rather limited by the low cosmic-ray intensity in the energy region concerned and by the difficulty of identification and energy determination of particles taking part in interactions. This is in sharp contrast to the case of studies with accelerators.

In cosmic rays nuclear interactions are induced by nucleons, charged pions, and sometimes by heavy nuclei. In most experiments charged pions and protons cannot be distinguished individually. In most cases the targets available are complex nuclei, such as those in photographic emulsions. Nuclear collisions are identified by the secondary particles produced. Therefore elastic collisions and inelastic ones with few secondary particles are apt to be missed. Much effort in cosmic-ray physics is devoted to disentangling such complexities and only after this has been achieved can the very nature of high-energy interactions be determined.

* A general summary of this subject was given by Rozental and Chernavsky (Rozental 54) in 1954 and, from a theoretical point of view, by Koba and Takagi (Koba 59) in 1958. Later developments were reviewed by Perkins (Perkins 60) and by Sitte (Sitte 61) in 1959 and by Fujimoto and Hayakawa (Fujimoto 67) in 1964.

3.1 Characteristic Quantities in High-Energy Interactions

High-energy events that give rise to the multiple production of particles have been observed directly by means of counter hodoscopes, cloud chambers, and photographic plates, and indirectly through studies of high-energy photons, nucleons, and muons and of extensive air showers (EAS). Observations with photographic plates have been the most extensive, and customarily the terminology derived from this method is used. In photographic emulsions the multiple particles produced in a high-energy interaction look like a shower (or like a jet if the particles are highly collimated) emerging from a point at which the interaction has taken place. The particles produced are divided into two groups according to their appearance; one consists of collimated thin tracks and the other of gray and black tracks that are less anisotropic. Particles belonging to the former group, which are called shower particles, are of relativistic energies and are mostly charged pions. Those belonging to the latter, which are called heavy particles, are of nonrelativistic energies and are mostly protons arising from the disintegration of the target nucleus. According to Powell et al. (Powell 59), the numbers of particles of the respective groups are designated by n_s and N_h. We characterize a jet shower by $N_h + n_{si}$, where i stands for p, n, or α, indicating that the primary particle is singly charged, neutral, or doubly charged; for example, an event of $2 + 16_p$ means that $N_h = 2$, $n_s = 16$, and the primary particle is a proton or a charged meson.

The symbol n_s gives us a rough idea of multiplicity, whereas N_h indicates the approximate number of nucleons taking part in the interaction. If N_h is large, say larger than 3, we regard the event as due to multiple collisions with nucleons in a target nucleus, so that features of the collision with a single nucleon are more or less masked. Consequently we are more interested in events of small N_h, which may be due to the collisions with free protons or to peripheral collisions with heavy nuclei.

If the incident energy is as high as 10^{13} eV or higher, even a head-on collision with a heavy nucleus often gives small N_h, and this reveals essential features of the collision with a single nucleon. At such a high energy most shower particles are emitted within a narrow cone like a jet. In most cases the shower particles can be divided into two distinct groups, one belonging to a very narrow cone and the other to a rather diffuse cone. They may be regarded as particles emitted forward and backward, respectively, in the center-of-mass system.

In addition to charged particles we often observe cascade showers initiated by high-energy photons that arise from the decay of neutral pions. The observation of cascade showers has an advantage in identifying pions among the secondary particles, whereas charged particles may contain kaons and baryons besides pions. The particles that are heavier than pions are collectively referred to as X-particles. Neutral particles that do not immediately decay are X-particles. The neutral X-particles produced by a primary collision can be detected by observing secondary interactions without visible parent tracks within a cone formed by secondary charged particles. However, such events are sometimes produced by uncorrelated neutral particles.

If a secondary charged particle is nonrelativistic or moderately relativistic, its mass can be obtained by measuring the grain density and multiple scattering. Otherwise the identification of each secondary particle is difficult, but the numbers of the various kinds of secondary particles are obtained only statistically. It is now known that pions are most abundant and kaons are second in abundance. Thus the high-energy nuclear collision is characterized by multiple production of mesons.

Among the many particles produced, one particle may be regarded as the descendant of the primary particle. In most cases this takes a great fraction of the incident energy, the fraction being called elasticity. The descendant particle is not always identical to the primary one; for example, in the case of proton incidence the descendant may be a neutron or an excited baryon. In the latter case a descendant particle may eventually emit secondary particles. The target particle is recoiled, and this may be considered as the descendant of the target particle in the coordinate system in which the incident particle is at rest. The large elasticity results in a rather small energy of the recoil particle, which is often observed as a gray track on photographic plates. The fraction of the energy that is not given to the descendant and recoil particles is called inelasticity. The inelastic part of the energy is shared with many secondary particles produced.

The energy of each particle that emerges after a collision is not always easy to determine, whereas its angle of emission is more easily measured. If both the energy and the emission angle are measured, the transverse component of its momentum can be determined. After analyzing various pieces of experimental information, Nishimura (Nishimura 56) suggested that the transverse momentum is about several hundred MeV/c and is essentially independent of primary and secondary energies. This was later verified by many experiments. Once this is established, the energy of a secondary particle can be estimated approximately from its emission angle alone.

Angular-distribution studies of secondary particles show that their angles of emission (and consequently their energies) are in most cases distributed in one or more groups (Ciok 58, Cocconi 58, Niu 58). This is interpreted in terms of the fireball model; each of these groups behaves like a fireball; secondary particles are emitted in analogy to thermal radiation. One can thus define the temperature of the fireball, by which the energy distribution of mesons is obtained.

In the fireball model a number of secondary particles are emitted with nearly the same energy, and the energy of individual particles is limited by the speed of the fireball. The latter becomes far smaller than the speed of a descendant baryon that may be excited to emit mesons. Then mesons from an excited baryon have energies much greater than others and may give important contributions to various phenomena (Peters 62). This is called the excited-baryon model and has been suggested independently by a number of authors (Akashi 62, Zatsepin 62).

The fireball model is often characterized by a small momentum transfer between two colliding particles, so that outgoing particles are emitted in the center-of-mass system with small angles. The smallness of momentum transfer is connected with that of transverse momenta and indicates that the impact parameter that is effective for meson production is large. Such a collision is called peripheral, and the momentum transfer is sometimes accounted for in terms of the exchange of a single pion (Dremin 60, Salzman 60). The momentum transfer between two fireballs also plays an important role in this model. A quantity that is closely related to the momentum transfer is the effective target mass—the mass of a fictitious particle that takes direct part in the collision (Birger 59).

The various concepts introduced above are regarded as characteristic of high-energy collisions and are employed in analyzing experimental results.

3.2 Nature of Primary and Secondary Particles

3.2.1 Identification of Primary Particles

A particle that induces an interaction is identified with one of the known particles by the following convention.

At very high altitudes in the atmosphere many of the interactions are produced by the heavy nuclei in primary cosmic rays. Except at high magnetic latitudes they are of relativistic energies, and their charges can be determined by the ionization processes they cause. The charge determination can be made with satisfactory accuracy for low atomic numbers, say $Z \leq 10$, but it becomes less accurate as Z increases. In most nuclear collisions a heavy nucleus disintegrates into fragments that emerge with emission angles that are smaller than those of mesons. The sum of the charges of these fragments provides the calibration of the charge determination by ionization processes. In some cases isotopic ratios are given statistically by taking account of properties of nuclear reactions, as will be discussed in Chapter 6.

At high altitudes most single-charge particles are protons, since few secondary particles are produced at such altitudes. As the altitude decreases, the number of secondary particles increases, and even protons are among those that have suffered nuclear interactions. Among them the electronic component can be distinguished from the nuclear active component, because the former produces an electronic cascade shower. The nuclear active component contains nucleons, charged pions, charged kaons, and neutral kaons, K_2^0; other particles are believed to be negligible because of their short lifetimes. Since kaons are believed to be fewer than other particles, most neutral particles are conventionally identified as neutrons. As many protons as neutrons are expected at energies at which the loss of energy by ionization is unimportant, because of the charge-exchange property in nuclear collisions. Therefore if one observes the ratio of charged to neutral particles responsible for nuclear interactions, the contribution of pions can be determined by the deviation of the charged-to-neutral ratio from unity. Experimental results described in Chapter 4 indicate that the majority of nuclear active particles in the lower atmosphere are nucleons.

Underground only muons and neutrinos survive. But not all charged particles that produce nuclear interactions are muons, since pions produced by the nuclear interactions of muons contribute as much as muons. These two agencies of nuclear interactions underground can be distinguished by the fact that pions are associated mostly with penetrating showers (Higashi 64).

Deep underground even muons are very few. The particles that can be detected are neutrinos and their secondary particles, mostly muons. These muons are distinguished from those produced in the atmosphere because the latter are practically negligible in the horizontal and upward directions.

Most of the above methods of identification depend on statistics, and individual identification is possible only in restricted cases. In some cases the information on secondary particles is helpful in identifying the primary particle.

3.2.2 *Identification of Secondary Particles*

Of the various secondary particles, charged ones are directly observable as ionizing tracks. Thus the number of charged shower particles, n_s, is a well-defined quantity, although some uncertainty arises if the tracks form a very narrow cone, in which individual tracks are indistinguishable from each other near the starting point and are masked by cascade showers far from the interaction.

The cascade showers are initiated by γ-rays from neutral-pion decays. In a homogeneous medium showers from individual γ-rays overlap with each other. An approximate value of the energy imparted to all neutral pions may be obtained by observing the cascade showers, but the number of them can be obtained only by counting electron pairs before they develop into showers. This, however, requires a statistical treatment that separates such pairs from the pairs that are produced in later stages of the cascade process. This makes it possible to determine the number of neutral pions, n_{π^0}, statistically. There are other neutral particles that decay into γ-rays—for example, $\Sigma^0 \rightarrow \Lambda^0 + \gamma$. However, they are believed to be much fewer than neutral pions.

In a dense medium many neutral secondary particles undergo nuclear interactions before they disintegrate. Such neutral particles, collectively called X^0, are observed as nuclear interactions without charged parent particles. If one finds N_n neutral interactions, scanning a distance D below a primary interaction, the number of the neutral particles produced is given by $N_n(l_n/D)$, where l_n is their interaction mean free path. In an appreciable number of cases it cannot be determined whether or not an interaction is associated with charged parent particles. They are arbitrarily shared equally between neutral and charged interactions. Furthermore it is physically meaningful to subtract a neutron that is a survivor of the primary nucleon. It is assumed that half the survival nucleons are neutrons. The above considerations allow us to estimate the number of neutral X-particles produced as

$$n_X{}^0 = (N_n + \tfrac{1}{2}N_?)\frac{l_n}{D} - \tfrac{1}{2}, \tag{3.2.1}$$

where $N_?$ is the number of unidentified interactions.

The counterpart of the neutral X-particle is the charged X-particle, which can be distinguished from the charged pion if the masses are measured. This may be possible for the particles that are emitted backward in the center-of-mass system. In practice only the X-π ratio is reliable—rather than the absolute numbers of X and π.

3.2.3 Neutral-to-Charged Ratio

The situation is the same for neutral pions. Their total number is hardly obtainable, but their number relative to the number of charged shower particles can be determined rather unambiguously if they are compared in the same solid angle region; this is based on the assumption that the angular distributions are identical. A further assumption about the conversion length of γ-rays allows us to obtain the neutral-to-charged ratio,

$$R = \frac{n_{\pi^0}}{n_s} \tag{3.2.2}$$

in which n_{π^0} and n_s are based on many jets of nearly the same primary energy. This quantity has been measured by many authors, as summarized in Table 3.1. The value of R is essentially energy independent, and their weighted average is

$$R = 0.40 \pm 0.03. \tag{3.2.3}$$

Table 3.1. Neutral-to-Charged Ratio, $R = n_{\pi^0}/n_s$

Primary Energy (eV)	R	Method	Reference
$(1 \text{ to } 2) \times 10^{10}$	0.50 ± 0.10	Cloud chamber	1
$\sim 2.5 \times 10^{10}$	0.48 ± 0.11	Emulsion	2
$(2 \text{ to } 15) \times 10^{10}$	0.40 ± 0.04	Cloud chamber	3
$\sim 5 \times 10^{11}$	0.33 ± 0.08	Emulsion	2
$(1 \text{ to } 5) \times 10^{12}$	0.46 ± 0.09	Emulsion	4
$(1 \text{ to } 5) \times 10^{12}$	0.40 ± 0.04	Emulsion	5
$\sim 1 \times 10^{13}$	0.50 ± 0.11	Emulsion	6

References:

1. G. Salvini and Y. Kim., *Phys. Rev.*, **85**, 921 (1952); **88**, 40 (1952).
2. R. R. Daniel *et al.*, *Phil. Mag.* **43**, 753 (1952).
3. S. Lal, Y. Pal, and R. Raghaven, *Journ. Phys. Soc. Japan*, **17**, Suppl. A–III, 393 (1962).
4. M. F. Kaplon, W. D. Walker, and M. Koshiba, *Phys. Rev.*, **93**, 1424 (1954).
5. F. A. Brisbout *et al.*, *Phil. Mag.*, **1**, 605 (1956); B. Edwards *et al.*, *Phil. Mag.*, **3**, 237 (1958).
6. M. Koshiba and M. F. Kaplon, *Phys. Rev.*, **97**, 193 (1955).

The ratio of neutral to charged interactions has been measured by observing secondary interactions in narrow cones of jets (Brisbout 56, Edwards 58, Lohrmann 58), as described in Section 3.2.2. In this case, too, data on interactions within a given angular region of many jets were collected. The number ratio of neutral to charged interactions is thus given by

$$Q = \frac{N_n + \frac{1}{2}N_? - \frac{1}{2}n_N(1 - e^{-D/l_N})}{N_{ch} + \frac{1}{2}N_? - \frac{1}{2}n_N(1 - e^{-D/l_N})}, \tag{3.2.4}$$

where N_{ch} is the number of charged interactions. Here the contribution of survival nucleons is subtracted by assuming the expected number of survival nucleons, n_N, and their interaction mean free path l_N, the latter being assumed as equal to l_n.

This gives the number ratio of neutral X-particles to charged shower particles on the assumption that the interaction mean free paths of all secondary particles are essentially equal; that is,

$$Q = \frac{n_X{}^0}{n_s}. \tag{3.2.5}$$

The numerical value of Q obtained for jets of primary-particle energies of about several TeV observed with emulsion is (Brisbout 56, Edwards 58, Lohrmann 58, Koshiba 63*a*)

$$Q = 0.25 \pm 0.04. \tag{3.2.6}$$

It is interesting that the relative abundances obtained in (3.2.6) from jets of primary energies greater than 1 TeV are greater than those at energies as low as 10 GeV; in the latter case $Q \approx 0.1$.

3.2.4 *Relative Abundance of Charged* X-*Particles*

Identification of secondary particles is possible by measuring the ionization-scattering relation for particles emitted backward in the center-of-mass system. These particles form the outer cone of a jet and have energies low enough to allow such measurements. The values of $n_{X^\pm}/n_s$ thus obtained are summarized in Table 3.2. At lower energies

Table 3.2. Relative Abundance of Charged X-Particles

Primary Energy (eV)	$n_{X^\pm}/(n_{\pi^\pm} + n_{X^\pm})$	Method	Reference
$(2.2 \text{ to } 200) \times 10^{10}$	0.24 ± 0.05	Cloud chamber	1
$(1 \text{ to } 70) \times 10^{12}$	0.25 ± 0.08	Emulsion	2
$(1 \text{ to } 100) \times 10^{12}$	0.16 ± 0.06	Emulsion	3
$\sim 2 \times 10^{12}$	0.25 ± 0.06	Emulsion	4

References:

1. L. F. Hansen and W. B. Fretter, *Phys. Rev.*, **118**, 812 (1960).
2. B. Edwards *et al.*, *Phil. Mag.*, **3**, 237 (1958).
3. E. Lohrmann and M. W. Teucher, *Phys. Rev.*, **115**, 636 (1959).
4. C. O. Kim, *Phys. Rev.*, **136**, *B*515 (1964).

an observation made with a cloud chamber has also given the same result.

It is worth remarking that in a cloud-chamber experiment involving 41 interactions in a carbon plate (Hansen 60), 13 X^--particles were observed, compared with 32 X^+. In 30-GeV p-Be and p-Al collisions the relative abundances of secondary particles were found to be approximately (Cool 61)

$$n(\pi^+) : n(K^+) : n(K^-) : n(\bar{p}) \simeq 1 : 0.15 : 0.05 : 0.01, \qquad (3.2.7)$$

also indicating a large positive excess of kaons. The relative abundance of hyperons is roughly 5 percent of all secondary particles at this energy. At several tens of GeV 49 Λ^0 and 18 $\Sigma^\pm$ were found in comparison with 42 K^0 (Manjavidze 62). If most charged X-particles were positive kaons, they would have positive strangeness and would have to be associated with hyperons due to the conservation of strangeness. This would also result in many more hyperons than $K_1{}^0$. However, the above result indicates that neutral-kaon decays are as abundant as hyperon decays, and for TeV jets in emulsion few hyperon decays have been observed in jets. This suggests that a considerable part of negative kaons are produced in pairs with positive kaons.

For particles emitted at narrow angles a different result is reported (Kim 64). For particles emitted within 5° in the backward direction in the center-of-mass system kaons are found to be as abundant as pions, and 10 protons were identified in 20 interactions. The latter seems to indicate that the descendants of incident protons are nearly equally divided between protons and neutrons. One hyperon decay was observed among these secondaries.

These results obtained for TeV jets are in sharp contrast to the relative abundances of particles in the accelerator energy region. The yields of various charged particles emitted within 8 mrad from a hydrogen target bombarded by a 23.1-GeV/c proton beam are shown in Fig. 3.1 (Dekkers 65). Here the yields of $\pi^\pm$, $K^\pm$, and $\bar{p}$ relative to the proton yield are plotted against momentum, since the proton is the survivor of the incident proton. The fact that relative yield decreases as momentum increases indicates a large elasticity for the collision; the absolute yield of protons increases with momentum, whereas that of other secondary particles decreases. It is interesting to note that the relative yields are rather insensitive to the angle of emission and to the mass number of the target nucleus. The general features of the relative yields may be interpreted to indicate that secondary-particle production in the accelerator energy region takes place substantially through resonance states (Lindenbaum 57, Sternheimer 61, Peters 62).

3.2.5 *Relative Abundances of Secondary Particles*

The experimental results discussed above allow us to evaluate the relative abundances of various secondary particles if conservation laws are further taken into consideration; for example, the isovector nature of pions results in

$$\frac{n_{\pi^0}}{n_{\pi^\pm}} = \tfrac{1}{2}. \tag{3.2.8}$$

From (3.2.3), (3.2.6), and (3.2.8), obtained for a given angular region of the forward cone, we thus obtain

$$\begin{aligned}
&\frac{n_{X^\pm}}{n_{\pi^\pm}+n_{X^\pm}} = 1-2R = 0.20 \pm 0.05,\\
&\frac{n_{X^\pm}}{n_{\pi^\pm}} = \frac{1-2R}{2R} = 0.25 \pm 0.06,\\
&\frac{n_{X^0}}{n_{\pi^\pm}} = \frac{Q}{2R} = 0.30 \pm 0.06,\\
&\frac{n_{X^0}}{n_{X^\pm}} = \frac{Q}{1-2R} = 1.2 \pm 0.4,\\
&\frac{n_{X^\pm}+n_{X^0}}{n_{\pi^\pm}+n_{\pi^0}+n_{X^\pm}+n_{X^0}} = 1-\frac{3R}{1+R+Q} = 0.27 \pm 0.05.
\end{aligned} \tag{3.2.9}$$

The first relation is consistent with those obtained for the backward cone in Table 3.2.

It should be mentioned that the comparison of the relative yields in (3.2.9) obtained for TeV jets with those at lower energies, as shown in Table 3.2 and Fig. 3.1, indicates a rather weak energy dependence. A smaller K/π ratio at accelerator energies is considered to be due to an energy dependence of the contribution of resonance states.* Apart from this, the relative yields of secondary particles appear to be essentially independent of energy between several tens of GeV and several TeV.

3.3 Interaction Mean Free Path

When a beam of intensity J passes through matter, the fraction of the beam that undergoes interaction in a thickness Δx is proportional

* M. Koshiba has suggested that most kaons in the TeV region are produced through a resonance state ϕ, whose yield increases with energy.

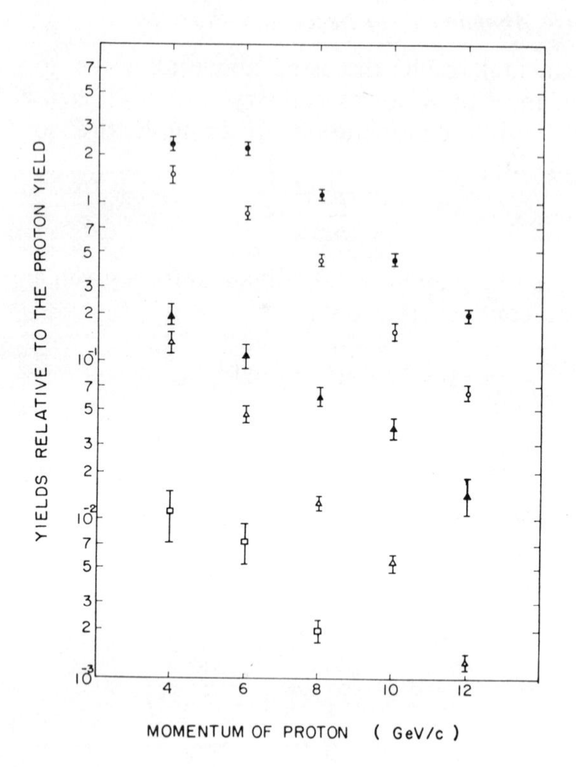

Fig. 3.1 Relative yields of secondary particles by p-p collisions at 23.1 GeV/c: ●—π^+/p; ○—π^-/p; ▲—K^+/p; △—K^-/p; □—$\bar{p}/p$.

to Δx as

$$\Delta J = -\left(\frac{\Delta x}{l}\right) J. \tag{3.3.1}$$

The coefficient l is defined as the interaction mean free path. If the thickness is large, the beam intensity, disregarding interactions after passing through x, is

$$J(x) = J(0)e^{-x/l}. \tag{3.3.2}$$

If the beam intensity is known, either (3.3.1) or (3.3.2) is used for determining the interaction mean free path.

The number of nuclear interactions that take place in a thickness Δx in g-cm^{-2} is proportional to the number of nuclei per gram, N/A, and the interaction cross section σ_A. The interaction mean free path is expressed as

Plate 1. A high-energy nuclear interaction produced by a neon nucleus in nuclear emulsion. The narrow cone consists of a breakup nucleus and charged secondary particles emitted forward in CMS; several thin tracks with considerable opening angles are those emitted backward in CMS; several heavy tracks emerge from the interaction point. Courtesy of M. Koshiba. See Chapter 3, Section 3.

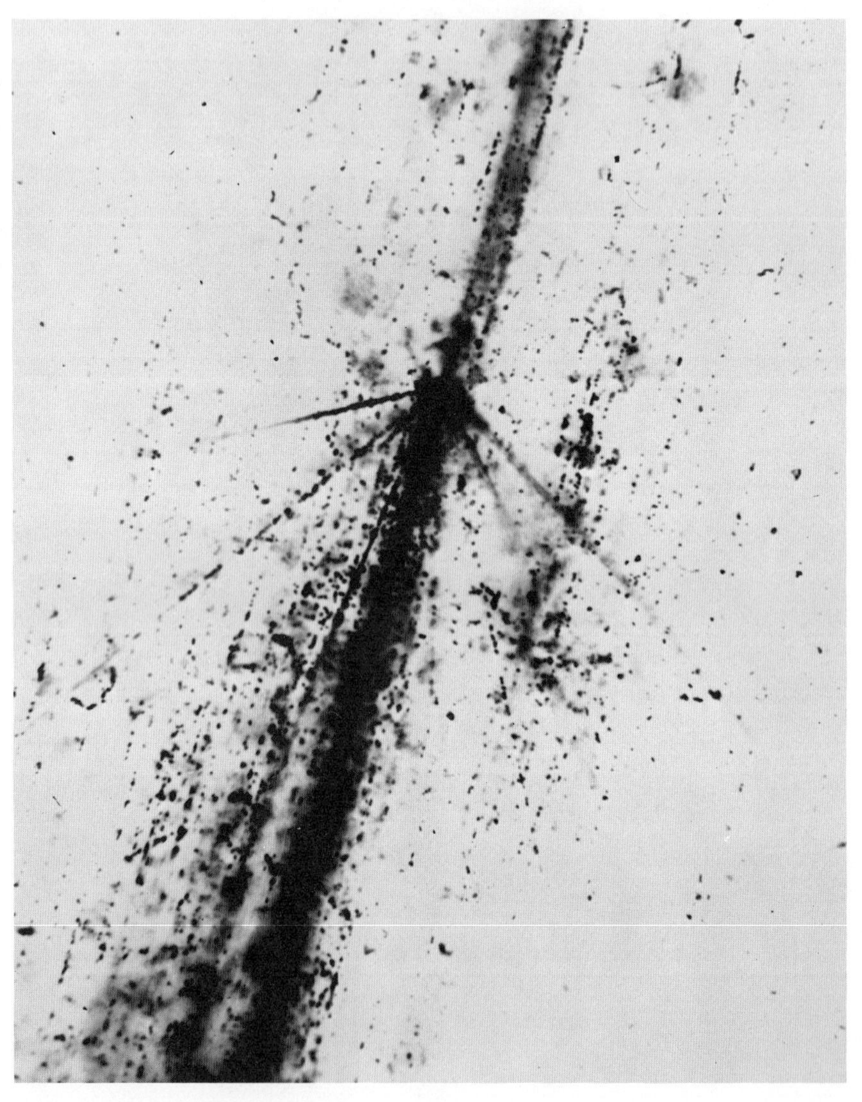

Plate 2. A high-energy interaction produced by a secondary particle in a jet shower, one of the events obtained by the International Collaboration of Emulsion Flight. Parallel thin tracks represent relativistic particles produced by a primary interaction. One of them produces a secondary interaction that shows typical features of the jet shower. Courtesy of K. Niu. See Chapter 3, Section 3.

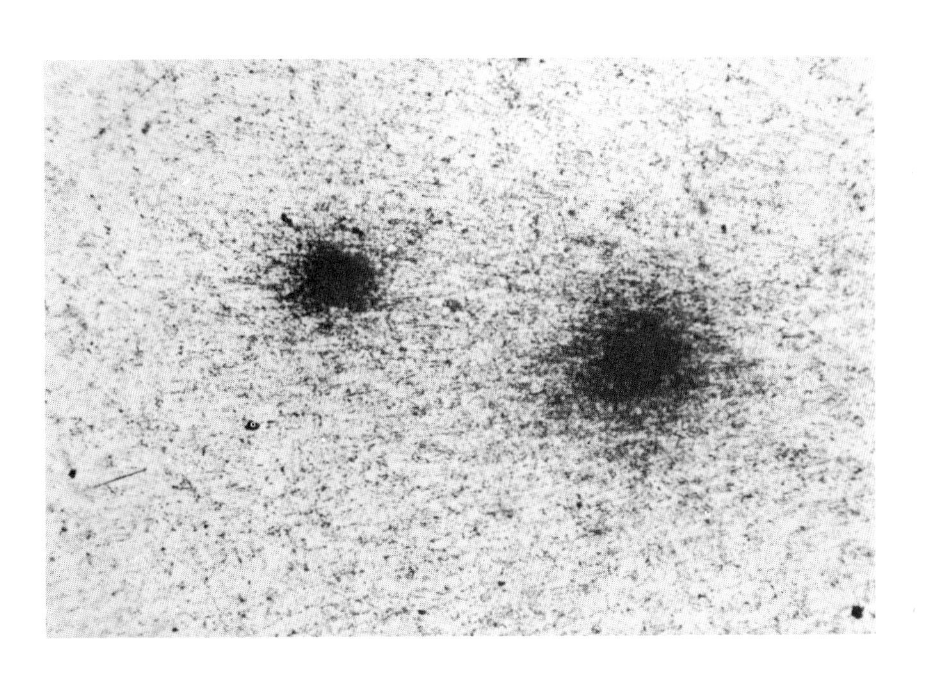

Plate 3. A pair of high-energy cascade showers observed in an emulsion plate in an emulsion chamber. These two are considered to be due to two γ-rays from a neutral pion. Courtesy of J. Nishimura. See Chapter 3, Section 3.

$$\frac{1}{l} = \frac{N}{A}\sigma_A. \tag{3.3.3}$$

The cross section σ_A is conventionally represented by the geometrical one, πR^2, where the nuclear radii R are given in Section 2.9.1. As discussed in Section 2.9, however, the transparency of a nucleus has to be taken into account, and the interaction cross section is supposed to deviate considerably from the geometrical one for light nuclei. However, experimental results with accelerators give interaction cross sections that are in close agreement with the geometrical ones.

This has been interpreted as being a result of the tapered density distribution of nucleons in a nucleus (Cronin 57). The charge distribution from electron scattering by nuclei is well approximated by

$$\rho(r) = \rho_0 u(r) = \frac{\rho_0}{\exp[(r-c)/a]+1}. \tag{3.3.4}$$

The density distribution of nucleons is considered to be essentially the same as this. If a particle passes through a nucleus with impact parameter b, the effective length corresponding to the optical depth is given by

$$t(b) = \rho_0 \bar{\sigma} \int_0^{\infty} u(\sqrt{s^2+b^2})\, ds, \tag{3.3.5}$$

where $\bar{\sigma}$ is the effective collision cross section with nucleons. Hence the interaction cross section is expressed as

$$\sigma_A = 2\pi \int_0^{\infty} \{1 - \exp[-2t(b)]\}\, b\, db. \tag{3.3.6}$$

The values of σ_A calculated for

$$c = 1.13\mathrm{A}^{1/3} \times 10^{-13}\ \mathrm{cm}, \qquad a = 4.23 \times 10^{-13}\ \mathrm{cm}, \qquad \bar{\sigma} = 33\ \mathrm{mb} \tag{3.3.7}$$

are compared with the geometrical cross sections in Fig. 3.2. The former is found to reproduce the $A^{2/3}$ dependence rather well—in good agreement, for example, with an experimental result obtained by 23-GeV protons (Ashmore 60). In more detail (3.3.6) is approximated by the $A^{0.69}$ dependence, as also shown in Fig. 3.2 (Williams 64). If $\bar{\sigma}$ is slightly different from 33 mb, (3.3.6) is represented by an empirical expression,

$$\sigma_A = 44A^{0.69}[1 + 0.039A^{-1/3}(\bar{\sigma} - 33) - 0.0009A^{-1/3}(\bar{\sigma} - 33)^2]\ \mathrm{mb}. \tag{3.3.8}$$

The reason for taking account of the $\bar{\sigma}$-dependence may be understood on the following physical grounds. The effective cross section $\bar{\sigma}$ is the cross section for an incident particle to give a part of its energy to the target nucleus. If the collision with a nucleon is inelastic or meson producing, this is certainly responsible for $\bar{\sigma}$. Even if the elementary collision is elastic, this contributes to $\bar{\sigma}$ when the recoil energy is high enough to excite the target nucleus. This is the case if the angle scattered is sufficiently large. Therefore only a part of elastic scattering with small angles does not contribute to $\bar{\sigma}$. Such a fraction of the elastic cross section is rather small up to a few GeV, but it increases with increasing incident energy. Near 1 GeV $\bar{\sigma}$ for incident pions is essentially equal to the total cross section, whereas $\bar{\sigma}$ for incident nucleons consists of the inelastic cross section and about one-third of the elastic one. Above 5 GeV $\bar{\sigma}$ for nucleons is essentially equal to the inelastic cross section.

Because the value of $\bar{\sigma}$ is subject to further quantitative investigation, the dependences of the interaction mean free paths on $\bar{\sigma}$ are shown in Fig. 3.3 according to (3.3.8) for two representative media—air and nuclear emulsion (Williams 64).

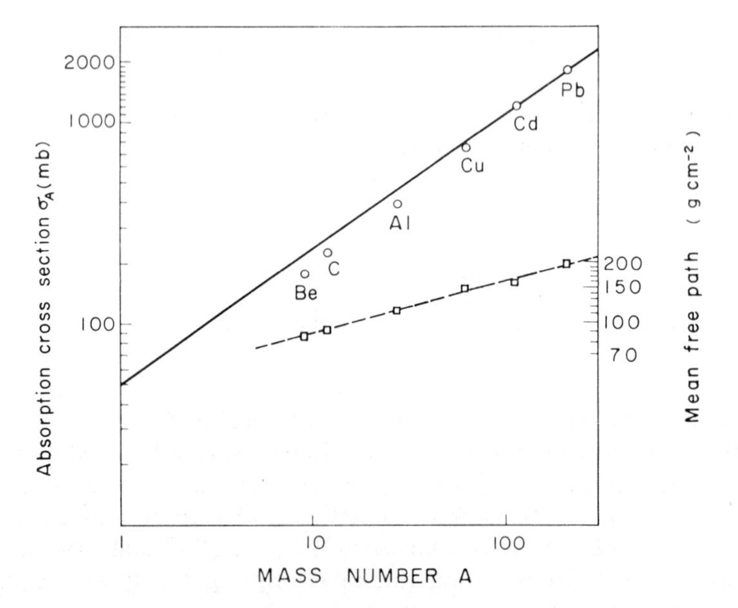

Fig. 3.2 Cross sections (○) and mean free paths (□) for the absorption of 24.2-GeV/c protons in various elements. The solid line represents the nuclear-absorption cross sections given by $\sigma_A = \pi(r_0 A^{1/3})^2$ with $r_0 = 1.26 \times 10^{-13}$ cm, and the dashed line represents the corresponding nuclear absorption mean free paths (Ashmore 60).

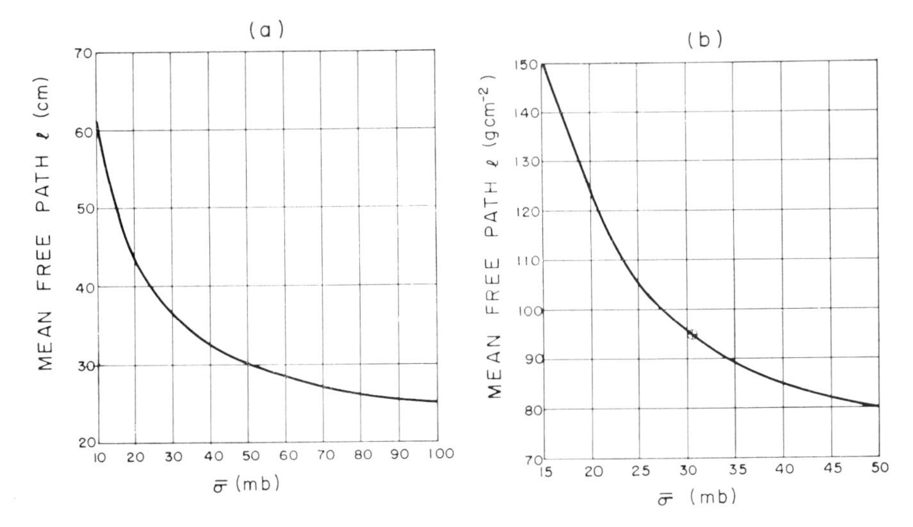

Fig. 3.3 Mean free paths for interactions as functions of the effective elementary cross section $\bar{\sigma}$ in Ilford G-5 nuclear emulsion of density 3.85 g-cm^{-2}, (*a*); and in air, (*b*). The experimental value in (*b*) is obtained by interpolation of the results in Fig. 3.2 (Williams 64).

In nuclear emulsions a number of experimental results are available, as shown in Table 3.3. The scatter in the experimental results indicates difficulties in experimental technique, such as in scanning efficiency. The weighted average for proton interactions of energies between 6 and 27 GeV is

$$l = 36.5 \pm 1.0 \text{ cm}. \tag{3.3.9}$$

This is interpreted, according to Fig. 3.3(a) by adopting

$$\bar{\sigma} = 31 \text{ mb}, \tag{3.3.10}$$

in good agreement with those obtained for other nuclei by counter experiments.

Table 3.3 also lists several experimental values of the pion mean free path in nuclear emulsions. This is essentially the same as the proton mean free path.

For cosmic rays the flux of incident particles is hard to determine accurately. Therefore the interaction mean free path obtained by means of (3.3.1) is subject to an error that is directly proportional to the uncertainty in the flux value. In nuclear emulsion the number of interactions and the total potential track length of incident particles are observed, but a sophisticated method for deriving the mean free path is

Table 3.3 Interaction Mean Free Path in Nuclear Emulsions

Energy (eV)	Particle	Mean Free Path (cm)	Reference
5.7×10^9	p	35.6 ± 2.4	V. Y. Rajopadhye, *Phil. Mag.*, **5**, 537 (1960)
6.2×10^9	p	38.2 ± 1.5	H. Winzeler et al., *Nuovo Cim.*, **17**, 8 (1960)
		34.7 ± 3.4	M. V. K. Appa Rao et al., *Proc. Ind. Acad. Sci.*, **A43**, 181 (1956)
8.7×10^9	p	35.0 ± 1.3	G. B. Zhdanov, *Sov. Phys. JETP*, **10**, 442 (1960)
9.0×10^9	p	36.9 ± 0.9	V. A. Kobzev et al., *Sov. Phys. JETP*, **14**, 538 (1962)
	p	35.9 ± 0.4	T. Visky et al., *Sov. Phys. JETP*, **14**, 763 (1962)
	p	37.1 ± 1.0	N. Bogachev et al., *DAN*, **121**, 615 (1958)
14×10^9	p	29.0 ± 1.0	C. Bricman et al., *Nuovo Cim.*, **20**, 1017 (1961)
20.5×10^9	p	35.6 ± 1.5	H. Meyer et al., *Nuovo. Cim.*, **28**, 1399 (1963)
23.5×10^9	p	36.6 ± 1.0	G. Cvijanovich et al., *Nuovo Cim.*, **20**, 1012 (1961)
25×10^9	p	32.6 ± 1.6	P. G. Bizzeti et al., *Nuovo. Cim.*, **27**, 6 (1963)
26.7×10^9	p	41.9 ± 3.6	Y. K. Lim, *Nuovo. Cim.*, **26**, 1221 (1962)
27×10^9	p	37.2 ± 70.4	Y. Baudinet-Robinet et al., *Nuclear Physics*, **32**, 452 (1962)
27×10^9	p	35.1 ± 1.5	H. Meyer et al., *Nuovo. Cim.*, **28**, 1399 (1963)
28×10^9	p	37.9 ± 1.2	P. L. Jain et al., *Nuovo. Cim.*, **21**, 859 (1961)
1.7×10^{11}	Breakup p	41 ± 10	E. Lohrmann et al., *Phys. Rev.*, **122**, 672 (1961)
$\sim 10^{12}$	Breakup p	26 ± 9	M. Koshiba, *Int. Conf. at Jaipur*, **5**, 293 (1963)
7×10^{12}	Primary p	27^{+14}_{-7}	R. R. Daniel et al., *Nuovo. Cim.*, Suppl. **1**, 1163 (1963)
4.2×10^9	π	38.7 ± 2.3	J. O. Clark, *Phil. Mag.*, **2**, 37 (1957)
4.3×10^9	π	33.7 ± 4.7	A. Margnes, *Nuovo Cim.*, **5**, 291 (1957)
7×10^9	π	35 ± 3	V. A. Belyakov et al., *J. Exp. Theoret. Phys.* **39**, 937 (1960)
7.5×10^9	π	38.6 ± 1.4	L. C. Grote et al., *Nuclear Phys.*, **24**, 300 (1961)
1.6×10^9	π	39.2 ± 1.0	A. Baldassare et al., *Aux-en-Province Conf.* 1961, 427
1×10^{11}	Secondaries	41 ± 8	A. Barkow et al., *Phys. Rev.*, **122**, 617 (1961)

needed. The mean free path thus obtained is subject to uncertainty, even if a considerable number of events are collected (Daniel 63). The method based on (3.3.2) may be regarded as more reliable because the absolute flux value is not used. In this method one measures the frequency of penetrating showers produced in a lower material layer as a function of thickness of an upper layer through which incident particles traverse without interactions. Around 1950 a number of such experiments were carried out, and the interaction mean free paths thus determined were not far from those obtained by more reliable methods using accelerator beams. However, the probability of missing penetrating showers is expected to be so appreciable that the result obtained with this method is subject to critical examination (Sitte 61).

A rather reliable source of cosmic-ray beams is available in secondary particles of nuclear interactions. The secondary particles provide a beam of a common origin and with a rather small energy spread. In particular the breakup products of a heavy nucleus are essentially parallel and monoenergetic, so that they form a beam of good quality.

There are a few such experimental results, as listed in Table 3.3. For the average energy of about 100 GeV the interaction mean free path for single-charge secondary particles is found to be 41 ± 8 cm. Most of the particles under consideration are regarded as pions, and accordingly this indicates an inappreciable energy dependence of the pion mean free path up to 100 GeV. At about the same energy the mean free path is measured for the breakup protons of heavy cosmic rays and is again consistent with that at lower energies. At about 1 TeV and higher fewer experimental data are available. This and other indirect evidence discussed in Chapters 4 and 5 do not show any drastic energy dependence of the interaction mean free path. This would indicate the constancy of the inelastic cross section of elementary particles at very high energy.

The interaction mean free paths of heavy nuclei are also available. Table 3.4 lists experimental results of the interaction mean free path of α-particles. This is definitely smaller than the proton mean free path, and all the experimental values up to 20 GeV per nucleon are consistent with the weighted mean value,

$$l_\alpha = 18.7 \pm 0.6 \text{ cm}. \tag{3.3.11}$$

Values at higher energies are not inconsistent with (3.3.11). It should be noted that for α-particles the use of incident beams is a rather reliable method because the scanning efficiency is high because of the heavy ionization of α-particles.

The interaction mean free paths of heavy nuelci are obtained by

Table 3.4 Interaction Mean Free Path for α-Particles in Nuclear Emulsions

Energy per Nucleon (GeV)	Source of Particles	Mean Free Path (cm)	Reference
0.09	Accelerated α	20 ± 1.7	G. Quareni et al., *Nuovo Cim.*, **1**, 1282 (1955)
0.09	Accelerated α	18.4 ± 0.9	D. Willoughby, *Phys. Rev.*, **101**, 324 (1956)
6	Primary α	20.5 ± 2.2	C. J. Waddington, *Phil. Mag.*, **45**, 1312 (1956)
8	Breakup α	18.4 ± 4	F. Hänni, *Helv. Phys. Acta* **29**, 281 (1956)
12	Primary α	17.5 ± 1.1	M. V. K. Appa Rao et al., *Proc. Ind. Acad. Sci.*, **A43**, 181 (1956)
20	Primary α	19.7 ± 2.2	M. Shapiro et al., *Bull. Am. Phys. Soc.*, **1**, 319 (1956)
20	Breakup α	20.2 ± 1.7	E. Lohrmann et al., *Phys. Rev.*, **115**, 636 (1959)
166	Primary and Breakup α	27 ± 7	E. Lohrmann et al., *Phys. Rev.*, **112**, 672 (1961)
Several thousand	Primary α	27^{+14}_{-7}	R. R. Daniel et al., *Nuovo Cim.*, Suppl. **1**, 1163 (1963)

observing nuclear interactions of heavy primary cosmic rays. Since they are not as abundant as α-particles, no experimental value has yet been obtained at high energies. Most available experiments are concerned with the mean free path averaged over a wide energy range above the geomagnetic cutoff. Table 3.5 gives an experimental result in which the energies of individual heavy nuclei are measured. This might be regarded as showing a slight energy dependence of the mean free path (Aizu 60). The results in two energy regions are in substantial agreement with other results in the respective energy regions, and those for a higher energy region agree with the results obtained for the geomagnetic cutoff of 7 GeV per nucleon (Lohrmann 59).

In interpreting the nucleus-nucleus cross section a simple geometrical consideration has been used (Bradt 50). If the mass numbers of two colliding nuclei are A_1 and A_2, the cross section is given by

$$\sigma = \pi r_0^2 (A_1^{1/3} + A_2^{1/3} - b)^2, \tag{3.3.12}$$

where b represents an overlapping effect of two nuclei. The values of r_0 and b are obtained as (Aizu 60).

$$r_0 = (1.0 \pm 0.3) \times 10^{-13} \text{ cm}, \qquad b = (-0.5 \pm 1.1) \times 10^{-13} \text{ cm} \quad \text{for} \quad \sim 0.5 \text{ GeV} \qquad (3.3.13a)$$

$$r_0 = (1.1 \pm 0.2) \times 10^{-13} \text{ cm}, \qquad b = (-0.7 \pm 1.6) \times 10^{-13} \text{ cm} \quad \text{for} \quad \sim 3 \text{ GeV} \qquad (3.3.13b)$$

The result above may, within statistical accuracy, indicate the energy independence of the cross section and the blackness of the nucleus.

More accurately one has to take into account the transparency effect and the tapered distribution of nucleons. Taking these effects into account Aizu et al. (Aizu 60) calculated the inelastic cross section in a way analogous to the derivation of (3.3.6) by assuming a Gaussian density distribution

$$\rho(r) = \frac{A}{(a\sqrt{\pi})^3} e^{-r^2/a^2}, \qquad a = (0.63A^{1/3} + 0.3) \times 10^{-13} \text{ cm}, \qquad (3.3.14)$$

and obtained the cross section

$$\sigma = \pi(a_1{}^2 + a_2{}^2)[\gamma + \ln \alpha - E_i(-\alpha)], \qquad (3.3.15)$$

Table 3.5 Interaction Mean Free Paths[a] of Heavy Nuclei[b]

Nucleus	Medium	Energy per Nucleon (GeV)			
		~0.7[c]	0.5[d]	~3[c]	2.5[d]
Light (beryllium)	Emulsion	18.7 ± 2.1	16.2	14.3 ± 1.2	15.6
Medium (nitrogen)	Emulsion	16.2 ± 1.3	14.0	13.2 ± 0.7	13.5
Heavy (magnesium)	Emulsion	12.1 ± 1.4	10.7	10.1 ± 1.5	10.6
Very heavy (chromium)	Emulsion	12.1 ± 1.4	10.7	8.4 ± 0.7	8.3
Light (beryllium)	Air		40.0		38.2
Medium (nitrogen)	Air		28.6		27.6
Heavy (magnesium)	Air		22.4		21.6
Very heavy (chromium)	Air		14.4		13.9

[a] Units in centimeters for interactions in emulsion and in g-cm^{-2} for interactions in air.
[b] From H. Aizu et al., *Prog. Theor. Phys.*, Suppl., No. 16, 54 (1960).
[c] Experimental.
[d] Calculated.

NOTE: Other references to experimental results may be found in the following:

P. S. Freier et al., *Phys. Rev.*, **84**, 322 (1951).
J. H. Noon et al., *Phys. Rev.*, **97**, 769 (1955).
P. H. Fowler et al., *Phil. Mag.*, **2**, 239 (1957).
V. Y. Rajopadhye et al., *Phil. Mag.*, **3**, 19 (1958).
R. Cester et al., *Nuovo Cim.*, **7**, 37 (1958).

where $\gamma = 0.577$ is Euler's constant and

$$\alpha \equiv \frac{\bar{\sigma} A_1 A_2}{\pi(a_1{}^2 + a_2{}^2)}.$$

$\bar{\sigma}$ is the effective nucleon-nucleon cross section discussed before. The calculated results for $\bar{\sigma} = 30$ and 37 mb at 0.5 and 2.5 GeV per nucleon, respectively, are compared with the experimental ones in Table 3.5. They are in good agreement at approximately 3 GeV, but the experimental mean free paths appear to be systematically longer than the calculated ones at approximately 0.7 GeV. The latter may be explained by invisible peripheral collisions in which only neutrons are emitted. In the same table the interaction mean free paths in air are also given on the basis of (3.3.15).

3.4 Transverse Momenta

Since the observation that the transverse momenta of secondary particles are rather small and independent of primary and secondary energies (Nishimura 56), many observations have been made with various methods. The first systematic attempt was to measure the energies and emission angles of neutral pions by means of emulsion chambers, in which individual decay γ-rays are spatially separated while they traverse a low Z-material (Minakawa 58, 59). With emulsion stacks the energy of each γ-ray was measured by the relative scattering of a pair of electrons, whereas that of a charged particle emitted backward in the center-of-mass system was determined by the multiple-scattering method (Edwards 58). Both experiments gave the same result for neutral pions, thus verifying Nishimura's prediction, whereas in the latter the transverse momenta of X-particles were found to be appreciably higher. At lower energies, however, no significant difference between the transverse momenta of pions and X-particles was observed (Hansen 60). This was also found to be the case at an accelerator energy (Morrison 61).

The distribution of transverse momentum, p_T, is rather narrow and may be represented for 23-GeV p-p collisions by (Cocconi 62)

$$f(p_T) 2\pi p_T \, dp_T = \exp\left(\frac{-p_T}{p_{T_0}}\right) 2\pi p_T \frac{dp_T}{p_{T_0}^2}, \tag{3.4.1}$$

$$p_{T_0} = 0.17 \text{ GeV}/c, \qquad \langle p_T \rangle = 2p_{T_0} = 0.34 \text{ GeV}/c. \tag{3.4.2}$$

It is also shown that the average value of transverse momenta, $\langle p_T \rangle$, is the same for all secondary particles. In a recent experiment with π^--p

collisions at 10 GeV/c, however, the value of $\langle p_T \rangle$ increased slowly with the mass of a particle produced, as shown in Table 3.6 (Bigi 62).

Table 3.6 Transverse Momenta of Particles Produced by 10-GeV/c π^--p Collisions

Secondary Particle	Average Transverse Momentum (GeV/c)
π^+	0.30 ± 0.01
K^0	0.39 ± 0.02
p	0.44 ± 0.05
Λ_0	0.46 ± 0.02
$\Sigma^\pm$	0.51 ± 0.04
Ξ^-	0.56 ± 0.08

The exponential behavior of the transverse-momentum distribution is also observed for protons elastically scattered by hydrogen, and the average transverse momentum is essentially the same as that in (3.4.2) (Orear 64):

$$\langle p_T \rangle = 0.30 \text{ GeV}/c. \tag{3.4.3}$$

The average value and the distribution of transverse momenta are essentially independent of energy, as shown by cosmic-ray evidence. Table 3.7 lists a number of experimental results on the average transverse momenta obtained in a wide range of energy. All but those of X-particles by Edwards *et al.* give essentially the same average value for all secondary particles.

The energy independence holds also for the distribution of transverse momenta of pions. The distributions at three different primary energies are compared in Fig. 3.4 (Hasegawa 65). All of them are well represented by an exponential function of the average value consistent with (3.4.2) and (3.4.3), in spite of the fact that the primary energy ranges from 25 GeV to 50 TeV.

In some experiments a few events of transverse momentum greater than 1 GeV/c are observed. The distribution of transverse momenta of γ-rays found in jets of primary energies around 10 TeV extends as far as 1.5 GeV/c, as shown in Fig. 3.5 (Fowler 63). The exponential distribution for pions results in the distribution for γ-rays represented by $Ei(-p_{T\gamma}/p_{T0})$. The observed distribution may be represented by the sum of two Ei functions with

$$\langle p_{T\pi^0} \rangle = 0.28 \text{ GeV}/c, \qquad \langle p_{T\pi^0} \rangle = 1.5 \text{ GeV}/c. \tag{3.4.4}$$

Table 3.7 Mean Values of Transverse Momenta (MeV/c)

Primary Energy	Particle[a]	Target	$\pi^\pm$	π^0	$X^\pm$	X^0	Reference
11.4 GeV	π^-	Hydrogen	339		411(p)	376(K^0)417 (n)	1
24 GeV	p	Emulsion		470 ± 35			2
24 GeV	p	Hydrogen	340 ± 14				3
20 to 150 GeV	N	Carbon		390 ± 25			4
~100 GeV	N	Carbon	307 ± 26		310 ± 44		5
~200 GeV	N	Lithium hydride	300 ± 20				6
1 to 70 TeV	N	Emulsion	410 ± 130	520 ± 160	800 ± 100	1400 ± 200	7
~2 TeV	N	Emulsion	300 ± 100		330 ± 100		8
~10 TeV	N	Carbon		390 ± 20			9
~50 TeV	N	Air		300 ± 40			10

[a] N stands for nuclear active particles, which consist mainly of nucleons.

References:

1. T. Ferbel and H. Taft, *Nuovo Cim.*, **28**, 1214 (1963).
2. Y. Pal and T. N. Rengarajan, *Phys. Rev.*, **124**, 1575 (1961).
3. D. R. O. Morrison, *Proc. Int. Conf. on Theoretical Aspects of Very High Energy Phenomena*, CERN, 61–22, 153 (1961).
4. S. Lal, Yash Pal, and R. Raghaven, *J. Phys. Soc. Japan*, **17**, Suppl. A–III, 393 (1962).
5. L. F. Hansen and W. B. Fretter, *Phys, Rev.*, **118**, 812 (1960).
6. V. V. Guseva et al., *J. Phys. Soc. Japan*, **17**, Suppl. A–III, 375 (1962).
7. E. Edwards et al., *Phil. Mag.*, **3**, 237 (1958).
8. C. O. Kim, *Phys. Rev.*, **136**, B515 (1964).
9. O. Minakawa, et al., *Nuovo Cim., Suppl.* **8**, 761 (1958); **11**, 125 (1959).
10. M. Akashi et al., *Progr. Theor. Phys. Suppl.*, **32**, 1 (1964).

This indicates that the transverse-momentum distribution could have an extended tail.

There are some indications of large transverse momenta in extensive air showers. In the central part of an extensive air shower double cores have been observed with a neon hodoscope at sea level (Miura 62, Oda 62) as well as with a cloud chamber at a mountain altitude (Miyake 63). Although the energy involved in each core can be determined only approximately (on the order of 10^{13} eV) the transverse momenta estimated from pairs of cores are as high as several GeV/c. The relative frequency of occurrence of such double cores is estimated as a few percent; this may be compared with the rareness of transverse momenta greater than 2 GeV/c in emulsion experiments, shown in Fig. 3.5.

The large transverse momentum observed in extensive air showers is not necessarily attributable to a meson produced but may be that of a survival nucleon. In the accelerator energy region the transverse momentum of the survival nucleon is found to be nearly equal to that of the secondary pion. For those produced from emulsion nuclei by 25-GeV protons (Pal 63) and those from beryllium and aluminum by

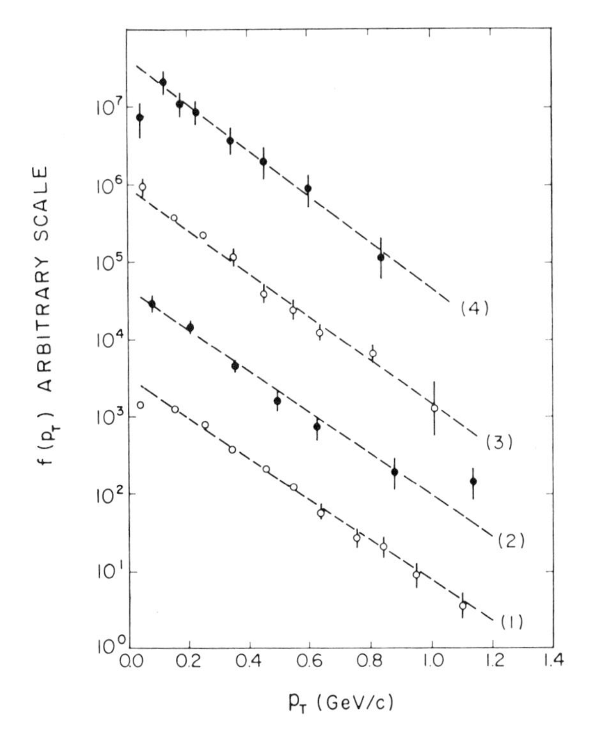

Fig. 3.4 Distribution of transverse momenta of pions. Numbers on curves represent the experimental data of (1) Morrison 61, (2) Hanson 60, (3) Guseva 62, and (4) Akashi 65; the references are given in Table 3.7. (Hasegawa 65).

30-GeV protons (Baker 61) the average values of the transverse momenta are obtained as

$$\langle p_{Tp} \rangle = 400 \text{ and } 460 \text{ MeV}/c, \tag{3.4.5}$$

respectively. However, there is an indication that the distribution of transverse momenta of survival protons differs from that of secondary pions. The former shows the most probable value to be appreciably smaller than that of the latter, thus having an extended tail (Pal 63). In fact, the distribution represented by the exponential form is less steep than that for pions (Baker 61).

An analysis has been made at about 1 TeV for breakup events of heavy nuclei in emulsion; namely, the measurement of transverse momenta of nucleons produced by heavy nucleus interactions.* Single-charge particles emitted at much smaller angles than others are

* H. Aizu and M. Koshiba, private communication, 1965.

defined as the survival protons, and neutral interactions in nearly the same angular region are defined as the survival neutrons. The distribution of the transverse momenta of the survival nucleons thus obtained is shown in Fig. 3.6. The distribution is exponential except at small p_T. The latter part is attributed to the evaporation nucleons of heavy nuclei after their interactions whose transverse momenta are expected

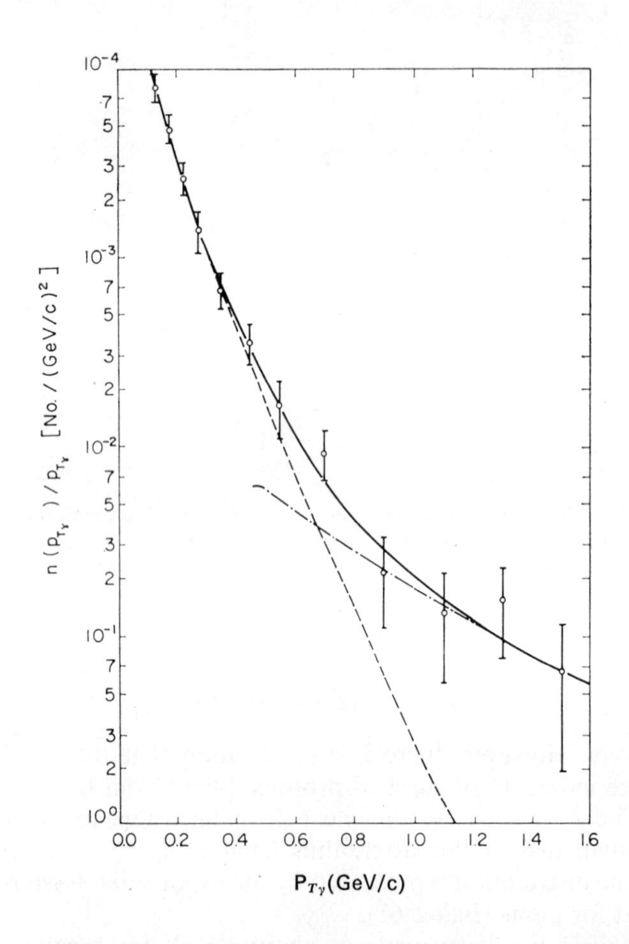

Fig. 3.5 Distribution of transverse momenta of γ-rays in jets. The dashed and dot-dashed curves represent the $p_{T\gamma}$-distribution expected from the exponential $p_{T\pi}$-distribution; namely, $n(p_{T\gamma}) \propto -Ei(-p_{T\gamma}/p_0)2\pi p_{T\gamma}$.
$- - -\ p_0 = 0.14$ GeV/c; $- \cdot - \cdot -\ p_0 = 0.75$ GeV/c;
—— sum of - - - and - · - · -. (Hasegawa 65).

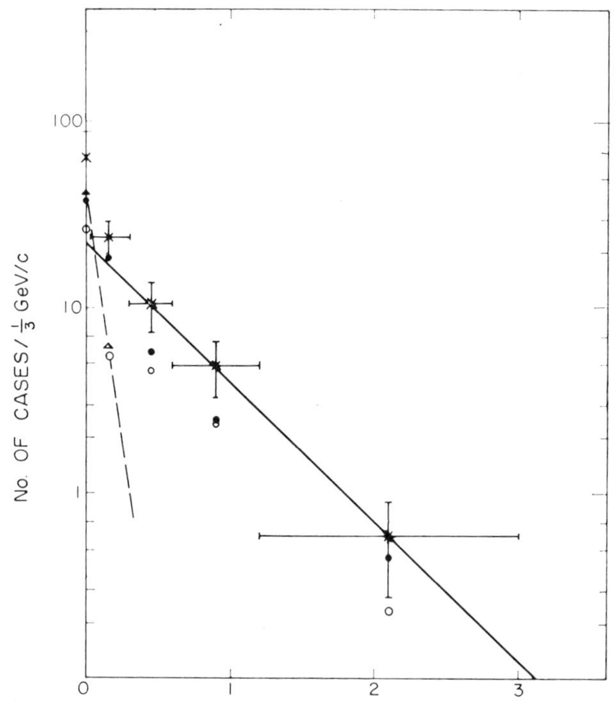

Fig. 3.6 Differential distribution of the transverse momenta of survival nucleons ● protons; ○ neutrons; × protons plus neutrons; △ uninteracted nucleons —— distribution for interacted nucleons, $\langle p_T \rangle = 580 {+380 \atop -170}$ MeV/c; - - - distribution for uninteracted nucleons, $\langle p_T \rangle \simeq 80$ MeV/c (after Koshiba).

to be as small as the Fermi momenta of nucleons in the nucleus. By subtracting this part, with an average transverse momentum of about 80 MeV/c, we find an exponential distribution with the average value of

$$\langle p_{Tp} \rangle = 580 {+\,380 \atop -\,170} \text{ MeV}/c. \tag{3.4.6}$$

Another experimental result is shown in Figure 3.24, but this is subject to an ambiguity in kinematics, as discussed in Appendix B.9. It should, however, be kept in mind that the study of survival nucleons has just begun, and further studies will be needed for a conclusive result.

3.5 Angular Distribution

Secondary particles produced by high-energy interactions are highly collimated in the forward direction because of a large value of the Lorentz factor γ_c for the transformation between the laboratory system and the center-of-mass system. From (B.14) we see that no particle is emitted backward if the velocity of a particle emitted in the center-of-mass system, $c\beta^*$, is smaller than the velocity of the center-of-mass system, β_c, where $\gamma_c = (1 - \beta_c{}^2)^{-1/2}$. In this case the maximum angle of emission corresponds to the angle of emission in the center-of-mass system, θ^*, given by

$$\cos \theta^* = \frac{-\beta^*}{\beta_c},$$

and its value θ_m is given by

$$\tan \theta_m = \frac{\beta^*/\gamma_c \beta_c}{\sqrt{1 - (\beta^*/\beta_c)^2}}. \tag{3.5.1}$$

Since this is an increasing function of β^*/β_c, the maximum angle of emission could give the maximum energy of secondary particles in the center-of-mass system.

The angle of emission of a particle of momentum p^* in the center-of-mass system is related to θ^* by

$$\sin \theta^* = \frac{p_T}{p^*}. \tag{3.5.2}$$

For the particle of largest emission angle this gives

$$\tan \theta_m = \frac{\beta^*}{\gamma_c \beta_c} \frac{p^*}{p_T}. \tag{3.5.1$'$}$$

Thus the measurement of θ_m enables us to obtain p^* or γ_c if one of these is known. However, this method is of little practical use because it relies on the information from a single secondary particle alone.

It is therefore preferred to use the emission angles of all measurable secondary particles. If both β_c and β^* are assumed to be nearly equal to unity, the angles of emission in the laboratory system and the center-of-mass system are related simply by (Castagnoli 53)

$$\log \tan \theta = -\log \gamma_c + \log \tan \frac{\theta^*}{2}, \tag{3.5.3}$$

as given in Appendix B. Hence the distribution of log tan θ gives an intuitive indication of the angular distribution in the center-of-mass system, which is topologically invariant under the Lorentz transformation. Several typical examples of the log tan θ plot are shown in Fig. 3.7.

Another way of representing the angular distribution is the so-called F-plot (Duller 54). A quantity $F(\theta)$ is so defined that $F(\theta)$ is proportional to the number of particles emitted with angles smaller than θ and is normalized to $F(\pi) = 1$. For the isotropic distribution in the center-of-mass system, as shown in (B.57), we have

$$\log \frac{F(\theta)}{1 - F(\theta)} = 2 \log \gamma_c + 2 \log \tan \theta. \tag{3.5.4}$$

If we plot $F/(1 - F)$ against tan θ on a log-log scale, the isotropic distribution gives us a straight line with slope 2. Any deviation from the straight line indicates departure from isotropy. If log $[F/(1 - F)]$ has the same form above and below the log tan θ axis, the emission is regarded as symmetric. Several typical examples of the F-plot are shown in Fig. 3.8.

Since the log tan θ plot and the F-plot are two ways of representing

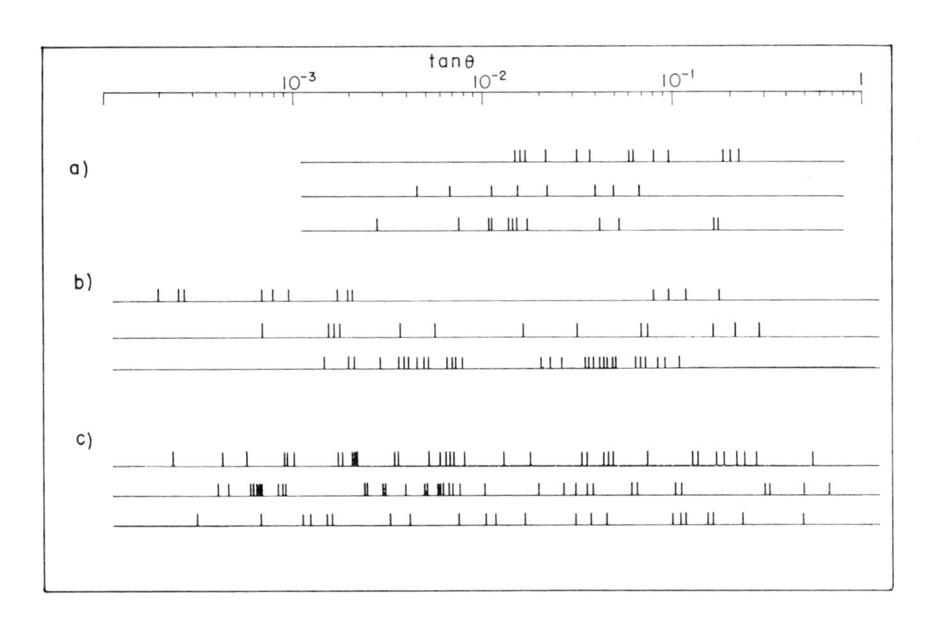

Fig. 3.7 Typical examples of log tan θ plots: examples of one lump, (*a*); examples of two lumps, (*b*); examples of four lumps, (*c*).

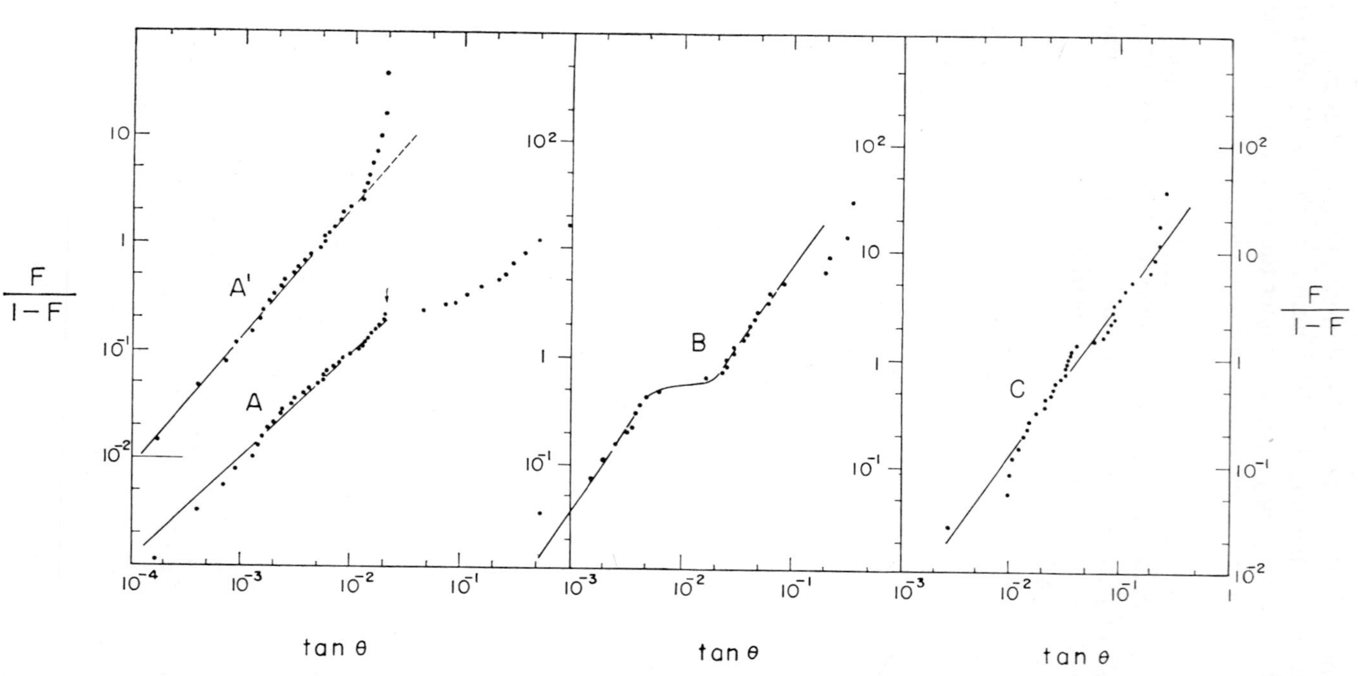

Fig. 3.8 Typical examples of F-plots and the median angles (θ_M) of secondary particles (after P. K. Malhotra and Y. Tsuzuki). $A(\theta =_M 9.5 \times 10^{-3}$ rad): points beyond the arrow are attributed to secondary interactions in the same nuclei; $A'(\theta_M = 5.0 \times 10^{-3}$ rad): F-plot constructed by removing particles beyond the arrow; the steepness at the largest angle may be due to inappropriateness of the removal process; $B(\theta_M = 1.2 \times 10^{-2}$ rad): F-plot representing the two-fireball model; $C(\theta_M = 4 \times 10^{-2}$ rad): F-plot representing the one-fireball model.

the same thing, we shall discuss mainly the former, which seems to be a little simpler.

By putting an observable quantity log tan θ as

$$x = \log \tan \theta, \tag{3.5.5}$$

we can evaluate the moments of x. The first moment, the average value of x, gives

$$\langle x \rangle = -\log \gamma_c \tag{3.5.6}$$

for the symmetric distribution. With γ_c thus determined, we can divide secondary particles into forward and backward groups in the center-of-mass system. Let the numbers of particles in respective groups be n_f and n_b and the sums of energies therein be E_f and E_b. Then we may define the asymmetric parameters as

$$a_n \equiv \frac{n_f - n_b}{n_f + n_b}, \qquad a_E \equiv \frac{E_f - E_b}{E_f + E_b}. \tag{3.5.7}$$

If the value of a_n or a_E deviates from zero by more than the standard deviation, this may be regarded as an indication of asymmetry.

Using a_E obtained from a calorimeter experiment, the asymmetry for the nucleon-nucleon collision around a few hundred GeV is found to be about 40 percent (Guseva 62). It is possible that asymmetry could arise from the incorrect assignment of γ_c. This ambiguity may be avoided by selecting events of known primary energies, such as those caused by the breakup secondaries of heavy nuclei (Koshiba 63*b*). For these events the n_f-n_b relation is obtained as shown in Fig. 3.9. In Fig. 3.9 curves are drawn indicating one, two, and three standard deviations. The value of a_n with significant deviation from zero is found to be 0.2 to 0.3. Although the result is not quantitative yet, this evidence demonstrates that about a quarter of observed interactions leads to asymmetric emission. A method suitable for asymmetric events is given in (B.63) to (B.69) (Yajima 65 *a,b*).

The second moment of x represents the degree of anisotropy. This is conveniently expressed by

$$\sigma \equiv \left[\frac{\langle (x - \langle x \rangle)^2 \rangle}{n_s}\right]^{1/2}. \tag{3.5.8}$$

For the isotropic distribution we have $\sigma = 0.39$. A greater value of σ indicates the concentration of particles in the forward and backward cones. It has been found for primary energies around 10^{12} eV that three-fourths of the jets studied have $\sigma > 0.6$, indicating two maxima in the distribution of x (Gierula 60). It is interesting to note that the values of σ for forward and backward particles are obtained as (Miesowicz 62)

$$\sigma_f = 0.40 \pm 0.05, \qquad \sigma_b = 0.43 \pm 0.05,$$

respectively, thus demonstrating isotropy in both groups. This is the case for $N_h \leq 5$, but for large N_h we have $\sigma_f < \sigma_b$ and $n_f < n_b$. This is considered to be due to multiple collisions in a nucleus.

The third way is to represent the angular distribution in the center-of-mass system as $\cos^m \theta^* \, d(\cos \theta^*)$ (Edwards 58). The value of m can be obtained by taking the average value of cosec θ^*, as given in (B.70),

$$\langle \operatorname{cosec} \theta^* \rangle = \frac{\int_0^{\pi/2} \operatorname{cosec} \theta^* \cos^m \theta^* \, d(\cos \theta^*)}{\int_0^{\pi/2} \cos^m \theta^* \, d(\cos \theta^*)}$$

$$= \sqrt{\pi} \, \frac{\Gamma(m + 3/2)}{\Gamma(m + 2/2)}. \tag{3.5.9}$$

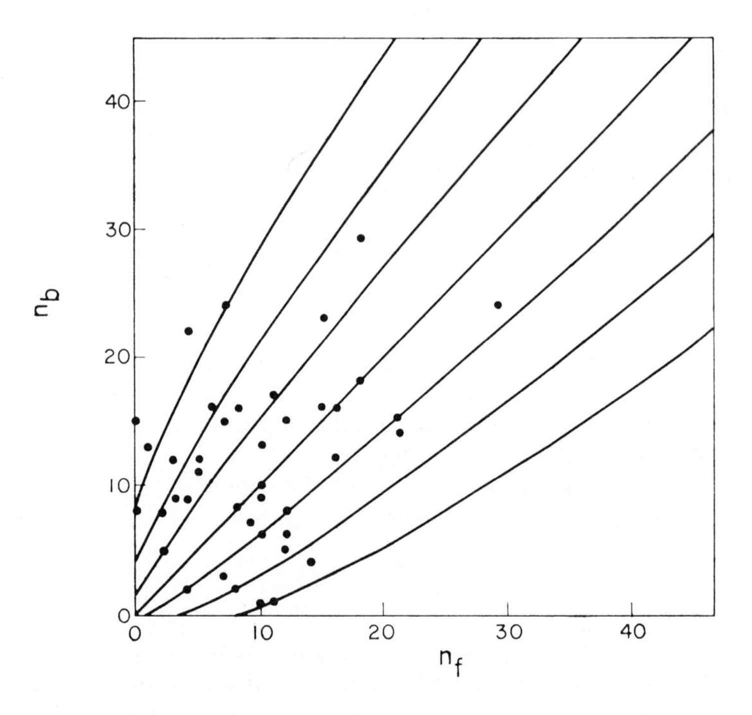

Fig. 3.9 Forward-backward asymmetry. The numbers of forward and backward shower particles, n_f and n_b, are plotted on the n_f-n_b diagram. Full curves indicate a line of $n_f = n_b$, and one, two, and three standard deviations from it (Koshiba 63b).

For the isotropic distribution we have $m=0$ and

$$\langle \text{cosec } \theta^* \rangle_{m=0} = \frac{\pi}{2} = 1.57. \tag{3.5.10}$$

whereas for large m (3.5.9) is approximated as

$$\langle \text{cosec } \theta^* \rangle \simeq \sqrt{\pi m/2} = 1.25\sqrt{m}. \tag{3.5.11}$$

The values of $\langle \text{cosec } \theta^* \rangle$ are shown as a function of m in Fig. 3.10. The approximation (3.5.11) is quite good for $m > 10$, but it deviates from (3.5.9) more than 30 percent for $m < 3$.

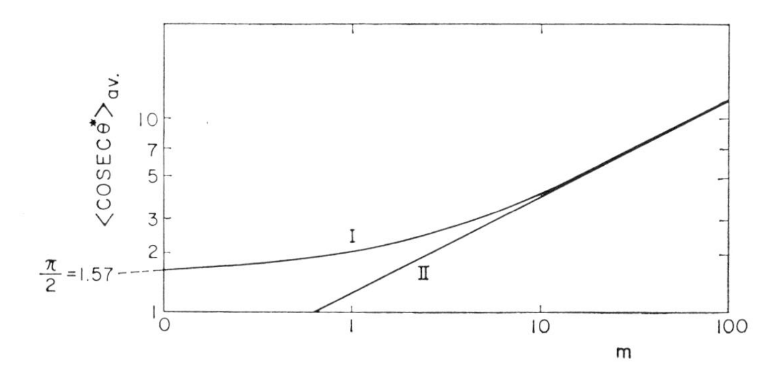

Fig. 3.10 The average value of $\langle \text{cosec } \theta^* \rangle$ versus m when the angular distribution is represented as $\cos^m \theta^*$. I: exact (3.5.9); II: approximate (3.5.11).

The third method may be regarded as disadvantageous because the angle θ^* can be obtained only after the center-of-mass system has been found, and the value of γ_c is hardly reliable for most events. The effect of the misassignment of γ_c is unimportant only if $|\Delta\gamma_c/\gamma_c| < 0.5$.

On the other hand, this method has an advantage, because of the rather direct physical meaning of m. As can be seen from (3.5.9), the value of m is determined mainly from the angular distribution at small θ^*. Then the angular distribution is approximated, in particular for large m, as

$$\cos^m \theta^* \simeq (1 - \tfrac{1}{2}\,\theta^{*2})^m \simeq 1 - \frac{m}{2}\,\theta^{*2}.$$

Hence the most probable angle of emission is given approximately by $\theta_m^* \simeq \sqrt{2/m}$. Here θ_m^* corresponds to the impact parameter responsible for interactions, and its value b is related to θ_m^* as

$$\theta_m^* \simeq \frac{\hbar}{p^* b},$$

where p^* is the momentum of a particle emitted. Since p^*b is simply the angular momentum of this particle, $\hbar L$, the value of m gives the angular momentum

$$L \simeq \frac{p^* b}{\hbar} \simeq \sqrt{m/2}. \tag{3.5.12}$$

This relation can be obtained also from (3.5.11) if $b \simeq \hbar / p_T$. If n secondary particles are emitted, the angular momentum of each particle adds up statistically to (Ezawa 59)

$$J \simeq \sqrt{n/2} L \simeq \sqrt{nm/4} \simeq \sqrt{3n_s\, m/8}. \tag{3.5.13}$$

In the last relation use has been made of $n \simeq (\frac{3}{2}) n_s$. Thus J may be regarded as the angular momentum of the system consisting of all secondary particles.

The values of m obtained experimentally are plotted agbinst γ_c in Fig. 3.11. They are smaller than 10 for $\gamma_c < 20$ but widely distribute for large γ_c. On the average, m behaves roughly as γ_c, indicating $p^* \propto \gamma_c^{1/2}$, according to (3.5.12).

3.6 Energy Distribution

The angular distribution is closely related to the energy distribution through (3.5.2) or in the laboratory system through

$$\sin \theta = \frac{p_T}{p} \tag{3.6.1}$$

with $p_T =$ constant. The angular distribution obtained from jets around 10^{12} eV indicates that the number of secondary particles with cosec θ larger than a certain value is proportional to $\sin \theta$,

$$N(> \operatorname{cosec} \theta) \propto \sin \theta. \tag{3.6.2}$$

This results in the momentum spectrum (Hasegawa 57)

$$f(p)\, dp \propto \frac{dp}{p^2}. \tag{3.6.3}$$

An analogous result was found for the momentum spectrum in the center-of-mass system (Edwards 58).

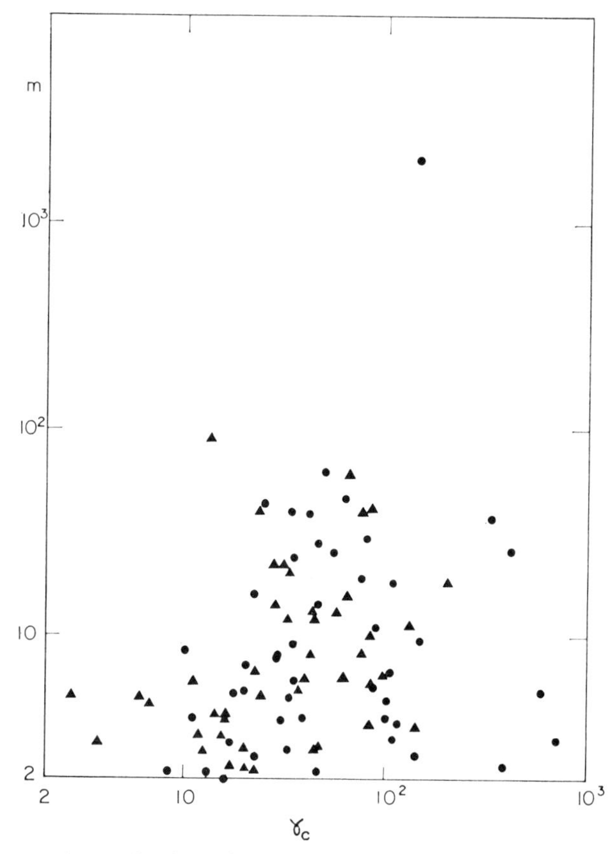

Fig. 3.11 Experimental values of m versus γ_c (collected by T. Kawamura). ●—$N_h = 0, 1$ ▲—$N_h = 2, 3$ $(m \geq 2)$.

More exactly one has to take account of the distribution of p_T, so that the angular and energy distributions are not uniquely related to each other. In fact, the momentum spectrum depends rather strongly on the angle of emission as was found for charged pions from 10- to 30-GeV protons incident on beryllium (Baker 61) and for γ-rays in the same energy region (Fidecaro 61). The result of the former is reproduced in Fig. 3.12. The differential number of pions produced per interacting proton in Fig. 3.12 is expressed empirically for a given emission angle θ as

$$\frac{d^2 n_\pi}{dp\, d\Omega} = Ap^2 e^{-4.8p/\sqrt{p_0}} e^{-2.6p\sqrt{p_0}\,\theta^2} + \frac{Bp^2}{p_0} e^{-10.4(p/p_0)^2} e^{-3.9p\theta} \tag{3.6.4}$$

where p_0 and p are the momenta in GeV/c of the incident proton and a

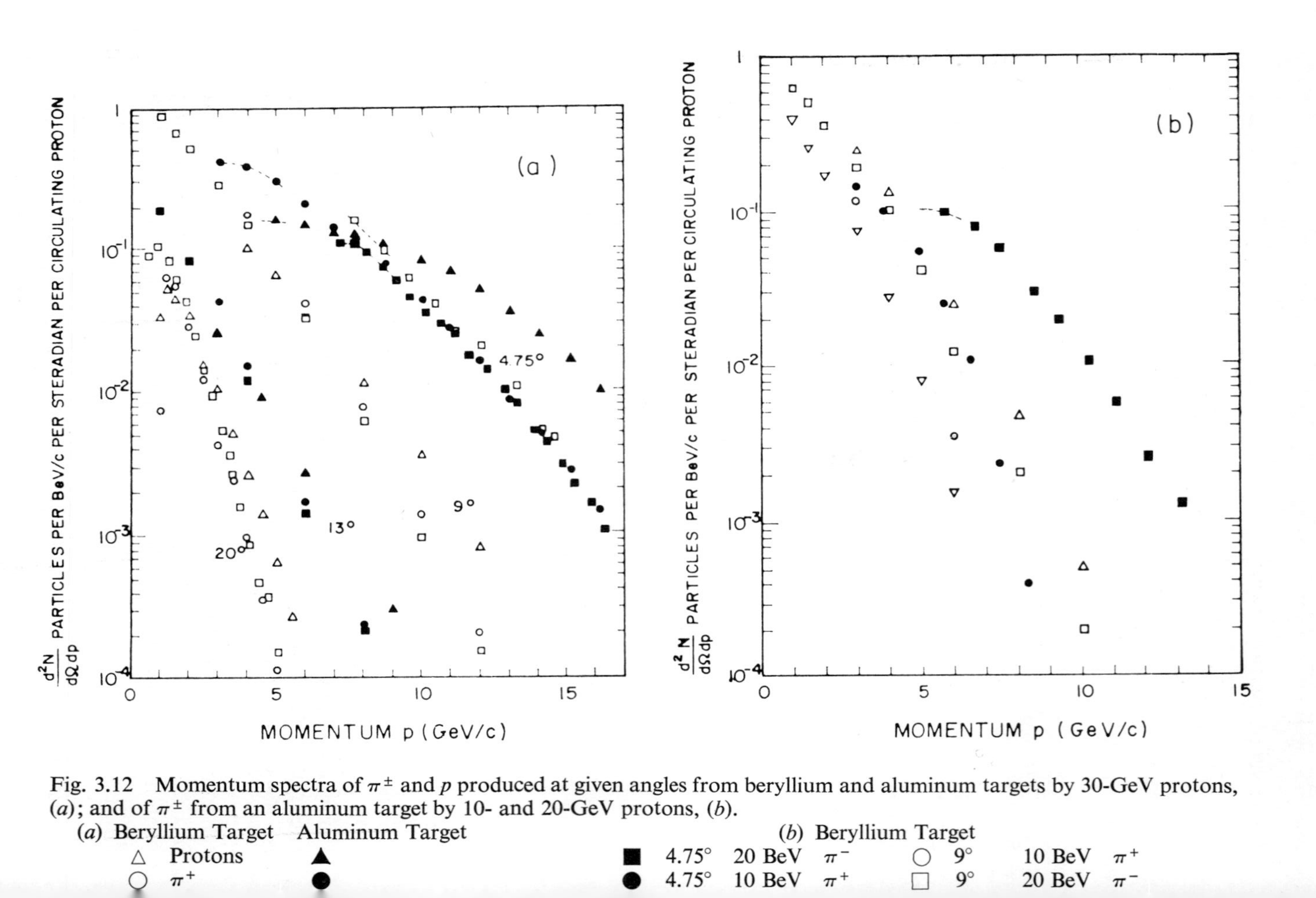

Fig. 3.12 Momentum spectra of $\pi^{\pm}$ and p produced at given angles from beryllium and aluminum targets by 30-GeV protons, (*a*); and of $\pi^{\pm}$ from an aluminum target by 10- and 20-GeV protons, (*b*).

(*a*) Beryllium Target Aluminum Target
△ Protons ▲
○ π^+ ●

(*b*) Beryllium Target
■ 4.75° 20 BeV π^- ○ 9° 10 BeV π^+
● 4.75° 10 BeV π^+ □ 9° 20 BeV π^-

secondary pion, respectively. The A and B are constants, $A(\pi^{\pm}) = 1.65$, $B(\pi^{+}) = 3.12$, and $B(\pi^{-}) = 1.04$. The integration of (3.6.4) over angle gives

$$\frac{dn_\pi}{dp} = \bar{A}\frac{p}{\sqrt{p_0}}\, e^{-4.8p/\sqrt{p_0}} + \frac{\bar{B}}{p_0}\, e^{-10.4(p/p_0)^2}, \qquad (3.6.4')$$

where $\bar{A}(\pi^{\pm}) = 2.0$, $\bar{B}(\pi^{+}) = 1.24$, and $\bar{B}(\pi^{-}) = 0.41$. The two peaks of the momentum distribution arising from the two terms above are not well separated but represent an increase of the peak width with incident energy. It should also be noted that the expressions above are concerned only with forward pions in the center-of-mass system, and the angular distribution does not hold for transverse momenta greater than 1 GeV/c.

Analogous expressions can be obtained for other secondary particles, such as protons and kaons, but they are not given here. It is open to question whether a distribution such as (3.6.4) holds at higher energies. Since no accelerator can as yet reach such high energies, we shall mention some qualitative results obtained from cosmic ray experiments.

Measurements of the energy spectrum at about 100 GeV have been made with cloud chambers (Hansen 60, Fujioka 61, Lal 62) as well as with emulsions (Matsumoto 63). These results show the energy distribution in the center-of-mass system to be expressed approximately by

$$f(E^*)\, dE^* \propto \frac{dE^*}{E^{*2}},$$

with the average energy of pions

$$\langle E_\pi^* \rangle \simeq 0.45 \text{ GeV}. \qquad (3.6.5)$$

In these experiments the determination of the center-of-mass system is made on the basis of the angular distribution. A different method has been developed by employing a calorimeter, with which the primary energy is measured, combined with a cloud chamber in which the magnetic rigidities of secondary particles are measured (Guseva 62, Dobrotin 63). The momentum spectrum of secondary particles thus measured is shown in Fig. 3.13. The spectrum may be expressed as the Planck distribution

$$f(p^*)\, dp^* = \frac{z^3 p^{*2}\, dp^*}{m_\pi{}^3 F(z)} \{\exp [z\sqrt{(p^*/m_\pi)^2 + 1} - 1]\}^{-1}, \qquad (3.6.6)$$

with

$$F(z) = z^2 \sum_{m=0}^{\infty} \frac{K_2[z(m+1)]}{m+1} \qquad \left(z = \frac{m_\pi}{T}\right),$$

where m_π is the pion rest energy and T is the temperature in units of energy. The distribution in Fig. 3.13 is fitted for $T = 0.65\ m_\pi$. It should, however, be remarked that a long tail extends to high values of momentum.

Some of these high-energy particles may be associated with secondary X-particles and surviving primary particles. In fact, the average energy of X-particles is found to be as high as (Hansen 60)

$$\langle E_X^* \rangle = 1.3 \pm 0.2\ \text{GeV}. \tag{3.6.7}$$

As the primary energy increases to about 1 TeV the angular distribution begins to show two peaks, and consequently the energy distribution in the center-of-mass system is shifted toward high energy. Since the emission angles in the center-of-mass system are also small at such high energies, the log tan θ plot gives the distribution of energies, in use of (3.5.2), (3.5.3), and (3.6.1). Hence the double maxima in the angular distribution give those in the momentum distribution—one corresponding to the forward group and the other, to the backward group. The average momenta of the respective groups can be obtained from the angular distribution. The distribution of momenta about the average is found to be analogous to (3.6.6) with the width essentially

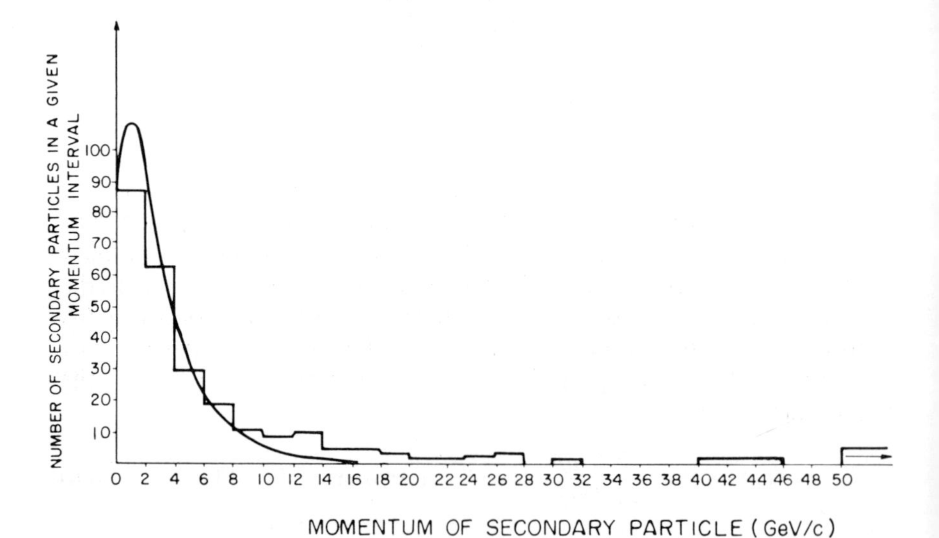

Fig. 3.13 Momentum spectrum of secondary particles for primary energies of about 300 GeV. The solid curve represents the spectrum expected from the Planck distribution with $T = 0.65 m_\pi$ (Dobrotin 63).

equal to (3.6.7) (Niu 58). This provides a basis for regarding each of these groups or lumps as an entity that is formed by particles trapped for a period long enough to reach a thermal equilibrium.

The shape of the energy distribution in the lump system may also be represented by the distribution of $E_\pi/\Sigma E_\pi$ in the laboratory system, where ΣE_π is the sum of the energies of forward or backward particles, because both E_π and ΣE_π are transformed to the lump system in a similar manner. The forward particles may be selected by observing high-energy γ-rays, from which the spectrum of $E_{\pi^0}/\Sigma E_{\pi^0}$ may be obtained. The spectra thus obtained with emulsion chambers in three energy regions are shown in Fig. 3.14.

Although statistics is not enough, Fig. 3.14 demonstrates that the spectrum for $\Sigma E_{\pi^0} = 10^{11}$ to 10^{12} eV is Planck-like, whereas that for $\Sigma E_{\pi^0} > 10$ TeV has a steep rise toward low energies and a rather flat tail at high energies.

This feature of the spectrum can also be expressed by the ratio E_1/E_2, where E_1 is the highest secondary energy and E_2 the second highest. In the Planck-like distribution E_1/E_2 is rather small, whereas in the distribution with a flat tail it is in most cases large. The plot of E_1/E_2 against ΣE_{π^0} shown in Fig. 3.15 indicates small E_1/E_2 for $\Sigma E_{\pi^0} \lesssim 1$ TeV and the wide scattering of E_1/E_2 as ΣE_{π^0} increases. This result

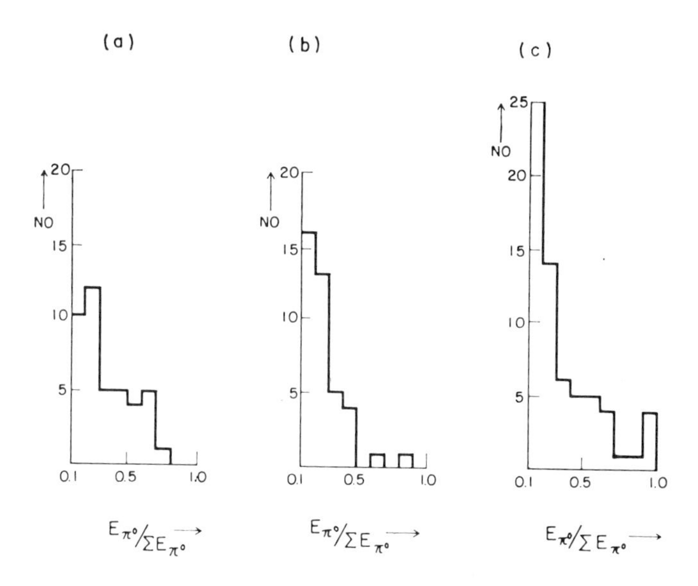

Fig. 3.14 Distributions of $E_{\pi^0}/\sum E_{\pi^0}$ observed with emulsion chambers. The values of $\sum E_{\pi^0}$ are 0.1 to 1 TeV, (*a*); 10 to 20 TeV, (*b*); and $\geq$20 TeV, (*c*) (after Fujimoto).

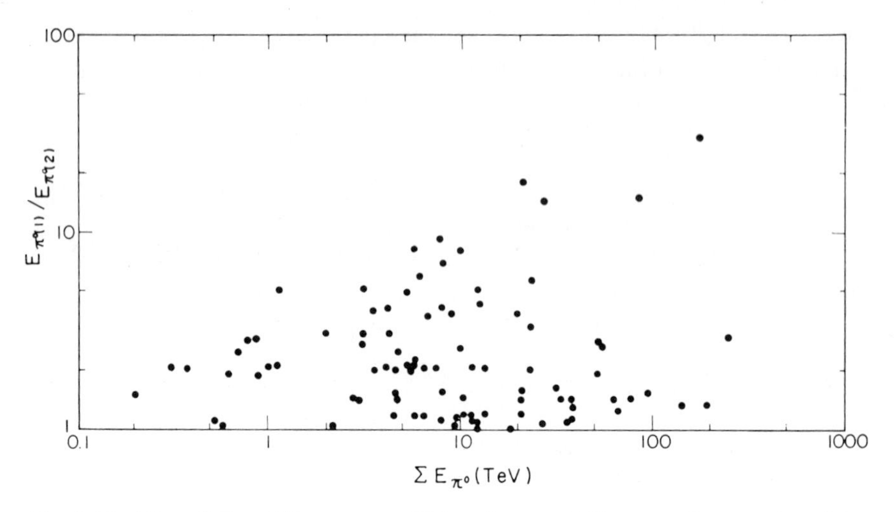

Fig. 3.15 Plot of $E_{\pi^0(1)}/E_{\pi^0(2)}$ versus $\sum E_{\pi^0}$; $E_{\pi^0(1)}$ and $E_{\pi^0(2)}$ are the energies of the pions of the greatest and the second greatest energies, respectively (after Fujimoto).

suggests that secondary particles are further divided, in addition to the division into two lumps, into those concentrated toward low energies and a few that may or may not form a further lump.

3.7 Determination of Primary Energy

In spite of its great importance the primary energy of a high-energy interaction cannot be determined very accurately. Several methods thus far proposed may give only rough values of the energies of individual events, although they seem to be applicable to the determination of the average energy of many events. The following are the methods often conventionally used. Methods 1, 2, and 3 are based on kinematical relations concerning the emission angles of secondary particles, as discussed in Section 3.5 and Appendix B (B.4 and B.9); methods 4, 5, and 6 refer to physical information given in Section 3.4 and elsewhere. Among them methods 1 through 4 have been widely employed in emulsion work.

1. The median-angle method. If the median angle of emission of secondary particles, $\theta_{1/2}$, can be obtained from observation, it can be uniquely related to the Lorentz factor γ_c of the center-of-mass system of colliding particles, as given in (B.58). Since the primary energy is given by $E_0 = 2\gamma_c^2$ for $\gamma_c^2 \gg 1$, a reasonable estimate of the primary

energy may be obtained as

$$E_{\text{med}} = \frac{2}{\tan^2 \theta_{1/2}}. \tag{3.7.1}$$

Actually, however, the emission of particles is not always symmetric in the forward and backward directions in the center-of-mass system, and the unambiguous determination of the median angle is difficult even in the symmetric case because the number of charged secondary particles is not great and the gap between the inner and outer cones is large at high energies.

2. The Castagnoli method (Castagnoli 53). As given in Section 3.5, the log-tan plot of emission angles allows us to deduce the value of γ_c and consequently the primary energy:

$$E_{\text{Cas}} = 2\gamma_c^{\,2}, \qquad \log \gamma_c = -\frac{1}{n_s} \sum_i \log \tan \theta_i, \tag{3.7.2}$$

where θ_i is the emission angle of the ith charged particle. This method is applicable to collimated jets and is widely used in emulsion work. However, the contribution of neutral particles is sometimes so important as to influence the result, especially in the case of small multiplicity.

3. The Yajima–Hasegawa method (Yajima 65a,b). The value of γ_c obtained by (3.7.2) is greatly affected by particles with small emission angles. This would result in a considerable underestimate of γ_c if a neutral particle is emitted with the smallest angle, and in an overestimate for the case of forward asymmetry. These drawbacks are to some extent reduced by using (B.65),

$$\gamma_c = \left(\frac{\sum_i \cot \theta_i}{\sum_i \tan \theta_i} \right)^{1/2}. \tag{3.7.3}$$

The effect of asymmetry is taken into account in (B.69) as

$$\gamma_c = \frac{1}{(1 + \mu^2/p_T^{\,2})^{1/2}} \left(\frac{1 - Z}{1 + Z} \right)^{1/2} \left(\frac{\sum_i \cos \theta_i}{\sum_i \tan \theta_i} \right)^{1/2}, \tag{3.7.4}$$

where μ is the pion mass. The asymmetry factor Z is defined by (B.68) but may in practice be evaluated as equal to a_n in (3.5.7).

4. The E_{ch}-method. Because of the constancy of transverse momentum, we may regard $p_T \operatorname{cosec} \theta_i$ as a reasonable value of the momentum of a charged secondary particle. Hence the energy given to charged secondary particles may be given by

$$E_{\text{ch}} = \sum_i p_T \operatorname{cosec} \theta_i, \tag{3.7.5}$$

in which only the emission angles of charged secondary particles are measured. The average total energy of all charged secondary particles is a fraction of the primary energy E_0. Denoting this fraction as K_{ch}, we obtain

$$E_0 = \frac{E_{ch}}{K_{ch}}. \tag{3.7.6}$$

The conventional values adopted for K_{ch} and p_T are

$$K_{ch} = 0.31, \qquad p_T = 0.4 \text{ GeV}. \tag{3.7.7}$$

The value of p_T here is taken as larger than that in Section 3.4 because of the sampling of data. This method is quite simple, but it seems to give a rather good estimate.

5. The ΣE_{π^0}-method. In some cases only the electromagnetic cascade arising from a nuclear interaction is observed. This is thought to be due mainly to neutral pions, and consequently the total energy spent to the cascade indicates the sum of the energies of neutral pions, ΣE_{π^0}. As before, we conventionally divide ΣE_{π^0} by a factor K_γ to get the primary energy:

$$E_0 = \frac{\Sigma E_{\pi^0}}{K_\gamma} \qquad (K_\gamma \simeq 0.16). \tag{3.7.8}$$

This method is not free from ambiguity caused by some of the secondary particles producing secondary interactions that in turn lead to additional cascades mixed with the primary cascade. Moreover, both methods 4 and 5 are subject to rather large fluctuations in the inelasticity factors, K_{ch} and K_γ.

6. The Calorimeter method (Grigorov 58, 59). If a material underneath the producer of jet showers is thick enough, all the energy brought in by a primary particle is absorbed therein. There is a variety of such devices, called calorimeters, that measure the energy dissipated by secondary particles. Since essentially all of the energy is converted into ionization, the device is so designed as to measure the ionization by means of ionization chambers, scintillation counters, and other analogous detectors. Such detectors can catch only a part of ionization because much energy is dissipated in heavy materials that are responsible for the development of nuclear and electromagnetic cascades. It is therefore necessary to know the conversion factor k that relate the measured ionization I to the primary energy:

$$E_{cal} = kI. \tag{3.7.9}$$

The value of k may be estimated by assuming that the ionization is proportional to the track length, but in the absence of a suitable method of calibration the absolute value thus obtained may be subject to a systematic error.

There are a number of modifications of methods 1–6 above, but none of them seems to give the primary energy of a single event to better than within a factor of 2.

The inaccuracy of the estimated primary energy may be best illustrated by applying one of the above methods to the breakup products of a high-energy heavy nucleus. Since the internal energy of the nucleon in a nucleus is small compared with the energy under consideration, all evaporation products from a single incident nucleus have essentially the same energy per nucleon. If the method of energy determination is appropriate, the energy of each product ought to be the same. As an example, the energies of breakup products produced by a heavy primary particle in emulsion have been estimated by the Castagnoli and E_{ch} methods; they are listed in Table 3.8 (Koshiba 63b). It is seen that E_{Cas} fluctuates more than E_{ch}. Part of the fluctuations may be due to those

Table 3.8 Estimated Interaction Energies (E_{Cas} and E_{ch}) of the Breakup Products of a Heavy Primary Particle[a]

Event[b]	E_{Cas} (TeV)	E_{ch} (TeV)	K_{ch} ($E_{ch}/\langle E_{Cas}\rangle$)
$(3+5+H)H$	2.32	0.26	—
$(11+44+H)H$	2.44	2.7	—
$(7+8)n$	0.044	0.037	0.024
$(0+3+H)H$	21.8	1.04	—
$(0+51+\mathrm{Li}+\mathrm{He})H$	5.04	5.53	—
$(8+6)p$	4.92	1.03	0.68
$(1+4)\mathrm{He}$	18.0	0.44	—
$(2+1)p$	0.105	0.003	0.002
$(3+45)n$	0.21	0.43	0.28
$(0+15)n$	1.43	0.51	0.33
$(0+44)\mathrm{Li}$	5.3	4.5	—
$(9+40)p$	0.98	0.52	0.34
$(2+4)n$	7.06	0.22	0.14
$(3+24)p$	0.92	0.90	0.59
$(16+23)n$	0.76	0.42	0.27
$(14+15)p$	0.139	0.35	0.22

[a] $\langle E_{Cas}\rangle = 1.53$ TeV.

[b] The events are classified in the customary notation: ($N_h + n_s +$ heavy fragments) incident particle. The symbol H stands for an unidentified heavy nucleus.

of inelasticity because some of the breakup products suffer from meson-producing collisions.

The observation of the breakup products of a heavy nucleus not only illustrates the accuracy of an energy-determination method but also provides a monoenergetic beam of nucleons with a well-defined energy. Although the energies estimated for individual particles spread over a considerable range, their average value may give an energy close to the true one. The average energy thus obtained is regarded as the energy per nucleon of each breakup product. In this sense breakup events are of great importance to the investigation of high-energy interactions.

Energy-determination methods apply mainly in the energy region between 10^{11} and 10^{15} eV. Above this region high-energy interactions are observed mostly through extensive air showers. The most reliable method of estimating the primary energy of an extensive air shower is equivalent to the calorimeter method; the amount of energy dissipated in the atmosphere and underground is estimated from experiments. However, experimental results available for this purpose are so limited that some particular quantities such as the number of electrons and the density of muons are referred to in estimating relative values of the primary energies. In some cases a specific model is assumed, which permits calculation of observable quantities for a given value of the primary energy and comparison of the calculated result with observations. Detailed description of such methods will be given in Chapter 5.

3.8 Collisions with Nuclei

One of the difficulties in energy determination arises from the fact that most available interactions involve complex nuclei. In this case both a survival nucleon and mesons produced by the first collision in a nucleus may undergo subsequent collisions in the same nucleus, thus resulting in the intranucleus cascade (Messel 52, 54). This seems to be the case for primary energies not greater than 10^{12} eV, since both multiplicities and N_h in nucleon-nucleus collisions are large in this energy region.

As energy increases the emission angles of particles become so small that only the fraction of nucleons that are contained in a tunnel "punched" by the incident particle are involved in successive collisions. Heavy tracks result from the evaporation process corresponding to the excitation energy produced by the presence of the tunnel (Heitler 53, McCusker 53, Harber-Schaim 54). More precisely, some of the secondary particles with relatively low energies are emitted with relatively large angles, and consequently the cross-sectional area of the tunnel increases as a nuclear cascade proceeds in the nucleus, the tunnel assuming a funnel shape (McCusker 53, Terashima 55). Apart from such detail the tunnel model consists essentially of a nuclear cascade developing in a cylindrical region. Hence the collisions in later generations are of low energy and the secondary particles produced therein are emitted with larger angles. The result of this would be that the primary energy estimated from the angular distribution, say by the Castagnoli method, should give the geometric mean of the energies of the particles that are responsible for individual collisions in the nucleus. The multiplicity observed is the sum of the multiplicities of the collisions (Fukuda 53).

In this model the particle most responsible for meson production in successive collisions may be regarded as the incident nucleon. Its descendant loses a fraction, K of its energy at each collision. If N collisions take place in a nucleus, the geometric mean of the energies of the survival nucleon is $(1 - K)^{(1+N)/2} E_0$, where E_0 is the energy of the incident nucleon. Therefore the primary energy in the tunnel model may be related to the Castagnoli energy by

$$E_{\text{tun}} = (1 - K)^{(1+N)/2} E_{\text{Cas}}. \tag{3.8.1}$$

Since the value of K is known to be about $\frac{1}{2}$, the primary energy estimated from the angular distribution by assuming the nucleon-nucleon collision may not be too far from the true value. This is is also expected from the fact that the major contribution to E_{Cas} or E_{ch} comes from secondary particles with small emission angles and arising mainly from the first collision.

The smallness of the difference would also hold in other methods of energy estimation, as can readily be seen from the discussion in the preceding section.

The validity of the successive-collision model may be questioned on the ground that the time interval between two collisions may be shorter than the time for which secondary particles are free. The former is $\tau_{\text{coll}} \simeq r/c\gamma_c$, where r is the internucleon distance, whereas the latter is $\tau_{\text{em}} \simeq \hbar/\varepsilon^*$, where ε^* is the average energy of emitted mesons in the center-of-mass system. If, for example, we take $r = 2\hbar/m_\pi c$ and $\varepsilon^* = 2m_\pi c^2 \gamma_c^{1/2}$, the ratio of these two times would be

$$\frac{\tau_{\text{coll}}}{\tau_{\text{em}}} \simeq \frac{4}{\gamma_c^{1/2}}. \tag{3.8.2}$$

Then the above situation would hold for $\gamma_c > 16$ or for incident energies greater than 500 GeV. If this were the case, no mesons should be emitted until the incident nucleon has left the target nucleus. Then N

nucleons in the nucleus would participate in the collision and form one target. Since the target has the mass NM, the Lorentz factor for the transformation to the center-of-mass system between an incident nucleon and N target nucleons is given as in (B.9) by

$$\gamma_{\text{comp}} \simeq \frac{2\sqrt{N}}{N+1}\gamma_c \quad \text{for} \quad \gamma_c \gg n. \tag{3.8.3}$$

Then the primary energy estimated from the angular distribution would give

$$E_{\text{comp}} \simeq 2\gamma_{\text{comp}}^2 \simeq \frac{4N}{(N+1)^2} E_0 , \tag{3.8.4}$$

where E_0 represents the true primary energy. This is called the composite-collision model (Cocconi 54) and would be valid if the collision time were smaller than τ_{em} given above. According to the fireball model, however, fireballs seem to be freed within a rather short time, although the emission of mesons from a fireball may take time. Again in this case the survival nucleon makes successive collisions as in the tunnel model. Therefore the whole collision seems to look like that in the tunnel model.

The difference may arise if a recoil nucleon or a slow fireball produced in a later stage of collision meets with the fireball produced in an earlier stage. This may be approximated by the fluid model (Belenkii 55, Amai 57), according to which the meson fluid propagates in a nucleus. This model is worked out in the coordinate system in which two colliding bodies are of equal velocity, thus being the same as the center-of-mass system of two colliding nucleons. The angular spread is governed by the lateral pressure and the thermal motion of the fluids. Since this is responsible for the transverse component of momentum, the primary energy estimated by E_{ch} in (3.7.5) may be valid as it stands. This also suggests that the methods of estimating the primary energy in nucleon-nucleon collisions could provide suitable means even in nucleon-nucleus collisions.

The above discussions indicate that gross features of the nucleon-nucleus collision are similar to those of the nucleon-nucleon collision if the primary energy is so high that collisions are concentrated in a narrow tunnel. It can, however, be said that the multiplicity of mesons is certainly higher in the nucleon-nucleus collision. The value of multiplicity expected again depends on the model assumed. However, the most distinctive feature, which is almost independent of the model, may be the number of target nucleons that participate in the collision, and this seems to be revealed by the number of gray tracks that are associated with a jet.

Most gray tracks are protons knocked out by collisions in a nucleus, but some of them are mesons, deutrons, and so on. The relative abundances of particles observed in gray tracks are shown in Table 3.9.

Table 3.9 Relative Abundances of Particles in Gray Tracks

Particle	Relative Abundances (%)	
	Reference 1	Reference 2
Proton	74	87 (Proton, Deuteron and triton)
Deuteron and triton	21	
Pion	4	6
Kaon	1	2
Unidentified	—	5

References:

1. C. Dahanayake et al., *Nuovo Cim.*, **1**, 888 (1955).
2. C. B. A. McClusker and L. S. Peak, *Nuovo Cim.*, **31**, 525 (1964).

According to these experimental results, three-fourths of gray tracks are due to protons.

If n_s shower particles and N_g gray tracks are observed, the number of charged secondary particles resulting from a single collision, n_s^0, may be estimated as follows. Since about equal numbers of protons and neutrons are contained in a target nucleus, the number of target nucleons participating in the collision with a nucleus is $2(\frac{3}{4})N_g$. In each collision n_s^0 particles are produced and, as a result, $2(\frac{3}{4})N_g n_s^0$ particles are created. This number should be equal to $n_s - \frac{1}{2}$, where the $\frac{1}{2}$ subtracted represents a survival proton. If no gray track is associated with a collision, the interpretation is that only one collision takes place with a knock-out neutron. Thus we are able to deduce the value of n_s^0 as

$$n_s^0 = \frac{n_s - \frac{1}{2}}{2(\frac{3}{4})N_g} \quad \text{for} \quad N_g \geq 1, \qquad (3.8.5)$$

$$n_s^0 = n_s - \tfrac{1}{2} \quad \text{for} \quad N_g = 0.$$

It should be noted that the value of n_s^0 given by (3.8.5) corresponds to the multiplicity at the incident energy given, for example, by (3.8.1).

Since the relations (3.8.5) as well as (3.8.1) hold only statistically, this is applicable when there are many events produced by essentially mono-energetic nucleons.

With the method described above applied to jets having a median E_{Cas} of 2.8 TeV (McCusker 64), the average number of gray tracks was found to be 1.6 ± 0.4 per jet, of which 1.2 ± 0.3 were expected from knock-out protons; hence the number of collisions would be 2.4 ± 0.6. This may be compared with the value calculated from the mean mass number A and radius $R = r_0 A^{1/3}$. The average path length through the nucleus is

$$\int_0^R 2(R^2 - x^2)^{1/2} \frac{2\pi x\, dx}{\pi R^2} = \tfrac{4}{3}R = \tfrac{4}{3} r_0 A^{1/3},$$

and for a composite medium, each component having the fractional area of cross section p_i, the average path length is $(\frac{4}{3}) r_0 \sum_i p_i A_i^{1/3}$. By dividing this by the total nucleon-nucleon cross section σ we obtain the average number of collisions,

$$n_c = \tfrac{4}{3} r_0 \sum_i \frac{p_i A_i^{1/3}}{\sigma} \tag{3.8.6}$$

With $r_0 = 1.2 \times 10^{-13}$ cm, $\sum_i p_i A^{1/3} = 3.8$ and $\sigma = 40$ mb, we have $n_c = 2.5$, in good agreement with the observed value of 2.4 ± 0.6. The observed values of n_s and N_g yield, from (3.8.5),

$$n_s^{\,0} = 10.5 \pm 1.0. \tag{3.8.7}$$

The dependence of the multiplicity of secondary particles on the number of collisions in a nucleus may be seen in Fig. 3.16 (Koshiba 63*a*). This is based on the secondary interactions with known incident energies estimated from E_{ch} in (3.7.5). They are divided into six groups, according to $E'_{\text{ch}} = 31.6$ to 316 GeV, $E'_{\text{ch}} = 316$ to 3160 GeV, where $E'_{\text{ch}} = E_{\text{ch}}/(1 + R + Q)$, and $N_h = 0, 1; 2 \leq N_h \leq 5; N_h \geq 6$. The selection of events is made for $n_s \geq 7$, and the multiplicity distribution for $n_s < 7$ is extrapolated by referring to a number of well-analyzed events. The values of $\langle n_s \rangle$ before correction, shown in Figure 3.16, indicate the average values of multiplicities with $n_s \geq 7$, whereas those after correction are the average ones of those including $n_s < 7$, whose fractions are also shown therein.

Figure 3.16 shows a trend that the multiplicity increases with N_h, the latter increasing with the number of collisions within a nucleus. Since N_h is larger than N_g, we cannot estimate the number of collisions from N_h. However, we are able to use an empirical method by assuming that no tertiary collisions take place in a nucleus. Events with $N_h = 0, 1$

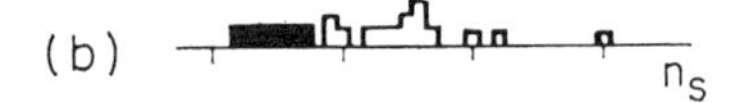

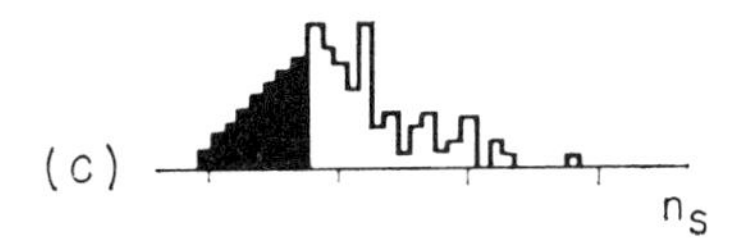

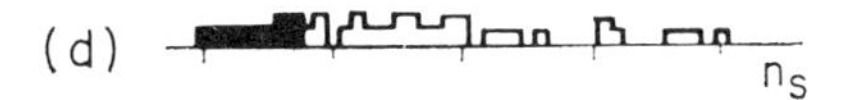

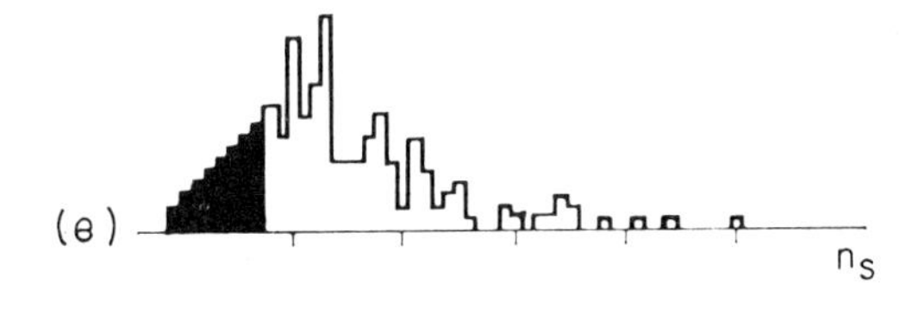

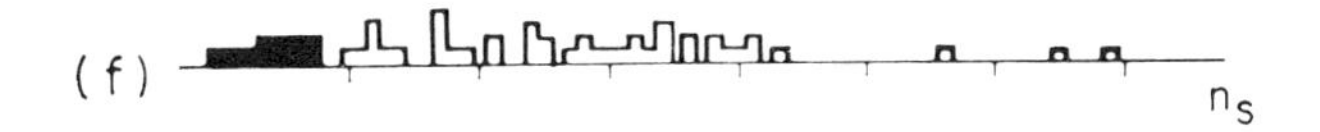

Fig. 3.16 Multiplicity n_s-N_h relation. For the corrected value of $\langle n_s \rangle$ the shaded parts are taken into account (Koshiba 63a).

	N_h	E_{ch} (GeV)	$\langle n_s \rangle$ Uncorrected	$\langle n_s \rangle$ Corrected	Percent of $n_s < 7$
(*a*)	0, 1	31.6 to 316	12.9	8.7	40
(*b*)	0, 1	316 to 3160	15.6	12.5	29
(*c*)	2 to 5	31.6 to 316	12.7	10.0	26
(*d*)	2 to 5	316 to 3160	20.8	16.4	21
(*e*)	≥ 6	31.6 to 316	16.4	13.8	17
(*f*)	≥ 6	316 to 3160	28.8	24.6	13

represent single collisions in which the multiplicity of charged particles is n_1. For $N_h \geq 2$ multiple collisions are assumed to take place; αn_1 secondary particles produced by the first collision, including both the survival nucleon and the mesons created, can take part in the secondary collisions, in which the contribution of neutral secondaries is included in α. In each secondary collision n_2 charged particles are produced; if the average energies of the primary and secondary particles are E_1 and E_2, respectively, and the multiplicity law is expressed as $E^{1/4}$, we have $n_2 = n_1(E_2/E_1)^{1/4}$. The value of E_1 is estimated by using the fraction of energy transferred to secondary charged particles, K_{ch}, as follows:

$$E_2 = \frac{K_{\text{ch}} E_1}{n_1}.$$

Accordingly, the value of n_s observed is expressed by

$$n_s = n_1 + \alpha n_1 n_2 = n_1 + \alpha n_1{}^2 \left(\frac{E_2}{E_1}\right)^{1/4} = n_1 + \alpha n_1^{7/4} K_{\text{ch}}^{1/4}. \quad (3.8.8)$$

The values of α are obtained for two groups of N_h as

$$\alpha = 2.7\% \quad \text{for} \quad N_h = 2 \text{ to } 5;$$
$$\alpha = 13.4\% \quad \text{for} \quad N_h > 5. \quad (3.8.9)$$

The smallness of α indicates that the survival nucleon contributes mainly to subsequent collisions, as is assumed in deriving (3.8.6).

The case of $N_h > 5$ and $316 < E'_{\text{ch}} < 3160$ corresponds to the one analyzed by McCusker and Peak (McCusker 64). In this case $(\frac{3}{2})N_g \simeq 2.4 \pm 0.6$ and $n_s{}^0 = 10.5 \pm 1.0$, approximately equal to $\alpha n_1 \simeq 1.9$ and $n_1 = 14$, respectively.

The mass-number dependence of the multiplicity is obtained also by averaging over N_h. This is expressed by

$$n_s \propto A^{0.14 \pm 0.03} \quad (3.8.10)$$

for incident protons of 20.5 and 27 GeV/c (Meyer 63). A similar weak mass-number dependence is also found at higher energies.

3.9 Inelasticity

To evaluate the primary energy the values of the inelasticity coefficients are often used, as in (3.7.7) and (3.7.8). Inelasticity is defined as the ratio of the energy imparted to all secondary particles to that brought by the incident particle. If the mass of the incident particle is M, the inelasticity coefficient in the laboratory system is expressed by

$$K_L \equiv \frac{\Sigma_i E_i}{(E_0 - M)}, \tag{3.9.1}$$

where E_i is the energy of a secondary particle that is produced. The subscript L, which indicates the laboratory system, is needed because the value of inelasticity so defined depends on the choice of the coordinate system, as shown in (B.80) to (B.83).

The value of K_L has been known from various experimental evidence, to deviate appreciably from unity. This has been observed particularly from the structure of extensive air showers and from the neutral-charged ratio of the nuclear active component in the atmosphere. This means that a considerable fraction of the incident energy is carried away by the survival particle; the energy of the latter is

$$E_0' - M = (1 - K_L)(E_0 - M). \tag{3.9.2}$$

The same situation should apply for the target particle, which is initially at rest and emerges with a recoil energy. In the mirror system, in which the incident particle is at rest and the target particle has energy E_0, the mirror inelasticity is given by (B.78) as

$$K_M \equiv \frac{\sum_i \gamma_L (E_i - \beta_L p_i \cos\theta_i)}{(\gamma_L - 1)M} \simeq \frac{\sum_i (E_i - p_i \cos\theta_i)}{M}. \tag{3.9.3}$$

The last expression, which holds in the extreme relativistic case, has an advantage, because the primary energy disappears. If the recoil particle is identified, and its energy, momentum, and emission angle are measured as E_0', p_0', and θ_0, respectively, we have

$$1 - K_M \simeq \frac{E_0' - p_0' \cos\theta_0}{M}. \tag{3.9.4}$$

The inelasticity in the center-of-mass system, is obtained from K_L and K_M as

$$K^* = \tfrac{1}{2}(K_L + K_M) = \frac{1}{2Z}(K_L - K_M) = \left(\frac{K_L K_M}{1 - Z^2}\right)^{1/2}, \tag{3.9.5}$$

where Z is the asymmetry factor defined in the center-of-mass system by

$$Z \equiv \frac{\sum_i p_i^* \cos\theta_i^*}{\sum_i E_i^*}. \tag{3.9.6}$$

A direct way of obtaining the inelasticity coefficient is to measure the energies of secondary components, which consist of charged particles,

γ-rays mainly from neutral pions, and other neutral particles. Hence the total inelasticity is decomposed into partial inelasticity coefficients:

$$K = K_{\text{ch}} + K_{\gamma} + K_n . \tag{3.9.7}$$

The partial inelasticity coefficients can be obtained rather directly, and the total inelasticity is obtained as their sum.

In Section 3.7 it is explained that K_{ch} is obtained from E_{ch} and the primary energy, the latter being equated to the mean Castagnoli energy of breakup nucleons from a heavy nucleus (Koshiba 63*a*). For individual events produced by breakup nucleons the values of

$$K_{\text{ch}} = \frac{E_{\text{ch}}}{\langle E_{\text{Cas}} \rangle} \tag{3.9.8}$$

are given for the example in Table 3.8. If individual Castagnoli energies are used in place of the average Castagnoli energy, the value of K_{ch} would be highly energy dependent. The values of K_{ch} in Table 3.8 obtained by (3.9.8) fluctuate little except for two events, whose incident energies could be much smaller than $\langle E_{\text{Cas}} \rangle$.

In the large emulsion stack, from which the data in Table 3.8 were obtained, cascade showers from individual interactions are observed. Comparing the energy of γ-rays estimated from the cascade showers arising from each interaction with E_{ch}, we obtain

$$\frac{K_{\text{ch}}}{K_{\gamma}} = \frac{E_{\text{ch}}}{\Sigma E_{\pi^0}} \simeq 2. \tag{3.9.9}$$

The value of K_n is known to be smaller than K_{γ}, and therefore we may conventionally assume that

$$K \simeq (\tfrac{3}{2}) K_{\text{ch}} . \tag{3.9.10}$$

If the primary energy is on the order of 100 GeV, the momenta of individual charged secondary particles can be measured by means of a magnet cloud chamber. Then we can obtain mirror inelasticity by applying (3.9.3). Furthermore, if the incident energy is measured, K_L can also be obtained. The comparison of K_L and K_M for charged secondaries has been made with a calorimeter (Guseva 62). The correlation between K_L and K_M shown in Fig. 3.17 demonstrates that about one-third of the cases are asymmetric with respect to the center-of-mass system. However, the average values of K_L and K_M are essentially the same, that is,

$$K_{\text{ch}} = \langle K_L \rangle = \langle K_M \rangle = 0.36 \pm 0.03. \tag{3.9.11}$$

It should be remarked that the primary energies determined by the

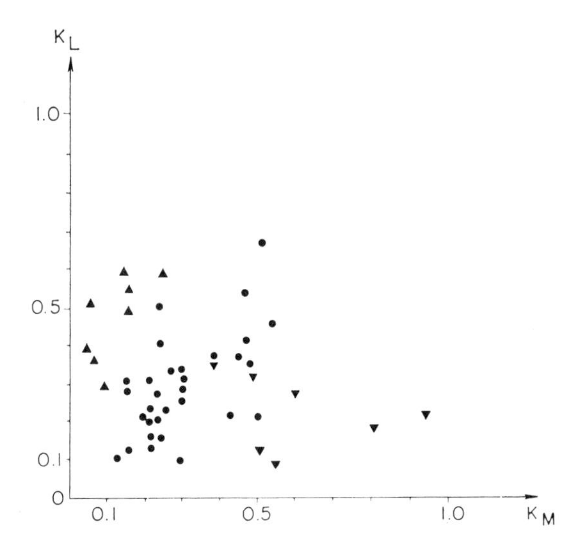

Fig. 3.17 Correlation between inelasticities in the laboratory and the mirror systems, K_L and K_M (Gusera 62).
▲—$K_L > K_M$; ▼—$K_L < K_M$; ●—$K_L \simeq K_M$.

calorimeter differ considerably from those obtained by the Castagnoli method. If the latter were used, the inelasticity would be found to decrease with increasing energy.

If the incident beam is monoenergetic, the measurement of the energies of survival particles through (3.9.2) provides a reliable means of obtaining inelasticity. In nuclear emulsions exposed to 25-GeV/c protons, about three-fourths of secondary particles emitted within 5° are positive and are identified with survival protons (Pal 63). Their momenta are measured by means of a pulsed magnetic field, and the average inelasticity for the collisions with emulsion nuclei is obtained.

These experimental results are listed in Table 3.10. There are many other data that are rather indirect and are regarded as less accurate. Most of them are not inconsistent with $K = \frac{1}{2}$.

However, there are indications that the inelasticity coefficient decreases with primary energy nearly in inverse proportion to E_0 (Hansen 60, Perkins, 60, Fujioka 61, Hasegawa 62). This could be due at least in part to the systematic error arising from the Castagnoli method of determining the primary energy, but it may imply a real energy dependence of K. In the air-shower energy region we see some indication that the inelasticity coefficient is slightly greater than $\frac{1}{2}$, as will be discussed in Chapter 5.

Table 3.10 Values of the Inelasticity Coefficients

Primary Energy	Target	K_{ch}	K	Reference
25 GeV	Emulsion		0.59 ± 0.07	1
~200 GeV	Lithium hydride	0.36 ± 0.03		2
1 to 20 TeV	Emulsion	0.31 ± 0.06	0.50 ± 0.07	3

References:

1. Y. Pal, A. K. Ray, and T. N. Rengarajan, *Nuovo Cim.*, **28**, 1177 (1963).
2. V. V. Guseva et al., *J. Phys. Soc. Japan*, **17**, *Suppl.* A–III, 373 (1962).
3. M. Koshiba et al., *Nuovo Cim.*, *Suppl.* **1**, No. 4, 1091 (1963).

Energy dependences, such as the above, could be due to the distribution of K. The probability of detection of extensive air showers is in favor of large inelasticity, because they are triggered by electrons whose number is roughly proportional to K_γ. Jets in emulsion are often found by cascade showers. If the primary energy is low, only a small fraction of events can give rise to cascade showers that are large enough to be detected. The correction for such a detection bias is possible only if the distribution of inelasticity is known.

For pions and nucleons with energies between 20 and 100 GeV colliding with carbon and copper the distribution of K_γ is approximately expressed as (Lal 65)

$$f(K_\gamma)\, dK_\gamma = A \exp\left[\frac{-B}{(1-K_\gamma)^2}\right] dK_\gamma . \tag{3.9.12}$$

This gives the distribution of $K_{ch} = 2K_\gamma$

$$f(K_{ch}) = A^2 \int_0^{K_{ch}} \exp\left[\frac{-B}{(1-x)^2}\right] \exp\left\{\frac{-B}{[1-(K_{ch}-x)]^2}\right\} dx. \tag{3.9.13}$$

This is in good agreement with that obtained at about 200 GeV, as shown in Fig. 3.18 (Guseva 62). The K-distribution at higher energies is not yet known, but it may also have an extended tail, as in the p_T-distribution.

3.10 Multiplicity-Energy Relation

The determination of multiplicity is subject to a number of difficulties. Firstly, most high-energy collisions take place with complex nuclei, although interest centers mainly on collisions with a single nucleon. As has

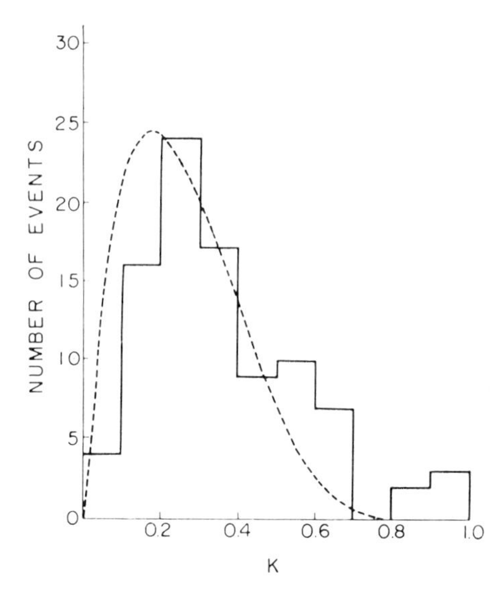

Fig. 3.18 Distribution of inelasticity. The solid curve represents $f(\frac{3}{2}K_{ch})$ given in (3.9.13).

been discussed in Section 3.8, however, the latter can be, in principle, distinguished from others by selecting events with $N_h = 0$ and 1. Secondly, variations in multiplicity are so large that many events must be collected. Further complexity arises from the fact that what is ordinarily observed is the multiplicity of either charged secondary particles or neutral pions. Moreover, the distribution of multiplicity is wider than Poisson distribution, and the dispersion of multiplicity is an important quantity for understanding the mechanism of multiple production. Thirdly, the selection of interactions is usually biased against events of small multiplicity as shown in Figure 3.16. Correction for this bias can be made only if the multiplicity distribution is known.

These points may result in systematic errors in the multiplicity-energy relation since these effects are energy dependent; for example, certain features of collision with a complex nucleus change with the primary energy. The nuclear-cascade model applies at relatively low energies, and the tunnel model at higher energies. The third point brings about a serious difficulty since in cosmic rays a monoenergetic beam of primary particles is not available. To correct for the selection bias we must *assume* a multiplicity-energy relation.

A more serious problem in obtaining the multiplicity-energy relation is the determination of the primary energy, as has been discussed in Section 3.7. Furthermore, this is not always independent of multiplicity;

for example, the primary energy determined by means of the E_{ch}-method with the average value of K_{ch} gives a strong correlation between energy and multiplicity.

In view of these difficulties we refer here to only those experimental results that are of relatively high reliability. They have two sources. The first is based on results obtained with monoenergetic beams. Most of these are taken from experiments with accelerators, but some are taken from cosmic-ray data on nuclear interactions produced by the breakup products of a heavy nucleus. Another set is based on the ICEF (International Collaboration of Emulsion Flight) data since many nuclear interactions are found by the same selection criteria and are analyzed in the same way.

Most of the experimental data collected in Table 3.11 were obtained with accelerator beams. Antiproton events were obtained with hydrogen bubble chambers, whereas π^- events were obtained either with hydrogen bubble chambers or with nuclear emulsions. Secondary particles in jets are believed to be mostly pions. Interactions produced by accelerator proton beams and by breakup nucleons of heavy nuclei are observed in emulsions, whereas those produced by cosmic-ray nuclear active particles are produced in lithium hydride and are observed with a calorimeter.

For interactions with emulsion nuclei events with $N_h \leq 1$ may be regarded as due to collisions with single nucleons. The multiplicities of such events are given in Table 3.11. Hence the multiplicities listed in Table 3.11 can be regarded as those for collisions with single nucleons. They are plotted in Figure 3.19 against the incident proton energy in the laboratory system; for the collisions of $\bar{p}$ and π the energies available in the center-of-mass system are taken to be equal to those for the p-N collisions. The multiplicity-energy relation thus obtained is approximately represented by a power law,

$$n_s \propto E^{1/4}. \tag{3.10.1}$$

However, caution should be used in interpreting the multiplicities at relatively low energies because they are concerned only with charged secondaries and are considerably affected by the conservation laws of charge and isospin; for example, the p-p collision would give more charged particles than the $\bar{p}$-p collision at the same energy available in the center-of-mass system. However, this effect does not seem to be too significant, according to the results in Table 3.11.

The number of events given above is so few that it is necessary to refer to other data of reasonable quality in order to improve the statistics. From the ICEF data we have selected 97 jets of $N_h \leq 3$ and $n_s \geq 6$, and for them we show n_s against E_{ch} in Fig. 3.20 (Kobayashi 64). The

Table 3.11 Average Multiplicity of Charged Relativistic Particles

Primary Particle	Energy	$\langle n_s \rangle$	Reference
$\bar{p}$	At rest	3.21 ± 0.12	N. Horowitz et al., *Phys. Rev.*, **115**, 472 (1959)
$\bar{p}$	1.05 GeV/c	3.3 ± 0.15	S. Goldhaber et al., *Phys. Rev.*, **121**, 1525 (1961)
$\bar{p}$	1.61 GeV/c	4.4 ± 0.1	J. Button et al., *Phys. Rev.*, **121**, 1788 (1961)
$\bar{p}$	1.99 GeV/c	4.6 ± 0.2	T. Elioff et al., *Phys. Rev., Letters*, **3**, 285 (1961)
π^-	4.5 GeV/c	2.0 ± 0.1	H. H. Aly et al., *Nuovo Cim.*, **28**, 1117 (1963)
π^-	11.4 GeV/c	3.8 ± 0.5	T. Ferbel and H. Taft, *Nuovo Cim.*, **28**, 1214 (1963)
π^-	16 GeV/c	4.1 ± 0.1	S. J. Goldsack et al., *Nuovo Cim.*, **23**, 94 (1962)
p	20.5 GeV/c	3.6 ± 0.3	H. Meyer et al., *Nuovo Cim.*, **28**, 1399 (1963)
p	26.7 GeV/c	4.4 ± 0.3	Y. K. Lim, *Nuovo Cim.*, **28**, 1214 (1963)
p	27 GeV/c	3.7 ± 0.3	H. Meyer et al., *Nuovo Cim.*, **28**, 1399 (1963)
N	~300 GeV	8 ± 1	N. A. Dobrotin and S. A. Slavatinsky, *Proc.* 1960 *High Energy Conf.*, p. 819
Secondary $\pi^{\pm}$	$0.47 {+0.09 \atop -0.06}$ TeV	12.7 ± 0.9	M. Koshiba et al., *Nuovo Cim. Suppl* **1**, 1091 (1963)
Breakup N	1.5 Tev	18 ± 1.5	M. Koshiba, *Proc. Int. Comf. on Cosmic Rays at Jaipur*, **5**, 293 (1963)
Breakup N	12.3 TeV	24 ± 4	M. Koshiba, ibid.

values of n_s are scattered over a wide range for a given value of E_{ch}, but they do depend significantly on the four-momentum transfer, as shown by the Japanese ICEF group (Fujioka 63). This will be discussed in the next section. As a result multiplicities are divided into four groups, according to the values of momentum transfer. In each group the multiplicity-energy relation is approximately represented by

$$n_s \propto \ln E, \tag{3.10.2}$$

with a coefficient that depends on momentum transfer.

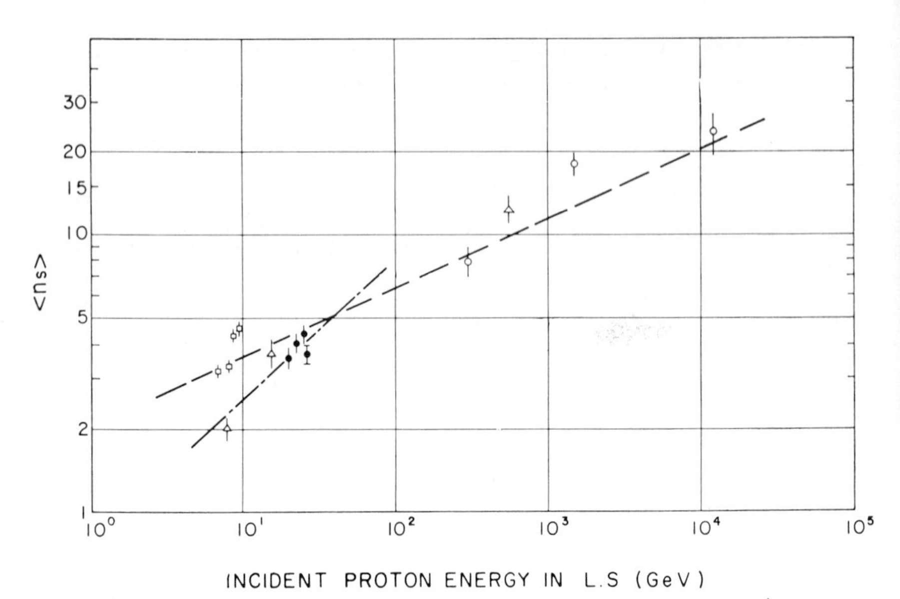

Fig. 3.19 Multiplicity-energy relation. Incident particles are nucleon (○), proton (●), pion (△), and antiproton (□). The laboratory energies for antiprotons are those corresponding to the *p-p* collisions of the same center-of-mass system energies available. The dashed line represents the $E^{1/4}$ law, and the dot-dashed line, the $E^{1/2}$ law.

The relation (3.10.2) interpreted in terms of the multi-fireball model in Section 3.16 results in that the dispersion of n_s, δn_s, increases with energy and consequently with $\langle n_s \rangle$. This is shown in Fig. 3.21, according to which the dispersion is significantly greater than that expected from the Poisson distribution.

At energies near and below 100 GeV a different multiplicity-energy relation has been found (Kaneko 58). This is represented by

$$n_s \propto KE^{1/2} \tag{3.10.3}$$

and is essentially confirmed by a calorimeter experiment, (Guseva 62).

3.11 Energy-Momentum Transfer

The dependence of multiplicity on energy-momentum or four-momentum transfer demonstrates the importance of the latter quantity in high-energy interactions. This quantity is defined as the energy-momentum transfer between two groups of outgoing particles. This depends on

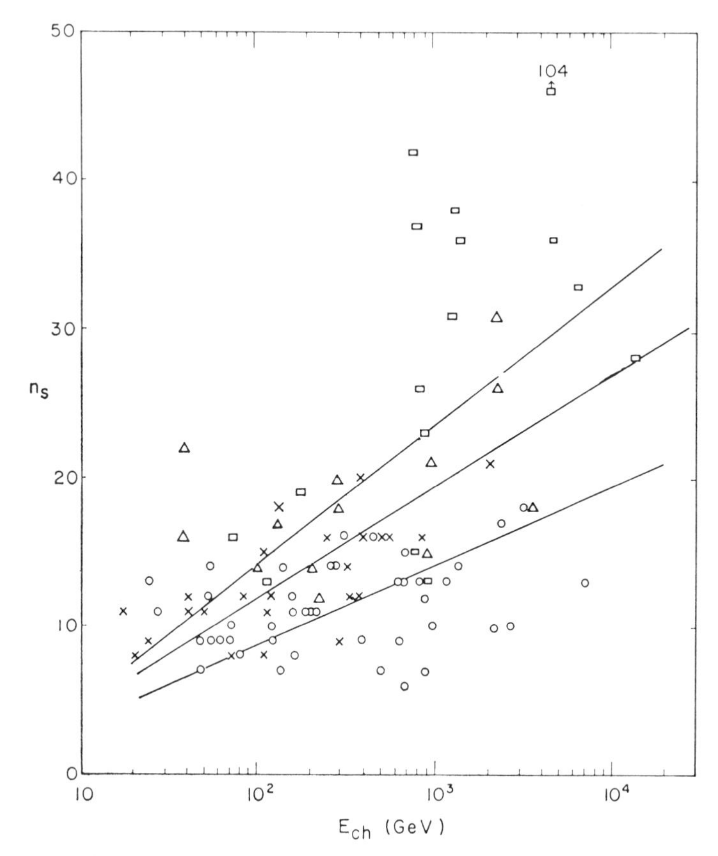

Fig. 3.20 Multiplicity-energy relation parametrized by four-momentum transfer Δ (Kobayashi 64).

○—$\Delta/M < 1.0$; ×—$1.0 \leq \Delta/M < 1.5$,
△—$1.5 \leq \Delta/M < 2.0$; □—$2.0 \leq \Delta/M$.

the manner in which the outgoing particles are divided into two groups, but it does not depend on the coordinate system. In the two-body collision this has a simple expression as given in (B.45), but in multiple production a simple expression is obtained only in the extremely relativistic case.

Let the mass, energy, and momentum of the incident particle be M_A, E_A, and $\mathbf{p}_A$, respectively, and those of the survival particle be represented by the subscript A'. The corresponding quantities for the target and the recoil particles are distinguished by the subscripts B and B', respectively. The secondary paricles produced are divided into f and b groups, respectively, representing those emitted forward and backward in the center-of-mass system. Kinematically, the division into these two

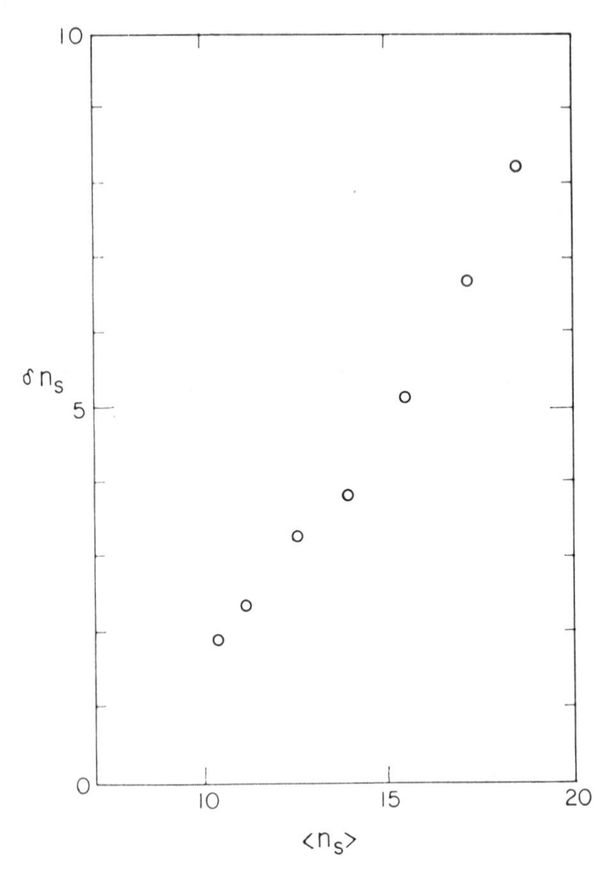

Fig. 3.21 Fluctuations of multiplicities, the dispersion of multiplicity versus the mean multiplicity (Kobayashi).

groups is arbitrary, but practically it implies a physical model. In the f-group the mass, energy, momentum, and emission angle of a secondary particle are represented by the subscript i, whereas in the b-group they are distinguished by the subscript j. The longitudinal and transverse components of momentum are represented by the subscripts L and T, respectively.

For the various groups of secondary particles the following equations hold:

$$E_f = \sum_i E_i = \sum_i (M_i^2 + \mathbf{p}_{iL}^2 + \mathbf{p}_{iT}^2)^{1/2}, \qquad \mathbf{p}_{fL} = \sum_i \mathbf{p}_{iL}, \; \mathbf{p}_{iT} = \mathbf{p}_{iL} \tan \theta_i,$$

$$E_b = \sum_j E_j = \sum_j (M_j^2 + \mathbf{p}_{jL}^2 + \mathbf{p}_{jT}^2)^{1/2}, \qquad \mathbf{p}_{bL} = \sum_i \mathbf{p}_{jL}, \; \mathbf{p}_{jT} = \mathbf{p}_{jL} \tan \theta_j. \tag{3.11.1}$$

The squared energy-momentum transfer is now defined as

$$\begin{aligned}\Delta^2 &= (\mathbf{p}_A - \mathbf{p}_{A'} - \mathbf{p}_f)^2 - (E_A - E_{A'} - E_f)^2 \\ &= (\mathbf{p}_B - \mathbf{p}_{B'} - \mathbf{p}_b)^2 - (E_B - E_{B'} - E_b)^2,\end{aligned} \tag{3.11.2}$$

where the second equality is due to energy-momentum conservation. This is a Lorentz-invariant quantity and can be evaluated in any coordinate system. Here we give the expression Δ^2 in the laboratory system, so that it can be evaluated by using directly observable quantities alone.

As shown in Appendix B, in expressions (B.101) through (B.106), Δ^2 is expressed as (Hasegawa 62)

$$\Delta^2 = \Delta_L{}^2 + \Delta_T{}^2, \tag{3.11.3}$$

where

$$\begin{aligned}\Delta_L{}^2 = &\left[\sum_i \mathbf{p}_{iT}\left(1 + \frac{M_i{}^2}{\mathbf{p}_{iT}^2}\right) \tan\theta_i + \frac{M_{A'}^2 + \mathbf{p}_{A'T}^2 - (1-K_L)M_A{}^2}{(1-K_L)E_A}\right. \\ &\times \left\{\sum_j \mathbf{p}_{jT}\left[\cot\theta_j + \frac{1}{4}\left(1 + \frac{M_j{}^2}{P_{jT}^2}\right)\tan\theta_j\right] + (E_{B'} + \mathbf{p}_{B'L} - M_B)\right\}\end{aligned} \tag{3.11.4}$$

in the relativistic approximation and

$$\Delta_T{}^2 = (\mathbf{p}_{A'T} + \mathbf{p}_{fT})(\mathbf{p}_{B'T} + \mathbf{p}_{bT}). \tag{3.11.5}$$

By taking account of the fact that the transverse momenta of outgoing particles are nearly the same and are greater than the pion mass, and that most of the secondary particles are pions, the expression (3.11.4) is further reduced to the following:

$$\Delta_L{}^2 \simeq \langle \mathbf{p}_T{}^2 \rangle \sum_i \tan\theta_i \sum_j \cot\theta_j. \tag{3.11.6}$$

The expression (3.11.5) may be evaluated in the following way. If in each group the transverse momenta of secondary particles are randomly distributed, we have

$$\mathbf{p}_{fT} \simeq \sqrt{n_f \langle \mathbf{p}_T{}^2 \rangle}, \qquad \mathbf{p}_{bT} \simeq \sqrt{n_b \langle \mathbf{p}_T{}^2 \rangle}, \tag{3.11.7}$$

where n_f and n_b are the numbers of secondary particles in the respective groups. Hence

$$\Delta_T{}^2 \simeq \mathbf{p}_{A'T}^2 + \sqrt{n_f n_b}\, \langle \mathbf{p}_T{}^2 \rangle. \tag{3.11.8}$$

Comparing $\Delta_L{}^2$ in (3.11.6) with $\Delta_T{}^2$ in (3.11.8), we find that the former is greater than the latter, if $\mathbf{p}_{A'T}$ is as small as the value of $\mathbf{p}_T$ for secondary particles. Then the value of Δ^2 is nearly equal to that of $\Delta_L{}^2$. The value of Δ_L^2 in (3.11.6) can be obtained by measuring the emission angles of secondary particles, but it depends on the manner of division into two groups. This is demonstrated by varying the division in Fig. 3.22. The distribution of Δ_L^2 thus obtained has a minimum for nearly

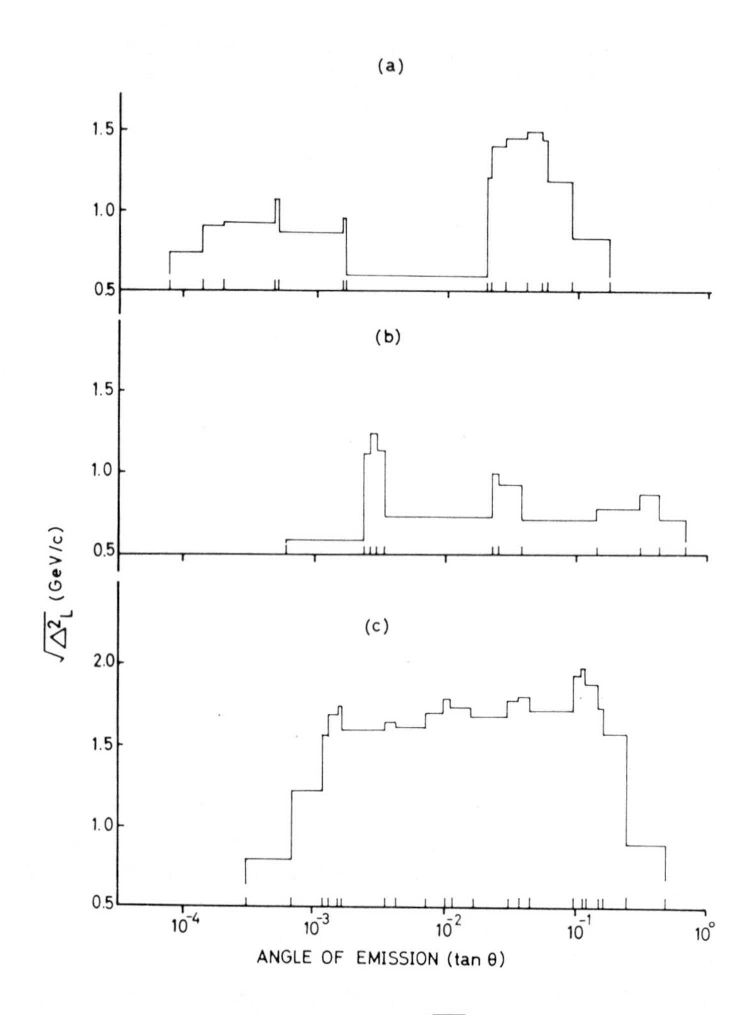

Fig. 3.22 Examples of $\sqrt{\Delta_L{}^2}$. The value of $\sqrt{\Delta_L{}^2}$ is given for an assumed manner of division into two groups. The following three typical events are analyzed by K. Yokoi, assuming $\langle p_T \rangle = 400$ MeV/c:
(*a*) $2 + 15p$. R. G. Glasser, D. M. Haskin, and M. Schein, *Phys. Rev.*, **99**, 1555 (1955);
(*b*) $1 + 12p$. A. G. Barkow et al., *Phys. Rev.*, **122**, 617 (1961);
(*c*) $0 + 22p$. B. Edwards et al., *Phil. Mag.*, **3**, 237 (1958).

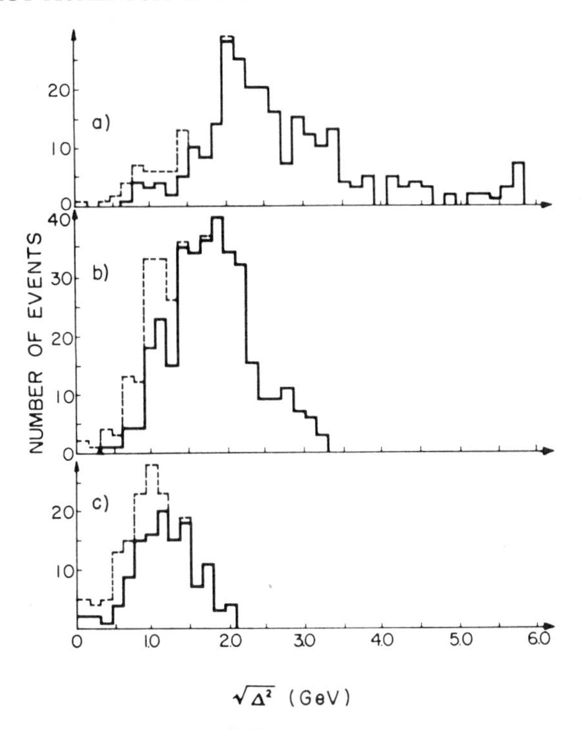

Fig. 3.23 The n_s-dependence of $\sqrt{\Delta^2}$ for all jets (Fujioka 63):
(*a*) $n_s > 17$, 14 jets, $\langle\gamma_c\rangle = 92.9$, $\langle n_s\rangle = 21.6$;
(*b*) $16 > n_s > 10$, 32 jets, $\langle\gamma_c\rangle = 45.1$, $\langle n_s\rangle = 13.6$;
(*c*) $n_s < 10$, 24 jets, $\langle\gamma_c\rangle = 135.3$, $\langle n_s\rangle = 8.3$.

equal division. This implies that the division into the forward and backward groups has a physical meaning; these two respectively form lumps of secondary particles called fireballs.

The value of $\sqrt{\langle\Delta^2\rangle}$ defined in this way is distributed around 2 GeV (Hasegawa 62), in sharp contrast with the small value obtained in the GeV region. The increase of $\sqrt{\langle\Delta^2\rangle}$ seems to become appreciable as energy increases above 10 GeV. In fact, the nuclear interactions induced by 30-GeV protons in nuclear emulsion give $\sqrt{\langle\Delta^2\rangle}$ close to 1 GeV.† Energy dependence becomes weak as energy increases but is still appreciable at about 1 TeV. According to a careful examination by Fujioka et al. (Fujioka 63), however, energy dependence takes place through dependence on multiplicity. In other words, the value of $\sqrt{\langle\Delta^2\rangle}$ depends directly on multiplicity, as shown in Fig. 3.23 as well

† M. Teranaka, personal communication, 1963.

as in Fig. 3.20. This suggests the great importance of energy-momentum transfer in multiple production.

There are many other sets of energy-momentum transfers that are Lorentz invariant. One of them is the energy-momentum transfer between the incident and the survival particles, and also that between the target and the recoil particles. They are, given in (B.91) and (B.89) respectively, by

$$\Delta_A{}^2 = (\mathbf{p}_A - \mathbf{p}_{A'})^2 - (E_A - E_{A'})^2 \simeq \frac{1}{1-K_L}(K_L{}^2 M_A{}^2 + \mathbf{p}_{A'T}^2) \tag{3.11.9}$$

$$\Delta_B{}^2 = (\mathbf{p}_B - \mathbf{p}_{B'})^2 - (E_B - E_{B'})^2 \simeq \frac{1}{1-K_M}(K_M{}^2 M_B{}^2 + \mathbf{p}_{B'T}^2). \tag{3.11.10}$$

The latter is directly expressed by means of the recoil kinetic energy as

$$\Delta_B{}^2 = \sqrt{2\, M_B(E_{B'} - M_B)}. \tag{3.11.11}$$

Since K_L and K_M may be as large as $\frac{1}{2}$, the value of Δ_A or Δ_B is on the order of 1 GeV for nucleons. This was found by Niu (Niu 58) by observing the recoil energy. If the transverse momentum of the survival nucleon were greater than 1 GeV/c, the value of Δ_A or Δ_B would be larger. If this were the case, $\Delta_T{}^2$ given in (3.11.8) would be as large as $\Delta_L{}^2$.

A few experimental data are available for Δ_A and Δ_B, but they are not considered to be conclusive, mainly because of the difficulty in identifying the survival and the recoil particles. For 25-GeV protons in nuclear emulsion both inelasticity and transverse momentum are measured (Pal 63), as discussed in Sections 3.4 and 3.9. The value of Δ_A that is obtained from them is about 0.9 GeV. For cosmic-ray events the distribution of recoil momenta, which are compared with Δ_B, is shown in Fig. 3.24.

3.12 Statistical Properties of Multiple Production

The very fact that many secondary particles are generated by high-energy collisions seems to suggest that the general features of multiple production are best understood by a statistical approach. This is analogous to nuclear reactions in the sense that a significant part of the energy is dissipated into many modes, exciting the nucleus as a whole,

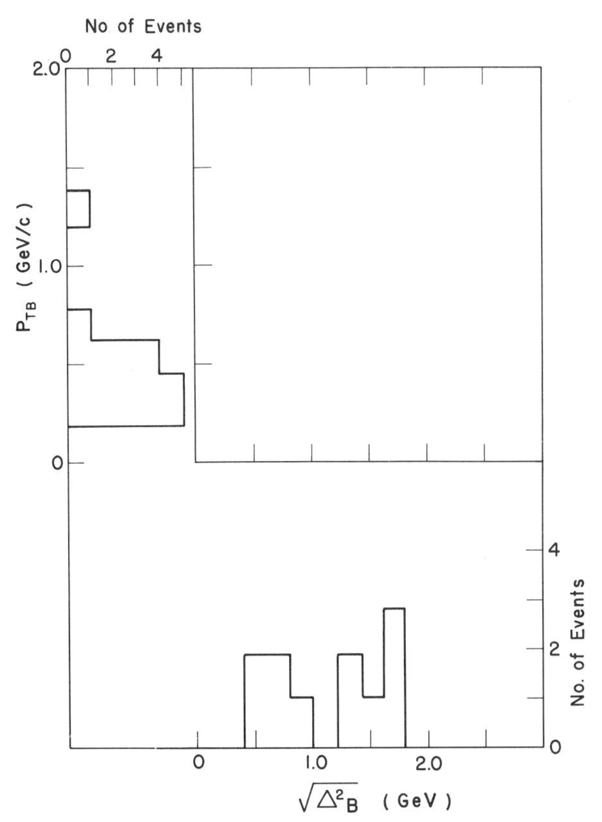

Fig. 3.24 Distributions of transverse momentum and momentum transfer of recoil protons. Twenty-two jets with $N_h \leq 3$ and 10^{11} eV $< E_0 \lesssim 10^{12}$ eV obtained by R. W. Huggett et al., *Proc. Int. Conf. on Cosmic Rays at Jaipur*, **5**, 3 (1963); $\langle p_{TB} \rangle = 0.44$ GeV/c, $\langle \sqrt{\Delta_B{}^2} \rangle = 1.1$ GeV.

and subsequently the nucleus emits particles through the evaporation process. In the excitation and evaporation processes detailed properties of interactions between individual particles are not revealed—only their statistical average. The importance of statistical considerations was first explicitly emphasized by Fermi (Fermi 50) in interpreting the multiplicity-energy relation. The statistical theory of multiple production was later elaborated in various respects.

In order for the statistical description to hold there are many possible final states $|f\rangle$ that may be formed from an initial state $|i\rangle$; for example, the resultant momentum of many outgoing particles can be formed in many ways by combinations of momenta of individual particles. The probability that an ensemble of all these final states, designated as F, be formed is expressed by

$$P(F) = \sum_f |\langle f|S|i\rangle|^2, \tag{3.12.1}$$

where $\langle f|S|i\rangle$ is the S-matrix, which represents the probability amplitude of forming a final state $|f\rangle$, and the summation extends over all possible states of F.

This involves integrals over the momenta of outgoings particles under the restriction of energy-momentum conservation. In order to express this explicitly $\delta(E_0 - \Sigma\, E_i)\delta(\mathbf{p}_0 - \Sigma\, \mathbf{p}_i)$ is factored out of the squared S-matrix and a part of Σ_f is expressed as $\int d\mathbf{p}_1, \cdots, d\mathbf{p}_n$. This part gives us the final-state density for emitting n particles of masses $m_1, \cdots, m_n$:

$$\rho_n(E_0, \mathbf{p}_0, m_1, \cdots, m_n) = \int \delta(E_0 - \Sigma\, E_i)(\mathbf{p}_0 - \Sigma\, \mathbf{p}_i)\, d\mathbf{p}_1, \cdots, d\mathbf{p}_n. \tag{3.12.2}$$

This is the density of states in a volume $(V/(2\pi)^3)^{n-1}$ and in the energy interval ΔE.

In strong interactions the conservation of isospin has to be taken into account. For a given isospin I and its z-component I_z, there are many possible sets of outgoing particles. The number of such possible sets $Q(I, I_z)$ is also factored out. This number is obtained by counting the number of sets formed by interchanging like particles. Since these sets are indistinguishable from each other, we have to divide Q by the number overcounted, which is equal to the permutation of all like particles $n_j!$. If a particle has spin S, there are $2S+1$ states that are degenerate. Hence the spin weights can also be factored out.

We finally reach the expression

$$P_n(F) = \frac{\Pi(2S_i + 1)}{\Pi n_j!}\, Q(I, I_z) \left[\frac{V}{2\pi^3}\right]^{n-1} \Delta E$$
$$\times \int |\langle \mathbf{p}_1, \cdots, \mathbf{p}_n|\mathbf{S}|i\rangle|^2\, \delta(E_0 - \Sigma\, E_i)\, \delta(\mathbf{p}_0 - \Sigma\, \mathbf{p}_i)\, d\mathbf{p}_1, \cdots, d\mathbf{p}_n, \tag{3.12.3}$$

where $\mathbf{S}$ is the dimensionless S-matrix that satisfies the conservation laws. If the integration over $\mathbf{p}_i$ is not performed, we obtain the energy and angular distribution of the ith particle as

$$P_n(F : \mathbf{p}_i)\, d\mathbf{p}_i =$$
$$C\, d\mathbf{p}_i \int |\langle \mathbf{p}_1, \cdots, \mathbf{p}_n\, |\mathbf{S}|i\rangle|^2\, \delta(E_0 - \Sigma\, E_i)\, \delta(\mathbf{p}_0 - \Sigma\, \mathbf{p}_i)\Pi'\, d\mathbf{p}_i \tag{3.12.4}$$

where Π' is the product without the ith factor and C represents the factors multiplied by the integral in (3.12.3).

It is practically impossible to evaluate the integral in (3.12.3) or (3.12.4), because the squared S-matrix has a complicated function. This can be evaluated approximately by applying the mean-value theorem as

$$P_n(F) = C|\langle \bar{p}_1, \cdots, \bar{p}_n|\mathbf{S}|i\rangle|^2 \, \rho_n(E_0, \mathbf{p}_0, m_1, \cdots, m_n). \quad (3.12.5)$$

The average squared matrix $|\langle \bar{p}_1, \cdots, \bar{p}_n|\mathbf{S}|i\rangle|^2$ is equal to the probability of finding n particles of momenta $\bar{p}_1, \cdots, \bar{p}_n$ in the interaction volume of two initial particles because of the reciprocal property of the S-matrix. Since the $n-1$ momenta are independent, the probability may be expressed as

$$|\langle \bar{p}_1, \cdots, \bar{p}_n|\mathbf{S}|\, i\rangle|^2 = (\Omega/V)^{n-1}, \quad (3.12.6)$$

where Ω is the interaction volume. If different kinds of particles are produced, the interaction volumes of the various kinds of particles may be different, depending on the mass and the coupling constant of each kind of particle. Hence (3.12.6) may be written better as

$$|\langle \bar{p}_1, \cdots, \bar{p}_n\, |\, \mathbf{S}\, |i\rangle|^2 = \prod_j \left(\frac{\Omega_j}{V}\right)^{n_j}, \qquad \Sigma\, n_j = n - 1. \quad (3.12.7)$$

In the limit of strong interactions, the interaction volume may be regarded as equal to the volume of a sphere with radius equal to the Compton wavelength but flattened by the Lorentz factor γ for two colliding particles in the center-of-mass system;

$$\Omega_j = \frac{4\pi}{3}\frac{1}{m_j{}^3\gamma} \equiv \frac{\Omega_j{}^{(0)}}{\gamma} \quad (3.12.8)$$

As the squared S-matrix is given by (3.12.6) or (3.12.7), we have to evaluate only the final-state density. The evaluation is quite involved if the conservation laws of momentum and angular momentum are taken into account. However, if the energy of each secondary particle is much smaller than the energy available, we obtain an approximate expression in the center-of-mass system as

$$\rho_n(E_0^*, 0) \simeq \left(\frac{\pi}{2}\right)^{n-1} E_0^{*\,3n-4} \frac{(4n-4)!\,(2n-1)}{(3n-4)!\,[(2n-1)!]^2} \quad (3.12.9a)$$

for the extremely relativistic case, and

$$\rho_n(W_0^*, 0) \simeq \frac{(2\pi)^{(3/2)(n-1)}}{[\frac{3}{2}(n-1)-1]!}\left(\frac{\Pi_{i=1}^n m_i}{\sum_{i=1}^n m_i}\right)^{3/2} W_0^{*(3/2)(n-1)-1} \quad (3.12.9b)$$

for the nonrelativistic case, where W_0^* is the kinetic energy available in the center-of-mass system.

In a coordinate system in which the total momentum does not vanish they are given by

$$\rho_n(E_0, \mathbf{p}_0) = \left(\frac{\pi}{2}\right)^{n-1} \frac{2n-1}{[(2n-1)!]^2} \frac{d^n}{dE_0{}^n} (E_0{}^2 - p_0{}^2)^{2n-2} \qquad (3.12.10a)$$

and

$$\rho_n(W_0, \mathbf{p}_0) = (2\pi)^{(3/2)(n-1)} \left(\frac{\Pi_{i=1}^n m_i}{\sum_{i=1}^n m_i}\right)^{3/2} \frac{(W_0 - (p_0{}^2/2)\sum_i m_i)^{(3/2)(n-1)-1}}{[\frac{3}{2}(n-1)-1]!}, \qquad (3.12.10b)$$

respectively.

If momentum conservation is neglected, the phase-space integral without $\delta(\mathbf{p}_0 - \Sigma_i \mathbf{p}_i)$ in (3.12.2) gives

$$\bar{\rho}_n(E_0) = \frac{(8\pi)^n E_0^{3n-1}}{(3n-1)!} \qquad (3.12.11a)$$

for the extremely relativistic case and

$$\bar{\rho}_n(W_0) = \frac{(2\pi)^{(3/2)n}\, \Pi_{i=1}^n m_i^{3/2} W_0^{(3/2)n-1}}{(\frac{3}{2}n-1)!}. \qquad (3.12.11b)$$

These are the well-known expressions derived by Fermi (Fermi 50).

The relation between (3.12.10) and (3.12.11) is approximated by assuming that the direction of momentum is randomly distributed (Ericson 61). Then the probability that the total momentum is p_0 is represented by a Gaussian function with dispersion

$$\sigma = \sqrt{\tfrac{1}{3}\sum_i p_i{}^2} \qquad (3.12.12)$$

as

$$(2\pi\sigma^2)^{-3/2} \exp\left[\frac{-p_0{}^2}{2\sigma^2}\right]. \qquad (3.12.13)$$

By integrating over angles in (3.12.2) therefore we obtain

$$\rho_n(E_0, p_0) \simeq (4\pi)^n \int \prod_i p_i{}^2\, dp_i (2\pi\sigma^2)^{-3/2} \exp\left[\frac{-p_0{}^2}{2\sigma^2}\right] \delta\left(E_0 - \sum_i E_i\right). \qquad (3.12.14)$$

The probability distribution depends on p_i only through σ, and this is nearly constant over the range of combinations of $\mathbf{p}_i$ that contribute to the integral. Hence (3.12.14) can be evaluated by the mean-value theorem as

$$\rho_n(E_0, p_0) = \left\langle (2\pi\sigma^2)^{-3/2} \exp\left[\frac{-p_0{}^2}{2\sigma^2}\right] \right\rangle \bar{\rho}_n(E_0)$$

$$\simeq (2\pi\langle\sigma^2\rangle)^{-3/2} \exp\left[\frac{-p_0{}^2}{2\langle\sigma^2\rangle}\right] \bar{\rho}_n(E_0), \qquad (3.12.15)$$

where

$$\langle\sigma^2\rangle = \tfrac{1}{3}\left\langle \sum_i p_i{}^2 \right\rangle = \frac{\sum_i (4\pi)^n \int \prod_i p_j{}^2\, dp_j\, p_i{}^2 \delta(E_0 - \sum_i E_i)}{\bar{\rho}_n(E_0)}$$

$$\simeq \tfrac{4}{3} \frac{E_0{}^2}{3n+1} \quad \text{or} \quad \tfrac{2}{3} \frac{\sum_i m_i}{n} W_0. \qquad (3.12.16)$$

It is shown that the approximation (3.12.15) is correct to the order of $1/n$.

Under this approximation the angular correlation of two secondary particles is given by

$$\left\langle \exp\left[\frac{-(p_i{}^2 + p_j{}^2)}{2\sigma^2}\right] \right\rangle \times \exp\left[\frac{-\langle p_i\rangle\langle p_j\rangle \cos\theta}{\sigma^2}\right]. \qquad (3.12.17)$$

Here σ^2 has to be taken for $(n-2)$ particles with total energy $(n-2)E_0/n$. Thus σ^2 in (3.12.17) is approximately estimated as

$$\sigma^2 \simeq \tfrac{4}{9} \frac{n-2}{n^2} E_0{}^2.$$

Now we give the multiplicity-energy relation in the simplest case where all the secondary particles are indistinguishable and are of extremely relativistic energies. Because of (3.12.7), (3.12.8), and (3.12.11*a*), the probability of having n particles is given as

$$P_n \simeq \frac{(2S+1)^n}{\pi^{2n}} Q(I, I_z)\Omega_0{}^n \frac{E_0^{*\,3n-1}}{n!(3n-1)!\gamma^{n-1}}. \qquad (3.12.18)$$

Since $\gamma = E_0^*/2M$, where M is the nucleon mass, the value of P_n has the maximum at

$$\bar{n} \simeq \left[\frac{(2S+1)2M\Omega_0}{3^3\pi^2}\right]^{1/4} E_0^{*1/2} \qquad (3.12.19)$$

for a sufficiently large value of n. This is the well-known multiplicity law proposed by Fermi (Fermi 50).

Next we evaluate the function $Q(I, I_z)$ in the same manner in which we obtained (3.12.13). If the random coupling of n classical vectors $\mathbf{I}_i$ gives a resultant $\mathbf{I}$, the probability of having $\mathbf{I}$ is obtained from (3.12.13) by replacing $p_0{}^2$ by $\mathbf{I}^2$ and $\mathbf{p}_i{}^2$ by $\mathbf{I}_i{}^2$. For isospin the z-component of $\mathbf{I}$, I_z, is also conserved. Therefore the probability distribution is obtained as

$$q(I, I_z)\, dI \simeq \pi^{-1/2}(2\sigma^2)^{-1/2} \exp\left[\frac{-I^2}{2\sigma^2}\right] \cdot 2I\, dI, \qquad (3.12.20)$$

with

$$\sigma^2 = \tfrac{1}{3} \sum_i I_i{}^2.$$

In quantum mechanics I^2 is replaced by $I(I+1)$.

If the charge of each secondary particle is specified, $I_z = \sum_i I_{iz}$ is well defined. Then instead of (3.12.20) we have

$$q(I, I_z) \simeq (2\sigma^2)^{-1}(2I+1) \exp\left\{\frac{-[I(I+1) - I_z{}^2]}{2\sigma^2}\right\}, \; I \geq |I_z|, \qquad (3.12.21)$$

with

$$\sigma^2 = \tfrac{1}{2} \sum_i [I_i(I_i+1) - I_{iz}]^2.$$

The conservation of angular momentum can be taken into account in the same spirit (Ericson 61), and the result is again in essential agreement with the results obtained by quantum-mechanical treatments (Koba 60, 61; Cerulus 61), which require involved computational work. Since angular momentum depends on the position vector $\mathbf{r}_i$ through $\mathbf{J}_i = \mathbf{r}_i \times \mathbf{p}_i$, the configurational-space integral cannot be separated from the momentum-space integral. Hence the phase-space integral under the restrictions of the total momentum $\mathbf{p}_0 = 0$ and of the total angular momentum $\mathbf{J}$ is expressed as

$$R_n(E_0, \mathbf{J}) = \int \cdots \int \Pi\, d\mathbf{p}_i\, d\mathbf{r}_i\, \delta\Big(\sum_i \mathbf{p}_i\Big)\delta\Big(E_0 - \sum_i E_i\Big)\delta\Big(\mathbf{J} - \sum_i \mathbf{r}_i \times \mathbf{p}_i\Big). \qquad (3.12.22)$$

By writing the expression above we are considering the angular momentum as a classical quantity. Accordingly the intrinsic spin is taken into account only as the statistical weight factor, and the orbital angular momentum alone is considered. Moreover, the interference between different partial waves is neglected, so that the forward-backward asymmetry does not appear.

Again, both momentum and angular momentum are assumed to be randomly oriented. Then the same procedure applies also in this case, and we obtain

$$R_n(E_0, \mathbf{J}) \simeq \left(\frac{2\pi}{3}\left\langle \sum_i p_i^2 \right\rangle\right)^{-3/2} \left(\frac{4\pi}{9}\langle r^2\rangle \left\langle \sum_i p_i^2 \right\rangle\right)^{-3/2}$$
$$\times \exp\left[\frac{-J^2}{\frac{4}{9}\langle r^2\rangle\langle\sum_i p_i^2\rangle}\right]\Omega^n \bar{\rho}_n(E_0), \qquad (3.12.23)$$

where Ω is the interaction volume.

Since the interaction volume is subject to the Lorentz contraction in the direction of flight of colliding particles, which is taken as the z-axis, it is convenient to divide $\mathbf{r}_i$ and $\mathbf{p}_i$ into their z-components and transverse components, the latter being $\mathbf{r}_{iT}$ and $\mathbf{p}_{iT}$, respectively. Correspondingly the total angular momentum is divided into

$$\mathbf{J}_z = \sum_i \mathbf{r}_{iT} \times \mathbf{p}_{iT}, \; \mathbf{J}_T = \sum_i (\mathbf{r}_{iT} \times \mathbf{p}_{iz} + \mathbf{r}_{iz} \times \mathbf{p}_{iT}).$$

If the contraction takes place by a factor α, we obtain

$$\langle r_{iz}^2\rangle = \alpha^2\langle r_{oz}^2\rangle = \tfrac{1}{3}\alpha^2\langle r_o^2\rangle,$$

where $\langle r_o^2\rangle$ is the mean-square radius of the uncontracted interaction volume. Thus (3.12.22) is modified to

$$R_n(E_0, \mathbf{J}) \simeq \left(\frac{2\pi}{3}\left\langle \sum_i p_i^2 \right\rangle\right)^{-3/2} \left(\frac{4\pi}{9}\langle r_o^2\rangle \left\langle \sum_i p_i^2 \right\rangle\right)^{-1/2}$$
$$\times \left(\frac{2\pi}{9}(1+\alpha^2)\langle r_o^2\rangle \left\langle \sum_i p_i^2 \right\rangle\right)^{-1}$$
$$\times \exp\left\{-\left[\frac{J_T^2}{1+\alpha^2} + \frac{J_z^2}{2}\right]\left(\langle \tfrac{2}{9} r_o^2\rangle \left\langle \sum_i p_i^2 \right\rangle\right)^{-1}\right\}$$
$$\alpha^n \Omega_o^n \bar{\rho}_n(E_0), \qquad (3.12.24)$$

where Ω_o is the uncontracted volume.

The effect of angular-momentum conservation reveals itself in the angular distribution of secondary particles. The angular momenta of individual particles tend to align with **J** and consequently they are emitted preferentially in the plane perpendicular to **J**. The direction of **J** is perpendicular to the beam direction but takes all directions equally in a plane perpendicular to the latter; this is due to the quantum-mechanical effect that the incident state is formed by the superposition of eigenstates that are represented by the total angular-momentum components perpendicular to the beam direction. Therefore the angular distribution is obtained by summing up all the probability amplitudes for specified directions of **J**. By the summation the amplitudes in the beam direction are simply added up, whereas those in other directions are subject to destructive interference between the states rotated by π. This results in an angular distribution peaked in the forward and backward directions (Ericson 58).

The angular-momentum conservation affects the multiplicity distribution through the Gaussian factor in (3.12.23) or in (3.12.24).

Because of (3.12.16), the Gaussian factor gives

$$n^{3/2} \exp\left(\frac{-27J^2 n}{16\,\langle r^2\rangle E_0{}^2}\right). \tag{3.12.25}$$

For $J \simeq 0$ the factor $n^{3/2}$ has a dominant effect in favor of large multiplicities, whereas for large J the exponential factor has a dominant effect in reducing the probabilities of large multiplicities.

The comparison with experimental results indicates that the statistical theory with angular-momentum conservation qualitatively explains some of the features of multiple production, such as the forward and backward peaks in angular distribution and the large multiplicity in antiproton-proton annihilation. However, no serious attempt has yet been made to examine the validity of statistical theory at high multiplicities, although this has been widely applied to the analysis of nuclear events in the accelerator energy region.

In comparison with experiment, it is often necessary to take the interaction volume larger than that given in (3.12.8), although angular-momentum conservation results in the reduction of the discrepancy to some extent. This may be due, at least in part, to the fact that the interaction volume is determined by outgoing particles rather than colliding particles. If this is the case, the Lorentz contraction of the interaction volume by the Lorentz factor of an incident particle might not give a proper answer. In order to avoid such an ambiguity the whole problem may be treated in a covariant way. Accordingly, the invariant phase-volume is introduced instead of (3.12.2).

A covariant modification of the momentum-space integral is given by

$$\omega_n = \int \delta^{(4)}\left(\tilde{p}_0 - \sum_{i=1}^{n} \tilde{p}_i\right) \prod_{i=1}^{n} \frac{d\mathbf{p}_i}{E_i} \tag{3.12.26}$$

where $\tilde{p}_i$ is the four-momentum of an outgoing particle and $\tilde{p}_0$ that of incident particles. This is expressed by (Kolkunov 62)

$$\omega_n \simeq 4 \cdot (4\pi)^{n-2}\left(\frac{|p_0|}{2n-\frac{3}{2}}\right)^{2n-2} \cdot (2n-\tfrac{3}{2})^2 e^{2n-3/2}. \tag{3.12.27}$$

If $n-2$ particles are identical, this is divided by $(n-2)!$ in order to get the probability of emitting n particles. Thus the most probable multiplicity is obtained as

$$\bar{n} \propto p_0^{2/3}. \tag{3.12.28}$$

This multiplicity law is identical to that obtained by introducing the invariant S-matrix and evaluating it by perturbation theory (Fukuda 50), since the multiplicity-energy relation is dictated by the invariant phase volume.

Experimental evidences have, however, shown that the average multiplicity increases more slowly with energy than (3.12.28). This implies that particles to be emitted are not independent but are strongly correlated. Hence they form a system that may be described by thermodynamics.

3.13 Thermodynamic Theory

If the correlation length is smaller than the dimension of the system under consideration, the system may be described by macroscopic theory, such as thermodynamics and hydrodynamics. The correlation length might be naïvely regarded as being equal to the mean free path of particles multiply produced in the system; namely,

$$l \simeq \frac{\Omega}{n\sigma}, \tag{3.13.1}$$

where σ is the collision cross section. However, neither the particle density n/Ω nor the cross section is well defined in the assembly of mesons because the particle-density operator is not positive definite, and the cross section can be defined only for the transition between two free particle states. Nevertheless, these two quantities have approximate meanings, and accordingly the mean free path may be used as an analogue of the correlation length.

In a sophisticated way the correlation length is derived on the basis of quantum statistical mechanics (Iso 59). When the energy density of mesons is high and their interactions are strong, the harmonic oscillation of a meson wave persists only as long as its De Broglie wavelength. Hence the correlation length is approximately equal to the De Broglie wavelength

$$l \simeq \frac{\hbar c}{T}, \tag{3.13.2}$$

where T is the average energy of elementary meson waves, which may be regarded as the temperature in energy units. This guarantees thermal equilibrium in a spatial region whose linear dimension is larger than l. In such a region we can define the energy density ε, the entropy density s, the local pressure p, and the local temperature T. The spatial differences in pressure and temperature are compensated by viscosity and thermal conductivity, respectively.

The viscosity and thermal conductivity coefficients, which are collectively called the transport coefficients, depend on the relaxation length, which is on the same order of magnitude as the correlation length. If the relaxation length were large and comparable to the linear dimension of the interaction volume Ω, the whole system would quickly reach thermal equilibrium. Thus we can introduce the internal energy and the entropy of the whole system respectively as

$$U = \varepsilon\Omega, \qquad S = s\Omega. \tag{3.13.3}$$

They are related as

$$\frac{dS}{dU} = \frac{1}{T}, \tag{3.13.4}$$

where T is the equilibrium temperature.

Now this system is assumed to consist of a gas of relativistic particles. The equation of state for this system is

$$p = \tfrac{1}{3}\varepsilon. \tag{3.13.5}$$

This leads to the Stefan-Boltzmann law

$$U = \frac{\pi^2}{30}\, ag\, \frac{T^4}{(ch)^3}\,\Omega \tag{3.13.6}$$

where g is the statistical weight arising from spin and isospin, and a is a numerical factor on the order of unity representing the relativistic effect. Then (3.13.4) with (3.13.6) gives

$$S = \frac{4\pi^2}{90}\, ag\left(\frac{T}{ch^2}\right)^3 \Omega. \tag{3.13.7}$$

The entropy represents the number of particles that are eventually emitted. Hence the multiplicity is given by

$$n \propto T^3\Omega \propto U^{3/4}\Omega^{1/4}. \tag{3.13.8}$$

If the interaction volume depends on energy as in (3.12.8), or $\Omega \propto \Omega^{(0)}U^{-1}$, the multiplicity-energy relation turns out to be

$$n \propto U^{1/2}. \tag{3.13.9}$$

This is the multiplicity law of Fermi (Fermi 50).

Although the multiplicity law explains the general features of experimental results, the theory implies the following consequences that do not at all agree with experimental facts. Firstly, the collision is totally inelastic, whereas the actual collision is distinctly elastic.

Secondly, particles emitted have an average energy that is equal to the temperature. The temperature depends on energy as

$$T \propto U^{1/2}, \tag{3.13.10}$$

and is considerably higher at high energies. Since the secondary particles are emitted isotropically, the transverse momentum increases with energy, giving high values. If angular-momentum conservation is taken into account, the angular distribution is peaked forward and backward, but the anisotropy thus attained is not enough to reduce transverse momentum.

Thirdly, the temperature is so high that as many massive particles, such as nucleons and antinucleons, have to be emitted as pions. The momentum distribution of particles of rest energy m is expressed as in (3.6.6), or

$$f(p)\,dp = \frac{gp^2\,dp}{T^3F(m/T)}\left[\exp\left(\frac{\sqrt{p^2+m^2}}{T}\right)-1\right]^{-1}, \tag{3.13.11}$$

where $F(x)$ is given in (3.6.6). Hence particles of $m < T$ are emitted in nearly equal abundances.

The last two difficulties arise from a high temperature of the system. At such a high temperature, however, the correlation length given in (3.13.2) is so small that the number of particles cannot be defined yet. In the initial stage of the collision the whole system does not reach thermodynamic equilibrium but behaves as a nonequilibrium system. The behavior of this system may be described by relativistic hydrodynamics (Landau 53).

Since the system immediately after the collision is contracted in the direction of incident particles, the pressure in this direction is predominant over that in other directions. Hence the system expands in the forward and backward directions. The solution of the hydrodynamic equation gives the distributions of energy and entropy that are concentrated near expanding fronts. In the course of expansion the transport mean free paths are so short that entropy is essentially conserved, but the temperature decreases. Finally, the temperature becomes so low that the correlation length is comparable to the linear dimension of the system. Then the number of particles can be defined and they escape as free particles.

Because of the isentropic expansion, the multiplicity law (3.13.9) remains unchanged, but the final temperature is low enough to avoid those two difficulties. Since the system may be as large as the pion Compton wavelength, the final temperature may be on the same order as the pion rest energy. Consequently, the transverse momenta of particles emitted are energy independent and as small as $m_\pi c$, and the production of heavier particles is less probable than pions (Koba 56). The number of particles of mass m is given by

$$n_m = \frac{4\pi}{h^3} \int \frac{(E^2 - m^2)^{1/2}\, E\, dE}{\exp(E/T) - 1}. \tag{3.13.12}$$

These consequences seem to be consistent with the experimental results described in Sections 3.2 and 3.4.

The multiplicity law based on thermodynamics depends on the equation of state in (3.13.5) and the energy dependence of the interaction volume in (3.12.8). These relations are dictated by the dynamics of interacting particles; for example, if the meson-nucleon interactions involve the derivative of the meson field, the factor $\frac{1}{3}$ in (3.13.5) may be modified. This results in an energy dependence of multiplicity other than according to (3.13.10). The multiplicity-energy relation may be indirectly checked by elastic scattering, as discussed below.

The entropy introduced in thermodynamical theory would play an important role in elastic scattering, too. The entropy is connected to the number of possible states, N, by

$$N = \exp(S). \tag{3.13.13}$$

Since the elastic process takes place as one of N possible processes, elastic scattering through the formation of a high-temperature collision complex has a cross section that is inversely proportional to N. Because elastic scattering at small angles is due mainly to the diffraction effect and diffraction scattering shows a steep angular distribution, only elastic scattering at large angles is attributed to the process under consideration.

In fact, the angular distribution of *p-p* scattering in the energy range between 10 and 30 GeV is found to be essentially flat at center-of-mass system angles greater than 70° (Cocconi 64). If the large-angle cross section is plotted against U/M as in Figure 3.25, we obtain an exponential dependence

$$\left(\frac{d\sigma}{d\Omega}\right)_{90^\circ} \propto \exp\left[-3.27\left(\frac{U}{M}\right)\right]. \tag{3.13.14}$$

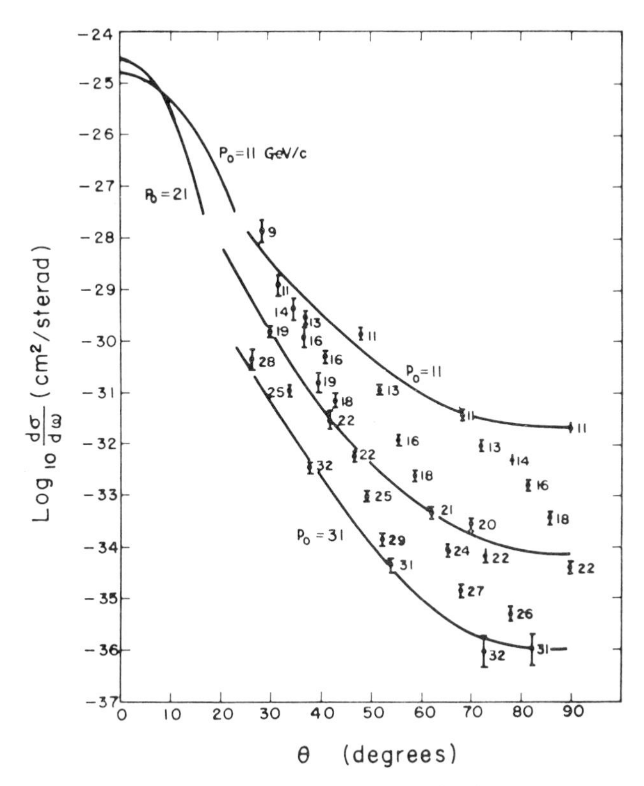

Fig. 3.25 Angular distributions of proton-proton elastic scattering. The numbers in experimental points represent the incident energies; curves represent the average angular distributions for the incident momenta $p_0 = 11$, 21, and 31 GeV/c, respectively. From (Cocconi 65).

The exponential law is consistent with the multiplicity law, $S \propto n \propto U$,

$$\left(\frac{d\sigma}{d\Omega}\right)_{90^\circ} = \frac{\sigma_c}{2\pi N} = \left(\frac{\sigma_c}{2\pi}\right) \exp(-S), \qquad (3.13.15)$$

where σ_c is the cross section of forming the collision complex. Thus the energy dependence of the elastic cross section would imply the multiplicity-energy relation through the entropy S.

Finally, we offer a brief comment on the validity of thermodynamic theory. The thermodynamic equilibrium does not hold on a time scale shorter than the correlation time

$$\tau_c = \frac{l}{c} \simeq \frac{\hbar}{T}. \qquad (3.13.16)$$

In the initial stage of the collision the energies of two colliding particles are exchanged in the impact time given by

$$\tau_i \simeq \frac{r}{c\gamma_c} \simeq \frac{\hbar}{m_\pi c^2 \gamma_c}, \tag{3.13.17}$$

where r is the force range, which is nearly equal to the pion Compton wavelength. Because the temperature in the initial stage is $T \simeq M\gamma_c^{1/2}$ where M is the nucleon mass, τ_i is greater than τ_c only for (Namiki 57)

$$\gamma_c \leq \left(\frac{M}{m_\pi}\right)^2. \tag{3.13.18}$$

This demonstrates that the macroscopic description is not applicable in the initial stage of very-high-energy collisions. Below a few TeV the macroscopic description would hold even at the initial stage. It is, however, necessary to consider the possibility that the effective force range may be smaller than the pion Compton wavelength since the four-momentum transfer observed is as large as, or larger than, the nucleon mass, and the cross section for the collision complex formation is on the order of 1 mb. If the force range is as small as the nucleon Compton wavelength, the validity of the macroscopic description is limited only in a later stage. It is therefore suggested that the dynamical description is necessary in the initial stage, whereas the thermodynamic and hydrodynamic description may be applied to later stages.

3.14 One-Fireball Model

As discussed in the preceding section the validity of the thermodynamic description may be questioned in two respects. Theoretically it is indicated that the local equilibrium that is required for the validity of the hydrodynamic model does not hold in the initial stage of a collision, so that the intermediate state of the collision cannot be described by one-collision complex alone. Experimentally it has been repeatedly emphasized that the incident particle gives only a part of its energy to the collision complex. The situation is analogous to the case of nuclear reactions.

The collision of a nucleon with a nucleus was thought to form a collision complex that would be described as a thermodynamic system. In the collision complex—the compound nucleus—the energy levels excited are so numerous that the energy brought by the incident particle is distributed over many modes. The distribution of the level density allows us to introduce the temperature of the compound nucleus, and particles are emitted by evaporation from the excited system. Although the evaporation theory can explain the main features of low-energy nuclear reactions, some of the emitted particles do not belong to the evaporated particles. They are interpreted as being due to direct collisions with nucleons in the nucleus. An appreciable fraction of the incident energy is taken away by such knocked-out nucleons, and the rest is spent to excite the compound nucleus, as discussed in Section 2.9. The fraction of energy used in the latter process corresponds to the inelasticity in very-high-energy collisions.

According to this analogy it is plausible to describe the very-high energy collision in the following way. An incident particle gives a fraction of its energy, K_L, to the collision complex, and the latter behaves as a thermodynamic entity. The same holds for the target particle in the mirror system, as described in Appendix B, Section 9.2; that is, a fraction of the energy, K_M, of the target particle is given to the collision complex. In the center-of-mass system, the collision complex receives a fraction of the incident energy,

$$K^* = \tfrac{1}{2}(K_L + K_M). \tag{3.14.1}$$

Hence the energy available for the collision complex should be multiplied by K^*, or

$$U = 2(E_0^* - M)K^* \tag{3.14.2}$$

in the case of the nucleon-nucleon collision with the center-of-mass-system energy E_0^*. The collision complex may be called the fireball, which implies an excited system at high temperature. If only one such fireball is formed, the collision process is said to be described by the one-fireball model, as shown in Fig. B.4.

By calling it model rather than theory, we mean that the whole process can be described mainly in kinematic terms, without referring to theoretical details. Hence important consequences of a model are readily obtained in an unambiguous way, so that they can be directly compared with experimental results. The theoretical basis of a model should be looked for after the model is found to be acceptable.

In the one-fireball model the mass and the translation energy of a fireball can be obtained as

$$M_F \simeq 2\gamma_c MK^*\sqrt{1 - Z^2} = 2\gamma_c M\sqrt{K_L K_M} \tag{3.14.3}$$

and

$$E_F^* \equiv \gamma_F^* M_F \simeq 2\gamma_c MK^* = \gamma_c M(K_L + K_M), \tag{3.14.4}$$

respectively, where the target particle is assumed to be a nucleon. Here Z is the asymmetry factor given by

$$Z = \beta_F^* = \frac{K_L - K_M}{K_L + K_M} = \frac{K_L - K_M}{2K^*} \tag{3.14.5}$$

and may, in practice but only approximately, be taken as being equal to a_n in (3.5.7),

$$Z \simeq a_n . \tag{3.14.6}$$

If the incident energy, or γ_c, is known, the values of K_L and K_M are obtained from the energies and emission angles of secondary particles, as discussed in Section 3.9 and Appendix B, (Section 9.2), and accordingly M_F and γ_F^* are evaluated.

The experimental support of the one-fireball model was most clearly given by the Soviet calorimeter experiment for incident energies on the order of 100 GeV (Guseva 62). With the experimental method described in Sections 3.6 and 3.9 it is possible to construct the center-of-mass system of secondary particles as well as to measure mirror inelasticity. The angular distribution of these secondary particles in their center-of-mass system is found to be essentially isotropic, and their energy distribution is well reproduced by the Planck distribution with a temperature as low as the pion rest energy.

This suggests that the behavior of the fireball may be described in thermodynamic terms, and accordingly the multiplicity may be given by (3.13.8) if U is replaced by (3.14.2). However, the interaction volume Ω is not well known. In order to avoid this ambiguity it is appropriate to express the multiplicity as

$$n \simeq \left(\frac{U}{T}\right) \propto K^* 2(\gamma_c - 1). \tag{3.14.7}$$

This results in the linear relation between n/K^* and $\gamma_c - 1$, and this had been, in fact, observed even earlier by Kaneko and Okazaki (Kaneko 58). Moreover, the absolute multiplicity is consistent with the low value of temperature, indicating the expansion of the fireball until the temperature becomes low enough to increase the correlation length to the dimension of the fireball. A slight anisotropy found as the forward-backward peaks may be accounted for in terms of the angular momentum of the fireball, as discussed in Section 3.12.

In the fireball model the distribution of transverse momentum of a pion is assumed to be the Planck distribution. This results in the emission-angle dependence of transverse momentum; the average transverse momentum for a given emission angle would be proportional to $\sin \theta^*$. However, no angular dependence has yet been observed.

Although problems concerning the angular dependence of transverse momentum seem to be open to future investigation, the one-fireball model explains many features of high-energy interactions. This model seems to be valid in the energy range between a few tens of GeV and a few hundred GeV.

3.15 Two-Fireball Model

As energy approaches 1 TeV the angular distribution of secondary particles is more and more peaked in the forward and backward directions. This means that the longitudinal momentum of a produced particle increases with incident energy, whereas transverse momentum is independent thereof.

This may be interpreted in terms of the fireball model in the following way. According to Section 3.12, the average value of the total angular momentum of the fireball increases with incident energy. Accordingly, the angular momenta of individual particles emitted increase, and they tend to align with the total angular momentum. The latter means that particles emitted from the fireball are divided into two groups—one going forward and the other backward. If particles belonging to one group are emitted in a small angle, their mutual interactions persist while they are in flight and may result in the formation of a fireball. In this extreme case, therefore, the collision ends up by the formation of two fireballs in addition to two survival particles (Ciok 58, Cocconi 58, Niu 58).

We distinguish these two fireballs going forward and backward by the subscripts f and b, respectively; their masses are M_f and M_b, and their total energies in the rest system of the primary fireball are $\gamma_f^+ M_f$ and $\gamma_b^+ M_b$. They are related to the mass of the primary fireball by

$$M_F = \gamma_f^+ M_f + \gamma_b^+ M_b. \tag{3.15.1}$$

The conservation of momentum also holds between two fireballs, as follows:

$$\beta_f^+ \gamma_f^+ M_f = \beta_b^+ \gamma_b^+ M_b. \tag{3.15.2}$$

If the two secondary fireballs are emitted in the direction of flights of the primary fireball, or if the transverse momenta of these fireballs are small, the velocities and the Lorentz factors are related to β_F^* and γ_F^* in a simple way. Then the asymmetry factor is given by

$$Z = \frac{\beta_f^* \gamma_f^* M_f - \beta_b^* \gamma_b^* M_b}{\gamma_f^* M_f + \gamma_b^* M_b} = \beta_F^*. \tag{3.15.3}$$

This is the same as in the case of the one-fireball model.

Since the asymmetry factor is known to be small, we shall give a number of kinematic relationships for the symmetric case. In this case we have

$$M_F = 2\gamma_f^* M_f \qquad (\gamma_f^* = \gamma_f^+). \tag{3.15.4}$$

The mass of a secondary fireball is related to energy-momentum transfer Δ as

$$M_f \simeq 2\gamma_f^* \Delta, \tag{3.15.5}$$

according to (B.107) and for the case of a small transverse component. If Δ is energy independent, the energy dependence of M_F given in (3.14.3) results in

$$\gamma_f^* \simeq \left(\frac{KM}{2\Delta}\right)^{1/2} \gamma_c^{\ 1/2}, \qquad M_f \simeq (2K\Delta M)^{1/2} \gamma_c^{\ 1/2}. \tag{3.15.6}$$

The former shows that the separation between the two humps in the log tan θ plot increases with energy, whereas the latter indicates the $E^{1/4}$ dependence of the multiplicity law.

It is still open to question whether or not the two-fireball model implies the existence of two collision complexes in the intermediate state. The characteristic features of the two-fireball model may also be accounted for in terms of the high spin of the primary fireball. The reality of the fireball will be seen by the emission-angle dependence of transverse momentum, as discussed in the preceding section. However, the formation of the collision complexes seems to be admissible since the relative momenta between secondary particles belonging to one lump are so small that they are contained in a relatively small phase-space volume and interact rather strongly with each other.

3.16 Multifireball Model

As energy further increases the longitudinal momentum distribution in the fireball system has an extended tail and may have a hump in the high-energy part. This has led Hasegawa (Hasegawa 61) to propose the multifireball model. It may also be said that the secondary fireball splits into two when it has a high spin.

Let us consider that the forward fireball splits into two, one going forward with velocity β_{ff}^{+} and mass M_{ff} and the other backward with velocity β_{fb}^{+} and mass M_{fb} in the rest system of the secondary fireball. In this system relationships similar to (3.15.1) and (3.15.2) hold:

$$M_f = \gamma_{ff}^{+} M_{ff} + \gamma_{fb}^{+} M_{fb}, \qquad \beta_{ff}^{+} \gamma_{ff}^{+} M_{ff} = \beta_{fb}^{+} \gamma_{fb}^{+} M_{fb}, \qquad (3.16.1)$$

where γ_{ff}^{+} and γ_{fb}^{+} are the Lorentz factors of tertiary fireballs. Their Lorentz factors in the center-of-mass system are given by

$$\gamma_{ff}^{*} = \gamma_f^{*} \gamma_{ff}^{+}(1 + \beta_f^{*} \beta_{ff}^{+}), \qquad \gamma_{fb}^{*} = \gamma_f^{*} \gamma_{fb}^{+}(1 - \beta_f^{*} \beta_{fb}^{+}), \qquad (3.16.2)$$

respectively. Again it can be shown that the asymmetry factor is the same as in (3.15.3).

For the sake of simplicity we again consider the symmetric case; namely, $M_{ff} = M_{fb}$, $\gamma_{ff}^{+} = \gamma_{fb}^{+}$ and $\beta_{fb}^{+} = \beta_{fb}^{+}$. As shown in Appendix B, a relation similar to (3.15.5) holds for the tertiary fireball:

$$M_{ff} = 2\gamma_{ff}^{+} \Delta. \qquad (3.16.3)$$

If Δ is independent of energy, we have

$$\gamma_{ff}^{+} \simeq \left(\frac{KM}{2\Delta}\right)^{1/4} \gamma_c^{\ 1/4}, \qquad M_{ff} \simeq (8KM\Delta^3)^{1/4} \gamma_c^{\ 1/4}. \qquad (3.16.4)$$

From (3.16.4) and (3.15.2) we obtain

$$\beta_{ff}^{+} \simeq 1 - \gamma_f^{*\,-1}. \qquad (3.16.5)$$

Because β_f^{*} is larger than β_{ff}^{+} and is nearly equal to unity, the two expressions in (3.16.2) are reduced, by virtue of (3.16.5), to

$$\gamma_{ff}^{*} \simeq 2\gamma_f^{*} \gamma_{ff}^{+} \propto \gamma_c^{\ 3/4}, \qquad \gamma_{fb}^{*} \simeq \gamma_{fb}^{+} \propto \gamma_c^{\ 1/4}. \qquad (3.16.6)$$

Hence the fast fireball gives very-high-energy secondary particles.

An alternative assumption was emphasized by Hasegawa (Hasegawa 61). The fireball may be an excited state of the meson, so that it has a definite mass, possibly equal to twice the nucleon mass in view of the fact that the boson consists of a fermion and its antiparticle (Sakata 56). This assumption can be incorporated with (3.15.6) and (3.16.4) if inelasticity depends on energy as

$$K \propto \gamma_c^{\ -1}. \qquad (3.16.7)$$

Important consequences of this assumption are that the velocity of the fireball is energy independent according to (3.15.5) and (3.16.3), and that multiplicity stays constant in an energy interval and increases stepwise when the number of fireballs increases. Although experimental evidence is not conclusive yet, the hypothesis itself is fascinating because it suggests a very highly excited state in a series of many resonance states. This is therefore called the H-quantum, where H stands for "heavy," and may imply a fundamental property of the elementary particle*.

Whatever the fundamental property of the fireball may be, the multi-fireball model results in the following multiplicity-energy relation (Frautschi 62). Since the mass of final fireballs is limited, multiplicity is proportional to the number of fireballs produced. If the splitting of fireballs continues k times, the number of fireballs is $N=2^k$. The mass of the primary fireball is expressed, according to (3.15.4), (3.15.5), (3.16.1), and (3.16.3) as

$$M_F \simeq \frac{M_f{}^2}{\Delta} \simeq \frac{M_{ff}^4}{\Delta^3} \simeq \left(\frac{M_k}{\Delta}\right)^N \Delta \qquad (N=2^k). \tag{3.16.8}$$

This gives

$$N \simeq \frac{\log(M_F)/\Delta}{\log(M_k/\Delta)} \propto \log(K\gamma_c). \tag{3.16.9}$$

Thus multiplicity increases with the logarithm of the incident energy. This is not inconsistent with the observed multiplicity law, as discussed in Section 3.10.

3.17 Excitation of Survival Particles

A group of particles with energies higher than the majority of secondary particles may also be interpreted as being due to the decay products of a survival particle that is excited to a resonance state (Nishimura 62, Peters 62, Zatsepin 62). Since the survival particle has a velocity much higher than that of the fireballs, the decay product of the former has a higher energy than that of the latter.

Again for the sake of simplicity we shall consider kinematic relations in the symmetric collision. If the survival particle is excited to mass M_x and leaves with energy $\gamma_x^* M_x$ in the center-of-mass system, the energy conservation is expressed by

$$\gamma_c M = \gamma_f^* M_x + \gamma_f^* M_f, \tag{3.17.1}$$

since energy transfer vanishes in the center-of-mass system of the sym-

* It has been argued by Koshiba et al. from their analysis of jet showers that the production of ϕ-mesons with a mass of 1019 MeV is dominant in the TeV region; the ϕ-meson decays into kaons with a large branching ratio.

metric collision. The partition of energy between the excited particle and the fireball is represented by the inelasticity coefficient K as

$$\gamma_x^* M_x = (1 - K)\gamma_c M, \qquad \gamma_f^* M_f = K\gamma_c M. \tag{3.17.2}$$

In the laboratory system the energy of the survival particle is given approximately by

$$\gamma_x M_x \simeq 2(1 - K)\gamma_c{}^2 M, \tag{3.17.3}$$

whereas that of the fireball is

$$\gamma_f M_f \simeq 2K\gamma_c{}^2 M. \tag{3.17.4}$$

If M_f is much greater than M_x, γ_x is distinctly separated from γ_f, unless K is close to unity.

In practice secondary pions from these parents are observed, and their separation has to be seen on the log tan θ plot. Pions from two fireballs distribute about $\log \gamma_c$ with dispersion

$$\sigma_f \simeq 0.5, \tag{3.17.5}$$

whereas those from the excited particle distribute about γ_x with dispersion

$$\sigma_x \simeq 0.29,$$

provided that the 3–3 resonance level of the nucleon is taken into account (Imaeda 63). Hence the separation can be clearly made if

$$\log \gamma_f - 2\sigma_x > \log \gamma_c + 2\sigma_f. \tag{3.17.6}$$

This is expressed as

$$(1 - K)\gamma_c \gtrsim 20. \tag{3.17.6$'$}$$

The same procedure should apply for the separation between the fireballs and the backward excited particle, and this is found to be experimentally easier for the identification of isolated particles that may arise from the excited particle than the use of the forward particles.

As a result the separation improves as inelasticity decreases; for example, among 167 events with $\gamma_c \geq 23$ and $N_h \leq 5$, about half the events with $K < 0.35$ are associated with the isolated particles, whereas the fraction of association decreases to about one-quarter for $K \geq 0.35$ (Imaeda 63). The average number of charged particles belonging to an excited particle is about 1.5 in the former case, whereas it is nearly unity in the latter case.

3.18 Nuclear Interactions of the Muon

One of the most important questions in elementary particle physics is whether or not the muon is identical with the electron in all respects except for a large difference in their masses. It was suspected that the muon would interact *anomalously* with nuclei (Fowler 58). By anomalous interaction we mean that the muon interacts with other particles through forces which are not involved in the electron's interactions. Normally the interactions of the muon should be accounted for in terms of the electromagnetic interactions of a spin $\frac{1}{2}$ particle and of the weak interactions responsible for muon decay and the nuclear capture of the muon. All these properties are shared with the electron, in spite of the fact that the muon is much heavier than the electron. If no other interactions are present, the nuclear interaction of the muon takes place through the electromagnetic field of the muon. In order to find any anomalous effect experiments of great precision have been attempted. In the framework of electromagnetism the g-factor of free muons and the photoproduction of muon pairs have been measured, but the results can be in perfect agreement with what is predicted by quantum electrodynamics; namely, the muon is a structureless Dirac particle. In the field of cosmic rays, scattering and pion production by nuclear collisions of muons have been studied for a long time (George 52).

3.18.1 Elastic Scattering

Most scattering events are caused by multiple scattering. If the scattering events one is looking for take place with a small cross section, the events of interest could be easily masked by the large-angle tail of multiple scattering. A serious difficulty lies in the fact that the angular distribution at large angles due to multiple scattering can be obtained only after involved analyses. Moreover, the effect of multiple scattering is sensitive to the energy of the incident particles concerned. Unless the energy is known with considerable accuracy, one can hardly subtract the contribution of multiple scattering with sufficient reliability. Most experiments that purported to show anomalous scattering seem to have suffered from these difficulties.

Defining the energies of incident muons and using the multiple-scattering formula given in (2.1.39), Fukui et al. (Fukui 59) found no anomalous effect. Experiments using accelerator muon beams have also been performed and have confirmed the cosmic-ray result. More quantitative experiments based on observing the scattering by protons with large momentum transfer have set the upper limits on muon size and the size difference between the muon and the electron. The upper limit of the charge radius thus found is as small as 3×10^{-14} cm, and no significant difference is found between the muon and the electron (Cool 65). The conclusion is consistent with that obtained from the g-factor and large-angle muon pair production.

3.18.2 Theoretical Considerations on Pion Production by Muons

Pion production by nuclear collisions of muons is considered to be due to photopion production by the virtual photons of muons. The number of virtual photons with energies between k and $k + dk$ that are associated with a muon of energy E is given by

$$n(E, k)\, dk = \frac{2}{\pi} \alpha \left[\ln\left(\frac{E}{k}\right) - 0.38 \right] \frac{dk}{k}. \tag{3.18.1}$$

Hence the cross section for energy transfer k in the muon-nucleon collision is expressed by

$$\frac{d\sigma_\mu(E, k)}{dk} = \sigma_\gamma(k) n(E, k), \tag{3.18.2}$$

where $\sigma_\gamma(k)$ is the photopion-production cross section. If $\sigma_\gamma(k)$ is energy independent, the cross section for energy transfer greater than E_c is

$$\sigma_\mu(E, > E_c) = \int_{E_c}^{E} \sigma_\gamma\, n(E, k)\, dk = \frac{\alpha}{\pi} \sigma_\gamma \ln^2 \left(\frac{E}{E_c}\right). \tag{3.18.3}$$

In the Williams-Weizsäcker (W-W) method described above energy-momentum conservation is not correctly taken into account. If we take energy-momentum conservation into account but neglect the effects of multipoles other than the electric dipole and of the longitudinal component of the electric field as in the case of the W-W method, the virtual-photon spectrum is modified to the following (Kessler 56):

$$n(E, k)\, dk = \frac{2}{\pi} \alpha \left[\ln\left(\frac{E}{m_\mu c^2}\right) - 1 \right] \frac{dk}{k}. \tag{3.18.4}$$

This gives for a constant photopion-production cross section

$$\sigma_\mu(E, > E_c) = \frac{2}{\pi} \alpha \sigma_\gamma \left[\ln\left(\frac{E}{m_\mu c^2}\right) - 1 \right] \ln\left(\frac{E}{E_c}\right). \tag{3.18.5}$$

This is in general greater than the one obtained with the W-W method in (3.18.3) and has a less steep spectrum of transferred energy E_c.

By including the longitudinal component and higher multipoles, Daiyasu et al. (Daiyasu 62) obtained a general expression for the muon-induced reaction. Let the energy and momentum of the incident muon be E and p, respectively, and those of the outgoing muon, E' and $\mathbf{p}'$ respectively. Energy and momentum transfers in the collision are expressed by

$$k = E - E', \qquad \mathbf{q} = \mathbf{p} - \mathbf{p}', \tag{3.18.6}$$

respectively. They are equal to the energy and momentum, respectively, of a virtual photon, and for four-momentum transfer we have

$$\Delta^2 = \mathbf{q}^2 - k^2 \neq 0, \tag{3.18.6'}$$

because the photon is not free but virtual. They are also equal to the energy and momentum, respectively, given to the system consisting of a recoil nucleon and the produced pions. Thus the differential cross section for given energy and momentum transfers is expressed by

$$\frac{d^2\sigma(E, k, \Delta)}{d\Delta^2\, dk} = \frac{\alpha}{8\pi^2} \frac{1}{p^2\Delta^4} \{L[(E^2 + E'^2)\Delta^2 - 2m^2{}_\mu k^2 - \tfrac{1}{2}\Delta^4] + L'(2m_\mu{}^2 - \Delta^2)\Delta^2\}. \tag{3.18.7}$$

Here L and L' are invariant functions of Δ^2 and k, representing the dynamic structure of interactions.

The term proportional to L represents the contribution of the transverse component, whereas the term proportional to L' represents that of the longitudinal one. These quantities appear in the photopion cross section as

$$\sigma_\gamma(k) = \frac{k}{4\pi} \left(L - L' \frac{\Delta^2}{k^2}\right)_{\Delta^2 = 0}. \tag{3.18.8}$$

Thus the L' term does not contribute to photopion production. This implies that pion production by muons or by electrons would explore yet unknown features that are not obtainable by photoreactions. They may represent the structure of the nucleon as well as that of the muon. However, a closer examination of (3.18.7) by reference to photopion production indicates that the contribution of the longitudinal component is not too important. Taking the L-term alone and assuming

$$L(k, \Delta^2) = L_0 \left(\frac{\Lambda^2}{\Delta^2 + \Lambda^2}\right)^2, \tag{3.18.9}$$

we find that the integral of the cross section (3.18.7) over Δ^2 and k lies between (3.18.3) and (3.18.5).

In (3.18.9) Λ^{-1} represents the size of the nucleon or the muon. Since muon size is as yet undetermined, as discussed in Section 3.18.1, it may be sensible to equate Λ^{-1} to the radius of the proton charge distribution, or

$$\Lambda^2 = 0.37(\mathrm{GeV}/c)^2. \tag{3.18.10}$$

The cross section (3.18.7) is expressed as a function of k and Δ, in which k can be obtained as the sum of the energies of secondary particles, and Δ^2 is a quantity that corresponds to ${\Delta_A}^2$, given for a nuclear collision in (3.11.10). In place of k and Δ it may be practical to introduce k and θ_μ, where θ_μ is the scattered angle of the muon. These two sets of variables are related to each other by

$$\Delta^2 = (\mathbf{p} - \mathbf{p}')^2 - (E - E')^2 \simeq 4EE' \sin^2 \frac{\theta_\mu}{2} = 4E(E - k) \sin^2 \frac{\theta_\mu}{2}, \tag{3.18.11}$$

where the relativistic approximations, $E^2 \simeq |\mathbf{p}|^2$ and $E'^2 \simeq |\mathbf{p}'|^2$ have been used. In (3.18.11) both k and θ_μ may be measurable, but the incident energy E cannot always be measured in cosmic-ray experiments. The incident energy can be eliminated by assuming the constancy of transverse momentum. Using $E' \simeq p_T' \operatorname{cosec} \theta_\mu$, we reduce the third expression of (3.18.11) to

$$\Delta^2 \simeq 2(k + p_T' \operatorname{cosec} \theta_\mu) p_T' \tan\left(\frac{\theta_\mu}{2}\right). \tag{3.18.12}$$

If transverse momentum p_T' is given, we need measure only k and θ_μ. Although the value of transverse momentum is expected to be small, no experimental check has yet been made. Therefore we tentatively assume that the transverse momentum of the outgoing muon is not different from that of a produced pion. This assumption is considered to be close to reality if one notes the angular distribution of outgoing muons and the balance of transverse momenta of all outgoing particles.

3.18.3 Experimental Results of Pion Production

The above discussion indicates that quantities to be observed in pion production by muons are analogous to those in the nucleon-nucleon and pion-nucleon collisions. In the early days most results obtained were restricted to the cross section and the average value of energy transfer. Experimental results comparable with those obtained for purely strong interactions are available in only a few cases. Here we refer to two representative works—one with cosmic rays performed at a depth of 50 meters water equivalent underground (Higashi 65) and the other with a negative-muon beam of 5 GeV/c obtained from a proton synchrotron (Jain 65).

In the underground cosmic-ray experiment, muons were selected as particles that were not associated with other nuclear active and penetrating particles; this selection procedure is essential in underground experiments because the interactions observed are nearly equally shared with those due to muons and those due to pions produced by muons in the upper layer of the earth—and features of the nuclear interactions produced by muons and pions are alike in many respects. The interactions observed took place in lead or iron plates in two cloud chambers. An interaction is characterized by the number of penetrating secondary particles, n_s, in which a survival muon is included, and by the number of heavily ionizing particles, N_h. The average values of n_s and N_h for lead and iron producers are given in Table 3.12. Essentially the same numbers are obtained for pion-induced showers. The frequencies of showers induced by muons and pions are also compared. We can see from Table 3.12 that no appreciable difference exists between the features of showers taking place in lead and in iron.

Table 3.12 Features of Muon-Induced Showers in Lead and Iron

Parameter	Lead	Iron
$\langle n_s \rangle$	3.6 ± 0.2	4.0 ± 0.3
$\langle N_h \rangle$	0.6 ± 0.1	0.6 ± 0.2
Frequency ratio of pion- to muon-induced showers	0.26 ± 0.11	0.35 ± 0.20

In the artificial muon experiment nuclear emulsion was exposed to a 5-GeV/c beam of negative muons with good purity. The interactions in emulsion were observed in the usual manner.

In both cases the cross section was found to be on the order of microbarns (10^{-30} cm^2) per nucleon. It should, however, be noticed that selection criteria in these two experiments were different; in the cosmic-ray experiment the cloud chamber were triggered by two or more particles that penetrated thick materials in cloud chambers and counter trays. Therefore only the interactions that gave energetic secondary particles were selected, and consequently the cross section thus derived should have been smaller than that obtained with emulsion. In the emulsion experiment, on the contrary, all inelastic events were, in principle, observed. Moreover, the cross section for the production of strange particles was distinguished from that for pion production. These cross sections were found to be

$$\sigma_\mu(\text{inelastic}) = 3 \pm 0.3 \ \mu b/\text{nucleon}, \qquad (3.18.13)$$
$$\sigma_\mu(\text{strange particles}) = 0.1 \pm 0.06 \ \mu b/\text{nucleon}$$

at 5 GeV. The cross sections for pion production may be compared as follows:

$$\sigma_\mu(\pi, \text{emulsion, 5 GeV}) = 0.7 \pm 0.14 \ \mu b/\text{nucleon},$$
$$\sigma_\mu(\pi, \text{lead, 50 mwe}) = 0.26 \pm 0.4 \ \mu b/\text{nucleon}, \qquad (3.18.14)$$
$$\sigma_\mu(\pi, \text{iron, 50 mwe}) = 0.36 \pm 0.8 \ \mu b/\text{nucleon},$$
$$\frac{\sigma_\mu(\pi, \text{lead, 250 mwe})}{\sigma_\mu(\pi, \text{lead, 50 mwe})} = 6.6 \pm 2.7.$$

The difference between the cross sections shows an effect of selection criteria. In the cloud-chamber experiment only the events that have transferred energies greater than 1 GeV were selected, whereas in the emulsion experiments most of the pions produced had kinetic energies of about 100 MeV. This would qualitatively explain the difference between the cross sections given in (3.18.14) but would hardly be quantitative enough to distinguish between possible assumptions such as the virtual-photon spectra discussed in the preceding subsection, because the absolute value as well as the energy dependence of the cross section are rather sensitive to the cutoff energy E_c. Nevertheless, we may say that the photopion-production cross section is as large as

$$\sigma_\gamma(k) \simeq 1 \times 10^{-28} \text{ cm}^2, \qquad (3.18.15)$$

essentially independent of energy up to a few tens GeV.

The energy of the virtual photon that is responsible for pion production can be estimated if the energies of all secondary particles as well as of a recoil nucleon are known. In the cosmic-ray experiment the energy of a charged secondary particle is estimated from its interaction; and that of a neutral pion, from a cascade shower. Since the recoil kinetic energy is evaluated as small, as will be shown below, the virtual-photon energy k is essentially equal to the sum of all secondary energies. On the basis of several events, for which secondary energies were measurable, the average value of k was found to be about 5 GeV. This is roughly one-third of the average energy of muons at a depth of 50 meters water equivalent, and the inelasticity coefficient may be as large as

$$K_\mu \simeq \tfrac{1}{3}. \qquad (3.18.16)$$

The smallness of inelasticity and the muon mass result, according to (3.11.10), in the value of Δ^2 being nearly equal to $p_T'^2$. If we tentatively assume $p_T' = 0.4$ GeV/c (as in the case of mesons produced in the nucleon-nucleon collision), the average value of Δ^2 is obtained from (3.18.12) as

$$\langle \Delta^2 \rangle^{1/2} \simeq 0.5 \text{ GeV}/c, \tag{3.18.17}$$

with a dispersion of about 0.35 GeV/c. This is not inconsistent with the value of Δ^2 derived from (3.11.10).

The transverse momentum of the outgoing muon is now compared with that of a produced pion. Although the number of pions that are suitable for the measurement of transverse momenta is small, it is found that the distribution of transverse momentum is approximately expressed by an exponential form with an average energy of

$$\langle p_{T\pi} \rangle \simeq 0.5 \text{ GeV}/c. \tag{3.18.18}$$

This is roughly equal to the transverse momentum of the outgoing muon and also is as large as the average longitudinal momentum of pions in their center-of-mass system. This suggests that the one-fireball model applies to pion production by muons.

In analogy to the discussion in Section 3.11 we introduce the four-momentum transfer of the target nucleon. Its longitudinal component can be estimated by referring to the mirror inelasticity by

$$\langle \Delta_{NL}^2 \rangle^{1/2} \simeq 0.3 \text{ GeV}/c. \tag{3.18.19}$$

If the transverse momentum of the recoil nucleon is also as large as 0.4 GeV/c, the value of $\Delta_N{}^2$ is essentially equal to that of Δ^2. This implies that the recoil energy is as small as 100 MeV.

Although the values of Δ^2 and $\Delta_N{}^2$ are smaller than those in nucleon-nucleon collisions, the value of p_T of a pion produced (as well as the angular distribution of pions) is analogous to that obtained in the nucleon-nucleon case. The composite angular distribution shown in Fig. 3.26 gives a form that is characteristic for pion emission from a fireball. We may therefore introduce the one-fireball model to describe pion production by muon-nucleon collisions (Fujimoto 67).

The mass of the fireball can be evaluated as in Appendix B, (Section 9.4). If an invariant scalar product, $\tilde{\Delta}_N \cdot \tilde{\Delta}$, is evaluated in the laboratory system by taking account of

$$k \simeq |\mathbf{q}|, \qquad \Delta_N{}^2 = 2MW_N' \simeq p_N'^2, \qquad \tilde{\Delta}_N \cdot \tilde{\Delta} \simeq k(p_N' - W_N'),$$

we obtain

$$M_F = (\tilde{\Delta}_N + \tilde{\Delta})^2 \simeq \Delta_N{}^2 + \Delta^2 + 2k(p_N' - W_N') \simeq 1.8 \text{ GeV}. \tag{3.18.20}$$

The mass of the fireball is equal to that in the case of the nucleon-nucleon collision.

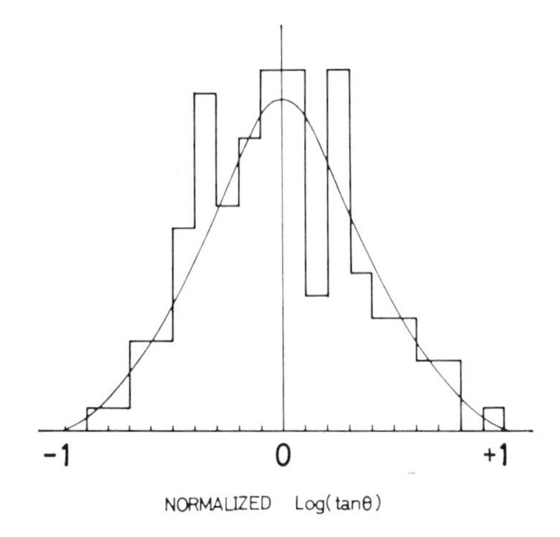

Fig. 3.26 Log tan θ distribution of pions produced by the nuclear interactions of muons (after T. Kitamura).

It is worthwhile to remark that the average features given in (3.18.18) through (3.18.20) are quite similiar to those in nucleon-nucleon and pion-nucleon collisions at several GeV. Insofar as we are concerned with such average behavior, the structure effects represented by L and L' in (3.18.7) will not be observable.

It has been attempted to obtain the structure effects by observing the angular distribution of outgoing muons and the distribution of Δ^2, both with an artificial muon beam of 5 GeV/c and with cosmic rays. Both results indicate the extended structure; that is, the results are in favor of a finite value of Λ^2 in L, but the value of Λ^2 derived by neglecting L' seems to be larger than that obtained from the elastic scattering of electrons. This would indicate a contribution of L' in (3.18.7), although the experimental results available are not sufficient to draw any quantitative conclusion.

3.19 Interactions of Neutrinos

3.19.1 Universal Fermi Interactions

The weak interactions of the muon are revealed in its spontaneous decay as well as in its capture by a nucleus. These interactions are represented by

$$\mu^{\pm} \rightarrow e^{\pm} + \nu_e(\bar{\nu}_e) + \bar{\nu}_\mu(\nu_\mu) \tag{3.19.1}$$

and

$$\mu^{-} + p \rightarrow n + \nu_\mu, \tag{3.19.2}$$

respectively. Together with the nuclear β-decay interaction

$$n \rightarrow p + e + \nu_e \tag{3.19.3}$$

these interactions are known to form the *universal Fermi interactions*, which are described by a combination of the vector and axial vector interactions. The coupling constant for the vector interaction, G_V, has a universal value,

$$G_V = 1.4 \times 10^{-49} \text{ erg-cm}^3 \simeq 1.0 \times 10^{-5}(\hbar c)\left(\frac{\hbar}{m_p c}\right)^2, \tag{3.19.4}$$

whereas that for the axial vector interaction, G_A, differs from G_V in the presence of strong interactions by

$$\left|\frac{G_A}{G_V}\right| \simeq 1.2. \tag{3.19.5}$$

These two interactions have opposite parities and opposite signs.

The coupling constant is not dimensionless as in strong interactions but has the dimension of length^2. This implies that there is a characteristic length within which the interaction energy exceeds the free-particle energy. If we write the characteristic length as $\hbar c/\Lambda$, Λ is the characteristic energy above which the perturbation calculation gives a cross section that exceeds the unitary limit. The value of Λ is obtained from (3.19.4) as

$$\Lambda \simeq \hbar c\sqrt{\hbar c/G} \simeq 300 \text{ GeV}. \tag{3.19.6}$$

Neutrino interactions are expected from (3.19.1) through (3.19.3) as

$$\nu_\mu(\bar{\nu}_\mu) + n(p) \rightarrow p(n) + \mu^-(\mu^+), \tag{3.19.7}$$

$$\nu_\mu + e^- \rightarrow \mu^- + \nu_e, \tag{3.19.8}$$

$$\nu_e(\bar{\nu}_e) + n(p) \rightarrow p(n) + e^-(e^+), \tag{3.19.9}$$

$$\bar{\nu}_e + e^- \rightarrow \mu^- + \bar{\nu}_\mu. \tag{3.19.10}$$

If the universal Fermi interaction holds between identical currents, such as $(e\nu_e)(e\nu_e)$, we expect elastic scattering:

$$\bar{\nu}_e + e \rightarrow \bar{\nu}_e + e. \tag{3.19.11}$$

Another elastic scattering,

$$\nu_\mu + p \rightarrow \nu_\mu + p, \tag{3.19.12}$$

would be possible if currents were composed of identical particles, such as $(\nu_\mu \nu_\mu)$ and (pp). The reaction (3.19.11) would have to be responsible for the energy-loss process in evolved stars with high central temperatures; a detailed analysis on the evolution of massive stars is not in favor of the neutrino pair process, $e^- + e^+ \to \nu_e + \bar{\nu}_e$ (Hayashi 62), whereas the evolution of stars toward white dwarfs would require this process. The reaction (3.19.12) was examined by using an artificial neutrino beam, but the occurrence of such events was found to be less than 5 percent of that of (3.19.7) (Bernardini 64).

In view of the universality of the weak Fermi interactions, especially of that of the vector coupling constant, an idea of introducing intermediate bosons has revived. According to the intermediate-boson hypothesis the weak interaction of four fermions does not take place directly but is mediated by a boson, which interacts semiweakly with a pair of fermions and may be called the weak boson, abbreviated as W. If the weak boson has a mass m_W and a coupling constant g_W, they are related to the Fermi coupling constant as

$$\frac{1}{\sqrt{2}} G_V = g_W{}^2 \left(\frac{\hbar c}{m_W}\right)^2. \tag{3.19.13}$$

Hence $g_W{}^2/\hbar c$ is dimensionless and as small as

$$\frac{g_W{}^2}{\hbar c} \simeq 0.7 \times 10^{-5} \left(\frac{m_W{}^2}{m_p}\right). \tag{3.19.14}$$

Once the weak boson was produced it would decay into two fermions, for example, through

$$W^+ \to \nu_\mu + \mu^+. \tag{3.19.15}$$

This could contribute to muons at great depths. The flux of neutrino-induced muons would be greater in this case than we would expect without the weak boson.

Muon production would be enhanced if a weak boson were produced in the intermediate state as follows:

$$\bar{\nu}_e + e^- \to W^- \to \bar{\nu}_\mu + \mu^-, \tag{3.19.16}$$

because this is a resonance process.

All these processes concerning the weak boson depend much on its mass. In the infinitely large mass the weak-boson theory becomes identical with the direct four-fermion interaction. A difference from what we expect from the latter could thus be attributed to the existence of the weak boson of finite mass.

The cross sections for these processes are proportional to G^2. Since G^2 is proportional to the fourth power of length, the cross section has to be proportional to the square of energy, as can be readily seen from dimensional analysis. The energy dependence should cease when the energy reaches the characteristic energy given in (3.19.6). Thus we are able to give an approximate cross section for neutrino interactions as

$$\sigma \simeq \left(\frac{G}{\hbar c}\right)^2 \left(\frac{E^*}{\hbar c}\right)^2 \frac{1}{1+(E^*/\Lambda)} \simeq \frac{(E^*/\Lambda)^2}{1+(E^*/\Lambda)^2} \cdot 5 \times 10^{-33}\ \text{cm}^2, \tag{3.19.17}$$

where E^* is the incident energy in the center-of-mass system.

The energy dependence in (3.19.17) yields a large difference between neutrino collisions with nucleons and with electrons. The neutrino energy in the laboratory system is

$$E_\nu \simeq \frac{E^{*2}}{2m_p} \quad \text{or} \quad \frac{E^{*2}}{2m_e}, \tag{3.19.18}$$

according to whether the target is a nucleon or an electron. The former is smaller by a factor of m_e/m_p than the latter. In other words, the cross sections in these cases differ by a factor of m_p/m_e at the same neutrino energy, provided $E^* \ll \Lambda$. Consequently neutrino interactions with electrons are harder to observe than those with nucleons. The relation (3.19.18) would be important also for the energy of resonance, because the resonance could take place at $E^* \simeq M_W c^2$.

3.19.2 Neutrino-Interaction Cross Sections

Elastic cross sections have been measured by using artificial neutrinos, and the results are in essential agreement with theory. The cross section for $\bar{\nu}_e + p \rightarrow n + e^+$ measured with neutrinos from reactors has been found to be

$$\sigma_{\text{el}}\,(\text{several MeV}) \simeq 1.1 \times 10^{-43}\ \text{cm}^2. \tag{3.19.19}$$

In this energy region the cross section increases with the square of energy and is expected to reach a value as large as 10^{-38} cm^2 at about 1 GeV.

In the GeV energy region neutrinos from accelerators are used. The elastic cross section for the process in (3.19.7) is found to be

$$\sigma_{\text{el}}\,(\text{several GeV}) \simeq 0.4 \times 10^{-38}\ \text{cm}^2. \tag{3.19.20}$$

The cross section seems to level off above a few GeV, as is expected from the effect of the nucleon form factor.

At about 1 GeV the inelastic process due to meson production occurs as frequently as the elastic process. The energy dependence of the inelastic cross section is expected to be

$$\sigma_{\text{in}} \simeq 0.4 \times 10^{-38} E_\nu \quad (\text{GeV})\ \text{cm}^2. \qquad (3.19.21)$$

In the inelastic event a muon gets about half the incident energy on the average, the rest being shared with the mesons produced and a recoil nucleon. This is compared with a value of 0.8 for the fractional energy transfer to a muon in the elastic process.

Cross sections for energies above 10 GeV can be obtained by means of cosmic-ray neutrinos. By the summer of 1965 several muons produced underground by neutrinos had been observed by each of two groups working in Indian (Achar 65) and South African (Reines 65) gold mines. The frequency of such muons seems to be not inconsistent with the cross sections given in (3.19.20) and (3.19.21). The existence of the weak boson would result in an additional flux of muons through:

$$\nu_\mu + N \rightarrow N' + \mu^- + W^+, \qquad W^+ \rightarrow \mu^+ + \nu_\mu. \qquad (3.19.22)$$

The W-meson production above may take place in the nuclear Coulomb field of nuclei, $\nu_\mu + N \rightarrow \nu_\mu + N' + W^- + W^+ \rightarrow N' + \mu^- + W^+$. Since this is a first-order process in g_W, the cross section is estimated to be as large as $10^{-37} Z^2$ cm^2 per nucleus. Hence this would give a considerable flux of neutrino-produced muons.

However, no conclusive evidence for the existence of the W-meson has yet been obtained. The neutrino experiments described above can rule out the W-meson of mass smaller than several nucleon masses.

REFERENCES

Achar 65 — Achar, C. V., et al., *Phys. Letters*, **18**, 196 (1965); **19**, 78 (1965).

Aizu 60 — Aizu, H., et al., *Prog. Theor. Phys. Suppl.*, No. 16, 54 (1960).

Akashi 62 — Akashi, M., et al., *J. Phys. Soc. Japan*, **17**, Suppl. A–III, 427 (1962); *Proc. Int. Conf. on High-Energy Physics*, CERN, (1962).

Amai 57 — Amai, S., et al., **17**, 241 (1957).

Ashmore 60 — Ashmore, A., et al., *Phys. Rev. Letters*, **5**, 576 (1960).

Baker 61 — Baker, W. F., et al., *Phys. Rev. Letters*, **7**, 101 (1961).

Belenkii 55 — Belenkii, S. Z., and Landau, L. D., *Fortschr. Physik*, **3**, 536 (1955).

Bernardini 64 — Bernardini, G., *Proc. Int. Conf. on High-Energy Physics at Dubna* (1964).

Bigi 62 — Bigi A., et al., *Proc. 1962 Int. Conf. on High-Energy Physics*, CERN, 247 (1962).

Birger 59 — Birger, N. G., and Smorodin, Yu. A., *J. Exp. Theor. Phys.* **36**, 1159 (1959); **37**, 1355 (1959).

Bradt 50	Bradt, H. L., and Peters, B., *Phys. Rev.*, **77**, 54 (1950).
Brisbout 56	Brisbout 56 F. A., et el., *Phil. Mag.*, **1**, 605 (1956).
Castagnoli 53	Castagnoli, C., et al., *Nuovo Cim.*, **10**, 1539 (1953).
Cerulus 61	Cerulus, F., *Nuovo Cim.*, **22**, 958 (1961).
Ciok 58	Ciok, P. *et al.*, *Nuovo Cim.*, **8**, 166 (1958).
Cocconi 54	Cocconi, G., *Phys. Rev.*, **93**, 1107 (1954).
Cocconi 58	Cocconi, G., *Phys. Rev.*, **111**, 1699 (1958).
Cocconi 62	Cocconi, G., Koester, L. J., and Perkins, D. H., UCRL–10022, 167 (1962).
Cocconi 64	Cocconi, G., *Nuovo Cim.*, **33**, 643 (1964).
Cool 61	Cool, R. C., *Proc. 1961 Int. Conf. on High-Energy Accelerators*, 15 (1961).
Cool 65	Cool, R. C., *et al.*, *Phys. Rev. Letters*, **14**, 724 (1965).
Cronin 57	Cronin, J., Cool, R. C., and Abashian, A., *Phys. Rev.*, **107**, 1121 (1957).
Daiyasu 62	Daiyasu, K., *et al.*, *J. Phys. Soc. Japan*, **17**, Suppl. A–III, 344, 357 (1959).
Daniel 63	Daniel, R. R. et al., *Nuovo Cim.*, *Suppl.* **1**, 1163 (1963).
Dekkers 65	Dekkers, D., et al., *Phys. Rev.*, **137**, B963 (1965).
Dobrotin 63	Dobrotin, N. A., et al., *Proc. Int. Conf. on Cosmic Rays at Jaipur*, **5**, 141 (1963).
Dremin 60	Dremin, I. M., and Chernavsky, D. S., *J. Exp. Theor. Phys.*, **38**, 229 (1960).
Duller 54	Duller, N. M., and Walker, W. D., *Phys. Rev.*, **93**, 215 (1954).
Edwards 58	Edwards, B., et al., *Phil. Mag.*, **3**, 237 (1958).
Ericson 58	Ericson, T. and Strutinski, V., *Nucl. Phys.*, **8**, 284 (1958); **9**, 689 (1958–9).
Ericson 61	Ericson, T., *Nuovo Cim.*, **21**, 605 (1961).
Ezawa 59	Ezawa, H., *Nuovo Cim.*, **11**, 745 (1959).
Fermi 50	Fermi, E., *Prog. Theor. Phys.*, **5**, 570 (1950).
Fidecaro 61	Fidecaro, M., *et al.*, *Nuovo Cim.*, **19**, 382 (1961).
Fowler 58	Fowler, G. N., and Wolfendale, A. W., *Progress in Elementary Particle and Cosmic Ray Physics*, **IV**, 105 (1958).
Fowler 63	Fowler, P. H., *Proc. Int. Conf. on Cosmic Rays at Jaipur*, **5**, 182 (1963).
Frautschi 62	Frautschi, S. C., *Nuovo Cim.*, **28**, 409 (1963).
Fujimoto 67	Fujimoto, Y., and Hayakawa, S., *Handbuch der Physik*, Bd 46/2, 115, Springer-Verlag, Berlin, 1967.
Fujioka 61	Fujioka, G., *J. Phys. Soc. Japan*, **16**, 1107 (1961).
Fujioka 63	Fujioka, G., et al., *Nuovo Cim.*, Suppl. **1**, 1143 (1963).
Fukuda 50	Fukuda, H., and Takeda, G., *Prog. Theor. Phys.*, **5**, 957 (1950).
Fukui 59	Fukui, S., Kitamura, T., and Watase, Y., *Phys. Rev.*, **113**, 315 (1959).
Fukuda 53	Fukuda, H., *Phys. Rev.*, **89**, 842 (1953).
George 52	George, E. P., *Progress in Cosmic Ray Physics*, **I**, 393 (1952).
Gierula 60	Gierula, J., *Proc. Int. Conf. on High-Energy Physics*, Rochester, 816 (1960).
Grigorov 58	Grigorov, N. L., Murgin, V. S., and Rapoport, I. D., *J. Exp. Theor. Phys. USSR*, **34**, 506 (1958).

Grigorov 59 — Grigorov, N. L., et al., *Proc. of Moscow Cosmic Ray Conference*, **I**, 130 (1959).

Guseva 62 — Guseva, V. V., et al., *J. Phys. Soc. Japan*, **17**, Suppl. A–III, 375 (1962).

Hansen 60 — Hansen, L. F., and Fretter, W. B., *Phys. Rev.*, **118**, 812 (1960).

Harber-Schaim 54 — Harber-Schaim, U., and Yekutieli, G., *Nuovo Cim.*, **11**, 172, 683 (1954).

Hasegawa 57 — Hasegawa, S., Nishimura, J., and Nishimura, Y., *Nuovo Cim.*, **6**, 697 (1957).

Hasegawa 61 — Hasegawa, S., *Progr. Theor. Phys.*, **26**, 150 (1961).

Hasegawa 62 — Hasegawa, S., INSJ 48, University of Tokyo (1962).

Hasegawa 67 — Hasegawa, S., and Yokoi, K., *Butsuri*, **20**, 586 (1965); *Proc. Phys. Soc. Japan.*

Hayashi 62 — Hayashi, C., Hoshi, R., and Sugimoto, D., *Prog. Theor. Phys.*, Suppl. No. 22, 1 (1962).

Heitler 53 — Heitler, W., and Terreaux, C., *Proc. Phys. Soc.*, **A66**, 929 (1953).

Higashi 64 — Higashi, S., et al., *Nuovo Cim.*, **32**, 1 (1964).

Higashi 65 — Higashi, S., et al., *Nuovo Cim.*, **38**, 107 (1965).

Imaeda 63 — Imaeda, K., and Kazuno, M., *Nuovo Cim.*, Suppl., **1**, 1197 (1963).

Iso 59 — Iso, C., Mori, K., and Namiki, M., *Prog. Theor. Phys.*, **22**, 403 (1959).

Jain 65 — Jain, P. L., and McNulty, P. J., *Phys. Rev. Letters*, **14**, 611 (1965).

Kaneko 58 — Kaneko, S. and Okazaki, S., *Nuovo Cim.*, **8**, 521 (1958).

Kessler 56 — Kessler, D., and Kessler, P., *Nuovo Cim.*, **4**, 601 (1956).

Kim 64 — Kim, C. O., *Phys. Rev.*, **136**, B515 (1964).

Koba 56 — Koba, Z., *Prog. Theor. Phys.*, **15**, 461 (1956).

Koba 59 — Koba, Z., and Takagi, S., *Fortschr. Physik*, **7**, 1 (1959).

Koba 60 — Koba, Z., *Nuovo Cim.*, **18**, 608 (1960).

Koba 61 — Koba, Z., *Acta Phys. Polonica*, **20**, 213 (1961).

Kobayashi 64 — Kobayashi, T., *et al.*, *Prog. Theor. Phys.*, **32**, 738 (1964).

Kolkunov 62 — Kolkunov, V. A., *Soviet Phys.*, JETP, **16**, 1025 (1962).

Koshiba 63a — Koshiba, M., et al., *Nuovo Cim.*, Suppl. **1**, No. 4, 1091 (1963*a*).

Koshiba 63b — Koshiba, M., *Proc. Int. Conf. on Cosmic Rays at Jaipur, 1963*, **5**, 293 (1963*b*).

Lal 62 — Lal, S., Pal, Y., and Raghaven, R., *J. Phys. Soc. Japan*, **17**, Suppl. **A–III**, 393 (1962).

Lal 65 — Lal, S., et al., *Phys. Letters*, **14**, 332 (1965).

Landau 53 — Landau, L. D., *Izv. Akad. Nauk USSR*, **17**, 51 (1953).

Lindenbaum 57 — Lindenbaum, S. J., and Sternheimer, R. M., *Phys. Rev.*, **105**, 1874 (1957).

Lohrmann 58 — Lohrmann, E., and Teucher, M. W., *Phys. Rev.*, **112**, 587 (1958).

Lohrmann 59 — Lohrmann, E., and Teucher, M. W., *Phys. Rev.*, **115**, 636 (1959).

Manjavidze 62 — Manjavidze, Z. Sh., and Roinishvili, N. H., *J. Phys. Soc. Japan*, **17**, Suppl. **A–III**, 420 (1962).

Matsumoto 63 — Matsumoto, S., *J. Phys. Soc. Japan*, **18**, 1 (1963).

McCusker 53 McCukser, C. B. A., and Roessler, F. C., *Nuovo Cim.*, **10**, 127 (1953).

McCusker 64 McCusker, C. B. A., and Peak, L. S., *Nuovo Cim.*, **31**, 525 (1964).

Messel 52 Messel, H., *Proc. Phys. Soc.* (London), **A65**, 465 (1952).

Messel 54 Messel, H., *Progress in Cosmic Ray Physics.*, **11**, 135 (1954).

Meyer 63 Meyer, H., Teucher, M. W., and Lohrmann, E., *Nuovo Cim.*, **28**, 1399 (1963).

Miesowicz 62 Miesowicz, M., *J. Phys. Soc. Japan*, **17**, Suppl. A–III, 458 (1962).

Minakawa 58 Minakawa, O., et al., *Nuovo Cim.*, Suppl. **8**, 761 (1958).

Minakawa 59 Minakawa, O., et al., *Nuovo Cim.*, **11**, 125 (1959).

Miura 62 Miura, I., and Tanaka, Y., *Proc. 1962 Int. Conf. on High-Energy Physics*, CERN, 637 (1962).

Miyake 63 Miyake, S., et al., *J. Phys. Soc. Japan*, **18**, 592 (1963).

Morrison 61 Morrison, D. R. O., *Proc. Int. Conf. on Throretical Aspects of Very High Energy Phenomena*, CERN, 61–22, 153 (1961).

Namiki 57 Namiki, M., and Iso, C., *Prog. Theor. Phys.*, **18**, 591 (1957).

Nishimura 56 Nishimura, J., *Soryū shiron Kenkyū*, **12**, 24 (1956).

Nishimura 62 Nishimura, J., *J. Phys. Soc. Japan*, **17**, Suppl. A-III, 432, 526 (*1962*).

Niu 58 Niu, K., *Nuovo Cim.*, **10**, 994 (1958).

Oda 62 Oda, M., and Y. Tanaka, Y., *J. Phys. Soc. Japan*, **17**, Suppl. A-III, 282 (1962).

Orear 64 Orear, S., *Phys. Rev. Letters*, **12**, 112 (1964).

Pal 63 Pal, Y., Ray, A. K. and Rengarajan, T. N., *Nuovo Cim.*, **28**, 1177 (1963).

Perkins 60 Perkins, D. H., *Progress in Elementary Particle and Cosmic Ray Physics*, **5**, 259 (1960).

Peters 62 Peters, B., *J. Phys. Soc. Japan*, **17**, Suppl. A–III, 552 (1962); *Proc. Int. Conf. on High-Energy Physics*, CERN (1962).

Powell 59 Powell, C. F., Fowler, P. H., and Perkins, D. H., *The Study of Elementary Particles by the Photographic Method*, Pergamon, New York, 1959.

Reines 65 Reines, F., *et al.*, *Phys. Rev. Letters*, **15**, 429 (1965).

Rosental 54 Rosental, L. L., and Chernavsky, D. S., *Usp. Fiz. Nauk*, **52**, 185 (1954).

Sakata 56 Sakata, S., *Progr. Theor. Phys.*, **16**, 686 (1956).

Salzman 60 Salzman, F., and Salzman, G. *Phys. Rev. Letters*, **5**, 377 (1960).

Sitte 61 Sitte, K., *Handbuch der Physik*, XLVI/1, 215, Springer-Verlag, Berlin (1961.

Sternheimer 61 Sternheimer, R. M., and Lindenbaum, S. J., *Phys. Rev.*, **123**, 333 (1961).

Terashima 55 Terashima, Y., *Progr. Theor. Phys.*, **13**, 1 (1955).

Williams 64 Williams, R. W., *Rev. Mod. Phys.*, **36**, 815 (1964).

Yajima 65a Yajima, N., and Hasegawa, S., *Progr. Theor. Phys.*, **33**, 184 (1965*a*).

Yajima 65b Yajima, N., et al., *Prog. Theor. Phys.*, Suppl. No. 33, 134 (1965*b*).

Zatsepin 62 Zatsepin, G. T., *J. Phys. Soc. Japan*, **17**, Suppl. A–III, 495 (1962).

CHAPTER 4

Behavior of Cosmic Rays in the Atmosphere and Underground

The behavior of cosmic rays in the atmosphere was investigated in detail mainly in the 1930s and 1940s with the intention of clarifying elementary particle interactions. These problems were solved except at high energies and are no longer central in cosmic-ray physics, but they do provide a basis for application to other fields. Therefore about half this chapter is inevitably classical and will serve as the background of newly developed fields in and around cosmic-ray physics. Very-high-energy cosmic rays, on the other hand, seem to imply something new.

4.1 Definitions of Intensities

By the *intensity* of cosmic rays we usually mean the number of particles that pass through unit area during unit time. When we refer to particles coming from a given direction the intensity is called *unidirectional intensity*.

The direction is usually assigned by the zenith angle θ and the azimuth angle φ as $j(\theta, \varphi)$. The *omnidirectional intensity* is obtained by integrating the unidirectional intensity over all directions as

$$J = \int j(\theta, \varphi) \sin \theta \, d\theta \, d\varphi; \qquad (4.1.1)$$

in terrestrial cases the integration is carried out over the upper hemisphere because particles travel mainly downward. We often refer to the vertical intensity, $j_{\perp} = j(\theta = 0)$, and express the zenith-angle dependence as

$$j(\theta, \varphi) = j_{\perp} \cos^n \theta, \qquad (4.1.2)$$

neglecting a weak azimuthal dependence. The value of n varies from one component to another, depending on altitude and energy.

If a component attenuates exponentially with the mean attenuation length L without changing its direction, the zenith-angle dependence at an atmospheric depth x is given by

$$j(\theta, x) = j_\perp(x=0) \exp\left(\frac{-x}{L \cos\theta}\right) \simeq j_\perp(x) \cos^{x/L}\theta, \qquad (4.1.3)$$

where isotropy at the top of the atmosphere is assumed.

A more general relation holds between J and $j_\perp$ provided that (*a*) cosmic rays impinge isotropically into the atmosphere, (*b*) all particles continue to propagate without changing their direction, and (*c*) the intensity change depends only on the thickness the rays traverse. Then the unidirectional intensity at zenith angle θ and depth x is equal to the vertical intensity at depth $x/\cos\theta$:

$$j_\theta(x) = j_\perp\left(\frac{x}{\cos\theta}\right). \qquad (4.1.4)$$

The omnidirectional intensity is expressed, on the assumption of azimuth-angle independence, as

$$J(x) = 2\pi \int_0^{\pi/2} j_\theta(x) \sin\theta \, d\theta = 2\pi \int_0^{\pi/2} j_\perp\left(\frac{x}{\cos\theta}\right) \sin\theta \, d\theta.$$

By differentiating this equation with respect to x we obtain

$$2\pi j_\perp(x) = J(x) - xJ'(x). \qquad (4.1.5)$$

This is the formula for obtaining the vertical intensity from the omnidirectional intensity and is called the *Gross transformation* (Gross 33). Since the variation of intensity with altitude is often measured with omnidirectional detectors such as ionization chambers, this relation is of practical use for obtaining the altitude dependence of unidirectional intensity.

The ionization chamber measures the number of ion pairs that are produced in unit time by cosmic rays in the gas contained in the chamber. The ionization current thus measured can be reduced to the rate of ion-pair production in the standard atmosphere; the number of ion pairs produced per second and per cm^3 in the standard atmosphere is called *ionization intensity*. This, designated by I, is related to the omnidirectional intensity as

$$J = sI. \qquad (4.1.6)$$

The s stands for *specific ionization* and represents the number of ion pairs produced by a cosmic-ray particle when it traverses 1 cm of the standard atmosphere. The value of s depends on the composition and energy of cosmic rays. The value of s is approximately 114 at sea level (Neher 52). On the other hand, $s = 60$ is found along tracks in cloud chambers. The discrepancy can be understood since the latter refers to primary ionization alone, whereas the former corresponds approximately to total ionization; in fact, this is consistent with the energy loss of 32 eV per ion pair.

Both flux intensity, j or J, and ionization intensity usually refer only to charged particles. Sometimes the intensities of neutral particles are measured by means of their conversion into charged particles; for example, thermal neutrons are detected with the aid of the $^{10}B(n, \alpha)^{7}\mathrm{Li}$ reaction. The cross section for this reaction is inversely proportional to the velocity of neutrons, v, and consequently the reaction rate is proportional to J/v; namely, to the *density* of thermal neutrons. Therefore the density is directly measurable for thermal and epithermal neutrons.

Detectors used for the observation of cosmic rays are sensitive to particles in a particular energy region, usually to those above a certain energy. The intensity may therefore be specified by the lower limit of energy E as $j(>E)$. By differentiating this or by measuring the intensity within a given energy range we can obtain the intensity of particles with energies between E and $E + dE$, $f(E)\,dE$. These two quantities are connected as

$$j(>E) = \int_E^{\infty} f(E')\,dE', \tag{4.1.7}$$

where $f(E)$ represents the *differential energy spectrum*, where $j(>E)$ is the *integral energy spectrum*. Although we refer here to the unidirectional intensity, both the differential and the integral energy spectra can, of course, be defined for the omnidirectional intensity and the density.

The energy spectrum is often expressed by a power law, with a negative power index that is nearly constant over a wide range of energy.

By energy we usually mean total energy. But we may also refer to kinetic energy, particularly when we are concerned with low-energy particles. In this case we distinguish kinetic energy by denoting it as W:

$$W = E - Mc^2, \tag{4.1.8}$$

where M is the mass of the particle concerned. At relativistic energies the difference between E and W is insignificant.

In some cases we refer to momentum, so that the *momentum spectrum* is used instead of the energy spectrum. In the extremely relativistic case momentum in units of eV/c is essentially equal to energy in eV, and consequently these two spectra are practically identical.

Momentum in eV/c is equal to rigidity in V for a single-charge particle. For multiple-charge particles such as heavy nuclei, however, they are different; rigidity R is connected with momentum p for a particle of charge Ze as

$$R = \frac{pc}{|Ze|}. \tag{4.1.9}$$

If we refer to momentum per nucleon for a heavy nucleus, (4.1.9) is rewritten as

$$R = \frac{A}{Z}\frac{pc}{e}. \tag{4.1.9'}$$

Hence the *rigidity spectrum* for a power law is related to the momentum spectrum as

$$R^{-\alpha-1}\,dR \propto \left(\frac{Z}{A}\right)^{\alpha} p^{-\alpha-1}\,dp. \tag{4.1.10}$$

4.2 Classification of Components

Cosmic rays are known to consist of various components. The components are defined depending on the method of observation. They have therefore phenomenological meaning. The component is a useful concept because the identification of a particle is often difficult from direct observation. The criteria for defining components are not always unique, as will be explained below.

Absorption. It has been known since the early 1930s that the absorption coefficient of cosmic rays changes rather abruptly as the thickness of an absorber increases, as described in Section 1.2. This observation is interpreted as due to the fact that cosmic rays consist of two components, one being rapidly absorbed and the other little absorbed. The former is called the *soft component*; and the latter, the *hard component*. These two components are conventionally distinguished by their penetrations, the soft component penetrating less than about 10 cm of lead and the hard component more.

The soft and hard components are respectively attributed mainly to electrons, which undergo bremsstrahlung, and to heavier particles, which do not. At sea level the latter consist mostly of muons.

Ionization Density. The ionization density of a track is most clearly revealed by the grain density in photographic emulsions and by the drop density of a track in cloud chambers. As is well known, the ionization density or the grain density decreases with increasing velocity of a particle and reaches a minimum value g_m. This is followed by the relativistic increase, but there is saturation at the plateau grain density g_p due to the density effect. Both the minimum grain density g_m and the plateau grain density g_p are defined for single-charge particles and provide standards of grain densities at different energies. Their absolute values depend not only on the kind of photographic plates but also on the conditions of development and other factors, whereas the relative values are nearly independent of these properties, $g_p/g_m \simeq 1.1$.

Tracks observed in photographic emulsions are conventionally classified according to their grain densities relative to g_m or g_p (Powell 59). The *thin track* has a grain density smaller than 1.4 g_m and is produced by a particle of relativistic or semirelativistic energy. The *gray track* is defined by a grain density of between 1.4 and 10 g_m, whereas the *black track* has a grain density greater than 10 g_m. The gray and black tracks are collectively called *heavy tracks*, because they are heavily ionizing and are produced mostly by particles heavier than mesons. The border values of the grain densities are not unique but are defined rather arbitrarily, although the above definition is the most common one. The energy ranges for protons and pions associated with these kinds of tracks are shown in Table 4.1.

Interactions with Matter. Although the above-mentioned ways of classifying components are based on the interactions of charged particles with matter, the production of secondary particles by interactions may also provide criteria for the classification of components. An interaction is often characterized by the generation of a shower that consists of a bundle of particles. There are two kinds of showers that are readily distinguishable from one another. One is generated frequently at rather small thicknesses of heavy materials; its secondary particles emerge

Table 4.1 Classification of Tracks by Grain Density

Track Type	Grain Density (g_m)	Kinetic Energy (MeV)	
		Proton	Pion
Thin	<1.4	>370	>55
Gray	1.4 to 10	25 to 370	4 to 55
Black	>10	<25	<4

with large angles and are rapidly absorbed. The other is generated with large mean free paths, and its secondary particles have a small angular divergence and great penetration.

The former is usually associated with the soft component, and the shower is reasonably accounted for in terms of the *cascade shower* of electrons and photons. Therefore the primary particles of such showers are identified as electrons and photons, and are collectively called the *electronic component*, or the E-*component*. The shower generated by this component is called the electronic, or cascade, shower.

The latter is associated with the hard component, and both its primary and secondary particles consist of nucleons and charged pions. The shower is a consequence of multiple meson production and/or a cascade process of these particles. Particles responsible for such showers —nucleons and charged pions—are called *nuclear active particles* and form the nuclear active component, or the *N-component*. A shower produced by this may be called the *N*-shower. Low-energy nuclear interactions do not give showers that consist of many particles coming down but give secondary particles that emerge almost isotropically, thus looking like *stars*. At high-energy nuclear interactions both *N*-showers and electronic showers occur together; these showers are called mixed showers and are associated with the radiative decay of the neutral pions produced.

Historically, "shower" is a name that is associated with cloud-chamber pictures, whereas "star" is associated with emulsion pictures. This is because material layers in cloud chambers were too thick for low-energy particles to come out, and photographic emulsions were not sensitive to high-energy shower particles. The *burst* is a phenomenon associated with a sudden increase of the ionization current of an ionization chamber and is regarded as due mainly to a shower or a star.

At sea level most cosmic-ray particles belong neither to the electronic nor to the nuclear active components. They penetrate a thick layer of matter without appreciable interactions; they form the *penetrating component* and are identified with muons. The only appreciable interactions of muons are the production of knock-on electrons, which may initiate knock-on showers, and at high energies bremsstrahlung and the creation of electron pairs with probabilities that are much smaller than the interaction probabilities of the electronic and *N*-components. Nuclear interactions of muons also take place with a probability that is much smaller than that of nuclear active particles.

It seems convenient to separate slow neutrons from the *N*-component because they are characterized by large scattering cross sections and also by particular capture processes.

Many other particles contained in cosmic rays are not important as far as the intensity observed with existing detectors is concerned. Neutrinos are nearly as abundant as muons, but they are difficult to observe. Hyperons and heavy mesons are so short-lived that their intensities are negligibly small compared with others.

Various ways of classifying particles and their interrelationships are summarized in Table 4.2.

Table 4.2 Classification of Cosmic-Ray Components

Absorption	Particle	Grain Density	Interaction
Hard component	Muon	Thin	Penetrating components
	Charged pion	Thin	Nuclear active component
	Proton	Thin	(N-component)
	Fast neutron	None	
Soft component	Proton	Gray, black	
	Charged pion	Gray	
	Muon	Gray	
	Electron	Thin	Electronic component
	Photon		(E-component)

4.3 Properties of Primary Cosmic Rays

Cosmic rays observed outside the earth's atmosphere are called *primary cosmic rays*, provided that the albedo effect is suitably subtracted. Their properties will be discussed in Chapter 6; here only a brief account is given in order to help the reader to understand later sections. Primary cosmic rays consist mainly of protons, and the flux of α-particles is a fraction of that of protons; other components are relatively unimportant as far as cosmic-ray phenomena in the atmosphere are concerned. If one is interested in the origin of cosmic rays, however, the composition of primary cosmic rays provides one of the most important pieces of information. In this chapter, however, we are not concerned with this point but leave this for Chapter 6.

For the discussions in this chapter, we have to know only the relative flux values,

$$j_p : j_\alpha : j_{\text{heavier}} = 1 : \tfrac{1}{7} : \tfrac{1}{60}, \tag{4.3.1}$$

that are observed at a given magnetic latitude. This means that nearly one-quarter of the primary nucleons are neutrons. There are also electrons, the flux of which is comparable to that of nuclei heavier than helium.

The composition of primary cosmic rays as described above is observed in a relatively low-energy region. In this region the primary particles are influenced by the geomagnetic field. At a given geomagnetic latitude and in a given direction only the particles that have rigidities higher than a certain value are able to arrive at the earth. With the aid of the geomagnetic effect the rigidity spectrum of primary cosmic rays can be obtained.

The rigidity spectrum in the geomagnetically sensitive region is subjected to modulation effects, as will be briefly discussed in Chapter 6. The intensity of primary cosmic rays is anticorrelated with solar activities. At the solar minimum the differential rigidity spectrum of protons has a maximum at about 1.6 GV, decreases toward high rigidity, and above 3 GV tends to (McDonald, 59, 60, 62)

$$j_p(R) = 5.0 \times 10^{-4} R^{-2.25} \quad \text{cm}^{-2}\,\text{sec}^{-1}\,\text{sr}^{-1}\,\text{MV}^{-1}, \qquad (4.3.2)$$

where R is measured in GV. The integral spectrum of protons is expressed by

$$j_p(>R) = 0.40 R^{-1.25} \quad \text{cm}^{-2}\,\text{sec}^{-1}\,\text{sr}^{-1} \qquad (4.3.3)$$

at high rigidities. At the solar maximum the intensity is reduced by a considerable factor. The reduction factor is smaller as rigidity increases, but it is still appreciable even above 15 GV.

The power spectrum given in (4.3.2) and (4.3.3) is conventional in the sense that the power index changes with energy; we can find a suitable power index in a rigidity range of interest. As energy increases the power index increases gradually. At an energy level of about 1 TeV the integral energy spectrum may be represented by

$$j_p(>E) = 1 \times E^{-1.6} \quad \text{cm}^{-2}\,\text{sec}^{-1}\,\text{sr}^{-1}, \qquad (4.3.4)$$

where E is the total energy in GeV.

The solar activity is responsible not only for the modulation of primary cosmic rays but also for the generation of energetic particles. Most solar particles are of relatively low energies and are observable only at high altitudes. In the case of intense solar particle production a sudden increase in cosmic-ray intensity is observed even at sea level.

4.4 Latitude Effects of Primary and Secondary Cosmic Rays

As mentioned in the preceding section, the geomagnetic effect determines the minimum rigidity of particles that can arrive at a given latitude. The minimum rigidity, or the *cutoff rigidity*, further depends on the direction of arrival. In most practical cases where the contribution of secondary cosmic rays is considered we refer to the cutoff energy for the vertical direction, because the vertical component gives the most important contribution. The vertical intensities measured at various latitudes are related to the cutoff energy, the primary spectrum, and the specific yield of secondary cosmic rays in the following way.

The vertical intensity of a component of secondary cosmic rays measured at latitude λ and depth x, $j_\perp(\lambda, x)$, is expressed as (Treiman, 52)

$$j_\perp(\lambda, x) = \sum_Z \int_{W_Z(\lambda)}^{\infty} j_Z(W_0) S_Z(W_0, x)\, dW_0, \tag{4.4.1}$$

where $j_Z(W_0)$ is the intensity of primary particles with atomic number Z and kinetic energy per nucleon W_0 in dW_0; $W_Z(\lambda)$ is the vertical cutoff energy, which depends on Z; $S_Z(W_0, x)$, called the *specific yield*, indicates the number of particles at x that are produced by a primary particle of energy W_0 and charge Z. The specific yield depends on the characteristics of the detector. If detectors are omnidirectional, the vertical intensity is derived from the omnidirectional intensity $J(x)$ by means of the Gross transformation (4.1.5) as

$$2\pi j_\perp(x) = \left(1 + \frac{x}{L}\right) J(x), \tag{4.4.2}$$

where L is the attenuation mean free path of the component under consideration.

Once the specific yield is known any change of the primary intensity and also that of the primary energy are in principle predictable from a change in the observed intensity.

The specific yield can be obtained from the latitude effect through the relation

$$\frac{\partial j_\perp(\lambda, x)}{\partial \lambda} = -\sum_Z \frac{dW_Z(\lambda)}{d\lambda} j_Z[W_Z(\lambda)] S_Z[W_Z(\lambda), x]. \tag{4.4.3}$$

If only protons were responsible for the observed intensity, the specific yield would be given by

$$S_p[W_p(\lambda), x] = -\frac{\partial j_\perp(\lambda, x)/\partial \lambda}{[dW_p(\lambda)/d\lambda]\, j_p[W_p(\lambda)]}, \tag{4.4.4}$$

in which the subscript p indicates quantities that pertain to primary protons. Actually, however, the contribution of primary α-particles is not negligible. Hence the average specific yield is given by

$$\bar{S}(\lambda, x) \equiv \sum_Z \frac{dW_Z}{d\lambda} \frac{j_Z S_Z}{\sum_Z (dW_Z/d\lambda) j_Z} = -\frac{\partial j_\perp/\partial \lambda}{\sum_Z (dW_Z/d\lambda) j_Z}. \tag{4.4.5}$$

The specific yield for a primary α-particle may be estimated by assuming that the collision of an α-particle produces four nucleons with energy equal to the energy per nucleon of the primary α-particle. Under this assumption it is expressed by

$$S_\alpha(x) = 4 \int \exp\left(\frac{-x'}{l_\alpha}\right) S_p(x - x') \frac{dx'}{l_\alpha}, \tag{4.4.6}$$

where l_α is the collision mean free path of α-particles. Because $S_p \propto \exp[-(x-x')/L]$ for large values of $x - x'$ and the integrand in (4.4.6) decreases rapidly as x' increases, (4.4.6) is approximately evaluated by

$$S_\alpha(x) \simeq 4\left(1 - \frac{l_\alpha}{L}\right)^{-1} S_p(x) \equiv A S_p(x). \tag{4.4.7}$$

By substituting (4.4.7) into (4.4.5) we obtain

$$\bar{S}(\lambda, x) = \frac{\dfrac{dW_p}{d\lambda} j_p(W_p) S_p(W_p, x) + A \dfrac{dW_\alpha}{d\lambda} j_\alpha(W_\alpha) S_p(W_\alpha, x)}{\dfrac{dW_p}{d\lambda} j_p(W_p) + \dfrac{dW_\alpha}{d\lambda} j_\alpha(W_\alpha)}$$

$$= \frac{\dfrac{dW_p}{d\lambda} S_p(W_p, x) + A f_\alpha(\lambda) \dfrac{dW_\alpha}{d\lambda} S_p(W_\alpha, x)}{\dfrac{dW_p}{d\lambda} + f_\alpha(\lambda) \dfrac{dW_\alpha}{d\lambda}} \tag{4.4.8}$$

where $f_\alpha(\lambda) \equiv j_\alpha(W_\alpha)/j_p(W_p)$.

As a useful example we give the specific yield concerning the neutron detector of Simpson (Simpson 53*a*, *b*). First, the average specific yield is obtained from the latitude effect as in (4.4.5). Second, the specific yield for a primary proton is obtained from (4.4.8), Finally, the specific yields for primary protons and α-particles are obtained separately with the aid of (4.4.7). The specific yields thus obtained for $x = 312$ g-cm^{-2} are shown in Fig. 4.1 as a function of the latitude rather than of the primary energy. Now we see that the difference between the latitudes at which shoulders occur in the specific-yield curves for a proton and an α-particle can be accounted for by the difference in the energies per nucleon at a given latitude. Hence the specific yield per nucleon can be given approximately as

$$S(W, 312) \begin{array}{ll} = 12(W - 0.83)\text{GeV}^{-1} & (W < 3 \text{ GeV}), \\ = 32 \text{ GeV}^{-1} & (W > 4 \text{ GeV}), \end{array} \tag{4.4.9}$$

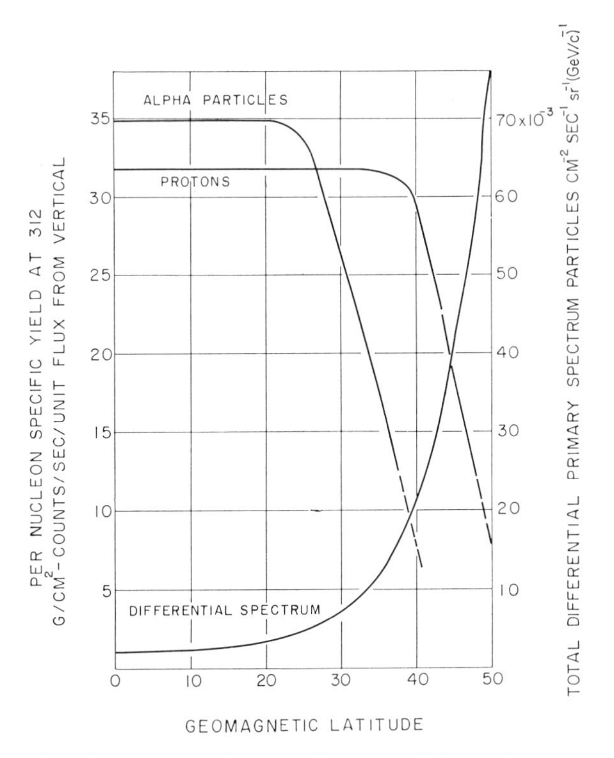

Fig. 4.1 Specific yields per nucleon at 312 g-cm^{-2} for primary protons and α-particles. The differential primary spectrum is also shown as a function of latitude (Simpson 35b).

where the kinetic energy per nucleon, W, is measured in GeV.

At $x = 680$ g-cm^{-2} the specific yield per nucleon is similarly obtained as

$$S(W, 680) \begin{array}{ll} = 0 & (W \leq 0.83 \text{ GeV}), \\ = 9.6 \ln \left(\dfrac{1+W}{1.83}\right) & (W > 0.83 \text{ GeV}) \end{array} \tag{4.4.10}$$

At greater atmospheric depths the specific yield has the same energy dependence as (4.4.10) but is reduced by a factor $\exp[-(x-680)/L]$ with $L \simeq 140$ g-cm^{-2}, because the latitude effect does not change with depth.

The specific yield at great depths continues to increase with energy, whereas that at 312 g-cm^{-2} levels off at about 4 GeV. This indicates that the development of the nuclear cascade is appreciable for $W > 4$ GeV and $x > 300$ g-cm^{-2}, so that the number of nucleons per incident nucleon increases with the thickness of the atmosphere as well as with incident energy.

The expression (4.4.1) is useful for decomposing the time variation of the intensity observed into three origins. The variation of (4.4.1) is expressed by

$$\delta j_{\perp}(\lambda, x) = \sum_{Z} \int_{w_Z(\lambda)}^{\infty} \delta j_Z(W_0)\, S_Z(W_0, x)\, dW_0 - \sum_{Z} j_Z(W_0)\, S_Z(W_0, x)\, \delta W_Z(\lambda)$$

$$+ \sum_{Z} \int_{w_Z(\lambda)}^{\infty} j_Z(W_0)\, \delta S_Z(W_0, x)\, dW_0 . \tag{4.4.11}$$

The first term on the right-hand side represents the interplanetary modulation, the second one is due to the change of the geomagnetic field, and the last term is attributed to atmospheric effects.

4.5 Composition of Heavy Nuclei in the Upper Atmosphere

The composition of heavy nuclei measured at a finite depth of the atmosphere is modified by interactions with air nuclei. The composition in free space is therefore obtained by correcting for this effect. The atmospheric correction is so important for obtaining the composition of primary cosmic rays that it has been investigated in detail, as will be shown below.

4.5.1 Solution of the Diffusion Equation

A heavy nucleus impinging on the atmosphere interacts with matter. The main consequences of the interactions are the breakup of nuclei and loss of energy by ionization. The deflection of primary and secondary particles due to the interactions is not of primary importance except at low energies. Hence the change of the intensity with atmospheric depth can be described by the one-dimensional diffusion equation,

$$\frac{\partial j_i(x, W)}{\partial x} = \sum_{k \geq i} \int_{W}^{\infty} dW'\, B_{ki}(W', W)\, j_k(x, W') - A_i(W)\, j_i(x, W)$$

$$+ \frac{\partial}{\partial W} [W_i(W)\, j_i(x, W)], \tag{4.5.1}$$

where $j_i(x, W)$ is the omnidirectional intensity of particles of type i at depth x with kinetic energy per nucleon W in dW. B_{ki} is the fragmentation probability per unit length of type k into type i, producing a fragmentation product of energy W in dW, and

$$A_i(W) = \int \sum_{k \leq i} B_{ik}(W, W')\, dW'$$

is the catastrophic absorption probability per unit length of a particle of type i. In the last term in (4.5.1), $W_i(W)$ is the ionization energy loss per unit length.

The fragmentation of a heavy nucleus takes place mainly through the emission of low-energy particles in the rest system of an incoming nucleus. Therefore the energy per nucleon is essentially conserved in the course of fragmentation, and the *fragmentation probability* is expressed as

$$B_{ki}(W', W) = \frac{\delta(W - W')P_{ki}(W)}{l_k(W)},$$

$$A_i(W) = \frac{1}{l_i(W)}, \tag{4.5.2}$$

where l_i is the *breakup collision mean free path* of particles of type i and P_{ki} is the fragmentation probability per collision. The diffusion equation (4.5.1) is then written as

$$\frac{\partial j_i(x, W)}{\partial x} = \sum_{k \geq i} \frac{P_{ki}(W)\, j_k(x, W)}{l_k(W)} - \frac{j_i(x, W)}{l_i(W)} + \frac{\partial}{\partial W}[W_i(W)\, j_i(x, W)]. \tag{4.5.1$'$}$$

At sufficiently high energies both the energy loss by ionization and the energy dependences of l and P may be neglected. Then the solution of (4.5.1′) is obtained as

$$j_i(x, W) = j_i(0, W) \exp\left(\frac{-x}{l_i'}\right) + \sum_{k > 1} F_{ki}\left[j_k(0, W) \exp\left(\frac{-x}{l_i'}\right) - j_k(x, W)\right], \tag{4.5.3}$$

where

$$F_{ki} = \frac{l_{ki}'}{l_k}\left(P_{ki} + \sum_{j=i+1}^{k-1} P_{kj} P_{ji} \frac{l_{ji}'}{l_j} + \cdots + P_{k,k-1} \right.$$
$$\left. + \cdots + P_{i+1,i} \frac{l_{k-1,k}' + \cdots + l_{i+1,i}'}{l_{k-1} + \cdots + l_{i+1}}\right), \tag{4.5.4}$$

$$l_{ki}' = \frac{l_k' l_i'}{l_i' - l_k'}, \qquad l_i' = \frac{l_i}{(1 - P_{ii})}.$$

If several types of nuclei are grouped, as is conventionally done because of the difficulty in distinguishing individual nuclei and because of their small intensities, the coefficients in the diffusion equation analogous to (4.5.1) are obtained by taking the weighted means of corresponding quantities of individual types; designating charge groups by $I, K, \ldots$, we have

$$\frac{1}{l_I} = \frac{\sum_{i \subset I} j_i / l_i}{\sum_{i \subset I} j_i},$$

$$P_{KI} = \frac{\sum_{k \subset K} P_{kI}\, j_k / l_k}{\sum_{k \subset K} j_k / l_k},$$

$$P_{kI} = \sum_{i \subset I} P_{ki},$$

$$\frac{1}{l'_I} = \frac{1 - P_{II}}{l_I} = \frac{\sum_{i \subset I} (1 - P_{iI})\, j_i / l_i}{\sum_{i \subset I} j_i}. \tag{4.5.5}$$

At low energies the loss of energy by ionization is no longer negligible. Assuming still the energy independence of P and l, we obtain the solution of (4.5.1′) as follows (Aizu 60):

$$j_i(x, W) = j_i(0, W) \frac{W_i(W_x{}^i)}{W_i(W)} e^{-x/l'_i}$$
$$+ e^{-x/l'_i} \int_0^x S_i(x', W^i_{x-x'}) \frac{W_i(W^i_{x-x'})}{W_i(W)} e^{x^i/l'_i}\, dx', \tag{4.5.6}$$

where

$$S_i(x, W) = \sum_{k>i} P_{ki} \frac{j_k(x, W)}{l_k};$$

$W_x{}^i$ is the energy of a particle with the residual range of $x + R_i(W)$, and $R_i(W)$ is the range of a particle of type i with energy W,

$$R_i(W_x{}^i) = R_i(W) + x.$$

When the depth x is much smaller than the fragmentation mean free path, l_i/P_{ki}, (4.5.6) can be approximated by

$$j_i{}^{(0)}(x, W) = j_i(0, W_x{}^i) \frac{W(W_x{}^i)}{W_i(W)} e^{-x/l'_i}. \tag{4.5.7}$$

4.5.2 Collision Mean Free Path and Fragmentation Probabilities

The collision cross section for two heavy nuclei of mass numbers A_1 and A_2 may be expressed in a simple way as in (3.3.12):

$$\sigma = \pi r_0{}^2 (A_1{}^{2/3} + A_2{}^{2/3} - b). \tag{4.5.8}$$

The values of r_0 and b are determined by taking into account the nuclear collisions of heavy nuclei in emulsions. Because of the energy dependence of the nucleon-nucleon cross sections, the values of r_0 and b are weakly energy dependent. With these values of r_0 and b the collision mean free paths of groups of heavy nuclei in air are obtained, as shown in Table 4.3.

Table 4.3 Collision Mean Free Paths of Heavy Nuclei

Energy (GeV/nucleon)	$r_0(10^{-13}$ cm)	$b(10^{-13}$ cm)	Group Z	Air (g-cm^{-2}) L 3–5	M 6–9	H 10–19	VH ≥ 20	Error
~ 0.5	1.0 ± 0.3	0.7 ± 1.6		33.4	28.6	22.4	14.4	$\pm 10\%$
~ 3	1.1 ± 0.2	0.5 ± 1.1		32.5	27.6	21.6	13.9	$\pm 5\%$

The fragmentation probabilities are very difficult to determine because no direct experiment has ever been made. One source of information comes from those interactions of heavy nuclei in emulsions that are associated with a small number of heavy tracks. Such interactions are supposed to be due mainly to collisions with light nuclei in emulsions, but some of them are due to those with silver and bromine nuclei associated with a few heavy tracks. The other comes from the interactions with airlike nuclei, such as carbon sandwiched between emulsions. The latter information is regarded as more reliable but its statistics are poorer than those of the former. The fragmentation probabilities for air are therefore obtained from the weighted average of the results from these two sources; according to O'Dell et al. (O'Dell 62) 1367 interactions with $N_h \leq 7$ and 206 interactions with airlike nuclei were averaged by doubling the statistical weight of the latter. Table 4.4

Table 4.4 Fragmentation Probabilities (%)

	Emulsion	Airlike Target	Weighted Average	Waddington	Aizu et al.
P_{LL}	15	13	15	13	15
P_{MM}	16	6	14	16	15
P_{ML}	26	33	27	21	25
P_{HH}	27	19	24	31	30
P_{HM}	34	23	31	33	33
P_{HL}	23	43	28	14	15

shows the fragmentation probabilities thus obtained as well as those adopted by Waddington (Waddington 62), based mostly on interactions in emulsions, and those adopted by Aizu et al. (Aizu 60).

The standard errors of the above figures based merely on statistics are rather small, on the order of 10 percent, but the systematic errors seem to be considerable, as indicated by the disagreement among different experiments.

4.5.3 Relative Abundances of Heavy Nuclei at the Top of the Atmosphere

The relative abundances of heavy nuclei in free space have to be deduced by taking account of the diffusion process discussed above. The technical details are so complicated (Engler 58, Aizu 60) that only essential points are described here.

The first method, which has been more popular, is based on extrapolation, making use of the solution of the diffusion equation (5.4.3). It is in principle possible if the collision mean free paths and the fragmentation probabilities are known. However, the values of the latter are not agreed on by all authors, and consequently the intensities at the top depend considerably on the fragmentation probabilities that are adopted; for example, the abundances of the light group ($Z = 3$ to 5) given by O'Dell et al. (O'Dell 62) and by Waddington (Waddington 62) are at variance with one another, and the disagreement is due mainly to the differences in P_{HL} and P_{ML}, as seen in Table 4.4.

The second method is based on the zenith-angle dependence of the intensities. If primary cosmic radiation is isotropic at the top of the atmosphere, the zenith-angle dependence can be reduced to the depth dependence, so that extrapolation to the top of the atmosphere may be possible. However, this assumption is not always valid because of the complicated geomagnetic effect. In order to take the geomagnetic effect into account the azimuth angle of a detector has to be kept constant. Even if this is done, effects of the shadow cone are very difficult to determine.

4.6 Genetic Relations Between Various Components

As discussed in the preceding section, heavy nuclei break up into nucleons in a small thickness of the atmosphere; both of which belong to a common component, the nuclear active component. The nuclear active particles are characterized by strong interactions with nuclei and result in nuclear disintegrations. Nuclear disintegration is almost always associated with pion production if the energy of a nuclear active particle is higher than 1 GeV.

The neutral pions that are produced decay immediately into photons, which initiate the cascade showers that form the electronic component. Most charged pions decay into muons and neutrinos; some of the muons decay into electrons and neutrinos, whereas the rest of them penetrate through the atmosphere. The decay electrons participate in the electronic component, and the neutrinos penetrate through the atmosphere and even the earth.

The decay rates of the pion and the muon are inversely proportional to their energies, because the mean lifetime of decay is proportional to the Lorentz factor. Let us consider an unstable particle of mass m and proper lifetime τ. The mean lifetime for total energy E is given by

$$\tau(E) = \frac{E}{mc^2}\tau = \gamma\tau. \tag{4.6.1}$$

The decay rate per unit length is therefore expressed by

$$\frac{1}{c\beta\tau(E)} = \frac{1}{c\beta\gamma\tau} = \frac{m}{p\tau}, \tag{4.6.2}$$

where $p = \beta\gamma mc$ is the momentum of the particle. In a medium of density ρ the mean free path for spontaneous decay l_d is given by

$$\frac{1}{l_d} = \frac{m}{\rho p\tau}. \tag{4.6.3}$$

If the decay mean free path is shorter than the interaction mean free path, the particle decays with a high probability. In the isothermal atmosphere the density depends on the pressure or the atmospheric depth. In the actual atmosphere the temperature decreases with altitude and tends to a nearly constant value in the stratosphere. Hence the scale height decreases as the altitude increases; the scale height defined by

$$H \equiv \frac{x}{\rho} \tag{4.6.4.}$$

is about 8.4 km at sea level and reaches 6.3 km for $x < 200$ g-cm^{-2}. The scale height versus the depth is shown in Fig. 4.2.

With the aid of the scale height the reciprocal decay mean free path (4.6.3) is expressed by

$$\frac{1}{l_d} = \frac{Hm}{xp\tau} = \frac{b}{xp}, \tag{4.6.5}$$

where b is the characteristic momentum for decay, or the *decay constant*,

$$b \equiv \frac{Hm}{\tau}. \tag{4.6.6}$$

The probability for the particle created at depth x_1 with momentum $p(x_1)$ to survive at depth x_2 is given by

$$w[x_1, x_2; p(x_1)] = \exp\left[-\int_{x_1}^{x_2} \frac{b(x)}{xp(x)}\, dx\right]. \tag{4.6.7}$$

If both b and p are independent of x, this is reduced to

$$w(x_1, x_2; p) = \left(\frac{x_1}{x_2}\right)^{b/p}. \tag{4.6.7$'$}$$

This means that a particle of momentum smaller than b decays with a high probability.

It is important for practical purposes to know the values of b for a number of unstable particles. Because most of them are produced at a level of about 100 g-cm^{-2}, the numerical values of b are evaluated for the scale height at the 100 g-cm^{-2} level:

$$H(100) = 6.3 \text{ km}. \tag{4.6.8}$$

Since the decay constants of hyperons and $K_1{}^0$ are as large as 100 TeV/c, they quickly decay before interactions unless their energies are higher than 100 TeV. For the muon we have

$$b_\mu = 1.0 \text{ GeV}/c. \tag{4.6.9}$$

Since the catastrophic collision of the muon is of negligible importance, the value of b_μ may be compared with the energy lost by ionization.

The decay constants of the charged pion and kaon are of most practical interest. Their values are

$$b_\pi = 115 \text{ GeV}/c, \qquad b_K = 850 \text{ GeV}/c. \tag{4.6.10}$$

Therefore they contribute to sources of muons at energy levels of about 10 GeV, whereas they participate in the nuclear component at about 1 TeV. Thus the genetic relations of cosmic rays are different at low and high energies.

At low energies the genetic relation is schematically expressed as follows:

$$\begin{array}{l}
N \to \pi^0 \longrightarrow \gamma \longrightarrow E \\
\quad\;\; \uparrow \\
N \to N \longrightarrow N \\
\quad\;\; \uparrow\!\downarrow \\
N \to \left\{\begin{matrix}\pi^\pm \\ K^\pm\end{matrix}\right\} \longrightarrow \mu,\ \nu
\end{array} \tag{4.6.11}$$

where the direct arrow from μ to E indicates the production of knock-on electrons, which in turn contribute to the electronic component. In this scheme the nuclear component consists mainly of nucleons.

At high energies the scheme is modified as follows:

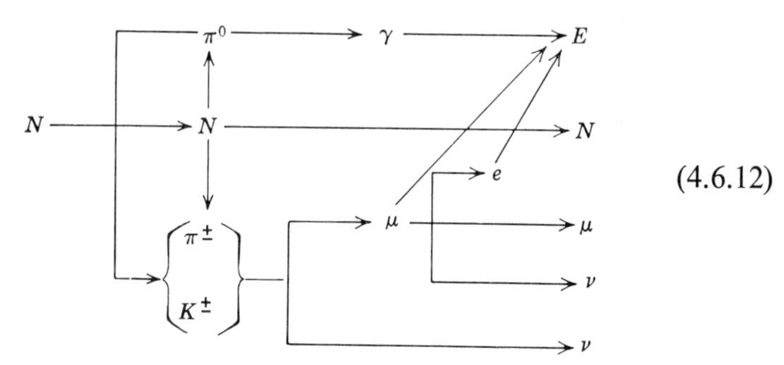

(4.6.12)

Here the muon decay is negligible, but the muons from charged pions and kaons are appreciable.

A number of general relations hold among related components. The charge symmetry in strong interactions affects the charge ratio of the penetrating component. The average value of the z-component of isospin for primary cosmic rays is

$$I_z = \frac{1}{2}\frac{j_p - j_n}{j_p + j_n} \simeq \frac{1}{4}. \tag{4.6.13}$$

In nuclear interactions associated with multiple pion production the probability that the survival nucleon is the proton or the neutron is known to be nearly $\frac{1}{2}$. Hence the z-component of isospin is transferred to pions and further to muons. If the numbers of positive and negative pions produced by a collision of a primary nucleon are n_+ and n_-, respectively, the z-component of isospin in a system of these pions is

$$I_z = n_+ - n_-. \tag{4.6.14}$$

Since I_z is conserved, we expect the *positive excess of penetrating* particles to be

$$\delta \equiv 2\frac{n_+ - n_-}{n_+ + n_-} = \frac{2I_z}{n_s} \simeq \frac{1}{2n_s}. \tag{4.6.15}$$

The observed positive excess is expected to be greater than this estimate because the contribution of positive kaons may become considerable.

The charge independence gives the *neutral-to-charged ratio* of pions,

$$\frac{n_0}{(n_+ + n_-)} = \frac{1}{2}. \tag{4.6.16}$$

This results in the relative intensities of the electronic component and muons as expected from (4.6.16). The contribution of kaons again affects the relative intensities, but it is complicated because the kaon is produced in association either with the hyperon, or with the antikaon, or with both due to the conservation of strangeness.

We have thus far not distinguished between two kinds of neutrinos —one associated with the muon, ν_μ; and the other, with the electron, ν_e. Each of them has its antiparticle, and the interactions of these four kinds of neutrinos are not the same. In elastic collisions with nucleons they are transformed into different particles, as follows:

$$\begin{aligned} \nu_\mu + n &\to p + \mu^-, \qquad & \bar{\nu}_\mu + p &\to n + \mu^+, \\ \nu_e + n &\to p + e, & \bar{\nu}_e + p &\to n + \bar{e}. \end{aligned} \tag{4.6.17}$$

They are produced from meson decays as follows:

$$\begin{aligned} \pi^+ &\to \mu^+ + \nu_\mu, \qquad & \pi^- &\to \mu^- + \bar{\nu}_\mu, \\ \mu^+ &\to \bar{e} + \nu_e + \bar{\nu}_\mu, & \mu^- &\to e + \bar{\nu}_e + \nu_\mu. \end{aligned} \tag{4.6.18}$$

Hence the flux of each kind of neutrinos is related to the flux of muons or of the electronic component.

4.7 Nucleons

In this section we are concerned mainly with the N-component of energies between several tens of MeV and several tens of GeV. Below several tens of MeV the range of nucleons is rather small because of their large nuclear-collision cross sections and because of the large ionization energy loss of protons, so that they are essentially in equilibrium with the N-component in the energy range concerned. Above several tens of GeV the nuclear collision of pions becomes comparable to, or more probable than, their decay in the atmosphere, so that pions are an important part of the N-component. In the energy range we are concerned with, therefore, the N-component consists mainly of protons and neutrons that take part in the nuclear cascade process caused by their successive collisions with air nuclei.

4.7.1 Attenuation Length in Air

The collision of a nucleon with a nucleus in air—nitrogen or oxygen— has been described in Sections 2.9 and 3.2. The reaction cross section is practically independent of energy above 1 GeV, but it decreases with decreasing energy, reaches a minimum at about 300 MeV, and increases toward low energy nearly inversely proportional to energy below 150 MeV. Since the energy dependence of the cross section is not too important for understanding the essential behavior of nucleons in the atmosphere, we may use the average cross section, which may be represented by the asymptotic value. Referring to Fig. 3.3, we choose the *collision mean free path* of nucleons in air as

$$l = 95 \text{ g-cm}^{-2}. \tag{4.7.1}$$

By reference to Section 2.9 we know that a relativistic nucleon produces one or more high-energy nucleons at each collision, and the secondary nucleons are well collimated in the forward direction. The relativistic nucleons may therefore be regarded as propagating in the atmosphere with little deflection. Since they attenuate in air as $\exp(-d/L)$ and air thickness d is related to atmospheric depth x as $d = x/\cos\theta$, where θ is the zenith angle, the unidirectional intensity at θ is expressed as in (4.1.3). If the energies of the nucleons concerned are appreciably smaller than the geomagnetic cutoff energy, the depth dependence tends to be flat at small x. The last expression of (4.1.3) represents the zenith-angle dependence that holds at small θ.

The omnidirectional intensity is obtained from (4.1.3) as

$$J(x) = 2\pi \int_0^{\pi/2} j(x, \theta) \sin\theta \, d\theta = 2\pi j_0 \left(\exp\frac{-x}{L} + \frac{x}{L} Ei\left(\frac{-x}{L}\right) \right), \tag{4.7.2}$$

where

$$-Ei(-x) \equiv \int_x^\infty e^{-z} \frac{dz}{z}. \tag{4.7.3}$$

The value of the term inside the parentheses in (4.7.2) deviates from the exponential function at large values of x/L.

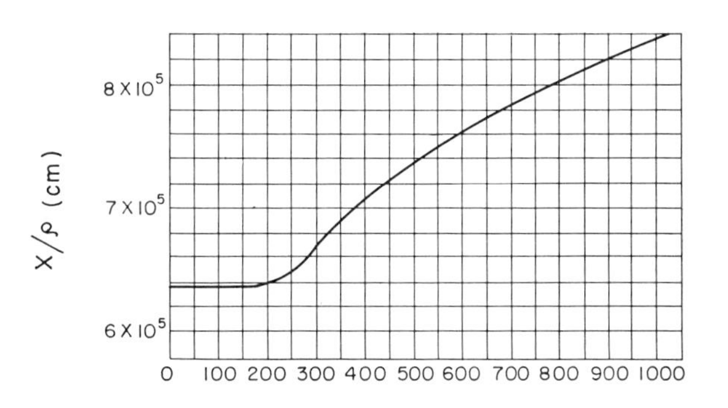

Fig. 4.2 The scale height, $H \equiv x/\rho$, as a function of x in the standard atmosphere; x is the atmospheric depth, ρ is the density of air (Rossi 48).

Relativistic nucleons are detected mainly by means of penetrating showers with emulsions, cloud chambers, ionization chambers, and counters. The energies of nucleons that are responsible for penetrating showers depend on the detecting system used and cannot be determined accurately. Tens of experiments have been made of penetrating showers under various conditions. The *attenuation lengths* thus obtained are scattered, but they are centered at about

$$L = 120 \text{ to } 125 \text{ g-cm}^{-2}; \tag{4.7.4}$$

this seems to decrease slightly as energy increases.

The attenuation mean free path L is greater than the collision mean free path l because of the contribution of secondary particles. The difference between L and l is accounted for in terms of the fraction of surviving nucleons, $f(\alpha)$, as follows:

$$\frac{1}{L} = \frac{1 - f(\alpha)}{l}. \tag{4.7.5}$$

The function $f(\alpha)$ depends on the power index α of the energy spectrum of the nucleons produced. If the energy spectrum is of a pure power shape and the energy-transfer probability is a function of the ratio of secondary and primary energy, $g(W'/W)\, dW'/W$, the function $f(\alpha)$ is obtained as follows:

$$f(\alpha) = \int_0^1 u^\alpha g(u)\, du \qquad \left(u \equiv \frac{W'}{W}\right). \tag{4.7.6}$$

Although the energy-transfer probability is not exactly of fractional shape, the relations (4.7.5) and (4.7.6) are useful for qualitative discussions. Since $g(u)$ is a decreasing function of u, $f(\alpha)$ decreases with increasing α. If the value of α increases gradually with energy, the value of L decreases correspondingly with increasing energy. This may be seen from the energy dependence of the attenuation length.

4.7.2 Energy Spectrum of Nucleons at Various Altitudes

The energy dependence of α is considered to be due partly to competition with pion production in nuclear collisions. As the pion production becomes predominant above 1 GeV, the fraction of energy transferred to nucleons decreases. Therefore the energy spectrum of nucleons at lower altitudes is expected to be steeper than that of primary protons. This is in fact the case, as shown in Fig. 4.3.

Figure 4.3 shows the momentum spectra of protons in the vertical direction measured at several altitudes above the knee of the latitude

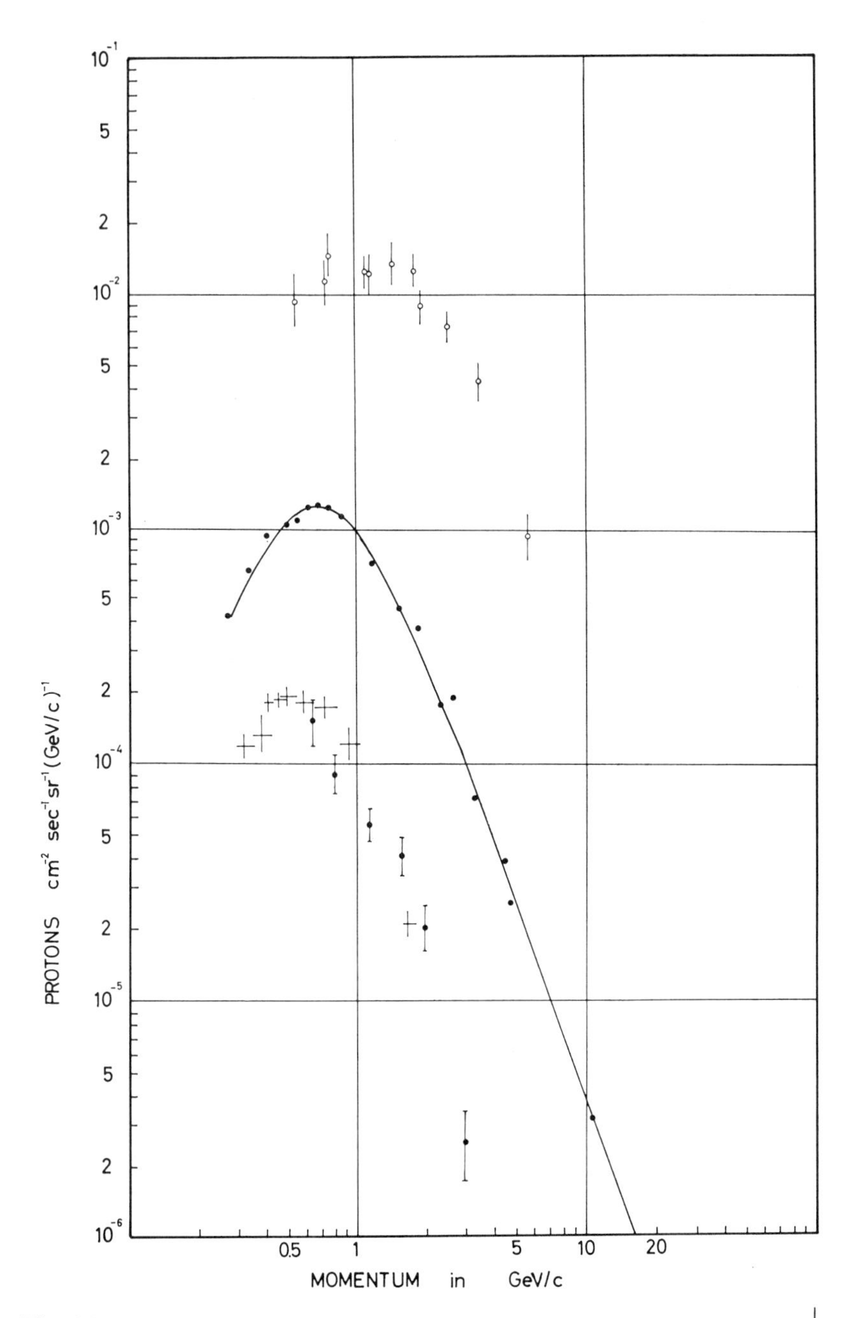

Fig. 4.3 Momentum spectra of protons at different altitudes: 310 g-cm^{-2} (○)—L. I. Baradzei et al., *J. Exp. Theor. Phys.* **36**, 1151 (1951); 700 g-cm^{-2} (●)—N. M. Kocharian, G. S. Saakian, and Z. A. Kirakosian, *J. Exp. Theor. Phys. USSR*, **35**, 1335 (1958); sea level (●)—M. G. Mylroi and J. G. Wilson, *Proc. Phys. Soc.* (London), **A64**, 404 (1951), and (+)—A. G. Meshkovskij and L. I. Sokolov, *J. Exp. Theor. Phys. USSR*, **33**, 542 (1957).

effect. An observation at the highest altitude (at an atmospheric depth of 50 g-cm^{-2}) was made with a balloon-borne emulsion experiment (Camerini 50). The energy of a proton was estimated from an empirical formula,

$$W = 1500\, n_s + 155\, N_h - 100 \text{ MeV}, \tag{4.7.7}$$

where n_s and N_h are the numbers of shower particles and heavy tracks, respectively. Since (4.7.7) gives only an approximate average energy, and the distinction between protons and pions is based on an assumed n/p ratio, the spectrum indicated by a solid curve gives us only a qualitative idea. The spectra at lower altitudes were obtained by magnet spectrometers and are therefore considered to be more reliable. The experimental data referred to in Fig. 4.3 are representative among many that are essentially consistent with each other.

The energy spectrum of protons can be fitted by the following semi-empirical expression.* Let the production spectrum of protons with kinetic energies between W and $W + dW$ at atmospheric depth x be

$$S_p(W, x)\, dW = l^{-1} S_0 (W + a)^{-\alpha - 1}\, dW\, e^{-x/L} \tag{4.7.8}$$

where a is a constant energy in units of GeV and l is an air thickness that may be equal to the nuclear-collision mean free path. The differential vertical intensity of protons, $j_p(W, x)$, obeys the diffusion equation

$$\frac{d}{dx} j_p(W, x) = -\left(\frac{1}{L}\right) j_p(W, x) + S_p(W, x). \tag{4.7.9}$$

This is solved as follows:

$$j_p(W, x) = S_0 \left(\frac{x}{l}\right) e^{-x/L} (W + a)^{-\alpha - 1}. \tag{4.7.10}$$

The best fit to the spectra in Fig. 4.3 is obtained for the values of the parameters

$$\frac{S_0}{l} = 1.0 \times 10^{-3} \quad \text{g}^{-1}\, \text{sec}^{-1}\, \text{sr}^{-1}, \qquad L = 125 \text{ g-cm}^{-2}, \tag{4.7.11}$$

$$\alpha = 1.8, \qquad a = 0.9 \text{ GeV}.$$

The momentum spectra

$$j_p(p, x) = j_p(W, x) \left(\frac{dp}{dW}\right)_{W = W(p)} \tag{4.7.10'}$$

* The results in this subsection were obtained by Y. Terashima.

obtained from (4.7.10) with (4.7.11) are shown by the solid curve in Fig. 4.3. The agreement is satisfactory for spectra at 700 g-cm^{-2} and sea level. The observed intensity at 310 g-cm^{-2} is higher than that expected from (4.7.10), and this may be attributed to the fact that protons of energies greater than 1 GeV are not quite latitude insensitive; in other words, the direct contribution of primary cosmic rays is appreciable at this altitude.

The parameter a implies several effects; namely, the steepening of the production spectrum with increasing energy, the loss of energy by ionization (which is important for low-energy protons), and the geomagnetic cutoff.

The first effect has a bearing on the energy dependence of the attenuation length. The second effect can be seen by comparing the intensities of protons and neutrons, since there is no ionization loss for neutrons. The *neutron-to-proton ratios* observed with emulsions (Camerini 50) are shown in Table 4.5, where the *n-p* ratios observed at two altitudes

Table 4.5 Neutron-Proton Ratio

n_s W (GeV)	0 ~ 0.5	1 ~ 2	≥ 2 ≥ 5
45 g-cm^{-2}	5.45 ± 0.87	0.91 ± 0.23	0.57 ± 0.19
690 g-cm^{-2}	7.10 ± 0.43	1.14 ± 0.15	0.76 ± 0.14
Expected from (4.7.10)	7.0	1.9	1.2

are given for the number of shower particles, n_s; the approximate primary energy corresponding to n_s is estimated from (4.7.7). If $a = 0$ is assumed for neutrons, the *n-p* ratios at 690 g-cm^{-2} expected from (4.7.10) are as given in the last row of Table 4.5.

At an atmospheric depth of 45 g-cm^{-2} the *n-p* ratios for $n_s = 1$ and $n_s \geq 2$ are less than unity because the direct contribution of primary nucleons is dominant. The *n-p* ratio of about $\frac{1}{2}$ for $n_s \geq 2$ indicates the contribution of primary heavy nuclei, due to which the *n-p* ratio in primary radiation is about $\frac{1}{4}$. Since most heavy primaries collide with air nuclei within 45 g-cm^{-2}, their contribution to the neutrons is considerable.

At 690 g-cm^{-2} the calculated *n-p* ratios are larger than the observed ones for $n_s = 1$ and $n_s \geq 2$. This indicates that either the value of a for neutrons is finite, or some primary particles persist even at this depth, or both.

The first alternative is demonstrated by examining the neutron spectrum at energies below 1 GeV. Indeed, the neutron spectrum in the energy range between 1 and 15 MeV observed through recoil protons in a cloud chamber filled with high-pressure hydrogen is less steep than

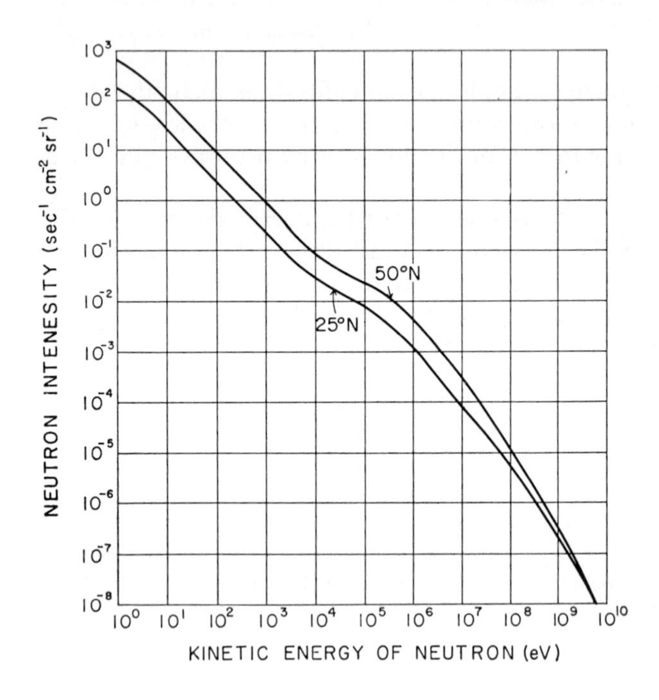

Fig. 4.4 Differential energy spectra of neutrons at geomagnetic latitudes 50°N and 25°N at mountain altitudes (Miyake 57).

the extrapolation of (4.7.10) with $a = 0$, and is represented by (Miyake 57)

$$j_n(W, x) = (1.2 \pm 0.5) \times 10^{-3}(W_0/W)^{1.25 \pm 0.10} \quad \text{cm}^{-2}\,\text{sec}^{-1}\,\text{st}^{-1}\,\text{MeV}^{-1},$$
$$x = 760\ \text{g-cm}^{-2}, \qquad W_0 = 1\ \text{MeV}. \tag{4.7.12}$$

If this is combined with the proton spectrum at high energies and the thermal-neutron flux, the neutron spectrum in a wide energy range looks like that shown in Fig. 4.4. If this is represented by a power spectrum, the power index should increase with energy as

$$j_n(W) \propto W^{-\alpha-1}, \qquad \alpha \simeq 0.1 + 0.3 \log \frac{W}{W_0}. \qquad W_0 = 1\ \text{MeV}. \tag{4.7.13}$$

4.7.3 *Latitude Effect*

The latitude dependence of a can be obtained by referring to the latitude effect of the proton intensity. Figure 4.5 shows the latitude effect of protons with momenta of about 1 GeV/c at the atmospheric

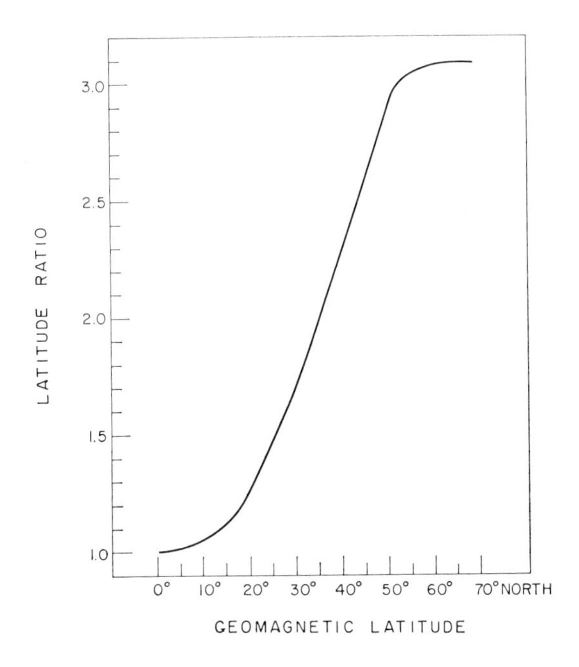

Fig. 4.5 Latitude effect of protons with momenta of about 1 GeV/c at 310 g-cm^{-2} (Conversi 50).

depth of 310 g-cm^2 (Conversi 50). This allows us to derive the values of a at several latitudes, as in Table 4.6. The geomagnetic cutoff also affects the altitude dependence of nucleon intensity. At low latitudes, where the geomagnetic cutoff energy is as high as 10 GeV or higher, the intensity

Table 4.6 Latitude Dependence of a in the Proton Spectrum

Latitude	$\geq 60°$	55°	30°	20°	0°
a (GeV)	0.85	0.90	1.10	1.35	1.54

of relativistic nucleons does not simply decrease with increasing atmospheric depth but levels off at high altitudes or even has a maximum. This indicates that the number of nucleons increases due to the nuclear cascade process in the atmosphere within the first few collision mean free paths, producing nucleons of energies below the cutoff. This further results in that the attenuation of nucleons is slower than that expected from the attenuation length (4.7.4) at lower altitudes. Such an effect becomes greater as the energy of nucleons decreases because a greater length is needed for nucleons of lower energies to be sufficiently supplied from the nuclear cascade process. In fact, the attenuation length measured with a standard neutron monitor of Simpson (Simpson 51) shows a significant latitude dependence; the neutron monitor is known to be sensitive to omnidirectional nucleons with energies of about 100 MeV. The attenuation lengths of the omnidirectional intensity thus measured are shown in Table 4.7.

Table 4.7 Latitude Dependence of the Attenuation Length of Nonrelativistic Nucleons

Latitude	Attenuation Length (g-cm^{-2}) for $200 \le x \le 600$ g-cm^{-2}
0°	212 ± 4
19°	206 ± 4
40°	181 ± 3
53°	157 ± 2
65°	157 ± 3

The latitude effect of nucleon intensity is nearly independent of atmospheric depth for x greater than 600 g-cm^{-2}. Consequently the attenuation length is independent of latitude (Simpson 53*a*):

$$L \simeq 140 \text{ g-cm}^{-2}. \tag{4.7.14}$$

This value is consistent with that derived from atmospheric-pressure effects.

4.8 Nuclear Active Particles at High Energies

As energy increases the contribution of charged pions to the N-component becomes considerable. If the ratio of the mean decay length l_d to the collision mean free path l_π,

$$\frac{l_d}{l_\pi} = \frac{px}{b_\pi l_\pi}, \tag{4.8.1}$$

is on the order of unity or larger, most pions do not decay but participate in the nuclear cascade process. Since b_π is as large as 100 GeV/c, as given by (4.6.10), the intensity of pions could be comparable to that of nucleons for $p \simeq 100$ GeV/c in the stratosphere and $p \simeq 10$ GeV/c near sea level.

4.8.1 Neutral-to-Charged Ratio and Energy Spectrum

From the above considerations one might expect that most of the nuclear active particles with energies of about 100 GeV observed at mountain altitudes would be pions, because the multiplicity of pions is greater than that of nucleons. If this were the case, the ratio of neutral to charged particles in the N-component would be small, because neutral particles consist only of neutrons, whereas charged ones consist of protons and pions. However, the *neutral-to-charged ratio* in Table 4.5 is rather high even for $n_s \geq 2$. This does decrease a little with increasing n_s. This tendency may be verified by referring to other experiments, shown in Table 4.8.

Table 4.8 Neutral-to-Charged Ratio (N/C) at High Energies

Atmospheric Depth (g-cm^{-2})	Energy Region (GeV)	N/C	Reference
710	80	0.77 ± 0.12	1
600	200 to 2000	0.74 ± 0.08	2
710	200 to 2000	0.65 ± 0.07	2
780	200 to 2000	0.56 ± 0.11	2
670	10	0.77 ± 0.04	3
800	20 to 150	0.69 ± 0.13	4

References:
1. K. Greisen and W. D. Walker, *Phys. Rev.*, **90**, 915 (1953).
2. L. A. Farrow, *Phys. Res.*, **107**, 1687 (1957).
3. M. Gervasi, G. Fidecaro, and L. Mezzetti, *Nuovo Cim.*, **1**, 300 (1955).
4. S. Lal, Yash Pal, and R. Reghaven, *Nuovo Cim.*, **28**, 1177 (1963).

The value of N/C in Table 4.8 cannot be predicted by the above consideration, but it suggests that an incident nucleon survives with a considerable fraction of energy after the nuclear collision.

From the above considerations we may, as a first approximation, assume that the quasi-elastic collisions of nucleons account for the gross feature of the N-component in the atmosphere. The numbers of protons and neutrons at depth x, $p(x)$, and $n(x)$, are then obtained as follows:

$$p(x) = \frac{1}{2}\left(\left\{\exp\left(-\frac{x}{L}\right) + \exp\left[-\left(\frac{2r}{l} + \frac{1-2r}{L}\right)x\right]\right\} p_0 + \left\{\exp\left(-\frac{x}{L}\right) - \exp\left[-\left(\frac{2r}{l} + \frac{1-2r}{L}\right)x\right]\right\} n_0\right),$$

$$n(x) = \frac{1}{2}\left(\left\{\exp\left(-\frac{x}{L}\right) - \exp\left[-\left(\frac{2r}{l} + \frac{1-2r}{L}\right)x\right]\right\} p_0 + \left\{\exp\left(-\frac{x}{L}\right) + \exp\left[-\left(\frac{2r}{l} + \frac{1-2r}{L}\right)x\right]\right\} n_0\right), \tag{4.8.2}$$

where r is the charge-exchange probability, and the attenuation length L and the collision mean free path l are related as in (4.7.5).

The ratio n/p is shown in Figure 4.6 for $r = 0.3$, 0.4, and 0.5; $L = 120$ g-cm^{-2}; and $l = 90$ g-cm^{-2}. The experimental values given in Table 4.8 are compared in the same figure. The rough agreement between the predicted and observed n/p ratios indicates a small contribution of charged pions to nuclear active particles. It is, however, indicated that

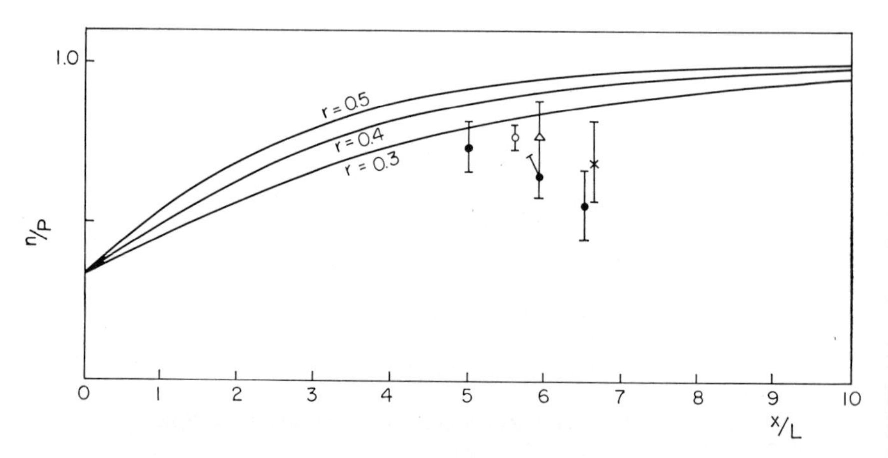

Fig. 4.6 Neutral-to-charged ratio of nuclear active particles. The curves represent the neutron-to-proton ratio given by (4.8.2), and the experimental values are taken from Table 4.8. △—Greisen; ●—Farrow; ○—Gervasi; ×—Lal. (See Table 4.8 for references.)

the value of N/C appears to decrease with increasing energy and atmospheric depth. This may suggest an increasing contribution of pions, as it should be.

The contribution of pions may also result in the steepening of the energy spectrum of nuclear active particles, because the energy is shared with many secondary particles, each of which has a relatively small energy. The energy dependence of the slope of the energy spectrum could also come from the energy dependence of the attenuation length. In order to see whether or not such effects are observed the experimental values of the attenuation length L and the power index of the integral energy spectrum α obtained by observing penetrating showers are summarized in Tables 4.9 and 4.10, respectively.

The experimental results given in Table 4.9 show that the attenuation length is essentially independent of energy above 10 GeV, and its value is equal to that given in (4.7.4); that is, $L = 120$ g-cm^{-2}. On the other hand, the energy spectrum tends to be steeper as energy increases, although the experimental data may not yet be conclusive.

A question may arise why the spectrum around 100 GeV is less steep than that at several GeV; in the latter $\alpha \simeq 1.8$ as given in (4.7.11) and Fig. 4.4. The steepness in the low-energy region could be due to the contribution of knock-on nucleons, which are of low energies and numerous. Since neither the method of energy determination nor the method of flux determination are entirely free from ambiguity, the results given in Table 4.10 cannot be taken too seriously. The energy spectrum around several TeV is found to be in disagreement with the last one given in Table 4.10.

4.8.2 Nuclear Active Particles with Energies above 20 TeV

In this energy region nuclear active particles are detected through the production of secondary particles, particularly through large cascade showers arising from secondary neutral pions. The total energy converted to γ-rays, ΣE_γ, is estimated by means of the emulsion-chamber technique as well as of the ionization pulses occurring in ionization chambers.

The fraction of energy given to all γ-rays, K_γ, distributes as $f(K_\gamma)$, as given in (3.8.12). In an emulsion-stack experiment the observed average inelasticity is

$$\langle K_\gamma \rangle = \left\langle \frac{\Sigma E_\gamma}{E_0} \right\rangle = \int_0^1 K_\gamma \, f(K_\gamma) \, dK_\gamma = 0.16 \pm 0.04. \qquad (4.8.3)$$

In the comparison between the energy spectra of the N-component and

Table 4.9 Observed Values of Attenuation Length

Atmospheric Depth Range (g-cm^{-2})	Energy Range (GeV)	L (g-cm^{-2})	Reference
300 to 1000	~10	118 ± 2[a]	1
650 to 850	30	123 ± 6	2
20 to 700	1000	129 ± 15	3
220 to 720	14 to 43	118.2 ± 0.4[a]	4
	43 to 75	121 ± 1[a]	4
	75 to 125	124 ± 1.5[a]	4
	125 to 260	124 ± 2.5[a]	4
	260	122 ± 4[a]	4

[a] Not corrected for the Gross transformation.

References:

1. J. H. Tinlot, *Phys. Rev.*, **74**, 1197 (1948).
2. V. F. Vishnesky, *Proc. Moscow Conf. on Cosmic Rays*, 1959, Vol. I, p. 188.
3. M. F. Kaplon et al., *Phys. Rev.*, **91**, 1573 (1953).
4. H. K. Ticho, *Phys. Rev.*, **88**, 236 (1952).

Table 4.10 Power Index, α, of the Energy Spectrum of Nuclear Active Particles

Energy Range (GeV)	Atmospheric Depth (g-cm^{-2})	α	Reference
50 to 200	760	1.32 ± 0.1	1
60 to 200	800	$1.30 {+0.26 \atop -0.29}$	2
200 to 2000	800	1.45 ± 0.15	3
2000 to 20,000	650	1.5 ± 0.1	4

References:

1. H. S. Bridge and R. H. Redikar, *Phys. Rev.*, **88**, 206 (1952).
2. S. Lal, Yash Pal, and R. Raghaven, *Nuovo Cim.*, **28**, 1177 (1963).
3. R. Raghaven et al., *J. Phys. Soc. Japan*, **17**, Suppl. A–III, 251 (1962).
4. O. I. Dovzenko et al., *Proc. Moscow Conf. on Cosmic Rays*, 1959, Vol. II, p. 134.

ΣE_γ, the energy scale of the former is obtained from that of the latter by multiplying the latter by a factor

$$\left\langle \frac{1}{K_\gamma} \right\rangle = \left[\int f(K_\gamma) K_\gamma^\alpha \, dK_\gamma \right]^{-1/\alpha} \simeq 10 \qquad (\alpha = 2). \tag{4.8.4}$$

Hence $\langle 1/K_\gamma \rangle^{-1}$ is different from $\langle K_\gamma \rangle$, and the former has to be used to obtain the spectrum of nuclear active particles from that of ΣE_γ.

The spectrum of ΣE_γ obtained by the Japanese-Brazilian emulsion group gives us the energy spectra of nuclear active particles at Mount Norikura (730 g-cm^{-2}) and at Mount Chacaltaya (550 g-cm^{-2}), as shown in Figure 4.7 (Akashi 64, 65). This gives

$$L = 110 \pm 10 \text{ g-cm}^{-2}, \qquad \alpha = 2.0 \pm 0.3. \tag{4.8.5}$$

By reference to other experimental data the altitude variation of the N-component intensity can be obtained from the frequency of γ-ray families with ΣE_γ greater than 2 TeV. The altitude dependence is represented by an exponential form, with the attenuation length equal to that given in (4.8.5), as shown in Fig. 4.8.

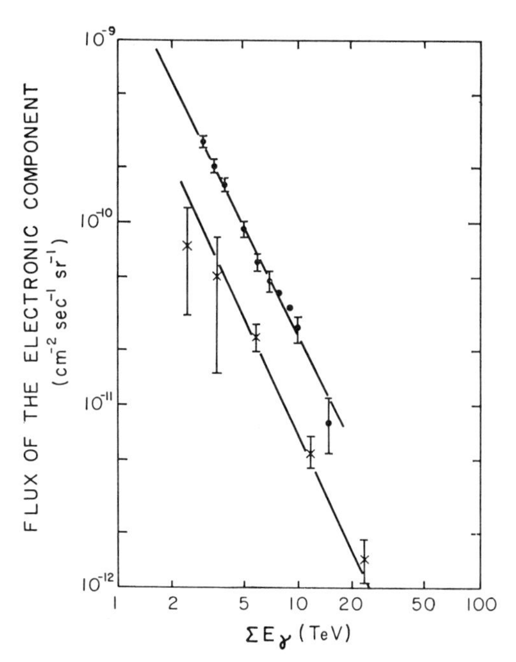

Fig. 4.7 The spectra of $\sum E_\gamma$ observed at Mount Chacaltaya (●) and Mount Norikura (×) (Akashi 64).

By extrapolating the N-component spectrum to the top of the atmosphere we would be able to obtain the primary spectrum. In view of the inaccuracy of the former data, however, the spectra of γ-rays and muons are also employed for this purpose. Several attempts have been made to derive the primary energy spectrum from the energy spectra of γ-rays, muons, and the N-component (Baradzei 61, Miyake 63, Brooke 64). Although they are different in detail, the methods of derivation adopted are essentially the same. Their results agree with each other in gross features, and the integral primary spectrum may be expressed simply as

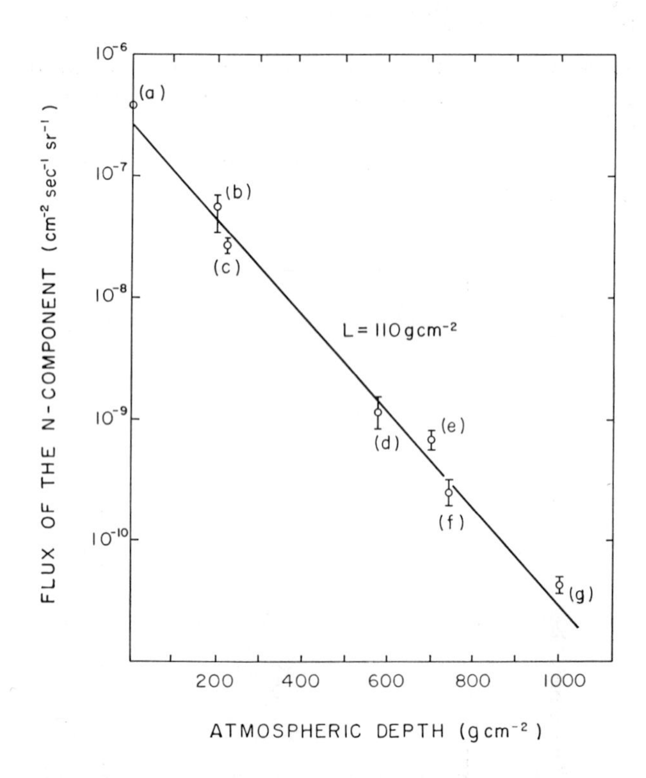

Fig. 4.8 The altitude dependence of the N-component flux with $\sum E_\gamma \geq 2$ TeV ($E_n \gtrsim 20$ TeV). (*a*) Equation 4.8.6; (*b*) L. T. Baradzei et al., *J. Phys. Soc. Japan*, **17**, Suppl. A–III, 433 (1961); *Proc. Int. Conf. on Cosmic Rays at Jaipur*, **5**, 283 (1963); (*c*) J. Duthie et al., *Nuovo Cim.*, **24**, 122 (1962); (*d*) and (*f*) M. Akashi et al., *Prog. Theor. Phys. Suppl.*, No. 32, 1 (1964); (*e*) S. Ya. Babetsky et al., *J. Phys. Acad. Sci. USSR*, **40**, 1551 (1961); (*g*) H. P. Babajan et al., *Proc. Acad. Sci. USSR*, **26**, 1559 (1962).

$$j_p(>E) = 1 \times E^{-1.6} \quad \text{cm}^{-2}\,\text{sec}^{-1} \quad \text{for} \quad 10\ \text{GeV} < E < 3 \times 10^4\ \text{GeV}, \tag{4.8.6}$$

where E is in GeV. The spectral index and the coefficient may have uncertainties of ± 0.05 and 50 percent, respectively.

At high energies the integral spectrum is estimated from extensive air showers to be (Clark 63)

$$j_p(>E) = (3.2 \pm 0.5) \times 10^{-10} \left(\frac{E}{10^6}\right)^{-2.1 \pm 0.1} \quad \text{cm}^{-2}\,\text{sec}^{-1}\,\text{sr}^{-1}, \tag{4.8.7}$$

for

$$10^6\ \text{GeV} < E < 4 \times 10^8\ \text{GeV}.$$

Although the spectra given in (4.8.6) and (4.8.7) may be subject to future revision, it is important to notice the fact that the primary spectrum bends between 10^{14} and 10^{15} eV. This should result in the bending of the N-component spectrum in the atmosphere; the break energy is supposed to decrease with increasing depth. The break should be further transferred to the spectra of γ-rays and muons. The energy at which the slope changes should decrease as altitude decreases. In order to illustrate how this happens we start with a simplified primary spectrum with a break at E_c, the spectral indices below and above E_c being α_1 and α_2, respectively. The values of these parameters are chosen as

$$E_c = 300\ \text{TeV}, \qquad \alpha_1 = 1.6, \qquad \alpha_2 = 2.1. \tag{4.8.8}$$

Further simplification is made by assuming a constant elasticity of nuclear collision and a constant interaction mean free path. Their values are chosen as

$$\zeta \equiv 1 - K = \tfrac{1}{2}, \qquad l = 80\ \text{g-cm}^{-2}, \tag{4.8.9}$$

respectively. Under these simplifying assumptions the altitude variation of the energy spectrum of survival nucleons can be mathematically treated (Hayakawa 64). The integral spectrum of the N-component at depth x is thus obtained as

$$j_n(>E, x) = j_{n0}^{(2)} \frac{\alpha_2}{\sqrt{2\pi q''(s)}} \left(\frac{E_c}{E}\right)^{\bar{s}} e^{q(\bar{s})}, \tag{4.8.10}$$

where $\bar{s}$ is a saddle point given by

$$\ln \frac{E_c}{E} - \frac{1}{s} + \frac{\zeta^s \ln \zeta}{l} x - \frac{1}{s - \alpha_1} + \frac{1}{\alpha_2 - s} = 0, \tag{4.8.11}$$

and

$$q(s) = -\ln \frac{s-(1-\zeta^s)x}{l} + \ln\left(\frac{1}{s-\alpha_1} + \frac{1}{\alpha_2 - s}\right),$$

$$q''(s) = -\frac{1}{s^2} + \frac{x}{l}\zeta^s(\ln \zeta)^2 + \frac{1}{(s-\alpha_1)^2} + \frac{1}{(\alpha_2 - s)^2}. \tag{4.8.12}$$

It may be appropriate to define the bending point as the energy at which

$$s = \frac{(\alpha_1 + \alpha_2)}{2}. \tag{4.8.13}$$

By substituting (4.8.13) into (4.8.11) we obtain the bending point as a function of depth x:

$$E_b(x) = E_c \exp\left[\zeta^{(\alpha_1+\alpha_2)/2}\left(\frac{x}{l}\right)\ln\zeta - \frac{2}{\alpha_1+\alpha_2}\right]. \tag{4.8.14}$$

The last term in the brackets represents an effect by which the bending point in the integral spectrum is shifted from that in the differential spectrum; that is,

$$E_b(0) = E_c \exp\left(-\frac{2}{\alpha_1+\alpha_2}\right).$$

The first term in the brackets in (4.8.14) represents the shift due to attenuation. The e-folding length for this is given by

$$L_b = \frac{-l}{\zeta^{(\alpha_1+\alpha_2)/2}\ln\zeta} \simeq 400 \text{ g-cm}^{-2}. \tag{4.8.15}$$

For $\zeta = 0.5$ and $(\alpha_1 + \alpha_2)/2 \simeq 2$ the bending energy is expressed simply by

$$E_b(x) \simeq 0.6 E_c e^{-x/L_b}. \tag{4.8.16}$$

The variation of the integral spectrum with depth is illustrated in Figure 4.9. The break looks rather smooth in the integral energy spectrum because of the smearing effect by integration. The break is further smoothed out as nucleons come down through the atmosphere in such a way that the spectrum gradually changes its slope. Then the bending point is defined as the energy at which the power index of the integral spectrum is equal to $(\alpha_1 + \alpha_2)/2$, as in (8.8.13).

This indicates that the shape of the N-component spectrum varies

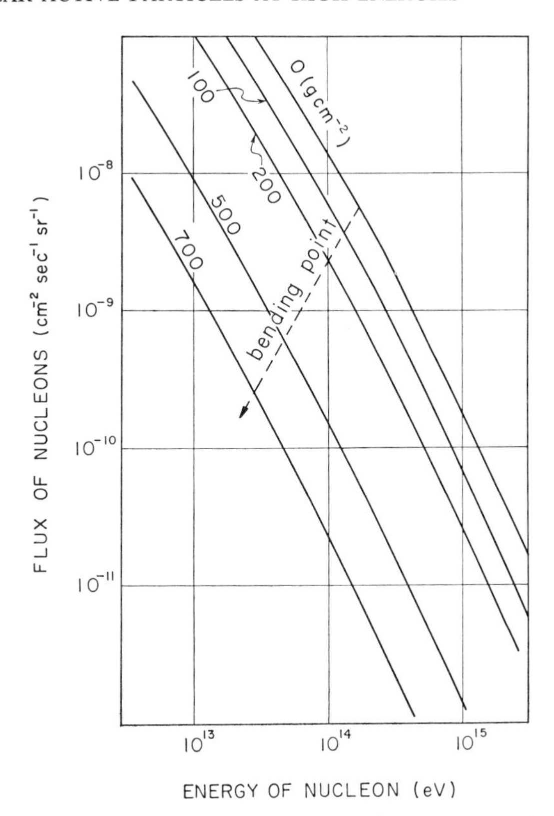

Fig. 4.9 Integral energy spectra of nucleons at different atmospheric depths. The change of the bending point with depth is indicated by a dashed arrow (Hayakawa 64).

rather slowly with atmospheric depth. The values of E_b are about 30 and 40 TeV at Mount Norikura and Mount Chacaltaya, respectively, and therefore the N-component spectra shown in Figure 4.7 may be regarded as representing the part above the bending point. In fact, the spectral index given in (4.8.5) is in agreement with α_2 rather than α_1.

4.8.3 *Contribution of Pions to Nuclear Active Particles*

The contribution of pions and kaons to the N-component can be calculated if their production rates are known. If the production rate of pions is expressed as $l^{-1}g_\pi(E)\exp(-x/L)$, the intensity of pions is obtained by solving the diffusion equation

$$\frac{dj_\pi(E, x)}{dx} = -\left(\frac{1}{l} + \frac{b_\pi cU(x)}{Ex}\right)j_\pi(E, x) + \frac{1}{l}\,g_\pi(E)e^{-x/L}, \quad (4.8.17)$$

where $U(x)=H(x)/H(100)$. This can be easily solved if $U(x)\simeq 1$ is assumed as constant; that is,

$$j_\pi(E,x)=\frac{1}{l}g_\pi(E)\int_0^x e^{-x'/L}e^{-(x-x')/l}\left(\frac{x}{x'}\right)^{b_\pi c/E}dx' \tag{4.8.18}$$

$$=\frac{x}{l}g_\pi(E)e^{-x/L}\Lambda\left(\frac{x}{l}-\frac{x}{L},\frac{b_\pi c}{E}\right),$$

with

$$\Lambda\left(y,\frac{b_\pi c}{E}\right)=\frac{b_\pi c}{E}\int_0^1 e^{-(1-t)y}t^{b_\pi c/E}\,dt. \tag{4.8.19}$$

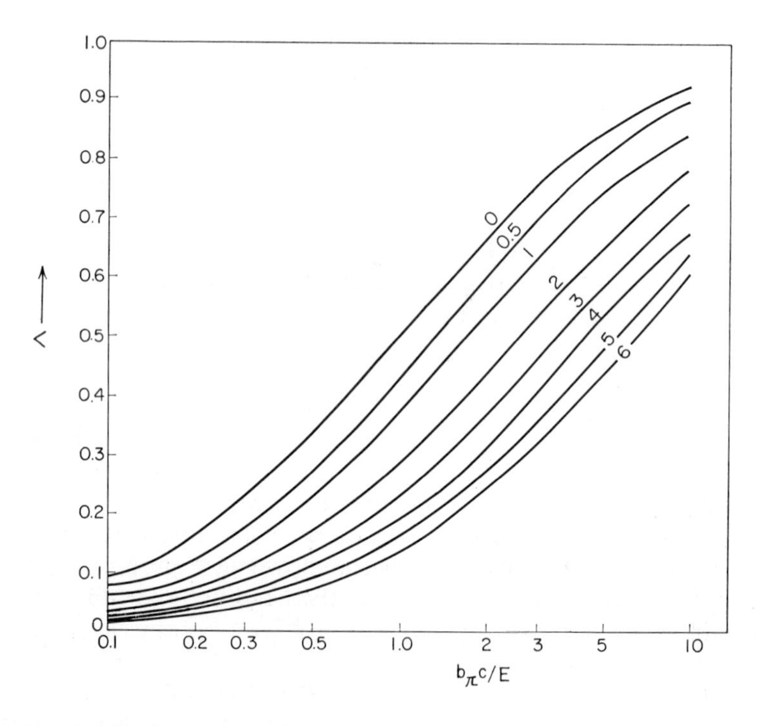

Fig. 4.10 The values of $\Lambda(y, b_\pi c/E)$ as defined in (4.8.19). The values of $y=(x/l)-(x/L)$ are attached to the curves (Subramanian 59).

The numerical values of Λ are shown in Fig. 4.10 (Subramanian 59).

At low energies, where $E \ll b_\pi c$, the factor $t^{b_\pi c/E}$ in the integrand of Λ has an appreciable value only when t is very close to unity. This yields

$$j_\pi(E, x) \simeq g_\pi(E) e^{-x/L} \frac{x}{l} \frac{E}{b_\pi c + E} \cdot \tag{4.8.20}$$

Since $g_\pi(E)$ is already much smaller than the nucleon intensity, a small factor $E/(b_\pi c + E)$ further reduces the pion intensity except at $x \gg l$. At high energies, where $E \gg b_\pi c$, the factor $t^{b_\pi c/E}$ is nearly unity for almost all values of t, and the pion intensity is evaluated as

$$J_\pi(E, x) \simeq \frac{L}{L - l} g_\pi(E)(e^{-x/L} - e^{-x/l}). \tag{4.8.21}$$

Since the pion contribution is expected to become appreciable for $E \gg b_\pi c$, this means that decays of charged pions with such high energies are negligible.

The expressions (4.8.20) and (4.8.21) show that the relative intensity of charged pions in the N-component increases with energy, and levels off as energy increases beyond $b_\pi c \simeq 115$ GeV. This could explain the steepening of the N-component spectrum with increasing energy, as seen in Table 4.10. The steepening at still higher energies may be attributed to the bending of the primary spectrum, as discussed in Section 4.8.2.

The relative intensity of charged pions in this higher energy region is estimated by observing jets produced in an emulsion chamber (Akashi 65). Neutral pions produced by a high-energy nuclear interaction give cascade showers in the emulsion chamber, whereas nuclear active particles simultaneously produced thereby may give secondary jets. The energies of each nuclear active particle can be estimated from the total energy of γ-rays in each secondary jet as $\Sigma E_\gamma / K_\gamma(N)$ or $\Sigma E_\gamma / K_\gamma(\pi^+)$, according to whether the nuclear active particle is a nucleon or a charged pion. Here $K_\gamma(N)$ and $K_\gamma(\pi^\pm)$ represent the fractions of energy given to γ-rays by the collisions of a nucleon and a charged pion, respectively. Since the energy spectrum of charged pions produced by the primary interaction can be estimated from that of neutral pions by observing cascade showers, we can compare the numbers of charged pions and nucleons in a given range of ΣE_γ. We thus obtain

$$\frac{j_\pi[\Sigma E_\gamma / K_\gamma(\pi)]}{j_N[\Sigma E_\gamma / K_\gamma(N)]} = 0.45 \pm 0.20. \tag{4.8.22}$$

If $K_\gamma(\pi)$ is equal to $K_\gamma(N)$, this gives the pion-nucleon ratio. The ratio given in (4.8.22) is consistent with the lowest value of the neutral-charged ratio listed in Table 4.8. This is also consistent with the pion-nucleon ratio obtained through the analysis of atmospheric γ-rays, as will be discussed in the following section.

4.9 Electronic Component from the Decay of Neutral Pions

4.9.1 Development of Cascade Showers in the Atmosphere

The production spectrum of charged pions, as given in Section 4.8.3, is directly related to that of neutral pions by

$$g_{\pi^0}(E_\pi) = \tfrac{1}{2} g_\pi(E_\pi), \tag{4.9.1}$$

in accordance with charge independence. Since the neutral pion decays immediately into two photons, the production spectrum of γ-rays, $g_\gamma(E_\gamma)\,dE_\gamma$, is connected with that of pions $g_{\pi^0}(E_\pi)\,dE_\pi$ as

$$g_\gamma(E_\gamma)\,dE_\gamma = 2\,dE_\gamma \int_{E_\gamma}^{\infty} \frac{1}{cp_\pi} g_{\pi^0}(E_\pi)\,dE_\pi = \frac{2}{\beta+1} g_{\pi^0}(E_\gamma)\,dE_\gamma, \tag{4.9.2}$$

where p_π is the momentum of a produced pion and can be put equal to E_π/c in the extremely relativistic case concerned. The last expression is obtained by assuming the power spectrum of integral spectral index β.

Hence the production rate of γ-rays with energies between E and $E + dE$ and at depths between x and $x + dx$ is given by

$$g_\gamma(E, x)\,dE\,dx = (\beta+1)^{-1}\bar{g}_\pi \left(\frac{E_0}{E}\right)^\beta e^{-x/L} \frac{dE}{E}\frac{dx}{L}, \tag{4.9.3}$$

where a single power law of the production spectrum and an exponential decline of the source intensity with attenuation length L is assumed as before. The term $\bar{g}_\pi$ is the production rate of charged pions with energies greater than E_0 in a column of the atmosphere.

The unidirectional intensity of electrons generated by the parent γ-rays is given by the one-dimensional cascade theory of approximation A as (Hayakawa 64)

$$\begin{aligned} j_e(E, x) &= L^{-1}(\beta+1)^{-1}\bar{g}_\pi \left(\frac{E_0}{E}\right)^\beta \frac{1}{E}\int_0^x \pi(E, x - x')e^{-x'/L}\,dx' \\ &= (\beta+1)^{-1}\bar{g}_\pi \left(\frac{E_0}{E}\right)^\beta \frac{1}{E}\left[a_1 \frac{e^{\lambda_1(\beta)x/X_0} - e^{-x/L}}{1+\lambda_1(\beta)L/X_0} + a_2 \frac{e^{\lambda_2(\beta)x/X_0} - e^{-x/L}}{1+\lambda_2(\beta)L/X_0}\right], \end{aligned} \tag{4.9.4}$$

where $X_0 = 37.1$ g-cm^{-2} is the radiation length in air, and λ_1 and λ_2

are the parameters familiar in cascade theory. The constants a_1 and a_2 obey the initial condition

$$a_1 + a_2 = 0. \tag{4.9.5}$$

The intensity of γ-rays is given by replacing a_1 and a_2 in (4.9.4) by $k_1(\beta)$ and $k_2(\beta)$, respectively. They are expressed by

$$k_1(\beta) = \frac{C(\beta)}{\sigma_0 + \lambda_1(\beta)} a_1, \qquad k_2(\beta) = \frac{C(\beta)}{\sigma_0 + \lambda_2(\beta)} a_2, \tag{4.9.6}$$

with the initial condition

$$k_1 + k_2 = 1. \tag{4.9.7}$$

The values of $\lambda_1(\beta)$, $\lambda_2(\beta)$, σ_0, and $C(\beta)$ are given in Appendix C, Table 2.

The integral intensity of the electronic component with energies greater than E is obtained as

$$j_{e\gamma}(>E, x) = \int_E^\infty [j_e(E', x) + j_\gamma(E, x)]dE' = \frac{1}{\beta + 1} G_\pi(E)\, P(\beta, x, L), \tag{4.9.8}$$

where

$$P(\beta, x, L) = N_1(\beta) \frac{e^{\lambda_1(\beta)x/X_0} - e^{-x/L}}{1 + \lambda_1(\beta)L/X_0} + N_2(\beta) \frac{e^{\lambda_2(\beta)x/X_0} - e^{-x/L}}{1 + \lambda_2(\beta)L/X_0}, \tag{4.9.9}$$

with

$$N_1(\beta) = a_1 + k_1(\beta), \qquad N_2(\beta) = a_2 + k_2(\beta), \qquad N_1 + N_2 = 1, \tag{4.9.10}$$

and $L^{-1}G_\pi(E)$ is the integral production spectrum of charged pions,

$$G_\pi(E) = \int_E^\infty g_\pi(E)\, dE = \beta^{-1}\bar{g}_\pi\left(\frac{E_0}{E}\right)^\beta. \tag{4.9.11}$$

The numerical values of $N_1(\beta)$ and $N_2(\beta)$ are given in Table 4.11.

Table 4.11 Numerical Values of $N_1(\beta)$ and $N_2(\beta)$

	β							
	1.6	1.8	2.0	2.1	2.4	2.6	2.8	3.0
$N_1(\beta)$	1.169	1.201	1.224	1.234	1.240	1.238	1.234	1.226
$N_2(\beta)$	−0.169	−0.201	−0.224	−0.234	−0.240	−0.238	−0.234	−0.226

The expression (4.9.9) is approximated for small x by

$$P(\beta, x, L) \simeq \frac{N_1(\beta)}{1+\lambda_1(\beta)L/X_0}\left[\frac{\lambda_1(\beta)}{X_0}+\frac{1}{L}\right]x+\frac{N_2(\beta)}{1+\lambda_2(\beta)L/X_0}\left[\frac{\lambda_2(\beta)}{X_0}+\frac{1}{L}\right]x. \tag{4.9.12}$$

For large x the terms proportional to $\exp(-x/L)$ become predominant, so that

$$P(\beta, x, L) \simeq \left[\frac{N_1(\beta)}{1+\lambda_1(\beta)L/X_0}-\frac{N_2(\beta)}{1+\lambda_2(\beta)L/X_0}\right]e^{-x/L}. \tag{4.9.13}$$

Thus the electronic-component intensity increases linearly at small depths, passes through a maximum, and then decreases exponentially as x increases; whereas the energy spectrum of γ-rays does not change at all as long as β is constant.

4.9.2 Comparison with Experiments

Taking these qualitative features into account we are able to determine the values of β and L by comparing (4.9.8) with observed data. Observations of the electronic component have been made by several groups, and their experimental conditions are listed in Table 4.12. The integral energy spectra of the electronic component observed thereby are shown in Fig. 10. They are expressed approximately by power shapes; the power indices obtained are given in Table 4.12. It is seen that each spectrum cannot exactly be represented by a single power law but shows gradual steepening as energy increases. It may therefore be appropriate to express the observed spectrum as consisting of two parts with different spectral indices—one below 1 TeV and the other above 1 TeV—as long as the energy range observed is wide enough. These spectra are not always consistent with each other. According to Akashi et al. (Akashi 64), however, the following values of the spectral indices are considered to be most reliable:

$$\beta \simeq 1.8 \text{ to } 2.0 \quad \text{below 1 TeV}, \qquad \beta \simeq 2.2 \text{ to } 2.3 \quad \text{above 1 TeV}. \tag{4.9.14}$$

The absolute flux of the electronic component and its altitude variation are shown in Fig. 4.11. For experiments at very high altitudes the depths are corrected by taking account of the zenith-angle distributions. The intensity-depth curve has an exponential form at great depths and a peak that is characteristic of the cascade shower at about 100

Table 4.12 Energy Spectrum of the Electronic Component at Various Depths

Atmospheric Depth (g-cm^{-2})	Apparatus	Energy Range (TeV)	Exponent of the Power	Energy Range of ΣE_γ (TeV)	Exponent of ΣE_γ	Group	Reference
9 (26[a])	Balloon emulsion stack	0.1 to 2	$1.9^{+0.3}_{-0.2}$			Chicago	1
22 (37)	Balloon emulsion chamber	0.3 to 5	1.75 ± 0.20	0.3 to 10	2.04 ± 0.22	Bristol-Bombay	2
30 (57)	Balloon emulsion chamber	0.1 to 2	2.0 ± 0.2			Japan	3
197	Airplane ion chamber	0.03 to 2	1.76 ± 0.11	0.1 to 5	1.92 ± 0.14	Moscow	4
220	Airplane emulsion stack	0.3 to 2	2.3 ± 0.2	0.3 to 2	2.5 ± 0.2	Bristol	5
		2 to 10	2.8 ± 0.3	2 to 10	3.5 ± 0.3		
310	Airplane ion chamber	0.03 to 2	1.83 ± 0.13	0.1 to 5	2.13 ± 0.17	Moscow	6
550	Mt. Chacaltaya emulsion chamber	0.5 to 10	2.2 ± 0.2	1.2 to 8	2.0 ± 0.5	Japan-Brazil	7
730	Mt. Norikura emulsion chamber	0.35 to 1	2.0 ± 0.2	2 to 23	2.1 ± 0.3	Japan	8
		1 to 10	2.3 ± 0.2				
730	Mt. Norikura scintillators	0.002 to 0.02	1.82			Kobe	9
1030 (sea level)	scintillators	0.002 to 0.02	1.86			Kobe	9

[a] The values in parentheses represent the effective depths.

References:

1. J. M. Kidd, *Nuovo Cim.*, **27**, 57 (1962); F. Abraham et al., ibid., **28**, 221 (1963).
2. P. H. Fowler, *Proc. Int. Conf. on Cosmic Rays at Jaipur*, **5**, 182 (1963).
3. O. Minakawa et al., *Nuovo Cim. Suppl.* **8**, 761 (1958).
4. L. T. Baradzei et al., *J. Phys. Soc. Japan*, **17**, Suppl. A–III, 433 (1961); *Proc. Int. Conf. on Cosmic Rays at Jaipur*, **5**, 283 (1963).
5. J. Duthie et al., *Nuovo Cim.*, **24**, 122 (1962).
6. L. T. Baradzei et al., *Proc. Int. Conf. on Cosmic Rays at Jaipur*, **5**, 283 (1963).
7. M. Akashi et al., *Proc. Int. Conf. on Cosmic Rays at Jaipur*, **5**, 326 (1936).
8. M. Akashi et al., *Progr. Theor. Phys. Suppl.*, **32**, 1 (1964).
9. T. Kameda, *J. Phys. Soc. Japan*, **15**, 1175 (1960); T. Kameda and T. Maeda, ibid., **15**, 1367 (1960).

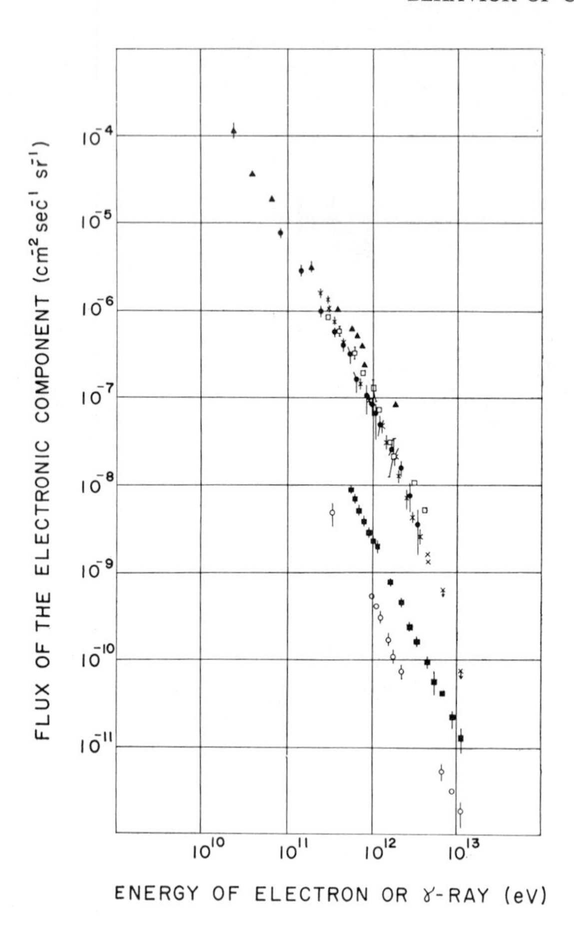

Fig. 4.11 Integral energy spectra of the electronic components. The references from which the data were taken are given in Table 4.12; the reference numbers for the respective curves are as follows: ●, 1; ▲, 6; □, 3; ×, 5; ■, 7; ○, 8 (Hayakawa 64).

g-cm^{-2}. The attenuation length in the exponential part is estimated to be

$$L = 110 \pm 10 \text{ g-cm}^{-2}. \tag{4.9.15}$$

In comparison with the experimental results the theoretical altitude dependence given by (4.9.8) is shown also in Fig. 4.12 for the following sets of parameters:

$$\beta = 2.0 \text{ for } E_\gamma < 1 \text{ TeV}, \qquad \beta = 2.3 \text{ for } E_\gamma > 1 \text{ TeV};$$
$$L = 100 \text{ and } 110 \text{ g-cm}^{-2}. \tag{4.9.16}$$

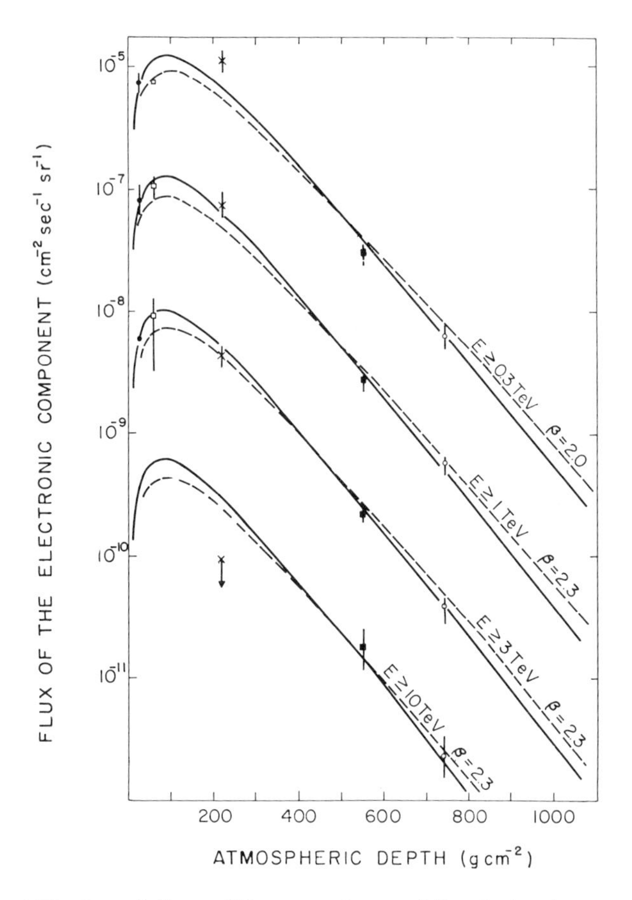

Fig. 4.12 Altitude variations of the γ-ray fluxes of the electronic component with energies $\geq$0.3, 1.0, 3.0, and 10 TeV. The full curves represent the cascade curves for $L = 100$ g-cm^{-2}; the dashed curves, for $L = 110$ g-cm^{-2}; legends the same as in Fig. 4.11 (Hayakawa 64).

The agreement at $x \lesssim 200$ g-cm^{-2} demonstrates the essential correctness of the picture adopted above.

4.9.3 *Bending of the Energy Spectrum*

The bending of the electronic-component spectrum remains to be discussed. This is considered to arise from the bending of the N-component spectrum and consequently from that of the production spectrum of neutral pions. The relation between the bending points of the observed spectrum of γ-rays and the production spectrum of neutral pions is somewhat complicated. Firstly, the bending point is shifted to the low-energy side by the decay of neutral pions. Secondly, observed γ-rays have experienced the atmospheric cascade starting from parent γ-rays, whereby the bending point is also shifted to the low-energy side. The amount of the shift, of course, depends on the spectral index β as well as the attenuation length L.

Let the production rate of the neutral pions be similar to (4.9.2) but let it consist of two parts of different spectral indices:

$$g_{\pi^0}(E, x) = \tfrac{1}{2}\bar{g}_\pi \left(\frac{E_1}{E}\right)^\beta \frac{dE}{E} p(x)$$

$$\beta = \begin{cases} \beta_1 & \text{for} \quad E \leq E_1, \\ \beta_2 & \text{for} \quad E \geq E_1. \end{cases} \tag{4.9.17}$$

The depth dependence, $p(x)$, is assumed to be the same as that of the nucleon spectrum given in Section 4.8. The integral energy spectrum of the electronic component generated from this source is expressed by

$$j_{e,\gamma}(>E, x) = \int_0^x dx' \int_0^\infty dE'[\Pi(E', E, x') + \Gamma(E', E, x')]\, g_\gamma(E', x - x')$$

$$= \bar{g}_\pi [2\pi q_1(\bar{s})]^{-1/2} \left(\frac{E_1}{E}\right)^{\bar{s}} e^{q_1(\bar{s})}, \tag{4.9.18}$$

where Π and Γ are the integral cascade functions of electrons and γ-rays, respectively. The saddle point $\bar{s}$ is obtained in a way that is analogous to (4.8.11); that is,

$$\ln \frac{E_1}{E} - \frac{1}{s} - \frac{1}{s+1} + \frac{\zeta^s \ln \zeta}{l} x$$

$$- \frac{\lambda'_1 - X_0 \zeta^s (\ln \zeta)/l}{\lambda_1 + X_0(1 - \zeta^s)/l} - \frac{1}{s - \beta_1} + \frac{1}{\beta_2 - s}\bigg|_{s = \frac{\beta_1 + \beta_2}{2}} = 0 \tag{4.9.19}$$

and

$$q_1(s) = -\ln\left[s(s+1) - (1 - \zeta^s)\frac{\lambda_0 x}{l}\right]$$

$$- \ln \frac{(1 - \zeta^s)\lambda_0}{l + \lambda_1(s)} + \ln\left(\frac{1}{s - \beta_1} + \frac{1}{\beta_2 - s}\right). \tag{4.9.20}$$

The bending point is given by

$$E_{b,e}(x) = E_b(x)\left(\frac{E_1}{E_c}\right) \exp\left\{-\left[\frac{2}{2+\beta_1+\beta_2} - \frac{\lambda_1' - X_0\,\zeta^{(\beta_1+\beta_2)/2}(\ln\zeta)/l}{\lambda_1 + X_0(1-\zeta^{(\beta_1+\beta_2)/2})/l}\right]\right\}$$

$$\simeq 0.4\,E_b(x)\left(\frac{E_1}{E_c}\right) \simeq 0.24\,E_1\,e^{-x/L_b}, \qquad (4.9.21)$$

where E_c is the bending point of the primary spectrum. In obtaining the numerical factor 0.4 in (4.9.21), we assume that $(\beta_1+\beta_2)/2 \simeq 2$ and that other parameters are taken to be the same as those in Section 4.8. The factor 0.4 comes from the energy shift by π^0-2γ decay and the energy degradation by the development of cascade showers. The former gives a factor 0.7; and the latter, a factor 0.55.

It may be worth remarking that the depth dependence of the bending point in the electronic-component spectrum is the same as that of the nuclear active component. This is due to an implicit assumption that the production mechanism of neutral pions in nuclear interactions is essentially independent of energy. However, the experimental results that are available are not extensive enough to permit a comparison with the altitude dependence of the bending point given in (4.9.21). If the bending were due to a change in the production spectrum with energy, the bending point would be independent of altitude.

4.10 High-Energy Muons in the Atmosphere

The intensity and the energy spectrum of the electronic component in the atmosphere allow us to obtain the production spectrum of neutral pions and consequently that of charged pions, as discussed in the preceding section. If the latter, $g_\pi(E)$, is given, the differential energy spectrum of charged pions at atmospheric depth x is expressed by (4.8.18).

4.10.1 Energy Spectrum of Muons from Charged Pions

The charged pion decays into a muon and a neutrino as in (4.6.18). In the rest system of the charged pion the energy and the momentum of the muon are given by

$$E_\mu^* = \frac{m_\pi{}^2 + m_\mu{}^2}{2m_\pi}\,c^2 = 109.8 \text{ MeV},$$

$$p_\mu^* = \frac{m_\pi{}^2 - m_\mu{}^2}{2m_\pi}\,c = 29.8 \text{ MeV}/c, \qquad (4.10.1)$$

respectively. Thus much of the pion rest energy is taken away by the muon, and the energy range allowed for the daughter muon is rather narrow.

A muon of energy E_μ can be generated from a pion of energy E_π lying in the range

$$E_\pi^- \equiv \frac{m_\pi}{m_\mu} \frac{E_\mu E_\mu^* - c^2 p_\mu p_\mu^*}{m_\mu c^2} \le E_\pi \le \frac{m_\pi}{m_\mu} \frac{E_\mu E_\mu^* + c^2 p_\mu p_\mu^*}{m_\mu c^2} \equiv E_\pi^+ . \qquad (4.10.2)$$

Since the pion decays at a rate of b_π/xp_π, as given by (4.6.5), the production rate of muons with energies between E_μ and $E_\mu + dE_\mu$ is found to be

$$\begin{aligned} l^{-1} g_\mu(E_\mu) e^{-x/\xi}\, dE_\mu \\ &= dE_\mu \int_{E_\pi^-}^{E_\pi^+} \frac{b_\pi}{xE_\pi} \frac{m_\pi}{2p_\pi p_\mu^*} j_\pi(E_\pi, x)\, dE_\pi \\ &= \frac{1}{l} \frac{m_\pi^2}{m_\pi^2 - m_\mu^2} e^{-x/L}\, dE_\mu \int_{E_\pi^-}^{E_\pi^+} g_\pi(E_\pi) \Lambda\left(\frac{x}{l} - \frac{x}{L}, \frac{b_\pi c}{E_\pi}\right) \frac{dE_\pi}{E_\pi}. \qquad (4.10.3) \end{aligned}$$

The range of the integral above is $E_\pi^+ - E_\pi^- = 2p_\mu p_\mu^* (m_\pi/m_\mu^2) = (m_\pi^2 - m_\mu^2) p_\mu c/m_\mu^2$ and is a small fraction of the muon energy concerned. Since the energy spectrum at high energies is considered to be monotonic, we are allowed to evaluate the integral with the aid of the mean-value theorem with the mean value

$$\bar{E}_\pi = \frac{m_\pi E_\mu}{m_\mu}. \qquad (4.10.4)$$

Thus (4.10.3) yields

$$g_\mu(E_\mu) \simeq \frac{m_\pi}{m_\mu} g_\pi\left(\frac{m_\pi}{m_\mu} E_\mu\right) \Lambda\left(\frac{x}{l} - \frac{x}{L}, \frac{m_\mu b_\pi c}{m_\pi E_\mu}\right). \qquad (4.10.5)$$

By integrating (4.10.3) with (4.10.5) over x we obtain the muon spectrum at depth x as

$$j_\mu(E_\mu, x) = \frac{m_\pi}{m_\mu} g_\pi\left(\frac{m_\pi}{m_\mu} E_\mu\right) M\left(\frac{m_\mu b_\pi c}{m_\pi E_\mu}, x\right), \qquad (4.10.6)$$

where

$$M(z, x) \equiv z \int_0^1 \frac{\{1 - \exp[(1 - t + (l/L)t)(-x/l)]\} t^z}{1 - t + (l/L)t}\, dt. \qquad (4.10.7)$$

At sea level this can be approximated, since $x \gg l$, by

$$M(z) = z\int_0^1 \frac{t^z}{1-t+(l/L)t}\,dt$$

$$= z\sum_{n=0}^{\infty} \frac{a^n}{z+1+n} \qquad (a = 1 - l/L). \tag{4.10.7'}$$

Asymptotic expressions of $M(z)$ are obtained for large and small z respectively as

$$M(z \gg 1) \simeq \frac{z}{z+1}\frac{L}{l} \simeq \frac{L}{l} \quad \text{for} \quad z \equiv \frac{m_\mu b_\pi c}{m_\pi E_\mu} \gg 1 \tag{4.10.8a}$$

and

$$M(z \ll 1) \simeq \frac{zL}{L-l}\ln\left(\frac{L}{l}\right) \quad \text{for} \quad z \equiv \frac{m_\mu b_\pi c}{m_\pi E_\mu} \ll 1. \tag{4.10.8b}$$

For intermediate values of z the numerical values of $M(z)$ are given in Fig. 4.13 (Subramanian 59) and may be approximately expressed by (Barrett 52)

$$M(z) \simeq \left[1 + \frac{L-l}{zl\ln(L/l)}\right]^{-1}. \tag{4.10.8c}$$

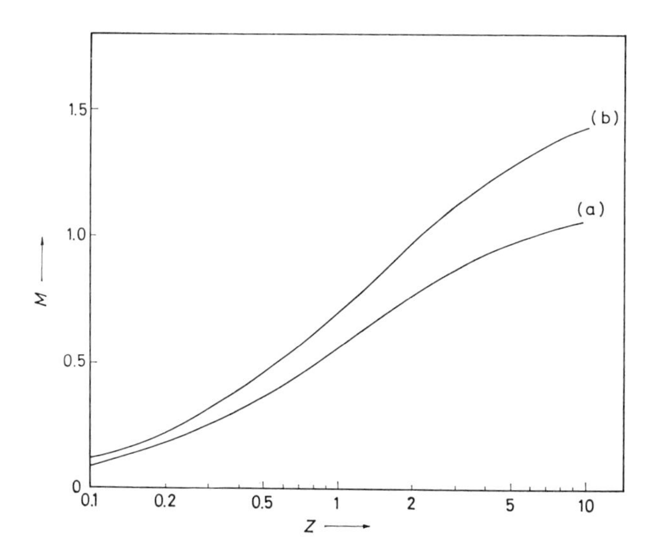

Fig. 4.13 The numerical values of $M(z)$, defined in (4.10.7′) (Subramanian 59):
(*a*) $l = 70$ g-cm^{-2}, $L = 120$ g-cm^{-2};
(*b*) $l = 100$ g-cm^{-2}, $L = 120$ g-cm^{-2}.

It is clear from (4.10.8) that at low energies $M(z)$ is essentially independent of z, and consequently the muon spectrum is obtained by shifting the argument of the pion-production spectrum—whereas at high energies a factor that is inversely proportional to energy is multiplied by the pion-production spectrum. This results in a gradual bend of the muon spectrum as we go across $E_\mu = (m_\mu/m_\pi)b_\pi c$. The lifetime of the pion was first estimated on this basis.

The use of the mean-value theorem in (4.10.5) results in a systematic underestimate of the muon intensity for the power spectrum. A better result can be obtained by carrying out the energy integration more exactly. Now the integration of (4.10.3) over x is performed first, and the x-dependence of the integral is neglected as in (4.10.7′). Hence the differential spectrum of muons at sea level is given by

$$j_\mu(E_\mu) = \int_{E_\pi^-}^{E_\pi^+} g_\pi(E_\pi) M(z) \frac{m_\pi}{2p_\mu^* p_\pi} \, dE_\pi, \tag{4.10.9}$$

where $z = b_\pi c/E_\pi$. Now $M(z)$ can be expressed as

$$M(z) = \left(\frac{K}{E_\pi}\right)^r, \tag{4.10.10}$$

where

$$r = -\frac{d \ln M(z)}{d \ln E_\pi} \simeq \left[1 + \frac{zl \ln (L/l)}{L - l}\right]^{-1} \tag{4.10.11}$$

is a parameter that is assumed to be independent of energy in the energy interval of integration in (4.10.9). Then the integral (4.10.9) can be easily evaluated for a power spectrum by

$$j_\mu(E_\mu) = \frac{1}{\beta + r + 1} \left[\frac{1 - (m_\mu/m_\pi)^{2(\beta+\gamma+1)}}{1 - (m_\mu/m_\pi)^2}\right] M(z_\mu) g_\pi(E_\mu), \tag{4.10.12}$$

where $z_\mu = b_\pi c/E_\mu$.

The difference between (4.10.6) and (4.10.12) is small but appreciable; the former is smaller by 10 to 20 percent than the latter, and the difference gradually increases with energy. The muon intensity given by (4.10.12) is overestimated by a few percent, but is good enough for practical use.

By integrating (4.10.12) over energy we obtain the integral muon flux at sea level as

$$j_\mu(>E_\mu) = \frac{\beta}{(\beta + r)(\beta + r + 1)} \left[\frac{1 - (m_\mu/m_\pi)^{2(\beta+r+1)}}{1 - (m_\mu/m_\pi)^2}\right] M(z) G_\pi(E_\mu). \tag{4.10.13}$$

This can be compared with the intensity of the electronic component given in (4.9.8). After eliminating $G_\pi(E)$ in (4.9.8) by reference to (4.10.3) we obtain the relation between the electronic component flux at depth x and the muon flux at sea level:

$$j_{e\gamma}(>E, x) = \frac{(\beta + r)(\beta + r + 1)}{\beta(\beta + 1)} \left[\frac{1 - (m_\mu/m_\pi)^2}{1 - (m_\mu/m_\pi)^{2(\beta+r+1)}}\right] \frac{P(\beta, x, L)}{M(z)}\, j_\mu(>E). \tag{4.10.14}$$

However, this cannot be directly compared with experimental results because an appreciable fraction of muons are expected to arise from kaons.

4.10.2 Contribution of Kaons

In order to estimate the contribution of kaons we refer to the lifetimes and the branching ratios of the charged and neutral kaons shown in Table 4.13.

Table 4.13 Lifetimes and Branching Ratios of the Decays of Kaons[a]

Particle	Lifetime (sec)	Decay Mode	Branching Ratio (%)
$K^\pm$	$(1.229 \pm 0.008) \times 10^{-8}$	$\mu^\pm + \nu$	63.2 ± 0.4
		$\pi^\pm + \pi^0$	21.3 ± 0.4
		$\pi^\pm + \pi^\pm + \pi^\mp$	5.52 ± 0.08
		$e^\pm + \nu + \pi^0$	4.9 ± 0.2
		$\mu^\pm + \nu + \pi^0$	3.4 ± 0.2
		$\pi^\pm + \pi^0 + \pi^0$	1.68 ± 0.05
$K_1{}^0$	$(0.881 \pm 0.010) \times 10^{-10}$	$\pi^+ + \pi^-$	68.5 ± 1.0
		$\pi^0 + \pi^0$	31.5 ± 1.0
$K_2{}^0$	$(5.77 \pm 0.59) \times 10^{-8}$	$e^\pm + \nu + \pi^\mp$	35.4 ± 2.7
		$\mu^\pm + \nu + \pi^\mp$	26.7 ± 2.6
		$\pi^+ + \pi^- + \pi^0$	13.6 ± 1.0
		$\pi^0 + \pi^0 + \pi^0$	24.8 ± 3.0
		2π	$(2.1 \pm 0.3) \times 10^{-3}$

[a] A. H. Rosenfeld et al., UCRL–8030, August 1965.

First, the decay of $K_2{}^0$ may be neglected, because its lifetime is considerably longer than that of the pion. Secondly, we take only the decay modes of large branching ratios that give major contributions to muons as well as γ-rays. They are

$$K^\pm \to \mu^\pm + \nu \qquad (2/3), \tag{4.10.15a}$$

$$K_1{}^0 \to \pi^+ + \pi^- \qquad (2/3), \tag{4.10.15b}$$

$$K_1{}^0 \to \pi^0 + \pi^0 \qquad (1/3), \tag{4.10.15c}$$

where the branching ratios adopted are shown in parentheses. The contributions of other modes are so small that they are estimated to be a correction to the contributions of the major modes.

The process (4.10.15*a*) is similar to the π-μ decay. The integral energy spectrum of muons from this decay mode is thus given, corresponding to (4.10.13), by

$$j_\mu^{(a)}(>E_\mu) = \frac{2}{3}\frac{\beta}{(\beta+r)(\beta+r_K+1)} \times \frac{1-(m_\mu/m_K)^{2(\beta+r_K+1)}}{1-(m_\mu/m_K)^2} M_{K^\pm}(z) G_{K^\pm}(E_\mu), \quad (4.10.16)$$

where the subscript K indicates quantities concerning the kaon and $G_K{}^\pm$ is the integral production spectrum for both kinds of kaons, K^+ and K^-.

In (4.10.15*b*) and (4.10.15*c*) the differential production spectrum of pions is obtained from that of kaons as

$$g_\pi(E_\pi) = \int_{E_{K^-}}^{E_{K^+}} \frac{m_K}{2p_\pi^* p_K} g_K(E_K) dE_K \simeq \xi(\beta) g_K(E_\pi), \quad (4.10.17)$$

where

$$\xi(\beta) = \frac{(1+\omega)^{\beta+1}-(1-\omega)^{\beta+1}}{2^{\beta+1}(\beta+1)^\omega}\left(\frac{m_K}{m_\pi}\right)^{2(\beta+1)} \quad \text{with} \quad \omega = \sqrt{1-2(m_\pi/m_K)^2}. \quad (4.10.18)$$

The muon flux that arises from (4.10.15*b*) is therefore given by

$$j_\mu^{(\beta)}(>E_\mu) = \frac{2}{3}\frac{\beta}{(\beta+r)(\beta+r+1)}\left[\frac{1-(m_\mu/m_K)^{2(\beta+r+1)}}{1-(m_\mu/m_K)^2}\right] \times M_\pi(z)\xi(\beta)G_{K_1{}^0}(E_\mu). \quad (4.10.19)$$

In deriving (4.10.19) it is implicitly assumed that the time delay in the $K_1{}^0$ decay is negligible; namely, $M_{K_1{}^0}(z)=1$. This assumption is found acceptable if we compare the values of $M(z)$ for various decay processes, as shown in Fig. 4.14.

The process (4.10.15*c*) contributes to the electronic component. Its flux is evaluated in a way that is analogous to (4.9.8) as

$$j_{e,\gamma}(>E, x) = \frac{2}{3}\frac{1}{\beta+1} P(\beta, x, L)\xi(\beta)G_{K_1{}^0}(E), \quad (4.10.20)$$

where $P(\beta, x, L)$ is given by (4.9.9).

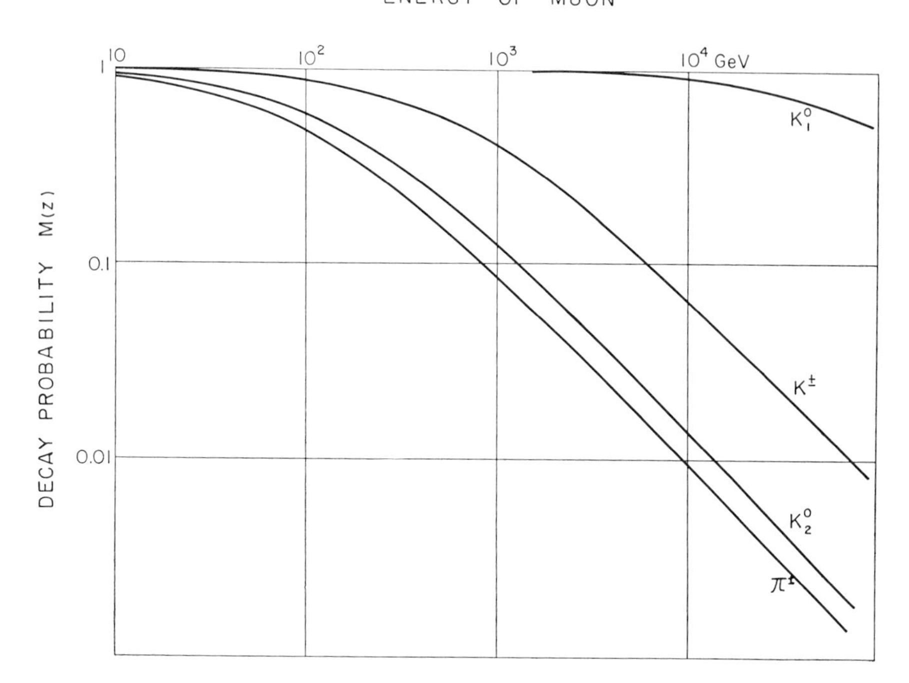

Fig. 4.14 Energy dependence of the decay probability of $\pi^\pm$, $K^\pm$, $K_1{}^0$, and $K_2{}^0$ with $l = 80$ g-cm^{-2} and $L = 110$ g-cm^{-2} (Hayakawa 64).

The restriction of decay modes to three in (4.10.15) is subject to a small error because other decay modes are not entirely negligible; for example, the $K^\pm \to \pi^\pm + \pi^0$ mode has a considerable branching ratio and may give an appreciable flux of muons at lower energies through the subsequent decay of $\pi^\pm$. If we take all the decay modes in Table 4.13 into account, the flux of the muons given by (4.10.16) and (4.10.19) should be corrected by a multiplicative factor $\varepsilon_\mu(E)$, and that of the electronic component given by (4.10.20) by a factor $\varepsilon_\gamma(E)$. The numerical values of these correction factors are approximately estimated and are shown in Fig. 4.15, assuming $g_{K^+} = g_{K^-} = g_{K_1{}^0} = g_{K_2{}^0}$.

4.10.3 Comparison Between the Intensities of the Electronic Component and Muons

The yield of kaons relative to pions can be estimated by comparing the intensities of muons and the electronic component because kaons may contribute appreciably to muons but only slightly to the electronic component.

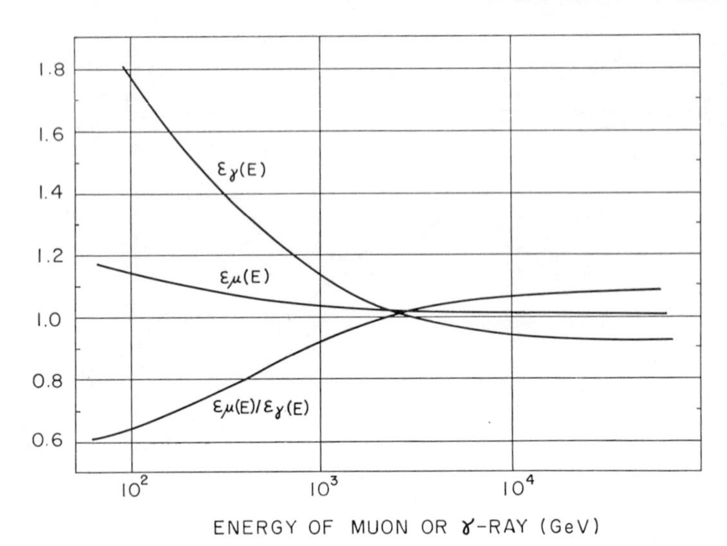

Fig. 4.15 Energy dependence of the correction factors $\varepsilon_\gamma(E)$ and $\varepsilon_\mu(E)$ applied to the muon and γ-ray fluxes, respectively, taking account of the kaon contributions (Hayakawa 64).

The contribution of kaons is represented by a parameter

$$f_K \equiv \frac{G_{K^\pm}(E)}{G_{\pi^\pm}(E)}, \tag{4.10.21}$$

the ratio of the integral production spectra corresponding to a given muon energy E. The intensities of muons arising from pions and kaons are in proportion to $G_{\pi^\pm}$ and $G_{K^\pm}$, respectively. The intensity of the electronic component arising from neutral pions is given in (4.9.8), and its relation to the intensity of muons arising from charged pions is shown in (4.10.14). Kaons also contribute to the electronic component, mainly through the decay of neutral kaons.

Taking all these processes into account we can express the intensity ratio of the electronic component to muons as follows:

$$\frac{j_{e\gamma}}{j_\mu} = \frac{j(\pi - e\gamma) + j(K - e\gamma)}{j(\pi - \mu) + j(K - \mu)} = \frac{j(\pi - e\gamma)}{j(\pi - \mu)} \cdot \frac{1 + j(K - e\gamma)/j(\pi - e\gamma)}{1 + j(K - \mu)/j(\pi - \mu)}. \tag{4.10.22}$$

The first factor in the third expression is given by (4.10.14) as

$$\frac{j(\pi - e\gamma)}{j(\pi - \mu)} = \frac{P(\beta, x, L)}{(\beta + 1)\eta_\pi(\beta) M_\pi(E_\mu)}, \tag{4.10.23}$$

where

$$M_\pi(E_\mu) = M_\pi(z_\mu), \qquad \eta_\pi(\beta) = \frac{\beta}{(\beta+\gamma)(\beta+\gamma+1)} \cdot \frac{1-(m_\mu/m_\pi)^{2(\beta+\gamma+1)}}{1-(m_\mu/m_\pi)^2}. \tag{4.10.24}$$

In the second factor of (4.10.22) the relative contributions of kaons and pions to muons are obtained by comparing (4.10.12) with the sum of (4.10.6) and (4.10.19) as

$$\frac{j(K-\mu)}{j(\pi-\mu)} = \frac{2}{3} f_K \frac{\eta_K(M_K + \frac{1}{2}\xi M_\pi)\varepsilon_\mu}{\eta_\pi M_\pi}, \tag{4.10.25}$$

where we have assumed $G_{K_1{}^0} = \frac{1}{2} G_{K^\pm}$. Likewise, we have, from (4.9.8) and (4.10.20),

$$\frac{j(K-e\gamma)}{j(\pi-e\gamma)} = \tfrac{1}{3} f_K \xi \varepsilon_\gamma. \tag{4.10.26}$$

Thus the ratio (4.10.22) is expressed as

$$\frac{j_{e\gamma}}{j_\mu} = \frac{P(\beta, x, L)}{(\beta+1)\eta_\pi M_\pi} Q(f_K, \beta, E_\mu), \tag{4.10.27}$$

where

$$Q(f_K, \beta, E_\mu) = \frac{1+\frac{1}{3} f_K \xi \varepsilon_\gamma}{1+\frac{2}{3} f_K (M_K + \frac{1}{2}\xi M_\pi)\varepsilon_\mu}. \tag{4.10.28}$$

In (4.10.27) the production ratio f_K is an unknown parameter. This can be determined by comparing (4.10.27) with experiment. For the comparison we refer to the energy spectrum of high-energy muons discussed in Section 4.12. Assuming the muon intensity, we are able to calculate the intensity of the electronic component by means of (4.10.27) with (4.10.28). For the latter we refer to the experimental result obtained with emulsion chambers at Mount Norikura at an atmospheric depth of 730 g-cm^{-2}. In Fig. 4.16 we compare the observed energy spectrum of the electronic component with that calculated from the muon spectrum for several values of f_K. Inspection of the figure leads us to conclude that we are able to rule out both $f_K = 0$ and $f_K \geq 1$. The most probable value of f_K may be (Hayakawa 64)

$$f_K = 0.4 \pm 0.2. \tag{4.10.29}$$

Figure 4.16 seems to suggest an increase of f_K with energy, but the accuracy of experiments and theoretical analyses is not good enough to establish this with certainty.

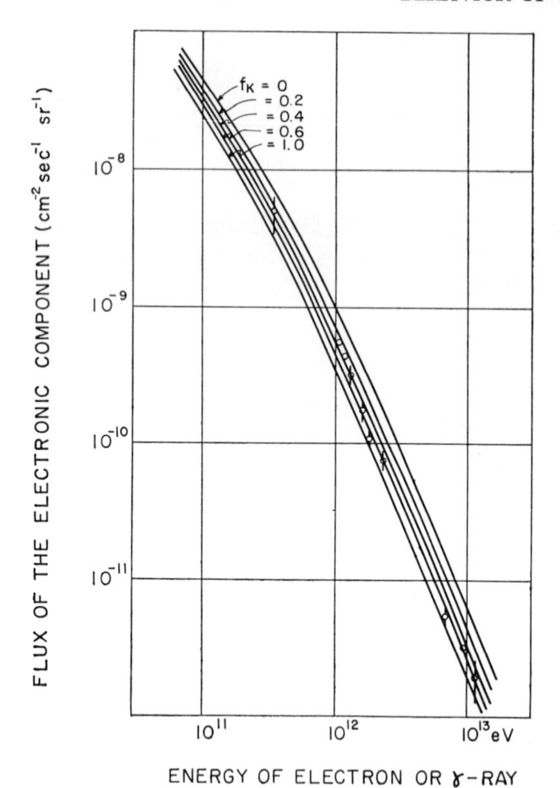

Fig. 4.16 Intregal spectrum of the electronic component expected at Mount Norikura from the muon spectrum given in Section 4.10 for a number of values of the K-π ratio f_K. Experimental values are those given in Fig. 4.11 (Hayakawa 64).

The value of f_K given in (4.10.29) can be compared with the K-π ratio, about 1/4, given in Chapter 3. In f_K the K-π ratio is given by fixing the energy of muons, whereas the K-π ratio in Chapter 3 is concerned with the relative yields in a nuclear interaction. In the latter the average energy of kaons produced is presumably higher than that of pions if both of them are emitted from a fireball of low temperature; the ratio of their average energies may be as large as $\sqrt{m_K^2 + p_T^2}/\sqrt{m_\pi^2 + p_T^2}$. In f_K in the present section the average energies of parent kaons and pions are estimated to be $2E_\mu$ and $(m_\pi/m_\mu)E_\mu$, respectively. Since the ratio of these energy values is comparable to that in nuclear interaction, the value of f_K here may be directly compared with the K-π ratio in Chapter 3.

It should, however, be noted that there is a difference in the primary energies. The K-π ratio given in Chapter 3 is based on the nuclear interactions produced at the primary energy of about 1 TeV, whereas f_K in this section is obtained for muon energies ranging from a fraction of TeV to 10 TeV. The latter corresponds to the primary energy range between 10 and 1000 TeV. Although the K-π ratio is found to be essentially independent of energy up to the primary energy of several TeV, it could be energy dependent at higher energy.

The K-π ratio is important not only in understanding the mechanism of strong interactions but also in estimating the neutrino flux, because the contribution of kaons to neutrinos is relatively greater than that of pions. There are several other phenomena in which the K-π ratio takes some part. They are the positive-negative ratio, the longitudinal polarization, and the zenith-angle distribution of muons. Since these quantities are measured with better accuracy at lower energies, they will be discussed mainly in the next section. In connection with the mathematical treatment in Section 4.10.1, however, the last problem will be briefly discussed in the following section.

4.11 Low-Energy Muons in the Atmosphere

The hard component observed in the lower atmosphere consists mostly of muons with energies smaller than 10 GeV. In this energy region the simplified treatment employed in the preceding section is not applicable, because of the following facts. First, the production spectrum of charged pions cannot be represented by a power law but has a maximum at an energy that depends on both the altitude and the latitude. Second, the energy loss and the decay of muons are no longer negligible but are of primary importance for the behavior of muons in the atmosphere. These facts demand the use of a more complicated mathematical analysis.

4.11.1 The Energy Spectrum and the Altitude Dependence of Low-Energy Muons

If we are concerned with muons of energies lower than 10 GeV, the time delay due to pion decay is negligible because the pion energy is low. Hence the function Λ in (4.10.3) is assumed to be unity, and the production spectrum of muons is given by

$$g_\mu(E_\mu) = \frac{m_\pi^{\ 2}}{m_\pi^{\ 2} - m_\mu^{\ 2}} \int_{E_\pi^-}^{E_\pi^+} g_\pi(E_\pi) \frac{dE_\pi}{p_\pi}, \tag{4.11.1}$$

where E_π^+ and E_π^- are given in (4.10.2). In this approximation the production spectra of pions and muons are related to each other in the following way. By differentiating (4.11.1) we obtain

$$\frac{dg_\mu(E_\mu)}{dE_\mu} = \frac{m_\pi}{2p_\pi^* p_\mu} \left[g_\pi(E_\pi^+) - g_\pi(E_\pi^-)\right]. \tag{4.11.2}$$

By choosing two points in the muon spectrum in such a way that $E_{\pi,1}^+ = E_{\pi,2}^-$, we can construct the pion spectrum by successively applying (Ascoli 50)

$$g_\pi(E_{\pi,2}^+) = g_\pi(E_{\pi,1}^-) + \frac{2p_\mu^*}{m_\pi}\left[\left(p_\mu \frac{dg_\mu(E_\mu)}{dE_\mu}\right)_1 + \left(p_\mu \frac{dg_\mu(E_\mu)}{dE_\mu}\right)_2\right]. \tag{4.11.3}$$

The production spectrum of pions can be obtained by reference to the energy spectrum of muons. It is convenient to use the range spectrum in place of the energy spectrum because the range is a unique function of energy, and we often consider the intensities of muons at various atmospheric depths. The range spectrum, the production rate of muons with range R in dR, is expressed by

$$G(R, \lambda)\, dR\, e^{-x/L} = g_\mu(E_\mu, \lambda)\, dE_\mu\, e^{-x/L}, \tag{4.11.4}$$

where the latitude dependence is indicated by a parameter λ, and the depth dependence is assumed to be exponential, as discussed in the preceding sections.

The muon thus produced decays with the probability per g-cm^{-2} of $b_\mu U(x)/p_\mu x$, where b_μ is defined analogously to b_π, as in (4.6.6), and its numerical value is given in (4.6.9). Hence the probability that the muon produced at x_1 with momentum $p(x_1)$ survives at x_2 is given by

$$w(x_1, x_2; p(x_1)) = \exp\left[-\int_{x_1}^{x_2} \frac{b_\mu U(x)}{p(x)x}\, dx\right], \tag{4.11.5}$$

where $U(x) = H(x)/H(100)$ as introduced in (4.8.17). The momentum lost while the muon traverses matter is expressed by the x dependence, $p(x)$. Since the momentum is uniquely related to x, we may write $p(x_2)$ in place of $p(x_1)$ in the *survival probability* w, whereby the momentum at the depth of observation is explicitly shown. At sufficiently high momenta the momentum loss is proportional to the thickness of matter traversed, so that

$$p(x) = p(x_1) - a(x - x_1) = p(x_2) + a(x_2 - x). \tag{4.11.6}$$

If, moreover, $U(x)$ is approximated as constant $\bar{U}$, the survival probability (4.11.5) can be expressed by

$$w(x_1, x_2; p(x_2)) = \left(\frac{x_1}{x_2} \frac{p(x_2)}{p(x_2) + a(x_2 - x_1)}\right)^{b_\mu \bar{U}/(p(x_2)+ax_2)}. \quad (4.11.7)$$

If $p(x_2) + ax_2 \gg b_\mu \bar{U}$, the survival probability w is practically unity; that is, the decay of muons is negligible. As momentum decreases the value of w decreases, indicating the decay effect. If $p(x_2)$ is expressed by the residual range R_2, the relation (4.11.6) becomes simpler:

$$R_2 = R_1 - (x_2 - x_1), \quad (4.11.8)$$

and the survival probability is expressed as $w(x_1, x_2; R_2)$. Its numerical values have been calculated by Olbert (Olbert 53) and are shown in Fig. 4.17.

For muons with energies of about 1 GeV at sea level the angular spread in the course of propagation is so small that we can adopt the one-dimensional model. Thus the vertical intensity of muons with

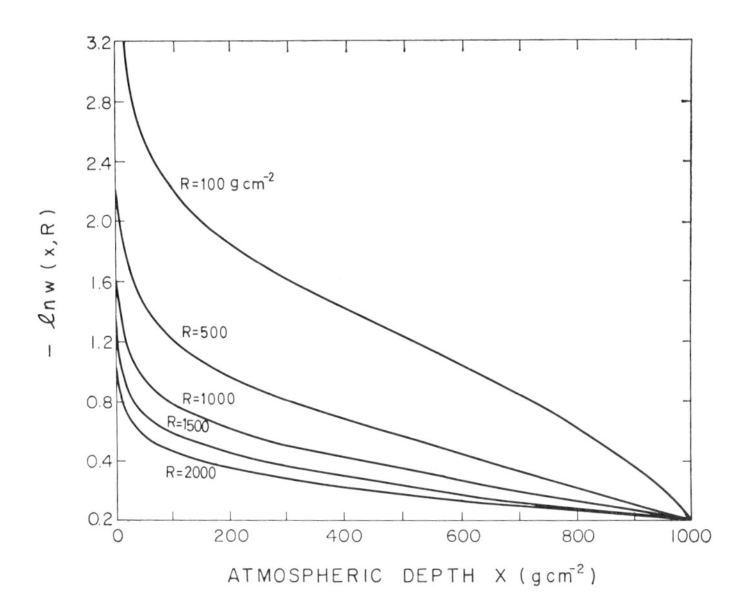

Fig. 4.17 Survival probability $w(x, R)$ of muons produced at atmospheric depth x and reaching sea level with various residual ranges R between 100 and 2000 g-cm^{-2} (Olbert 53).

range R in dR at depth x is given by (Olbert 54)

$$j_{\perp}(x, R; \lambda) = \int_0^x G(R + x - x'; \lambda)\, e^{-x'/L} w(x', x; R)\, dx'. \quad (4.11.9)$$

For a residual range R that is considerably greater than x the production spectrum G varies against x' much more slowly than other factors in the integrand of (4.11.9), so that we can approximate (4.11.9) as

$$j_{\perp}(x, R; \lambda) = G(R + x - x_m; \lambda) \int_0^x e^{-x'/L} w(x', x; R) dx'. \quad (4.11.10)$$

Here x_m is defined by

$$x_m(R) = \frac{\int_0^x x'\, e^{-x'/L} w(x', x; R)\, dx'}{\int_0^x e^{-x'/L} w(x', x; R)\, dx'} \quad (4.11.11)$$

The integrals above can be evaluated by approximating the survival probability as

$$w(x', x; R) \simeq v(R) \left(\frac{x'}{x}\right)^{u(R)}; \quad (4.11.12)$$

$v(R)$ and $u(R)$ can be determined for normal atmospheric conditions as shown in Fig. 4.18. This is found to be a good approximation for $x \simeq 1000$ g-cm^{-2}, $x' < 500$ g-cm^{-2}, and $R > 2000$ g-cm^{-2}. Since we are interested in the case of $x \gg L$, the integrals in (4.11.10) and (4.11.11) can be extended to infinity because of the steep decrease of $\exp(-x'/L)$. Thus we obtain

$$x_m(R) = [1 + u(R)]L \quad (4.11.11')$$

and

$$j_{\perp}(x, R; \lambda) = G(R + x - x_m; \lambda) L\Gamma(1 + u) w(L, x; R), \quad (4.11.13)$$

where $\Gamma(x)$ is the gamma function.

The production spectrum G can now be obtained from the momentum spectrum at great depths, say at sea level. There are several sets of experimental data (Pine 59, Ashton 60, Pak 61, Brooke 62) that are in good agreement with each other, and the momentum spectrum obtained by them at sea level and at high altitudes is shown in Fig. 4.19. The spectrum can be expressed by a power law $p^{-\alpha-1}$, with $\alpha = 1.64 \pm 0.05$ between 4 and 100 GeV/c. The spectrum becomes flatter as energy decreases and reaches a maximum at about 0.5 Gev/c.

In this energy range, where the power spectrum is a good approximation, momentum p is proportional to range R, and both $w(R)$ and $\Gamma(1+u)$ are slowly varying functions of R. Hence $G(R')$ with $R =' R + x - x_m$ can also be expressed by a power function as

$$G(R', \lambda) = \frac{G_0 R_0{}^{\alpha}}{(a + R')^{\alpha+1}} \cdot \tag{4.11.14}$$

Here a constant a indicates the flattening of the muon spectrum at small ranges, and its value cannot therefore be determined from the high-momentum part of the spectrum.

The value of a is determined by substituting (4.11.14) into (4.11.9) and comparing it with an experimental result at $\lambda = 50°$, $R = 100$ g-cm^{-2}, and $x \simeq 230$ to 1030 g-cm^{-2} (Conversi 50). The altitude dependence of the muon intensity with a residual range of 100 g-cm^{-2} air equivalent is shown in Fig. 4.20.

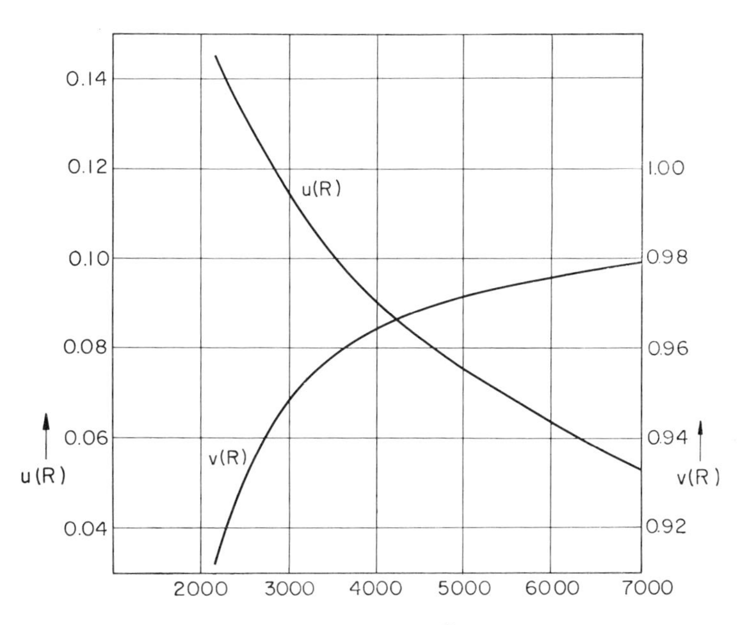

Fig. 4.18 The empirical functions $v(R)$ and $u(R)$ describing the survival probability w of muons produced at x and arriving at sea level with residual ranges $R > 2000$ g-cm^{-2} given in (4.11.12) (Olbert 54).

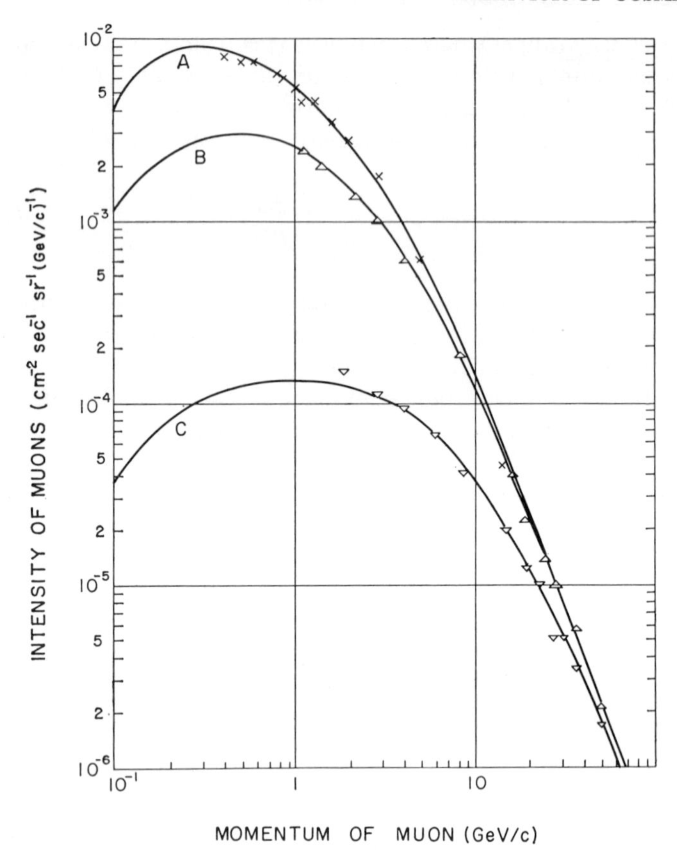

Fig. 4.19 Differential momentum spectra of muons at sea level and at a mountain altitude: at 3200 meters, vertical—A; at sea level, vertical—B; at sea level, 68°—C.

Thus Olbert determined the value of the parameters in (4.11.14) to be

$$G_0 = 1.3 \times 10^{-3} \quad \text{g}^{-1}\,\text{sec}^{-1}\,\text{sr}^{-1}, \qquad R_0 = 1000 \text{ g-cm}^{-2},$$
$$a = 520 \text{ g-cm}^{-2}, \qquad \alpha = 1.58, \tag{4.11.15}$$

for $\lambda = 50°$. The production spectrum (4.11.14) with (4.11.15) is found to reproduce the observed momentum spectrum at sea level and at a mountain altitude (Kocharian 56), as shown in Fig. 4.19, as well as the altitude dependence in Fig. 4.20.

4.11.2 Longitudinal Polarization of Muons

The shape of the production spectrum can also be obtained from the longitudinal polarization of muons. (Hayakawa 57, Fowler 58, Goldman 58). Because of the violation of parity conservation, the muon

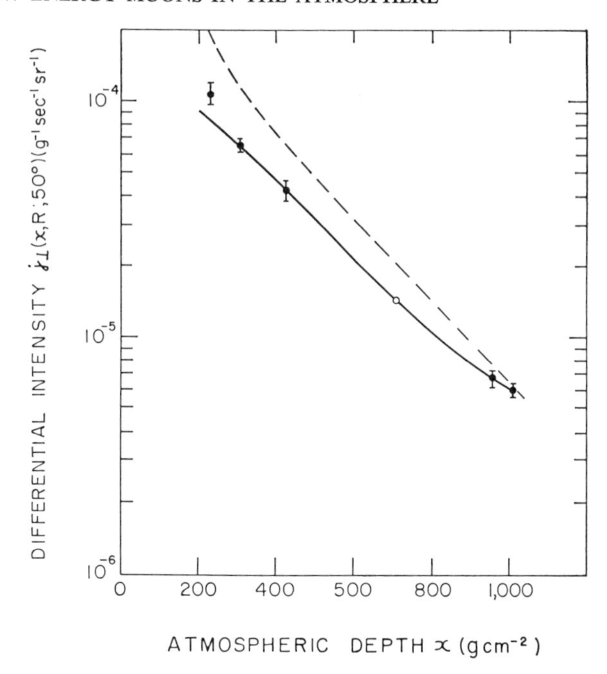

Fig. 4.20 Differential vertical intensity of muons with residual ranges $R_0 = 100$ g-cm^{-2} (air equivalent) as a function of the atmospheric depth x at 50°N geomagnetic latitude (Olbert 54). Closed circles indicate the measurements of Conversi (Conversi 50); open circle, the measurements of Kraushaar (Kraushaar 49). The dashed line represents Sand's measurements (Sands 50) on muons with residual ranges between 5 and 83 g-cm^{-2}; the solid line represents calculations by means of (4.11.10).

produced by the π-μ decay possesses the longitudinal component of polarization, and the polarization in the rest system of the parent pion is known to be complete. In the laboratory system a part of the muons coming downward are those produced upward in the rest system of the parent pion, and consequently their polarization with respect to the direction of flight is reversed. If we consider all muons coming downward, therefore, their polarization is not complete. As shown in Appendix B, (B.29), the polarization of a negative muon of energy E_μ and momentum p_μ produced by a pion of energy $\gamma m_\pi c^2$ is given by

$$P = \frac{E_\mu E_\mu^*}{p_\mu p_\mu^* c^2} - \gamma \frac{m_\mu^2 c^2}{p_\mu p_\mu^*}. \tag{4.11.16}$$

The polarization of a positive muon is negative but the dependence of the sign of polarization on the sign of charge has nothing to do with our problem as far as the up-down asymmetry of decay electrons is concerned because of the *CP* invariance. Decay negatons from negative muons at rest are preferentially emitted in the direction opposite to the sense of polarization of muons, whereas decay positons are preferentially emitted in the same direction as the sense of polarization. Consequently decay electrons are preferentially emitted backward independently of their signs.

In virtue of the above property the polarization of muons is measured by observing the up-down ratio R of decay electrons. With a measured value of R the degree of polarization is given by

$$\bar{P} = C\frac{R-1}{R+1}, \tag{4.11.17}$$

where C is a constant factor that depends on experimental conditions, such as the spectral range of decay electrons to be detected. The value of $\bar{P}$ thus measured is the average of (4.11.16) over the energy spectrum of muons measured and that of parent pions responsible for these muons.

For a given energy of muons the polarization to be observed is obtained by averaging (4.11.16) over the pion spectrum as

$$\bar{P} = \frac{\displaystyle\int_{\gamma_-}^{\gamma_+} P g_\pi(\gamma)\frac{d\gamma}{\sqrt{\gamma^2-1}}}{\displaystyle\int_{\gamma_-}^{\gamma_+} g_\pi(\gamma)\frac{d\gamma}{\sqrt{\gamma^2-1}}}, \tag{4.11.18}$$

where the pion spectrum is expressed as $g_\pi(\gamma)\,d\gamma$, and $\gamma^\pm = E_\pi^\pm/m_\pi c^2$ is given in (4.10.2). If the pion spectrum is of a power shape, $g_\pi(\gamma) \propto \gamma^{-\alpha-1}$, and $\gamma^2 \gg 1$, the expression (4.11.18) can be approximately evaluated for $\alpha > 0$ as

$$\bar{P} \simeq 1 + \left(\frac{E_\mu E_\mu^*}{p_\mu p_\mu^*} - 1\right)\left[1 - \frac{\alpha+1}{\alpha}\,\frac{1-(\gamma_-/\gamma_+)^\alpha}{1-(\gamma_-/\gamma_+)^{\alpha+1}}\right]. \tag{4.11.19}$$

The polarization given above is for muons at production. In the course of their coming down through the atmosphere and an absorber they lose energy and are scattered, so that polarization with respect to the direction of momentum changes. But the depolarization thus produced is found to be negligible unless a strong magnetic field is applied. Because the energy at production can be estimated from the thickness of matter that the muons have traversed, the value of α obtained from the measured polarization should indicate the power index of the pion spectrum in the energy range concerned if the contribution of kaons is negligible.

Since $\beta_\mu \simeq 1$, $\beta_\mu^* = c^2 p_\mu^*/E_\mu^* \simeq 0.27$, and $\alpha \simeq 1.65$, the polarization given by (4.11.19) yields

$$\bar{P}_\pi \simeq 0.31. \tag{4.11.20}$$

The polarization of muons from $K_{\mu 2}$ can be obtained in an analogous way. Since $\beta_\mu^* \simeq 0.91$ in this case, the polarization is obtained as $\bar{P}_{K(\mu 2)} = 0.54$ for the same value of α. For other decay modes we are able to calculate the polarizations by taking account of the two-stage decays through pions and the *V-A* coupling theory in the three-body decays. The polarizations thus obtained are listed in Table 4.14

Table 4.14 The Production Spectra and Polarization of Muons from Kaons

Decay Mode	Production Spectrum of Muons ($AE^{-2.65}\,dE$)	Polarization $\bar{P}$	Production Spectrum of Positive Muons ($AE^{-2.65}\,dE$)
$K_{\mu 2}$	0.2294	0.943	$0.2294r$
$K_{\pi 2}^{\pm}$	0.0640	0.327	$0.0640r$
$K_{\pi 3}^{\pm}(\tau)$	0.0210	0.327	$0.0070(r+1)$
$K_{\pi 3}^{\pm}(\tau')$	0.0023	0.327	$0.0023r$
$K_{\mu 3}^{\pm}$	0.0082	-0.273	$0.0082r$
$K_{\pi 2}^{0}$	0.1682	0.327	0.0841
$K_{\pi 2}^{0}$	0.0152	0.327	0.0076
$(K_{e3}^{0} + K_{\mu 3}^{0})$[a]	0.0660	0.327	0.0330
$K_{\mu 3}^{0}$ [b]	0.0330	-0.273	0.0165

[a] Includes muons from $K_{\mu 3}^{0}$ via two-stage decay only.
[b] Directly produced muons only.

(Osborne 64). The same table gives the relative yields of muons from all decay modes, including the decay of $K_2{}^0$, for the differential pion-production spectrum of $E^{-\alpha-1}\,dE$, with $\alpha = 1.65$.

In comparing the calculated polarization with the observed one it should be noticed that the polarization is obtained by measuring the up-down asymmetry of decay electrons from muons stopping in an absorber. Since most negative muons are captured by nuclei and do not give decay electrons, in practice one is observing the polarization of positive muons. It is therefore necessary to average the polarization by taking account of the positive excess of charged kaons. If the relative yield of positive kaons is defined as

$$r_K \equiv \frac{g_{K^+}(E)}{g_{K^\pm}(E)}, \tag{4.11.21}$$

the yields of positive muons from the respective decay modes are obtained by assuming $g_{K^\pm}(E) = g_{K_{1,2}^0}(E)$, as given in the last column of Table 4.14. Thus the polarization of positive muons from kaons is given by

$$\bar{P}_K \simeq \frac{0.04 + 0.24 r_K}{0.15 + 0.31 r_K} \simeq 0.54, \tag{4.11.22}$$

where the last number is obtained for $r_K = 0.6$.

The comparison between $\bar{P}_\pi$ and $\bar{P}_K$ demonstrates that the contributions of kaons increases the polarization of muons. Conversely, the observation of muon polarization could make it possible to obtain the K/π ratio. By summarizing a number of observed results at sea level, Osborne (Osborne 64) has given the mean values of polarization at several muon momenta, as in Fig. 4.21. It should be noted that the observed values lie between $\bar{P}_\pi$ and $\bar{P}_K$ shown in the same figure. Inspection of Fig. 4.21 leads us to conclude that there is an appreciable contribution by kaons to sea-level muons of energies between 1 and 10 GeV. It can also be seen that the K/π ratio depends little on energy.

If we adopt the mean polarization in the GeV region as

$$\bar{P} = 0.38 \pm 0.03, \tag{4.11.23}$$

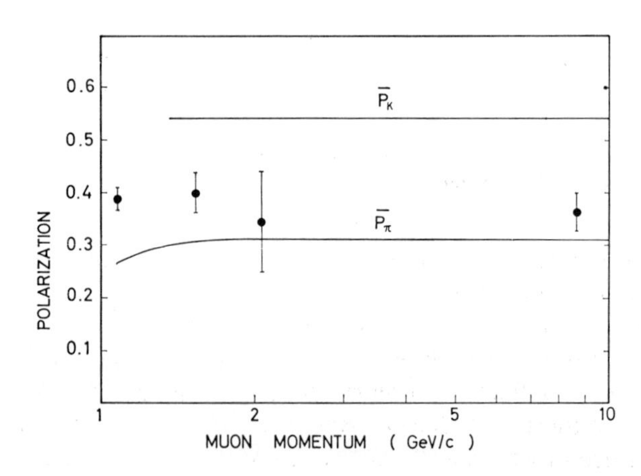

Fig. 4.21 Polarization of positive muons. Solid lines indicated by $\bar{P}_\pi$ and $\bar{P}_K$ are the values of polarization if all muons come exclusively from pions and kaons, respectively. Experimental values are those adopted by Osborne (Osborne 64).

the K/π ratio responsible for positive muons is estimated to be

$$f_K^+ = \frac{\bar{P} - \bar{P}_\pi}{\bar{P}_K - \bar{P}} = 0.4 \pm 0.2. \tag{4.11.24}$$

This appears to be consistent with the value given in (4.10.29) and further supports a weak energy dependence of the K/π ratio. However, the following should be taken into consideration.

4.11.3 Positive-to-Negative Ratio of Muons

The K/π ratio given in (4.11.24) applies to those parent particles that eventually give positive muons, whereas the value in (4.10.29) is the ratio of charged kaons to charged pions. In the former the contribution of neutral kaons amounts to about 40 percent, and consequently the K/π ratio to be compared with (4.10.29) may be a little smaller, thus being in favor of the K/π ratio decreasing as energy decreases, as seen in Fig. 4.16. For further discussion we have to consider the positive-to-negative ratio of muons in more detail.

The positive-to-negative ratio has long been studied, and it is sometimes expressed as the positive excess defined as

$$\text{positive excess} \equiv 2\,\frac{j_{\mu^+} - j_{\mu^-}}{j_{\mu^+} + j_{\mu^-}}. \tag{4.11.25}$$

The value of the positive excess has been found to be as large as 20 percent, and this is interpreted to be due to the predominance of protons in primary cosmic rays, as has been mentioned in Section 4.6.

The measurement of the positive-to-negative ratio has been extended to high energy, and the ratio stays at about 1.2 nearly independent of energy, as shown in Fig. 4.22. The weighted average of these experimental results is found to be*

$$\frac{j_{\mu^+}}{j_{\mu^-}} = 1.251 \pm 0.003. \tag{4.11.26}$$

If all the muons are assumed to come from pions, the relative yield of positive pions is

$$r_\pi = \frac{j_{\mu^+}}{(j_{\mu^+} + j_{\mu^-})} \simeq 0.56. \tag{4.11.27a}$$

If, on the other hand, they are assumed to come exclusively from kaons, the relative yield of positive kaons defined in (4.11.21) is

$$r_K \simeq 0.60. \tag{4.11.27b}$$

* Y. Kamiya, private communication.

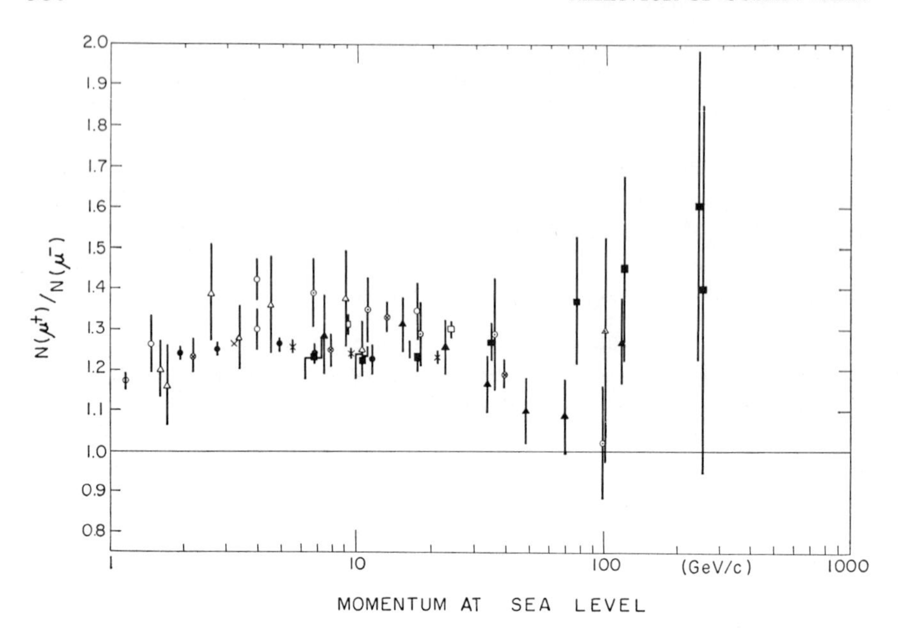

Fig. 4.22 Positive-negative ratio of muons at sea level. ●—B. G. Owen and J. G. Wilson, *Proc. Phys. Soc.* (London), **A64**, 417 (1951); ○—D. E. Caro et al., *Australian J. Sci. Rev.*, **7**, 423 (1954); △—J. R. Moroney and J. K. Parry, *Australian J. Sci. Rev.*, **7**, 423 (1954); ⨯—W. Pak et al., *Phys. Rev.*, **121**, 905 (1961) and private communication; ×—L. Filosofo et al., *Nuovo Cim.*, **12**, 809 (1954); ⊙—J. B. R. Holmes et al., *Proc. Phys. Soc.* (London), **78**, 505 (1961); ■—P. J. Hayman and A. W. Wolfdendale, *Nature*, **195**, 166 (1962); □—R. B. Brode and M. J. Weber, *Phys. Rev.*, **99**, 610 (1955); ▲—S. Kawaguchi et al., *Proc. Phys. Soc.*, in press; △—M. J. Campell et al., *Nuovo Cim.*, **28**, 885 (1963) prepared by Y. Kamiya.

The near equality of r_π and r_K indicates that the positive-to-negative ratio is rather insensitive to the K/π ratio, unless the observed value is far from the one given in (4.11.26). The value of f_{K^+}, the K^+/π^+ ratio, given in (4.11.24) by derivation from the polarization of muons is regarded to be nearly equal to f_K, the $K^\pm/\pi^\pm$ ratio.

The positive-to-negative ratio is related to the multiplicity of mesons produced, as shown in (4.6.15). The observed value in (4.11.26) indicates the multiplicity of $(3/2)n_s \simeq 4$. If the meson multiplicity increases with energy, the positive-to-negative ratio has to decrease. The fact that the positive-to-negative ratio is essentially independent of energy could be attributed to either of the following reasons. First, the relative intensity of heavy nuclei in primary cosmic rays may decrease with energy, so that the value of I_z in (4.6.13) approaches $\frac{1}{2}$ as energy increases. Such an indication will be discussed in Chapter 6. Second, the K/π ratio may increase with energy, and the positive excess of muons from kaons would then be greater. However, the kaon contribution is known to modify the positive-to-negative ratio only slightly, since most kaons seem to be produced with antikaons rather than with hyperons. If the second possibility were important, we could expect an appreciable zenith-angle dependence of the positive-to-negative ratio, because of the zenith-angle dependence of the K/π ratio.

4.11.4 Zenith-Angle Dependence

We have thus far considered muons that are incident mainly in the vertical direction, although most of the positive-to-negative ratios at high energies are obtained for inclined muons. Muons of inclined direction are of particular interest because the parent mesons have a greater chance to decay. Hence the zenith-angle dependence of their intensity cannot be given by (4.1.4). The intensity of high-energy muons at zenith angle θ is obtained from their vertical intensity by the substitutions $x \rightarrow x/\cos\theta$ and $b_\pi \rightarrow b_\pi/\cos\theta$. The former results in the reduction of the muon intensity at low energies because of the increasing energy loss and decay of muons, whereas the latter makes the muon intensity larger at high energies. As a result a parameter z introduced in $M(z, x)$ in (4.10.6) through (4.10.8) has to be replaced by

$$z \rightarrow \frac{z}{\cos\theta} = \frac{m_\mu b_\pi c}{m_\pi E_\mu \cos\theta}. \tag{4.11.28}$$

In the case of kaon decays an analogous substitution should be made. Accordingly, the function $M(z)$ is modified to

$$\begin{aligned} M\left(\frac{z}{\cos\theta}\right) &\simeq \frac{z}{z+\cos\theta}\frac{L}{l} \simeq \frac{L}{l} \quad \text{for} \quad \frac{z}{\cos\theta} \gg 1, \\ &\simeq \frac{z}{\cos\theta}\frac{L}{L-l}\ln\frac{L}{l} \quad \text{for} \quad \frac{z}{\cos\theta} \ll 1 \end{aligned} \tag{4.11.29}$$

at low- and high-energy limits, respectively. In other words, intensity at low energies is independent of the zenith angle because all pions decay into muons, whereas intensity at high energies increases with the zenith angle because the decay probability increases with the zenith angle.

As shown in Fig. 4.14, the value of $M(z/\cos\theta)$ depends on the kind of parent particles. The value of $M_\pi(z/\cos\theta)$ increases with θ more rapidly than $M_K(z/\cos\theta)$. The contribution of kaons results in a less steep zenith-angle dependence than that expected without kaons. Other quantities, such as the positive-to-negative ratio and polarization, should depend on the zenith angle, although their zenith-angle dependences have not been detected with sufficient reliability.

In addition, the decay of muons becomes more and more important as the zenith angle increases, because the path length of muons in the atmosphere is considerable. We shall discuss these effects below.

For muons with large zenith angles the energies at production are so great that their energy-loss rate can be assumed to be constant. Accordingly the survival probability is expressed by (4.11.7), with x replaced by $x/\cos\theta$ and b_μ by $b_\mu/\cos\theta$:

$$w(x_1, x; p, \cos\theta) \simeq \left[\frac{x}{L}\frac{p\cos\theta}{a(x-L)+p\cos\theta}\right]^{b_\mu/(p\cos\theta+ax)}, \tag{4.11.30}$$

where the production depth x_1 is taken to be $L/\cos\theta$. For $p\cos\theta + ax \gg b_\mu$ the survival probability (4.11.30) can be approximated by

$$w(x, p, \cos\theta) \simeq 1 - \frac{b_\mu}{p\cos\theta + ax} \ln\left[\frac{x}{L}\frac{a(x-L)+p\cos\theta}{p\cos\theta}\right]. \tag{4.11.30'}$$

Thus the θ-dependence of the survival probability is rather weak, and consequently the zenith-angle distribution of the hard component is essentially represented by the integral range spectrum $(R/\cos\theta)^{-\alpha}$. The zenith-angle distribution of $\cos^\alpha\theta$ is in good agreement with the observed zenith-angle distribution

$$j_\theta(>p) \simeq j_\perp(>p)\cos^2\theta \tag{4.11.31}$$

if we take into account $\alpha \simeq 1.64$ and the weak θ-dependence of w. At low energies, on the contrary, the θ-dependence of the survival probability becomes increasingly important as energy decreases, thus giving a steeper $\cos\theta$ dependence; for slow muons an observed zenith-angle dependence is represented by

$$j_\theta(\text{slow } \mu) \simeq j_\perp(\text{slow } \mu)\cos^3\theta. \tag{4.11.32}$$

If we fix the energy of muons, an interesting fact may be noted in that the differential intensity of muons with energies greater than several tens of GeV increases with increasing zenith angle (Budini 53). This is because the probability of pion decay increases as the zenith angle increases. The increase in muon intensity with θ continues until the survival probability becomes appreciably smaller than unity. The deviation of w from unity begins, say, at

$$\frac{b_\mu}{p\cos\theta}\ln\frac{x}{L} \simeq \tfrac{1}{10}. \tag{4.11.33}$$

or

$$\cos\theta \simeq 20\,\frac{b_\mu}{p}. \tag{4.11.33$'$}$$

A quantitative result of the zenith-angle dependence of w is shown for $L = 120$ g-cm^{-2} in Fig. 4.23 (Allen 61).

The effects of the π-μ decay represented by $M(z)$ in (4.11.29) include the following. Since $M(z)$ is an increasing function of z, this increases with decreasing $\cos\theta$, as shown in (4.11.29). On the other hand, muon energy at production is expressed as $E_\mu(x) + ac(x-L)/\cos\theta$ and

$$z(\theta) = \frac{m_\mu}{m_\pi}\,\frac{b_\pi c}{E_\mu(x)\cos\theta + ac(x-L)} \geq z(0)\,. \tag{4.11.34}$$

Because of (4.11.29) and the fact that $b_\pi \gg a(x-L)$, it follows that $M(z)$ is essentially flat for $z(0) > 1$ or $E_\mu(x) < b_\pi c$, whereas $M(z)$ increases with θ for $z(\theta) < 1$ for $E_\mu(x)\cos\theta \geq b_\pi c$.

The product of the survival probabilities of pions and muons, $w(\theta)M(\theta)$, gives us the zenith-angle distribution of muons of a particular energy. Figure 4.24 shows the zenith-angle distributions thus obtained for several energies, in which $l = L = 120$ g-cm^{-2} is assumed for

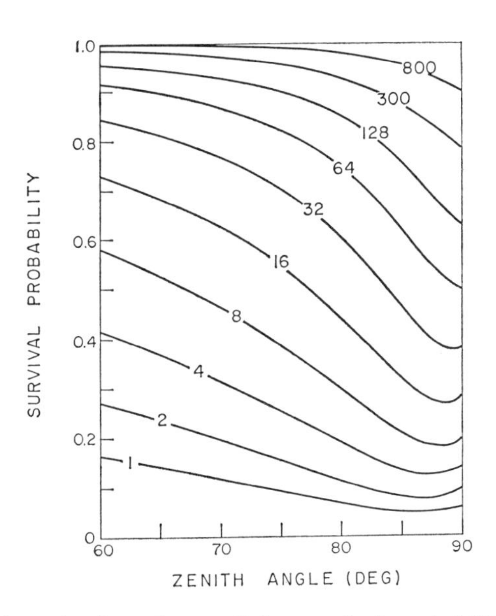

Fig. 4.23 Zenith-angle dependence of the survival probability of muons. The numbers indicate the energies in GeV of muons at sea level (Allen 61).

simplicity and the pion-production spectrum with $\alpha = 1.67$ is taken into account (Smith 59). The zenith-angle distribution of the integral intensity is derived from this and is shown in Fig. 4.25.

Another way of expressing the zenith-angle distribution is by considering the momentum spectra at various zenith angles. These are shown in Fig. 4.26 for $l = L = 120$ g-cm^{-2} and the pion spectrum of $0.15\, E_\pi^{-2.55}\, dE_\pi$ cm^{-2} sec^{-1} sr^{-1} with E_π in GeV (Allen 61). As the zenith angle increases the flat portion of the spectrum becomes broader due to the energy loss of muons in the atmosphere. At large zenith angles the low-momentum part of the spectrum is suppressed because of the muon-decay effect, as indicated by dotted lines in Fig. 4.26. At high momenta the intensity is found to increase with zenith angle. The momentum spectra thus predicted are in good agreement with observed spectra (Pak 61, Allen 61).

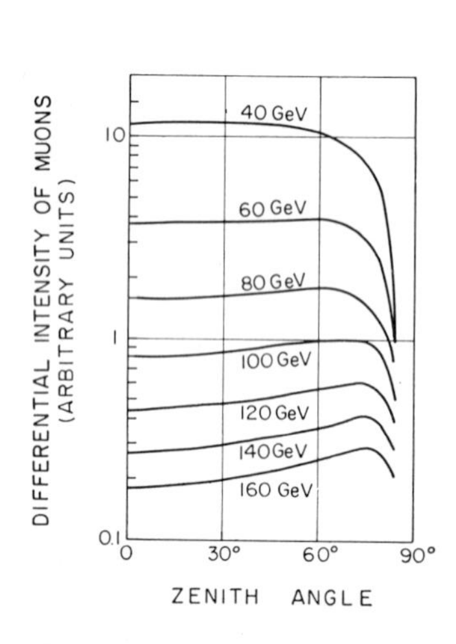

Fig. 4.24 Zenith-angle distribution of the intensity of muons with given energies.

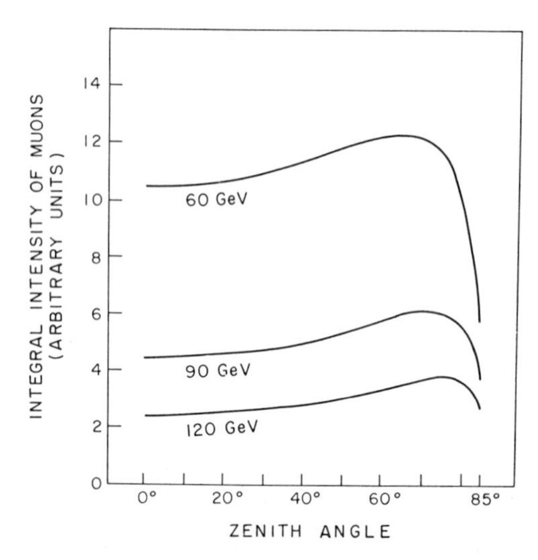

Fig. 4.25 Zenith-angle dependence of the intensities of muons with energies greater than 60, 90, and 120 GeV (Smith 59).

For practical purposes the angular spread of muons has to be taken into account at large zenith angles, where the zenith-angle distribution is very steep. There are two sources of angular spread of muons in the course of their propagation through the atmosphere.

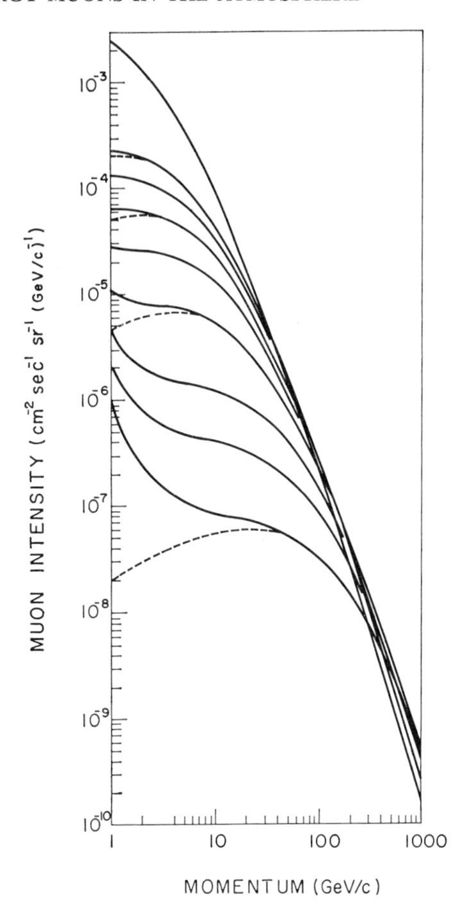

Fig. 4.26 Momentum spectra of muons at various zenith angles, 0°, 60°, 65°, 70°, 75°, 80°, 85°, 87.5° and 90° from the top to the bottom. The dotted curves represent spectra corrected for the decay effect of muons (Allen 61).

The first source is multiple scattering in the atmosphere. The mean-square angle of multiple scattering is given by (Olbert 54).

$$\langle\theta_s^2\rangle = \frac{E_s^2}{X_0}\int_{pv(x)}^{pv(L)} \frac{d(pv)}{p^2v^2[-d(pv)/dx]} \simeq \frac{E_s^2}{X_0\langle -d(pv)/dx\rangle}\left[\frac{1}{pv(x)} - \frac{1}{pv(L)}\right] \tag{4.11.35}$$

for a muon produced at depth L with momentum $p(L)$ and velocity $v(L)$ and reaching depth x with $p(x)$ and $v(x)$. With $E_s = 21$ MeV,

$X_0 = 37$ g-cm^{-2}, and $-d(pv)/dx \simeq 2.0$ MeV/g-cm^{-2} we obtain

$$\langle \theta_s^2 \rangle^{1/2} \simeq \sqrt{5.8 \text{ MeV}/pv(x)}, \qquad (4.11.35')$$

where $pv(x) \ll pv(L)$ is assumed; for example, this is about 2° for $p = 5$ GeV/c and considerably modifies the zenith-angle distribution for $\theta \gtrsim 80°$. This makes the zenith-angle distribution less steep and the differential intensity greater at low momenta. In Fig. 4.26 the corrected momentum spectra are shown by solid lines.

The second source is deflection by the geomagnetic field. The radius of curvature, r, of a single-charge particle with momentum p is

$$r = \frac{pc}{300 B_\perp} = 3.3 \times 10^9 \frac{p}{100 \text{ GeV}/c} \frac{0.1 \text{ gauss}}{B_\perp}, \qquad (4.11.36)$$

where $B_\perp$ is the magnetic-field strength (in gauss) perpendicular to the momentum. The path length of a muon in the atmosphere with zenith angle θ is approximately given, if deflection is neglected, by

$$l = \frac{\bar{H}}{\cos\theta} \ln \frac{X_0}{L \cos\theta} = \frac{6.0 \times 10^5}{\cos\theta} (2.2 - \ln \cos\theta), \qquad (4.11.37)$$

where $\bar{H} = 6.0 \times 10^5$ cm is the average scale height. For $\theta = 80°$, l is as large as 10^7 cm. Since this is a few percent of the earth radius, the curvature of the earth's surface is not negligible.

The deflection angle is given by

$$\phi = \frac{l}{r}, \qquad (4.11.38)$$

which is several degrees for a muon of several tens of GeV. The sense of deflection is different for positive and negative particles. For particles coming from the east a positive muon has a longer path length than a negative one, whereas for those coming from the west a positive muon has a shorter path length than a negative one. For particles traversing the short path the path length given in (4.11.37) may be corrected by replacing θ by $\theta - \phi$. For particles traversing the long path length θ may be replaced by $\theta + \phi$.

Magnetic deflection brings about the following effects: (*a*) The energy at production is higher for particles traversing the long path and (*b*) the survival probability is smaller for those traversing the long path. Both effects result in the reduction of the intensity for the long path (Kamiya 63).

An interesting consequence of geomagnetic deflection is its effect on the positive-to-negative ratio of muons and its azimuth dependence. As can be expected from the discussions above, a considerable amount of the negative excess is observed for muons coming from the east because a positive muon has a long path in this direction. The azimuth dependences of the positive excess at sea level for $\theta = 78°$ are shown in Fig. 4.27 for three energy ranges (Kamiya 62).

4.11.5 Behavior of Slow Muons

Since slow muons have high probabilities of decay into electrons, they provide an important source of electrons in the lower atmosphere. In a dense material, however, they lose their energies so rapidly that

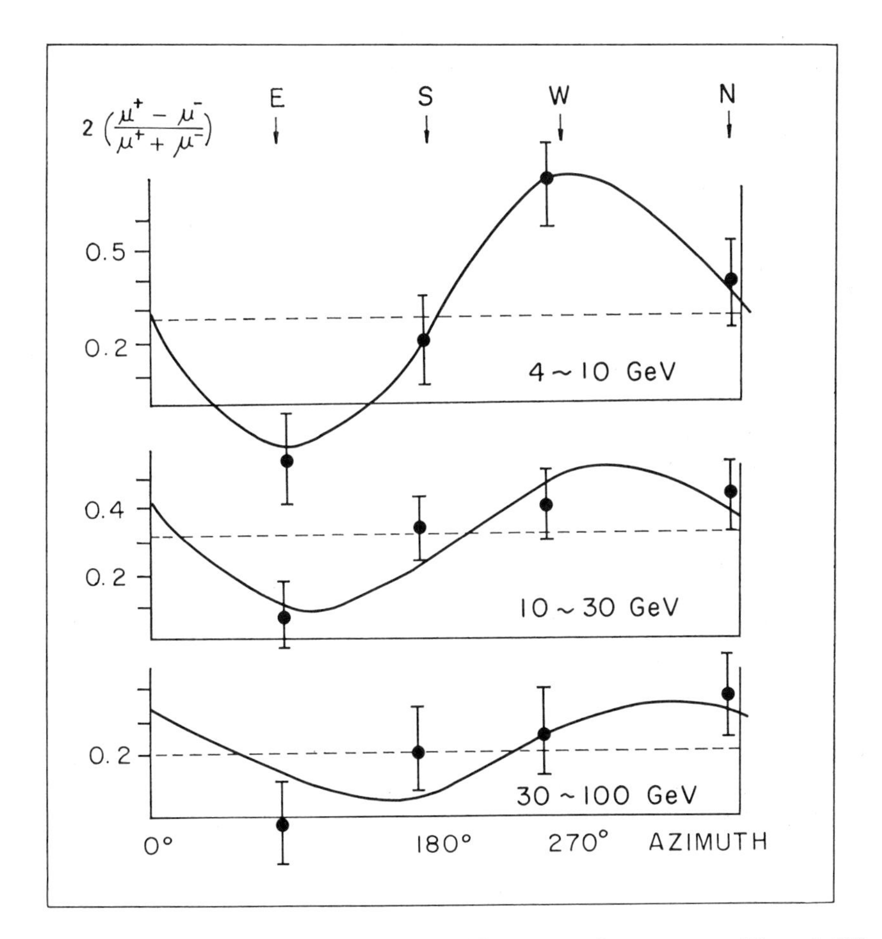

Fig. 4.27 Azimuth dependence of the positive excess of muons at zenith angle 78° at sea level. The energy intervals of sea-level muons are indicated (Kamiya 62).

few of them decay in flight. The time necessary for a muon to be slowed down to thermal energy is estimated to be less than 10^{-11} sec in solid and liquid media. A thermal muon is subject to the thermal oscillations of atoms and is attracted or repelled by the Coulomb field of a nucleus. The positive muon has a negligible probability of approaching a nucleus, whereas the negative muon tends to be trapped thereby. The probability of trapping is found to be independent of Z, thus being proportional to the atomic concentration in chemical compounds (Sens 58). The muon thus trapped falls into lower atomic levels by radiative and Auger transitions, emitting X-rays and Auger electrons, respectively. In a liquid, such as liquid hydrogen, an atom that has trapped a muon diffuses through the medium and meets another ordinary atom, forming a molecule in which the muon is responsible for molecular binding.

In the ground state of the muonic atom the radius of its K-orbit is so small that the muon spends most of the time inside the nucleus. Hence the density of the muon is proportional to Z_{eff}^3 rather than Z^3, and the capture probability is proportional to Z_{eff}^4 rather than Z^4. The effective charge, Z_{eff}, for uniform nuclei with radii of $1.20 \times 10^{-13} A^{1/3}$ cm is found to be (Tennent 60)

$$Z_{\text{eff}} = Z\left[1 + \left(\frac{Z}{42.0}\right)^{1.47}\right]^{-1/1.47}. \tag{4.11.39}$$

A slight modification of Z_{eff} due to the shape of individual nuclei is significant for light nuclei, but the amount of modification is no greater than 2 percent.

The disappearance rate of a bound muon is the sum of the decay rate, $1/\tau_d$, and the nuclear capture rate, $1/\tau_c$. The capture rate is proportional to Z_{eff}^4 and depends on the neutron excess $(N-Z)/A$, where $N = A - Z$, as well as on the states available for the recoil neutron arising from

$$\mu^- + p \rightarrow n + \nu_\mu .$$

These dependences are represented by the parameters δ and γ, respectively, so that the rate of capture by a nucleus with A and Z is expressed by (Primakoff 55)

$$\frac{1}{\tau_c}(A, Z) = \gamma\left[\frac{1}{\tau_c}(1, 1)\right] Z_{\text{eff}}^4 \left[1 - (A - Z)\frac{\delta}{2A}\right]. \tag{4.11.40}$$

Good agreement with experimental capture rates (Sens 59) can be obtained with

$$\frac{\gamma}{\tau_c}(1, 1) = 188\ \text{sec}^{-1}, \qquad \delta = 3.15. \tag{4.11.41}$$

For more detailed discussions of the capture rate we have to consider the nuclear structures of initial and final nuclei.

The decay rate of a bound muon is not always equal to that of a free muon. The decay rate is expected to be reduced, first, by the decreases in available energy by the binding energy and, second, by muon motion, which causes the motional narrowing of the decay width. These effects are rather small, and the difference between the decay rates of the free and bound muons is calculated as (Groetzinger 51)

$$\frac{1}{\tau_\mu} - \frac{1}{\tau_d} = \frac{(Z/137)^2}{\tau_\mu}, \tag{4.11.43}$$

which is found to be in reasonable agreement with experiment, except for some nuclei such as iron. The increased decay rate may be accounted for in terms of the distortion of the outgoing electron wave by the nuclear Coulomb field.

The energy absorbed by a nucleus is mostly taken away by a neutrino, and the rest is left as nuclear excitation. The distribution of excitation energy is quite sensitive to the nuclear model, particularly to the reduced mass of a nucleon in nuclear matter and the momentum distributions of protons and neutrons therein. The excitation energy seems to extend to 40 MeV or so and to center at about 20 MeV. This is then followed by the emission of neutrons, protons, and photons. Most of the excitation energy goes to neutrons, but the absolute number of

Table 4.15 The Average Multiplication of Neutron Emission following Muon Capture

Element	Observed Multiplicity
Carbon	0.83 ± 0.06
Sodium	1.0 ± 0.4
Magnesium	0.6 ± 0.2
Aluminum	0.95 ± 0.17
Calcium	0.4 ± 0.4
Tin	1.54 ± 0.12
Silver	1.60 ± 0.18
Iodine	1.7 ± 0.4
Lead	2.14 ± 0.13, 1.64 ± 0.16
Bismuth	2.32 ± 0.17

neutrons emitted is very difficult to determine. The average multiplicities of neutrons thus far measured are not always in agreement with each other, as seen in Table 4.15 (Tennent 60). The emission of protons is found to take place with a probability of about 0.1 (Morinaga 53).

4.12 Muons Underground

The intensity and energy spectrum of muons discussed in the preceding sections have been compared with experimental results, as in Fig. 4.19. The energy spectrum has been measured with reasonable accuracy up to a few hundred GeV by means of magnet spectrographs (Hayman 62). The spectrum has been obtained by referring to the absorption of muons as well. For low-energy muons artificial absorbers such as lead can be used to measure the range spectrum. At energies higher than several GeV the thickness required for absorbers is so large that it is necessary to use natural absorbers, such as rock and water. This has stimulated the investigation of underground muons. Before going into this problem we shall discuss another method of obtaining the energy spectrum of muons.

4.12.1 Bursts Produced by Muons

A muon radiates a photon of high energy by bremsstrahlung in a heavy material, and the photon initiates a large cascade shower. The cascade shower that is observed with an ionization detector is a kind of burst, and it can be distinguished from other kinds of bursts because it is observable even at a large thickness. Bursts initiated by muons were investigated for the purpose of determining the properties of the muon, such as its spin and anomalous magnetic moment. At present, however, this problem is of little interest because there are better methods for this purpose. However, we are interested in muon-induced bursts in order to obtain the energy spectrum of muons. In order to exclude bursts produced by nuclear active particles, observations of the bursts at a considerable depth underground are preferable.

A muon of energy E radiates a photon of energy W with the probability $\psi(E, W)\, dE$ in a burst producer. Photon emission may take place at any thickness of the producer, and consequently the size of a burst produced by a photon of energy W distributes as $P(W, S)\, dS$, where the burst size S is ordinarily defined as the number of electrons passing through the detector but may be any quantity proportional to the amount of ionization produced therein. If the producer is as thick as, or slightly thicker than, the radiation length of the muon, the product $\int \psi(E, W)\, P(W, S)\, dW$ depends only weakly on thickness. Hence the frequency of bursts per unit area and solid angle with sizes between S and $S + dS$ is given by

$$f(S, \theta)\, dS = dS \int_0^\infty dE \int_0^E dW j(E, \theta)\, \psi(E, W)\, P(W, S). \quad (4.12.1)$$

By observing the burst spectrum $f(S)$ we can in principle deduce the energy spectrum of muons, $j(E)$. Since (4.12.1) involves many integrals, and both ψ and P are slowly varying functions of their arguments, it is not easy to obtain the energy spectrum from the burst size distribution. In particular, the absolute value of a muon energy corresponding to a given size is very difficult to determine.

4.12.2 Energy Loss of High-Energy Muons

A classical but still reliable method of obtaining the energy spectrum is the measurement of the muon intensity-depth relation. If the rate of muon energy loss is given, we can obtain the range-energy relation and consequently the energy spectrum. The rates of energy loss by electromagnetic interactions are discussed in detail in Chapter 2. Here, however, we repeat essential points that are important for the present purpose.

Energy Loss by Ionization. At energies of up to 1 TeV loss by ionization, excitation, and Čerenkov radiation predominates in the energy loss of muons. The rate of energy loss of a muon of momentum p, total energy E, and velocity $c\beta$ by these processes per g-cm^{-2} is expressed by

$$-\left(\frac{dE}{dx}\right)_{\text{ion}} = \frac{L}{\beta^2}\left[B + 0.69 + 2\ln\left(\frac{p}{m_\mu c}\right) + \ln\left(\frac{E'_m}{m_\mu c^2}\right) - 2\beta^2 - \delta\right], \quad (4.12.2)$$

where m_μ is the muon mass and

$$E'_m = \frac{p^2 c^2}{E + m_\mu{}^2 c^2 / 2m_e} \quad (4.12.3)$$

is the maximum transferable energy to an electron. The last term in the brackets, δ, represents the density effect and is given by

$$\delta = 4.606 \log_{10}\left(\frac{p}{m_\mu c}\right) - C \quad \text{for} \quad p > D m_\mu c. \quad (4.12.4)$$

The parameters L, B, C, and D in (4.12.2) and (4.12.4) are constants that are characteristic of the medium traversed. Although the ionization loss rate per g-cm^{-2} depends little on the medium, a slight medium dependence is significant for the derivation of the energy spectrum from the intensity-depth relation. In practice muons have been observed underwater as well as under rock. Since water is homogeneous, the depth that is measured under water is customarily taken as the standard value. The depth under rock is thus measured in meters water equivalent (mwe), which is roughly in units of 100 g-cm^{-2}. More accurately, however, the difference between the energy-loss rates in water and in rock has to be distinguished. For the latter we refer to rock in the Kolar Gold Field in India, where the deepest observations have been made (Miyake 64).

Loss by Pair Creation

The rate of energy loss by direct pair creation given in Section 2.5 is expressed by

$$-\left(\frac{dE}{dx}\right)_{\text{pair}} = b_p E. \tag{4.12.5}$$

Loss by Bremsstrahlung. The rate of energy loss by bremsstrahlung is given in the cases of no screening and complete screening, respectively, by

$$-\left(\frac{dE}{dx}\right)_b^{(1)} = b_b^{(1)} E\left[\ln\left(\frac{E}{m_\mu c^2}\right) - d^{(1)}\right], \tag{4.12.6a}$$

and

$$-\left(\frac{dE}{dx}\right)_b^{(2)} = b_b^{(2)} E. \tag{4.12.6b}$$

The values of parameters in the energy-loss rates above are tabulated for water and rock in Table 4.16. By comparing these energy-loss rates we can see that ionization loss constitutes the major part up to several hundred GeV and that the other processes are negligible below 100 GeV. In this energy region medium dependence is so weak that depth measured in meters water equivalent is meaningful. Above 1 TeV, on the other hand, losses by direct pair creation and bremsstrahlung constitute the main part, and the rate of energy loss by these processes is proportional to Z^2/A rather than to Z/A in the ionization process. Hence depth measured in muon radiation length is more significant.

Summing up these energy-loss processes, we can express the energy-loss rate by

Table 4.16 Values of Parameters in the Energy-Loss Rates of Muons

Parameter	Units	Sea Water	Rock[a]
Atomic number Z		7.43	12.8
Mass number A		14.79	25.8
Density	g-cm^{-3}	1.025	3.02
L	MeV/g-cm^{-2}	0.0853	0.0740
B		18.35	16.77
C		3.47	4.21
D		10^2	10^3
b_p	g^{-1} cm^2 (E in MeV)	1.1×10^{-6}	1.8×10^{-6}
$b_b^{(1)}$	g^{-1} cm^2	1.2×10^{-7}	2.1×10^{-7}
$d^{(1)}$		0.13	0.31
$b_b^{(2)}$	g^{-1} cm^2	1.3×10^{-6}	2.2×10^{-6}
a	MeV/g-cm^{-2}	1.75	1.84
$b_p + b_b$	g^{-1} cm^2	2.4×10^{-6}	4.0×10^{-6}
c	MeV/g- cm^{-2}	0.085	0.076

[a] Kolar Gold Field in India.

$$-\left(\frac{dE}{dx}\right)_{\text{total}} = a + bE + c \ln\left(\frac{E'_m}{m_\mu c^2}\right) \quad \text{MeV/g-cm}^{-2}. \tag{4.12.7}$$

Here b includes the effects of pair creation and bremsstrahlung with complete screening, and its values for water and rock at energies of several TeV are given together with the values of a and c in Table 4.16.

Loss by Nuclear Interactions. Electromagnetic interactions are not all the causes of energy loss. The nuclear interactions of the muon through the virtual-photon cloud surrounding the muon may also contribute. For muons with energies of up to a few tens of GeV the nuclear-interaction cross section can be accounted for, as discussed in Section 3.17, in terms of the energy-independent photonuclear cross section of $\sigma_\gamma = (1.4 \pm 0.3) \times 10^{-28}$ cm^2 per nucleon and the Williams-Weizsäcker formula. Pion production in this energy region is interpreted essentially by the one-fireball model with inelasticity of $K \simeq \frac{1}{2}$. If this holds, the energy-loss rate is given by

$$\begin{aligned} -\left(\frac{dE}{dx}\right)_N &= \frac{2}{\pi}\alpha K N \sigma_\gamma E \ln\frac{E}{E_c} \\ &\simeq 2.0 \times 10^{-7} E \ln\frac{E}{E_c} \quad \text{MeV/g-cm}^{-2}, \end{aligned} \tag{4.12.8}$$

where E is measured in MeV and E_c is on the order of the muon rest energy. No reliable information is yet available for higher energies. If, however, the fireball model such as in the nucleon-nucleon collision holds also in the muon-nucleon collision, the expression (4.12.8) may be extended to higher energy. The energy-loss rate thus obtained may be comparable to the rate of loss by bremsstrahlung, and may be taken into account as a part of b in (4.12.7). However, no positive evidence for nuclear-interaction loss has yet been observed.

The value of b in (4.12.7) implies further inaccuracy if (4.12.7) is applied in a wide energy range, because the screening effect in pair creation and bremsstrahlung is not complete at about 1000 GeV. Hence the value of b varies with energy. If the value of b changes by an amount Δb, the depth measured in muon radiation length should be changed by an amount Δt:

$$\frac{\Delta t}{t} = \frac{\Delta b}{b}. \tag{4.12.9}$$

The ionization loss ε per radiation length should be correspondingly changed by an amount of $\Delta \varepsilon_1$:

$$\frac{\Delta \varepsilon_1}{\varepsilon} = \frac{-\Delta t}{t} = \frac{-\Delta b}{b}. \tag{4.12.10}$$

The rate of energy loss by ionization given by the first and third terms of (4.12.7) also has an energy dependence. This results in the change of the critical energy by $\Delta \varepsilon_2$. Hence the resultant change in the critical energy is

$$\Delta \varepsilon = \Delta \varepsilon_1 + \Delta \varepsilon_2 = -\left(\frac{\varepsilon}{b}\right) \Delta b + \Delta \varepsilon_2. \tag{4.12.11}$$

This affects the intensity of muons by an amount Δi:

$$\frac{\Delta i}{i} = -\frac{n\, \Delta \varepsilon}{\varepsilon}, \tag{4.12.12}$$

where n is the power index of the range spectrum. The energy dependences of $\Delta b/b$ and $\Delta \varepsilon_2/\varepsilon$ are shown in Fig. 4.28 (Hayakawa 64).

Taking these considerations into account we are able to use the energy-loss rate given in (4.12.7) in order to obtain the range-energy relation. It must, however, be remarked that (4.12.7) gives the average energy-loss rate, and the actual loss rate fluctuates considerably around this value. The fluctuation effect is of great importance in bremsstrahlung, because the fraction of energy lost by this process is a significant part of the muon energy. The fluctuations affect the intensity-depth curve a

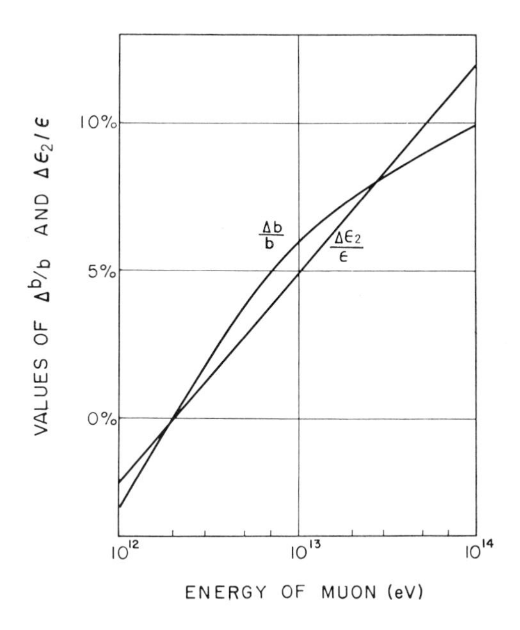

Fig. 4.28 Energy dependences of parameters in the energy-loss rate given in (4.12.9) through (4.12.11) (Hayakawa 64).

great deal, since the energy spectrum falls off steeply at high energy. Although the distribution of the energy-loss rate about its average value is nearly symmetric, particles with energy losses smaller than average penetrate down to a greater depth and constitute a great part of the intensity at this depth, whereas particles sustaining larger energy losses make up only a minor part of the intensity at a lesser depth. Thus the fluctuation effect gives an intensity larger than what we would expect without the fluctuations.

4.12.3 Intensity-Depth Relation

At low energies, say below 200 GeV, the energy-loss rate may be approximately expressed by

$$-\frac{dE}{dx}=\bar{a}, \tag{4.12.13}$$

since the term bE in (4.12.7) is negligibly small and the logarithmic term is slowly varying in energy. The effective energy-loss rate $\bar{a}$ around 100 GeV may be assumed to be

$$\bar{a}\simeq 2.5\ \text{MeV/g-cm}^{-2} \tag{4.12.13$'$}$$

Thus the range-energy relation is simplified to

$$x = \frac{E}{\bar{a}}. \tag{4.12.14}$$

The intensity-depth relation for vertical muons is therefore expressed by

$$i_{\perp}(x) = j_{\perp}(> \bar{a}x), \tag{4.12.15}$$

where $i_{\perp}$ is the vertical intensity and $j_{\perp}(>E)$ is the integral energy spectrum of the vertical component at sea level.

The relation (4.12.15) demonstrates how the intensity-depth relation represents the energy spectrum. The bending of the energy spectrum at about 100 GeV due to the π-μ decay is revealed by that of the intensity-depth curve at several hundred meters water equivalent. If the intensity-depth curve is represented by a power function

$$i(x) \propto x^{-n}, \tag{4.12.16}$$

the power index n is related to the power index of the integral energy spectrum β by

$$\begin{aligned} n &\simeq \beta \qquad (x < x_b), \\ n &\simeq \beta + 1 \qquad (x > x_b), \end{aligned} \tag{4.12.17}$$

where $x_b \equiv (m_\mu/m_\pi)(b_\pi c/\bar{a})$ is the depth around which the intensity-depth curve bends. The actual intensity-depth curve deviates from that expected from (4.12.17) because of the energy dependence of the energy loss, the contribution of kaons, and so forth.

Muons at large zenith angle θ traverse a thickness $x/\cos\theta$. Hence the unidirectional intensity at θ is obtained in analogy to (4.12.15) as

$$i(x, \theta) = j_\theta\left(> \frac{\bar{a}x}{\cos\theta}\right). \tag{4.12.18}$$

The zenith-angle dependence can be expressed also by a power function:

$$i(x, \theta) = i_{\perp}(x) \cos^m \theta. \tag{4.12.19}$$

If the energy spectrum at sea level does not depend on the zenith angle, the relation (4.12.18) leads to $m = n$. However, the spectrum depends on the zenith angle, as discussed in Section 4.11.4. The transition from the zenith-angle-independent part to the dependent part takes place at $E\cos\theta = (m_\mu/m_\pi)b_\pi c$. Since the energy E is related to the thickness traversed by $E = \bar{a}x\cos\theta$, the transition point is $x = x_b$, independent of the zenith angle. For $x < x_b$ the sea level spectrum is essentially independent of θ, whereas for $x > x_b$ it depends on θ as $1/\cos\theta$. Thus we obtain the zenith-angle dependence as

$$\begin{aligned} m &= n = \beta \qquad & (x < x_b), \\ m &= n - 1 = \beta \qquad & (x < x_b). \end{aligned} \tag{4.12.20}$$

The zenith-angle dependence does not change as the depth crosses x_b.

By taking the zenith-angle dependence into account we can extend the energy spectrum of muons to higher energies than are obtained from only vertical muons. At such high energies the approximation (4.12.13) is no longer valid, and we have to take the bE term in (4.12.7) into account.

At high energies, where bE is no longer negligible, the energy-loss rate may be approximated by

$$-\frac{dE}{dx} = \bar{a} + bE. \tag{4.12.21}$$

Hence the range-energy relation is given by

$$x = \frac{1}{b} \ln\left(\frac{\bar{a} + bE}{\bar{a}}\right), \qquad E = \frac{\bar{a}}{b}(e^{bx} - 1). \tag{4.12.22}$$

At depths of $x > 1/b$ the average energy of a muon reaching x is given by

$$E = \left(\frac{\bar{a}}{b}\right) e^{bx}. \tag{4.12.22'}$$

If the integral energy spectrum is represented by $E^{-(\beta+1)}$, the intensity-depth relation is

$$i(x, \theta) = \frac{1}{\cos\theta} j_{\perp}\left(> \frac{\bar{a}}{b} e^{bx/\cos\theta}\right) \propto \frac{1}{\cos\theta} \exp\left[-(\beta+1)b \frac{x}{\cos\theta}\right]. \tag{4.12.23}$$

Thus the intensity decreases exponentially, and the angular distribution becomes steep with increasing depth according to

$$m \simeq (\beta + 1)bx - 1. \tag{4.12.24}$$

Since $-(\beta + 1)bx$ is the slope of the intensity-depth curve, $m = n - 1$ still holds at such great depths.

At great depths the range of fluctuations is of great importance. The fluctuation problem was solved by Nishimura (Hayakawa 64, Nishimura 67), who expressed the depth in units of muon radiation length. The intensity at depth t, $i(t)$, deviates from that without fluctuations, $i_0(t)$, by

$$f(t) \equiv \frac{i(t) - i_0(t)}{i_0(t)}. \tag{4.12.25}$$

The values of $f(t)$ are given in Table 4.17 and in Fig. 4.29. Hence the intensity-depth curve is less steep than that without fluctuations. Furthermore, as the integral power index of the muon spectrum increases the deviation between them becomes larger.

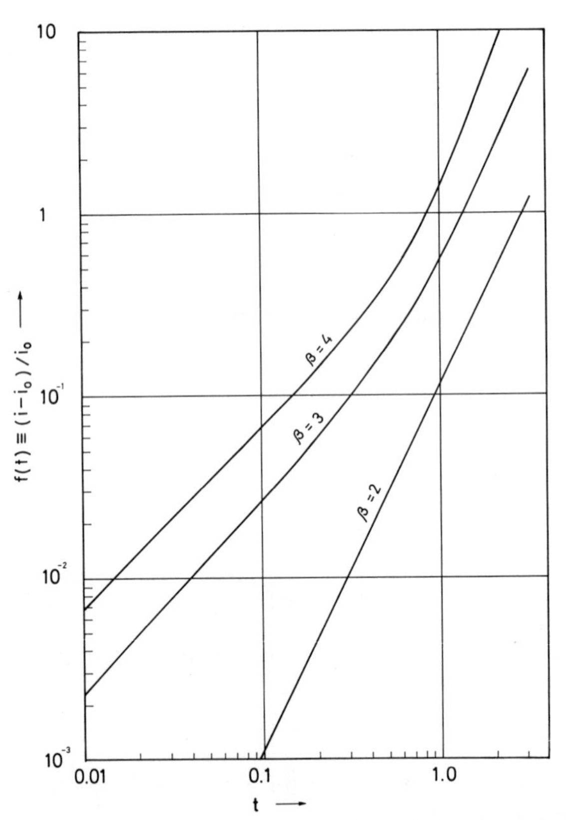

Fig. 4.29 Degree of range fluctuations of muons, $f(t) \equiv (i - i_0)/i_0$, given in Table 4.17. The values of the integral power index ($\beta = 2$, 3, and 4) are attached to the curves. Abscissa: thickness of matter in units of muon radiation length; ordinate: $(i - i_0)/i_0$.

Table 4.17 Degree of Fluctuations $f(t) \equiv [i(t) - i_0(t)]/i_0(t)$

Spectral Index β	Depth t (units of muon radiation length)						
	0.01	0.1	0.6	0.8	1.0	1.5	2.0
2	0.000	0.0011	0.04	0.075	0.12	0.27	0.46
3	0.0024	0.027	0.23	0.38	0.55	1.26	2.21
4	0.007	0.070	0.52	0.89	1.36	3.63	7.89

The qualitative features discussed above are obtained from experimental results. Because the values of parameters in the energy-loss rate are different for water and rock, the intensity-depth relations in these two media are shown separately.

A number of attempts have been made to observe the muon intensity deep in lakes and the sea. The deepest point reached is as deep as 3000 meters (Barton 61). Here, however, we refer only to a systematic observation down to 1380 meters near the Pacific coast of Japan (Higashi 65*a*). The intensities in both the vertical and the inclined directions are indicated in Fig. 4.30, where the normalization of the intensity was made at 20 meters below sea level. It is seen that the slope is essentially

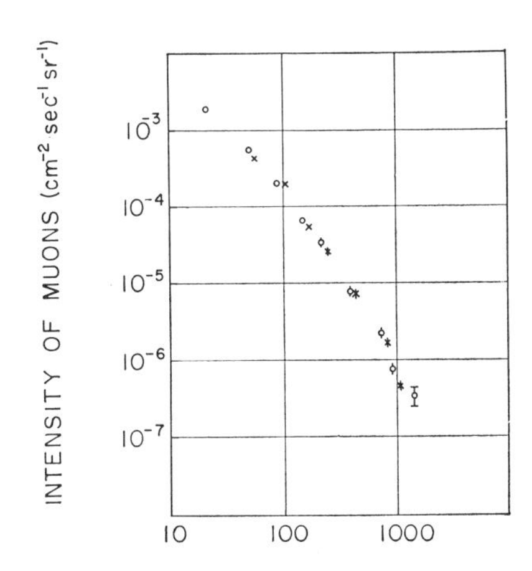

Fig. 4.30 Intensity-depth curve of muons in sea water (Hayakawa 64). ⌽—vertical direction; ✱—inclined direction; ⊙—normalization point.

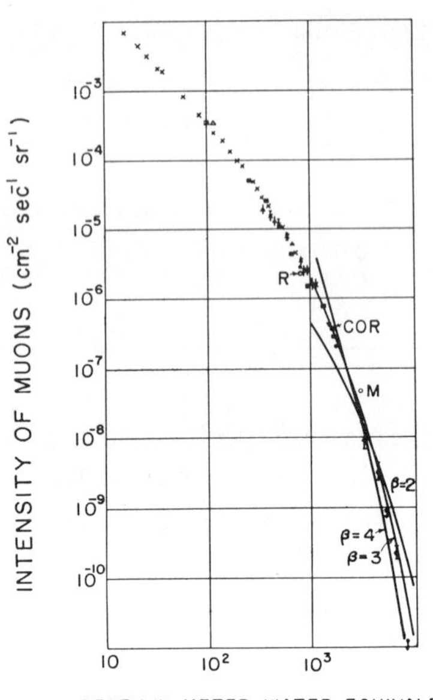

Fig. 4.31 Intensity-depth curve of muons in rock. The curves are obtained from the power energy spectra of muons with power indices $\beta = 2$, 3, and 4 by taking account of the fluctuation effect (Hayakawa 64). ●—S. Miyake et al., *Nuovo Cim.*, **32**, 1505 (1964); COR—P. H. Barrett et al., *Rev. Mod. Phys.*, **24**, 133 (1952); ▽—L. M. Bollinger, Ph.D. Thesis, Cornell University, 1951; ■—J. Clay and A. von Gemert, *Physica*, **6**, 497 (1939); ▲—J. C. Barton, *Phil. Mag.*, **6**, 1271 (1961); ○R—C. A. Randal and W. E. Hazen, *Phys. Rev.*, **81**, 144 (1951); △—B. V. Sreekantan and S. Naranan, *Proc. Ind. Acad. Sci.*, **36**, 97 (1952); ○M—Y. Miyazaki, *Phys. Rev.*, **76**, 1733 (1949); ×—V. C. Wilson, *Phys. Rev.*, **53**, 337 (1938).

constant as far as several hundred meters of water and becomes gradually steeper as the depth increases.

Figure 4.31 shows the intensity-depth curve observed under rock by various workers. Expressing the rock thickness in meters water equivalent, we notice its similarity to the intensity-depth curve in Fig. 4.30. Beyond 1000 meters water equivalent the fluctuation effect is found to be significant; the curves drawn in the figure represent those derived theoretically for given values of the power index of the integral energy spectrum. Comparison with experimental results indicates that the value of the power index is

$$\beta = 3.3 \pm 0.2. \qquad (4.12.26)$$

The deepest point reached is 8400 meters water equivalent at Kolar

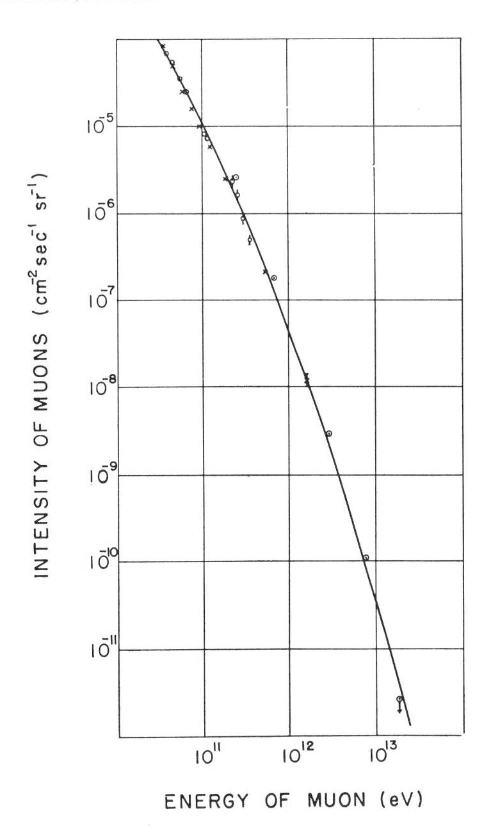

Fig. 4.32 Integral energy spectrum of muons (Hayakawa 64). ⊙—S. Miyake et al., *Nuovo Cim.*, **32**, 1505 (1964); ○—S. Higashi et al., *Kagaku* (*Japan*), **35**, 461 (1965); ×—P. T. Hayman and A. W. Wolfendale, *Proc. Phys. Soc.*, **80**, 710 (1962); J. L. Osborne et al., ibid., **84**, 911 (1964).

Gold Field in India (Miyake 64). It is worthwhile remarking that no count was obtained during 2 months. This demonstrated that the neutrino experiment would be feasible at this depth.

These two intensity-depth curves are transformed into the energy spectrum of muons, as shown in Fig. 4.32. The agreement between the two is good if the energy scale in the measurement under rock is reduced by 10 percent. The energy spectrum thus derived is also in good agreement with that measured with a magnet spectrograph (Hayman 62) up to several hundred GeV. The spectrum extends up to 10 TeV, and gradual steepening of the slope is seen at about 1 TeV.

The energy spectrum of muons thus obtained is used in Fig. 4.16 for comparison with the energy spectrum of the electronic component.

The physical implications of the comparison have been discussed in Section 4.10.

The observation of muon intensities under water and rock has further implications. The consistency of the intensity-depth relations in two different media and with the directly measured energy spectrum verifies the energy-loss rate given in (4.12.7) together with the values of the parameters given in Table 4.16. In particular this puts the upper limit on the energy-loss rate by nuclear interactions and consequently that of photonuclear reactions at high energies as

$$b_N \lesssim 0.9 \times 10^{-6}\ \text{g}^{-1}\ \text{cm}^2, \qquad \sigma_{\gamma N} \lesssim 3 \times 10^{-28}\ \text{cm}^2 \text{ per nucleon.} \tag{4.12.27}$$

This indicates that the photopion-production cross section does not increase with energy.

4.13 Neutrino Flux

4.13.1 Qualitative Features of Neutrino Interactions

At great depths underground no cosmic-ray particles other than neutrinos are able to survive. Since the neutrino has no charge and a negligibly small magnetic moment, its electromagnetic interactions produce no observable effect; the magnetic moment would cause the heating of the earth's interior by solar neutrinos if the anomalous part were too large. Its interactions are considered as being mediated exclusively by the weak interaction, as discussed in Section 3.19. The cross section is expected to be very small, as given in (3.19.19) through (3.19.21). The interaction mean free path corresponding to the cross section of 10^{-38} cm^2 is about 10^{14} g-cm^{-2}—much greater than the diameter of the earth. Hence the neutrino can pass through the earth without appreciable loss.

Although the interaction is very weak, the neutrino flux is so strong that secondary particles produced by neutrinos may be of a detectable intensity. Let the flux of neutrinos be j_ν and the interaction mean free path be l_ν. The frequency of interactions taking place in matter is given by j_ν/l_ν, which is as small as 10^{-16} g^{-1} sec^{-1}, since $j_\nu \simeq 10^{-2}$ cm^{-2} sec^{-1}. Even if we were able to detect interactions in a material of 100 tons, we would expect only one interaction in several years. The effective volume of a detector can be increased if secondary particles of long ranges are observed. In this case the effective thickness of the detector is equal to the range. If the range is 3×10^3 g-cm^{-2}, a detector with an area as large as 30 meters^2 corresponds to a detector of 1000 tons. Hence we would expect to record a secondary particle in several months. Most muons observed at very great depths may be those produced by neutrinos underground. In fact, the intensity of muons at the deepest point described in Section 4.12 is smaller than 10^{-10} cm^{-2} sec^{-1} sr^{-1}, and the flux of neutrinos with energies greater than 10 GeV is estimated to be about 10^{-3} cm^{-2} sec^{-1} sr^{-1}. Because of the steep zenith-angle dependence of muons and the increase of neutrino intensity with zenith angle, muons in the horizontal direction are considered to be mainly those that are produced by neutrinos.

4.13.2 Evaluation of the Neutrino Flux

In order to determine whether muons are of atmospheric origin or the secondary products of neutrino interactions, and to determine the neutrino interaction cross section, we have to know the flux of neutrinos as accurately as possible.

If the neutrino is produced in association with the muon by a two-body decay, there is a simple relation between the fluxes of neutrinos and muons. Let us take the π-μ decay as an example. The neutrino energy in the rest system of the parent pion is given by reference to (4.10.1) as

$$E_\nu^* = p_\nu^* c = \frac{m_\pi{}^2 - m_\mu{}^2}{2m_\pi} c^2 = 29.8 \text{ MeV}. \tag{4.13.1}$$

If a pion of energy E_π and momentum p_π emits a neutrino at angle θ, the energy of the neutrino in the laboratory system is given by

$$E_\nu = \frac{E_\nu^* m_\pi c^2}{E_\pi - p_\pi c \cos\theta}. \tag{4.13.2}$$

Since the velocity of the neutrino is equal to the velocity of light and is greater than the velocity of the parent pion, the neutrino may be emitted in the backward direction in the laboratory system. For a fixed neutrino energy, therefore, the energy of the parent pion is given for $\theta < \pi/2$ by

$$E_\pi = \frac{1}{\sin^2\theta}\left\{\frac{E_\nu^*}{E_\nu} - \cos\theta\left[\left(\frac{E_\nu^*}{E_\nu}\right)^2 - \sin^2\theta\right]^{1/2}\right\} m_\pi c^2. \tag{4.13.3}$$

This is obviously valid for

$$\sin\theta \leq \frac{E_\nu^*}{E_\nu}. \tag{4.13.4}$$

At the maximum angle, $\theta_{max} = \sin^{-1}(E_\nu^*/E_\nu)$, the pion energy is maximum, whereas it is minimum at $\theta = 0$, insofar as we are concerned with neutrino energies greater than E_ν^*. Hence the energies of the parent pions lie in the region

$$E_\pi^- \equiv \left(\frac{E_\nu^*}{2E_\nu} + \frac{E_\nu}{2E_\nu^*}\right) m_\pi c^2 \leq E_\nu \leq \frac{E_\nu}{E_\nu^*} m_\pi c^2 \equiv E_\pi^+. \qquad (4.13.5)$$

This is comparable with (4.10.2) for muons. For $E_\mu \simeq E_\nu \gg m_\mu c^2$, both (4.10.2) and (4.13.5) may be approximated by

$$E_\pi^-(\mu) \equiv E_\mu \leq E_\pi \leq \left(\frac{m_\pi}{m_\mu}\right)^2 E_\mu \equiv E_\pi^+(\mu), \qquad (4.13.6a)$$

$$E_\pi^-(\nu) \equiv \frac{m_\pi^2}{m_\pi^2 - m_\mu^2} E_\nu \leq E_\pi \leq \frac{2m_\pi^2}{m_\pi^2 - m_\mu^2} E_\nu \equiv E_\pi^+(\nu). \qquad (4.13.6b)$$

The neutrino flux can therefore be obtained from the muon flux by replacing the limits on the integral in (4.10.3) for muons by (4.13.6*b*).

Since the range of the integral for neutrinos is larger than that for muons, application of the mean-value theorem will bring about an incorrect result since this theorem is questionable even in the case of muons. It is better to replace E_μ in the muon spectrum by $E_\nu m_\pi^2/(m_\pi^2 - m_\mu^2)$. A difference in the upper limits arising from this procedure may be evaluated in an approximate way. Thus we obtain the neutrino spectrum

$$j_\nu(E_\nu, \pi) = j_\mu\left(\frac{m_\pi^2}{m_\pi^2 - m_\mu^2} E_\nu\right) + \frac{m_\pi^2}{m_\pi^2 - m_\mu^2}\left(2 - \frac{m_\pi^2}{m_\mu^2}\right)\left(-\frac{dj_\mu}{dE_\mu}\right) E_\mu', \qquad (4.13.7)$$

where

$$E_\mu' = \frac{m_\pi^2}{m_\mu^2} \frac{m_\pi^2}{m_\pi^2 - m_\mu^2} E_\nu. \qquad (4.13.8)$$

Since the muon spectrum is a decreasing function of energy, this demonstrates that the intensity of neutrinos is smaller than that of muons.

The situation is slightly different in the case of $K_{\mu 2}$ decay because the mass of the kaon is much greater than that of the muon. Now the lower limits on the integrals are nearly equal for muons and neutrinos, whereas the upper limit for muons is much larger than that for neutrinos. Because of the steeply falling spectrum we may evaluate the neutrino spectrum as

$$j_\nu(E_\nu, K) = j_\mu(E_\nu, K) - j_\mu(2E_\nu, K) + E_\nu\left(-\frac{dj_\mu}{dE_\mu}\right) E_\mu', \qquad (4.13.9)$$

where

$$E'_\mu = \frac{m_K^{\,2}}{m_\mu^{\,2}} E_\nu . \qquad (4.13.10)$$

Since the first two terms on the right-hand side of (4.13.9) give a slight overestimate, the last term for correction may be discarded. The neutrino flux given in (4.13.9) is smaller than the muon flux by only a small amount. If this is compared with (4.13.7), we may see that kaons are more efficient than pions as producers of neutrinos.

Although the method described above might be regarded as advantageous in view of the direct connection of the neutrino flux with the observed muon flux, it is not always practical because the muon flux employed may be reduced by energy loss and decay in the atmosphere. The method may be of practical value for vertical neutrinos with energies greater than 10 GeV and for horizontal ones with energies greater than 100 GeV. Since in neutrino experiments horizontal neutrinos with energies of about 10 GeV are important, it is not practical to evaluate the neutrino flux by referring to the observed muon flux. Moreover, the contribution of neutrinos produced by muon decay is appreciable. For horizontal neutrinos the contributions of pion, kaon, and muon decays are comparable to each other between 10 and 100 GeV, and the contribution of muons is the greatest below 10 GeV.

More quantitatively it is necessary to take into account the curvature of the earth's surface and the geomagnetic effect, because a considerable part of the horizontal neutrinos are produced by primary cosmic rays of various incident angles at remote places; for example, many low-energy neutrinos are produced near the poles, and neutrinos produced in the equatorial region would be subject to a considerable east-west effect. Unless these complicated effects are taken into account a precise evaluation of the neutrino flux is not possible. Since no such quantitative calculation has yet been made, we here present a semi-quantitative result obtained by Cowsik et al. (Cowsik 66), with suitable corrections in the choice of parameters.

The method of calculation is analogous to that in Sections 4.10 and 4.11, and the result is believed to be correct with an uncertainty of several tens of percent for neutrino energies between 2 and 200 GeV. In this analysis the neutrinos associated with the muon, ν_μ and $\bar{\nu}_\mu$, arise mainly from the decay modes

$$\begin{aligned} &\pi^\pm \to \mu^\pm + \nu_\mu(\bar{\nu}_\mu), && K^\pm \to \mu^\pm + \nu_\mu(\bar{\nu}_\mu), \\ &\mu^\pm \to e^\pm + \nu_e(\bar{\nu}_e) + \bar{\nu}_\mu(\nu_\mu), && K_2^{\,0} \to \pi^\mp + \mu^\pm + \nu_\mu(\bar{\nu}_\mu), \end{aligned} \qquad (4.13.11a)$$

and those associated with the electron, ν_e and $\bar{\nu}_e$, are produced mainly by

$$\mu^{\pm} \to e^{\pm} + \nu_e(\bar{\nu}_e) + \bar{\nu}_\mu(\nu_\mu), \qquad K_2{}^0 \to \pi^{\pm} + e^{\pm} + \nu_e(\bar{\nu}_e). \quad (4.13.11b)$$

The contribution of muon decay to muon-neutrinos becomes less than 10% for energies greater than 10 GeV in the vertical direction, whereas in the horizontal direction it is the principal contributor up to 10 GeV. This is because in muon decay the fraction of energy that is given to a neutrino is larger than that in the pion-muon decay, and it remains as large as one-quarter even at 100 GeV. In the vertical direction the kaon contribution is the greatest if $f_K = 0.4$, whereas in the horizontal direction it is only slightly larger than the pion contribution up to 100 GeV. Above this value the contribution of kaons predominates. The intensity ratio of ν_μ to $\bar{\nu}_\mu$ is nearly proportional to the positive-to-negative ratio of muons—except at low energies, where the contribution of muon decay is important. The ratio becomes less than unity below 20 GeV in the horizontal direction because of the contribution of muon decay.

Most electron-neutrinos come from muon decay, but the decay of $K_2{}^0$ into electrons makes a not negligible contribution because the lifetime of $K_2{}^0$ is shorter than that of the muon. In fact, the former contribution exceeds the muon contribution above 10 GeV in the vertical direction, whereas in the horizontal direction these two become comparable only at 200 GeV. The $\nu_e/\bar{\nu}_e$ ratio is near unity and is almost independent of energy.

The energy spectra of four kinds of neutrinos in the vertical and horizontal directions are shown in Fig. 4.33. It can be readily seen that the flux of muon-neutrinos is much greater than that of electron-neutrinos, especially at high energies. Therefore the neutrino interactions to be detected would be produced mostly by muon-neutrinos.

The flux of neutrinos thus estimated is quite large. Among various components of cosmic rays with energies greater than 1 GeV only muons have a flux larger than that of neutrinos at sea level, whereas the neutrino component is the strongest one underground. Although the neutrino flux is large, the detection of neutrinos is extremely difficult because of the weakness of their interactions.

4.13.3 Secondary Particles Produced by Neutrinos

The unidirectional intensity of secondary particles is expressed in the one-dimensional approximation by

$$j_s(\theta) = \int dE_\nu \int_{E_{\text{th}}}^{E_\nu} dE_s \, R(E_s) N\sigma(E_\nu, E_s) j(E_\nu, \theta), \qquad (4.13.12)$$

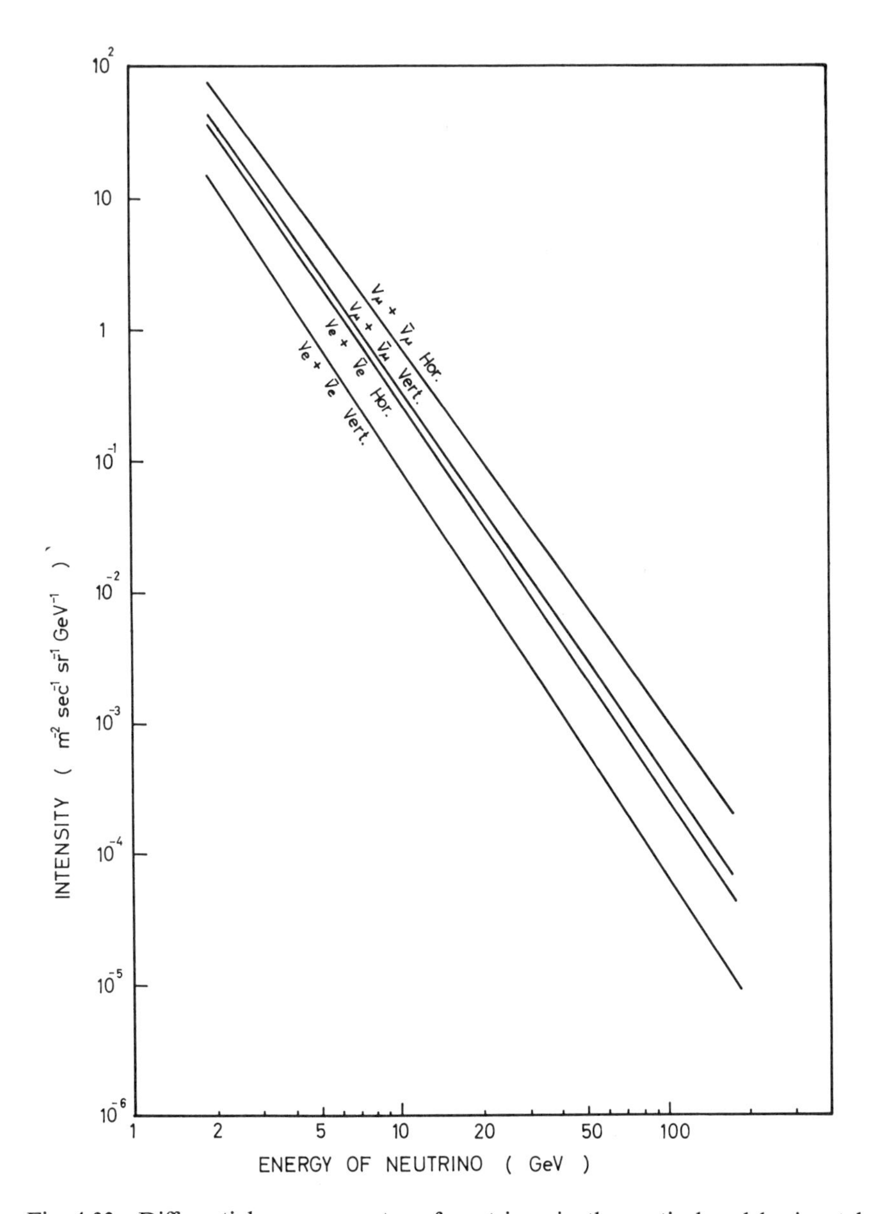

Fig. 4.33 Differential energy spectra of neutrinos in the vertical and horizontal directions.

where $\sigma(E_\nu, E_s)\, dE_s$ is the cross section per nucleon for a neutrino of energy E_ν to produce a secondary particle with energy E_s in dE_s; E_{th} is the threshold energy for detecting secondary particles.

The effective range $R(E_s)$ is given for a secondary muon by

$$R(E_\mu) = \frac{E_\mu}{\bar{a}}, \tag{4.13.13}$$

where $\bar{a}$ is defined as in (4.12.13) and is roughly equal to 2 MeV/g-cm^{-2} at about 10 GeV. Together with muons, pions and other strongly interacting particles may also be produced, but their effective range is as large as the attenuation length of nuclear active particles; that is, on the order of 100 g-cm^{-2}. The electron may be produced by an electron-neutrino and initiates a cascade shower. The penetration depth of the shower given in (2.6.23) may be regarded as the effective range, so that

$$R(E_e) \simeq 5.0 \times \ln\left(\frac{E_e}{30\ \text{MeV}}\right) \text{g-cm}^{-2}, \tag{4.13.14}$$

where the radiation length in rock is taken as about $X_0 \simeq 24$ g-cm^{-2}. Because of the logarithmic dependence on energy, the effective range of the electron increases only slowly with energy and is exceeded by that of the muon at 0.8 GeV. If we are interested in the interactions of neutrinos with energies greater than 10 GeV, therefore, the detection of secondary muons provides the most feasible method.

The flux of secondary muons can be estimated if the cross section $\sigma(E_\nu, E_\mu)\, dE_\mu$ is known. We refer to the discussion in Section 3.19 and assume

$$\sigma(E_\nu, E_\mu) = \sigma_0 \left(\frac{E_\nu}{E_0}\right)^n \delta(E_\mu - fE_\nu). \tag{4.13.15}$$

By substituting this into (4.13.12) and carrying out the integration over E_ν we obtain the flux of secondary muons with energies greater than E_μ, as follows:

$$j_\mu(> E_\mu, \theta) = N\sigma_0 f^{-(n+1)} \int_{E_\mu}^{\infty} \frac{E_\mu^{n+1}}{E_0{}^n \bar{a}}\, j_\nu\!\left(\frac{E_\mu}{f}, \theta\right) dE_\mu. \tag{4.13.16}$$

If $n = 1$ as given in (3.19.21), the contribution of neutrinos increases with energy as long as the power index of the integral energy spectrum of neutrinos is smaller than 2. This seems to be the case for horizontal neutrinos with energies smaller than 100 GeV.

Hence we may for simplicity assume the neutrino spectrum to be

$$j_\nu(E_\nu) = j_{\nu 0}\,\frac{dE_\nu}{E_\nu} \times \begin{cases} \left(\dfrac{E_c}{E_\nu}\right)^{\alpha_1} & (E_\nu \le E_c), \\ \left(\dfrac{E_c}{E_\nu}\right)^{\alpha_2} & (E_\nu \ge E_c). \end{cases} \tag{4.13.17}$$

If $\alpha_1 < n+1 < \alpha_2$, the flux of secondary muons is expressed by

$$j_\mu(>E_\mu) = N\sigma_0\, j_{\nu 0}\,\frac{E_c}{f^{n+1}\bar{a}}\left(\frac{E_c}{E_0}\right)^n \left\{\frac{f^{\alpha_1}}{n+1-\alpha_1}\left[1-\left(\frac{E_\mu}{E_c}\right)^{n+1-\alpha_1}\right] + \frac{f^{\alpha_2}}{\alpha_2-n-1}\right\}. \tag{4.13.18}$$

If we choose $E_0 \ll E_c$, this is approximately reduced to

$$j_\mu(>E_\mu) \simeq N\sigma_0\, j_{\nu 0}\,\frac{(1/f)^{n+1-\alpha_1}}{n+1-\alpha}\left(\frac{E_c}{E_0}\right)^n \frac{E_c}{\bar{a}}, \tag{4.13.18$'$}$$

provided that $n+1-\alpha_1 < \alpha_2 - n - 1$. Thus the secondary muon flux depends rather critically on E_c, the energy at which the neutrino spectrum bends. This implies that the neutrino interaction cross section deduced from muon intensity is influenced by the detailed shape of the neutrino spectrum. If the energy dependence of the cross section were much weaker, such ambiguity would disappear.

If we refer to the energy spectrum of horizontal neutrinos $(\nu_\mu + \bar{\nu}_\mu)$ given in Fig. 4.33, we have

$$j_{\nu 0}\left(\frac{E_c}{E_0}\right)^{\alpha_1} = 3.6 \times 10^{-2} \quad \text{cm}^{-2}\,\text{sec}^{-1}\,\text{sr}^{-1} \qquad (\alpha_1 = 1.9), \tag{4.13.19}$$

where $E_0 = 1$ GeV. For $n = 1, f = \frac{1}{2}$, and $\sigma_0 = 4 \times 10^{-39}$ cm^2. The contribution of neutrinos with energies smaller than E_c is given, according to (4.13.17), by

$$j_\mu(>E_\mu) = 4.3 \times 10^{-13}\left[\left(\frac{E_c}{E_0}\right)^{0.1} - \left(\frac{E_\mu}{E_0}\right)^{0.1}\right] \quad \text{cm}^{-2}\,\text{sec}^{-1}\,\text{sr}^{-1}. \tag{4.13.20}$$

A detector used for the observation of secondary muons may be as large as 40 m^2 sr (Achar 65, Reines 65), and this gives a detection frequency of several events per year.

In the experiment at Kolar Gold Field in India, however, 15 muons were observed during 7 months. Since the detector was placed at a depth of 6800 meters water equivalent underground, most muons with relatively small zenith angles were considered to be produced in the atmosphere. Thus about one-half of 10 single-muon events were attributed to atmospheric muons.

This and other ambiguities arising mainly from the difficulty of estimating the neutrino flux make it difficult to obtain the neutrino interaction cross section to better than within a factor of 2.

4.14 Secondary Particles Underground

In experiments underground it is necessary to take account of the contributions of secondary particles produced by muons. Although we might think that cosmic-ray components deep underground would be only muons and neutrinos, various secondary components are produced by these penetrating components. These can produce interactions as frequently as the penetrating components; for example, high-energy nuclear interactions associated with multiple meson production are produced almost equally by muons and pions, the latter being the secondary component. When we observe radioactivity produced by solar neutrinos, the production of radioactive nuclides by muons and their secondary particles has to be subtracted. For such purposes it is important to estimate the flux intensities of various secondary components.

Since the secondary components of interest are of rather short range, they may be considered as being in equilibrium with the parent component, say, muons. Let us consider that the muons with the energy spectrum $j_\mu(E_\mu, x)\, dE_\mu$ produce secondary particles of energies between E_0 and $E_0 + dE_0$ at depths between x and $x + dx$ with the probability $P(E_\mu, E_0)\, dE_0\, dx$. The secondary particle may initiate a multiplicative process or may be absorbed, so that it gives an average number in the energy interval between E and $E + dE$, $n(E_0, E, x)\, dE$ after traversing a thickness x. Then the intensity of the secondary particles at x with energies E in dE is

$$j(E, x)\, dE = \int_{E^-}^{\infty} dE_\mu \int_E^{E_m} dE_0 \int_0^x dx' j_\mu(j_\mu(E_\mu, x - x')\, P(E_\mu, E_0)\, n(E_0, E, x'), \tag{4.14.1}$$

where E_m is the maximum energy transfer corresponding to the muon energy E_μ, and E^- is the minimum muon energy corresponding to the secondary-particle energy E. Since $n(x')$ in the integrand varies much more rapidly with x' than $j_\mu(x - x')$, the integral over x' may be evaluated by making use of the mean-value theorem as

$$j(E, x)\, dE = \int_{E^-}^{\infty} dE_\mu \int_{E}^{E_m} dE_0\, j_\mu(E_\mu, x - \bar{x})\, P(E_\mu, E_0) Z(E_0, E), \quad (4.14.2)$$

where $Z(E_0, E)$ is the track length defined as

$$Z(E_0, E) \equiv \int_0^{\infty} n(E_0, E, x)\, dx. \quad (4.14.3)$$

If the depth x is sufficiently large, the argument $x - \bar{x}$ in $j_\mu(x - \bar{x})$ can be equated to x.

In virtue of this simplification, we are able to estimate the intensities of the various secondary components in the following section.

4.14.1 Electronic Component

Electrons and photons produced by muons initiate cascade showers as described in Section 2.6 and Appendix C. Since elements in earth are of relatively low atomic number, analytic solutions may be accurate enough for the present purpose. The track length of electrons in a cascade shower is given in approximation B by

$$Z(E_0, E) = 2.3 X_0 \frac{E_0}{\varepsilon_0{}^2} \left[-\frac{d\phi(\varepsilon)}{d\varepsilon} \right], \simeq X_0 \frac{E_0}{2.3[E + (\varepsilon_0/2.3)]^2}, \quad (4.14.4)$$

where $\varepsilon = 2.3E/\varepsilon_0$ and $\phi(\varepsilon)$ is given in (2.6.16). The radiation length X_0 and the critical energy ε_0 depend on the composition of matter under consideration. In rock those for silica as given in Table 2.7 may be used as being representative:

$$X_0 = 27 \text{ g-cm}^{-2}, \qquad \varepsilon_0 = 47 \text{ MeV}, \quad (4.14.5a)$$

whereas in water

$$X_0 = 36.4 \text{ g-cm}^{-2}, \qquad \varepsilon_0 = 73.0 \text{ MeV}. \quad (4.14.5b)$$

If we are interested not in the energy spectrum but in the total flux of electrons, we should use the integral track length

$$Z(E_0) = \left(\frac{E_0}{\varepsilon_0} \right) X_0 . \quad (4.14.6)$$

Electrons and photons underground are generated mainly by the knock-on, bremsstrahlung, direct pair-creation, and neutral-pion-production processes. Each of these processes will be discussed below.

Knock-on Process. As shown in Section 2.1.2, the knock-on probability for a relativistic muon is given by

$$P_k(E_\mu, E)\, dE = K \frac{dE}{E^2} \left[1 - \frac{E}{E_m} + \frac{1}{2} \left(\frac{E}{E_\mu} \right)^2 \right], \quad (4.14.7)$$

where

$$K = 2\pi\left(\frac{ZN}{A}\right)\frac{e^4}{mc^2 \cdot 1\ \text{MeV}} = 0.153\,\frac{Z}{A}\,g^{-1}\ \text{cm}^2. \qquad (4.14.7')$$

The definition of K above implies that the energy of an electron is measured in units of MeV. The term E_m is the maximum energy transfer given in (4.12.3) and is here expressed by

$$E_m = \frac{E_\mu{}^2 - (\mu c^2)^2}{E_\mu + \mu^2 c^2/2m} \simeq \frac{E_\mu{}^2}{E_\mu + \mu^2 c^2/2m}, \qquad (4.14.8)$$

where m and μ are the masses of the electron and the muon, respectively, and m^2 is neglected in comparison with μ^2.

If we are interested in the case of E_μ, $E_m \gg \mu c^2$, E, the number of electrons with energies E in dE associated with a muon of energy E_μ is obtained as

$$\begin{aligned}\bar{n}_{e,k}(E_\mu, E)\,dE &= \int_E^{Em} dE_0\,P_k(E_\mu, E_0)\,Z(E_0, E)\\ &= KX_0\,\frac{dE}{2.3[E + (\varepsilon_0/2.3)]^2}\left[\ln\left(\frac{E_m}{E}\right) - 1 + \left(\frac{E_m}{E_\mu}\right)^2\right].\end{aligned} \qquad (4.14.9)$$

Likewise, the total number of such electrons is

$$\begin{aligned}\bar{N}_{e,k}(E_\mu) &= \int_0^{E_m} dE_0\,P(E_\mu, E_0)\,Z(E_0)\\ &\simeq \frac{KX_0}{\varepsilon_0}\left[\ln\left(\frac{E_m}{\varepsilon_0}\right) - 1 + \left(\frac{E_m}{E_\mu}\right)^2\right].\end{aligned} \qquad (4.14.10)$$

If we use the numerical values in (4.14.5*a*) and (4.14.7′), this is approximately estimated as

$$\bar{N}_{e,k}(E_\mu) \simeq 0.10 \log\frac{E_m}{\varepsilon_0}. \qquad (4.14.10')$$

Hence a considerable fraction of muons are accompanied by electrons initiated by knock-on processes.

The energy spectrum of electrons can be obtained by integration over E_μ. Since the function $\bar{n}_{e,k}(E_\mu, E)$ is a slowly varying function of E_μ, the integral may be approximated as

$$\begin{aligned}j_{e,k}(E) &\simeq \langle \bar{n}_{e,k}(E_\mu, E)\rangle\, j_\mu^{(0)}\left(\frac{2mc^2}{E}\right)^{\beta/2}\\ &\simeq \frac{KX_0 j_\mu^{(0)}}{2.3[E + (\varepsilon_0/2.3)]^2}\left(\frac{2mc^2}{E}\right)^{\beta/2},\end{aligned} \qquad (4.14.11)$$

where the muon spectrum is represented as $j_\mu(E_\mu) = \beta j_\mu^{(0)}(\mu c^2/E_\mu)^\beta E_\mu{}^{-1}$. Hence the electron spectrum depends on energy approximately as $E^{-2-\beta/2}$. For $\beta \simeq 1.8$ this is similar to the muon spectrum.

These electrons often appear as showers. The shower frequency may be roughly estimated by choosing the minimum energy of electrons as $fS\varepsilon_0$, where S is the shower size and f is a numerical factor on the order of 10, slowly increasing with shower size.

Bremsstrahlung. The bremsstrahlung of the muon has a cross section that is much smaller than that for the knock-on process, but the energy transferred to a photon is so great that the track length of a cascade shower initiated by this photon may compensate for the smallness of the cross section at high muon energies.

The bremsstrahlung cross section in (2.4.1) gives the radiation probability of muons as

$$P_r(E_\mu, E)\,dE = \left(\frac{m}{\mu}\right)^2 \frac{1}{X_0 \ln(191\, Z^{-1/3})} F\left(E_\mu, \frac{E}{E_\mu}\right)\frac{dE}{E}. \tag{4.14.12}$$

Since the radiation process gives an appreciable contribution only at high muon energies, we have to use only the energy distribution $F(E_\mu, E/E_\mu)$ in the case of complete screening, as given in (2.4.13d).

The differential energy spectrum of electrons associated with muon bremsstrahlung is therefore obtained as

$$\begin{aligned} \bar{n}_{e,r}(E_\mu, E)\,dE &= \int_E^{E_\mu} P(E_\mu, E_0)\, Z(E_0, E)\, dE_0\, dE \\ &\simeq \frac{(m/\mu)^2\, dE}{2.3[E + (\varepsilon_0/2{,}3)]^2} E_\mu. \end{aligned} \tag{4.14.13}$$

This is as small as $(m/\mu)^2 E_\mu / K X_0 \ln(E_m/E)$ times the same quantity in the knock-on process. This means that the contribution of bremsstrahlung exceeds that of the knock-on process for

$$E_\mu \gtrsim 2\left(\frac{\mu}{m}\right)^2 \ln\left(\frac{E_m}{E}\right) \text{MeV} \simeq 10^{11} \ln\left(\frac{E_m}{E}\right) \text{eV}.$$

In correspondence to (4.14.10) we have

$$\bar{N}_{e,r}(E_\mu) = \int_0^{E_\mu} P(E_\mu, E_0)\, Z(E_0)\, dE_0 = \left(\frac{m}{\mu}\right)^2 \left(\frac{E_\mu}{\varepsilon_0}\right). \tag{4.14.14}$$

Hence a muon with energy as high as $4 \times 10^4 \varepsilon_0$ or higher is almost always accompanied by electrons.

The energy spectrum of electrons due to bremsstrahlung is given by

$$\begin{aligned} j_{e,r}(E) &= \int_E^\infty j_\mu(E_\mu)\bar{n}_{e,r}(E_\mu, E)\,dE_\mu \\ &\simeq \frac{\beta}{(\beta-1)}\left(\frac{m}{\mu}\right)^2 \frac{\mu c^2}{2.3[E+(\varepsilon_0/2.3)]^2}\left(\frac{\mu c^2}{E}\right)^{\beta-1} j_\mu^{(0)}. \end{aligned} \tag{4.14.15}$$

At great depths, therefore, the electron spectrum continues to rise toward low energy with a power index of $\beta + 1$, whereas the muon spectrum tends to be flat.

Direct Pair Creation. The cross section for direct pair creation given in Section 2.5.1 indicates that the energy spectrum of electrons produced is approximately $1/E$ for $E \lesssim (m/\mu)E_\mu$, whereas it is $1/E^3$ for $(m/\mu)E_\mu \lesssim E \lesssim E_\mu$. Hence the contribution to the electronic component comes mainly from the low-energy region. Taking only the low-energy region into account, we may give an approximate probability for direct pair creation as

$$P_p(E_\mu, E)\,dE = \frac{7}{9\pi}\frac{\alpha}{X_0}\ln\left(\frac{E_\mu}{\mu c^2}\right)\frac{dE}{E} \qquad E \lesssim \frac{m}{\mu}E_\mu, \qquad \alpha = \tfrac{1}{137}. \tag{4.14.16}$$

This is greater than the radiation probability by a factor of $(\frac{1}{9}\pi)(\mu/m)^2\alpha\ln(E_\mu/\mu c^2)$, but the energy transfer is limited to small values. These two effects compensate for each other, and, as a result, this gives nearly the same contribution as bremsstrahlung.

The differential energy spectrum of electrons associated with direct pair creation is thus obtained as

$$\begin{aligned} \bar{n}_{e,p}(E_\mu, E) &= \int_E^{(\mu/m)E_\mu} P_p(E_\mu, E_0)\, z(E_0, E)\, dE_0 \\ &\simeq \frac{7}{9\pi}\alpha\frac{m}{\mu}\frac{E_\mu}{2.3[E+(\varepsilon_0/2.3)]^2}\ln\left(\frac{E_\mu}{\mu c^2}\right). \end{aligned} \tag{4.14.17}$$

In correspondence to (4.14.14) we have

$$\bar{N}_{e,p}(E_\mu) = \frac{7}{9\pi}\alpha\frac{m}{\mu}\frac{E_\mu}{\varepsilon_0}\ln\left(\frac{E_\mu}{\mu c^2}\right). \tag{4.14.18}$$

This is greater than $\bar{N}_{e,r}$ in (4.14.14) by a factor of

$$\frac{\bar{N}_{e,p}}{\bar{N}_{e,r}}\left(\frac{Z}{9\pi}\right)\alpha\left(\frac{\mu}{m}\right)\ln\left(\frac{E_\mu}{\mu c^2}\right) \simeq 0.37\ln\left(\frac{E_\mu}{\mu c^2}\right). \tag{4.14.19}$$

The same factor should be multiplied by (4.14.15) to obtain the energy spectrum. However, the contribution of direct pair creation can be evaluated only approximately, because the cross section is known less accurately.

Neutral-Pion Production. The mean free path for pion production is given by reference to Section 3.18 as

$$l = \frac{1}{N\sigma\,(\mu - \pi)} \simeq \frac{5 \times 10^6}{\ln\,(E_\mu/\mu c^2)}\ \text{g-cm}^{-2}. \tag{4.14.20}$$

The energy spectrum of pions produced is not well known but may be approximated by a delta function because the fireball model may possibly hold, as

$$P(E_\mu, E)\,dE = l^{-1} 2n_0\,\delta\!\left(\frac{E - K_\gamma E_\mu}{2n_0}\right) dE, \tag{4.14.21}$$

where $K_\gamma \simeq \frac{1}{6}$ is the inelasticity for neutral-pion production and n_0 is the average multiplicity of energetic neutral pions, each of which decays into two photons.

Hence the number of electrons associated with neutral-pion production is obtained as

$$\bar{N}_{e,\pi}(E_\mu) = \int P(E_\mu, E_0)\,Z(E_0)\,dE_0 = \left(\frac{X_0}{l}\right) K_\gamma \left(\frac{E_\mu}{\varepsilon_0}\right). \tag{4.14.22}$$

If this is compared with $\bar{N}_{e,r}$ in (4.14.14), we have

$$\frac{\bar{N}_{e,\pi}}{\bar{N}_{e,r}} = \frac{X_0}{l}\left(\frac{\mu}{m}\right)^2 K_\gamma \simeq 4 \times 10^{-2} \ln \frac{E_\mu}{\mu c^2}. \tag{4.14.23}$$

This indicates that the contribution of neutral-pion production to the electronic component is smaller by a factor of 10 than that of direct pair creation.

Summary. In order to obtain the number of electrons associated with a muon it is more quantitative to refer to the energy-loss rate given in Section 4.12, since the total track length given in (4.14.6) is proportional to the energy dissipated eventually by the ionization of electrons. For process j we have therefore

$$\bar{N}_{e,j} = -f_j\left(\frac{dE}{dx}\right)_j \left(\frac{X_0}{\varepsilon_0}\right). \tag{4.14.24}$$

Here f_j represents a fraction of energy that is responsible for the electrons concerned. For ionization the energy loss due only to the knock-on process should be taken into account. For pion production $f_j = 1/3.5$ is assumed because only neutral pions are considered to be

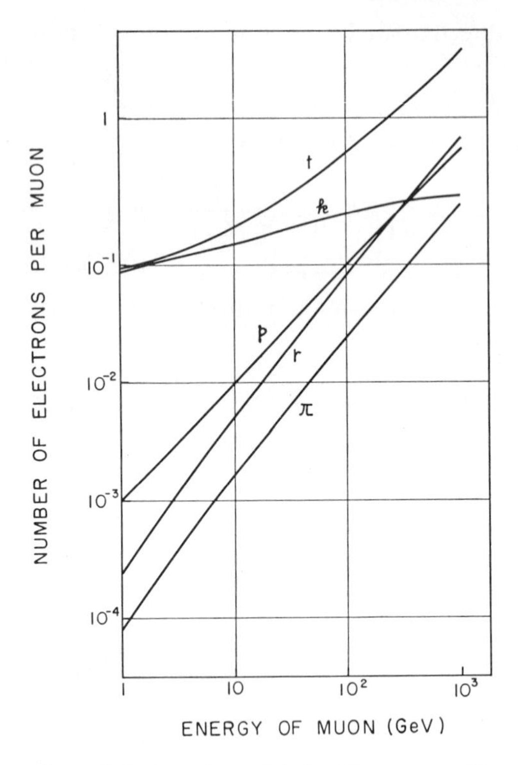

Fig. 4.34 The number of electrons associated with a muon. Curves k, p, r, and π represent the number of electrons per muon due to the knock-on, pair creation, bremsstrahlung, and pion-production processes, respectively. Curve t represents their sum.

responsible for the electrons concerned. The values of $\bar{N}_{e,j}$ as well as their sum in rock are plotted against muon energy in Fig. 4.34.

4.14.2 N-*Component*

As mentioned in Section 3.18, the nuclear interactions that take place underground are partly produced by nuclear active particles. Since nuclear active particles are produced by muons with the interaction mean free path l_μ and attenuate with the mean free path L, their intensity relative to muon intensity is estimated to be L/l_μ. Although this is very small, the interaction mean free path of nuclear active particles, l_n, is so small that they produce nuclear interactions at a rate proportional to $L/l_\mu l_n$. This is comparable to the corresponding quantity for muon interactions, $1/l_\mu$. Thus we expect that nuclear interactions produced by nuclear active particles are as frequent as those produced by muons.

In order to estimate the flux of nuclear active particles more quantitatively we have to take account of the energy and angular distributions at their production. Since little quantitative information is available on these points, we consider only two representative cases with the aid of simplified models.

Mesons Responsible for Penetrating Showers. In high-energy nuclear interactions of muons nucleons receive a rather small fraction of incident energy, and most of them are considered to be of relatively low energy. Among secondary nuclear active particles of high energies we need therefore take account only of mesons. Since mesons are produced probably through the fireball stage, we may assume a simple model that gives the production probability (4.14.21); namely, a fraction K_{ch} of incident energy is given to n_s energetic mesons of unique energy. Analogously to (4.14.21), the production probability is expressed by

$$P(E_\mu, E)\,dE = l_\mu^{-1}\, n_s\, \delta\left(E - \frac{K_{ch}E_\mu}{n_s}\right) dE. \tag{4.14.25}$$

If these mesons are produced with negligible angular divergences, the energy spectrum of unidirectional mesons is obtained as

$$\begin{aligned} j_\pi(E)\,dE &= L\int j_\mu(E_\mu)\, P(E_\mu, E)\, dE_\mu\, dE \\ &= \left(\frac{L}{l_\mu}\right)\left(\frac{n_s^2}{K_{ch}}\right) j_\mu\left(\frac{n_s E}{K_{ch}}\right) dE. \end{aligned} \tag{4.14.26}$$

The integral spectrum is

$$\begin{aligned} j_\pi(>E) &= \left(\frac{L}{l_\mu}\right) n_s\, j_\mu\left(>\frac{n_s E}{K_{ch}}\right) \\ &\simeq \left(\frac{L}{l_\mu}\right)\left(\frac{K_{ch}^\beta}{n_s^{\beta-1}}\right) j_\mu(>E), \end{aligned} \tag{4.14.27}$$

where the last expression is derived by assuming a power spectrum.

This gives us the following frequency ratio of penetrating showers produced by pions and muons:

$$\frac{l_\pi^{-1}\, j_\pi(>E)}{l_\mu^{-1}\, j_\mu(>E)} \simeq \frac{L}{l_\pi}\,\frac{K_{ch}^\beta}{n_s^{\beta-1}}. \tag{4.14.28}$$

If we take $K_{ch} = 1/3$ and $n_s = 4$, as given in Section 3.18, and assume $L/l_\pi = 1.5$, as given for nucleons in air in Section 4.7.1, the ratio is obtained as

$$\frac{l_\mu j_\pi}{l_\pi J_\mu} \simeq 0.5, \tag{4.14.28'}$$

where $\beta = 1$ is assumed because of the slope of the energy spectrum at the average energy.

A pion produced with energy E deviates from the trajectory of the parent muon due to its angular divergence and multiple scattering. The angular deviations due to these causes are $p_T c/E$ and $(E_s/E)\sqrt{L/X_0}$, respectively. Since $p_T c$ is much greater than E_s, the former is mainly responsible for the angular deviation. The mean distance between the pion and the muon is therefore expected to be about $(p_T c/E)(L/\rho)$ in a medium of density ρ. For $E > 1$ GeV this is as small as, or smaller than, a few tens of centimeters in rock. Hence most energetic mesons are correlated with a muon, and pion interactions may therefore be distinguished from muon interactions by observing whether they are associated with penetrating particles or not. The frequency ratio obtained with this criterion for penetrating showers produced in lead was found to be (Higashi 65*b*)

$$\frac{l_\mu j_\pi}{l_\pi j_\mu} = 0.26 \pm 0.11. \tag{4.14.29}$$

The difference from the theoretical estimate in (4.14.28′) is possibly due to the difference between the nuclear-interaction mean free paths in rock and in lead. If (4.14.29) is corrected for this effect by multiplying a factor $[A(\text{Pb})/A(\text{rock})]^{1/3} \simeq 2$, the agreement between them is reasonably good.

A remark may be necessary concerning photopion production by real photons produced by muons. Since the intensity of such photons is comparable to that of electrons as given in Section 4.14.1, the knock-on process is the main source of photons at small depths; for example, at a depth of 50 meters water equivalent, the muon spectrum becomes flat at energies below 10 GeV, and the lower limit on the integral of a power spectrum may be replaced by $E_0 = \bar{a}x \simeq 10$ GeV, where $\bar{a}x$ is the energy lost by ionization while a muon traverses a thickness x. Hence the intensity of photons associated with muons is approximately expressed by

$$j_\gamma(>E) \simeq 0.57\, KX_0 \frac{\ln(E_0/E)}{[E + (\varepsilon_0/2.3]^2}\, j_\mu(x). \tag{4.14.30}$$

This is compared with the intensity of virtual photons. If Heitler's spectrum given in Section 3.18 is used, this is approximately expressed by

$$j_{v\gamma}(E) \simeq \frac{2}{\pi} \alpha \ln\left(\frac{E_0}{E}\right) \frac{1}{E} j_\mu(x). \tag{4.14.31}$$

Their ratio is

$$\frac{j_\gamma(E)}{j_{v\gamma}(E)} \simeq \frac{0.57\pi K X_0}{2\alpha} \frac{E}{[E + (\varepsilon_0/2.3)]^2} \simeq \frac{2.5 \times 10^2}{E\,(\text{MeV})}, \tag{4.14.32}$$

where the last expression holds for $E \gg \varepsilon_0/2.3$. This demonstrates that the contribution of real photons is relatively small if one is interested in large energy transfer—say, $E > 1$ GeV. This is the case as long as nuclear active particles are detected by means of penetrating showers. On the contrary, real photons play a main role in low-energy interactions.

Neutrons. Nuclear interactions produce many kinds of particles other than energetic mesons. The charged particles of low energies quickly lose their energies, and, with the exception of negative pions, they do not produce appreciable nuclear interactions. Neutrons have a relatively long mean free path and take part in nuclear interactions.

Neutrons are produced from various sources; that is, directly by muons through their virtual photons, by nuclear active particles of high energies, and by γ-rays—the latter two being the secondary products of muons. The first and the third are due to photoneutron processes, in which the giant resonance caused by relatively low-energy photons plays a main role. Since the photon spectrum increases as energy decreases, the contribution of the photoneutron processes is dominant over that of energetic nuclear active particles. The relative importance of the two photoneutron processes can be compared by reference to (4.14.32). Since the giant resonance takes place at a photon energy of about 20 MeV, the contribution of real photons is found to be dominant. Because neutron production by muons is of interest in itself, however, this process will be discussed first.

Since the spectrum of virtual photons is essentially the same as the bremsstrahlung spectrum, the photoreactions caused by bremsstrahlung may be referred to. Section 2.9.5 discusses nuclear reactions induced by photons and describes experimental results on photostars in nuclear emulsion irradiated by bremsstrahlung. From the emulsion results we can estimate the production cross sections and the multiplicities of neutrons in rock and in lead.

The photoreaction is caused mainly by giant resonance and photopion production. The mass-number dependences of the cross sections for these processes are approximately $A^{4/3}$ and A, respectively, as given in (2.9.16) and (2.9.19). The average neutron multiplicity for lead is estimated to be 2 for the giant resonance and three times the average prong number for the photopion process in emulsion, whereas the average neutron multiplicity in rock is nearly unity for the giant resonance and equal to that in emulsion for the photopion process. This holds for a muon of about 1-GeV energy. As energy increases, the photopion cross section increases approximately in a logarithmic way, whereas the contribution of the giant resonance is essentially energy independent.

Referring to the experimental information on photostars in emulsion as given in Figs. 2.60 and 2.61, and taking the above considerations into account, we are able to derive semiempirical cross sections and neutron multiplicities as given below. Here we give the products of the cross sections and multiplicities in lead and rock versus muon energy.

$$\sigma_m(\text{Pb}) \simeq 0.4 + 0.6[\ln(10E) - 1] \times 10^{-28}\ \text{cm}^2, \tag{4.14.33a}$$

$$\sigma_m(\text{rock}) \simeq 0.1 + 0.2[\ln(10E) - 1] \times 10^{-28}\ \text{cm}^2, \tag{4.14.33b}$$

where E is muon energy in GeV. The semiempirical relation for lead is compared with an experimental result in Fig. 4.35. The essential agreement encourages us to use (4.14.33b) for estimating neutron intensity underground.

According to (4.14.32) the photopion process underground is due mainly to the virtual photons of muons, whereas the giant resonance is due mainly to real photons. The contribution of the virtual photon to neutron production is given in (4.14.33b). For real photons we need take only the giant resonance into account. The resonance energy is given in (2.9.15) and may be taken as about $E_g \simeq 28$ MeV for rock. Since this is only slightly larger than $\varepsilon_0/2.3 \simeq 20$ MeV, we are able to evaluate the photon-absorption rate per nucleus, by referring to (2.9.16), as

$$\begin{aligned} q_g(x) &= \int \sigma_{\text{abs}}\, j_\gamma(E)\, dE \simeq \int \sigma_{\text{abs}} \frac{dE}{E} E_g\, j_\gamma(E_g) \\ &= 0.21 A^{4/3} K X_0 \frac{\ln(E_0/E_g) E_g}{[E_g + (\varepsilon_0/2.3)]^2}\, j_\mu(x) \times 10^{-27}\ \text{sec}^{-1} \\ &\simeq 3.8 \times 10^{-28} \ln(8x)\, j_\mu(x)\ \text{sec}^{-1}, \end{aligned} \tag{4.14.34}$$

where x is the depth in meters water equivalent. Since the average number of neutrons emitted thereby is unity, this gives the neutron-production rate per nucleus.

The neutron-production rate by the photopion process is obtained from the second term on the right-hand side of (4.14.33b) as

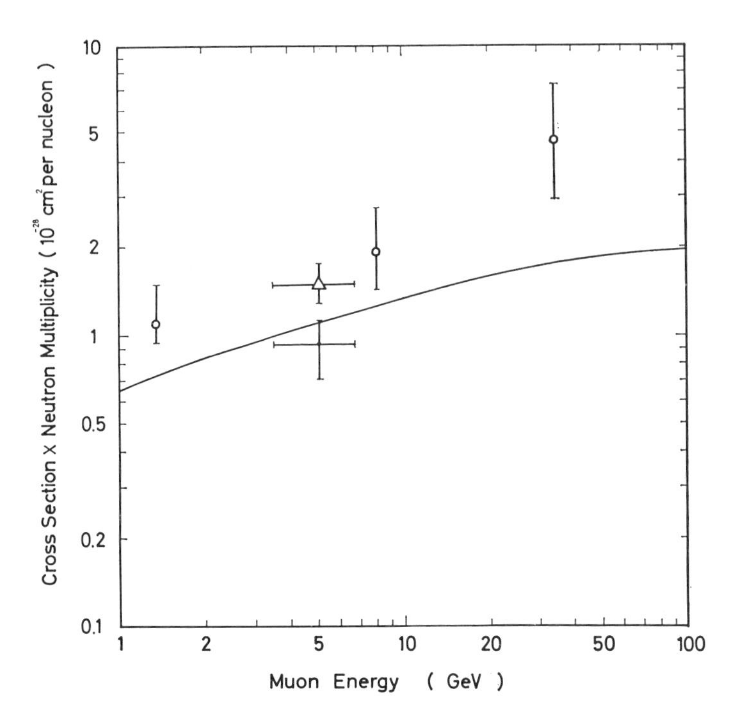

Fig. 4.35 Cross section for neutron production by muons in lead. The curve represents the cross section times the neutron multiplicity given in (4.14.37*a*) against the muon energy. Experimental values are taken from (triangle symbol) —J. de Pagter and R. D. Sard, *Phys. Rev.*, **118**, 1353 (1960), not corrected for the contribution of knock-on showers; (open circle symbol)—M. A. Meyer et al., *J. Phys. Soc. Japan*, **17**, Suppl. A–III, 342 (1962), not corrected for the contribution of knock-on showers; (filled circle symbol) — J. de Pagter and R. D. Sard, *Phys. Rev.*, **118**, 1353 (1960), corrected for the contribution of knock-on showers.

$$q_\pi(x) = \int m\sigma_{\mu F}\, j_\mu(E)\, dE \simeq 0.2[\ln (8x) - 1] \times 10^{-28} j_\mu(x) \sec^{-1}. \tag{4.14.35}$$

Most of the neutrons produced are isotropic, so that the omnidirectional intensity has practical meaning. This is given by

$$\begin{aligned} J_n(x) &= LQ(x) \\ &\simeq 1 \times 10^{-3} \ln (8x) J_\mu(x), \end{aligned} \tag{4.14.36}$$

where $L \simeq 150$ g-cm^{-2} is the attenuation length of neutrons and

$$Q(x) = \frac{N}{A} \int \left[q_\pi\left(\frac{x}{\cos\theta}\right) + q_g\left(\frac{x}{\cos\theta}\right)\right] 2\pi \, d\cos\theta. \tag{4.14.37}$$

The neutrons thus produced participate in nuclear reactions underground at a rate of $Q = J/L$. Nuclear reactions are also produced directly by muons and indirectly by photons, negative pions, and other particles. The rate of nuclear reactions is on the order of $Q(x)$.

4.15 Low-Energy Electronic Component in the Lower Atmosphere

4.15.1 Intensity of Electrons

Electrons of energies lower than several GeV behave analogously to those underground; they are secondary products and are nearly in equilibrium with their parent components at altitudes lower than 10 km. Near sea level the parents are mostly muons, whereas the contribution of nuclear active particles increases rapidly with altitude. The latter has been discussed in Section 4.9; the intensity of the electronic component given in (4.9.8) with (4.9.11) and (4.9.13) can account for experimental ones of energies higher than 10 GeV.

Below 1 GeV (4.9.8) does not hold exactly, because the pion-production spectrum begins to deviate from a single power law. Nevertheless, it is not far from reality to use (4.9.8) with a smaller value of β (down to 100 MeV), because the deviation is not too large and the contribution to the electron flux is roughly proportional to the energy of a produced pion. Since 100 MeV is nearly equal to the critical energy in air, the flux of electrons with energies below 100 MeV is nearly equal to the flux of electrons with higher energies. In this way the total flux of electrons arising from the N-component may be expressed as

$$j_{eN}(>0, x) = 2\frac{0.437}{(\beta+1)\beta}\bar{g}_\pi\left(\frac{E_0}{0.1}\right)^\beta \left[\frac{1}{1+\lambda_1(\beta)L/X_0} - \frac{1}{1+\lambda_2(\beta)L/X_0}\right] e^{-x/L}$$
$$\simeq 6\times 10^{-4} \exp\left[\frac{(x_0 - x)}{L}\right], \tag{4.15.1}$$

where we have used $L = 125$ g-cm^{-2}, and $\beta = 1.8$ between 10 and 100 GeV and $\beta = 1.6$ between 100 MeV and 10 GeV.

The electron intensity in (4.15.1) decreases exponentially with increasing depth and becomes very small at low altitudes. On the other hand, the knock-on electrons of muons give a considerable contribution, in the same way as those underground. The energy spectrum of the electrons due to the knock-on process is therefore given by (4.14.11).

The total flux of these electrons can be evaluated by taking the integral track length as

$$j_{e,k}(>0, x) = \langle \bar{N}_{e,k}(E_\mu) \rangle \int_{E^-}^{\infty} j_\mu(E_\mu, x)\, dE_\mu . \tag{4.15.2}$$

In practice an electron detector has a threshold energy, so that electrons with energies below the threshold are not detectable. If the threshold is 5 MeV, we have $E^- \simeq 230$ MeV. This implies that muons responsible for knock-on electrons belong to the hard component. Thus the electron intensity due to the knock-on process is proportional to the intensity of the hard component. In evaluating the average track length in (4.15.2) we take into account that the average energy of muons is about 1 GeV and the maximum energy transfer corresponding to it is about 100 MeV. Thus we obtain

$$\begin{aligned} j_{e,k}(>0, x) &\simeq \frac{KX_0}{\varepsilon_0} \left[\ln \left(\frac{E_m}{E} \right) - 1 \right] j_\mu(>E^-, x) \\ &\simeq 0.070\, j(\text{hard } \mu, x). \end{aligned} \tag{4.15.3}$$

There is another source of electrons that is unimportant underground. This is the spontaneous decay of muons. We shall avoid the complexity associated with kinematic problems, and give here an approximate method of evaluating the intensity of decay electrons.

Since the average energy of decay electrons is one-third the muon energy,

$$\langle E_e \rangle \simeq \frac{E_\mu}{3}, \tag{4.15.4}$$

the intensity of muons that give electrons of energies between E_e and $E_e + dE_e$ is approximately given by

$$j_\mu(E_\mu)\, dE_\mu = j_\mu(3E_e) 3dE_e . \tag{4.15.5}$$

The decay rate of a muon with energy E_μ is

$$\frac{b_\mu}{p_\mu x} \simeq \frac{b_\mu c}{3E_e x}, \tag{4.15.6}$$

where $b_\mu c \simeq 1.3$ GeV is the muon-decay constant near sea level. The energy dependence of the decay rate results in the energy spectrum of electrons being steeper by a power index of 2 than the muon spectrum at energies greater than the critical energy of the cascade shower. The intensity of decay electrons may be estimated, in an analogous way to that due to neutral-pion decays, as

$$j_{e,d}(>0, x) \simeq \frac{X_0}{x}\frac{b_\mu c}{3\varepsilon_0}\, j_\mu(>3\varepsilon_0, x)$$

$$\simeq 0.20\left(\frac{x_0}{x}\right) j(\text{hard } \mu, x), \tag{4.15.7}$$

where $x_0 = 1030$ g-cm^{-2} is the depth of sea level.

Summing up these contributions, we obtain the intensity of electrons as

$$j_e(x) = j_{e,k} + j_{e,d} + j_{e,N} \simeq \left[0.07 + 0.20\left(\frac{x_0}{x}\right)\right] j_\mu(x) + 6 \times 10^{-4} \exp\left[\frac{(x_0 - x)}{L}\right]. \tag{4.15.8}$$

The contributions of respective processes and their sum are shown in Fig. 4.36.

4.15.2 *Analysis of the Soft Component*

This is compared with the observed intensity of the soft component (Rossi 48). The measurement of the absolute cosmic-ray intensity at sea level was made at latitudes higher than the knee of the latitude effect. The vertical and omnidirectional intensities of cosmic rays that can penetrate a counter wall 2.3 g-gm^{-2} thick are found to be

$$j_{\perp,t} = 1.14 \times 10^{-2}\ \text{cm}^{-2}\ \text{sec}^{-1}\ \text{sr}^{-1}, \tag{4.15.9a}$$

$$J_t = 2.41 \times 10^{-2}\ \text{cm}^{-2}\ \text{sec}^{-1}, \tag{4.15.9b}$$

respectively. Those of the hard component defined as capable of penetrating 10 cm of lead are

$$j_{\perp,h} = 0.82 \times 10^{-2}\ \text{cm}^{-2}\ \text{sec}^{-1}\ \text{sr}^{-1}, \tag{4.15.10a}$$

$$J_h = 1.66 \times 10^{-2}\ \text{cm}^{-2}\ \text{sec}^{-1}, \tag{4.15.10b}$$

respectively. Their differences give the respective intensities of the soft component:

$$j_{\perp,s} = 0.32 \times 10^{-2}\ \text{cm}^{-2}\ \text{sec}^{-1}\ \text{sr}^{-1}, \tag{4.15.11a}$$

$$J_s = 0.75 \times 10^{-2}\ \text{cm}^{-2}\ \text{sec}^{-1}. \tag{4.15.11b}$$

The intensity of the soft component thus obtained is significantly greater than the electron intensity given in (4.15.8); the soft-hard ratio

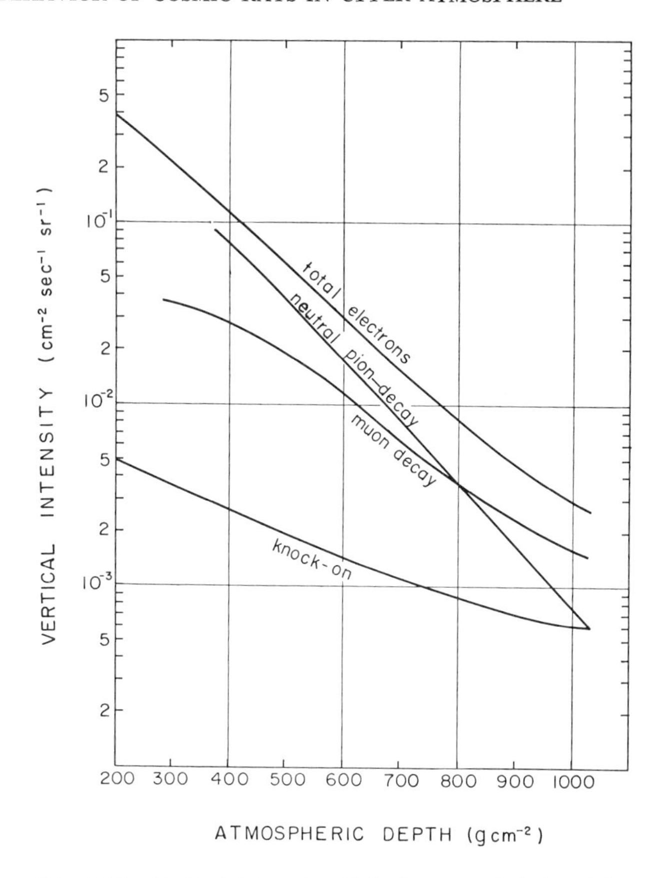

Fig. 4.36 Vertical intensity of electrons and their origins.

of the vertical flux, $j_{\perp,s}/j_{\perp,h} = 0.38$, should be compared with the electron-muon ratio, $j_e/j_\mu \simeq 0.30$. The difference may be attributed to soft muons and soft nuclear active particles. Referring to the discussions given in Sections 4.7 and 4.11, we are able to obtain their intensities (also given in Figure 4.36). The sum of all the contributions is consistent with the intensity of the soft component (Rossi 49, Hayakawa 52).

4.16 Behavior of Cosmic Rays in the Upper Atmosphere

At high altitudes the equilibrium of secondary components with their parents does not hold, and their behavior is dictated by complicated interactions with air and spontaneous decays. The intensities of various components as functions of altitude were once of interest because they revealed characteristic features of interactions and decays, but nowadays their main importance is in the sense that atmospheric cosmic rays cause a harmful background in the studies of primary cosmic rays at balloon altitudes. The discussions on heavy nuclei in Section 4.5 and on the high-energy electronic component meet this need. We are here concerned with the bulk of cosmic rays, mainly of relatively low energies, and shall estimate the intensities and the energy spectra of several components.

4.16.1 Simplified Model

For the sake of simplicity we shall discuss the intensities of various components of relativistic energies, for which the one-dimensional approximation is permissible (Okuda 65). They may be calculated under the following assumptions and approximations:

1. Primary cosmic rays are isotropic at the top of the atmosphere; that is, the azimuth effect due to the geomagnetic field is neglected.
2. The angular spreads in the production of secondary particles are neglected; that is, the one-dimensional treatment is assumed.
3. The deflection of secondary particles by the geomagnetic field is also neglected.
4. The heavy primary particles are assumed to behave in nuclear collisions like an assembly of free nucleons, but their correct collision mean free paths are used.
5. The atmosphere of the earth is approximated by an isothermal model in which the scale height is assumed to be constant, with a value of 6.5 km.

Although the above assumptions may be considered to be oversimplified, the obtained results reproduce the essential behavior of cosmic rays in the upper atmosphere.

Assumption 1 is not acceptable if we are concerned with the azimuthal distribution and the charge ratio. Hence the results of the simplified model are of practical use only for the intensities averaged over azimuth angle. Assumption 2 barely holds for the production of pions with energies smaller than 1 GeV, since their transverse momenta are not negligible compared with their longitudinal momenta. In other interactions and decay processes this is acceptable with reasonable accuracy. Hence the model fails if we are concerned with particles whose energies are too low. Other assumptions are less questionable than the above two.

Once this model is adopted, the calculation of the intensities at a given depth and zenith angle is straightforward. The methods described in Sections 4.8 through 4.11 are used with appropriate simplification. There is one modification in that the electronic component arises not only from neutral pions but also from muon decays. The latter are taken into account with the same approximation as in Section 4.15. If we assume that the intensity of protons after passing through a thickness x is

$$j_p(>E, x) = 0.4E^{-1.15} \exp(-x/110) \quad \text{cm}^{-2}\,\text{sec}^{-1}\,\text{sr}^{-1}, \quad (4.16.1)$$

where E is measured in GeV and x in g-cm^{-2}, the intensities of other components can be evaluated by numerical calculation.

4.16.2 Results and Comparisons with Experiments

The altitude dependences of the vertical intensities of muons at several energies are shown in Fig. 4.37, in which the atmospheric depth is expressed in units of the attenuation length of the nuclear active component, 110 g-cm^{-2}.

The same parameters for vertical electrons and γ-rays of energies greater than 100 MeV are given in Fig. 4.38. The γ-rays come almost exclusively from the γ-decays of neutral pions; those that are due to the cascade process starting from electrons are unappreciable. At small depths the electrons come mainly from the decay of charged mesons, whereas below a certain depth the electrons due to the cascade process starting from neutral-pion-decay γ-rays dominate over the former. Because the altitude dependences of muon intensities at different energies are nearly parallel to each other, the same should hold for electrons and γ-rays. The energy spectra of electrons and γ-rays may be expressed by power laws of $E^{-\alpha-1}\,dE$, with $\alpha = 2.0$ for electrons and $\alpha = 1.64$ for γ-rays. The value of α is expected to decrease as energy decreases; for example, muon intensities at energies below 1 GeV as given in Fig. 4.37 may be higher than the true values.

This is, in fact, seen by comparing the calculated intensity-altitude relation with the observed relation. (Clark 52).

The comparison is also made for electrons and γ-rays in Fig. 4.38. The experiments referred to are concerned with electrons and γ-rays with energies much lower than 1 GeV. Since, however, the altitude dependences are essentially dictated by the nuclear-collision mean free path and the cascade development, their energy dependence practically disappears in the first approximation. Hence we compare the intensity-depth curves for energies greater than 100 MeV with the experimental values by extrapolating the calculated energy spectra of γ-rays and electrons to energies as low as 100 MeV.

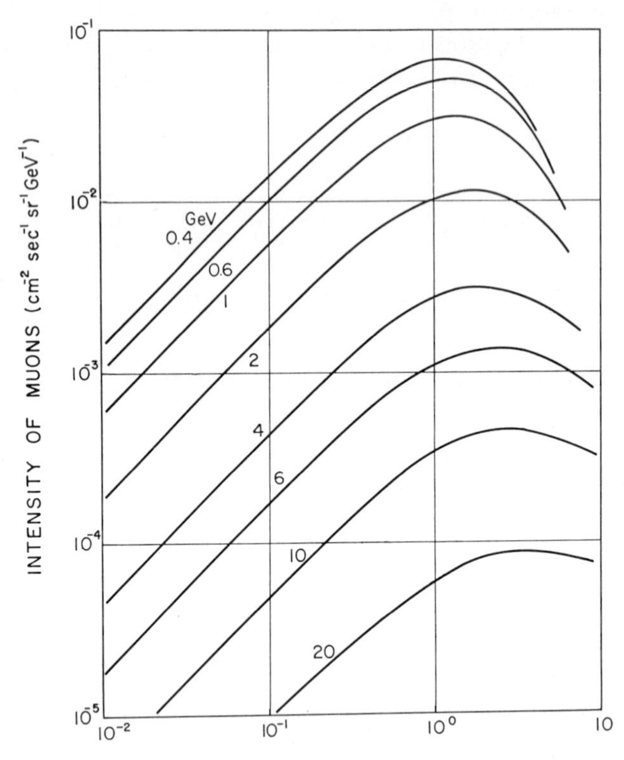

Fig. 4.37 Vertical intensities of muons versus atmospheric depth, in units of the nuclear attenuation length, 110 g-cm^{-2}. The energies of muons are attached to respective curves (Okuda 65).

The energies of the electrons in the experiment of Meyer and Vogt (Meyer 62) were greater than 100 MeV. The deviation of the experimental values from the calculated curve at small depths is interpreted as due to primary electrons. The energies of γ-rays in the experiments of Cline, Kraushaar and Clark, and Duthie et al. (Cline 61, Kraushaar 62, Duthie 63) are several tens of MeV. All give similar altitude dependences and are parallel to the calculated value, but the absolute values are not consistent with each other.

The fair agreement between the experimental results and the simple theory demonstrates that the essential features of the behavior of cosmic rays in the upper atmosphere can be accounted for in terms of the simplified model. If we rely on this model, we can derive the zenith-angle dependences for electrons and γ-rays with energies of 1 GeV at several atmospheric depths. According to this calculation the intensity at a small depth increases rapidly with zenith angle, and the zenith-angle dependence varies considerably with depth. These facts are important in making the comparison with experiment, in which a detector has a finite angle of acceptance.

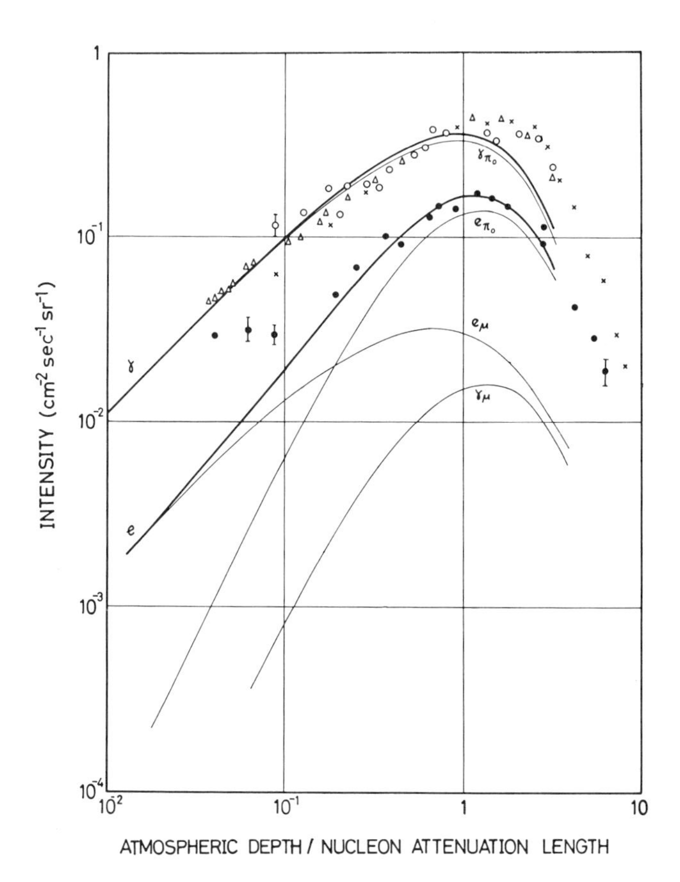

Fig. 4.38 Intensities of electrons (e) and γ-rays (γ) of energies above 100 MeV versus atmospheric depth, in units of the nucleon attenuation length, 110 g-cm^{-2}. The contributions of π^0-2γ decays (γ_{π^0}, e_{π^0}) and π-μ-e decays (γ_μ, e_μ) to γ-rays and electrons are shown separately (Okuda 65). Experimental results shown are taken from ●—P. Meyer and R. Vogt, *Phys. Rev. Letters*, **8**, 387 (1962), multiplying by a factor of 1.2 (>100 MeV); △—J. G. Duthie et al., *Phys. Rev. Letters*, **10**, 364 (1963), multiplying by a factor of 0.8 (>60 MeV); ○—W. L. Kraushaar and G. W. Clark, *Proc. Fifth Interamerican Seminar on Cosmic Rays*, Vol. 1, XVI–1 (1962), multiplying by a factor of 3 (>50 MeV); ×—M. A. Cline, *Phys. Rev. Letters*, **7**, 109 (1961), multiplying by a factor of 3 (>70 MeV).

4.17 Neutrons in the Atmosphere

Another important component of secondary cosmic rays consists of neutrons, because they take part in geophysical phenomena; neutrons escaping into outer space provide one of the sources of geomagnetically trapped particles, and the resulting nuclear interactions give radioactive nuclides such as ^{14}C. Since energetic neutrons have been discussed in Section 4.7, we are concerned mainly with slow neutrons—say, with energies below 10 MeV.

Slow neutrons are produced mainly by the evaporation-like process as a consequence of the nuclear interactions of nuclear active particles with higher energies. The energy spectrum of such neutrons is empirically expressed (as discussed in Section 2.9) by

$$S(E)\,dE \propto E \exp\left(\frac{-E}{T}\right) dE, \tag{4.17.1}$$

where $T = 1$ MeV is chosen on the basis of experimental as well as theoretical information. The behavior of these neutrons in the atmosphere can be described by diffusion theory, and the diffusion equation can be solved with the aid of multigroup diffusion theory (Hess 61). To this we add those neutrons of higher energies that are produced by knock-out processes, although they do not significantly affect the energy spectrum below 1 MeV. They contribute mainly to the absolute flux in such a way that one-half of them are degraded to less than 10 MeV—mostly between 3 and 10 MeV. Since the ratio of evaporation to knock-on sources is estimated as about 4, only one-eighth of the slow neutrons arise from the knock-on source.

The solution of the diffusion equation is compared with the observed intensities and spectra at atmospheric depths between 200 g-cm^{-2} and sea level at geomagnetic latitude 44°N (Hess 59). The altitude dependence for depth $x > 200$ g-cm^{-2} is represented by $\exp(-x/L)$,

$$L \simeq 155 \text{ g-cm}^{-2}; \tag{4.17.2}$$

this is somewhat greater than the nucleon attenuation length at about 100 MeV, as given in (4.7.14).

The energy spectra calculated for various depths are shown in Fig. 4.39. The spectrum at 700 g-cm^{-2} can be compared with that in Fig. 4.4. The altitude dependences of neutron fluxes at various energies are given in Fig. 4.40. It should, however, be noted that the thermal-neutron flux measured with enriched and normal ^{10}B plastic scintillators gives a

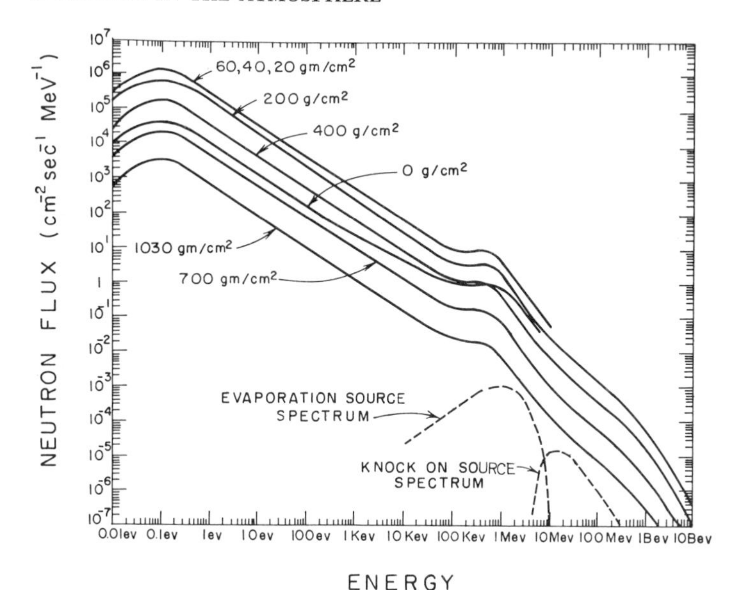

Fig. 4.39 The equilibrium neutron flux versus energy at different depths in the atmosphere for geomagnetic latitude 44°N. The energy spectra for 200 to 1030 g-cm^{-2} are experimental values; for depths less than 200 g-cm^{-2} the spectra are calculated (Hess 61). The shape of the two neutron source spectra are also shown (Hess 61).

value that is smaller by a factor of about 2 (Boella 63); namely, at 200 g-cm^{-2} at geomagnetic latitude 46.5°

$$J(<0.4 \text{ eV}) = 3.0 \pm 0.5 \text{ cm}^{-2} \text{ sec}^{-1}. \tag{4.17.3}$$

This indicates that the absolute flux is subject to some uncertainty.

By reference to the neutron flux given in Figs. 4.39 and 4.40, the total production rate of neutrons in a column of the atmosphere and the rates of their dissipation by various processes are summarized in Table 4.18. The former is 6.2 neutrons cm^{-2} sec^{-1}—5.0 by evaporation and 1.2 by knock-on. Among them 3.97 are lost by $^{14}N(n, p)^{14}C$ reaction, 1.20 are captured by other processes, and 1.03 leak out of the atmosphere.

The global average values can be obtained by multiplying by a factor of 0.75. Thus the global average values of the neutron-production rate and the ^{14}C production rate are found to be

$$\begin{aligned} \bar{S} &= 4.6 \text{ cm}^{-2} \text{ sec}^{-1}, \\ \bar{Q}(^{14}\text{C}) &= 2.9 \text{ cm}^{-2} \text{ sec}^{-1}, \end{aligned} \tag{4.17.4}$$

respectively. The leakage rates are estimated as 0.4 cm^{-2} sec^{-1} at the equator and 1.8 cm^{-2} sec^{-1} at the pole.

Table 4.18 Dissipation Rates of Neutrons at Latitude 44° (cm^{-2} sec^{-1})

Process	<1 MeV	1 to 10 MeV	>10 MeV	Total
$^{14}N(n, p)^{14}C$ reaction	3.86	0.11	0	3.97
Other capture processes	0.18	0.49	0.53	1.20
Leakage out of the atmosphere	0.62	0.35	0.06	1.03
Strike earth	—	—	—	<0.01
Total	4.66	0.95	0.59	6.20

The neutron-production rate given in Table 4.18 corresponds to about 20 neutrons produced per primary nucleon with energies of several GeV. This implies that most neutrons are produced by secondary nucleons with energies of about 100 MeV, because neutron multiplicity from light nuclei is small.

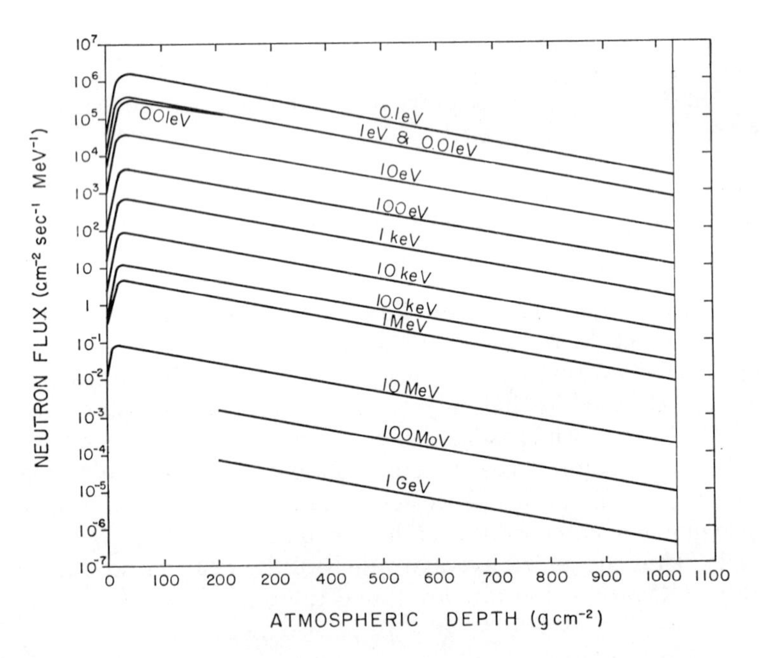

Fig. 4.40 The equilibrium neutron flux versus altitude at different neutron energies for geomagnetic latitude 44°N. The data for altitudes above the 200 g-cm^{-2} level are calculated; the rest are experimental (same data as Fig. 4.39) (Hess 61).

The flux of neutrons is roughly proportional to the rate of nuclear reactions and consequently to the fluxes of other nuclear active particles produced by the same reactions. These quantities can be estimated from a knowledge of the ranges and the multiplicities of the respective components. The flux of component i with range L_i is given by

$$J_i = m_i L_i q, \tag{4.17.5}$$

where m_i is the multiplicity and q is the nuclear reaction rate in unit volume. From this we can easily see how small the flux of low-energy charged particles is because their range is much smaller than the diffusion length of neutrons.

4.18 Summary of the Intensities of Various Components

By comparing the intensities of various components at different altitudes we can check the consistency of their values and transformation processes. Here we give the intensities of charged components versus the atmospheric depth at geomagnetic latitude 45°, since those at lower latitudes are given in Fig. 1.7.

At this latitude we refer to the compilation made by Rossi (Rossi 48), but a minor modification is made in the intensity of the soft component. Here the lower limit of energy is taken to correspond to the penetration of 2.3 g-cm^{-2} brass—instead of 5 g-cm^{-2} adopted by Rossi—because scintillation counters with thin shields are widely used. Since no experimental data are available for direct comparison, our results may be subject to an error on the order of 10 percent.

Apart from this modification, the procedures for obtaining the intensities at different altitudes are the same as those employed by Rossi. From the hard component we subtract the hard nuclear active component whose intensity is assumed to be represented by a simple exponential law. In this way the intensity of hard muons can be obtained. By referring to the energy spectra of muons and protons we can estimate the intensities of soft muons and protons. These are subtracted from the intensity of the soft component to give the intensity of electrons. The electron intensity is decomposed into three sources, as discussed in Fig. 4.36. These procedures may be permitted at lower altitudes. At high altitudes the results given in Section 4.16 are referred to. Thus we can draw the intensity-depth curves of the various components as shown in Fig. 4.41.

A similar analysis has been made of the intensity-depth relation at a

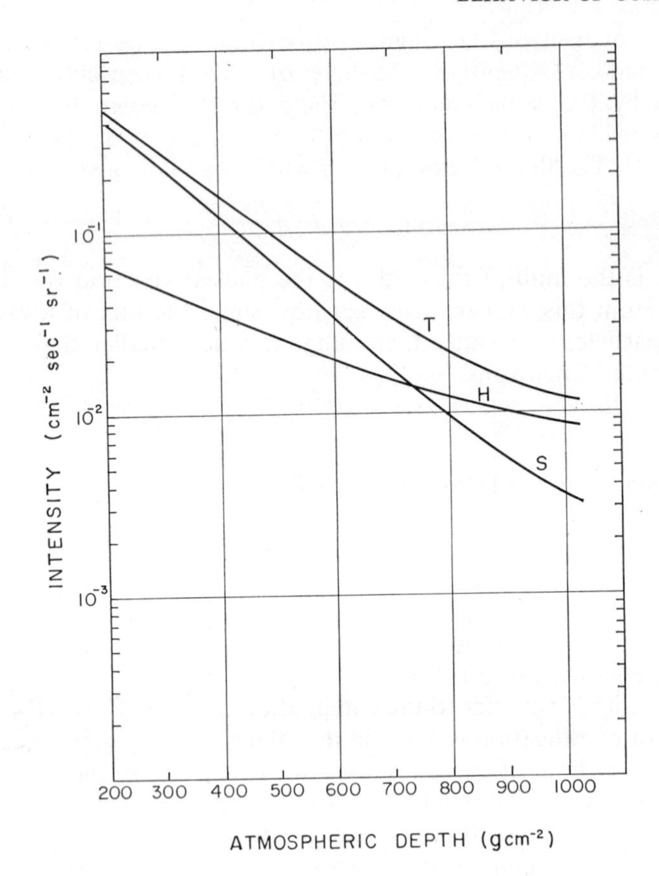

Fig. 4.41 (a) Vertical intensities versus atmospheric depth of the total (T), hard (H), and soft (S) components at geomagnetic latitude 45°.

high latitude; namely, 60.5°N. The result at the solar minimum is shown in Fig. 4.42 (Komori 62).

By referring to these analyses we can estimate the energy of cosmic rays dissipated in the atmosphere. Charged particles lose their energy by ionization. Although some of them are converted into neutral particles, they again turn into charged particles, their energy thus being eventually lost by ionization. The energy dissipated by ionization in the atmosphere is directly determined by ionization measurements (Neher 52). Part of the cosmic rays penetrate the whole atmosphere and escape out of the atmospheric ionization. The amount of such energy can be obtained from the total track length of underground cosmic rays, which is proportional to the energy lost by ionization underground.

There are two other causes of energy dissipation. One of them is the energy spent for nuclear binding. Most particles and photons that are produced by nuclear disintegrations dissipate their energy by ionization, but some neutrons escape out of the atmosphere. Neutrinos emitted in β-decays from disintegration products also escape from ionization. More important is the difference between the binding energies of a target nucleus and its disintegration products. Since the fission of a light nucleus has a negative Q value, this is a rather important source of energy dissipation. For air nuclei the binding energy per nucleon is about 11.5 MeV. Because the emission of an α-particle spends a rather small amount of energy, and because the number of protons emitted is about half that of neutrons, the binding energy per neutron production is estimated to be about 25 MeV.

Another source of dissipation is by neutrinos produced in meson decay. The resultant energy dissipation is estimated by referring to the

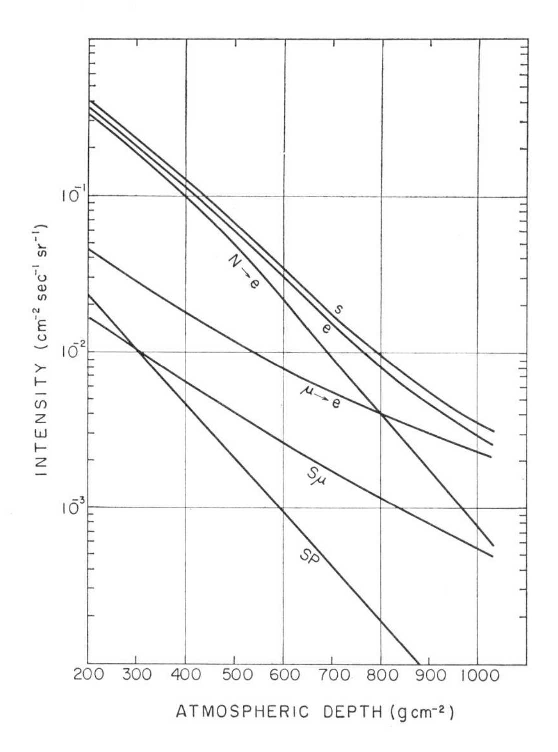

Fig. 4.41 (b) Vertical intensities versus atmospheric depth of the soft component (S) and its subcomponents; $S = e + s\mu + sp$, e (electrons) $= N \rightarrow e$ (electrons from π^0) $+ \mu \rightarrow e$ (electrons from the knock-on and decay processes of muons).

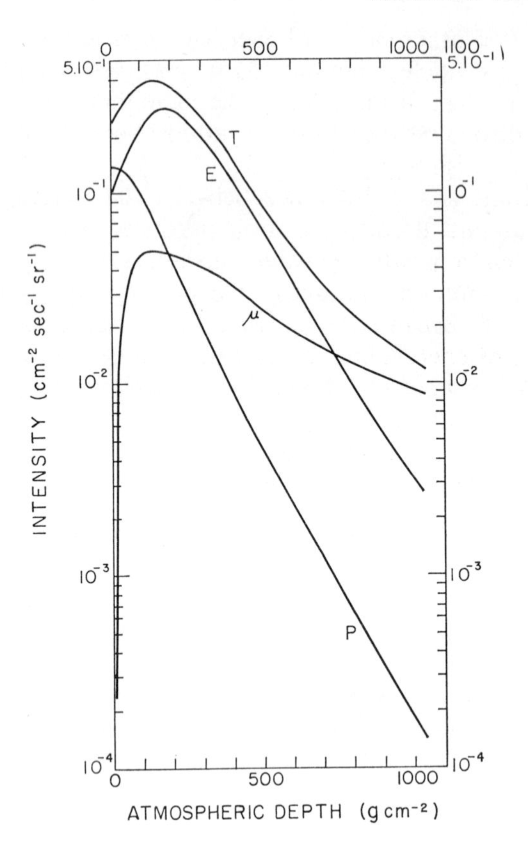

Fig. 4.42 Vertical intensities of various components versus atmospheric depth at Saskatoon in the solar minimum year (Komori 62). *T*—total, *E*—electrons, μ—muons, *p*—protons.

intensity and energy spectrum of muons. Thus the energy given to neutrinos in the decays of pions and muons can be evaluated.

In this way we are able to derive the distribution of energy dissipation. Table 4.19 lists the distribution at the geomagnetic latitude 50°, partly referring to the analysis by Puppi (Puppi 56). The total energy dissipated is in fair agreement with the amount of energy brought in by primary cosmic rays.

If we know the production spectrum of pions, it may be more appropriate to write the balance sheet in a slightly different way. Since muons arise from charged pions and the electronic component from neutral pions and muons, the energy given to pions covers most of the energy dissipated by secondary cosmic rays. In addition to this, we need

Table 4.19 Distribution of Energy Dissipation at Latitude 50°

Process	Dissipation (MeV-cm^{-2} sec^{-1} sr^{-1}
Ionization in the atmosphere	730
Residual energy at sea level	40
Nuclear disintegration	150
Neutrinos	230
Total	1150

Table 4.20 Incident and Dissipated Energies (MeV cm^{-2} sec^{-1} sr^{-1}). Errors Indicated are Those Arising from Theoretical Estimates and Experiments

		Station			
		Saskatoon	Minneapolis	San Angelo	Guam
Geomagnetic Coordinates[a]	latitude	60.5°N	55.2°N	41.4°N	3.6°N
	longitude	311.9°	330.9°	317.3°	212.9°
Cutoff energy[a] for protons (GeV)		0.18	0.55	3.92	14.9
Incident energy:					
Protons		1000 ± 25	889 ± 25	565 ± 17	346 ± 9
He-nuclei		204 ± 4	200 ± 4	178 ± 4	78 ± 4
L-elements		6 ± 2	6 ± 2	4 ± 2	1
M-elements		48 ± 2	47 ± 1	28 ± 1	21 ± 1
H-elements		40 ± 1	40 ± 1	21 ± 1	5
Sum		1300 ± 30	1180 ± 30	796 ± 18	452 ± 10
Dissipated energy:					
Ionization loss of protons		131 ± 3	129 ± 3	76 ± 2	16
$\pi^{\pm}$		423 ± 15	416 ± 14	359 ± 13	208 ± 8
π^{0}		278 ± 25	265 ± 24	227 ± 21	157 ± 13
Nuclear disintegrations		314 ± 68	301 ± 68	139 ± 31	53 ± 12
Sum		1150 ± 80	1110 ± 80	801 ± 40	434 ± 20

[a] Based on "The Geomagnetic Field," J. J. Quenby and W. R. Webber, *Phil Mag.*, **4**, 90 (1959).

estimate only the energies dissipated by the ionization loss of protons and by nuclear binding.

A detailed comparison between the energy dissipated and the incident energy has been attempted by Komori (Komori 62). The incident energy is estimated from the energy spectra of primary particles at the solar minimum, with correction for the albedo effect. The result is shown in Table 4.20. Here the method of evaluating the energy spent for nuclear binding is different from that described above, but the result turns out to be almost the same because the excessively high neutron-production rate adopted by Komori has been compensated for by our high value for the nuclear-binding energy per neutron production. The agreement between the incident and dissipated energies is satisfactory.

REFERENCES

Achar 65	Achar, C. V., et al., *Phys. Letters*, **18**, 196; **19**, 78 (1965).
Aizu 60	Aizu, H., et al., *Prog. Theor. Phys. Suppl.*, No. 16, 54 (1960).
Akashi 64	Akashi, M., et al., *Prog. Theor. Phys.*, Suppl. No. 32, 1 (1964).
Akashi 65	Akashi, M., et al., *Proc. Int. Conf. on Cosmic Rays at London*, **2**, 878 (1965).
Allen 61	Allen, J. E., and Apostolakis, A. J., *Proc. Roy. Soc.* (London), **A265**, 117 (1961).
Ascoli 50	Ascoli, G., *Phys. Rev.*, **79**, 812 (1950).
Ashton 60	Ashton, F., et al., *Nature*, **185**, 354 (1960).
Baradzei 61	Baradzei, L. T., et al., *Proc. Int. Conf. on Cosmic Rays at Jaipur*, **5**, 283 (1963).
Barrett 52	Barrett, P., et al., *Rev. Mod. Phys.*, **24**, 133 (1952).
Barton 61	Barton, J. C., *Phil. Mag.*, **6**, 1271 (1961).
Brooke 62	Brooke, G., et al., *J. Phys. Soc. Japan*, **17**, Suppl. A–III, 311 (1962).
Brooke 64	Brooke, G., et al., *Proc. Phys. Soc.* (London), **83**, 853 (1964).
Boella 63	Boella, G., et al., *Nuovo Cim.*, **29**, 103 (1963).
Budini 53	Budini, P., and Molière, G., *Kosmische Strahlung*, ed. by W. Heisenberg, Springer-Verlag, Berlin, 1953.
Camerini 50	Camerini, U., et al., *Phil. Mag.*, **41**, 413 (1950).
Clark 52	Clark, M. A., *Phys. Rev.*, **87**, 87 (1952).
Clark 63	Clark, G., et al., *Proc. Int. Conf. on Cosmic Rays at Jaipur*, **4**, 65 (1963).
Cline 61	Cline, M. A., *Phys. Rev. Letters*, **7**, 109 (1961).
Cowsik 66	Cowsik, R., Pal, Y., and Tandon, S. N., *Proc. Indian Acad. Sci.*, **43**, 217 (1966).
Conversi 50	Conversi, M., *Phys. Rev.*, **79**, 750 (1950).
Duthie 63	Duthie, J. G., et al., *Phys. Rev. Letters*, **10**, 364 (1963).
Engler 58	Engler, A., Kaplon, M. F., and Klarmann, J., *Phys. Rev.*, **112**, 597 (1958).
Fowler 58	Fowler, J. M., Primakoff, H., and Sard, R. D., *Nuovo Cim.*, **9**, 1027 (1958).

Groetzinger 51 Groetzinger, G., Berger, M. J., and McClure, G. W., *Phys. Rev.*, **81**, 969 (1951).

Gross 33 Gross, B., *Z. Physik*, **83**, 214 (1933).

Goldman 58 Goldman, I. I., *J. Exp. Theo. Phys. USSR*, **34**, 1017 (1958); *Soviet Phys. JETP*, **7**, 702 (1958).

Hayakawa 52 Hayakawa, S., *Kagaku* (Japan), **22**, 278 (1952).

Hayakawa 57 Hayakawa, S., *Phys. Rev.*, **108**, 1533 (1957).

Hayakawa 64 Hayakawa, S., Nishimura, J., and Yamamoto, Y., *Prog. Theor. Phys.*, Suppl. No. 32, 104 (1964).

Hayman 62 Hayman, P. T., and Wolfendale, A. W., *Proc. Phys. Soc.* (London), **80**, 710 (1962).

Hess 59 Hess, W. N., et al., *Phys. Rev.*, **116**, 445 (1959).

Hess 61 Hess, W. N., Canfield, E. H., and Lingenfelter, R. E., *J. Geophys. Res.* **66**, 665 (1961).

Higashi 65a Higashi, S., et al., *Kagaku* (Japan), **35**, 461 (1965).

Higashi 65b Higashi, S., et al., *Nuovo Cim.*, **38**, 107 (1965).

Kamiya 62 Kamiya, Y., et al., *J. Phys. Soc. Japan*, **17**, Suppl. A–III, 315 (1962).

Kamiya 63 Kamiya, Y., *J. Geomag. Geol.*, **14**, 191 (1963).

Kocharian 56 Kocharian, N. M., et al., *J. Exp. Theor. Phys. USSR*, **30**, 243 (1956).

Komori 62 Komori, H., *J. Phys. Soc. Japan*, **17**, 620 (1962).

Kraushaar 62 Kraushaar, W. L., and Clark, G. W., *Proc. Fifth Interamerican Seminar on Cosmic Rays* (La Paz), Vol. LXVI–1, 1962.

McDonald 59 McDonald, F. B., and Webber, W. R., *Phys. Rev.*, **115**, 194 (1959).

McDonald 60 McDonald, F. B., *J. Geophys. Res.*, **65**, 767 (1960).

McDonald 62 McDonald, F. B., *J. Phys. Soc. Japan*, **17**, Suppl. A–II, 428 (1962).

Meyer 62 Meyer, P., and Vogt, R., *Phys. Rev. Letters*, **8**, 387 (1962).

Miyake 57 Miyake, S., Hinotani, K., and Nunogaki, K., *J. Phys. Soc. Japan*, **12**, 113 (1957).

Miyake 63 Miyake, S., *J. Phys. Soc. Japan*, **18**, 1226 (1963).

Miyake 64 Miyake, S., Narasimham, V. S., and Ramana Murthy, P. V., *Nuovo Cim.*, **32**, 1505 (1964).

Morinaga 53 Morinaga, H., and Fry, W. F., *Nuovo Cim.*, **10**, 308 (1953).

Neher 52 Neher, H. V., *Progress in Cosmic Ray Physics*, Vol. 1, 243 (1952).

Nishimura 67 Nishimura, J., *Handbuch der Physik*, Bd. 46/2,1 Springer-Verlag, Berlin, 1967.

O'Dell 62 O'Dell, F. W., Shapiro, M. F., and Stiller, B., *J. Phys. Soc. Japan*, **17**, Suppl. A–III, 23 (1962).

Okuda 65 Okuda, H., and Yamamoto, Y., *Rep. Ionosph. Space Res.*, **19**, 322 (1965).

Olbert 53 Olbert, S., *Phys. Rev.*, **92**, 454 (1953).

Olbert 54 Olbert, S., *Phys. Rev.*, **96**, 1400 (1954).

Osborne 64 Osborne, J. L., *Nuovo Cim.*, **32**, 816 (1964).

Pak 61 Pak, W., et al., *Phys. Rev.*, **121**, 905 (1961).

Pine 59 Pine, J., Davisson, R. J., and Greisen, K., *Nuovo Cim.*, **14**, 118 (1959).

Powell 59 — Powell, C. F., Fowler, P. H., and Perkins, D. H., *The Study of Elementary Particles by the Photographic Method*, Pergamon, London, 1959.

Primakoff 55 — Primakoff, H., *Proc. of the 5th Rochester Conference*, 174 (1955).

Puppi 56 — Puppi, G., *Progress in Cosmic Ray Physics*, Vol. 3, 338 (1956).

Reines 65 — Reines, F., et al., *Phys. Rev. Letters*, **15**, 429 (1965).

Rossi 48 — Rossi, B., *Rev. Mod. Phys.*, **20**, 537 (1948).

Rossi 49 — Rossi, B., *Rev. Mod. Phys.*, **21**, 104 (1949).

Simpson 51 — Simpson, J. A., *Phys. Rev.*, **83**, 1175 (1951).

Simpson 53a — Simpson, J. A., and Fagot, W. C., *Phys. Rev.*, **90**, 1068 (1953*a*).

Simpson 53b — Simpson, J. A., Fonger, W., and Treiman, S. B., *Phys. Rev.*, **90**, 934 (1953*b*).

Sens 58 — Sens, J. C., et al., *Nuovo Cim.*, **7**, 536 (1958).

Sens 59 — Sens, J. C., *Phys. Rev.*, **113**, 679 (1959).

Smith 59 — Smith, J. A., and Duller, N. M., *J. Geophys. Res.*, **64**, 2297 (1959).

Subramanian 59 — Subramanian, A., and Verma, S. D., *Nuovo Cim.*, **13**, 573 (1959)

Treiman 52 — Treiman, S. B., *Phys. Rev.*, **86**, 917 (1952).

Tennent 60 — Tennent, R. M., *Progress in Elementary Particles and Cosmic Ray Physics*, Vol. 5, 364 (1960).

Waddington 62 — Waddington, J., *J. Phys. Soc. Japan*, **17**, Suppl. A–III, 63 (1962).

CHAPTER 5

Extensive Air Showers

In the preceding chapter secondary particles produced in the atmosphere were assumed to be incoherent, except for γ-ray families; that is, two or more particles with a common origin do not hit a detector. As in the case of γ-ray families, however, two or more coherent particles are sometimes observed if the energy concerned is high enough to produce many particles in a narrow spatial region. Such a phenomenon is nothing but the shower developed in the atmosphere, which is called the *air shower*. If the energy of a primary particle is as high as 10^{14} eV or higher, the total number of particles in a shower is very large, and the distance at which two or more particles are coherent may extend as far as several hundred meters. Thus the air shower having a large coherence distance is called the *extensive air shower*, or *EAS* for short.

The observation of EAS provides a unique method of detecting ultra-high-energy particles and the interactions produced by them. This is due to the extensiveness of EAS. Because the spread of coherent particles is large, we can increase the effective area of detection to as large as 1 km^2 by using many small detectors scattered over the area. The observation of ultra-high-energy particles is thus made feasible.

The physical significance of the observation of EAS is twofold. One is the investigation of interactions of particles at very high energies, which cannot be reached by any other means available today. The other is to obtain astronomical information through ultra-high-energy particles. Leaving the latter to the next chapter, we shall here be concerned mainly with the former.

The investigation of high-energy interactions by means of EAS is as indirect as that by observation of secondary particles in the atmosphere and underground. Even with an elaborate analysis of various phenomena the information thus obtained remains in most cases qualitative rather than quantitative. Nevertheless, painstaking phenomenological studies of EAS have been made both experimental and theoretical since no other means can provide information on interactions at such high energies.

5.1 Components and Their Genetic Relations

The primary particles that generate EAS are mainly nuclear particles. Let us consider that a high-energy proton of energy E_0 impinging on the atmosphere collides with air nuclei with a mean free path l. Then the proton survives as a proton or a neutron with energy $(1 - K)E_0$, and the rest of the energy KE_0 is given to secondary particles. Most of the secondary particles are pions, whose multiplicity is n. Few charged pions decay before they collide with air nuclei, because their energies are so high that their lifetime in flight is considerable; neutral pions, on the other hand, immediately decay into γ-rays.

The survival nucleon, the charged pion, and other nuclear active secondary particles produce further secondary particles in their collisions with air nuclei. Thus nuclear active particles that participate in such chains of nuclear interactions form a component that is called the *nuclear active component*, or the *N-component*. The behavior of the N-component in EAS is described by the N-cascade; namely, the number of nuclear active particles increases with depth due to the multiple production of nuclear active particles. It reaches a maximum when the increase is just compensated for by the dissipation of nuclear active particles as their energies become too low to produce further particles. After this the number of nuclear active particles decreases with the *attenuation length* L_n, which depends weakly on atmospheric depth.

Not all charged pions can participate in the N-cascade because some of them decay into muons if their energies are not too high. Thus the number of muons increases as the N-cascade develops. These muons are highly penetrating because they may be lost only by decay and ionization energy loss, both of which are of little importance for muons of moderately high energies. Due to these features, muons are clearly distinguished from nuclear active particles, thus forming a component that is called the *μ-component*. The number of muons also decreases at great depths, because their decay and energy loss are appreciable, and few muons are supplied by the N-component. The attenuation length of the μ-component, L_μ, can thus be defined; its value is much larger than L_n.

The γ-rays that are produced by nuclear interactions initiate cascade showers of the *electronic component*, or the *E-component*. The whole behavior of the E-component is described by a complicated superposition of individual electronic cascades. At small depths cascades initiated by a few γ-rays of high energies predominate, but at greater depths those initiated by many γ-rays of moderate energies compete with the former. If the latter are dominant, the attenuation length of the E-component L_e is nearly equal to L_n, whereas in the other case L_e is greater than L_n.

The EAS consists of these three components as well as neutrinos that escape observation as in the case for the bulk of cosmic rays. These particles cover a nearly circular area about an axis, and the number of particles that belong to the respective components in EAS are denoted by N_n, N_μ, and N_e, in which N_e normally represents the number of electrons alone, without including photons, but N_n includes both protons and neutrons. Of these N_e is the largest, except in the case of nearly horizontal EAS. The values of these parameters N are not directly measured—only the number of particles falling on a limited number of detectors is counted. Therefore it is impossible to obtain the total number of particles without knowing the density distribution.

5.2 Lateral Structure

The density of particles in EAS can be measured directly with a number of detectors arranged over a certain area. The *density* of a component at a distance r from the axis of EAS is given by

$$\Delta(r) = Nf(r), \tag{5.2.1}$$

where $f(r)$ is the *lateral structure function* normalized as

$$\int_0^\infty f(r)2\pi r\, dr = 1;$$

N is the number of particles in EAS, and is called the *size*. Here $f(r)$ is assumed for simplicity to be independent of the azimuth angle, but its azimuth dependence has to be taken into account for inclined EAS and in the case where deflection by the geomagnetic field is appreciable.

If the variation of $f(r)$ is negligible within the effective area of a detector S, the density can be obtained from the number of particles detected thereby, n, as

$$\Delta = \frac{n}{S}. \tag{5.2.2}$$

The *frequency* of EAS with densities Δ in $d\Delta$ has been measured with single detectors, such as an ionization chamber (Lapp 43) and a cloud chamber (Brown 49).

In general, however, many detectors are used, thus increasing the accuracy of the density measurement. Three different methods have been used for observing the density with multiple detectors.

5.2.1 *Density Spectrum*

In the early days around 1950 many EAS observations were made with several detectors situated about 10 meters apart. Suppose an EAS containing N particles falls on an area in which m detectors with the same effective area S are involved. Then the probability of the m-fold coincidence is given by

$$\prod_{i=1}^{m} \{1 - \exp[-\Delta(r_i)S]\},$$

where r_i is the distance of the ith detector from the shower axis. If the frequency of EAS with size N in dN whose axes fall on an area dA is expressed as

$$\phi(N)\,dN\,dA = \phi_0 k\left(\frac{N_0}{N}\right)^k N^{-1}\,dN\,dA, \tag{5.2.3}$$

the m-fold coincidence rate is given by

$$\begin{aligned} C_m(S) &= \phi_0(N_0 S)^k k \int dA \int d(NS)(NS)^{-k-1} \prod_{i=1}^{m} [1 - e^{-NSf(r_i)}] \\ &= \phi_0(N_0 S)^k I(k, m), \end{aligned} \tag{5.2.4}$$

where

$$I(k, m) \equiv k \int dA \int x^{-k-1} \prod_{i=1}^{m} [1 - e^{-f(r_i)x}]\,dx$$

is a function depending on the geometrical configuration of detectors. In the second expression of (5.2.4) we have implicitly assumed that the structure function $f(r)$ is independent of N.

If m detectors are so close to each other that they observe practically the same density, we can define the density in the spatial region covered by the detectors. With the frequency of having the density Δ in $d\Delta$,

$$h(\Delta)\,d\Delta = h_0 k\left(\frac{\Delta_0}{\Delta}\right)^k \Delta^{-1}\,d\Delta, \tag{5.2.5}$$

the m-fold coincidence rate is expressed by

$$C_m(S) = h_0(\Delta_0 S)^k k \int (\Delta S)^{-k-1}(1 - e^{\Delta S})^m\,d(\Delta S) = h_0(\Delta_0 S)^k J(k, m),$$

where

$$J(k, m) \equiv k\Gamma(-k) \sum_{j=0}^{m} (-1)^{j+1} j^k \binom{m}{j}. \tag{5.2.6}$$

Either form (5.2.4) or (5.2.6) the power index of the *density spectrum* k can be obtained by varying the area of a detector S (Cocconi 46). This can also be obtained by varying the number of detectors, m, but the reliability of the latter method is smaller, because $I(k, m)$ or $J(k, m)$ depends on the arrangement of the detectors.

5.2.2 Decoherence Curve

In the earliest days of EAS studies the coincidence rate of two detectors was measured as a function of the distance between them (Auger 38). In fact, this led to the discovery of EAS, as described in Chapter 1. Suppose two detectors are separated by a distance d and each of them measures particles of density Δ or greater. If one detector observes a density Δ and the other a density greater than Δ, the axis of EAS must have fallen on an arc, as shown in Figure 5.1. The area element along the arc is given by

$$dA = 2\cos^{-1}\left(\frac{d}{2r}\right) r\, dr.$$

Considering that two detectors are equivalent, the coincidence rate that two detectors will observe densities Δ or greater is found to be (Blatt 49, Williams 48)

$$C(\Delta, d) = \int_{d/2}^{\infty} 4\cos^{-1}\left(\frac{d}{2r}\right) \Phi\left[\frac{\Delta(r)}{f(r)}\right] r\, dr, \tag{5.2.7}$$

where $\Phi(N) = \int_N^{\infty} \phi(N)\, dN$, called the *integral size spectrum*. The curve C against d is called the *decoherence curve* and gives information on the structure function. However, the structure function is given by

$$\Phi\left(\frac{\Delta}{f}\right) = \frac{1}{4\pi} \int_r^{\infty} \left[\left(\frac{d}{2}\right)^2 - r^2\right]^{-1/2} \frac{\partial^2 C}{\partial d^2}\, dd \tag{5.2.8}$$

only through the second derivative of the decoherence curve. Hence the structure function f and the size spectrum Φ can be obtained only with limited accuracy.

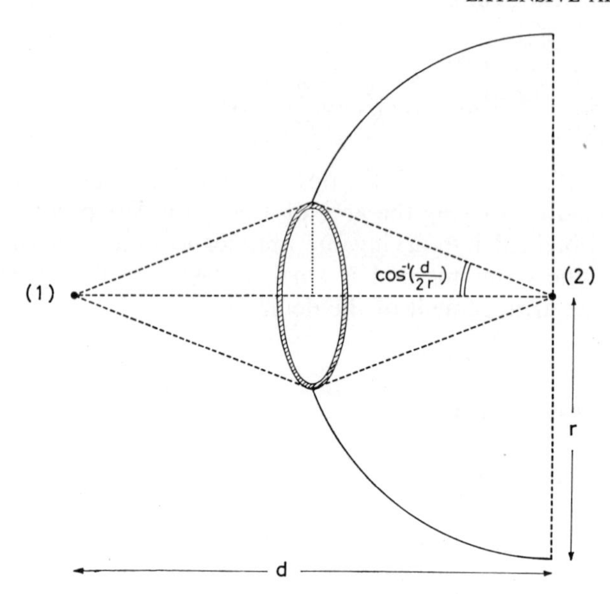

Fig. 5.1 Location of the axis in the case of double coincidence. Numbers 1 and 2 indicate two detectors. The shaded area indicates region that the shower axis may hit.

5.2.3 *Core Selection*

In the two methods discussed above the axis of EAS is not located in spite of the fact that the distance from the axis is used in analyzing experimental data. If the position of the axis is known, the quality of information will increase considerably. The axis of EAS is characterized by a small region called the *core*, in which both the densities and the average energies of the *E*- and *N*-components are higher than those in the surrounding region.

In an early attempt (Cocconi 49) the axis was located by several core selectors, in each of which several Geiger-Müller counters were discharged by penetrating showers produced in a thick lead layer covering the counter system. In order to distinguish these penetrating showers from those caused by single high-energy nuclear active particles, coincidence with an uncovered counter tray several meters away was required. By observing the densities of electrons and muons with other detectors located far from the core selectors, it was possible to obtain the density distributions of the *E*- and μ-components.

Although the core-selector method has an advantage in locating the axis, the probability that the core hits one of the core selectors is small because of their limited area; moreover, the accuracy of the core position is not satisfactory enough because the core selector may be triggered by energetic nuclear active particles that are not close to the core. Such disadvantages are overcome by the correlated hodoscope method (Dobrotin 56), in which the densities of particles at various positions are measured with many detectors scattered over a large area.

A typical example of the densities observed with the correlated hodoscopes is shown in Figure 5.2. The number of particles observed with these detectors make it possible to locate the axis as the central peak of density and to obtain the density distribution by reference to the axis thus located as well as to the total number of particles. Observations carried out at an altitude of 3860 meters and at sea level give the lateral distributions,

$$f(r) \propto \begin{cases} r^{-1} \exp(-r/r_0) & \text{for} \quad r < 100 \text{ m}, \\ r^{-n} & \text{for} \quad r > 100 \text{ m} \end{cases} \tag{5.2.9}$$

for $N \simeq 10^5$. The values of r_0 and n in (5.2.9) depend on the altitude as

$$\begin{aligned} r_0 &\simeq 80 \text{ m}, & n &\simeq 3 & &\text{at 3860 m}, \\ r_0 &\simeq 55 \text{ m}, & n &\simeq 2.6 & &\text{at sea level.} \end{aligned}$$

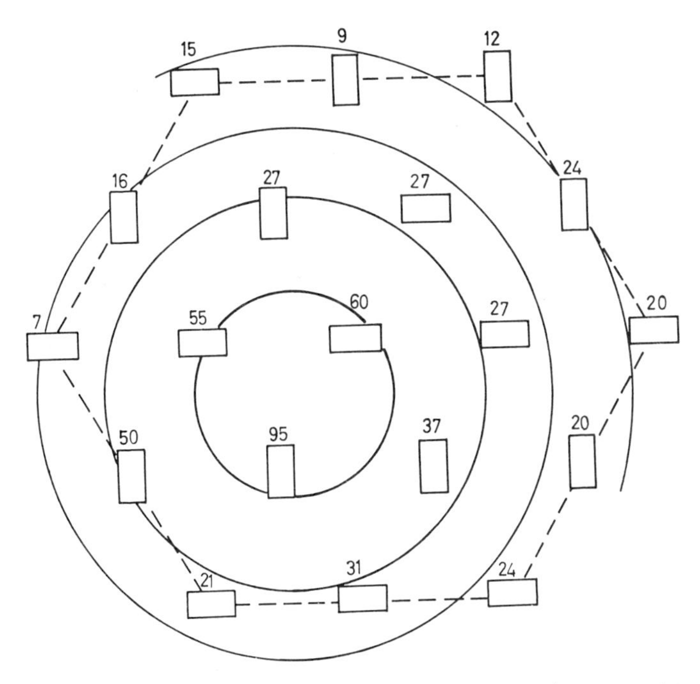

Fig. 5.2 Arrangement of correlated hodoscopes. The number of particles falling on each detector is represented by the figure (Dobrotin 56).

The particles considered above are mostly electrons, because the detectors used for deriving the lateral distribution are sensitive to electrons, and the number of electrons is larger by orders of magnitude than that of other particles. Hence the lateral distribution and the size given in (5.2.9) should be considered as those of electrons. In addition to the E-component, both the N- and μ-components can be observed with detectors shielded with thick materials. These two components can be distinguished from each other by determining whether a detected particle is associated with a penetrating shower or not. The number of particles belonging to each of these components is about 1 percent of the total size, excluding the μ-component in the peripheral region. The lateral distribution of the N- component is approximated by r^{-1} within 10 meters, and that of the μ-component is much flatter.

Although the correlated-hodoscope method is powerful in investigating various properties of EAS, the direction of arrival cannot directly be obtained with this method. If an EAS is inclined, the lateral distribution is elongated along the direction of the EAS, and the effective area of detectors is reduced. In analyzing experimental data, therefore, these effects have to be taken into account. If the direction of arrival is known, the analytical procedure is much simplified. Moreover, the zenith-angle dependence of EAS frequency and the distribution of arrival directions provide valuable information on the structure of EAS and the origin of cosmic rays.

The observation of arrival direction was made by means of the fast-timing technique (Bassi 53). For an EAS falling with zenith angle θ two particles separated by d arrive on a horizontal plane with a time difference of $d \sin \theta/c$, as shown in Fig. 5.3, provided that all particles in an EAS are contained in a thin disk. Since the thickness of the front is observed to be 10^{-9} sec or smaller, the zenith angle θ can be measured with an accuracy of several degrees if the span of detectors is on the order of 10 meters.

With more than 10 scintillation detectors set on concentric circles with the largest radius of 230 meters, EAS as large as 10^8 were observed within a year (Clark 57, 58). It was possible to determine the arrival direction within 5°, to locate the core within 10 percent of the distance from the center of the detector array, and to determine the size within 20 percent. For the latter two the lateral distribution of form (5.2.9) with $r_0 = 74$ meters was used. In addition to the scintillation detectors, Geiger–Müller counter hodoscopes with heavy shields were located near and far from the center of the array, so that the lateral distribution of muons has been observed.

This apparatus is suitable for obtaining the overall features of EAS; in particular, the arrival direction and size. These pieces of information are of great value in the investigation of the origin of cosmic rays. If a different aspect is aimed at, the arrangement of detectors should be modified.

In order to investigate more detailed features of EAS—to obtain information on the properties of high-energy interactions—it is necessary to observe quantities that are sensitive to parameters characterizing the interactions. For this purpose the structure of the EAS core should be investigated in more detail by setting detectors close together near the center of the array. When a core hits the central part, the lateral and energy distributions of particles and the composition thereof must be observed with various detectors concentrated in the central part. A typical arrangement for this purpose is shown in Fig. 5.4 (Fukui 60).

In this arrangement a dozen scintillation counters served to determine the size, the position of the axis, and the arrival direction as well. The lateral distribution was obtained with better accuracy because of the concentrated distribution of these detectors. The distribution in the core was observed with a neon hodoscope of large area. Muons were detected by a shielded hodoscope as well as by scintillation counters underground. Nuclear active particles were observed with a cloud chamber as well as with a transition meter in which N-cascades develop. Furthermore, total-absorption Čerenkov counters were employed for the measurement of the energy flow of the E-component.

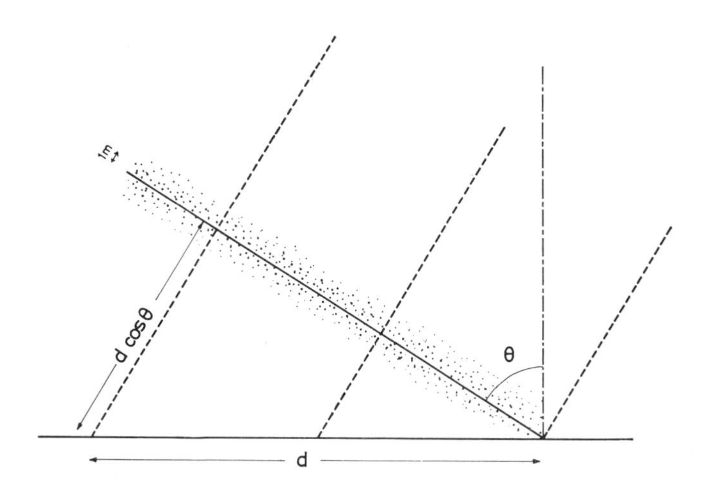

Fig. 5.3 The front structure of EAS (Bassi 53).

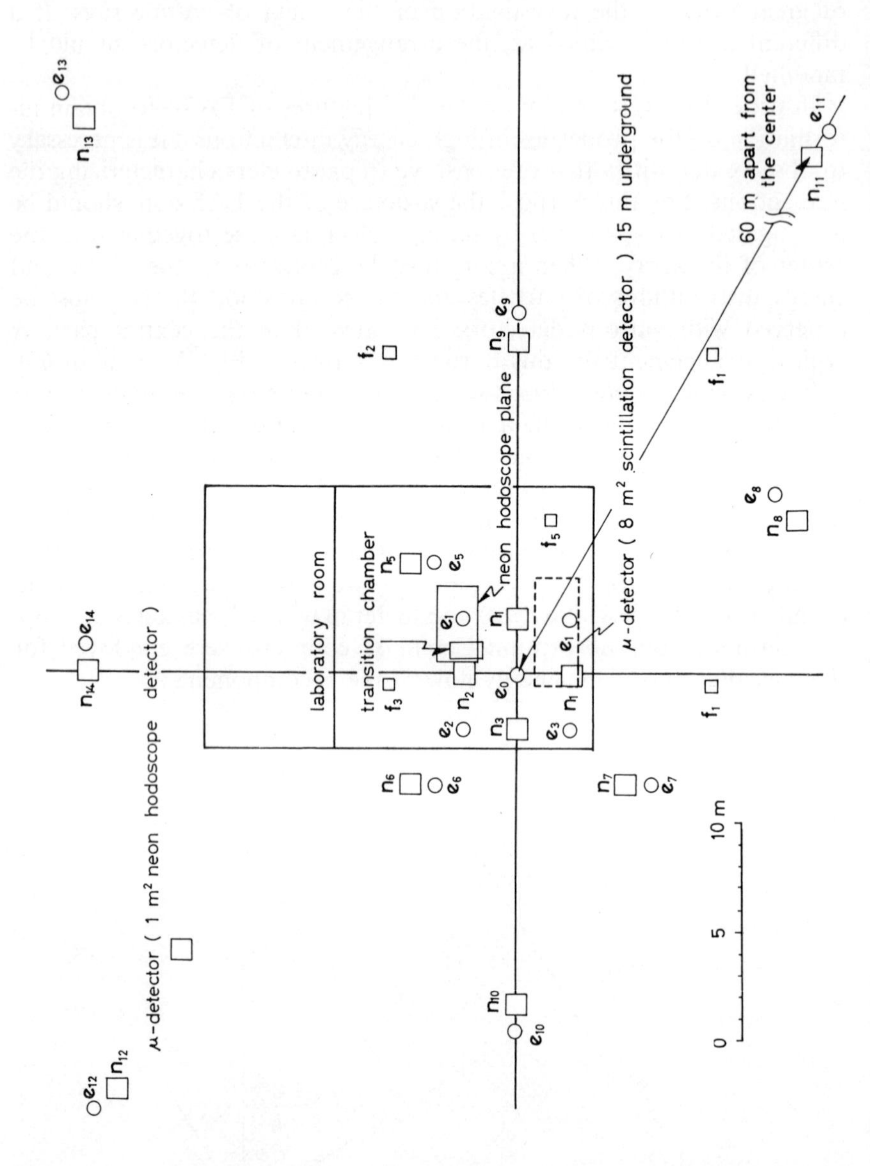

Fig. 5.4 Typical arrangement of detectors for investigating the structure of EAS (Fukui 60). Abbreviations: *n*—plastic scintillator for measuring the number of particles; *e*—lead-glass Čerenkov detector for measuring energy flow; *f*—plastic scintillator for fast timing. The transition chamber is a sandwich counter for detecting nuclear active particles.

5.3 Longitudinal Structure

5.3.1 Methods of Observation

The methods described above are concerned with the observation of a cross section of EAS. The longitudinal structure of EAS can be obtained by making observations at various atmospheric depths. This is possible with the same apparatus by observing EAS of different zenith angles because the air thickness traversed increases with zenith angle. For observations at small depths, however, it is necessary to compare observations at various mountain altitudes with different apparatus or to use airplanes with simple detector systems (Kraybill 49). Even with such methods results are available only at discrete levels.

The observation of Čerenkov light produced in the atmosphere by extremely relativistic particles in EAS provides a means of investigating the longitudinal structure of EAS (Chudakov 60). Čerenkov light is emitted mainly by electrons of several tens of MeV and greater, and its intensity is proprtional to the total energy dissipated in the atmosphere. An advantage of this method is a good angular resolution, suitable to the study of anisotropy of primary cosmic rays, if any.

New methods other than those described above are under development. If the shower size is very large—too large to be detected with conventional methods because of the very small frequency—the effective detecting area will be increased by observing scintillation light produced by EAS particles at high altitudes or the scattering of electromagnetic waves by an ionized column produced by EAS, as will be discussed in Section 5.8.5. The development of such new methods is motivated by the search for EAS of very large sizes that may be due to primary cosmic rays of energies greater than 10^{20} eV, if any.

5.3.2 Shower-Size-Fluctuations

The size of EAS depends at least on the primary energy E_0 and the atmospheric depth x, thus being represented by $N(E_0, x)$. This is not a definite function but is subject to considerable *fluctuations.* Among the various causes of fluctuations those of the first interaction point are considered to give rise to the largest effect (Kraushaar 58, Miyake 58), because in subsequent interactions the number of participating particles is so great that fluctuations in individual interactions are more or less compensated for.

Under this assumption the shower curve that starts at x_0 from a primary particle of energy E_0 has no fluctuations; the size is thus expressed as $N(E_0, x - x_0)$. An EAS of size N observed at x may be one that started at small x_0 with large E_0 or that started at large x_0 with small E_0, as shown in Fig. 5.5. The smallest energy, E_1, for a given size N is given by $N(E_1, x - x_0 = x_m) = N$, where x_m is the depth at which the size is maximum. The largest energy, E_2, is given by $N(E_2, x) = N$; namely, by EAS starting at $x_0 = 0$. The latter equation has two solutions, x_1 and x_2—one before the maximum and the other after.

Now let us consider EAS of varying primary energy, so that x_1 and x_2 are functions of E_0. Extensive air showers of sizes greater than N can be produced by primary particles with energies greater than E_1, provided that their starting points satisfy the following conditions. For $E_0 \geq E_2$ only an EAS with $x - x_0 < x_1(E_0)$ is of a size smaller than N. For $E_1 \leq E_0 \leq E_2$ only those with $x_1 < x - x_0 < x_2$ are of sizes greater than N. The probability that the first interaction takes place within this range is given by

$$\int_{x-x_2}^{x-x_1} e^{-x_0/l} \frac{dx_0}{l} = e^{-(x-x_2)/l} - e^{-(x-x_1)/l}.$$

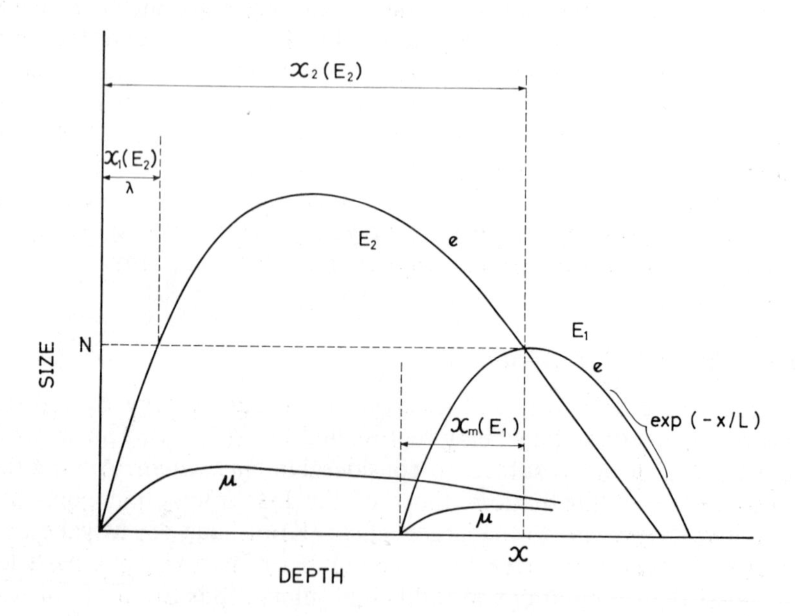

Fig. 5.5 Schematic shapes of shower curves in two limiting cases.

Thus the frequency of EAS of sizes greater than N is expressed by (Miyake 58)

$$\Phi(N) = \int_{E_2(N)}^{\infty} [1 - e^{-(x-x_1)/l}] j_0(E_0)\, dE_0$$

$$+ \int_{E_1(N)}^{E_2(N)} [e^{-(x-x_2)/l} - e^{-(x-x_1)/l}] j_0(E_0)\, dE_0, \qquad (5.3.1)$$

where $j_0(E_0)\, dE_0$ is the unidirectional differential energy spectrum of primary particles. The frequency without fluctuations is obtained by dropping the second term on the right-hand side of (5.3.1) and neglecting $\exp[-(x - x_1)/l]$ in comparison with unity. If ($\exp[-(x - x_1)/l]$) is neglected, the expression (5.3.1) is considerably simplified (Kraushaar 58).

In this approximation (5.3.1) can be readily integrated, under the assumption that

$$N(E_0, x) = N_0\left(\frac{E_0}{E_c}\right)^{\beta} \exp(-x/L) = N_c(x)\left(\frac{E_0}{E_c}\right)^{\beta}, \qquad (5.3.2)$$

$$j(E_0) = \frac{j_0\, \alpha(E_c/E_0)^{\alpha}}{E_0}, \qquad (5.3.3)$$

as

$$\Phi(N) = j_0\left(\frac{N_c}{N}\right)^{\alpha/\beta}\left\{1 + \frac{1}{1 - (\beta L/\alpha l)} \exp\left[\left(\frac{\alpha}{L\beta} - \frac{1}{l}\right)(x - x_m) - 1\right]\right\}. \qquad (5.3.4)$$

The contributions of the above two terms to the size spectrum are shown in Fig. 5.6. The second term in the braces represents the effect of fluctuations. The fluctuation effect is large if α/β is great compared with L/l. In this case large-size EAS are produced by those that start at great depths with small primary energies rather than those that start at the top with large primary energies.

The attenuation length of the shower frequency can be obtained by taking the logarithmic derivative of (5.3.1). The function $\Phi(N)$ appears to depend on x only through the factor $\exp(-x/l)$ and E_2. However, the dependence on E_2 does not appear because the two integrands on the right-hand side of (5.3.1) are equal at E_2. Hence the *frequency attenuation length* Λ is given by

$$-\frac{1}{\Lambda} = \frac{d \ln \Phi}{dx} = -\frac{1}{l}\left[1 - \frac{\int_{E_2}^{\infty} j_0(E_0)\, dE_0}{\Phi}\right]. \qquad (5.3.5)$$

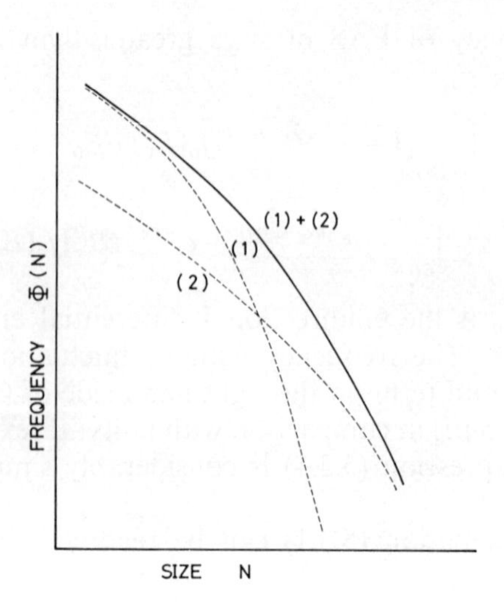

Fig. 5.6 Contributions of the two terms in (5.3.4) to the size spectrum:

$(1) = j_0(N_c/N)^{\alpha/\beta}$

$(2) = j_0(N_c/N)^{\alpha/\beta}\left\{\frac{1}{1-(\beta L/\alpha l)} \exp\left[\left(\frac{\alpha}{L\beta}-\frac{1}{l}\right)(x-x_m)-1\right]\right\}$

Under the same approximation that leads to (5.3.4), this is reduced to

$$\Lambda = l\left\{\frac{1+[1-(\beta L/\alpha l)]}{\exp\,[(\alpha/L\beta - 1/l)(x-x_m)-1]}\right\}. \tag{5.3.6}$$

If α/β is much smaller than L/l—namely, if the fluctuation effect is negligible—then (5.3.6) is further reduced to

$$\Lambda \simeq \frac{\beta}{\alpha} L. \tag{5.3.7}$$

This can also be derived directly if

$$\Phi(N) = \int_{E_2}^{\infty} j_0(E_0)\, dE_0 .$$

Hence the value of Λ lies in the range

$$l < \Lambda < \frac{\beta}{\alpha} L. \tag{5.3.8}$$

This indicates that the directly observable quantity Λ can be with difficulty related to the theoretically significant quantities.

If we know the value of Λ, however, the size spectrum can be expressed in a simple way. By substituting (5.3.6) into (5.3.4) we obtain

$$\Phi(N) = j_0 \left(\frac{N_c}{N}\right)^{\alpha/\beta} \frac{\Lambda}{\Lambda - l}. \tag{5.3.9}$$

This represents the way in which the shape of the size spectrum deviates from that of the primary energy spectrum. In the case of small fluctuations the value of Λ is nearly equal to $(\beta/\alpha)L$, independently of the size, so that the deviation arises only from the energy-size relation given by (5.3.2). If the primary spectrum is so steep that the fluctuations become important, Λ approaches l. As an extreme case we assume that the primary spectrum has a cutoff at E_c. Then the size spectrum beyond N_c falls off approximately as $(N_c/N)^{L/l}$, as shown in (5.3.15), until N reaches the maximum size corresponding to E_c.

5.3.3 *Analysis of Longitudinal Structure*

The above arguments are so general that they hold for any component in EAS. We shall hereafter specify the longitudinal structure that is applicable to the E- and μ-components. Since the μ-component has a large attenuation length, the fluctuation effect may be assumed to be negligible. By observing these two components, therefore, we can obtain a deeper insight into the longitudinal structure, separating out the fluctuation effect to some extent (Fukui 60).

The size-attenuation law that is given in (5.3.2) is specified as

$$N = N_m \exp\left[\frac{-(x - x_m - x_0)}{L}\right] \tag{5.3.10}$$

for x sufficiently larger than $x_0 + x_m$, where N_m is the size at the shower maximum; x_m is the atmospheric thickness in which a shower develops into maximum size and is related to the primary energy as follows:

$$x_m = B \ln \frac{E_0}{\varepsilon} + C, \tag{5.3.11}$$

where B and C are constants independent of E_0. The maximum size is proportional to the primary energy as follows:

$$N_m = QE_0. \tag{5.3.12}$$

The same relations apply for the μ-component but with different values of parameters, which are distinguished by the subscript μ. The trajectories on the N-N_μ plane represent the way in which EAS develop, as schematically drawn in Fig. 5.7.

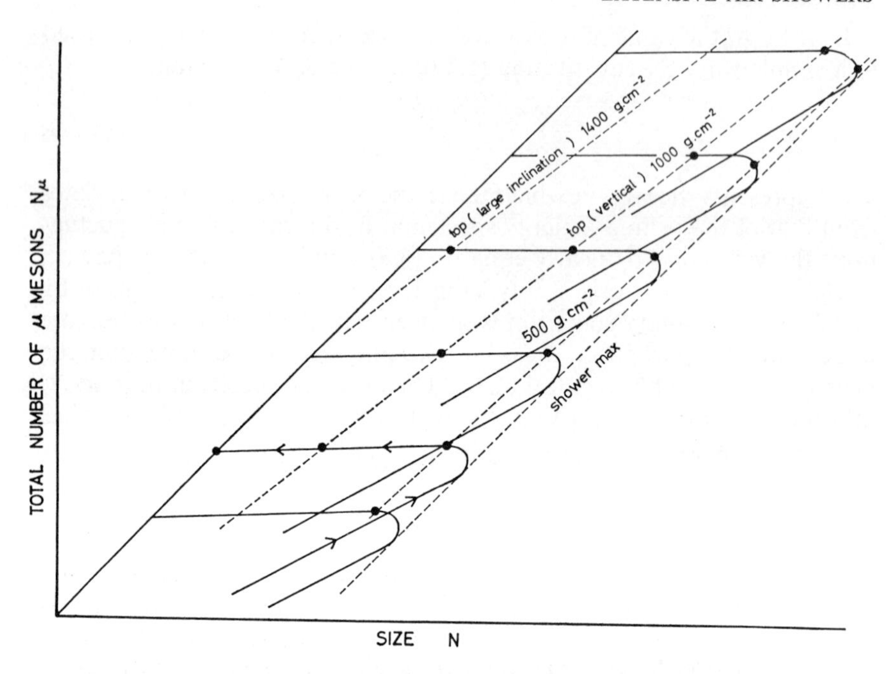

Fig. 5.7 N-N_μ diagram representing the development of EAS.

Muons of sufficiently high energies are produced mainly at high altitudes. Then the attenuation factor in (5.3.10) is nearly equal to unity (as shown in Fig. 5.5), so that

$$N_\mu \simeq N_{m\mu} \simeq Q_\mu E_0 . \tag{5.3.13}$$

Because of this relation, E_0 in (5.3.10) is eliminated, and x_0 is determined to be

$$x_0 = (x - C + L) \ln\left(\frac{N}{Q\varepsilon}\right) - (L + B) \ln\left(\frac{N_\mu}{Q_\mu \varepsilon}\right). \tag{5.3.14}$$

The frequency of EAS produced by primary particles with energy E_0 in dE_0 and at depth x_0 in dx_0 is given by

$$\begin{aligned} &j(E_0)\, dE_0 \exp\left(\frac{-x_0}{l}\right) \frac{dx_0}{l} \\ &\quad = j[E_0(N, N_\mu)] \exp\left[\frac{-x_0(N, N_\mu)}{l}\right] \frac{1}{l} \left|\frac{\partial(E_0, x_0)}{\partial(N, N_\mu)}\right| dN\, dN_\mu \\ &\quad \propto N^{-(1+L/l)}\, dN\, N_\mu^{-(\alpha+1-(L+B)/l)}\, dN_\mu . \end{aligned} \tag{5.3.15}$$

The limits of N and N_μ are determined by $x_0 \geq 0$ and $x \geq x_0 + x_m$.

These conditions are, because of (5.3.12), (5.3.13), and (5.3.14), reduced to

$$\varepsilon Q\left(\frac{N_\mu}{Q_\mu \varepsilon}\right)^{1+B/L} e^{-(x-C)/L} \leq N \leq \frac{Q}{Q_\mu} N_\mu \qquad (5.3.16a)$$

for fixed N_μ, and to

$$\frac{Q_\mu}{Q} N \leq N_\mu \leq Q_\mu \varepsilon\left(\frac{N}{Q\varepsilon}\right)^{L/(L+B)} e^{(x-C)/(L+B)} \qquad (5.3.16b)$$

for fixed N. By integrating (5.3.15) either over N_μ or N we obtain the size spectra of respective components as follows:

$$\phi(N)\, dN \propto N^{-(\alpha+1-B/l)}\, dN, \qquad (5.3.17a)$$

$$\phi(N_\mu)\, dN_\mu \propto N_\mu^{-(\alpha+1)}\, dN_\mu. \qquad (5.3.17b)$$

In this analysis it is important to notice that the size spectrum of one component for a fixed size of the other shows a characteristic dependence on zenith angle. Since x increases with zenith angle, the lower limit of the N-spectrum is shifted toward low N for inclined showers, whereas the upper limit of the N_μ-spectrum is shifted toward high N_μ, as illustrated in Fig. 5.8. It should, however, be remarked that at large zenith angles the effect of muon attenuation is no longer negligible, so that (5.3.13) is not valid any more. It is also important to notice that (5.3.15) and (5.3.17) enable us to determine four parameters; namely, α, l, L, and B.

The above analysis takes part also in the attenuation of frequency. The frequency at zenith angle θ is related to the vertical frequency Φ_v according to (5.3.5) as follows:

$$\Phi(\theta) = \Phi_v \exp\left[\frac{-(x-x_v)}{\Lambda}\right], \qquad (5.3.18)$$

where x_v is vertical depth and x is depth along the inclined direction; their relationship is $x/x_v = \sec\theta$. The relation (5.3.18) is well verified by experiment (Fukui 60), as shown in Fig. 5.9. The same relation holds also for the differential frequency $\phi(N)\, dN$ if suitable correction is made for the triggering efficiency, which depends on θ (Clark 58). This is nothing but the comparison of two areas, 1 and 2 in Fig. 5.8*b*, because they are the shower frequencies of given N integrated over N_μ-distributions for two different zenith angles. If, however, they are compared for the same N_μ-range, we are observing EAS in a corresponding range of primary energies, and the difference between their frequencies represents the difference in probability for the primary particles

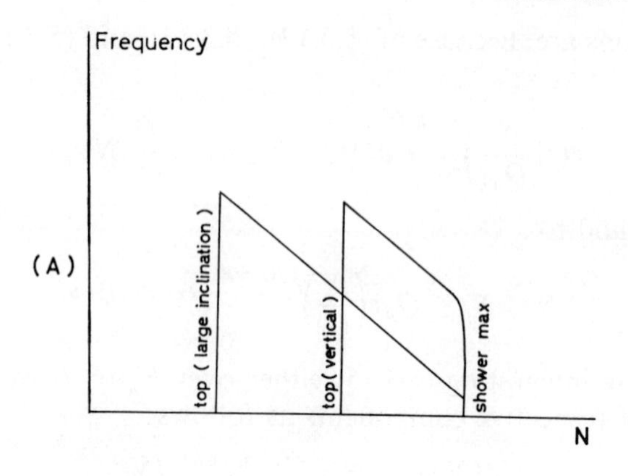

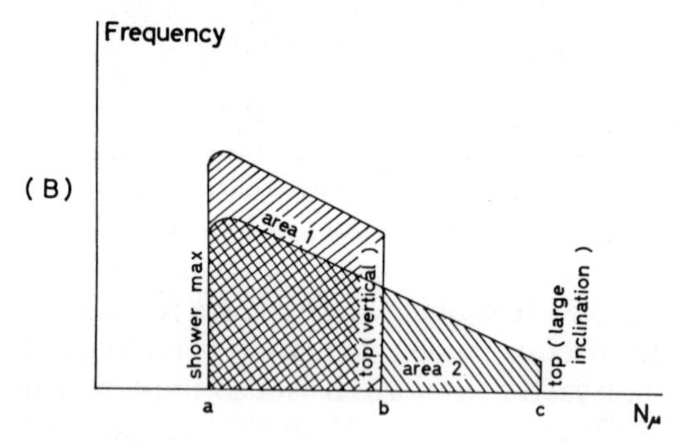

Fig. 5.8 N- and N_μ-spectra for a fixed size of the other component:

(*a*) N_μ fixed, $N^{-(1+L/l)}$

(*b*) N fixed, $N_\mu^{-(\alpha+1-(L+B)/l)}$,

$$\frac{\text{area 1}}{\text{area 2}} = e^{-x_0/L}\,\frac{\text{cross-hatched region}}{\text{area 1}}$$
$$= e^{-x_0/l}.$$

of the same energy to initiate EAS at two different depths but with the same path length in the atmosphere, as schematically shown in Fig. 5.10. From the zenith-angle dependence of the shower frequency shown in Fig. 5.11, therefore, we can obtain the interaction mean free path l as

$$l = \frac{x_{02} - x_{01}}{\ln(\phi_1/\phi_2)}. \tag{5.3.19}$$

Having obtained the value of l, we can get the value of L from the N-distribution with fixed N_μ, shown in Fig. 5.12. The value of B is

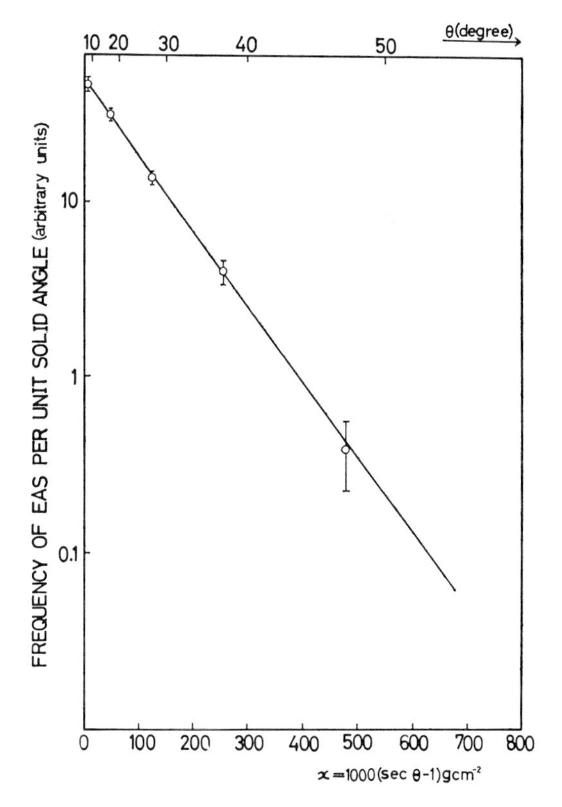

Fig. 5.9 Zenith-angle distribution of shower frequency per unit solid angle of EAS with $N \geq 10^5$ (Fukui 60).

most directly obtained from the upper edge of the distribution of N_μ against N. This can be expressed according to (5.3.16*b*) as

$$N \propto N_\mu^{1+B/L}. \tag{5.3.20}$$

This relation is indicated by dashed lines in Fig. 5.7.

The experimental values thus obtained are summarized as follows:

$$\begin{aligned}
\Lambda &= 102 \pm 15 \text{ g-cm}^{-2} \quad &&\text{for} \quad N \geq 10^5, \\
l &= 90 \pm 15 \text{ g-cm}^{-2} &&\text{for} \quad 10^5 \lesssim N \lesssim 10^6, \quad 10^3 \lesssim N_\mu \lesssim 10^4, \\
1 + \frac{L}{l} &= 2.75 \pm 0.25 &&\text{for} \quad N_\mu \simeq 10^4, \\
1 + \frac{B}{L} &= 1.3 \pm 0.1
\end{aligned} \tag{5.3.21}$$

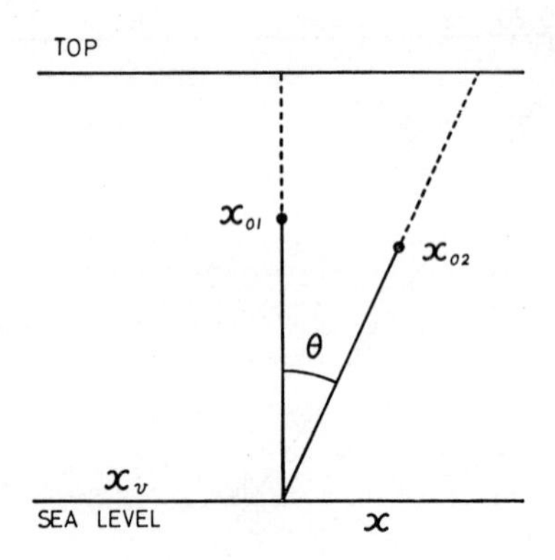

Fig. 5.10 Extensive air showers of the same size with different zenith angles; $x_v - x_{01} = x - x_{02}$.

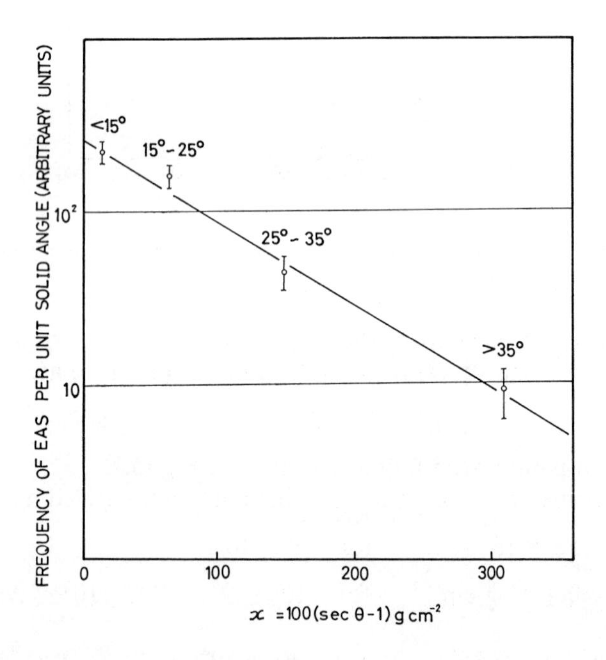

Fig. 5.11 Zenith-angle distribution of shower frequency per unit solid angle of EAS with fixed N and N_μ. The solid line represents exp $(-x/l)$ with $l = 90$ g-cm^{-2} (Fukui 60).

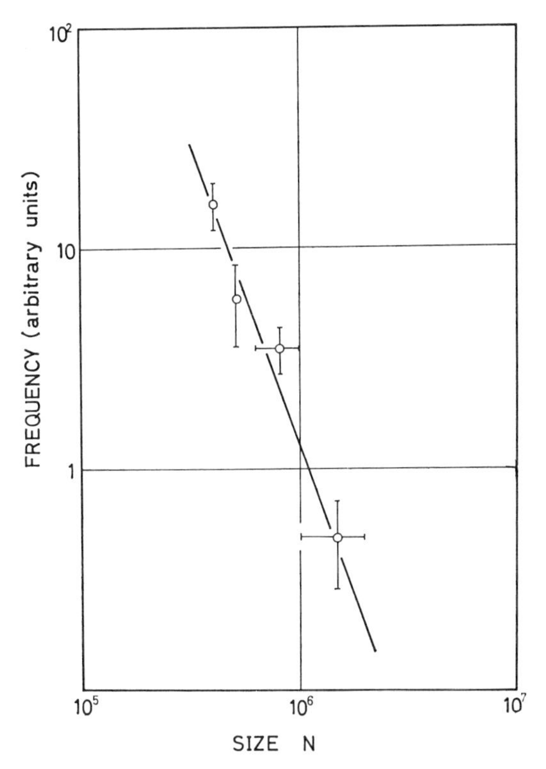

Fig. 5.12 N-spectrum with fixed N_μ (Fukui 60). The solid line represents $N^{-(1+L/l)}$.

From these relations we can deduce that

$$\begin{aligned} L &= 158 \pm 35 \text{ g-cm}^{-2}, \\ B &= 47 \pm 19 \text{ g-cm}^{-2}, \end{aligned} \tag{5.3.22}$$

The internal consistency of this analysis can be verified by the N-spectrum without fixing N_μ, and by the N_μ-spectrum with fixed N. The former has a break at about $N = 10^6$; whereas the latter is nearly flat, as shown in Fig. 5.13.

The experimental values of the power indices are estimated to be

$$\begin{aligned} \alpha - \frac{B}{l} &= 1.4 \pm 0.1 \quad \text{for} \quad N < 10^6 \\ &= 2.0 \pm 0.2 \quad \text{for} \quad N > 10^6 \end{aligned} \tag{5.3.23}$$

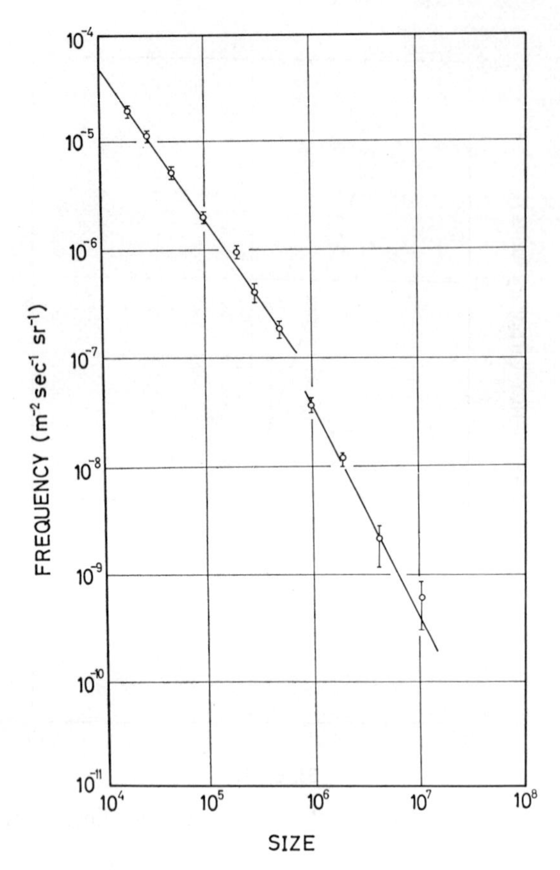

Fig. 5.13 N-spectrum with N_μ not fixed. The solid lines represent power laws of the integral exponents of 1.4 ± 0.1 and 2.0 ± 0.2 in a lower and a higher size regions, respectively (Fukui 60).

for the N-spectrum, and

$$\alpha - \frac{(L+B)}{l} = -1.0 \pm 0.3 \tag{5.3.24}$$

for the N_μ-spectrum. The value of L/l given by these relationships is too large in comparison with the one given in (5.3.21). Moreover, the N_μ-distribution with fixed N becomes flatter as N increases, which is the opposite of the trend expected from (5.3.20).

These observations indicate that the model adopted here is too simple. A considerable error seems to arise from neglecting the attenuation of muon size and consequently its fluctuations. It is quite possible that both L and B are larger than those given in (5.3.22).

Nevertheless, the values of parameters obtained above seem to give the right orders of magnitude and may be referred to in further discussions. Firstly, the value of l is essentially equal to the geometrical mean free path for the nuclear radius of $1.2 \times 10^{-13} A^{1/3}$ cm and is consistent with that given in Section 3.3 on the basis of low-energy experiments. Secondly, the fact that B is appreciably smaller than l implies that the development of EAS to the maximum is due to the mixture of electron and nuclear cascades because, if either one of them plays a main role, the value of B is equal to the mean free path of either process; that is, 37 or 90 g-cm^{-2}.

5.3.4 *Attenuation of EAS in the Atmosphere*

The preceding discussions apply for a modern method by which both the size and the incident direction can be measured. With the classical method, however, the density of EAS is measured without knowing the incident direction. This requires a complicated analysis to obtain the vertical frequency.

The omnidirectional frequency at the vertical depth x, $\overline{\Phi}(x)$, is related to the unidirectional frequency $\Phi(x, \theta)$ as follows:

$$\overline{\Phi}(x) = 2\pi \int_0^{\pi/2} \Phi(x, \theta) \sin \theta \, d\theta, \tag{5.3.25}$$

where $\Phi(x, \theta)$ is expressed approximately by

$$\Phi(x, \theta) = \Phi\left(\frac{x}{\cos \theta}, 0\right) = \cos^n \theta \, \Phi(x, 0), \tag{5.3.26}$$

which is the zenith-angle distribution. By introducing (5.3.26) into (5.3.25) we obtain

$$\Phi(x, 0) = \frac{1}{2\pi} \left[(n+1)\overline{\Phi}(x) - x \frac{d\overline{\Phi}}{dx}\right]. \tag{5.3.27}$$

This is the *Gross transformation* for EAS.

Now we need to find out only the value of n. The relation (5.3.26) represents the relation between the unidirectional frequency at the vertical x for the zenith angle θ and the vertical frequency at the vertical depth $x/\cos \theta$. The change of the vertical depth from x to $x/\cos \theta$ results in the increase of the air density by a factor of $\cos \theta$. As a consequence the lateral spread of particles decreases, whereas the area of detectors remains as it is. Looking at this effect in relation to EAS, we observe the same shower with large detectors. The shrinkage of the lateral spread results in the density to be observed increasing as $\cos^{-2a} \theta$, where a is nearly equal to unity. Then the frequency of observing densities greater than a given value of Δ increases as $\cos^{-2ak}\theta$, where k is the power index of the integral density spectrum. On the other hand, the decrease of the lateral spread decreases the area on which the core falls, thus reducing the frequency by a factor of $\cos^{2a}\theta$. The third effect is the change in the effective area of detectors for an inclined direction. The effective area changes as $\cos^{b}\theta$, where $b=0$ for isotropic detectors, $b=1$ for perfectly flat ones, and $b=\frac{1}{2}$ for separated cylindrical detectors of infinite length. This leads to an increase in frequency by $\cos^{-bk}\theta$. Finally, the separation between detectors normal to the shower axis depends on θ. From the decoherence curve this gives a contribution of $\cos^{c}\theta$ to the frequency. Summing up all the above effects, we have

$$n = 2a(k-1) + bk - c. \tag{5.3.28}$$

For most of the detector systems used in classical experiments the values of a, b, and c are estimated to be (Greisen 56)

$$a = 0.9, \qquad b = 0.35, \qquad c = 0.2.$$

A systematic measurement of the altitude variation was performed with a triangular arrangement of Geiger-Müller counter trays (Imai 61). The result, supplemented by earlier data, is shown in Fig. 5.14. Two curves for the average densities of 45 and 180 per square meter are nearly parallel. The attenuation lengths of omnidirectional EAS for $x < 500$ g-cm^{-2} are obtained as

$$\begin{aligned} \bar{\Lambda} &= 120 \pm 4 \text{ g-cm}^{-2} \quad \text{for} \quad \Delta = 45 \text{ m}^{-2}, \\ &= 136 \pm 10 \text{ g-cm}^{-2} \quad \text{for} \quad \Delta = 180 \text{ m}^{-2}. \end{aligned} \tag{5.3.29}$$

The Gross transformations for a curve of $\Delta = 45$ m^{-2} are shown by the dotted curves in Fig. 5.14, with $k = 1.5$ and 1.7. It is worthwhile to note that the altitude variations of the omnidirectional and the vertical frequencies are nearly the same for $x \gtrsim 700$ g-cm^{-2}. Therefore the value of $\bar{\Lambda}$ given in (5.3.29) can be regarded as being equal to the frequency-attenuation length.

From the observed altitude variation it is possible to obtain the zenith-angle distribution from (5.3.26). The distributions that are expected from the altitude variation are compared with experimental results, obtained with the fast-timing method at sea level (Bassi 53), cloud chambers at mountain altitudes, and discharge chambers at an airplane altitude, as shown in Fig. 5.15. The good agreement at all altitudes verifies the correctness of the altitude variation and the value of n given by (5.3.28).

The altitude variation is revealed also by the barometric effect. The barometric coefficients measured by many authors distribute about

$$\beta = -11\% \text{ (cm Hg)}^{-1} \tag{5.3.30}$$

for sizes between 10^4 and 10^6 at sea level as well as at mountain altitudes (Greisen 56, Bennet 62).

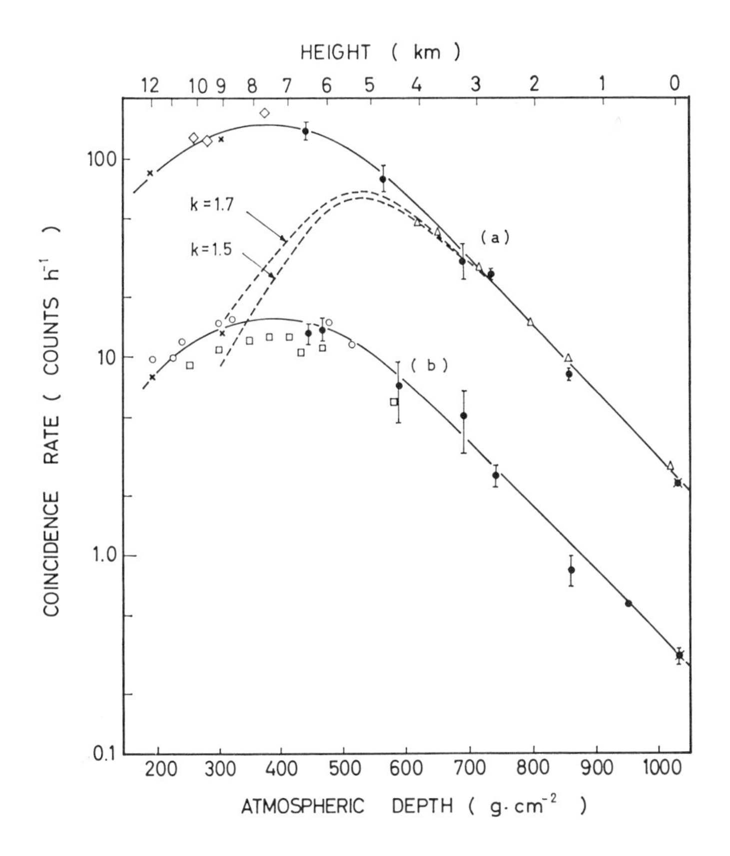

Fig. 5.14 Altitude variations of EAS frequencies. ●—T. Imai et al., *Sci. Pap. IPCR*, **55**, 42 (1961); △—N. Hilberry, *Phys. Rev.*, **60**, 1 (1941); ○—H. L. Kraybill, *Phys. Rev.*, **76**, 1092 (1949); □—A. L. Hodson, *Proc. Phys. Soc.* (London), **A66**, 49 (1953); ◇—H. L. Kraybill, *Phys. Rev.*, **93**, 1360, 1362 (1954); ×—R. A. Antonov et al., *Proc. Moscow Conf. on Cosmic Rays*, **II**, 96 (1960).
The density Δ and the frequency attenuation length $\overline{\Lambda}$ are related as follows:

(*a*) $\Delta = 45 \text{ m}^{-2}$, $\overline{\Lambda} = 120 \pm 4 \text{ g-cm}^{-2}$

(*b*) $\Delta = 180 \text{ m}^{-2}$, $\overline{\Lambda} = 136 \pm 10 \text{ g-cm}^{-2}$

The dashed curves represent the Gross transformations of (*a*) with $k = 1.5$ and 1.7.

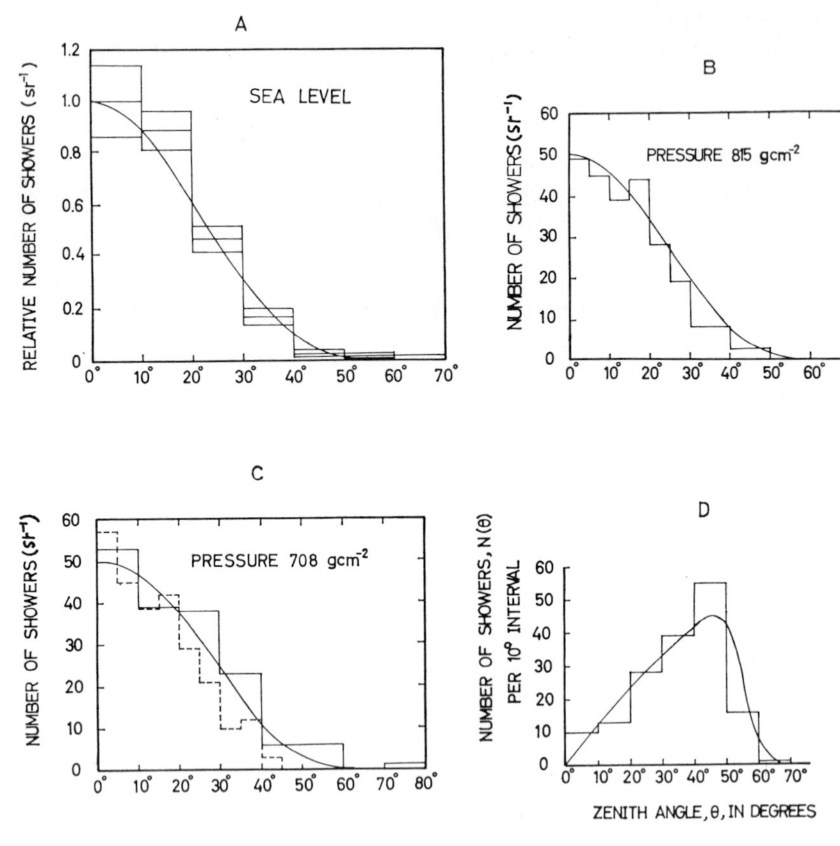

Fig. 5.15 Zenith-angle distributions. A. Sea level, $\Delta \gtrsim 20\ \text{m}^{-2}$, P. Bassi, G. Clark, and B. Rossi, *Phys. Rev.*, **92**, 441 (1953); B. 815 g-cm^{-2}, $\Delta \gtrsim 400\ \text{m}^{-2}$, J. Daudin, *J. Phys. Radium*, **6**, 302 (1945); C. 708 g-cm^{-2}; solid-line histogram: $\Delta \gtrsim 300\ \text{m}^{-2}$, W. W. Brown and A. S. MacKay, *Phys. Rev.*, **76**, 1034 (1949); dashed-line histogram: $\Delta \gtrsim 200\ \text{m}^{-2}$, W. E. Hazen, R. W. Williams, and C. A. Randall, *Phys. Rev.*, **93**, 578 (1954); D. 330 g-cm^{-2}, $\Delta \gtrsim 20\ \text{m}^{-2}$, K. Kamata, *Sci. Pap. IPCR*, **55**, 130 (1961).

A summary of experimental values of the frequency-attenuation length is given in Table 5.1.

5.4 Size Distribution

As shown in (5.2.5), the density spectrum of EAS is represented by a power law. However, the power index k is not strictly constant but increases slowly with density; it is expressed by

$$k = a + b \ln\left(\frac{\Delta}{\Delta_0}\right) \quad \text{for} \quad 1 < \Delta < 10^4 \text{ m}^{-2}, \qquad \Delta_0 = 1 \text{ m}^{-2}. \tag{5.4.1}$$

Thus the integral density spectrum is represented by

$$H(\Delta) = h_0 \left(\frac{\Delta_0}{\Delta}\right)^{a+(1/2)b \ln(\Delta/\Delta_0)} \tag{5.4.2}$$

Table 5.1 Frequency Attenuation Length

Size	Altitude	Λ(g-cm^{-2})	Method	Reference
10^4 to 10^6	Sea level	120	Barometric effect	Greison 56 Bennet 62
$\sim 5 \times 10^5$	500 to 1000 g-cm^{-2}	120 ± 4	Altitude variation	Imai 61
$\sim 2 \times 10^6$	500 to 1000 g-cm^{-2}	136 ± 10	Altitude variation	Imai 61
10^5 to 10^6	Sea level	102 ± 15	Zenith angle	Fukui 60
10^6 to 10^8	460 meters	107 ± 11	Zenith angle	Clark 60
1×10^7 to 3×10^7	Sea level	111 ± 7	Zenith angle	Bennet 62
3×10^7 to 1×10^8	Sea level	137 ± 13	Zenith angle	Bennet 62
5×10^7	820 g-cm^{-2}	140	Zenith angle	Linsley 62
1×10^8	820 g-cm^{-2}	170	Zenith angle	Linsley 62
2×10^8	820 g-cm^{-2}	210	Zenith angle	Linsley 62
5×10^5 to 2×10^6	530 g-cm^{-2}	97 ± 13	Zenith angle	Suga 63

The values of a, b, and h_0 obtained at sea level and at mountain altitudes of about 700 g-cm^{-2} are shown in Table 5.2 (Greisen 56).

Table 5.2 Values of the Parameters in the Density Spectrum

Altitude	a	b	h_0(sec^{-1})
Sea level	1.32	0.038	0.210
$\sim$700 g-cm^{-2}	1.26	0.048	1.75

On the basis of the density spectrum (5.4.2), the integral size spectrum is expressed as follows:

$$\Phi(N) = C\left(\frac{N_0}{N}\right)^{-[A+(1/2)B \ln(N/N_0)]} \tag{5.4.3}$$

For $N_0 = 10^6$ and $10^4 < N < 3 \times 10^7$ the values of A, B, and C are given in Table 5.3 (Greisen 56). For vertical EAS, C should be replaced by C_v.

Table 5.3 Values of the Parameters in the Size Spectrum

Altitude	A	B	$C(\text{m}^{-2}\ \text{sec}^{-1})$	$C_v(\text{m}^{-2}\ \text{sec}^{-1}\ \text{sr}^{-1})$
Sea level	1.53	0.039	3.5×10^{-8}	5.5×10^{-8}
~700 g-cm^{-2}	1.50	0.050	5.3×10^{-7}	6.0×10^{-7}

The size spectrum (5.4.3) can be approximated by a single power law at about $N = N_1$ as follows:

$$\Phi(N) \simeq C\left(\frac{N_1}{N_0}\right)^{(1/2)B\ln(N_1/N_0)}\left(\frac{N}{N_0}\right)^{-\gamma}, \qquad \gamma = A + B\ln\left(\frac{N_1}{N_0}\right) \tag{5.4.4}$$

If the power-law fit is made for $N < 10^6$ and $N > 10^6$ separately, the values of γ are found to be (Kraushaar 58, Fukui 60)

$$\begin{aligned} \gamma &= 1.4 \pm 0.1 \quad \text{for} \quad N < 10^6, \\ &= 1.9 \pm 0.2 \quad \text{for} \quad N > 10^6. \end{aligned} \tag{5.4.5}$$

For $N < 10^6$ the value of γ is in essential agreement with (5.4.4) and with the values of A and B in Table 5.3, whereas for $N > 10^6$ the value of γ is appreciably greater. A number of representative experimental results for the size spectrum are summarized in Table 5.4 and Fig. 5.16.

Table 5.4 Size Spectrum

Altitude	Size	γ	$\Phi(N)(10^6/N)^{-\gamma}$ $(\text{m}^{-2}\ \text{sec}^{-1}\ \text{sr}^{-1})$	Reference
Sea level	10^4 to 10^6	1.4 ± 0.1	7×10^{-8}	1
Sea level	10^6 to 10^7	2.0 ± 0.2	5×10^{-8}	1
Sea level	10^6 to 10^8	1.90 ± 0.10	$(3.48 \pm 0.53) \times 10^{-8}$	2
Sea level	3×10^6 to 3×10^8	1.88 ± 0.15	2.6×10^{-8}	3
Sea level	6×10^6 to 6×10^9	1.84 ± 0.06	$(3.58 \pm 0.28) \times 10^{-8}$	4
820 g-cm^{-2}	5×10^7 to 10^{10}	1.7	$\sim 3 \times 10^{-8}$	5

References:

1. S. Fukui et al., *Prog. Theor. Suppl.* No. 16, 1 (1960).
2. G. Clark et al., *Phys. Rev.*, **122**, 637 (1961).
3. T. E. Cranshaw et al., *Phil. Mag.*, **3**, 377 (1958).
4. J. Delvaille et al., *J. Phys. Soc. Japan*, **17**, Suppl. A–III, 76 (1962).
5. J. Linsley, *Proc. Int. Conf. on Cosmic Rays at Jaipur*, **4**, 77 (1963).

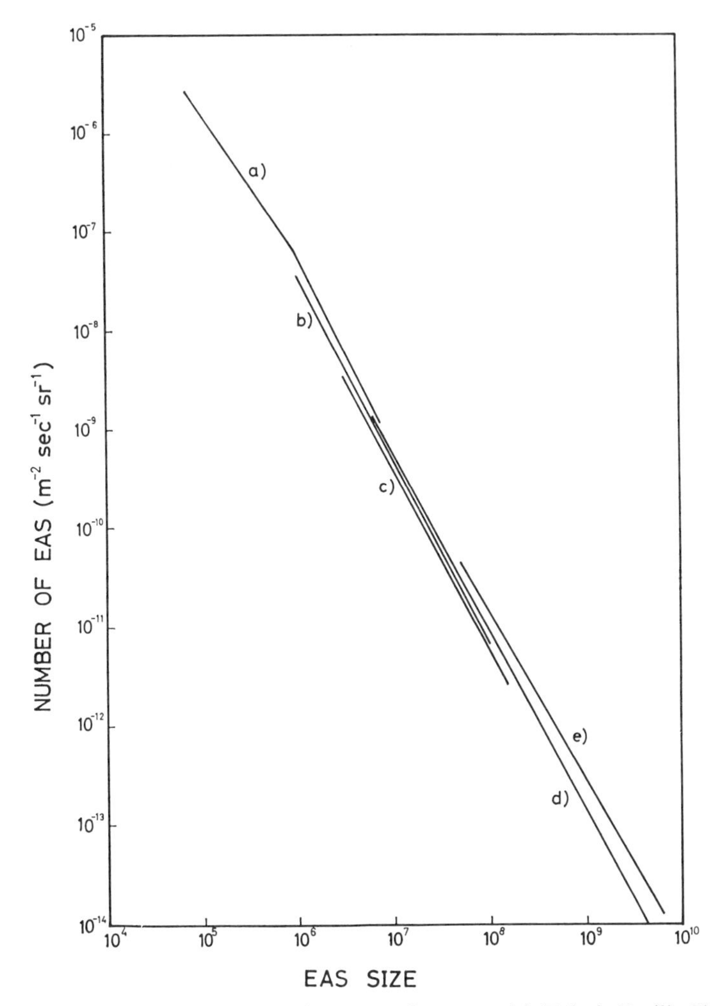

Fig. 5.16 Integral size spectrum of EAS. References: (*a*) Fukui 60, (*b*) Clark 61, (*c*) Cranshaw 58, (*d*) Delvaille 62, (*e*) Linsley 63.

It should, however, be remarked that the estimation of size is not always free from systematic error. The size could be overestimated at large N, so that the value of γ for $N > 10^8$ would be reduced to 1.7 in some experiments (Delvaille 62). If this is the case, the size spectrum is flattened at large sizes after the steepening for N between 10^6 and 10^8.

It has been conventionally assumed that the primary energy spectrum can be obtained from the size spectrum at sea level by putting

$$E_0 \simeq QN \tag{5.4.6}$$

where $Q \simeq 10^{10}$ eV for N of about 10^5. However, the linearity does not necessarily hold at a given level. The linear relation may be acceptable for EAS of the maximum development. This is the case for $N > 10^8$, and the value of Q is approximately 2×10^9 eV.

5.5 Electronic Component

5.5.1 A Qualitative Description of Shower Theory

The behavior of electrons and photons in the atmosphere is well described in terms of cascade shower theory (Kamata 58, Nishimura 67). In this theory the thickness of air is measured in units of the *radiation length* X_0 and the energy in units of the *critical energy* ε_0. It is also convenient to measure the lateral spread in *Molière units*:

$$r_M \equiv \frac{E_s}{\varepsilon_0} X_0, \tag{5.5.1}$$

where $E_s = 21$ MeV is the characteristic energy for multiple scattering. It is more accurate to replace E_s by the *scattering constant* $K = 19.3$ Mev, but we keep the traditional definition (5.5.1) because the difference between E_s and K is small.

The numerical values of these units are given for air as

$$\begin{aligned} X_0 &= 37.1 \text{ g-cm}^{-2} \qquad (287\, P^{-1}[T/273]) \text{ m}, \\ \varepsilon_0 &= 81 \text{ MeV}, \\ r_M &= 9.6 \text{ g-cm}^{-2} \qquad (74.3\, P^{-1}[T/273]) \text{ m}, \end{aligned} \tag{5.5.2}$$

where P is the pressure in atmospheres and T is the temperature in degrees Kelvin.

Let us consider a cascade shower initiated by a photon of energy W_0. The number of electrons with energies greater than E at depth t is approximately expressed, as also described in 2.6.1, by (Greisen 56)

$$N(W_0, E, t) \simeq \frac{0.135}{y^{1/2}} \exp\,[\lambda_1(s)t + sy], \qquad y \equiv \ln\left(\frac{W_0}{E}\right), \tag{5.5.3}$$

where s is the *shower age* and s and t are related as given in (2.6.13) and

$$t \simeq \frac{2ys}{(3 - s)}. \tag{5.5.4}$$

Other relations are obtained by changing E_0 in 2.6.1 into W_0

In actual EAS photons are produced by the N-component with the generation spectrum $g(W_0, t)\,dW_0$. Hence the number of electrons is given by

$$N_e(t) = \int dW_0 \int_0^t dt'\, g(W_0, t')N(W_0, t-t')$$
$$\equiv f(s)e^{\lambda_1(s)t}\, G(s, t), \tag{5.5.5}$$

where

$$G(s, t) = \int dW_0 \int_0^t dt'\, g(W_0, t') \exp\,[-\lambda_1(s)t' + sy]. \tag{5.5.6}$$

In the approximation leading to (5.5.3) and (2.6.16) the function $f(s)$ is independent of s. Therefore the age parameter is determined by

$$\lambda_1'(s)t + \frac{\partial \ln G(s, t)}{\partial s} = 0. \tag{5.5.7}$$

The size-attenuation length is given by

$$-\frac{1}{L} = \lambda_1(s) + \frac{\partial \ln G(s, t)}{\partial t}, \tag{5.5.8}$$

with the value of s obtained from (5.5.10).

In order to derive an accurate value of s as well as of L, we need a detailed knowledge of the N-component. However, an approximate value of s can be obtained by taking into account the fact that $\partial \ln G/\partial t$ in (5.5.8) is positive. This implies that

$$\lambda_1(s) \leq -\frac{1}{L} \quad \text{or} \quad -\lambda_1(s) \geq \frac{1}{L}. \tag{5.5.9}$$

Since the observed value of L is about 200 g-cm^{-2} or smaller, (5.5.9) gives

$$-\lambda_1(s) \gtrsim 0.18 \quad \text{or} \quad s \gtrsim 1.2. \tag{5.5.10}$$

This demonstrates that the main contribution to EAS comes from those cascade showers that have passed their maxima; thus the electrons in an EAS result from the sum of the electrons in small cascade showers, as schematically shown in Fig. 5.17. The envelope of individual cascade showers forms the electron number versus depth curve and evidently passes through points past the maxima of the cascade showers in the declining phase of the EAS.

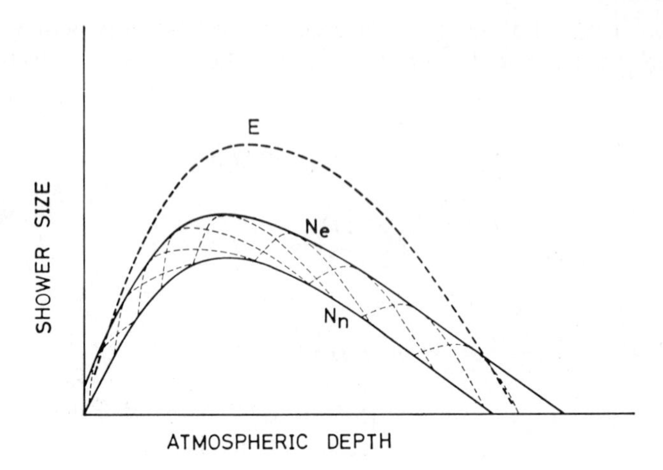

Fig. 5.17 Schematic view of EAS development. N_n: N-component; N_e: E-component, the sum of many electronic showers indicated by dashed curves; E: an electronic shower initiated by a high-energy photon.

In such a case we may evaluate (5.5.5) in an approximate way as

$$N_e(t) = \int dW_0 \int_0^t dt' \, g(W_0, t-t')C(W_0, t')$$

$$\simeq \int dW_0 \, g(W_0, t-\bar{t})Z(W_0), \tag{5.5.11}$$

where

$$Z(W_0) = \int_0^\infty C(W_0, t) \, dt = \frac{W_0}{\varepsilon_0}. \tag{5.5.12}$$

Here $\bar{t}$ is the depth at which the contribution is maximum and may be given, because of the value of s in (5.5.10) and (5.5.4), by

$$\bar{t} \simeq \tfrac{3}{2}y. \tag{5.5.13}$$

The approximation leading to (5.5.11) implies that the E-component is in equilibrium with the N-component, so that the size-attenuation length of EAS is equal to that of the N-component.

In order to make this approximation valid, the value of $\bar{t}$ should not be too large—say, $\bar{t} \lesssim 10$. Then $Z(W_0)$ is as small as 10^3. However, there is an appreciable probability that photons of $W_0 \gtrsim 10^{14}$ eV are produced. Such photons produce large cascade showers that are near the maxima in an observing level; for example, a photon of $W_0 = 10^{14}$ eV starting at $t = 5$ gives 10^5 electrons at the maximum, $t \simeq 19$.

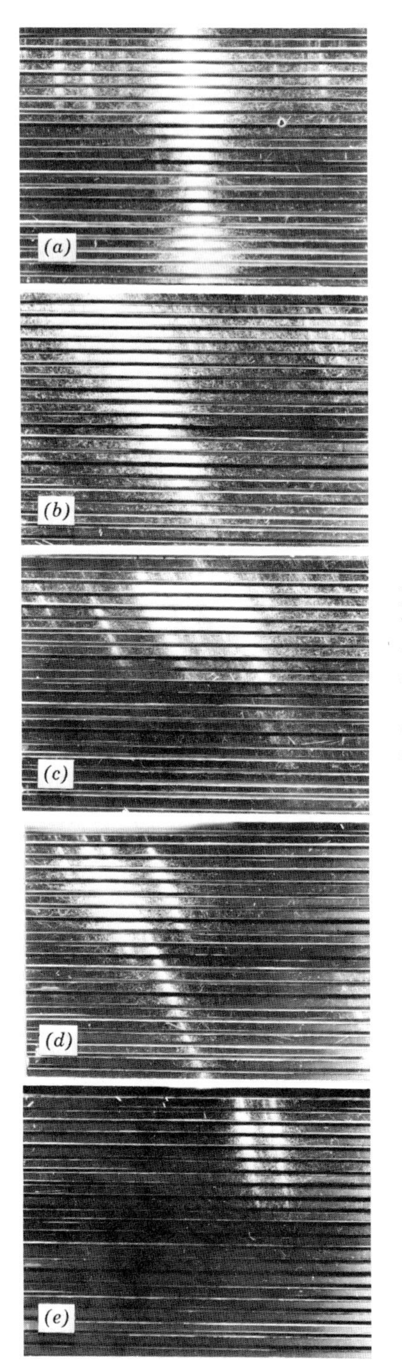

Plate 4. Core structures of EAS observed with a cloud chamber (200 by 130 by 65 cm^3) at Mount Norikura. These cores are associated with EAS of sizes around 10^5 and of different ages: (*a*) $s \simeq 0.4$; (*b*) $s \simeq 0.6$; (*c*) $s \simeq 0.8$; (*d*) $s \simeq 1.4$; (*e*) $s \simeq 1.6$. Courtesy of S. Miyake. See Chapter 5, Section 5.2.

Thus at a mountain altitude (say, $t = 20$), an appreciable fraction of EAS contain such large cascade showers, and their ages may be smaller than unity.

5.5.2 *Lateral Structure and Its Age Dependence*

The determination of age may be made by measuring the energy spectrum of electrons. As seen from (5.5.3) the integral energy spectrum of electrons in a single cascade is given by $(W_0/E)^s$. In studying the energy spectrum it is important to note that the energy of an electron depends strongly on the distance from the axis, because the lateral spread increases as the energy of an electron decreases. This indicates that the lateral distribution depends on age.

According to the theory of multiple scattering the rms angle of deflection in one radiation length is given approximately by $\theta \simeq K/E$, and consequently the rms lateral spread is

$$r \simeq \left(\frac{K}{E}\right) X_0 \simeq \left(\frac{\varepsilon_0}{E}\right) r_{\mathrm{M}}. \tag{5.5.14}$$

By substituting E thus related to r in y given in (5.5.3) we can modify (5.5.4) to

$$s \simeq \frac{3t}{t + 2y_0 + 2x}, \qquad y_0 \equiv \ln\left(\frac{W_0}{\varepsilon_0}\right), \qquad x = \ln\left(\frac{r}{r_{\mathrm{M}}}\right). \tag{5.5.15}$$

This shows that age depends on distance from the axis. In practice, however, the r-dependence of s is rather weak for $-1 < x < 1$ because $t + 2y_0$ is significantly greater than 10. Only at very small distances from the axis is the r-dependence important.

If combined with the energy spectrum, the relation (5.5.14) gives an approximate form of the lateral distribution. The number of electrons with energies between E and $E + dE$ is proportional to $E^{-s-1}\, dE$. This is transformed into the lateral distribution as

$$E^{-s-1}\, dE \propto r^{-(2-s)} r\, dr. \tag{5.5.16}$$

At the shower maximum therefore, the lateral distribution is represented as $1/r$ near $r = r_M$; it is less steep at small distances but steeper at large distances. For $r > r_{\mathrm{M}}$ the energy of electrons is smaller than ε_0, so that they cannot travel through one radiation length. This results in a distribution that falls off more steeply than that given in (5.5.16). An accurate expression of the lateral distribution was calculated by Nishimura and Kamata (Nishimura 51, 52; Kamata 58), and is called the *N-K* function. Leaving detailed discussions on the lateral-structure function to

Appendix C, we here give an approximate expression of the N-K function introduced by Greisen (Greisen 56), which is called the N-K-G function. Its expression and numerical values are given in (2.8.21) and Table 2.10, respectively.

In most experiments the lateral distribution is measured at distances between a few meters and about a hundred meters. Since $x \lesssim 1$ in this region, the distribution is essentially represented by the first factor, x^{s-2}, in (2.6.21). At distances smaller than several meters the value of s found at $x \sim 1$ no longer holds. It should be increased to an appreciable extent because of the x-dependence of s given in (5.5.15), the distribution thus being flattened. This is in fact observed by experiment, as shown in Fig. 5.18 (Fukui 61).

The lateral distributions of many EAS are drawn in Fig. 5.18, normalized at size $N = 10^5$. They are not always represented by a unique structure function in a distance range from a few meters to 80 meters, but by the N-K functions of $s = 0.6$ to 1.4. Within several meters of the axis the distributions are measured with the aid of a neon hodoscope; and distribution tends to be flatter and fluctuates in a wide range. The distributions corresponding to $s = 0.6$, 1.0, and 1.4 exhibit flattening in qualitative agreement with the x-dependence of s predicted by (5.5.15).

The large fluctuations may be due at least in part to the difference in the primary energy. According to (5.5.15) the age at $x = 1$ depends only on y_0/t, thus being insensitive to the primary energy. At small x, however, the contribution of $2x$ in the denominator is large if y_0 and consequently t are small. Thus the flattening with decreasing x is more pronounced as the primary energy is smaller. The lateral-structure function calculated for given W_0/ε_0 in Fig. 5.19 clearly shows this tendency (Nishimura 67).

In deriving the age distribution we have to consider the fact that the efficiency of detection depends on age. A requirement for EAS detection is that the density of particles in each detector exceeds a certain level. Since the density given in (5.2.1) depends on s, the distance from the axis r is a function of N, Δ, and s. Hence the effective detecting area depends on s as $\pi r^2(N, s)$ for a given triggering level of Δ. The detection areas for the triggering level of $\Delta = 20\ \mathrm{m}^{-2}$ are calculated as shown in Fig. 5.20 (Kamata 63). The age distributions observed at sea level (Fukui 61), at Mount Norikura (Miyake 63*b*), and at Mount Chacaltaya (Escobar 63) are corrected for detection efficiency, and the results are shown in Fig. 5.21. This shows that the distribution at Mount Chacaltaya is narrower than at lower altitudes, thus indicating that the fluctuation effect is less important at high altitudes.

At large distances the lateral distribution is dictated by the last factor

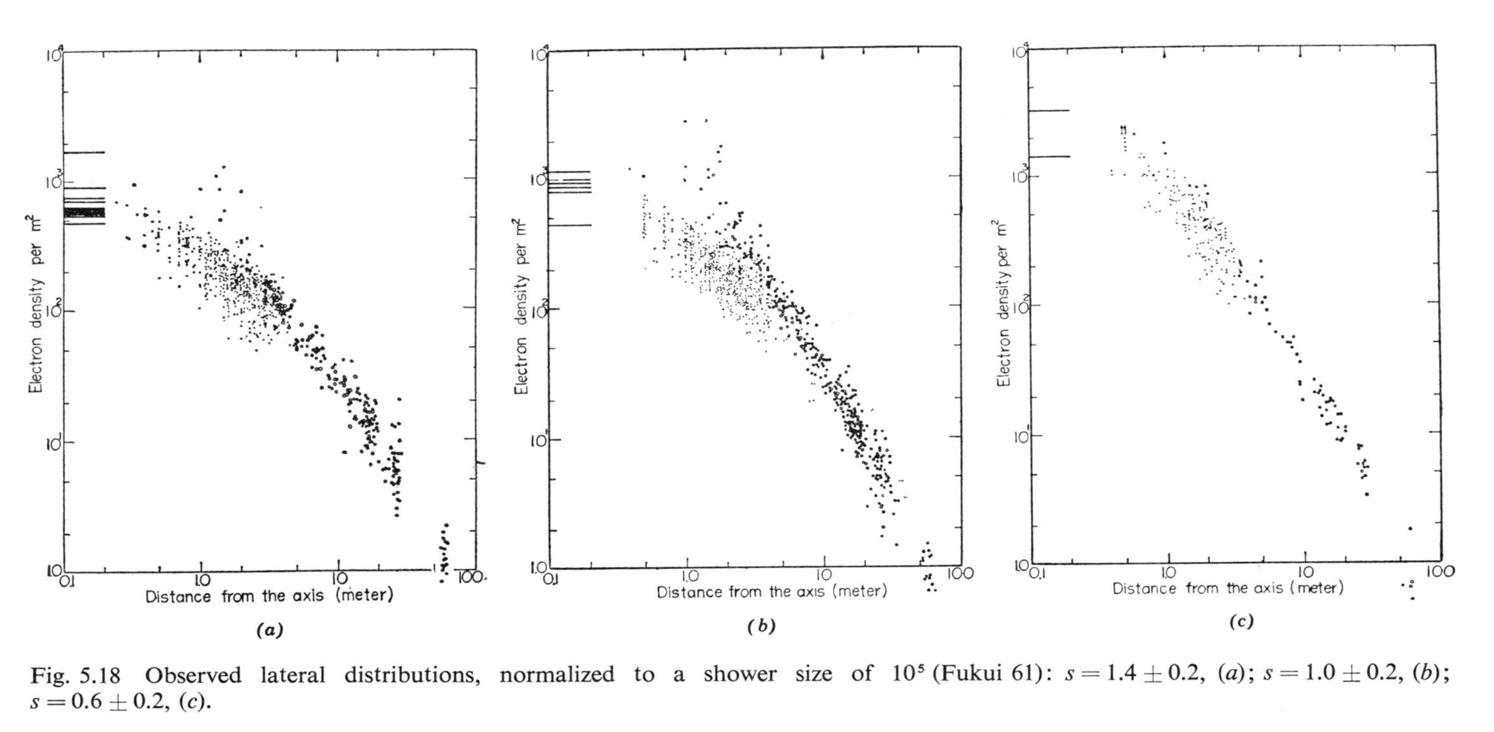

Fig. 5.18 Observed lateral distributions, normalized to a shower size of 10^5 (Fukui 61): $s = 1.4 \pm 0.2$, (*a*); $s = 1.0 \pm 0.2$, (*b*); $s = 0.6 \pm 0.2$, (*c*).

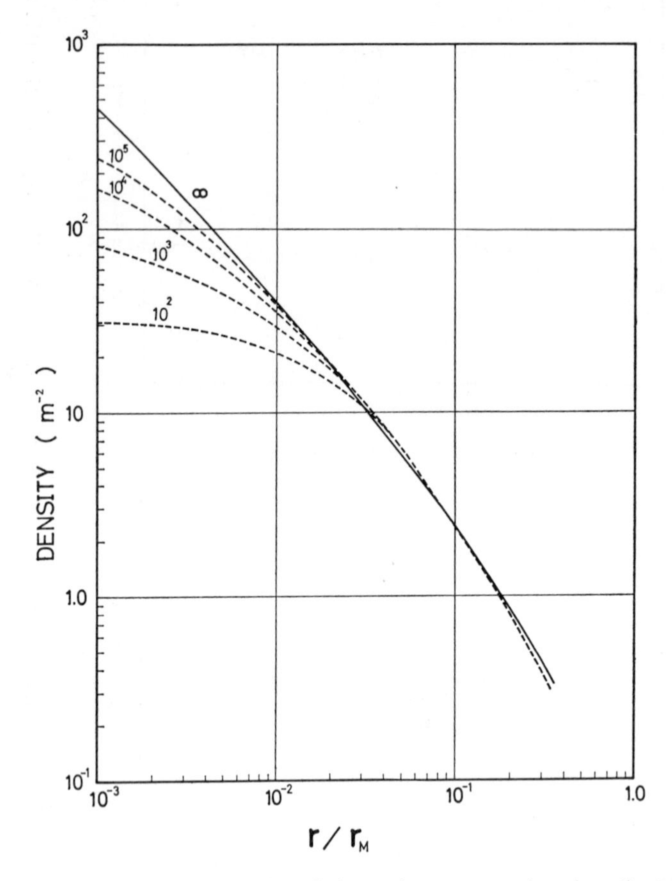

Fig. 5.19 Lateral distributions for finite primary energies ($s = 1$). Numbers attached to the curves of W_0/ε_0 (Nishimura 67).

in (5.5.20). An observation carried out at the Massachusetts Institute of Technology Volcano Ranch Station at an atmospheric depth of 820 g-cm^{-2} indicates that the power index of this factor varies with size and zenith angle (Linsley 62*a*). It is more appropriate to modify (2.6.21) to

$$f(x) = C(s, q)\left(\frac{r}{r_M}\right)^{s-2}\left(\frac{r}{r_M}+1\right)^{-(q+s-2)}, \tag{5.5.17}$$

where

$$C(s, q) = \frac{\Gamma(q+s-2)}{2\pi\Gamma(s)\,\Gamma(q-2)}.$$

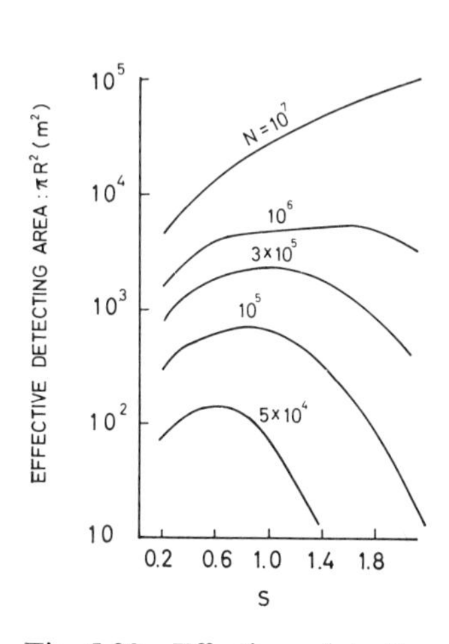

Fig. 5.20 Effective detecting area of EAS of $\Delta = 20\ \text{m}^{-2}$ versus shower age s (Kamata 63).

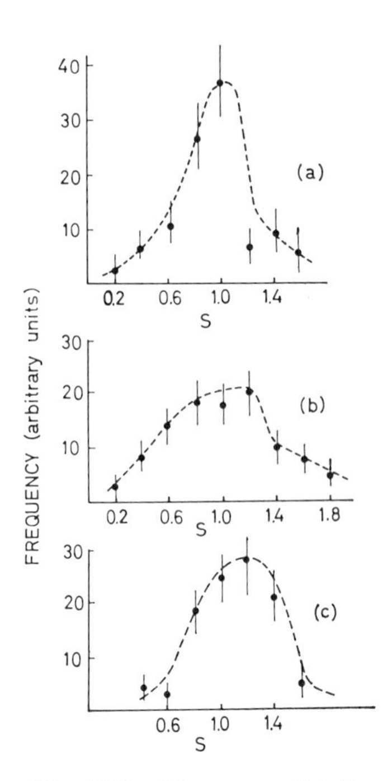

Fig. 5.21 The age distributions of EAS with $10^5 \leq N \leq 10^6$ at different altitudes (Kamata 63): Mount Chacaltaya, (*a*); Mount Norikura, (*b*); sea level, (*c*).

For $r/r_M \gg 1$ the lateral distribution is represented by $(r/r_M)^{-q}$, and the value of q is approximately 3 for $N \sim 10^9$, increasing with size but decreasing as the zenith angle increases.

5.5.3 Energy Spectrum of Electrons

As mentioned before, the energy spectrum of electrons gives information that is nearly equivalent to the lateral distribution. However, there have been only a few experiments on the energy spectrum, by means of both total-absorption Čerenkov counters (Fukui 60) and multiplate cloud chambers (Kameda 62, Toyoda 62). Here we refer to the results of the latter, in which the energy of an electron or a photon is determined by counting the number of tracks between lead plates through which a cascade shower is developed.

The energy spectra of the E-component were measured by varying the size, distance from the axis, and energy range. According to (5.5.4) shower age increases with the lower limit of the electron energy E. This tendency was in fact observed. For $E \geq 5$ GeV the lateral distribution fits the N-K function of $s = 2$; whereas for $E \geq 1.5$ GeV, $s = 1.6$ gives good fit independent of size and altitude, as shown in Fig. 5.22.

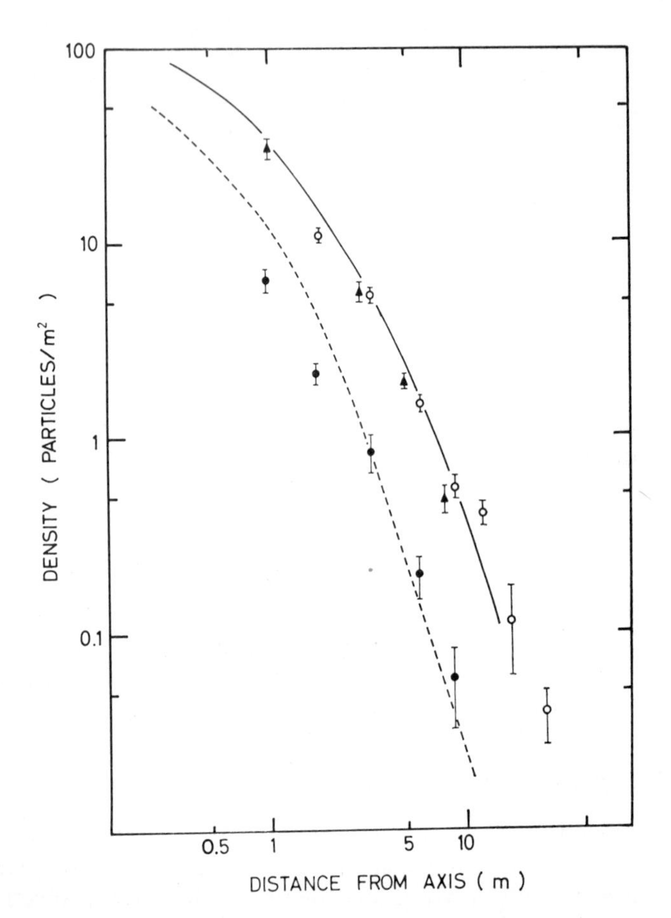

Fig. 5.22 Lateral distribution of the E-component with an average size of 5×10^5 (Kameda 63). ○—$E > 1.5$ GeV at sea level; ▲—$E > 1.5$ GeV at Mount Norikura; ●—$E > 5$ GeV at sea level. Solid curve: N-K function with $s = 1.6$; dashed curve: N-K function with $s = 2.0$.

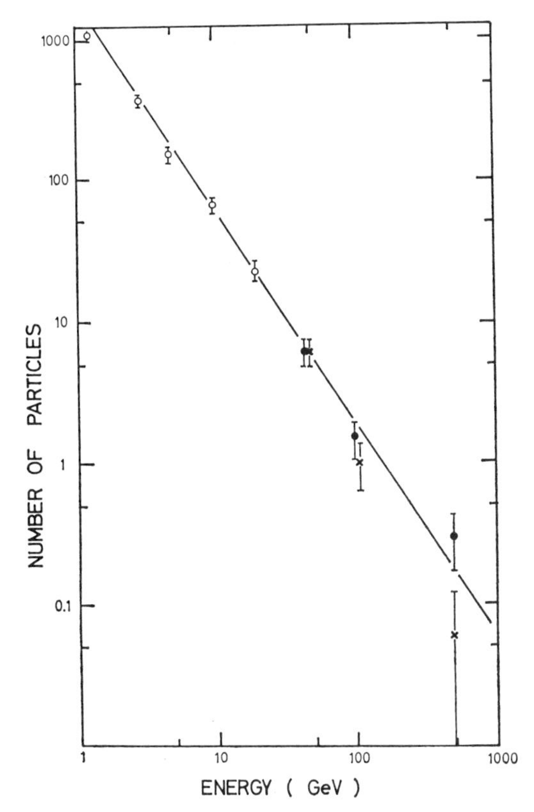

Fig. 5.23 Integral energy spectrum of the E-component (Kameda 63). ●— $5 \times 10^6 > N \geq 7 \times 10^4$; ×—$6 \times 10^4 \geq N \geq 2 \times 10^4$; ○—mean. The solid line represents the integral spectrum of $E^{-1.5}$.

The integral energy spectrum shown in Fig. 5.23 is represented by a power law,

$$N_e(>E) = 2 \times 10^3 \left(\frac{N}{10^5}\right) E^{-n}, \qquad n = 1.5 \pm 0.1, \qquad E \text{ in GeV}, \tag{5.5.18}$$

which is in reasonable agreement with that expected from cascade theory.

The energy spectrum at a given distance from the axis is more difficult to measure because the number of electrons and photons in a narrow distance range is too small to give good statistics. Nevertheless, the increase of energy with decreasing distance is qualitatively verified, as expected from (5.5.14). Quantitatively, however, the average energy of electrons decreases with distance more slowly than $1/r$, as shown in Fig. 5.24 (Fukui 60, Escobar 63).

If this is integrated over distance, the average energy of the E-component in EAS can be obtained. The experimental values at sea level and at Mount Chacaltaya are (Tanahashi 65)

$$\begin{aligned} \langle E_e \rangle &= 230 \pm 50 \text{ MeV at sea level}, \\ &= 360 \pm 50 \text{ MeV at 5200 m}, \end{aligned} \tag{5.5.19}$$

almost independent of size from 3×10^4 to 3×10^6.

5.5.4 *Structure of the Core*

At a distance as small as 1 meter, the average energy of the E-component is about 10 GeV and the energy spectrum is flatter (Kameda 62):

$$N_{eb}(>E) = (2.5 \pm 0.5)\left(\frac{N}{10^5}\right)E^{-(1.0 \pm 0.2)}. \tag{5.5.20}$$

This spectrum appears to be connected with (5.5.18), but the electrons responsible for this part are concentrated near the axis and form bundles of electrons and photons. These bundles are likely to be due to atmospheric cascade showers that are initiated near the observing level by neutral pions of energies greater than 1 TeV.

Such bundles have been observed by a neon hodoscope of large area (Oda 62). Extensive air showers of sizes greater than 5×10^4 are associated with bundles of electrons that have a spread of about 10 cm with a frequency of about 5 percent. Since the average density of electrons in such EAS does not exceed 10^3 m^{-2} according to Fig. 5.21, a bundle of higher electron density can be distinguished by means of the neon hodoscope; for example, a bundle of 300 electrons is detected and the energy contained in the bundle is measured as about 10 TeV by means of a Čerenkov counter and a lead-shielded scintillation counter underneath the neon hodoscope. If this is a cascade shower that is initiated by a single neutral pion, it should have been produced at an altitude of a few kilometers. The age parameter for such a shower is about $\frac{1}{2}$, consistent with the steep lateral distribution.

It is worthwhile to remark that about one-third of such bundles consist of *multiple cores* separated by 1 meter or so. If each of the multiple cores is initiated by a single neutral pion, the transverse momentum between these pions may be as high as several GeV/c, an order of magnitude greater than the average transverse momentum observed in a TeV

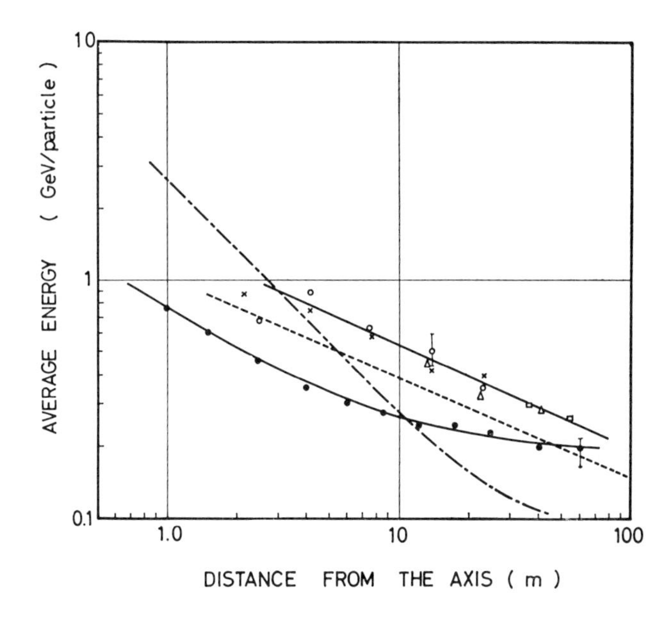

Fig. 5.24 Average energy of electrons and photons versus distance from the axis. Mount Chacaltaya: ○—$1 \times 10^5 \leq N < 5 \times 10^5$; ×—$5 \times 10^5 \leq N < 1 \times 10^6$; △—$1 \times 10^6 \leq N < 5 \times 10^6$; □—$5 \times 10^6 \leq N < 1 \times 10^7$. Sea-level values are indicated by solid circles. The dashed line is obtained from the data at Mount Chacaltaya by reducing the distance scale by a factor of $\frac{1}{2}$, in order to permit the two distributions at Mount Chacaltaya and at sea level to be directly compared for the same air density. The dot-dash curve represents what is expected from the N-K function with $s = 1.0$ (Tanahashi 65).

region. Similar events have been observed also with a large cloud chamber at Mount Norikura (Miyake 63*b*). It is also found that two associated cores are not of the same age, indicating that they started at different stages of EAS development.

An extended study of the core structure has been made with a spark chamber of large area (Shibata 65). The several events of double cores that have been found so far were of high transverse momenta, such as ≥ 20 and 47 GeV/c. Double cores with transverse momenta greater than 5 GeV/c are associated with 2 percent of EAS with $N \simeq 10^5$. This corresponds to the probability of 0.1 percent for secondary particles having transverse momenta greater than 5 GeV/c. This seems to be not inconsistent with the transverse-momentum distribution of survival nucleons, but it is an open question whether or not their distribution as discussed in Section 3.4 has a tail that is long enough to account for the events of very large transverse momenta, greater than 10 GeV/c.

5.6 Nuclear Active Component

The behavior of nuclear active particles in EAS can be described in terms of the N-cascade. In comparison with the E-cascade, an unambiguous result can hardly be obtained in the N-cascade, mainly because of the complexity of elementary interactions. We are therefore forced to limit ourselves to a qualitative treatment of the N-cascade.

5.6.1 Lateral and Energy Distributions

Since nuclear active particles that are produced in nuclear interactions are emitted with small transverse momenta, those of high energies form a bundle of small spread. Hence nuclear active particles far from the core may be regarded as those emitted from the bundle with relatively low energies. If their energies are not lower than several hundred MeV, the ionization loss of charged nuclear active particles is negligible compared with absorption by nuclear collisions. Thus we assume that the observed energy of a particle is equal to its energy at production.

Let us consider a particle of momentum p produced at height z. When this arrives at the observing level, it spreads as

$$r=\left(\frac{p_T}{p}\right)z \tag{5.6.1}$$

where p_T is the transverse momentum. Since the distribution of transverse momentum is narrow, we assume that all particles have a constant transverse momentum of $p_T = 0.4$ GeV/c. A particle produced at z survives without interactions with the probability of $\exp(-z/l)$, where l is the effective interaction mean free path for the air density near the observing level. On the other hand, the number of nuclear active particles that are responsible for the production of the observed particles increases with height as $\exp(z/L_n)$, where L_n is the effective attenuation length. Thus the number of particles that are produced at z in dz and arrive at r in dr is proportional to

$$\begin{aligned} \exp(x/L_n)\exp(-x/l)\,dz &= \exp(-r/r_0)\left(\frac{p}{p_T}\right)r^{-1}r\,dr, \\ r_0 &= \frac{lL_n}{L_n-l}\frac{p_T}{p} \simeq l\frac{p_T}{p}. \end{aligned} \tag{5.6.2}$$

In deriving the last expression we have assumed $L_n \gg l$. Since l is about 1 km, the value of r_0 is as large as 100 meters for most nuclear active particles. The $1/r$ distribution thus expected is in fair agreement with experiment at distances not far from the core.

In the lowest energy region nuclear active particles are detected at 3300 meters above sea level by their neutron production (Danilova 63). In the kinetic energy range of $W = 0.2$ to 3 GeV, the lateral distribution is represented by (5.6.2) with $r_0 = 70$ meters at 0.2 GeV and 50 meters at 3 GeV. The energy dependence of r_0 weaker than (5.6.2) may be interpreted as due to the fact that energy loss is so effective for low-energy particles that the value of l decreases with decreasing energy. The absolute value of $r_0 = 50$ meters is somewhat smaller than that predicted by (5.6.2). At higher energies the value of r_0 decreases inversely proportional to energy.

The *integral energy spectrum* is represented by a power law $W^{-0.45 \pm 0.15}$ but becomes E^{-1} above 10 GeV (Dobrotin 57). As energy further increases the spectrum becomes gradually steeper. The energy spectra of nuclear active particles measured at sea level by means of the burst-size spectra obtained with a shielded lead-glass Čerenkov counter and with a scintillator at 5 meters underground are shown for several shower sizes in Fig. 5.25 (Tanahashi 65). Similar energy spectra were obtained at Mount Chacaltaya, at 5200 meters above sea level.

The energy spectrum may reveal the nature of nuclear interactions at very high energies. As discussed in Chapter 3, we can divide the secondary particles produced by an interaction into a survival particle and others produced through fireballs and excited states. Since the energy of the former is much higher than the energy of a particle belonging to the latter, the energy spectrum may have a secondary hump separated from the bulk of the particles. This feature may be preserved after several generations of nuclear interactions in the atmosphere. The relative contributions of these two components may be seen by referring to the simple model discussed below.

Let us assume that the survival nucleon receives a constant fraction, $1 - K$, of the incident energy. After n interactions the energy of the survival nucleon is reduced to

$$E_N(n) = (1 - K)^n E_0, \tag{5.6.3}$$

where E_0 is the energy of a primary nucleon. This gives the probability that the survival nucleon at depth x has energy greater than E_N :

$$P(>E_N) = \sum_{k=0}^{n} \frac{(x/l)^k}{k!} \exp\left(\frac{-x}{l}\right). \tag{5.6.4}$$

This, however, gives a spectrum that is appreciably flatter than the

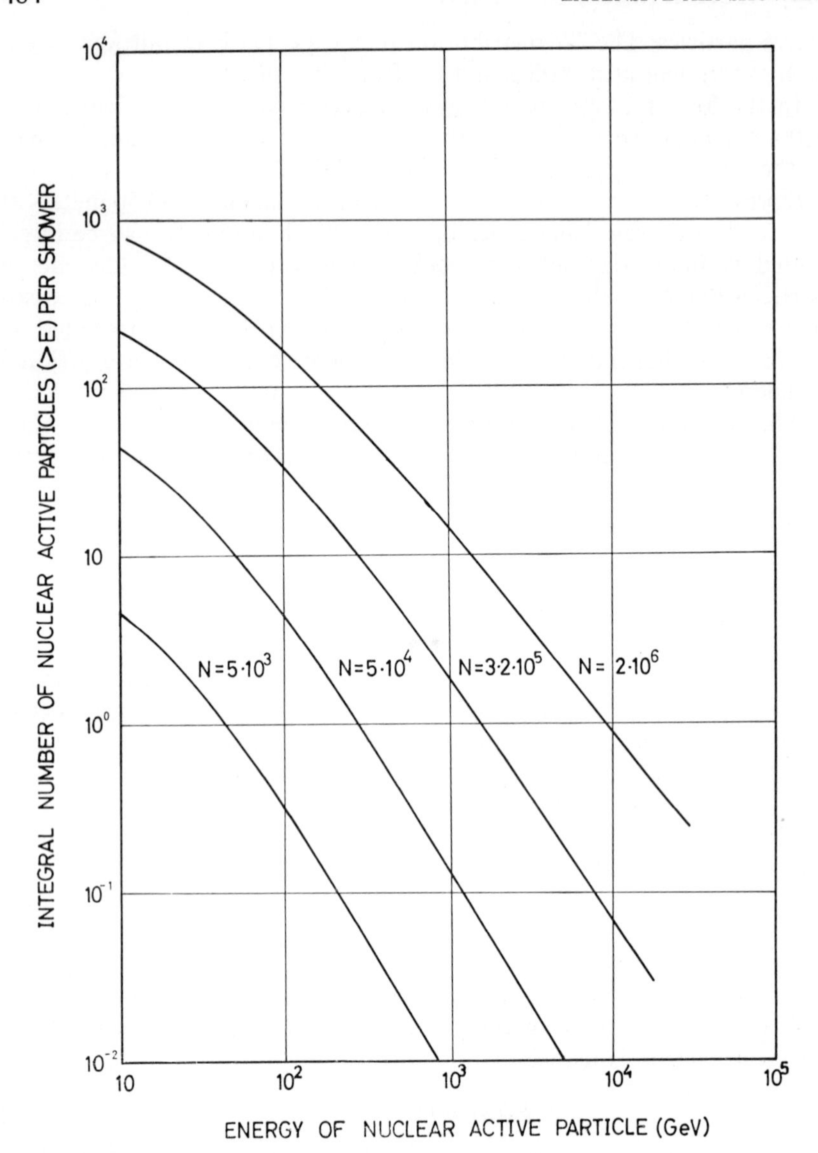

Fig. 5.25 Integral energy spectrum of nuclear active particles in EAS at sea level. The numbers attached are the average sizes of EAS (Tanahashi 65).

observed one for any set of K and l. This is due to the fact that most low-energy nuclear active particles arise from particles other than survival nucleons. In the lower half of the atmosphere these two groups are not well separated but form a continuous spectrum, as is seen in Fig.

5.25. The absence of a gap or a dip in the spectrum indicates that the energies of the two components overlap and that elasticity in the nuclear interaction is not too large.

Since the separation between two components is not possible, the following method may be of practical use for estimating the value of inelasticity K. Let us consider nuclear active particles after n generations. These particles passed through k stages of survival particles and $n-k$ stages of other particles. Of the latter, the neutral pions immediately decay into γ-rays and escape from the nuclear active component; some low-energy charged mesons also decay and escape. Hence only a fraction, $f(E_0, n-k)$, of the energy participates in the nuclear cascade process. If most charged mesons do not decay, the value of f tends to $1/(1+R) \simeq 0.7$, where R is the ratio of neutral to charged secondary particles, as given in Section 3.2. This is the case if E_0 is large and k is small. Accordingly the energy carried by nuclear active particles is given by

$$E_{\text{na}}(n) = \sum_k \binom{n}{k} (1-K)^{n-k} (fK)^k E_0 . \tag{5.6.5}$$

The summation can be reduced by taking the average value of $f(E_0, n-k)$, $\bar{f}$, in

$$E_{\text{na}}(n) = [(1-K) + \bar{f}K]^n E_0 . \tag{5.6.6}$$

This should be compared with $E_N(n)$ in (5.6.3).

In practice one observes nuclear active particles at a given atmospheric depth x. Since the number of generations follows the Poisson distribution, the average energies of survival nucleons and of all other nuclear active components in many showers are obtained as

$$\bar{E}_N = \sum_n E_N(n) \frac{(x/l)^n}{n!} e^{-x/l} = \exp\left(\frac{-Kx}{l}\right) E_0 , \tag{5.6.7}$$

$$\bar{E}_{\text{na}} = \sum_n E_{\text{na}}(n) \frac{(x/l)^n}{n!} e^{-x/l} = \exp\left[-(1-\bar{f})K \frac{x}{l}\right] E_0 . \tag{5.6.8}$$

The value of $\bar{E}_N$ is considered to be equal to the average energy of most energetic nuclear active particles observed for many EAS events, whereas the value of $\bar{E}_{\text{na}}$ is obtained by integrating the energy flow over the energy spectrum. In the latter the energy carried by particles with energies lower than 10 GeV escapes observation; however, it is estimated to be less than 20 percent, because of the flattening of the spectrum as energy decreases. Thus the values of $\bar{E}_N$ and $\bar{E}_{\text{na}}$ at sea level are obtained as given in Table 5.5 (Tanahashi 65). The same table gives the values of $\bar{E}_{\text{na}}$ at Mount Chacaltaya.

Table 5.5 Energies of the Most Energetic Nuclear Active Particle and the Energy Flow in EAS

Shower Size (N)	3.3×10^4	2.1×10^5	1.3×10^6
Sea level			
Energy of the most energetic particle ($\bar{E}_N$ in TeV)	0.6 ± 0.2	3.2 ± 1.2	20 ± 10
Energy flow of nuclear active particle ($\bar{E}_{\text{na}}$ in TeV)	2.5 ± 0.8	19 ± 7	120 ± 40
$\bar{E}_{\text{na}}$ at Mount Chacaltaya ($x = 530$ g-cm^{-2})		80 ± 40	600 ± 300

The ratio of $\bar{E}_N$ to $\bar{E}_{\text{na}}$ in Table 5.5 is approximately 1/6. If this is explained by (5.6.7) and (5.6.8), then we have

$$\frac{\bar{E}_N}{\bar{E}_{\text{na}}} = \exp\left[-\bar{f}K\frac{x}{l}\right] \simeq \tfrac{1}{6}, \quad \text{or} \quad \bar{f}K \simeq \tfrac{1}{6}. \tag{5.6.9}$$

In the last expression $l \simeq 90$ g-cm^{-2} given in (5.3.21) is taken into account. The value of K may be obtained from (5.6.7) by taking $E_0 \simeq 14\,N$ GeV and $K \simeq 0.6$. This gives $\bar{f} \simeq 0.3$—much smaller than the asymptotic value 0.7. This would imply that more than half the energy of charged secondary particles would be lost from the nuclear active component.

However, the conclusions are only qualitative. If the distribution of K, the energy dependence of f, and other effects are taken into account, the results may change considerably.

5.6.2 *Relative Intensity of Nuclear Active Particles*

Since the direct measurement of the total number of nuclear active particles is difficult, its value relative to the total number of charged particles is often measured. This, of course, depends on the lowest energy of the nuclear active particles measured. Conventional N-detectors have threshold energies of several GeV, and the intensity of nuclear active particles relative to the total size is about 1 percent for $N = 10^4$ to 10^5. This has a weak altitude dependence in the lower atmosphere; at sea level (Fujioka 55)

$$\frac{N_n}{N} \simeq 0.5\% \quad \text{for} \quad N \simeq 10^6 \tag{5.6.10a}$$

and at a mountain altitude (3860 meters) (Dobrotin 57)

$$\frac{N_n}{N} \simeq 1\% \quad \text{for} \quad N \simeq 10^5, \tag{5.6.10b}$$

although the threshold energies of the N-detectors are not the same.

With the same N-detector, the altitude dependence of the N-E ratio is inappreciable at low altitudes; however, it becomes great above 4000 meters (Cool 51, Hodson 53, Piccioni 53). The altitude dependence of the N-intensity is compared with that of the E-intensity in Fig. 5.26. This demonstrates the parallelism of the N- and E-altitude dependences at depths greater than 700 g-cm^{-2}, whereas the deviation increases with altitude. Figure 5.26 also shows the altitude dependence of the intensity of nuclear active particles that are not associated with EAS. This increases exponentially as the atmospheric depth decreases, whereas particles associated with EAS seem to have an intensity maximum. This is considered to be due to the development of the N-cascade.

The N-E ratio decreases with increasing size. For the minimum energy of 0.2 GeV at a 3330-meter elevation a significant size dependence was obtained (Danilova 63):

$$\frac{N_n}{N} = (1.2 \pm 0.4) \times 10^{-2} \left(\frac{N}{10^6}\right)^{-0.34 \pm 0.01} \quad \text{for} \quad 3 \times 10^3 < N < 10^7. \tag{5.6.11}$$

A similar result was obtained for vertical showers of the minimum nucleon energy of 1 GeV at 2200 meters (Chatterjee 63):

$$\frac{N_n}{N} = 2.1 \times 10^{-3} \left(\frac{N}{10^6}\right)^{-0.35 \pm 0.05} \quad \text{for} \quad 10^5 < N < 10^7. \tag{5.6.12}$$

For showers with zenith angles greater than 40° the ratio is reduced by a factor of 0.6, but the size dependence is essentially the same as that for vertical ones.

5.7 The μ-Component

The behavior of the μ-component in EAS has been discussed in Section 5.3 in connection with longitudinal development. The number of muons referred to was obtained for an average energy of 10 GeV and a lateral distribution of up to 100 meters from the shower axis. More precisely, however, the total number of muons depends on the lateral distribution at larger distances, as well as on the energy spectrum. These dependences bring about significant effects, as will be discussed below.

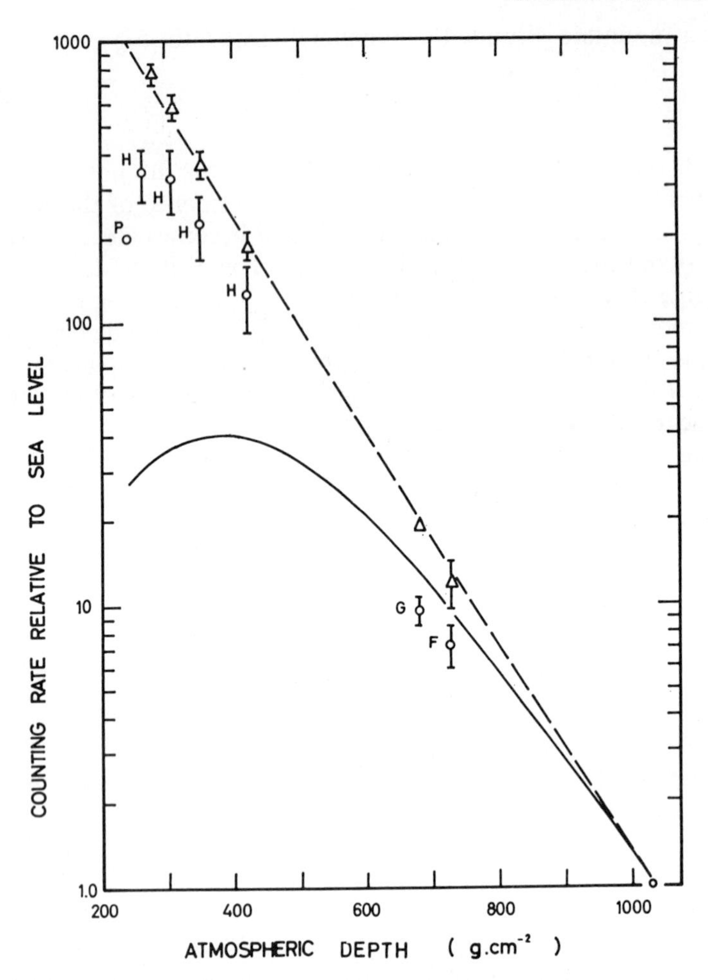

Fig. 5.26 Altitude variation of the penetrating shower frequency. ○—associated with EAS; △—not associated with EAS; *P*—Cool 51, Piccioni 53; *H*—Hodson 53; *G*—E. R. George and A. C. Jason, *Proc. Phys. Soc.* (London), **A63**, 108 (1950); *F*—W. B. Fretter, *Phys. Rev.*, **76**, 511 (1949). The dashed curve represents the altitude variation of unassociated penetrating shower frequency, $\exp(-x/L)$, $L = 130$ g-cm^{-2}. The solid curve represents the altitude variation of EAS frequency (Greisen 56).

5.7.1 Lateral Distribution of Muon Density

In order to illustrate the contribution of muons at large distances an experimental observation of lateral distribution at an altitude of 3860 meters is shown in Fig. 5.27 (Vavilov 57). The muons measured have

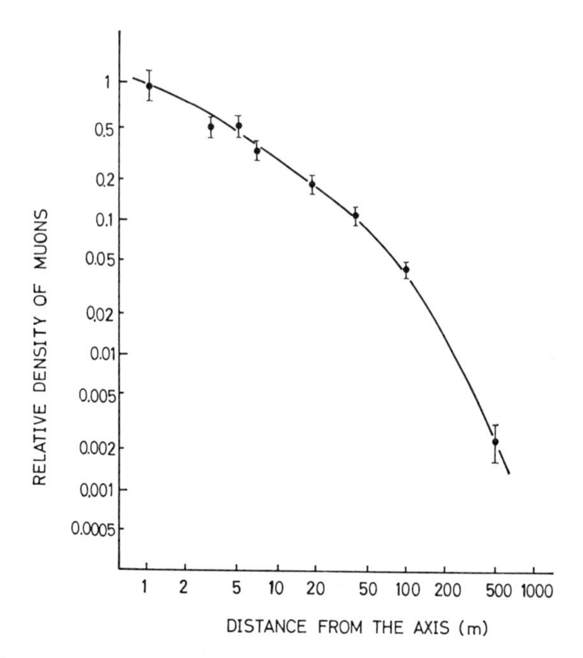

Fig. 5.27 The lateral distribution of the density of muons with energy greater than 440 MeV, associated with EAS of sizes between 5×10^4 and 10^6 (Vavilov 57).

energies greater than 0.44 GeV, and their average energy lies between 3.5 and 3.9 GeV. The lateral distribution is found to become steep beyond 100 meters from the axis, being proportional to r^{-n} with $n = 1.8 \pm 0.2$ for distance r between 100 and 500 meters. This means that the contribution from greater distances is predominant.

Muons produced at high altitudes spread out of the shower axis by multiple scattering and geomagnetic deflection. Since the distance through which muons travel is about 10 km, the lateral spread by these effects is as large as several tens of meters. The magnetic deflection gives rise to an anisotropy about the axis, larger in the east-west direction, and is quite important for inclined showers, in which the path lengths of muons are large. More important than these two causes is the angular divergence of parent pions at their production, because the transverse momentum of a pion produced is greater than the characteristic momentum for multiple scattering and the momentum of a particle with unit radius in the geomagnetic field. The lateral spread due to angular divergence is on the order of

$$r_\mu \simeq \left(\frac{p_T}{p}\right) z, \tag{5.7.1}$$

which is as large as 1 km for muons of several GeV. The fact that the transverse momentum is mainly responsible for the lateral spread of muons would suggest a lateral distribution that is analogous to (5.6.2). In the case of muons the value of r_0 is so large that lateral distribution not far from the shower axis may be represented by an inverse power of distance as $r^{-\alpha}$. The value of α is not always equal to unity because of the longitudinal structure of EAS, energy loss, and effects of muon decay.

The lateral distribution of muons observed by an Indian group (Sreekantan 63) at an atmospheric depth of 800 g-cm^{-2} is well represented by an inverse power law at distances of 10 and 100 meters from the axis. This depends significantly on shower size and the lateral distribution of electrons, as this is the case for nuclear active particles. The results are summarized in Table 5.6.

Table 5.6. Power Indices of the Lateral Distributions ($r^{-\alpha}$)

Component	$<10^6$		$>10^6$	
E	1.1	$1.7-1.9$	1.1	$1.7-1.9$
N	1.0 ± 0.2	1.8 ± 0.15	1.5 ± 0.2	1.65 ± 0.15
μ	1.1 ± 0.15	1.5 ± 0.15	0.95 ± 0.15	1.1 ± 0.15

The correlation between the value of α of the E-component and those of the N- and the μ-components can be understood in the following way. The former represents the age of a shower. For a young shower $\alpha(E)$ is large, and most shower particles are produced near the observation level. Hence both the N- and the μ-components spread to a lesser degree. The size dependence is rather difficult to explain. A shower of large size may have experienced many interactions that contribute to particles at the observation level. This smears out the age dependence of the later distributions of the N- and the μ-components. A contribution to muons is made by particles that are produced at high altitudes, and their lateral distribution becomes flatter.

The *lateral distribution* of muons becomes steeper as the distance further increases; however, it is less steep than that of electrons, as represented by (Greisen 60)

$$\Delta_\mu(N, r) = 18\left(\frac{N}{10^6}\right)^{3/4} r^{-3/4}\left(1 + \frac{r}{320}\right)^{-2.5} m^{-2}, \tag{5.7.2}$$

where r is the distance from the axis in meters.

Since the density of muons is measured relative to that of electrons, it is practical to express the lateral distribution of muons by the muon-electron ratio as a function of distance. The results obtained at sea level and mountain altitudes are shown in Fig. 5.28*a* and *b*, respectively.

It should be remarked that the experiments referred to in Fig. 5.28 measured all penetrating particles consisting of muons and nuclear active particles. The flattening of the ratio at small distances is considered to be due to the contribution of the N-component (Fujioka 55). At large distances the penetrating particles consist exclusively of muons, and the muon-electron ratio as a function of distance can be represented by a simple power law (Linsley 63b). At about 1 km from the axis the density of muons is comparable to that of electrons. This behavior is approximately expressed by

$$\frac{\Delta_\mu}{\Delta_e} \simeq 0.05\left(\frac{r}{100}\right)^{0.74} \tag{5.7.3}$$

for $r \geq 100$ meters at the atmospheric depth of 820 g-cm^{-2}. By integrating the muon density over distance we find the total number of muons relative to electrons to be about 10 and 5 percent at sea level and at mountain altitudes, respectively (Cocconi 61).

5.7.2 *Relative Intensity of Muons*

The intensity of muons relative to electrons depends on the size, the lower limit of muon energies, and the zenith angle of arrival (as in the case of the N-E ratio); it depends more strongly on lateral distance, as given in (5.7.3). Actually, however, the lateral-distance dependence is weaker than that given in (5.7.3) for $r/r_M < 1$. Therefore the μ-E ratio as measured near the shower axis is only approximate. This is about 0.8 percent, nearly independent of the atmospheric depth between sea level (McCusker 50, Fujioka 55) and 600 g-cm^{-2} (Greisen 50, Sitte 50).

The dependence of the μ-E ratio on shower size is measured at 700 g-cm^{-2}, as in Table 5.7, where the N-E ratios are also shown. The

Table 5.7. Relative Intensities of the N- and the μ-Components at 700 g-cm^{-2} Near the Axis[a]

N_e	$N_\mu/N(\%)$	$N_n/N(\%)$
3×10^4 to 2×10^5	0.86 ± 0.08	0.63 ± 0.08
2×10^5 to 10^6	0.85 ± 0.07	0.56 ± 0.06
$>10^6$	0.74 ± 0.43	0.43 ± 0.06

[a] H. L. Kasnitz and K. Sitte, *Phys. Rev.*, **94**, 977 (1954).

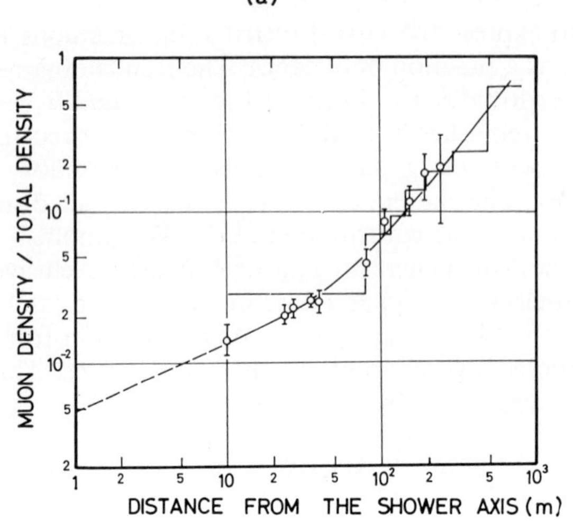

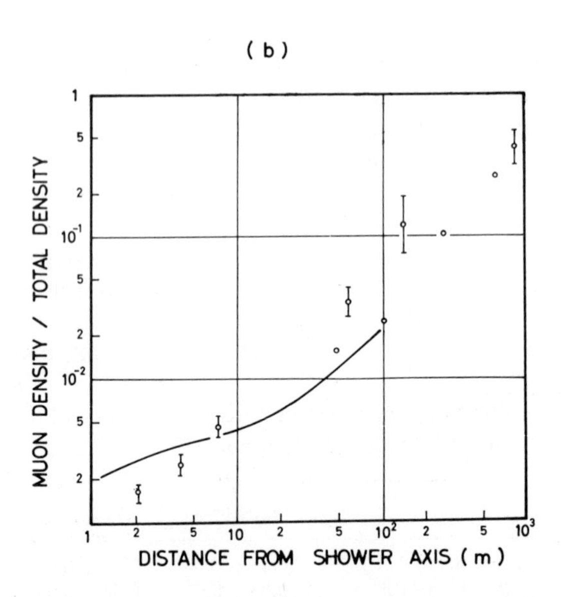

Fig. 5.28 The ratios (muon density to total density) at various distances from the shower core at sea level, (*a*); and at mountain altitudes (3000 to 4000 meters), (*b*) (Cocconi 61).

size dependence is weak, but it indicates a decrease of the μ-E ratio as the size increases, in qualitative agreement with the prediction given by (5.3.20) with (5.3.21).

The measurement of the μ-E ratio is subject to a bias due to the correlation of the lateral distributions of these two components. If the shower size is small, the triggering system of EAS is biased for showers of a less steep lateral distribution of electrons. In such showers the lateral distributions of muons is also less steep, as seen in Table 5.6, and the chance of observing a low density of muons is relatively high. It is therefore expected that the μ-E ratio at small sizes given in Table 5.7 could be lower than the true value.

A less biased measurement is carried out at 800 g-cm^{-2} by integrating the densities of electrons and muons up to 50 meters from the axis (Chatterjee 63). The results show a steeper size dependence as well as dependence on the minimum muon energy.

$$\frac{N_\mu(>0.5\text{ GeV})}{N_e} = 0.64\left(\frac{10^6}{N_e}\right)^{0.56\pm0.04}\% \tag{5.7.4a}$$

$$\frac{N_\mu(>1\text{ GeV})}{N_e} = 0.46\left(\frac{10^6}{N_e}\right)^{0.40\pm0.04}\% \tag{5.7.4b}$$

The size dependences above are obtained for EAS of zenith angles smaller than 40°. For inclined EAS the μ-E ratio is greater, as expected.

The μ-E ratio thus far discussed is concerned only with the average values of muon and electron numbers. It is, however, known, as discussed in Section 5.3, that the number of muons fluctuates considerably for a fixed value of the electron number. The density of muons for a fixed size observed at sea level is distributed over a wide range, as shown in Fig. 5.29. This is represented by a Pólya-Eggenberger distribution (Matano 63):

$$P(\Delta) = \frac{\bar{\Delta}(\bar{\Delta}+\alpha)\cdots[\bar{\Delta}+\alpha(\Delta-1)]}{\Delta!(1+\alpha)^{\Delta+\bar{\Delta}/\alpha}} \qquad (\alpha = 5). \tag{5.7.4}$$

This can be expected if the logarithm of the number of muons, $\log N_\mu$, distributes uniformly between $\log(\sqrt{7}\,\bar{N}_\mu)$ and $\log(\bar{N}_\mu/\sqrt{7})$; the distributions of the density expected from the number distributions are shown by dashed curves in Fig. 5.29. The observed distribution agrees reasonably well with the expected one for $\bar{N} = 2 \times 10^5$ and 10^6, but it is narrower for $\bar{N} = 7 \times 10^6$. The narrowing of the distribution with increasing size is shown by the size dependences of the upper and the lower edges of the N_μ-distribution:

$$N_{\mu,\max} \propto N^{0.65}, \qquad N_{\mu,\min} \propto N^{0.9}. \tag{5.7.6}$$

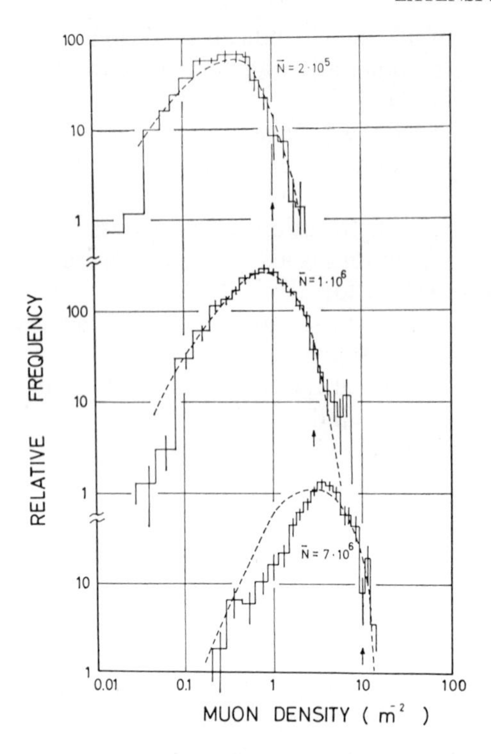

Fig. 5.29 Muon-density distribution for fixed total size (Matano 63). Data are selected from EAS with zenith angles smaller than 30° and with the average sizes of 2×10^5, 1×10^6, and 7×10^6 (attached to the graphs). The dashed curves represent those expected for a square-shaped distribution of N_μ for fixed N, as described in the text. Arrows indicate that the right-hand sides are due to heavy primaries.

The number of muons relative to electrons is related to the primary particle that initiates an EAS. An EAS of large N_μ/N may be attributed to a heavy primary particle; whereas that of small N_μ/N, to a γ-ray. The arrows in Fig. 5.29 indicate that events on the right-hand side of the arrows may be tentatively attributed to heavy primaries, and they are 6 percent of all events.

The narrowing of the distribution of N_μ/N is considered to be due partly to the change in the composition of primary particles with energy. The N_μ/N distribution for $N \geq 5 \times 10^7$ observed at Mount Chacaltaya is well represented by a Gaussian distribution with a dispersion of 25 percent, as shown in Figure 5.30 (Toyoda 65). This is interpreted in terms of a single kind of primary particle, probably protons alone (Linsley 62*b*). A dotted curve represents the distribution that is expected if the composition is the same as at low energies. This suggests that heavy primaries are absent at energies above 10^{17} eV per nucleon. Its significance will be discussed in Chapter 6.

The existence of γ-primaries may be seen from the existence of *few muon events*. The distribution of N_μ/N for $5 \times 10^5 < N < 5 \times 10^6$ observed at distances from the axis greater than 25 meters at Mount Chacaltaya is similar to that in Fig. 5.29, but at the smallest part we find a peak, as shown in Fig. 5.31*a*. The frequency of such showers is 5×10^{-4} of the total frequency of EAS of the same size. At larger sizes, $N \geq 5 \times 10^6$, the peak is not seen, as shown in Fig. 5.31*b*.

Whether or not the low muon peak belongs to a different kind of EAS may be determined by observing the zenith-angle distribution, which

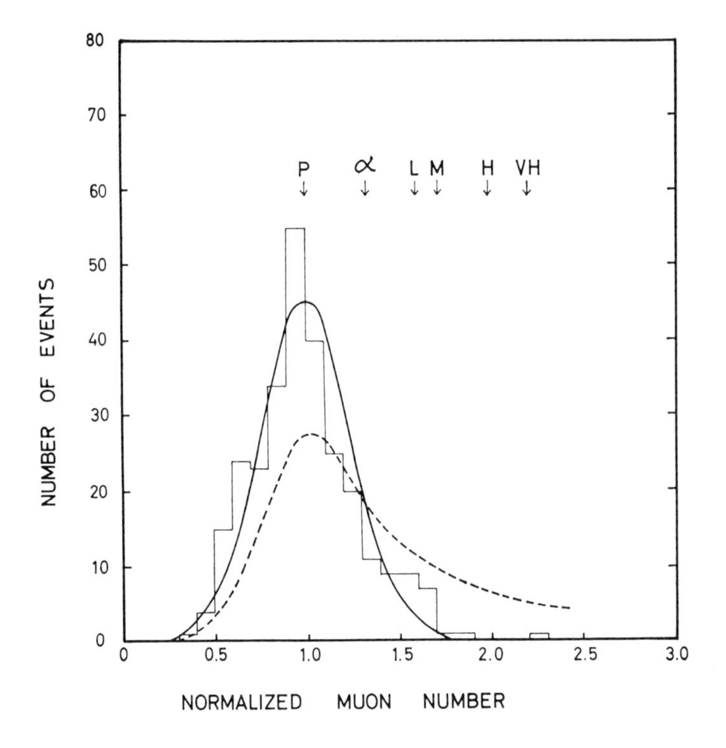

Fig. 5.30 The distribution of muon size (Toyoda 65). Data are taken from 280 EAS with $N \geq 5 \times 10^7$. The solid curve represents a Gaussian distribution, expected if primary particles consist of protons alone. The dashed curve represents a distribution expected if the composition of primary cosmic rays is the same as at low energies; p, α, etc., and arrows indicate the positions of peaks of the N_μ-distributions due to the various primary particles.

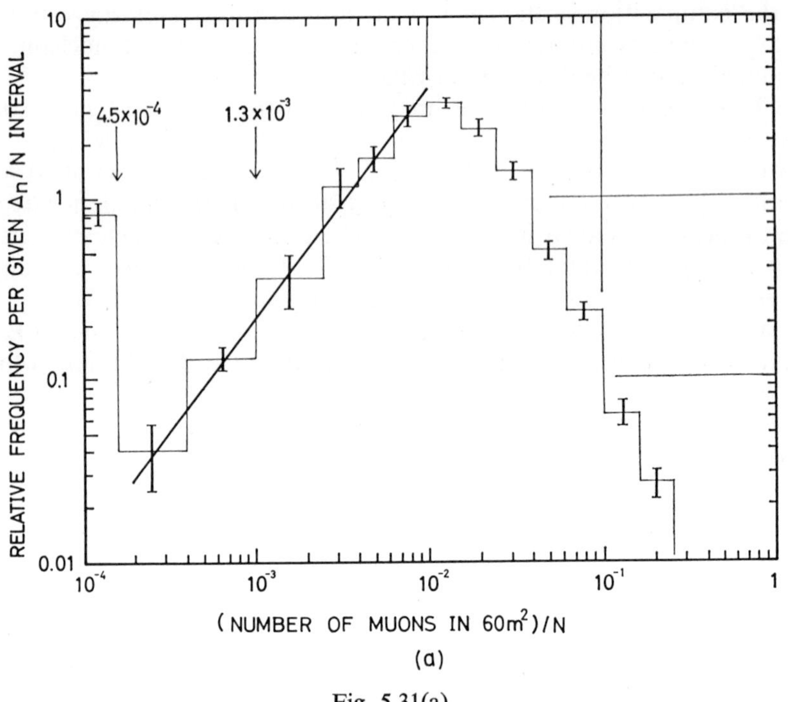

Fig. 5.31(a)

represents the attenuation of the shower frequency. For the low muon showers, the maximum of the frequency is broader and appears at a lower altitude than that of ordinary showers, and the attenuation length after the maximum is smaller (Suga 63):

$$\Lambda(\text{low } \mu) = 67 \pm 10 \text{ g-cm}^{-2}, \qquad \Lambda(\text{ordinary}) = 97 \pm 13 \text{ g-cm}^{-2} \tag{5.7.7}$$

This would support the γ-ray primary of the low-muon shower. If this is taken for granted, the relative intensity of primary γ-rays is estimated to be 2×10^{-4} in the energy range between 10^{15} and 10^{16} eV, and it is less than 5×10^{-5} at energies above 10^{16} eV.

5.7.3 Energy Spectrum of Muons

The energy dependence of the μ-E ratio given in (5.7.4) indicates that the energy spectrum of muons is a slowly decreasing function of energy at about 1 GeV. The absorption of muons associated with EAS at mountain altitudes gives the integral energy spectra at 3260 meters (Sitte 50) and at 2200 meters (Hinotani 63):

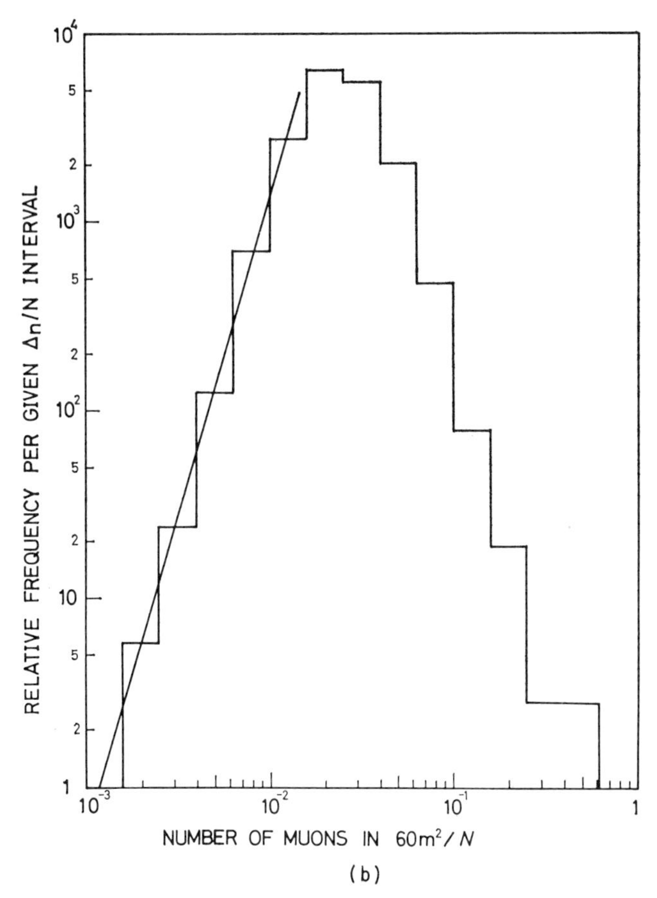

Fig. 5.31 Muon-density distribution for fixed size (Toyoda 65). The muon density is represented by muons falling in an area of 60 square meters at distances from the axis greater than 20 meters. Curve (*a*) is based on 8400 events of $5 \times 10^5 < N < 5 \times 10^6$; the vertical arrows indicate that the relative frequencies of showers low in muons on the left-hand sides of them are 4.5×10^{-4} and 1.3×10^{-3}, respectively. Cuıve (*b*) is based on 19,000 events of $N > 5 \times 10^6$.

$$N_\mu(>E) \propto E^{-0.5}, \qquad 0.6 \text{ GeV} \leq E \leq 1.3 \text{ GeV at 3260 m} \tag{5.7.8a}$$

$$\propto E^{-0.6}, \qquad 1 \text{ GeV} < E < 2 \text{ GeV at 2200 m}, \tag{5.7.8b}$$

respectively.

The energy spectrum at higher energies can be obtained from the attenuation of muon numbers with depth. This can be derived from the muon size spectra at different depths,

$$\Phi_\mu(d) = K(d) N_\mu^{-\gamma_\mu(d)}. \tag{5.7.9}$$

The muon size with a given frequency at different depths gives the muon number attenuation,

$$N_\mu(d; \Phi_\mu) = \left[\frac{K(d)}{\Phi_\mu}\right]^{1/\gamma_\mu(d)}. \tag{5.7.10}$$

The values of K and γ_μ are summarized in Table 5.8. From these data and direct measurements of the energy spectrum we can obtain the

Table 5.8. Underground Muons Associated with EAS

Depth (mwe)	Median Distance (meters)	K (m^{-2} sec^{-1} sr^{-1})	γ	N_μ/N		$N_\mu(r < 60$ m$)/N$	Reference
				$N = 10^6$	$N = 10^5$	$N = 10^6$	
0	300	1.6	1.5	1.0×10^{-1}	1×10^{-1}	0.9×10^{-1}	Equation 5.7.1,2
60	40	0.8	2.2	2.0×10^{-3}	4×10^{-3}	2.0×10^{-3}	1
1580	9	0.007	2.4	1.4×10^{-4}	3×10^{-4}	1.4×10^{-4}	2

References:

1. E. P. George, J. W. MacAnuff, and J. W. Sturgess, *Proc. Phys. Soc.* (London), **A66**, 345 (1953).
2. P. H. Barrett et al., *Rev. Mod. Phys.*, **24**, 133 (1952).

energy spectrum of muons up to several hundred GeV from (Greisen 60)

$$N_\mu(>E, N) = 1.7 \times 10^5 \left(\frac{2}{E+2}\right)^{1.37} \left(\frac{N}{10^6}\right)^{0.75}, \tag{5.7.11}$$

where E is measured in GeV. This gives the average muon energy of 5.4 GeV.

The energy spectrum depends on the distance from the axis, as is represented by Fig. 5.32 (Greisen 60). The observed spectra at different distances are approximately expressed by

$$\Delta_\mu(r, > E) = \frac{14.4\, r^{-0.75}}{(1 + r/320)^{2.5}} \left(\frac{N}{10^6}\right)^{0.75} \left(\frac{51}{E+50}\right) \left(\frac{3}{E+2}\right)^{0.14 r^{0.37}} \text{m}^{-2}, \tag{5.7.12}$$

where r is the distance from the axis in meters and E is in GeV. In fact, the energy spectrum derived from the observation of muons that were detected by counters as much as 100 meters apart at two depths underground is steeper than that given in (5.7.11), being represented by $E^{-1.5}$ (Chatterjee 65).

The softening of muons with distance results in the decrease of the lateral spread with depth, because muons at greater distances are more

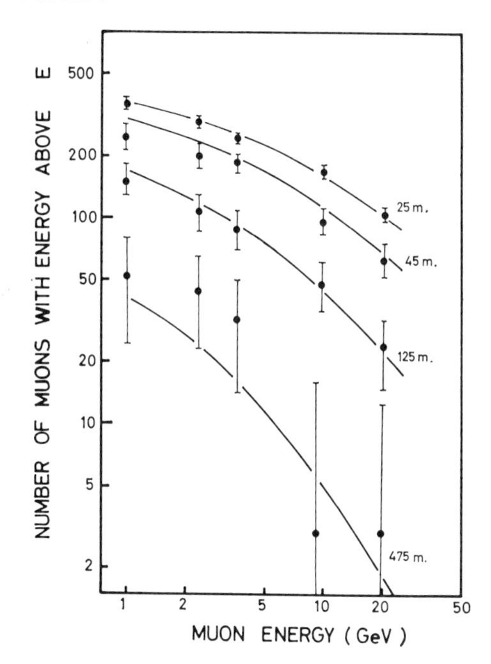

Fig. 5.32 Integral energy spectrum of muons at various distances from the axis (Greisen 60). The distances are indicated on the curves. The solid curves represent (5.7.11).

rapidly absorbed. A few experiments have evaluated the median distance of muons from the axis; that is, the distance within which half the muons fall. The results are also given in Table 5.8.

5.7.4 *Muon Showers*

Occasionally a bundle of muons, which are penetrating and parallel to each other, is observed with a detector underground. This event was considered peculiar because the frequency of muon showers of such high densities could not be as large as observed (Higashi 57). Further progress in the study of the multiple penetrating particles at 50 meters water equivalent underground has, however, shown that a substantial part of such muon showers can be interpreted in terms of the surviving muons in EAS (Higashi 62). Figure 5.33 gives a comparison of the density spectra of muons underground and at sea level. They are essentially parallel to each other, and the power index of the integral spectrum is $-(2.5 \pm 0.4)$. The other properties of muons and the accompanying

EAS are normal. Therefore the multiple-penetrating-particle events are due to high-density muon-showers associated with EAS.

However, we cannot necessarily rule out the possibility that the muon bundle reveals a muon-generation process that is as yet unknown. In fact, the frequency of such bundles at distances of less than 1 meter from

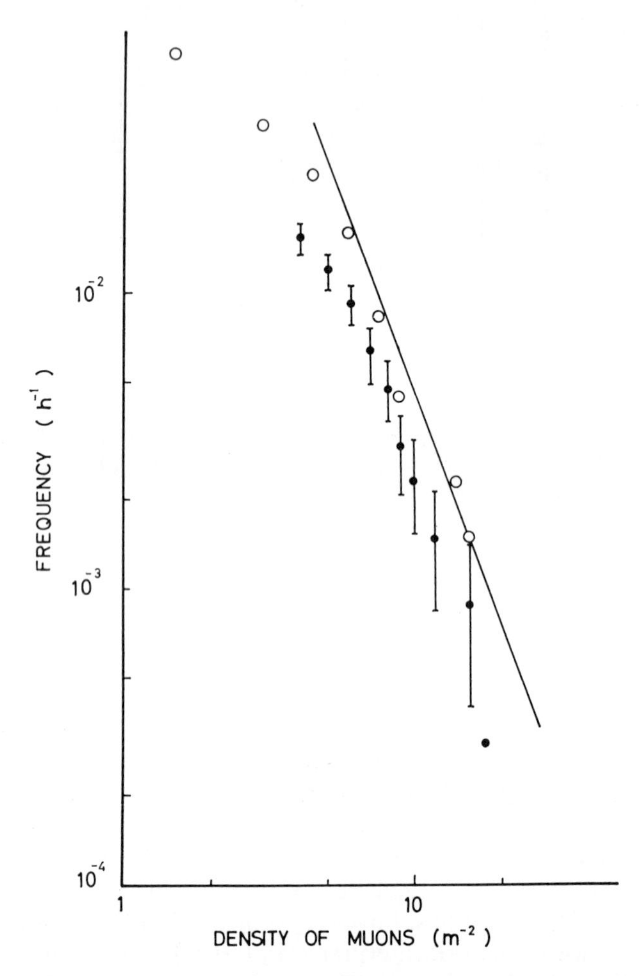

Fig. 5.33 Integral density spectrum of multiple penetrating particles (Higashi 62). ●—multiple penetrating particles; ○—all muons in EAS with $N \geq 2 \times 10^6$. The solid line represents the integral density spectrum of muons expected for EAS.

Plate 5. A bundle of muons associated with an EAS, observed with the same cloud chamber as in Plate 4. Six parallel tracks penetrating all lead plates without noticeable interactions are muons forming a bundle. Showers produced by nuclear active particles are seen near the right edge. Courtesy of S. Miyake. See Chapter 5, Section 7.4.

the axis is appreciably higher than that expected from the extrapolation of the lateral distribution known for greater distances (Vernov 62). As an extreme example, five muons are found in a bundle of diameter as small as 10 cm (Miyake 63a). It is not always unlikely that a considerable number of muons of very high energies exist in EAS.

If muons of very high energies exist in EAS, they can survive down to a great thickness of matter after all other particles are absorbed. At a great thickness of matter, therefore, the EAS consists of muons and their secondary products. Such muons are observed at sea level at large zenith angle.

The detection of such muons would be very difficult if muons alone were observed. Actually high-energy muons are associated with many secondary particles, as discussed in Section 4.14, and the electron showers initiated by muons serve to increase the probability of muon detection.

It may be worthwhile to note that nearly horizontal showers were observed to show very high energy muons and a great amount of energy transferred to the electronic component (Matano 65). One of the examples shows that the shower size is 9×10^4, the lateral distribution shows a young feature, and a nuclear interaction of 30 to 50 GeV is associated. Since the shower must have traversed an atmospheric thickness of 12 kg-cm^{-2}, only muons and neutrinos of high energies could have survived to produce the shower. The energy of such a muon is estimated to be greater than 3×10^{14} eV, and a substantial fraction of its energy is transferred to the electronic component.

5.8 Information on Primary Cosmic Rays

The properties of EAS depend not only on interactions of various particles with air but also on primary cosmic rays. These two are not clearly separated but reveal themselves in various features of EAS. In this section we discuss the energy spectrum and the composition of primary cosmic rays.

5.8.1 Energies of Primary Particles

The energy spectrum of primary cosmic rays at very high energies is derived from the size spectrum of EAS, assuming a primary energy-size relation. The latter has been customarily obtained from a theoretical calculation based on a model of EAS. Because the result depends considerably on the model assumed and the model deviates from the real

EAS by an unpredictable extent, we can hardly rely on the model calculation of EAS. It is therefore advisable to use the energy-balance method, as applied to the bulk of cosmic rays in Section 4.18.

From the altitude variation of the shower frequency given in Fig. 5.14 we can derive the altitude variation of size $N(x)$, as discussed in Section 5.3. Since the energy of all charged particles is dissipated by ionization at a rate of ε per g-cm^{-2}, the integral $\varepsilon \int N(x)\, dx$ over the atmosphere gives the energy dissipated by ionization in the atmosphere. For $N = 5 \times 10^5$ at sea level Greisen (Greisen 56) obtained

$$\varepsilon \int N(x)\, dx = (11.5 \pm 2.9) N(\text{sea level})\ \text{GeV}. \tag{5.8.1}$$

The energy of particles striking the ground surface is not included in the track-length integral (5.8.1). Electrons at sea level have the average energy given in (5.5.19), and nearly the same amount of energy escapes below sea level as photons. Hence about 0.5 GeV per electron is lost by the electronic component underground.

The energy of muons is also dissipated underground. An exact estimate of the energy dissipated is difficult because both the energy spectrum and the relative density of muons vary with the distance from the shower axis, as discussed in Section 5.7. The average energy of muons may, however, be inferred to be several GeV, and the total number of muons is about 10 percent of the total electron number. Because some muons decay in the atmosphere and dissipate most of their energy into neutrinos, the contribution of muons to the energy dissipation is 1 GeV per electron.

The muons are considered to arise mainly from π-μ decays. Neutrinos arising thereby must be taken into account. Since a neutrino takes 0.27 of the muon energy, the neutrino contribution is estimated to be 0.3 GeV per electron.

The energy of nuclear active particles per electron is as small as 0.1 GeV, as shown in Table 5.9. Most of the energy dissipated by the N-component is not directly detectable. This is mainly lost by nuclear disintegrations—about 200 MeV per disintegration. The energy dissipated thereby in the atmosphere can be estimated by the track-length method by referring to the altitude variation in Fig. 5.29. This gives about 0.5 GeV per electron.

Summing up the contributions from various sources summarized in Table 5.9, we obtain

$$\begin{aligned} E_0 &= (14 \pm 3) N(\text{sea level})\ \text{GeV} \\ &= (7 \pm 2) \times 10^{15}\ \text{eV} \end{aligned} \tag{5.8.2}$$

Table 5.9 Energy Dissipated in EAS

Mode	Energy Dissipated (GeV) Altitude Size Sea Level 5×10^5	Energy Dissipated (GeV) Altitude Size 3860m 3.5×10^5
Ionization/N (GeV)	11.5 ± 2.9	9.7 ± 2.9
E-component/N (GeV)	0.5	2.1 ± 0.5
M-component/N (GeV)	1 }	$2.5^{+2.5}_{-0.6}$
Neutrinos/N (GeV)	0.3 }	
N-component/N (GeV)	0.5	$0.9^{+0.9}_{-0.6}$
Sum/N	14 ± 3	17^{+5}_{-3}
$E_0(10^{15}$ eV)	7 ± 2	$0.60^{+0.18}_{-0.11}$
Frequency ($>E_0$) (10^{-10} cm^{-2} sec^{-1} sr^{-1})	0.10 ± 0.02	5.3 ± 1.1
Reference	Greisen 56	Nikolsky 62

for $N = 5 \times 10^5$. The frequency of vertical showers with sizes greater than 5×10^5 is given by

$$\Phi_v(N \geq 5 \times 10^5) = (1.0 \pm 0.2) \times 10^{-11} \text{ cm}^{-2} \text{ sec}^{-1} \text{ sr}^{-1}. \quad (5.8.3)$$

The results in (5.8.2) and (5.8.3) give one point in the primary energy spectrum.

Since the energy spent by ionization in the atmosphere takes the greatest part in the balance sheet, the result depends most sensitively on its estimate. A more direct method is provided by the observation of Čerenkov light produced by EAS. Čerenkov light is emitted mainly by electrons of energies greater than the threshold energy—which is 20 to 100 MeV, depending on the air density—and the flux of Čerenkov light is proportional to the track length. The intensity of Čerenkov light observed at Pamir (Chudakov 60) was used in estimating the track length above the observation level. After adding the energy lost for ionization by electrons of energies below the Čerenkov threshold by reference to cascade theory, the total energy spent by ionization is found to be 9.7 ± 2.9 GeV per electron for a shower size of 3.5×10^5 (Nikolsky 62).

The energies shared with various components at the observation level are estimated in a way similar to Greisen's, and the results are listed also in Table 5.9. The total energy of EAS of $N = 3.5 \times 10^5$ at Pamir is thus obtained as

$$E_0 = (17^{+5}_{-3})N(3860\text{ m})\text{ GeV} = (6.0^{+1.8}_{-1.1}) \times 10^{14}\text{ eV}. \quad (5.8.4)$$

The frequency of EAS with sizes greater than 3.5×10^5 at Pamir is given by

$$\Phi_v(N > 3.5 \times 10^5, 3860\text{ m}) = (5.3 \pm 1.1) \times 10^{-10}\text{ cm}^{-2}\text{ sec}^{-1}\text{ sr}^{-1}. \quad (5.8.5)$$

The results in (5.8.4) and (5.8.5) give another point in the primary spectrum.

5.8.2 *Energy Spectrum of Primary Particles*

The primary energy spectrum has been often derived from the size spectrum by referring to the size-energy relation given in Table 5.9, as briefly mentioned toward the end of Section 5.4. However, the size-energy relation is not as simple as given in (5.4.6); it is greatly affected by the fluctuations discussed in Section 5.3. A less ambiguous relation is expected if the size near the shower maximum is compared with the primary energy; the maximum size is proportional to the primary energy; and a proportionality constant is insensitive to the fluctuations and the models of EAS assumed over a wide range of sizes.

The maximum size is obtained from the size spectrum at different depths $\Phi(N, x)$. The value of x may be varied by changing the zenith angle θ, according to $x = x_v/\cos\theta$. If the altitude at which the EAS are observed is so high that x_v is small, the shower maximum can be determined from the zenith-angle distribution. This can be performed by plotting the size $N(\Phi, x)$ against x for given frequencies, as shown in Fig. 5.34, which is based on EAS observations in Bolivia (Bradt 65). Thus one obtains the size spectrum at maximum development $\Phi(N_{max})$. The relation between N_{max} and the primary energy E_0 can be obtained at two checkpoints in Table 5.9 as

$$E_0 = 2\,N_{max}\text{ GeV}. \quad (5.8.6)$$

This gives the primary energy spectrum (Bradt 65) as

$$j(>E_0) = (2.0 \pm 0.4) \times 10^{-14}\left(\frac{E}{10^{17}}\right)^{-2.2\pm0.15}\text{ cm}^{-2}\text{ sec}^{-1}\text{ sr}^{-1} \quad (5.8.7)$$

for 8×10^{14} Ev $< E < 4 \times 10^{17}$ eV.

EAS initiated by primary particles of still higher energies have maxima near sea level. The size spectrum for sizes greater than 5×10^7 observed at an atmospheric depth of 820 g-cm^{-2} in New Mexico is considered as representative of the primary spectrum. With use of the relation

$$E_0 = 2N\text{ GeV}, \quad (5.8.8)$$

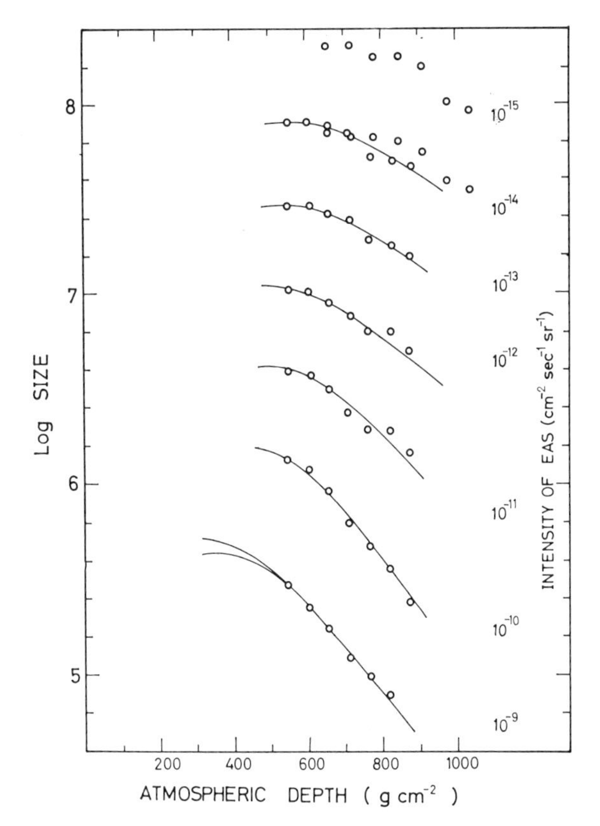

Fig. 5.34 Longitudinal development curves of EAS obtained from constant-intensity cuts on the plots of the size spectra obtained for various zenith angles or effective atmospheric depths. The smooth curves are drawn to fit the data points (Bradt 65). The numbers attached represent the intensity in cm^{-2} sec^{-1} sr^{-1}.

the integral primary spectrum for energies between 10^{17} and 3×10^{19} eV has a power index (Linsley 63*b*):

$$\frac{d \ln j(>E)}{d \ln E} \simeq 1.7. \tag{5.8.9}$$

This means that the spectrum in the highest energy region is less steep than that around 10^{16} eV.

It is remarkable that the largest size recorded is 5×10^{10}, and this corresponds to a primary energy of 10^{20} eV (Linsley 63*a*). Several showers with $N > 10^{10}$ have thus far been recorded, and the primary spectrum seems to extend to 10^{20} eV, with a power index that is close to the one given in (5.8.9).

The energy spectra given in (5.8.7) and (5.8.9) are of great significance for the origin of cosmic rays as well as for the structure of EAS. The power index of the spectrum in a low-energy region is about 1.6—much smaller than that for 8×10^{14} ev $< E < 4 \times 10^{17}$ eV. This causes the change in the size-spectrum slope that takes place at sizes between 10^5 and 10^6. The steepening of the spectrum is considered to be due to the cutoff of Galactic cosmic rays, because cosmic rays of higher rigidities would hardly be trapped in the Galaxy. Above 10^{17} eV cosmic rays of metagalactic origin become dominant, and the observed spectrum represents that of metagalactic cosmic rays.

5.8.3 *Abundance of Heavy Primaries*

It has been suspected that the spectrum of primary cosmic rays is cut off at such a rigidity that particles of rigidities higher than the cutoff value are no longer trapped in the Galaxy. Since the energy of a heavy nucleus with a given rigidity is greater than that of a proton, EAS of the greatest sizes could be produced almost exclusively by heavy primaries (Peters 60). The argument is based on the fact that the shower size is approximately proportional to the total energy of an incident particle. However, this is not exactly the case, as has been explored by the Tokyo air shower group (Fukui 60). A quantity that represents the primary energy is the muon size rather than the total size, as discussed in Sections 5.3 and 5.7. In general the presence of heavy primaries can be observed if the total size and muon size depend differently on the primary energy.

Let us assume that their dependences on the energy of primary protons are expressed by

$$N \propto E_0{}^{\beta_e}, \qquad N_\mu \propto E_0{}^{\beta_\mu}. \tag{5.8.10}$$

If an EAS is initiated by a nucleus of mass number A, this is nearly equivalent to an EAS initiated by A nucleons near the top of the atmosphere. Hence the sizes due to a nucleus of energy E_0 are given by

$$N \propto \left(\frac{E_0}{A}\right)^{\beta_e} A, \qquad N_\mu \propto \left(\frac{E_0}{A}\right)^{\beta_\mu} A. \tag{5.8.11}$$

Thus the mass-number dependences are $1 - \beta_e$ and $1 - \beta_\mu$ in the respective cases.

The relation (5.8.10) gives the N_μ-N relation:

$$N_\mu \propto N^{\beta_\mu/\beta_e} A^{1-(\beta_\mu/\beta_e)}. \tag{5.8.12}$$

The value of β_μ/β_e is about $\frac{3}{4}$, as described in Section 5.7. Therefore a heavy nucleus favors a large muon size. As a result the size of a muon due to a primary iron nucleus is about three times greater than that due to a proton. This leads one to interpret the upper edge of the muon size distribution for fixed N in terms of heavy primaries.

The contribution of heavy primaries to EAS would be quite large if their relative abundance were as large as at low energies. According to our convention the relative abundances are defined by the rigidity or the energy-per-nucleon scale. In the latter case the intensity of heavy nuclei of mass number A with energies greater than E_0 is expressed by

$$j_A\left(>\frac{E_0}{A}\right)=f_A\, j_p\left(\frac{E_0}{AE_c}\right)^{-\alpha},$$

where j_p is the intensity of protons with energies greater than E_c. The relation (5.8.10) gives the size spectrum of EAS arising from heavy nuclei of mass number A:

$$\Phi_A(N)=f_A A^{\alpha}\,\Phi_p(N), \tag{5.8.13}$$

where $\Phi_p \propto N^{-\alpha/\beta_e}$ is the size spectrum of EAS due to protons. The factor A^{α} would result in the heavy primaries contributing as much as protons if f_A were equal to the value at low energies, as given in Section 6.3.

The distribution of muon size discussed in Section 5.7.2 indicates that the value of $\sum_{A\geq 4} f_A A^{\alpha}$ is as small as several percent for 10^{15} eV $<$ $E_0 < 10^{16}$ eV. Section 6.3 discusses indications that the relative abundance of heavy primaries decreases as energy increases, as well as their implications in the origin of cosmic rays.

At $E_0 > 10^{17}$ eV the distribution of muon size is so narrow that EAS of such high primary energies are possibly produced exclusively by protons. This is interpreted to mean that primary cosmic rays of such high energies are of metagalactic origin and that heavy nuclei are destroyed by photodisintegrations in their collisions with starlight photons. In this energy region cosmic rays are unable to be trapped in the Galaxy, and the intensity of Galactic cosmic rays decreases rapidly as rigidity increases beyond a cutoff rigidity.

It has been argued by the Sydney group (Bray 64) that a great fraction of EAS of sizes greater than 10^6 are multicores and are initiated by heavy primaries. Although the relative frequencies of the multicore events observed by other groups are smaller, and it is not clearly understood why a heavy nucleus should produce such a pair of cores that is so widely spread, the following arguments may be referred to. A very-high-altitude collision of heavy nucleus with an air nucleus favors greater lateral spread. If a breakup nucleon undergoes n nuclear collisions, the average transverse momentum may be as large as $\sqrt{n}\, p_T$, where p_T is the transverse momentum of a survival nucleon. If p_T is as large as 1 GeV/c, as discussed in Section 4.4, the average lateral spread is a fraction of a meter for a primary nucleus with a total energy of 3×10^{15} eV. The experimental results obtained by the Sydney group could be interpreted as due to heavy primaries with total energies as large as 3×10^{15} eV. This could be an indication of a Galactic cutoff at a certain rigidity, so that most of the Galactic cosmic rays with total energies greater than the cutoff rigidity times Ze are heavy nuclei.

If the composition of primary cosmic rays changes as energy increases, the structure of EAS may change as the size increases. However, this problem is still open to future investigation.

5.8.4 Primary γ-Rays

As discussed in Section 5.7.2, EAS that are low in muons may be initiated by γ-rays. An EAS initiated by a γ-ray contains muons that are produced mainly by photopion production, but its cross section is much smaller than the cross section for pion production by nuclear active particles. The number of muons that are associated with an EAS is approximately proportional to the total track length of photons with energies greater than several GeV. For an EAS initiated by a primary γ-ray of energy W_0, the track length of photons with energies greater than W is $X_0(W_0/W)$. If the mean free path for photopion production is l, the number of muons in a γ-ray-initiated EAS would be as large as

$$N_{\mu,\gamma} \simeq \frac{X_0}{l}\frac{W_0}{W} \simeq 10^{-3}\frac{W_0}{W}, \tag{5.8.14}$$

since $X_0/l \simeq 10^{-3}$. The number of electrons to be observed depends on shower age. In most cases of interest it is about one-tenth of the electron number at the shower maximum and is slightly smaller than W_0/W. This means that

$$\left(\frac{N_\mu}{N}\right)_\gamma \lesssim 10^{-2}. \tag{5.8.15}$$

This is in contrast to $N_\mu/N \simeq 10^{-1}$ for ordinary EAS. Thus small muon size has been considered to indicate a γ-ray-origin for an EAS (Maze 60, Suga 62*a*).

With this method the Polish-French collaboration group obtained the relative intensity of γ-rays as being about 0.6 percent for $N \simeq 6 \times 10^5$ (Gawin 63), whereas the Bolivian Air Shower Joint Experiment gave a value of 0.02 percent for 10^{15} eV $< E < 10^{16}$ eV at Mount Chacaltaya (Toyoda 65). It is remarkable that the latter failed to detect few-muon EAS for $N > 5 \times 10^6$, thus indicating that the relative intensity of γ-rays is smaller than 5×10^{-5} for primary energies greater than 10^{16} eV.

The discrepancy between the two results mentioned above indicates that the determination of the relative intensity of primary γ-rays based on the N_μ/N ratio is not always unambiguous. A less ambiguous method can be used if one takes account of the possible anisotropy of primary γ-rays. If strong γ-rays are emitted from particular celestial sources, EAS may show peaks in such directions. Since the absolute intensity of such γ-rays is expected to be quite small, a good angular resolution is required for detecting anisotropic primary γ-rays. The fast-timing method described in Section 5.2.3 can achieve an angular resolution of $\pm 5°$. Detection of Čerenkov light emitted in the atmosphere by extremely relativistic electrons has a better angular resolution, say $\pm 2°$, and one can push down the upper limit of the γ-ray intensity by an order of magnitude (Chudakov 62). However, no positive evidence for γ-ray sources has been obtained yet.

5.8.5 Indirect Methods of Detecting Large EAS

The detection of EAS through Čerenkov light differs from conventional methods of detection in that it detects not particles but radiation produced in the atmosphere by particles. There are several other methods that are indirect in the above sense. These indirect methods are not necessarily suitable to investigation of the detailed structure of EAS, but they are useful for observing the gross features of large EAS, such as size and direction of arrival.

The intensity of Čerenkov light is approximately proportional to the size and accordingly to the primary energy. It has been suggested that a coherent effect may enhance Čerenkov radiation if the charge excess takes place in the front of EAS (Askaryan 61). The charge excess of δ gives the intensity of Čerenkov radiation proportional to $(\delta N)^2$ in virtue of the coherent effect, the intensity thus being enhanced by $\delta^2 N$ in comparison with the incoherent Čerenkov radiation. The coherence may rise if the wavelength of radiation is greater than the thickness of the shower front, the latter being about 2 meters (Bassi 53). In fact, a strong radio-wave emission has been observed at 44 MHz (Jelley 65) and at 100 MHz (Porter 65) for EAS of sizes greater than 5×10^6.

An EAS provides not only an emitter but also a scatterer of radio pulses, since it forms a column of dielectrics in the atmosphere. The detection of EAS by means of radar pulses was suggested many years ago (Blackett 41), but its feasibility has been discussed recently on a more refined analysis (Suga 62*b*).

For an EAS of size 10^{10} the ionization density is as large as 10^4 cm^{-3}. Electrons thus formed are lost mainly by attachment, with a mean lifetime of 10^{-7} sec, whereas ions can survive as long as several minutes against recombination. Hence molecular ions are responsible for radio echo. If n ions are formed per meter of an air-shower column located at a distance R, the echo power to be received is

$$2.5 \times 10^{-32} \frac{n^2\lambda^3}{R^3} \left(\frac{m}{M}\right)^2 \frac{1}{(4\pi r_0/\lambda)^2 + 1} \frac{1}{(\lambda \nu_c/\pi c)^2 + 1} PG^2 \text{ watt,} \tag{5.8.16}$$

where λ is the wavelength of a radio pulse, P is the power of the emitter in watts, and G is the antenna gain of an emitter and a receiver. The echo power depends on the ion mass M as $(m/M)^2$, where m is the electron mass; on the wavelength relative to the lateral spread r_0; and also on the collision frequency ν_c, which is on the order of 10^9 sec^{-1}. For $N = 10^{10}$, $P = 100$ kW, $r_0 = 100$ meters, $\lambda = 300$ meters, and $R = 100$ km the receiving power is as large as $(10^{-22}$ to $10^{-23})G^2$ watt. It is therefore feasible to detect EAS of sizes greater than 10^{10} falling on a wide area by integrating the echo power for a certain period.

Another method of observing large EAS is suggested; that is, the detection of light excited by EAS (Suga 62*b*). A primary particle of very high energy impinging on the atmosphere in a nearly horizontal direction develops into an EAS in the upper atmosphere. If the collision frequency is not much greater than the lifetime of a level excited, the energy spent for excitation is effectively converted into light. This may be the case for allowed transitions at an atmospheric depth as small as several tens of g-cm^{-2} or less. In order to avoid the background noise arising from airglow, the first negative band of $N_2^+(\lambda\, 3914$ Å) is considered as a candidate because this is not thermally excited, and its stationary intensity caused possibly by precipitated electrons and cosmic rays is not greater than several $R(1R = 10^6$ photons cm^{-2} sec^{-1}) at medium and low latitudes. The energy spent for excitation by fast charged particles is about 2 percent of the total energy loss, and accordingly a considerable fraction of EAS energy is emitted as the 3914-Å line. Since the light output forms a pulse, as the EAS crosses the field of view of a detector, EAS of sizes greater than 10^9 should be detectable by this method.

The indirect methods described in this subsection are in the development stage; but they will become fruitful in the near future, and the maximum energy of cosmic rays, if it exists, will be determined. New methods other than the above will also become feasible for future investigation of EAS.

REFERENCES

Askaryan 61 Askaryan, G. A., *J. Exp. Theor. Phys. USSR.*, **41**, 616 (1961).

Auger 38 Auger, P., Maze, R, and Grivet-Meyer, T., *Compt. Rend.*, **206**, 1721 (1938).

Bassi 53 Bassi, P., Clark, G., and Rossi, B., *Phys. Rev.*, **92**, 441 (1953).

Bennet 62 Bennet, S., et al., *J. Phys. Soc. Japan*, **17**, Suppl. A–III, 196 (1962).

Blackett 41 Blackett, P. M. S., and Lovell, A. C. B., *Proc. Roy. Soc.*, **117**, 183 (1941).

Blatt 49 Blatt, J. M., *Phys. Rev.* **75**, 1584 (1949).

Bradt 65 Bradt, H., et al., *Proc. Int. Conf. on Cosmic Rays at London*, **2**, 715 (1965).

Bray 64 Bray, A. D., et al., *Nuovo Cim.*, **32**, 827 (1964).

Brown 49 Brown, W. W., and McKay, A. S., *Phys. Rev.*, **76**, 1034 (1949).

Chatterjee 63 Chatterjee, B. K., et al., *Proc. Int. Conf. on Cosmic Rays at Jaipur*, **4**, 227 (1963).

Chatterjee 65 Chatterjee, B. K., et al., *Proc. Int. Conf. on Cosmic Rays at London*, **2**, 627 (1965).

Chudakov 60 Chudakov, A. E., et al., *Proc. Moscow Cosmic Ray Conference*, **2**, 50 (1960).

Chudakov 62 Chudakov, A. E., et al., *J. Phys. Soc. Japan*, **17**, Suppl. A–III, 106 (1962).

Clark 57 Clark, G. et al., *Nature*, **180**, 353, 406 (1957).

Clark 58 Clark, G., et al., *Nuovo Cim. Suppl.* **8**, 623 (1958).

Cocconi 46 Cocconi, G., Loverdo, A., and Tongiorgi, V., *Phys. Rev.*, **70**, 846 (1946).

Cocconi 49 Cocconi, G., Cocconi-Tongiorgi, V., and Greisen, K., *Phys. Rev.*, **76**, 1020 (1949).

Cocconi 61 Cocconi, G., *Handbuch der Physik*, Vol. 46/1, Springer-Verlag, Berlin, 1961, p. 215.

Cool 51 Cool, R. L., and Piccioni, O., *Phys. Rev.*, **82**, 306 (1951).

Danilova 63 Danilova, T. V., and Nikolsky, S. I., *Proc. Int. Conf. on Cosmic Rays, Jaipur*, **4**, 221 (1963).

Delvaille 62 Delvaille, J., Kendziorski, F., and Greisen, K., *J. Phys. Soc. Japan*, **17**, Suppl. A–III, 76 (1962).

Dobrotin 56 Dobrotin, N. A., et al., *Nuovo Cim., Suppl.* **3**, 635 (1956).

Dobrotin 57 Dobrotin, N. A., et al., *Nuovo Cim., Suppl.* **8**, 612 (1957).

Escobar 63 Escobar, I., et al., *Proc. Int. Conf. on Cosmic Rays at Jaipur*, **4**, 168 (1963).

Fujioka 55 Fujioka, G., *J. Phys. Soc. Japan*, **10**, 245 (1955).

Fukui 60 Fukui, S., et al., *Prog. Theor. Phys. Suppl.*, **16**, 1 (1960).

Fukui 61 Fukui, S., *J. Phys. Soc. Japan*, **16**, 604 (1961).

Gawin 63 Gawin J., et al., *Proc. Int. Conf. on Cosmic Rays at Jaipur*, **4**, 180 (1963).

Greisen 50	Greisen, K., Walker, W. P., and Walker, S. P., *Phys. Rev.*, **80**, 535 (1950).
Greisen 56	Greison, K., *Progress in Cosmic Ray Physics*, **III**, 1 (1956).
Greisen 60	Greisen, K., *Annual Review of Nuclear Science* **10**, 63 (1960).
Higashi 57	Higashi, S., et al., *Nuovo Cim.*, **5**, 597 (1957).
Higashi 62	Higashi, S., et al., *J. Phys. Soc. Japan*, **17**, Suppl. *A*–III, 209 (1962).
Hinotani 63	Hinotani, K., et al., *Proc. Int. Conf. on Cosmic Rays at Jaipur*, **4**, 277 (1963).
Hodson 53	Hodson, A. L., *Proc. Phys. Soc.* (*London*), **A66**, 65 (1953).
Imai 61	Imai, T., et al., *Sci. Pap. IPCR*, **55**, 42 (1961).
Jelley 65	Jelley, J. V., et al., *Nature*, **205**, 327 (1965).
Kamata 58	Kamata, K., and Nishimura, J., *Prog. Theor. Phys. Suppl.*, **6**, 93 (1958).
Kamata 63	Kamata, K., Murakami, K., and Kawasaki, S., *Proc. Int. Conf. on Cosmic Rays at Jaipur*, **4**, 214 (1963).
Kameda 62	Kameda, T., Toyoda, Y., and Maeda, T., *J. Phys. Soc. Japan*, **17**, Suppl. A–III, 270 (1962).
Kraushaar 58	Kraushaar, W., *Nuovo Cim., Suppl.* **8**, 649 (1958).
Kraybill 49	Kraybill, H. L., *Phys. Rev.*, **76**, 1092 (1949).
Lapp 43	Lapp, R. E., *Phys. Rev.*, **64**, 129 (1943).
Linsley 62a	Linsley, J., Scarsi, L., and Rossi, B., *J. Phys. Soc. Japan*, **17**, Suppl. A–III, 91 (1962).
Linsley 62b	Linsley, J., *Phys. Rev. Letters*, **9**, 123 (1962).
Linsley 63a	Linsley, J., *Phys. Rev. Letters*, **10**, 146 (1963).
Linsley 63b	Linsley, J., *Proc. Int. Conf. on Cosmic Rays at Jaipur*, **4**, 77 (1963).
Matano 63	Matano, T., et al., *Proc. Int. Conf. on Cosmic Rays at Jaipur*, **4**, 129 (1963).
Matano 65	Matano T., et al., *Phys. Rev. Letters*, **15**, 594 (1965).
Maze 60	Maze, R., and Zawadski, A., *Nuovo Cim.*, **17**, 625 (1960).
McCusker 50	McCusker, C. B. A., *Proc. Phys. Soc.* (London), **A63**, 1240 (1950).
Miyake 58	Miyake, S., *Prog. Theor. Phys.*, **20**, 844 (1958).
Miyake 63a	Miyake, S., et al., *J. Phys. Soc. Japan*, **18**, 465 (1963).
Miyake 63b	Miyake, S., et al., *J. Phys. Soc. Japan*, **18**, 592 (1963).
Nikolsky 62	Nikolsky, S. I., *Proc. Fifth Interamerican Seminar on Cosmic Rays*, **2**, 48 (1962).
Nishimura 51	Nishimura, J., and Kamata, K., *Prog. Theor. Phys.*, **6**, 262, 628 (1951).
Nishimura 52	Nishimura, J., and Kamata, K., *Prog. Theor. Phys.*, **7**, 185 (1952).
Nishimura 67	Nishimura, J., *Handbuch der Physik*, Vol. 46/2, Springer-Verlag, Berlin, 1967, p. 1.
Oda 62	Oda, M., and Tanaka, Y., *J. Phys. Soc. Japan*, **17**, Suppl. A–III, 282 (1962).
Peters 60	Peters, B., *Proc. Moscow Cosmic Ray Conference*, **3**, 157 (1960).
Piccioni 53	Piccioni, P., and Cool, R. L., *Phys. Rev.*, **91**, 433 (1953).
Porter 65	Porter, N. A., et al., *Phys. Letters*, **19**, 415 (1965).
Shibata 65	Shibata, et al., *Proc. Int. Conf. on Cosmic Rays at London*, 1965.
Sitte 50	Sitte, K., *Phys. Rev.*, **78**, 721 (1950).
Sreekantan 63	Sreekantan, B. V., *Proc. Int. Conf. on Cosmic Rays at Jaipur*, **4**, 143 (1963).
Suga 62a	Suga, K., et al., *J. Phys. Soc. Japan*, **17**, Suppl. A–III, 128 (1962).

Suga 62b — Suga, K., *Proc. Fifth Interamerican Seminar on Cosmic Rays*, **2**, 49 (1962).

Suga 63 — Suga, K., et al., *Proc. Int. Conf. on Cosmic Rays at Jaipur*, **4**, 9 (1963).

Tanahashi 65 — Tanahashi, G., *J. Phys. Soc. Japan*, **20**, 883 (1965).

Toyoda 62 — Toyoda, Y., *J. Phys. Soc. Japan*, **17**, 415 (1962).

Toyoda 65 — Toyoda, Y., et al., *Proc. Int. Conf. on Cosmic Rays at London*, **3**, 708 (1965).

Vavilov 57 — Vavilov Iu. N., Evstingneev, Iu. F., and Nikol'skii, S. I., *J. Exp. Theor. Phys. USSR*, **32**, 1319 (1957). (*Soviet Phys. JETP*, **5**, 1078 (1957)).

Vernov 62 — Vernov, S. N., et al., *J. Phys. Soc. Japan*, **17**, Suppl. A–III, 213 (1962).

Williams 48 — Williams, R. W., *Phys. Rev.*, **74**, 1689 (1948).

CHAPTER 6

Origin of Cosmic Rays*

We have thus far discussed the interactions of cosmic rays with matter, together with the morphology of cosmic ray phenomena on earth. There is another aspect of cosmic ray physics—the behavior of cosmic rays in outer space. The emphasis in cosmic-ray studies has gradually moved toward this aspect in recent years, and progress has been very rapid, especially since space probes have become available. This is particularly the case for phemomena that take place in the magnetosphere and interplanetary space, because space probes can directly touch this part of space. The situation seems to be somewhat different for phenomena outside the solar system, since we still have to get information transmitted over long distances in space; nevertheless, new channels have become available by means of observation at the top of and outside the atmosphere. The new components that have thus become observable are closely connected with the interactions of cosmic rays with celestial matter.

Accordingly, the behavior of cosmic rays in celestial space may be discussed with that on earth. Since cosmic ray interactions with matter reveal themselves when particles travel over long distances, we are less concerned with cosmic rays in the magnetosphere and interplanetary space (unless related to the subjects to be discussed) and more concerned with the astrophysical aspect. This implies that the present chapter is devoted to high-energy astrophysics, for which high-energy particles and radiation are responsible. Thus our discussions are not restricted to the behavior of relativistic particles but are extended to non-relativistic particles, radio emission, X-rays, and so forth.

* A substantial part of this chapter is based on S. Hayakawa, *Lectures on Astrophysics and Weak Interactions*, Brandeis Summer Institute in Theoretical Physics, Vol. II, Brandeis University, Waltham, Mass., 1963, pp. 1–164, as well as on articles in *Prog. Theor. Phys. Suppl.*, Nos. 30 and 31, 1964.

6.1 Astronomical Information

Cosmic radiation is one of the important means of obtaining astronomical information since it brings fragments of matter from inaccessible regions. However, the quality and the quantity of astronomical data thus far obtained from cosmic rays are poorer than those from optical astronomy and can hardly compete with those from radio astronomy. At the present stage of cosmic-ray studies, the cosmic-ray data are regarded as supplementing the information obtained by optical and radio means, and the latter is referred to in the discussion of the origin of cosmic rays. The present section is devoted to a brief description of the astronomical information that is necessary for cosmic-ray studies.

6.1.1 Structure of the Galaxy

Stars in our Galaxy are mostly concentrated in a *disk* with a thickness of a few hundred parsecs, but a small fraction of them are located away from the disk and form globular clusters. These stars and the family of high-velocity stars have orbits inclined to the disk, so that they seem to belong to a spherical system; they are usually population II type stars. Most of the stars in the disk system belong to population I. The density of stars in the disk is about 0.1 $(pc)^{-3}$ near the solar system but increases to about 10^2 $(pc)^{-3}$ in the nucleus. The total number of them in the Galaxy is on the order of 10^{11}. Bright stars (O and B types) are found to form the *arms* in spiral galaxies.

The sun is a star of standard size. Its mass and luminosity are

$$M_\odot = 1.99 \times 10^{33} \text{ g}, \qquad L_\odot = 3.90 \times 10^{33} \text{ erg-sec}^{-1},$$

respectively. It is located near the central plane of the disk at a distance of about 10 kpc from the Galactic center; in reference to the arm structure its position is near the inner edge of the Orion arm.

The arm structure is better understood by the distribution of gases. Most of the gases are known to be formed from neutral atoms and molecules. The region occupied mostly by neutral hydrogen atoms is called an H I cloud and is observed with the aid of a characteristic hyperfine line with a wavelength of 21 cm. The H I clouds occupy about one-tenth of the disk volume, and the average atomic density and the temperature therein are about 10 cm^{-3} and 100°K, respectively. Typical rms velocities are on the order of 10 km-sec^{-1}. Nine-tenths of the Galactic-disk volume is occupied by ionized gases, called the H II regions. The average particle density and the temperature are thought to be approximately 0.1 cm^{-3} and 10^4 °K, respectively. There is an indication that hydrogen molecules may be as abundant as hydrogen atoms; the abundance of the former is estimated to lie between 10 times and one-tenth of that of the latter. There are other types of molecules, among which free radicals such as OH have been spectroscopically identified in H I clouds, although their density is as small as 10^{-7} cm^{-3} or smaller.

The Doppler shifts of the 21-cm line indicate the rotation of the Galaxy. The angular velocity depends on the radial distance, as shown in Fig. 6.1, exhibiting a differential rotation tending to wind up the arms around the Galactic nucleus. The present shape of the arm structure would result if, about 4×10^8 years ago, the Galaxy had had two straight arms and had begun to rotate with an angular velocity equal to the present value. This would suggest that the arms are formed as a stationary pattern of waves excited and sustained in the interstellar gas differentially rotating in the disk.

A large radial velocity is observed for the gases flowing out from the nucleus. The rate of mass loss is about 1 $M_\odot$ per year, so that the nucleus would become empty within 10^8 years if no matter were supplied. It is inferred that matter may be supplied from the halo, thus suggesting the circulation of matter from the nucleus to the disk, then to the halo and back to the nucleus again.

The nucleus seems to contain a shell structure; for example, there appears to be a ring at a radius of 0.5 kpc from the center with angular velocity equal to 265 km-sec^{-1} and a hydrogen density of 1 cm^{-3}. The hydrogen density first decreases and then increases to a high value as the distance from the center decreases. A complicated structure is also

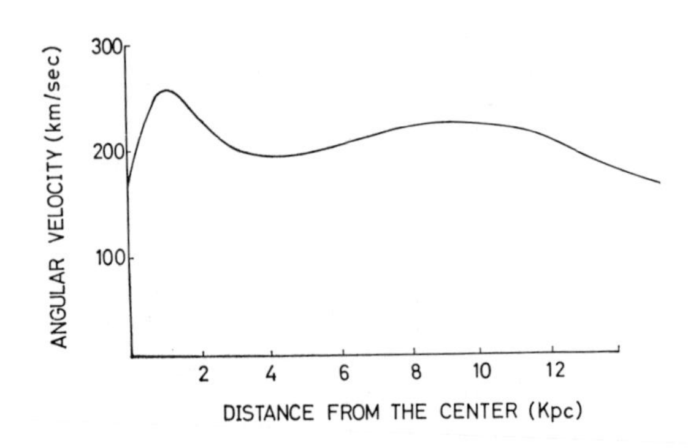

Fig. 6.1 Radial dependence of the angular velocity in the disk.

seen from the radio intensity distribution. One can thus see the active nature of the Galactic core.

There are dust grains in interstellar space, as indicated by the interstellar reddening and the polarization of starlight, which are believed to be due to absorption and scattering. These are considered to be caused by grains with an average size of a few tenths of a micron. A different possibility has been proposed—that macromolecules with a chain length of 10 to 100 Å are responsible for these processes. If grains are responsible, their mass density may be as small as 10^{-26} g-cm^{-3}, whereas the mass density would be much smaller if the latter is the case.

In either case the contribution by these bodies to the total mass of interstellar matter is of minor importance. The total mass, shared between neutral and ionized hydrogen, is a few percent of the whole mass of the Galaxy, and the average density is on the order of 10^{-24} g-cm^{-3}. To this the still uncertain mass of molecular hydrogen should be added. The spatial distribution of the density of atomic hydrogen in the disk is shown in Table 6.1. In the halo hydrogen is supposed to be

Table 6.1 Spatial Distribution of Matter Densities in the Galactic Disk

		Density			
			Gas		
R (kpc)	Mass within R (10^9 $M_\odot$)	Stars (10^{-23} g-cm^{-3})	(10^{-23} g-cm^{-3})	(atoms cm^{-3})	Period of Revolution (10^6 years)
0.01	0.03	24000	100	400	0.5
0.02	0.07	7800	10	40	1.0
0.1	0.92	710	0.5	2	3.1
0.5	8.3	42	0.2	1	12
3	16	69	0.1	0.4	96
10	60	1.0	0.2	1	280

mostly ionized, but its density is known only very approximately, probably as low as 10^{-26} g-cm^{-3} or lower.

The halo is a nearly spherical system extending on both sides of the Galactic disk. Globular clusters are found in this region. Another possible feature of the halo is its continuous-spectrum radio emission. The shape of the halo seen by radio emission is spheroidal. Its long radius is about 12 kpc and lies in the disk, and its short radius is about three-fifths as large. However, a considerable fraction of radio astronomers doubt the existence of the radio halo.

The radio emission would, if it exists, indicate the presence of magnetic fields in the halo, because it is likely to be due to synchrotron radiation from relativistic electrons. The intensity of the radio continuum is stronger toward the disk, thus indicating the presence of magnetic fields in the disk as well. This is also supported both by the stability of spiral arms against gravitational instability and by the polarization of starlight. The latter is interpreted as due to scattering by grains or chain molecules oriented by a magnetic field. The near uniformity of the polarization along an arm suggests that the magnetic field in an arm is rather uniform, probably lying along the arm with small irregularities of about 10 percent. Both the stability and the polarization suggest a magnetic-field strength on the order of 1 gamma, whereas the Zeeman splittings of the 21-cm absorption line in the direction of several radio sources indicate fields ranging from a few gamma to zero, depending on the direction measured. The magnetic-field strengths in the arms and the halo are closely related to cosmic rays and will be discussed later.

6.1.2 Abundances of the Elements

The relative abundances of elements in the Galaxy are determined from various sources. They are surprisingly homogeneous if one considers that the objects examined are considerably different from one another.

The abundances in the earth's crust are obtained by mineralogical and geological studies, and those in the whole earth are supplemented by geophysical means. Although a considerable error may be introduced by the averaging process over inhomogeneous samples, the earth provides a reliable source in obtaining the abundances of all elements, including the rarest ones and all stable and long-lived isotopes.

Meteorites are the only extraterrestrial objects to which chemical analysis can be applied. They are more homogeneous than the samples obtained from the earth; the relative abundances obtained from meteorites are similar to those in the earth, suggesting a common origin. The differences can be attributed to chemical fractionation and the effects of cosmic ray irradiation.

Both the earth and meteorites are lacking in the volatile elements such as hydrogen and helium. They are considered to have escaped long ago, because the gravitational forces are insufficient to prevent loss due to thermal motions. This is verified by the solar abundances of such volatile elements which are very large, whereas those of non-volatile ones are nearly the same as in the earth and meteorites. In fact, helium was discovered spectroscopically first on the sun, and hydrogen is the most abundant element in the sun. Moreover, a large abundance of hydrogen is observed in the atmosphere of large planets such as Jupiter, but not on small planets, (e.g., Mercury, Venus, and Mars). Except for such volatile elements and lithium, the relative abundances in meteorites (Suess 56) are in essential agreement with those in the sun (Aller 61), as seen in Table 6.2.

Outside the solar system the abundances in stars are observed to be in fair agreement with the solar abundances. In stars like the sun, however, some elements such as the rare gases cannot be observed, because the surface temperature is too low to excite these atoms. Their abundances are therefore determined only in high-temperature stars of O, B, and A types—and in planetary nebulae.

Table 6.2 gives a comparison of the logarithms of abundances obtained from these three sources; those in the sun and high-temperature stars are normalized to hydrogen, and those in the sun and in meteorites are normalized to silicon. They agree with each other within a factor of 3 except for lithium. This indicates that matter is well "mixed" in the galaxy and that the cosmic abundances make sense.

By looking at Table 6.2 we can notice the following characteristic features:

1. Hydrogen is far more abundant than other elements, and helium is next most abundant.
2. Lithium, beryllium, and boron are far less abundant than neighboring elements. Lithium is underabundant in the sun.
3. Carbon, nitrogen, and oxygen follow hydrogen and helium, and are more abundant than the heavier elements.
4. The abundances between $A = 20$ and $A = 100$ decrease with increasing A roughly exponentially except at peaks of $4N$ nuclei and the iron group.
5. Elements heavier than $A = 100$ have nearly equal abundances except at the double peaks that appear near the magic numbers of neutrons.

The general features above are interpreted in the following way (Burbidge 57, Hayashi 62). In the early epoch of the Galaxy matter consisted mostly, if not entirely, of hydrogen. When the matter forms stars, hydrogen burns up to helium in their central region by the p-p chain or the C-N-O cycle. When hydrogen is depleted and the central temperature increases as high as 10^8 °K, helium begins to burn by the 3α reactions and subsequent α-capture, and turns into carbon and oxygen. A further increase in temperature results in heavy-ion reactions such as C-C, and produces neon and heavier elements. The α-capture

Table 6.2 Abundances of the Elements (Logarithmic Scale)

Atomic Number and Element	Abundance		
	Sun	High-Temperature Stars	Meteorites
1 H	12.00	12.00	
2 He		11.21	
3 Li	0.96		3.50
4 Be	2.36		2.70
5 B			2.80
6 C	8.72	8.30	
7 N	7.98	8.18	
8 O	8.96	8.78	
9 F		(6.55)	3.98
10 Ne		8.72	
11 Na	5.44		6.14
12 Mg	7.40	7.95	7.46
13 Al	6.20	6.22	6.48
14 Si	7.50	7.45	7.50
15 P	5.34	5.49	5.20
16 S	7.30	7.49	6.49
17 Cl		6.22	4.82
18 A		6.92	
19 K	4.66		5.00
20 Ca	6.15		6.19
21 Sc	2.80		2.95
22 Ti	4.68		4.89
23 V	3.70		3.84
24 Cr	5.02		5.39
25 Mn	4.90		5.34
26 Fe	6.57		7.28
27 Co	4.64		4.76
28 Ni	5.91		5.94
29 Cu	3.50		3.83
30 Zn	3.52		3.76
31 Ga	2.51		2.56
32 Ge	2.49		3.31
33 As			
34 Se			
35 Br			
37 Rb	2.48		2.31
38 Sr	2.70		2.78
39 Y	3.20		2.45
40 Zr	2.65		3.24
41 Cb	2.30		1.40
42 Mo	2.30		1.88
44 Ru	1.82		1.82
45 Rh	1.37		1.35
46 Pd	1.27		1.61
47 Ag	1.04		1.04
48 Cd	1.66		1.78
49 In	1.28		0.91
50 Sn	2.05		1.62
51 Sb	0.42		0.58
52 Te			0.70
53 I			1.68
55 Cs			1.61
56 Ba	2.50		2.44
57 La			1.82
58 Ce			1.86
59 Pr			1.48
60 Nd			2.02
62 Sm			1.54
63 Eu			0.95
64 Gd			1.70
65 Tb			1.22
66 Dy			1.80
67 Ho			1.26
68 Er			1.70
69 Tm			0.96
70 Yb	2.28		1.68
71 Lu			1.18
72 Hf			1.24
73 Ta			1.00
74 W			2.61
75 Re			0.20
76 Os			1.48
77 Ir			0.99
78 Pt			1.68
79 Au			0.65
80 Hg			−0.72
81 Tl			0.54
82 Pb	1.33		0.59
83 Bi			0.82
90 Th			0.02
92 U			−0.25

processes following the photodissociation of neon into helium are also responsible for the synthesis of $4N$ nuclei. Neutrons may be produced, for example, by the $^{21}Ne(\alpha, n)^{24}Mg$ reaction, and they are successively captured to produce heavier odd nuclei. Such neutron-capture processes are also responsible for the generation of elements heavier than iron, thus producing one of the double peaks. Iron is the most stable nucleus and may be formed in equilibrium under high temperature. When all elements are turned into the iron group, no nuclear fuel is left and an increase in pressure results in the decomposition of iron nuclei into lighter ones, α, p, and n. Rapid reactions then take place by the captures of α, p, and n to form heavier nuclei. The rapid neutron-capture processes may be responsible for the synthesis of elements heavier than iron and possibly heavier than lead. Proton capture seems to be able to produce proton-rich nuclides.

In the course of the element systhesis sometimes violent processes occur. One such example is the helium flash, which takes place in a degenerate core in which the 3α reaction is triggered. Another example is the explosion of the core after the implosion associated with the decomposition of iron nuclei. These processes blow up the outer envelopes of stars and are likely to be identified with supernovae. In addition to such explosive processes, the injection of matter may occur for red giant stars with extensive envelopes. All these injection processes contribute matter to interstellar gases and enrich the heavy element abundances in interstellar matter.

Lithium, beryllium, boron, and deuterium are not produced by the thermonuclear reactions and capture processes such as described above. They may also be associated with the efficient convection that exists between the inner and surface regions of many stars, so that some elements existing on the surface are destroyed and the elements synthesized in the inner region are transported to the surface before they are destroyed. In fact, the abundance anomaly is often observed for stars that seem to be convective.

All the above discussion is still qualitative and consequently is open to future quantitative studies. However, the following summary may give the present status of thermonuclear processes (Hayashi 62). Table 6.3 shows important nuclear reactions from hydrogen to iron, together with their main products, temperatures at which reactions occur, and energies released thereby. It should be noticed that the temperatures for the various reactions hardly overlap at all, so that one nuclear burning process corresponds to each stage of stellar evolution. The rates of energy generation by respective processes are shown in Fig. 6.2. This also demonstrates that a single process is dominant in a given

Table 6.3. Nuclear Burning from Hydrogen to Iron

Fuel	Main Reactions	Main Products[a]	Temperature (°K)	Energy Release (10^{17} erg-g^{-1})
Hydrogen	$4\,^1H \rightarrow {}^4He$	^4He	$(1 \sim 4) \times 10^7$	60
Helium	$3\,^4He \rightarrow {}^{12}C$, $^{12}C(\alpha, \gamma)^{16}O$	^{12}C, ^{16}O	$(1 \sim 3) \times 10^8$	$5.8 \sim 8.6$[b]
Carbon	$2\,^{12}C \rightarrow {}^{20}Ne + {}^4He$, $^{23}Na + {}^1H$	^{20}Ne, ^{24}Mg	$(6 \sim 7) \times 10^8$	~ 4
Neon	$^{20}Ne(\gamma, \alpha)^{16}O$ $^{20}Ne(\alpha, \gamma)^{24}Mg$	^{16}O, ^{24}Mg	1.1×10^9	~ 2
Oxygen	$2\,^{16}O \rightarrow {}^{28}Si + {}^4He$ $^{31}P + {}^1H$	^{24}Mg, ^{28}Si, ^{32}S	1.3×19^9	~ 4
Sulfur	$^{32}S(\gamma, \alpha)^{28}Si$ $^{24}Mg(\alpha, \gamma)^{28}Si$	^{24}Mg, ^{28}Si	1.6×10^9	
Magnesium	$^{24}Mg(\gamma, \alpha)^{20}Ne$[c]	^{28}Si	1.8×10^9	~ 3
Silicon	$^{28}Si(\gamma, \alpha)^{24}Mg$[d] $^{28}Si + \alpha + p + n \rightarrow$	^{56}Fe	2.0×10^9	

[a] Unprocessed elements are included.
[b] The two values correspond to the cases where the final products are all carbon and all oxygen respectively.
[c] Followed by neon and oxygen burning.
[d] Followed by magnesium, neon, etc., burnings to form iron.

temperature range. Figure 6.2 also shows the rates of energy generation by neutrino processes; they do not contribute to the luminosity of stars but, if they are effective, only to the evolution of stars.

6.1.3 *Evolution of Stars*

The density of interstellar gases is not uniform, and its irregularities form clouds. When the density in a cloud increases and the temperature decreases by radiation, the cloud is subject to gravitational instability and splits into smaller units. If this process proceeds, the cloud turns into a cluster of protostars—and eventually into a cluster of stars as the protostars contract into stars.

In an early stage of the Galaxy gases are supposed to form a spherical system, and the distribution of such clusters may have been spherical; some of them remain as globular clusters. Stars belonging to population II are believed to be those formed in this stage and contain few heavy elements.

Clouds in the spherical system have a considerable chance of colliding and losing their angular momenta, so that they fall into a plane that is perpendicular to the rotation axis and thus form the Galactic

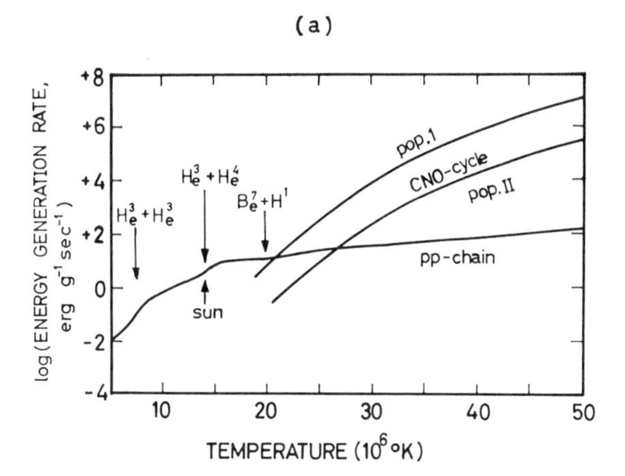

Fig. 6.2*a* Energy-generation rates of hydrogen burning versus temperature for a density of 10^2 g-cm^{-3} and a hydrogen concentration of 0.5. The C-N-O concentration is 0.016 for population I and 0.00064 for population II. The temperatures at which $^3He + {^3He}$, $^3He + {^4He}$, and $^1H + {^7Be}$ reactions participate in the *p-p* chain are indicated.

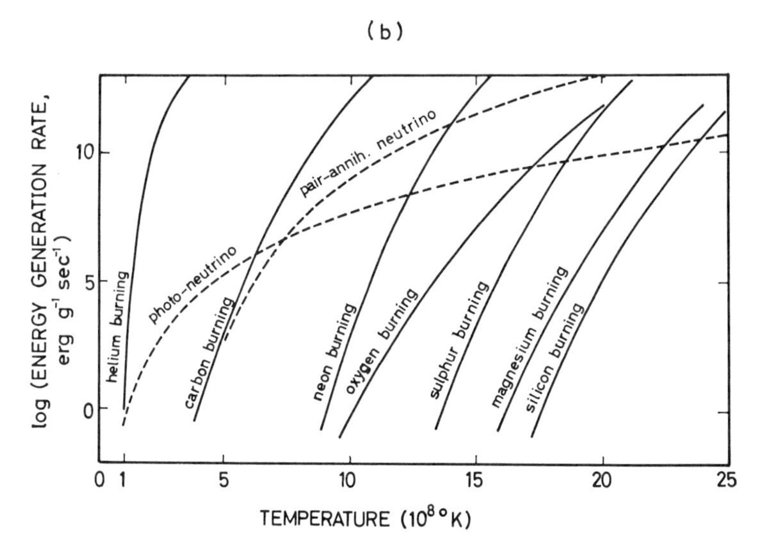

Fig. 6.2*b* Energy-generation rates of helium burning to silicon burning, and neutrino energy-loss rates, versus temperature. The concentration of each nuclear fuel is taken as 100 percent; the density as 10^4 g-cm^{-3} for helium burning and 10^5 g-cm^{-3} for all other curves (Hayashi 62).

disk. In this course a fractionation process may take place in such a way that gases that have fallen in the disk are relatively enriched in heavy elements. Stars formed from these gases belong to population I and are rich in heavy elements, in accordance with observation.

When a star is formed, it contracts by gravitation, converting the gravitational energy into radiation rather rapidly. The luminosity thus increases, and the star flares up. Then the contraction is gradually slowed down, and consequently the luminosity decreases. In this stage the star is wholly convective, and deuterium and lithium may be subject to nuclear reactions in the central part; this explains the underabundance of lithium in the sun. At a certain stage a radiative zone appears, and the star becomes wholly radiative. When hydrogen burning starts in the central region, the star reaches the main sequence. The evolutionary tracks in the Hertzsprung-Russell (H-R) diagram are shown in Fig. 6.3 for a number of representative values of stellar masses. The time scale for a star to reach the main sequence is 10^7 to 10^8 years, depending on its mass.

Stars stay on the main sequence for a considerable time, approximately 10^{10} $(M_\odot/M)^3$ years, until hydrogen is depleted in the core. Thus a helium core is formed, and further contraction results in an increase of central temperature. At this stage the envelope of such a

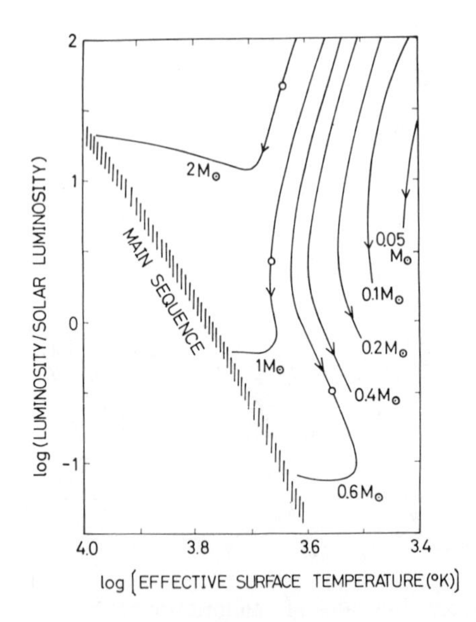

Fig. 6.3 Evolutionary tracks of contracting population I stars in the Hertzsprung-Russel diagram. The open circles denote the ends of the wholly convective stages. Mass values of stars are attached to the respective curves (Hayashi 62).

star expands, so that the star moves to the right on the H-R diagram. Then the hydrogen burning begins to take place mainly at the surface of the helium core. If the central temperature further increases, helium burning suddenly sets in as the helium flash for less massive stars, and then the star comes back toward the main sequence. They come to the left side of the main sequence after ejecting their envelopes by the helium flash, whereas those with hydrogen enevelopes remain in the red-giant branch. Massive stars burn helium less violently.

When helium is depleted, stars expand again until carbon starts to burn. The evolutionary tracks in these stages are shown in Fig. 6.4.

When stars are red giants and red supergiants, they have convective envelopes, and elements generated in the inner parts may be transported to their surfaces. These stars are also unstable and eject mass steadily. Mass ejection also takes place at the helium flash and at the

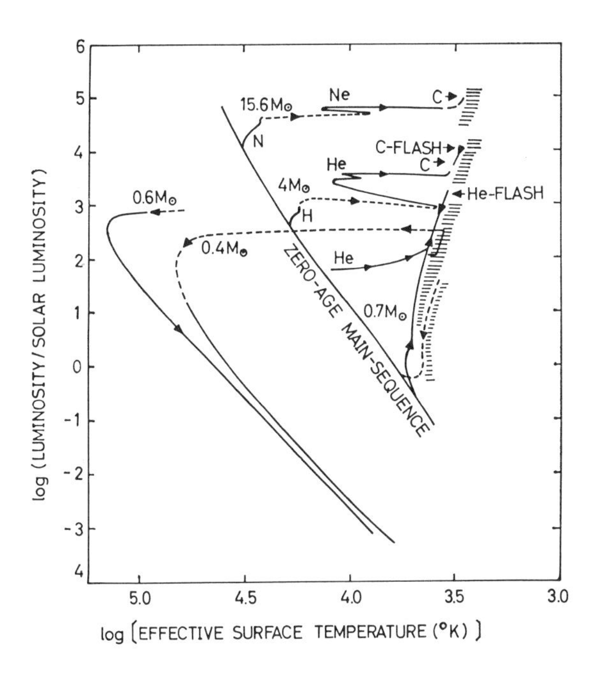

Fig. 6.4 Evolutionary tracks in the phase of helium burning together with the phases of hydrogen and carbon burning for stars of 15.6 $M_\odot$, 4 $M_\odot$ and 0.7 $M_\odot$ stars of 15.6 $M_\odot$ and 4 $M_\odot$ belongs to population 1, whereas those of 0.7 $M_\odot$ and contracting 0.4 $M_\odot$ to population II. A star of 0.6 $M_\odot$ is composed of metal alone. Phases of hydrogen, helium and carbon burning are indicated as H, He and C, respectively. A dashed curve approaching the main sequence represents the sun. The shaded areas represent forbidden regions for quasi-static equilibrium (Hayashi 62).

collapse of the iron core, which may correspond to supernovae of types I and II, respectively. Finally stars turn into white dwarfs or neutron stars and eventually into invisible dwarfs. The ejected mass contributes to interstellar matter, but the present theory is unable, as yet, to predict completely the abundances of elements thus ejected, because one can neither know which parts of the stars are ejected nor how nuclear reactions take place in association with the violent ejection.

6.1.4 Galaxies

Going far out of the Galaxy we find an enormous number of nebulae that resemble our Galaxy. These nebulae are in fact galaxies receding from us with velocity proportional to distance d,

$$v = Hd, \tag{6.1.1}$$

where $H \simeq 10^7$ cm sec^{-1} (Mpc)$^{-1}$ is the Hubble constant; $R = c/H \simeq 3 \times 10^3$ Mpc $\simeq 10^{28}$ cm is the so-called Hubble radius of the universe, beyond which no observation is possible.

$$\tau = \frac{1}{H} \simeq 3 \times 10^{17} \text{ sec} \simeq 10^{10} \text{ years} \tag{6.1.2}$$

is the *Hubble age*, the time for the universe to have passed from a point to the present radius if the universe were to have expanded with the recession velocity given in (6.1.1) though correction is needed.

The galaxies are classified according to their shapes, E (elliptic), S (spiral), SB (barred spiral), and Ir (irregular). The first three are nearly equally numerous, but Ir galaxies made up only about 10 percent of the total. The E galaxies are subdivided into E_n, where $n/10$ is the ellipticity. The S and SB galaxies have spiral arms and their openness increases in the order SO, Sa, Sb, and Sc—or SBO, SBa, SBb, and SBc. Our Galaxy belongs to Sb. It may be better to distinguish between large and small E's. Most distant E's are larger than S and SB, but some E's, such as M 32, are much smaller and resemble the nucleus of S. The size of Ir galaxies lies between those of S and small E. Since small E's at great distances cannot be seen, E hereafter means large E if not otherwise specified.

Roughly speaking, E and SO are spherical, and their average mass is $8 \times 10^{11}\ M_\odot$; whereas S and SB are flat, and their average mass is $3 \times 10^{10}\ M_\odot$. Both our Galaxy and the Andromeda Nebula (M 31) belong to the Sb class but are exceptionally massive, several times $10^{11}\ M_\odot$. The luminosity is not proportional to the mass but approximately to $M^{3/4}$. This is consistent with the spectral types; E and SO are of the dwarf type, and the spectral type shifts to the giant type as the mass decreases.

Several hundreds of these galaxies form a cluster of galaxies. It is also suggested that a number of clusters form a system called the supergalaxy.

About 10^{-4} of galaxies are observed as *radio galaxies*; their radio emission is more than 1000 times stronger than that of our own Galaxy. They are optically bright, on the average $M_{\text{pg}} = -20.5 \pm 0.8$, and their size is also large, about 100 kpc. Most of them consist of pairs of optical objects, and the radio sources do not always coincide with optical sources; in general the latter extend wider than the former. Since the polarization of radio waves is often observed, radio emission is believed to be due to synchrotron radiation. The spectral index of the radio emission lies around 0.7. The optical emission of some radio galaxies is due also to synchrotron radiation.

There are a number of famous radio galaxies; for example, Vir A is identified with M 87 (NGC 4486), a large E-type galaxy. It contains a jetlike structure that emits strongly polarized light. It is located at a distance of 11 Mpc, and the total radio luminosity is about 2×10^{41} erg-sec^{-1}. Cyg A is the strongest radio source, in spite of its distance of 170 Mpc. Hence its absolute radio luminosity is enormous, 4.4×10^{44} erg-sec^{-1}, with which only a few radio galaxies such as 3C 295 can compete. It consists of double optical systems and of large radio-emitting regions. Cen A (NGC 5128) is rather close to us, 4 Mpc, and consists of double sources of nearly equal strength. The Faraday rotation of radio polarization suggests that $n_e \simeq 3 \times 10^{-4}$ cm^{-3} and $H \simeq 5 \times 10^{-7}$ gauss in our local cluster. If $n_e H$ is averaged on a larger scale, however, its value seems to be smaller than 3×10^{-14} cm^{-3} gauss, as seen from the polarization of 3C 295.

The optical features of 3C 295 are starlike but bluer than stars. Galaxies like this are therefore called *quasistellar objects*. These objects may be optically identified by their blueness, and several dozens of such objects have thus far been found by this means. In some of these objects spectral lines are identified as being greatly red-shifted. The origin of the red shift may be considered to be cosmological; that is, due to the expansion of the universe. If this is taken for granted, the quasistellar objects are the farthest galaxies thus far observed, and their optical diameter is as small as several light years.

The quasistellar object may be considered as an explosive stage of a galaxy that will in turn evolve into an active radio galaxy such as Cyg A. The existence of other explosive galaxies of smaller scale, such as M 82, may suggest that the explosion would be a common galactic feature that takes place at a stage of galactic evolution. Apart from the explosion, in which nuclear energy release is presumed to take significant part, the evolution of galaxies is essentially dictated by gas dynamics in gravitational fields. Large spherical galaxies have smaller angular momenta, and gases in them are less turbulent, so that gases rapidly condense into stars in spherical systems. If, on the other hand, a gas cloud has a large angular momentum and is turbulent, it becomes flat, and stars are formed rather slowly. Thus the gross features of various types of galaxies are qualitatively understood.

Galaxies are immersed in intergalactic space, which is believed not to be empty, but information on intergalactic space is very poor. It is generally said that the density of matter is as small at 10^{-5} cm^{-3}, but it could vary by a factor of 10 either way. Properties of intergalactic space will be investigated by means of cosmic rays.

6.2 General Information on the Origin of Cosmic Rays

6.2.1 Observed Facts and Their Direct Consequences

There are several experimental facts that are of primary importance in considering the origin of cosmic rays.

1. Primary cosmic rays consist mainly of *nuclear particles*, whose atomic electrons are completely stripped off. The relative intensity of electrons is a few percent for given rigidity, and other components are weaker still. The abundances of the nuclear particles are similar to the cosmic abundances of elements—but not quite the same; hydrogen is the most abundant, helium is next, and heavier ones are less and less abundant (except for iron). However, similarity does not hold for those that are significantly less abundant than neighboring elements in the cosmic abundances; the abundance of lithium, beryllium, and boron in cosmic rays is comparable to that of carbon in cosmic rays, whereas the former are smaller by many orders of magnitude than the latter in stellar and interstellar matter.

2. Most cosmic-ray particles are of relativistic energy. Their energy spectrum is approximately represented by a power law with a negative power index; the value of the power index changes only within 50 percent, whereas the energy changes by a factor of 10^{10}. The highest energy of a particle thus far detected is as large as 10^{20} eV, greater than 1 joule.

3. The directional distribution of cosmic-ray particles is essentially isotropic; anisotropy, if any, is not greater than 0.1 percent for the bulk of cosmic rays, and it could be a few percent or less at very high energies.

4. The intensity of cosmic rays, as detected by radioactivity dating of meteorites and other methods, has been essentially constant, within a factor of 2 or so, in the past 10^8 years, except for possible abrupt changes over short periods.

5. The energy density of cosmic rays in interstellar space is as large as 1 eV cm^{-3}. This is comparable to the energy densities of starlight, turbulent motion of instellar gas, and magnetic fields.

6. The sun produces relativistic particles in association with solar flares. Their composition is essentially the same as the solar abundances of elements. Few solar particles have energies greater than 10 GeV, and the energy released as solar particles is smaller by many orders of magnitude than the radiation energy.

From these facts we are able to draw the following picture of the origin of cosmic rays:

1. The acceleration of cosmic rays is not so catastrophic as to destroy nuclei, as indicated above by the existence of heavy nuclei in primary cosmic rays. The nearly equal energy densities of cosmic rays, turbulent motion, and magnetic fields suggest energy exchanges among these three modes; in other words, cosmic rays may be accelerated by the motion of a magnetic field arising from the coupling of a magnetic field with turbulent motion.

2. The isotropy and the constant intensity of cosmic rays, as described in (3) and (4) above, suggest that cosmic rays are stored in a spatial region for long periods and are sufficiently stirred therein. Galactic magnetic fields are considered to be responsible for both storage and stirring as well as acceleration. The regions in which cosmic rays are stored may form a hierarchy: the solar system, other cosmic-ray sources, the disk of our Galaxy, our Galaxy including the halo, other galaxies, clusters of galaxies, and the whole universe. The behavior of cosmic rays in the various stages of the hierarchy will be discussed separately in later subsections.

3. Acceleration and storage result in a power energy spectrum, provided that the rate of energy gain is proportional to energy; that is, $dE/dt = aE$. A particle of initial energy E_0 gains energy E after time t, as given by

$$E = E_0 e^{at}. \tag{6.2.1}$$

If the mean lifetime of particles to be stored is τ, the probability that the particle remains until time t in the stored region is $\exp(-t/\tau)$. Since t is related to energy as in (6.2.1), this represents the probability that the particle has energy greater than E:

$$\exp\left(\frac{-t}{\tau}\right) = \left(\frac{E}{E_0}\right)^{-1/a\tau}. \tag{6.2.2}$$

4. The existence of nuclides that are rare in stellar and interstellar matter is interpreted as being due to the fragmentation of heavier nuclides while they propagate in interstellar space. The fragmentation takes place in collisions with matter after cosmic-ray nuclei are accelerated to energies that are high enough for nuclear transformations. The relative intensity of such nuclides indicates that the thickness of matter traversed by cosmic rays is a few grams per square centimeter. This is consistent with the long storage of cosmic rays in a region of low density of matter.

5. The fact that electrons are fewer than protons is due either to the mass dependence of acceleration, or to the mass dependence of energy loss, or both. The energy loss associated with synchrotron radiation is more effective for electrons than for protons. This process should be intimately connected with nonthermal radiation. Electrons can also be produced by energetic nuclear collisions through the production of mesons and their decays. The same collisions can produce γ-rays as well through the decay of neutral pions. The intensity of γ-rays is smaller than that of such electrons, which again supports the theory that charged particles are stored in magnetic fields.

6.2.2 Relation Between Cosmic Rays and Electromagnetic Radiation

The near equality of the energy densities of cosmic rays and starlight might be regarded as a close relation between them. If, however, we consider that cosmic rays wander in magnetic fields and are stored in a spatial region for a long period—whereas starlight propagates rectilinearly—a large difference between their rates of energy generation becomes apparent. Let us denote the linear dimension of the stored region by L and the mean lifetime of cosmic rays stored therein by τ. If their energy densities, which we denote by ε, are the same, the rates of energy generation in these two modes are compared as follows:

$$\frac{\varepsilon}{\tau} \ll \frac{\varepsilon c}{L}; \tag{6.2.3}$$

since the time for starlight to pass through this region, L/c, is much shorter than τ, the rate of energy generation for starlight, the right-hand side of (6.2.3), is much greater than that for cosmic rays.

This is also the case for the sun. Most of the energy of the sun is emitted as visible light. The radiation intensities in the radio and *X*-ray regions are higher than those expected from blackbody radiation. They are due to thermal emission from the corona, which has a much higher temperature than the photosphere. On certain occasions radio emission is enhanced, and its spectrum is no longer thermal. A significant fraction of such events is associated with solar particle emission, and they are therefore believed to be related to high-energy particles.

It is now generally accepted that some types of solar radio emission are caused by magnetic bremsstrahlung or synchrotron radiation, which is emitted by relativistic electrons while they move in a magnetic field, like relativistic electrons in a synchrotron. The solar radio outburst of type IV is believed to belong to this category, since this has a continuous spectrum and is emitted from a wide spatial region extending high above the solar surface.

Since synchrotron radiation is a very efficient mechanism of radio emission, it is natural to assume that this is mainly responsible for nonthermal radio emission and that nonthermal radio sources are strong sources of relativistic electrons and consequently of protons as well. Many radio sources have nonthermal spectra; in some of them—for example, the Crab Nebula and the jetlike part in the central region of the elliptic galaxy M 87—optical emission is also accounted for in terms of synchrotron radiation because of its linear polarization. It is certain that very-high-energy electrons exist in such objects. This leads us to assume that supernova remnants, with which most Galactic radio sources are believed to be identified, are important sources of cosmic rays in our Galaxy, and that radio galaxies supply a significant part of metagalactic cosmic rays.

In addition to localized sources, radio waves are emitted from wide spatial regions. The general Galactic radio emission indicates that Galactic cosmic rays contain a considerable intensity of electrons. A strong intensity of isotropic radio waves in the centimeter region suggests their metagalactic origin; it is speculated that they could be greatly red-shifted thermal radiation emitted in the remotest past.

Electromagnetic radiation such as the above fills up space. Its spectrum has a peak in the visible region in our Galaxy, and the photon density is as high as 0.1 to 1 cm^{-3}. The density of microwave photons in the metagalaxy is as large as 10^3 cm^{-3}. These photons collide with high-energy electrons with the Compton cross section. As a result the electrons lose energy, and some of the photons gain energy. The former is added to the energy loss by synchrotron radiation, whereas the latter may produce X-rays of a continuous spectrum.

Rather strong intensities of X-rays are observed from several localized regions along the Galactic plane as well as uniformly in all directions. One of the sources is located in the central region of the Crab Nebula. Although the mechanism of X-ray emission is open to question, it may well be related to cosmic rays—possibly through the collisions of energetic electrons with matter and photons, and by sharing at least a part with the acceleration of cosmic rays.

The above discussions suggest that radio waves and X-rays should be considered important parts of our subject, the origin of cosmic rays. It is almost needless to remark that electromagnetic radiation of still higher energy belongs to cosmic rays.

6.2.3 Cosmic Rays in the Solar System

It has been suspected, as described in Chapter 1, that the sun is a principal producer of the observed cosmic rays. In fact, the sun produces high-energy particles, and indirectly the solar wind causes the acceleration of particles in its interaction with the magnetic fields of the earth and possibly of Jupiter. These particles can be stored and stirred by interplanetary magnetic fields and be made isotropic. The storage time is estimated as

$$\tau \simeq \frac{L^2}{lv} \simeq 10^8 \text{ sec} \tag{6.2.4}$$

if the linear dimension of the solar system is taken as $L \simeq 10^{15}$ cm, the mean scattering length by magnetic irregularities as $l \simeq 10^{12}$ cm, and the mean velocity of the particles as $v \simeq 10^{10}$ cm-sec^{-1}. Even if the magnetic storage were complete, as in the case of the dipole magnetic field, absorption by planets should limit the mean lifetime to as little as 10^4 years.

If we adopt a moderate value of the lifetime given in (6.2.4), the contribution of solar particles to cosmic rays is found to be not negligible at all. The number of solar particles emitted in a strong solar flare is as large as 10^{33}. Several solar events were observed in the last solar cycle, and most of them were associated with the solar flares in the western quadrant of the near-side hemisphere. More than 10 such events may have taken place over the entire solar surface. This would give the solar production rate as

$$q_s \simeq 10^{26} \text{ sec}^{-1}. \tag{6.2.5}$$

Accordingly the density of solar particles in the solar system is estimated as

$$n_s \simeq q_s \frac{\tau}{L^3} \simeq 10^{-11} \text{ cm}^{-3}. \tag{6.2.6}$$

This is about one-tenth of the cosmic-ray density observed.

Indeed, a component of cosmic rays was found to behave in parallel with solar activities; namely, its intensity was considerable at the solar maximum but inappreciable in a quiet period (Vogt 62). However, this component is limited to a low-energy region. Most cosmic rays, on the contrary, behave in the opposite way. The 11-year variation of cosmic rays clearly shows a decrease in intensity as solar activity increases. The solar modulation of cosmic rays with the solar cycle is interpreted as being due to a diffusion-convection effect in irregular magnetic fields flowing outward with the solar wind.

The relative contributions of the solar and extrasolar components can be observed from their energy spectra. The solar particles associated with flares have an energy spectrum that is much steeper than the spectrum of steady cosmic rays, and their energies rarely reach 100 GeV. A particle of energy higher than 10 TeV cannot be trapped by the interplanetary magnetic field because the strength of the latter is only about several gammas ($1\ \gamma = 10^{-5}$ gauss).

Further evidence is obtained from the difference between the atomic abundances of solar particles and steady cosmic rays, as briefly mentioned in Section 6.2.1. These facts allow us to conclude that most of the cosmic rays are not of solar origin, only a fraction of the low-energy components.

6.2.4 Cosmic Rays in Our Galaxy

If stars such as the sun were mainly responsible for the generation of cosmic rays in the Galaxy, the rate of cosmic-ray production would be

$$q_{\text{stars}} \simeq 10^{37}\ \text{sec}^{-1}, \tag{6.2.7}$$

based on (6.2.5) and about 10^{11} stars in the Galaxy. If cosmic rays were stored in the disk of volume $V_{\text{disk}} \simeq 10^{67}$ cm^3, the mean storage time would have to be

$$\tau \simeq \frac{V_{\text{disk}}\, n_G}{q_{\text{stars}}} \simeq 10^{20}\ \text{sec},$$

since the density of cosmic-ray particles in the Galaxy, $n_G \simeq 10^{-10}$ cm^{-3}. This is much greater than the age of the universe. It is therefore concluded that ordinary stars play only a minor part in the generation of cosmic rays.

The mean storage time can be estimated from the thickness of matter traversed by cosmic rays. As mentioned in Section 6.2.1, the relative intensity of relativistic light nuclei—lithium, beryllium, and boron—can be accounted for in terms of the fragmentation of heavier nuclei that have traversed a thickness of

$$X \simeq 3 \text{ g-cm}^{-2}. \tag{6.2.8}$$

This is proportional to the time spent by cosmic rays traveling from their source to the earth. If cosmic rays were trapped in the disk, in which the density is $\rho_{\text{disk}} \simeq 10^{-24}$ g-cm^{-3}, the travel time would be

$$\tau_{\text{disk}} \simeq \frac{X}{c\rho_{\text{disk}}} \simeq 10^{14} \text{ sec}. \tag{6.2.9}$$

On the other hand, the mean trapping time in the disk can be estimated, according to the diffusion model as in (6.2.4). Taking $L \simeq 10^{21}$ cm to be the disk thickness and $l \simeq 3 \times 10^{19}$ cm equal to the mean distance between interstellar clouds, the value of τ is as small as 10^{12} sec. The mean trapping time could be increased to 10^{14} sec only if the value of l were as small as 0.1 pc.

It is therefore natural to ask if cosmic rays are stored in the halo. The volume of the halo is $V_{\text{halo}} \simeq 3 \times 10^{68}$ cm^3, and the mass density therein is probably not higher than 10^{-26} g-cm^{-3}. If this is the case, matter is mostly concentrated in the disk, and the mean density averaged over the whole Galaxy, including the halo, may be about 3×10^{-26} g-cm^3. This results in

$$\tau_G \simeq \frac{X}{c\rho_G} \simeq 3 \times 10^{15} \text{ sec}. \tag{6.2.10}$$

This is long enough to permit isotropy of cosmic rays, whereas $\tau_{\text{disk}} \simeq 10^{14}$ sec would give a greater anisotropy, which is barely consistent with observation.

The storage time can be directly obtained from the energy spectrum of electrons produced by nuclear collisions in interstellar space (Hayakawa 58a, Fujimoto 64). Nuclear collisions of cosmic rays with interstellar matter produce mesons, which eventually decay into electrons. These electrons are stored in the same way as protons, but they lose their energy by synchrotron radiation and the inverse Compton effect. The energy lost is proportional to the travel time but does not directly depend on the density of matter. Since, moreover, the rate of energy loss is proportional to the square of the energy, the energy spectrum of electrons becomes steeper as the electron energy exceeds a value above which the half lifetime against energy loss is smaller than the electron travel time. The electron spectrum of such a feature could be observed through the radio spectrum; unfortunately the bending energy corresponds to such a high frequency that thermal radio emission masks the nonthermal one.

The cosmic-ray flux and the density of matter in the disk and the halo could be obtained from the intensity of γ-rays, which also arise from meson decays. The observation of γ-rays, has been attempted by many people, but no positive evidence was obtained until the summer of 1967.

Since the experimental knowledge about electrons and γ-rays is not yet definite, it is not possible to decide whether cosmic rays are stored in the entire Galaxy, including the halo, for periods as long as 10^8 years or in the disk for several million years. In either case the ratio of the stored volume to the mean lifetime is on the order of 10^{53} cm^3 sec^{-1}, and the rate of cosmic-ray production is

$$q_G = \frac{n_G V_G}{\tau_G} \simeq 10^{43} \text{ sec}^{-1}. \tag{6.2.11}$$

This gives the rate of energy generation,

$$\langle E \rangle q_G \simeq 10^{41} \text{ erg-sec}^{-1}. \tag{6.2.12}$$

This may be compared with the maximum rate of energy release, 10^{46} erg-sec^{-1}, which is obtained if all nuclear energy available is released at a constant rate over the age of the Galaxy.

The large rate of energy generation as given in (6.2.12) may be attributed to violent phenomena. The most violent phenomenon in the Galaxy is the supernova explosion. A supernova remnant generates nonthermal energy at as high a rate as 10^{38} to 10^{39} erg-sec^{-1}. If the supernova explosions take place on the average once every 100 years and its remnant activity lasts for 10^4 years, it is not impossible to account for the energy-generation rate (6.2.12) in terms of supernovae. The supernova origin of cosmic rays is supported by the nonthermal radio emission discussed in Section 6.2.2 as well as by the overabundance of heavy elements in cosmic rays. The latter point is connected with the evolution of stars, according to which the supernova is the last stage of evolution, and the relative abundance of heavy elements increases as evolution proceeds.

This does not necessarily rule out other sources of cosmic rays. There are many peculiar objects that may be active enough to generate cosmic rays. Some novae may behave like supernovae in the acceleration of particles but on a smaller scale. Peculiar stars and magnetic variable stars of type A show overabundances of heavy elements. High-energy particles may be responsible for the production of such elements, although the other process seems to be more likely. Stars of the T-Tauri type are known to have an overabundance of lithium, which may be a spallation product of high-energy collisions. Supergiants and red giants are so unstable that matter is ejected with a supersonic velocity. Nucleus stars of planetary nebulae belong to an advanced stage of evolution and are active. These candidates may be responsible for the acceleration of particles, although the reality of these possibilities is open to question.

6.2.5 Metagalactic Cosmic Rays

Since nuclear absorption plays a minor part in the lifetime of cosmic rays in our Galaxy, most of the cosmic rays should escape from the Galaxy. Hence cosmic rays are ejected into intergalactic space at a rate of 10^{43} sec^{-1}, as given in (6.2.12). The cosmic-ray-generation rates of other galaxies may be different from this value, depending on the properties of the galaxy, but (6.2.12) would give a reasonable order of magnitude.

Cosmic rays in intergalactic space may also be subject to scattering by intergalactic magnetic fields. The latter are expected to be weak, probably as weak as, or weaker than, 10^{-7} gauss—but strong enough to scatter cosmic-ray particles while they travel in a region of large linear dimensions. If the mean scattering length is $l \simeq 10^{23}$ cm, the diffusion distance of cosmic rays within the age of the universe, $\tau_U \simeq 3 \times 10^{17}$ sec, is estimated, according to (6.2.5), to be as large as 10 Mpc. Since this is smaller than the dimension of a supergalaxy, an inhomogeneity of the cosmic-ray density, if any, may exist in the scale of a supergalaxy.

Apart from the possible inhomogeneity, the average density of cosmic rays can be estimated in the following way. Let the average generation rate of cosmic rays per galaxy be q_g and the mean density of galaxies be n_g. The cosmic-ray density in intergalactic space is given by

$$n_{ig} = q_g n_g \tau_U \simeq 10^{-14}\ \mathrm{cm}^{-3}. \tag{6.2.13}$$

The numerical values obtained are $q_g = 10^{43}\ \mathrm{sec}^{-1}$, $n_g = 2 \times 10^{-75}\ \mathrm{cm}^{-3}$, and $\tau_U = 3 \times 10^{17}$ sec. With the reservation of uncertainty of an order of magnitude, this is smaller by four orders of magnitude than the Galactic intensity.

Direct evidence for the intergalactic intensity of cosmic rays could be obtained from the observation of isotropic γ-rays. The collisions of cosmic-ray nuclei with intergalactic matter and photons give neutral pions, which in turn decay into γ-rays, as in the case of Galactic γ-rays. Since γ-rays of energies smaller than the threshold of pair creation by collision with intergalactic photons are not appreciably absorbed, such γ-rays produced as far as the Hubble radius of the universe can contribute to the observed intensity. Because the Hubble radius is far greater than the radius of the Galaxy, the intensity of the intergalactic γ-rays would have to exceed that of the Galactic γ-rays if the intensity of intergalactic cosmic rays were comparable to that of the galactic cosmic rays. The upper limit on the intensity of γ-rays thus far observed is in favor of the intergalactic cosmic-ray intensity being appreciably smaller than the Galactic intensity.

The distribution between Galactic and intergalactic cosmic rays is meaningful only if there is effective storage of cosmic rays in the Galaxy. For cosmic rays of very high energies—say, energies greater than 10^{17} eV—however, the radius of curvature in Galactic magnetic fields is so large that they are not effectively stored, and metagalactic cosmic rays can freely enter the Galaxy. Moreover, cosmic rays of such high energies could hardly be accelerated in our Galaxy, but they may be generated in the active galaxies having strong radio emission, which have large dimensions and strong magnetic fields. If the rate of cosmic-ray generation in an active radio galaxy is proportional to its radio power, the intensity of very-high-energy cosmic rays can be accounted for in terms of the acceleration of cosmic rays up to 10^{20} eV or greater (Fujimoto 64). The less steep energy spectrum of primary cosmic rays above 10^{17} eV supports this idea.

The active galaxies may be responsible also for the generation of γ-rays and X-rays. However, their contribution would be comparable to the contribution of an ordinary galaxy because of the relative number of the active galaxies. Intergalactic starlight and microwave photons are responsible not only for the absorption of high-energy γ-rays by pair creation but also for the generation of isotropic X-rays by Compton collisions with intergalactic electrons of relativistic energies.

6.3 Composition of Primary-Cosmic-Ray Nuclei

As was described in Section 6.2, primary cosmic rays consist mainly of naked nuclei. Their relative abundances may vary with energy, depending on the sources that are responsible for the generation of cosmic rays in particular energy regions and also on acceleration and modulation operating in space between the sources and the earth. It is therefore legitimate to consider the composition in separate energy regions. This is practical from an experimental point of view because the technique and the accuracy of determination of the composition depend on energy. For heavy nuclei it is convenient to speak of energy per nucleon, instead of the energy of a nucleus.

Below a few GeV modulation in interplanetary space is important, and for very-low-energy particles observations are possible outside the earth's magnetosphere. Between a few GeV and a few tens of GeV interplanetary modulation is moderate, and observations are easier than in other energy regions, particularly because the geomagnetic effect is applicable. Above this energy region the intensity of primary cosmic rays is so low that very little information is available. Above 10^{14} eV extensive air showers provide qualitative data.

6.3.1 Composition in the GeV Region

In the energy region between a fraction of a GeV and several tens of GeV heavy primaries have been observed mainly at balloon altitudes. Their relative abundances are derived after the atmospheric correction, as described in Section 4.5. Helium nuclei are found to be about one-seventh of the protons above a given cutoff rigidity, and others are only about 1 percent of the protons. They are so few that one customarily groups several elements into one group, as shown in Table 6.4. This

Table 6.4 Classification of Nuclear Particles in Primary Cosmic Rays

Component	Z	Intensity (m^{-2} sr^{-1} sec^{-1})[a] Rigidity $\geq$4.5 GV	Relative Intensity[1] for given W/per Nucleon	Cosmic Abundance[2]
Protons	1	610 ± 30	100	100
α-particles	2	88 ± 2	5.00 ± 0.2	15
Light nuclei	3 to 5	1.60 ± 0.40	0.090 ± 0.03	5.7×10^{-7}
Medium nuclei	6 to 9	5.70 ± 0.28	0.33 ± 0.02	0.15
Heavy nuclei	$\geq$10	1.94 ± 0.25	0.11 ± 0.02	0.014
Very heavy nuclei	$\geq$20	0.39 ± 0.06[3]	0.03 ± 0.01	0.00071

[a] The intensities here refer to the values in a solar quiet period except for the value of *VH* which was observed in early 1959.

References:

1. C. J. Waddington, *J. Phys. Soc. Japan*, **17**, Suppl. A–III, 63 (1962).
2. A. G. W. Cameron, *Astrophys. J.*, **129**, 672 (1959).
3. K. A. Neelakanten and P. G. Shukla, *J. Phys. Soc. Japan*, **17**, Suppl. A–III, 20 (1962).

grouping is in accordance with the origin of the elements. The light group (L) consists of the elements lithium, beryllium, and boron, which are much less abundant than others in stars and ambient gas, and cannot be generated by thermocuclear reactions but presumably can be by bombardment with high-energy particles. The medium weight group (M) with $6 \leq Z \leq 9$ consists of elements that participate in the C-N-O cycle and may also be connected with the helium-burning process. The heavy group (H) with $Z \geq 10$ consists of elements that are generated mainly by heavy ion reactions, which may be followed by neutron capture processes and also by rapid proton and α-particle reactions. The heavy group may be subdivided into the ordinary heavy group and the very heavy group (VH), because otherwise it covers too wide a range of the charge spectrum. This subdivision seems to be meaningful because there is a valley in the charge spectrum at $Z = 16$ to 19 on the one hand, and very heavy nuclei belong to the iron peak, which is caused by the equilibrium process at very high temperature, or to the silicon burning process also under a high temperature. Nuclei with $Z > 30$ are grouped into the very, very heavy group (VVH).*

The abundances of individual nuclei in the light and medium groups have been observed with considerable accuracy, whereas for heavy nuclei the determination of Z within $\Delta Z < 1$ is so difficult that the accuracy of the estimated abundances is considered to be less than this. The percentage abundances of nuclei heavier than helium shown in Table 6.5 should be read in the light of the above remark.

Among the data shown in Table 6.5, the data from references 1, 2, and 4 are based on observations in a solar active period, whereas the data from reference 3 are based on three observations in a solar quiet period. Data in reference 2 were obtained with Čerenkov counter, whereas the other results were obtained with emulsions. Abundances given in reference 5 represent the weighted average values of available data, among which the greatest weight is put on the data in reference 4, which were obtained from an emulsion exposure at an altitude that was very high compared with that in the other experiments.

Tables 6.4 and 6.5 indicate the following features of the composition of heavy primaries.

1. Although solar activity has a considerable effect on the intensity of heavy primary nuclei, it has practically no effect on the composition (Waddington, 60, 62).

2. The relative abundances are nearly independent of rigidity, although there may be some indication of an increase of L/M and L/H with decreasing rigidity (Aizu 60).

3. The light group is far overabundant in comparison with the cosmic abundances.

* The abundances of VVH nuclei have been observed to be similar to those expected from the rapid neutron capture process.

Table 6.5 Relative Abundances of Heavy Primaries

Z	Group and Element	Rigidity Interval: 1.3 to 2.7 GV[1]	>2 GV[2]	>4.5 GV: 3	4	5	Cosmic Abundances 6
	L group:						
3	Lithium	4.4		5.2	5.3	3.9	2.5×10^{-4}
4	Beryllium	6.6	6.7	4.3	2.3	1.7	4.9×10^{-5}
5	Boron	9.3	10.1	11.9	7.4	11.6	5.9×10^{-5}
	M group:						
6	Carbon	25.5	28.6	25.1	30.1	26.0	22.9
7	Nitrogen	12.8	13.3	14.9	9.7	12.4	5.9
8	Oxygen	20.0	17.9	14.5	19.4	17.9	61.7
9	Fluorine	1.2	4.0	4.0	2.4	2.6	4.0×10^{-3}
	H_3 group:						
10	Neon	2.4	16.6	4.3	23.4	23.9	2.0
11	Sodium	0.5	for H	2.7	for H	for H	0.11
12	Magnesium	4.8		4.7			2.2
13	Aluminum	1.2		0.9			0.23
14	Silicon	2.4		1.6			2.5
15	Phosphorus	0.8		0.2			2.5×10^{-2}
	H_2 group:						
16	Sulfur	1.1		0.0			0.95
17	Chlorine	0.8		0.1			6.5×10^{-3}
18	Argon	0.3		0.5			0.37
19	Potassium	0.3		0.0			7.3×10^{-3}
	H_1 group:						
20	Calcium	1.8		0.9			0.12
≥ 21		4.1		3.9			0.30

References:

1. H. Aizu et al., *Prog. Theor. Phys. Suppl.* No. 16, 54 (1960).
2. F. B. McDonald and W. R. Webber, *J. Phys. Soc. Japan*, **17**, Suppl. A–II, 428 (1962).
3. J. Waddington, *Progress in Nuclear Physics*, **8**, 1 (1960); earlier works are cited there.
4. F. W. O'Dell, M. M. Shapiro, and B. Stiller, *J. Phys. Soc. Japan*, **17**, Suppl. A–III, 23 (1962).
5. C. J. Waddington, *J. Phys. Soc. Japan*, **17**, Suppl. A–III, 63 (1962).
6. A. G. W. Cameron, *Astrophys. J.*, **129**, 672 (1959).

4. The relative abundances in the medium and heavy groups are qualitatively similar to the cosmic abundances; for example, even nuclei are more abundant than odd ones, although the difference between them is smaller in cosmic rays than in celestial elements.

5. Carbon is more abundant than oxygen, whereas the opposite holds in the cosmic abundances.

6. Heavy nuclei are relatively more abundant in cosmic rays than in celestial matter.

7. In the heavy group the abundance of nuclei with $Z = 16$ to 19 is significantly smaller than that of other nuclei.

Table 6.6 Intensities of the Three Subgroups in the Heavy Group

Subgroup	Z	Intensity (m^{-2} sec^{-1} sr^{-1})[1] Rigidity ≥ 4.5 GV	Relative Intensity[a] [2] Energy ≥ 0.7 GeV per Nucleon
H_3	10 to 15	1.67 ± 0.19	17.0
H_2	16 to 19	$\lesssim 0.13$	1.9
H_1	≥ 20	0.69 ± 0.30	3.8
H	≥ 10	2.40 ± 0.32	22.7

[a] The total abundance of nuclei with $Z \geq 3$ is normalized as 100.

References:

1. R. R. Daniel and N. Durgaprasad, *J. Phys. Soc. Japan*, **17**, Suppl. A–III, 15 (1962).
2. M. Koshiba et al., *Phys. Rev.*, **131**, 2692 (1963).

The last point is revealed in Table 6.6 (Daniel 62), although another experiment gave the H_2/H_1 ratio to be 1/2 for nuclei of energies greater than 700 MeV per nucleon (Koshiba 63). The intensity given in (1) of Table 6.6 is a little higher than that given in Table 6.4, but such a difference is rather common among various observations.

6.3.2 *Composition at High Energies*

The composition of heavy nuclei at high energies has been measured in a geomagnetically senstitive region (Waddington 62). No essential difference from the composition at lower energies has been found. At about 10^{12} eV per nucleon an early experimental result indicates nearly the same composition as in the GeV region (Naugle 56), but the ratio of heavy nuclei to protons seems to be about 60 percent of its value at lower energies (H. Aizu 64). Air showers rich in muons with energies above 10^{14} eV per nucleon are inferred as evidence for the existence of heavy nuclei, and such EAS show anisotropy. If heavy nuclei are the sole components that are responsible for the anisotropy, their intensity in the energy-per-nucleon scale should be only a fraction of a percent of the total primary intensity (Hasegawa 62*a*). At energies higher than 10^{17} eV per nucleon the fluctuations of EAS indicate that primary cosmic radiation consists of a single kind of particle, probably protons (Linsley 62), as discussed in Sections 5.7 and 5.8.

Although experimental information is meager, there seems to be an indication that the relative abundance of heavy nuclei decreases as energy increases and vanishes at energies greater than 10^{17} eV. It should, however, be mentioned that there are a number of experimental results for or against the above trend; multicore EAS events are interpreted as being due to heavy primaries with energies of about 10^{16} eV per nucleon, and their frequency suggests an increase in the ratio of heavy nuclei to protons in this energy region compared with that in the GeV region (McCusker 63). The energy dependence of the composition has an important bearing on the origin of cosmic rays, as will be discussed in later sections.

6.3. *Isotopic Abundances*

The abundances of isotopes in primary cosmic rays are also useful in such tasks as identifying cosmic-ray sources, studying the propagation of cosmic rays; for example, from the abundances of isotopes such as ^{2}H and ^{3}He, which are believed to be rare in the universe, we can estimate the path length of cosmic rays in interstellar matter, and from the abundance ratio such as ^{13}C/^{12}C we can select the types of nuclear synthesis in cosmic-ray sources. From the observation of radioactive isotopes such as ^{3}H and ^{10}Be we can obtain information about the lifetime of cosmic rays in the vicinity of the earth and in the Galaxy.

The discrimination of isotopes is possible by conventional means of determining the mass of nonrelativistic particles. With nuclear emulsion the ionization-range method is used for determining the masses of nuclei that stop in the emulsion. With a counter telescope the product of the ionization energy loss and the energy of a particle is measured as in low-energy nuclear physics. In the relativistic region methods based on the ionization process are useless, but nuclear reactions provide a useful method (Hasegawa 61).

In discussing the derivation of isotopic abundances by any of the above means a remark may be necessary on the difference between the rigidity and the energy-per-nucleon scales. The relative abundances depend on which of the scales is used. Let us consider isotopes of charge Ze and mass numbers A and A'. If both of them are fragmentation products or if the lighter one is a fragmentation product of the heavier one, their kinetic energies per nucleon are essentially equal: $W = W'$. Their rigidities are

$$R = \frac{A}{Ze}\sqrt{(W+M)^2 - M^2}, \qquad R' = \frac{A'}{Ze}\sqrt{(W'+M)^2 - M^2}, \qquad (6.3.1)$$

where M is the nucleon rest energy. For $W = W'$ we have

$$R' = \frac{A'}{A} R, \qquad \frac{dR'}{dW} = \frac{A'}{A} \frac{dR}{dW}. \tag{6.3.2}$$

Hence the isotopic ratio in the energy-per-nucleon scale is related to that in the rigidity scale as

$$\frac{(dj/dW)_{A'}}{(dj/dW)_A} = \frac{A'}{A} \frac{(dj/dR')_{R'}}{(dj/dR)_R} \tag{6.3.3}$$

If the rigidity spectra of different nuclei are alike and of a power shape, $dj/dR \propto R^{-\alpha-1}$, the isotopic ratio is expressed by

$$\frac{(dj/dW)_{A'}}{(dj/dW)_A} = \left(\frac{A}{A'}\right)^{\alpha} \frac{(dj/dR)_{A'}}{(dj/dR)_A}. \tag{6.3.4}$$

Since α is positive at relativistic energies, the isotopic ratio in the energy-per-nucleon scale is greater than that in the rigidity scale, In the nonrelativistic region, on the other hand, the value of α is negative, and the former is smaller than the latter. The shape of the spectrum at low energies varies considerably with solar activity, and consequently the difference between these two isotopic ratios changes with time. This provides an important means of understanding the mechanism of solar modulation (Hildebrand 66). This is particularly the case if we compare the intensities of ^{4}He, ^{3}He, and ^{2}H; the last two are considered to be the fragmentation products of ^{4}He and the values of Z/A of ^{4}He and ^{2}H are the same.

The rigidity-energy relation allows us to determine the isotopic ratio if we measure the energy per nucleon at a magnetic latitude. For a given cutoff rigidity the cutoff energy per nucleon is higher for a lighter isotope. The isotopic ratio of helium is thus measured at a cutoff rigidity of 14.6 GV, but the result is found to be rather sensitive to the energy spectrum (Balasubrahmanyan 63).

Keeping the above remarks in mind, we now present experimental results on the isotopic ratios. Attempts have thus far been made to measure the isotopic ratios of hydrogen, helium, beryllium, and carbon but results on beryllium seem to be far from conclusive.

The isotopic ratio of deuterons to protons has been obtained only for kinetic energies of less than 100 MeV per nucleon. The differential energy spectrum of deuterons is found to be approximately proportional to W^2, and the isotopic ratios at a given energy per nucleon and rigidity are (Fan 66):

$$\begin{aligned}(^2\mathrm{H}/^1\mathrm{H})_{60\ \mathrm{MeV/nucleon}} &= 0.05 \pm 0.01 \\ (^2\mathrm{H}/^1\mathrm{H})_{0.7\ \mathrm{GV}} &= 0.005 \pm 0.001\end{aligned} \tag{6.3.5}$$

A more important quantity is the $^2H/^4He$ ratio, since these two nuclides have the same charge-to-mass ratio and their ratio is free from modulation effects. At 60 MeV per nucleon the ratio is about 0.15.

The $^3He/(^3He + ^4He)$ ratio has been measured since 1957, and results prior to 1965 have been summarised by Biswas et al.* The results are rather divergent, due partly to solar modulation effects but probably more to experimental reasons. Here we refer only to some of the latest results that were obtained near the solar minimum. The results given in Table 6.7 demonstrate that the isotopic ratio is energy dependent, as in the case of the $^2H/^4He$ ratio, approximately W^2 below 100 MeV per nucleon. Above 100 MeV per nucleon the ratio seems to be approximately energy independent; the values obtained by Biswas et al. may be regarded as representative.

The isotopic ratio of carbon is estimated by observing in nuclear emulsion the charge-retention collision of carbon nuclei with energies greater than 500 MeV per nucleon (Hasegawa 63). The cross section for this collision is obtained as 76 ± 20 mb, whereas the charge-retention cross section of ^{12}C is estimated by reference to proton and heavy ion collisions with carbon targets as 30 ± 5 mb. The cross section for an outer neutron of ^{13}C to be stripped off is theoretically estimated to be about 80 mb. Using these cross sections we obtain

$$\frac{^{13}C}{^{12}C} \simeq 1. \tag{6.3.6}$$

The overabundance of ^{13}C is supported also by examining carbon-boron reactions. It is reasonable to assume $\sigma(^{12}C \to B) \simeq \sigma(^{12}C \to C)$, whereas $\sigma(^{13}C \to B)$ is smaller than $\sigma(^{13}C \to C)$ because in the latter process the stripping of an outer neutron takes considerable part. In fact, nine $C \to B$ events have been observed in comparison with fourteen $C \to C$ events.

6.3.4 *Fragmentation of Heavy Nuclei in the GeV Region in Interstellar Space*

The composition of heavy nuclei in primary cosmic rays is not the same as that at cosmic-ray sources but is affected by nuclear interactions that occur in their propagation through interstellar space. The nuclear interactions result in the transformation of heavy nuclei into lighter ones, and as a result some nuclei that are practically absent at the source may be produced. As has been pointed out first by Bradt and

* S. Biswas et al., *Proc. Int. Conf. on Cosmic Rays in London*, **9**, 368 (1965).

Table 6.7 Summary of Experimental Values of $^3He/(^3He + ^4He)$

λ	Date	Atmospheric Depth (g-cm^{-2})	Kinetic Energy Interval (MeV per Nucleon)	$\frac{^3He}{^3He + ^4He}$	Rigidity Interval (GV)	$\frac{^3He}{^3He + ^4He}$	Type of Detector	Reference
73°	June 15, 1963	3.1	115 to 210	0.11 ± 0.03	0.85 to 1.12	0.32 ± 0.07	Emulsion	1
69.5°N	May 12, 1965	3.5 to 4.5	80 to 150	0.19 ± 0.035	0.6 to 1.0	0.39 ± 0.09	Scintillator	2
Outside magnetosphere	May 29, 1965		~ 50	0.02 ± 0.01			Solid counter	3
	September 20, 1965		~ 100	0.07 ± 0.013				

References:

1. S. Biswas et al., *Proc. Int. Conf. on Cosmic Rays at London*, **9**, 368 (1965).
2. D. J. Hoffman and J. R. Winckler, *Phys. Rev. Letters*, **16**, 109 (1966).
3. C. Y. Fan et al., *Phys. Rev. Letters*, **16**, 813 (1966).

Peters (Bradt 50), the light nuclei—lithium, beryllium, and boron—in primary cosmic rays should, if at all, have been produced by the nuclear collisions of heavier nuclei with interstellar matter, because their cosmic abundances are far smaller than those of neighboring elements. If we know the partial cross sections for producing light nuclei at such collisions, we can deduce the thickness of matter traversed by cosmic rays. The partial cross sections are estimated by referring to the nuclear disintegrations of some nuclei by energetic protons, since interstellar matter consists mostly of hydrogen. In this way Hayakawa et al. (Hayakawa 58*a*) have estimated the thickness to be about 3 g-cm^{-2}, and Aizu et al. (Aizu 60) have made an estimate of 4 g-cm^{-2}, both being subject to errors of about 20 percent.

Recently Badhwar et al. (Badhwar 62) attempted a more careful investigation, which will be referred to below. They subdivided the H-group into three (H_1, H_2, and H_3) and took into account the partial cross sections to give individual L and M nuclei. In doing so they were able to check the internal consistency of the assumed value of thickness by comparing the L/M ratio and the relative abundances in the L- and M-groups with observed ones. The thickness of matter thus obtained is

$$X = 2.5 \text{ g-cm}^{-2}, \tag{6.3.7}$$

hardly lying outside the range of X between 2 and 3 g-cm^{-2}. The procedure to get this result will be described below.

Let the intensity of the ith component at thickness x be $J_i(x)$. It undergoes nuclear collision with a mean free path l_i, so that its intensity decreases at a rate of $1/l_i$, whereas it increases due to the fragmentation of heavier components with probabilities P_{ji}. Hence the diffusion equation is expressed—if the energy change is neglected—by

$$\frac{dJ_i(x)}{dx} = \frac{1}{l_i} J_i(x) + \sum_{j \geq i} \frac{1}{l_j} P_{ji} J_j(x), \tag{6.3.8}$$

where the summation is carried out for the ith component and heavier ones. This is exactly the same as the diffusion equation of heavy nuclei in the atmosphere, as discussed in Section 4.5, and the solution is expressed in the same way as (4.5.3). For x smaller than $l'_{ij} = l_i l_j / |l_i - l_j|$ the solution can be expressed approximately as

$$J_i(x) \simeq J_i(0)\left(1 - \frac{x}{l'_i}\right) + \sum_{j=1}^{i-1} \frac{P_{ji}}{l_j} J_j(x) x, \tag{6.3.9}$$

where

$$l'_i = \frac{l_i}{1 - P_{ii}}.$$

If i stands for the L-group, we have $J_L(0) = 0$ and

$$\frac{J_L(x)}{\sum_{j=H_1}^{M} J_j(x)} = \left[\frac{\sum_{j=H_1}^{M} (P_{jL}/l_j) J_j(x)}{\sum_{j=H_1}^{M} J_j(x)}\right] x \tag{6.3.10}$$

Hence the value of X can be obtained from the observed intensities of the various components, provided that the collision mean free paths and the fragmentation probabilities are known.

The mean free paths and the cross sections of the various components are given in Table 6.8. Here the H_1- and H_3-groups are represented by

Table 6.8 Cross Sections and Mean Free Paths of Heavy Nuclei in Hydrogen

Component	Representative Nucleus	Inelastic Cross Section (mb)	Interaction Mean Free Path in Hydrogen (g-cm^{-2})
H_1 group	^{52}Cr	635	2.63
H_3 group	^{27}Al	407	4.10
Fluorine	^{19}F	320	5.22
Oxygen	^{16}O	287	5.82
Nitrogen	^{14}N	262	6.37
Carbon	^{12}C	225	7.43

^{52}Cr and ^{27}Al, respectively, and the H_2-group is neglected because of its very small intensity. The fragmentation probabilities of these groups are given in Table 6.9.

Table 6.9 Fragmentation Probability

	Secondary				
Primary	H_1	H_2	H_3	M	L
$H_1(Z = 20$ to $26)$	0.44	0.16	0.14	0.09	0.12
$H_2(Z = 15$ to $19)$					
$H_3(Z = 10$ to $15)$			0.28	0.32	0.27
$M(Z = 6$ to $9)$				0.32	0.48

Some of the fragments that are produced are not stable but eventually decay into elements of different Z. If the fragmentation probabilities are obtained by reference to the direct products of nuclear interactions, they cannot be used for our problem, because the heavy nuclei we observe have traveled for a long time, probably much longer than the lifetimes of ordinary β-decays. The fragmentation probabilities given in Table 6.9 have been corrected for the decay effect, so that all unstable nuclei except ^{7}Be are assumed to have turned into stable ones. For comparison P_{iL}, uncorrected for the decay effect, are shown in Table 6.10.

Table 6.10 Comparison of Fragmentation Probabilities P_{iL}

i	H_1	$H_2 + H_3$	H_3	M	Reference
P_{iL} corrected for decays	0.12		0.27	0.48	1
P_{iL} uncorrected for decays	0.14		0.32	0.48	1
P_{iL} uncorrected for decays	0	0.15		0.40	2

References:

1. G. D. Badhwar et al., *Prog. Theor. Phys.*, **28**, 607 (1962).
2. H. Aizu et al., *Prog. Theor. Phys. Suppl.*, No. 16, 54 (1960).

A comparison is also made for the fragmentation probabilities obtained from hydrogenlike collisions in emulsions (Aizu 60). Although these would have to coincide with those uncorrected for decays, the discrepancy between them is significant. The values of P_{iL} adopted by Aizu et al. are smaller than those given by Hayakawa (Hayakawa 58*a*) and Badhwar et al. (Badhwar 62), and consequently the thickness traversed turns out to be greater. The discrepancy is due mainly to statistics, but one of the reasons for this may be the fact that the fragmentation probabilities given by Aizu et al. give some weight to lower energy particles observed by them; the probability of producing nuclei whose mass numbers are far from those of the target nucleus decreases as energy decreases.

A few comments on the decay effect should be added. The fragmentation effect indicates the thickness of matter traversed but not the absolute length or time of travel in interstellar space. If the average age of cosmic rays were shorter than the mean lifetimes of some unstable nuclei, we would have to use the fragmentation probabilities uncorrected for their decay. Therefore the presence or absence of the decay effect would tell us the age of cosmic rays. The longest-lived suitable nucleus is ^{10}Be, whose mean lifetime is about 4×10^6 years. Since the age of cosmic rays is about 10^6 or 10^8 years, the presence or the absence of ^{10}Be could be used to distinguish between these two alternatives. An unstable nuclide that could survive in cosmic rays may be ^{7}Be, because it can decay into ^{7}Li only by electron capture. The mean free path for picking up a free electron to a K-orbit by a relativistic nucleus of charge Z and of energy per nucleon E is given by (Hayakawa 58a)

$$l\,(\text{electron pickup}) \simeq 6 \times 10^8 Z^{-5} \frac{E}{Mc^2}\ \text{g-cm}^{-2}, \qquad (6.3.11)$$

where M is the nucleon mass. Since this is much larger than the average thickness traversed, ^{7}Be is expected to exist among primary cosmic rays—unless the energy is very low, so that the pickup cross section is large. However, this does not modify P_{iL} because the decay product belongs also to the L-group.

If we are concerned with the abundances of individual nuclides, we have to know the partial cross sections for giving them. The cross sections adopted on the basis of experimental cross sections by Badhwar et al. (Badhwar 62) and an empirical formula given in Section 2.10.4 (Rudstam 62) are reproduced in Table 6.11.* The partial cross sections

Table 6.11 Partial Cross Sections

Target Total Cross Section (mb)	C 225.0	N 262.0	O 287.0	Al(H_3) 407.0	Cr(H_1) 635.0
Product nucleus:					
H_1					287.0
$H_2 + H_3$[a]				115.5	201.0
F[a]				13.8	7.1
O[a]				64.9	28.3
N[a]			112.5	15.5	15.8
C[b]		59.9	45.6	35.4	8.9
^{11}C	23.4	9.5	10.3	4.9	0.65
^{10}C	0.67	1.4	1.6	1.2	0.15
B[b]	48.4	34.4	24.4	34.0	23.4
^{10}Be	6.8	6.7	3.0	5.5	5.8
^{9}Be	15.9	14.8	6.3	12.0	8.3
^{7}Be	10.6	10.5	4.7	8.5	5.9
Li[b]	40.0	39.6	15.7	42.1	29.1
^{6}He	0.6	0.7	0.8	1.3	4.0
αp	78.5	82.7	56.2	54.7	0.0

[a] Including parent nuclides that eventually decay into it.
[b] Stable nuclides.

* A recent experiment has given the cross section for ^{10}Be production smaller by an order of magnitude. This implies that the determination of cosmic-ray age through the observation of ^{10}Be is technically difficult.

giving in the line designated as αp represent those for the complete breakup into α-particles and nucleons directly and indirectly through ^{8}Be, but they do not include production by knockout and evaporation processes; the latter takes a main part in the production of α-particles and nucleons, especially from heavy targets.

The fragmentation process will be first analyzed for groups H_1, H_2, H_3, M, and L by assuming the initial absence of the L-group. By using (6.3.9) as well as Tables 6.8 and 6.9, we obtain the thickness of interstellar matter traversed by relativistic heavy nuclei as $X = 2.5$ g-cm^{-2}, with a possible error not exceeding 30 percent, as given in (6.3.7).

The value of X is checked by calculating the relative abundances of individual nuclei given in Table 6.5(4) and the partial cross sections given in Table 6.11. The results are shown in Table 6.12; the relative

Table 6.12 Relative Abundances[a] of Heavy Nuclei Observed and Expected with $X = 2.5$ g-cm^{-2} (Normalized for Carbon)

Group or Element	Observed[b]	Calculated	Source	Cosmic[c]
H_1	23.4	23.7	27.0	1.9
H_2	54.7	3.6	0.0	5.8
H_3		49.0	51.0	30.7
O + F	72.6	69.5	73.0	269
N	32.3	29.8	17.0	25.8
C	100.0	100.0	100.0	100.0
B	24.6	24.7	0.0	2.6×10^{-4}
Be	7.7	8.7	0.0	2.1×10^{-4}
Li	17.1	15.0	0.0	1.1×10^{-3}

[a] Normalized for carbon.
[b] From Reference 4 in Table 6.5.
[c] From Reference 6 in Table 6.5.

abundances of light- and medium-weight nuclei are correctly given with $X = 2.5$ g-cm^{-2}. The agreement between the calculated and observed compositions implies the following points:

1. The composition of the L-group is reproduced if all unstable nuclides except those undergoing electron capture are assumed to decay.
2. The source abundance of the H_2-group is taken to be zero. This is in favor of the value of X being as small as in (6.3.7).
3. The even-odd effect is seen in the H_3-group, as shown in Table 6.5; whereas the fragmentation of H_1 nuclei should give an almost vanishing even-odd effect if correction is made for the decay effect. This implies, together with the existence of the H_2-valley, that heavy nuclei in cosmic

rays are not entirely the fragmentation products of the heaviest nuclei but are mostly those that are accelerated.

The considerable abundances of rare isotopes are also interpreted as being due to fragmentation in interstellar space. Because the even-odd ratio of heavy nuclei is about 1/4, the relative abundance of a rare isotope is expected to be as large as a few tens percent. In the GeV region the ratio of ^{13}C to ^{12}C would be about 0.1 if the abundance of ^{13}C at the origin were negligible and ^{13}C were exclusively a fragmentation product. However, the ^{13}C-^{12}C ratio given in (6.3.6) is as great as unity. If this is accepted, the isotopic abundances of carbon cannot be explained by the fragmentation effect alone.

6.3.5 *Fragmentation at High Energies*

As described in Section 6.3.2, the atomic abundances of primary cosmic rays seems to change as energy increases beyond 1 TeV, although experimental evidence is not conclusive as yet. The energy dependence of the composition is not unreasonable, as will be discussed below.

If cosmic ray particles gain energy slowly with time, as represented by (6.2.1), the length of time a particle has existed increases with increasing energy. Accordingly the thickness of matter traversed is energy dependent:

$$x(E) - x(E_0) = X \ln\left(\frac{E}{E_0}\right). \tag{6.3.12}$$

The consequences of this energy dependence are easily predicted from (6.3.12) to result in (*a*) a relative increase in the intensities of L- and H_2-groups; (*b*) a decrease in the intensity of the H_1-group; and (*c*) the diminishing of the even-odd ratio, the isotopic ratios, and so forth. If (6.3.12) were valid, $x(E)$ would become as large as the collision mean free path of α-particles at 10^{12} eV per nucleon, and few heavy nuclei would survive in primary cosmic rays. As a result, the intensity of heavy nuclei relative to that of protons would be reduced by a factor of 2 or so. The experimental information would indicate a weaker energy dependence than predicted by (6.3.12), thus suggesting that a considerable part of acceleration has taken place rapidly at the sources.

At energies higher than 10^{17} eV most cosmic rays are likely to be of metagalactic origin. Then cosmic-ray particles may have traveled as far as the radius of the universe, about 10^{28} cm. Although the path length is very large, the corresponding thickness of matter may be smaller than 1 g-cm^{-2}, since the density of matter in intergalactic space is presumed at most to be on the order of 10^{-5} cm^{-3}. However, the density of starlight may be not too small—it may be as large as 10^{-2} cm^{-3}. This results in the photodisintegration of cosmic-ray nuclei while they travel through intergalactic space (Gerasimova 61).

In the rest system of an impinging nucleus the energy of a starlight photon is given by

$$\tilde{\varepsilon} = \varepsilon(\gamma + \sqrt{\gamma^2 - 1}\cos\theta) \simeq \varepsilon(1 + \cos\theta)\gamma, \qquad (6.3.13)$$

where γ is the Lorentz factor of the high-energy nucleus, ε is the energy of the starlight photon in the laboratory system, and θ is the angle between the nucleus and the photon with respect to the head-on collision. Since the energy threshold for photoreactions is 10 to 20 MeV, heavy nuclei of energies greater than 10^{16} eV per nucleon have considerable probabilities of disintegration. The cross sections averaged over the giant resonance and above the pion threshold are roughly expressed by (see Section 2.10.5)

$$\sigma_g \simeq 0.4\, A^{4/3} \text{ mb}, \qquad \sigma_d(E > m_\pi c^2) \simeq 0.1\, A \text{ mb}, \qquad (6.3.14)$$

respectively. Thus the mean free path for heavy nuclei to disintegrate in metagalactic space is about 4×10^{28} cm for $A = 4$. Therefore there are practically no metagalactic heavy nuclei at energies greater than 10^{16} eV per nucleon.

The microwave photons that have been recently discovered are responsible for the photodisintegration of heavy nuclei with energies greater than 10^{19} eV. Since their density is as high as 10^3 cm^{-3}, even Galactic heavy nuclei should be eliminated within a relatively short period. These considerations are supported by the experimental evidence given in Section 6.3.2 for the absence of heavy nuclei with energies greater than 10^{17} eV. If the composition of heavy nuclei at such high energies were equal to that at low energies, we would expect their photodisintegration by sunlight in the solar system. The products of a photodisintegration could form coherent air showers separated by 1 km or so, and the frequency of such showers is expected to be on the order of 1 km^{-2} sr^{-1} $month^{-1}$ (Zatsepin 51, Gerasimova 60).

*6.3.6 Composition in the Nonrelativistic Region**

In a number of places in this section we have briefly mentioned that the composition of heavy nuclei may be different in the nonrelativistic

* Experimental data are rapidly accumulating. Although more quantitative description is possible, it is a little too early to finalize the experimental information at this stage. The qualitative picture given here remains valid even if recent data are taken into consideration, except for the existence of very low energy nuclei as given in a footnote towards the end of Section 6.4.2.

region; the abundances of the L-group shown in Table 6.5 for the rigidity regions of 1.3 to 2.7 and > 4.5 GV, as well as the relative abundances of the H_1- and H_2-groups given in Table 6.6 for two different rigidities do not seem to be the same.

The energy dependence of composition at low energies is expected from the following three effects:

1. The loss of energy by ionization increases as energy decreases.
2. The fragmentation cross sections depend rather strongly on energy at nonrelativistic energies.
3. The path length traversed by a particle may be greater because the deflection in magnetic fields could be more effective at low rigidities.

If the third effect is significant, interstellar fragmentation becomes more important, and as a consequence both the L/M ratio and the H_2/H_1 ratio increase as energy decreases. This seems to have been indicated in Tables 6.5 and 6.6, as remarked above. The same, however, can be also accounted for in terms of the energy dependence of the fragmentation cross sections. Since the fragmentation cross sections for some nuclei are known from laboratory experiments at low energies, we shall first discuss the second effect for light nuclei, on which the first effect is insignificant.

The fragmentation probabilities in hydrogen at energies of about 200 MeV per nucleon are obtained by reference to a number of laboratory as well as cosmic-ray experiments, as given in Table 6.13 (Badhwar 63*a*, Dahanayake 64). The difference between the results of these two groups tells us how uncertain the fragmentation probabilities are; in

Table 6.13 Fragmentation Probabilities in Hydrogen at about 200 MeV per Nucleon

Primary	$p + n$	^{2}H	^{3}H	^{3}He	^{3}H + ^{3}He		^{4}He	
					1	2	1	2
^{4}He	1.8	0.48	0.03	0.57	0.60	0.3		0.05
C	2.7	0.30	0.18	0.11	0.29	0.20 (C, O)	0.68	0.40 (C, O)
O	2.1	0.47	0.05	0.09	0.14		0.56	
Al	2.2	0.31	0.06	0.07	0.12	0.15 (Al, Ni)	0.80	0.80 (Al, Ni)
Ni	2.2	0.25	0.05	0.06	0.11		0.60	

References:

1. G. D. Badhwar and R. R. Daniel, *Prog. Theor. Phys.*, **30**, 615 (1963).
2. C. Dahanayake et al., *J. Geophys. Res.*, **69**, 3681 (1964).

particular the difference in $P(^4\text{He} \rightarrow {}^3\text{He})$ by a factor of 2 brings about a great effect on the implication of the isotopic ratio of helium nuclei.

The abundances of hydrogen and helium isotopes calculated by using the fragmentation probabilities given in Table 6.13 (reference 1) and $X = 2.5$ g-cm^{-2} are given in Table 6.14. The isotopic ratios thus obtained for a given energy per nucleon are

$$\frac{^2\text{H}}{^1\text{H}} \simeq 0.01, \qquad \frac{^3\text{He}}{^4\text{He} + {}^3\text{He}} \simeq 0.07. \tag{6.3.15}$$

Table 6.14 Isotopic Abundances of Cosmic-Ray Hydrogen and Helium[a]

Nucleus	Calculated[b]		Observed[c] at ~200 MeV/A	Cosmic Abundances
	At Sources	At the Earth		
^{1}H	140	200	200	2.7×10^3
^{2}H	0	2.0	$\lesssim 10$	<0.1
^{3}He	0	2.2	3	~0.1
^{4}He	25	30	32	410

[a] All values are normalized to the carbon abundance.
[b] The source composition is given in the same manner as in Table 6.12. The composition at the top of the atmosphere is calculated for $X = 2.5$ g-cm^{-2} by using the fragmentation probabilities given in Table 6.13 (Reference 1).
[c] The helium-carbon ratio is taken from V. K. Balasubrahmanyan et al., *J. Geophys. Res.*, **71**, 1771 (1966), and the hydrogen-helium ratio (=5.7) from V. K. Balasubrahmanyan et al., *Phys. Rev.*, **140B**, 1157 (1965).

The latter is in fair agreement with the experimental results given in Table 6.7. Since the fragmentation probabilities are considerably energy dependent and the available results are not reliable enough, we can only say that the thickness of matter favor a larger value at about 100 MeV per nucleon. It should also be noted that the derivation of the thickness of matter depends on the shape of the unmodulated spectrum.

In comparing the calculated results with observed ones one had to be cautious about the solar modulation effect. Both the absolute intensity and the energy spectrum of cosmic rays are greatly affected by solar modulation, especially at low energies. The modulation effects for a given energy per nucleon depends on Z/A and consequently is different for different isotopes. This also suggests that acceleration depends on Z/A. Therefore it is not always adequate to compare the energy dependence of the relative abundances of nuclei with different Z/A values. In order to avoid this complexity the ratios of L/M and He/C are compared as follows.

The relative abundances at energies of about 100 MeV per nucleon observed with satellites and at energies greater than 600 MeV per nucleon observed with a balloon are compared in Table 6.15 (Balasu-

Table 6.15 Energy Dependences of the Relative Abundances of Heavy Nuclei

Ratio	Energy per Nucleon	
	100 MeV	>600 MeV
Light to medium nuclei	0.29 ± 0.07	0.30 ± 0.03
Carbon to helium	0.023 ± 0.005	0.036 ± 0.004

brahmanyan 66). The results would indicate the energy independence of L/M but a decrease of C/He as energy decreases. However, the energy-independent L/M ratio is accidental. A number of experimental data on the L/M ratio (plotted against energy in Fig. 6.5) clearly show an increase in the L/M ratio in the nonrelativistic region, and the L/M ratio at 100 MeV per nucleon seems to indicate that it decreases again as energy decreases below 300 MeV per nucleon. This demonstrates that fragmentation probabilities are highly energy dependent in the nonrelativistic region.

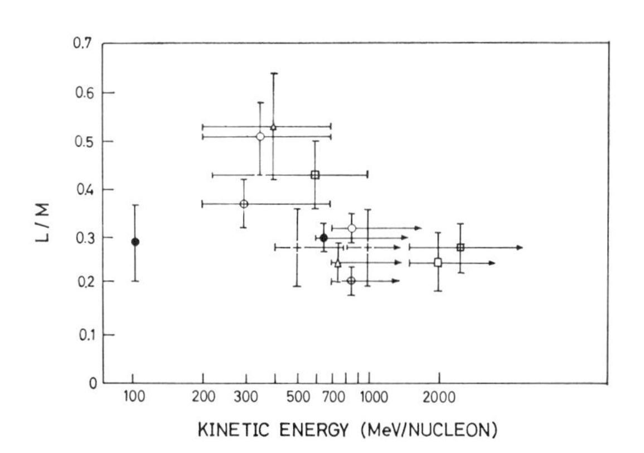

Fig. 6.5 Summary of L/M ratios. Arrow denotes integral measurements. ●—Balasubrahmanyan 66; +—Balasubrahmanyan 64; △—G. D. Badhwar, S. N. Devanathan, and M. F. Kaplon, *J. Geophys. Res.*, **70**, 1005 (1965); ○—Koshiba 63; ⊞—F. Foster and A. Debenedetti, *Nuovo Cim.*, **28**, 1190 (1963); □—O'Dell 62 (Table 6.5); ⊕—Aizu 60.

The energy dependence of the C/He ratio can be partly accounted for in terms of energy loss by ionization. Although the presence of this effect is indisputable, a quantitative result is hardly obtainable, because this depends sensitively on the shape of the energy spectrum at the sources. Figure 6.6 shows the Z-dependence of the energy spectrum for a rather steep source spectrum (Appa Rao 63). The effect is very large for the very heavy group; a steep spectrum would give the thickness to be traversed smaller than 2 g-cm^{-2}, as shown in Fig. 6.7. On the other hand, the ^{3}He/(^{3}He + ^{4}He) ratio decreases with decreasing energy (Fan 66). This and the absence of F would indicate that shortening of the path length toward low energy.

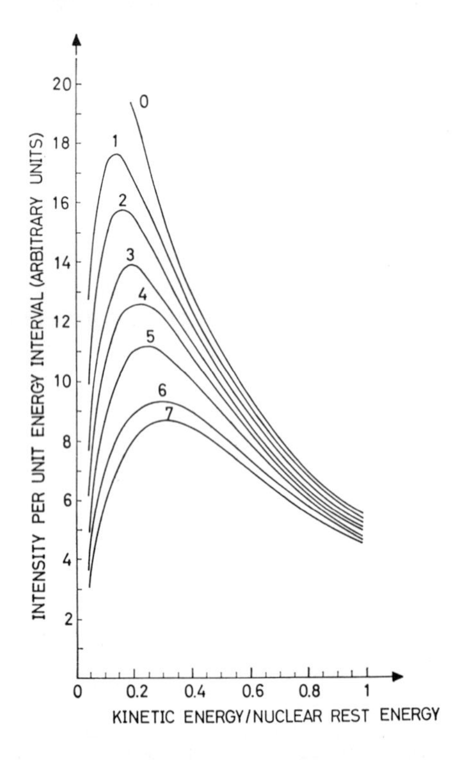

Fig. 6.6 The differential energy spectra obtained for the given nuclear species when the source spectrum (labeled 0) traverses 2 g-cm^{-2} of hydrogen undergoing loss by ionization only. Each component is normalized to the arbitrarily normalized source spectrum. The abscissa is kinetic energy per nucleon in units of the nucleon rest mass; (1) H or ^{4}He, (2) ^{9}Be, (3) ^{12}C, (4) ^{19}F, (5) ^{27}Al, (6) ^{40}Ca, (7) ^{56}Fe (Appa Rao 63).

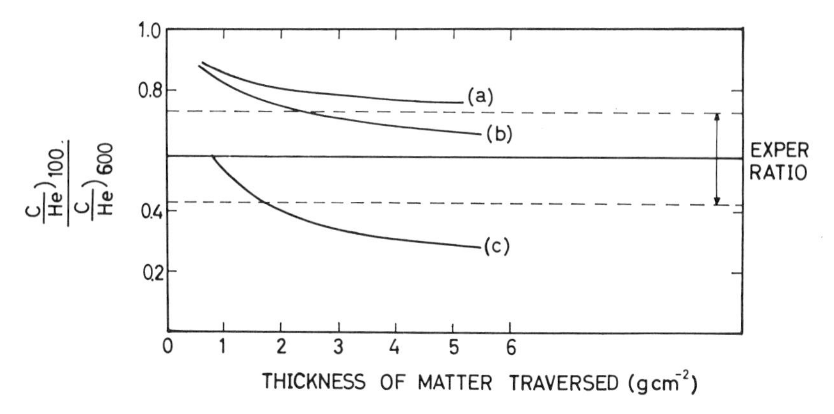

Fig. 6.7 Calculated variation of the ratio of $j_C(100 \text{ MeV})/j_{He}(100 \text{ MeV})$ to $j_C\,(>600 \text{ MeV}/A)/j_{He}(>600 \text{ MeV}/A)$ plotted as a function of the interstellar matter traversed. The calculation allows for energy loss by ionization for the nuclei in the interstellar medium but neglects nuclear interactions. Solid curves correspond to the cases of source spectra. (*a*) $j(E) = \text{const}$; (*b*) $j(E) \propto E^{-2.5}$; (*c*) $j(W) = W^{-2.5}$ (Balasubrahmanyan 66).

6.3.7 *Composition of Cosmic Rays at Sources*

With correction for fragmentation in interstellar space, the composition of cosmic rays at sources is given in Tables 6.12 and 6.14. This is compared with the so-called cosmic abundances (Cameron 59). The difference between them is obvious, as already noted in Section 6.3.1 before the correction for fragmentation.

Some remarks may be necessary before going into a detailed discussion. First, the value of $X \simeq 2.5$ g-cm^{-2} represents thickness X traversed after particles gain energy that is high enough for nuclear disintegrations. However, it does not include the thickness traversed in the course of initial acceleration at the source. If particles traverse a large thickness of matter in their acceleration from 10 to 100 MeV per nucleon, the relative abundances of light nuclei should be modified in the following way. Since the cross sections for the production of ^{11}C and ^{7}Be from ^{12}C bombarded by protons have pronounced peaks at several tens of MeV— whereas the cross section for the production of lithium is less energy dependent—the ratios of boron to lithium and beryllium to lithium would have to be much greater than the observed ones. This indicates that the thickness traversed while particles are accelerated from 20 to 100 MeV per nucleon is negligibly small (Badhwar 63*a*).

The second remark is concerned with the abundance ratio of protons to heavy nuclei. Since acceleration and modulation depend on Z/A, as has been already mentioned, this ratio does not necessarily coincide with the cosmic abundance ratio. However, the difference seen in Table 6.14 seems to be too large to be accounted for in these terms.

In view of the above remark, it is permissible to compare the relative abundances of heavy nuclei at the source of cosmic rays with those of general galactic matter because for heavy nuclei $Z/A \simeq 1/2$. In either case the abundance generally decreases with increasing mass number, but the rate of decrease is less rapid for cosmic rays than for galactic elements, as can be readily seen from Tables 6.12 and 6.14. In order to demonstrate this point a number of representative abundance ratios are compared in Table 6.16.

Table 6.16 Difference Between the Compositions of Cosmic Ray Nuclei and Galactic Elements

Ratio	Cosmic Rays	Galactic Elements
p/M	10^2	10^3
He/C	25	410
O/C	0.7	2.7
$^{13}C/^{12}C$	~1	0.011
M/H_1	7	150
H_3/H_1	2	16
H_2/H_1	<0.1	3

In addition to this general trend, there are several other abundance ratios that show distinct differences between these two cases. In the first place, the oxygen-carbon ratio is inverted for cosmic rays, so that carbon is most abundant among cosmic-ray nuclei, whereas oxygen is most abundant in galactic elements.* In the second place, the isotopic ratio of carbon, $^{13}C/^{12}C$, is nearly unity in cosmic rays, whereas it is as small as a few percent in galactic elements. In the third place, the H_2/H_1 ratio is smaller than unity in cosmic rays, whereas it is greater in galactic elements. There are two alternatives in explaining the difference in the general trend: (1) the composition of elements at cosmic-ray sources is the same as that of galactic elements, but the acceleration efficiencies favor heavy elements, and (2) cosmic-ray sources are limited to particular regions in which the composition of elements is anomalous.

* The oxygen-carbon ratio is found to be greater than unity, and the even-odd ratio is large in the nonrelativistic region (Fichtel 66).

The first alternative is suggested in connection with an acceleration mechanism, according to which an ion begins to be accelerated in a partially ionized state, so that at an early stage the rate of Fermi acceleration relative to the loss of energy by ionization increases with increasing mass. This is because the acceleration rate is proportional to mass, whereas the rate of energy loss is proportional to the square of the charge, the latter being equal for all ions at the beginning (Korchak 58). If this were the case, neither the H_2-valley or the small oxygen-carbon ratio would exist. We are therefore inclined to be in favor of the second alternative. Further evidence against the first alternative is provided by the composition of solar particles, as will be discussed in the next subsection.

The second alternative may be supported by the fact that there are a number of celestial objects in which the composition of elements is considerably different from the average composition. It is well known that heavy elements are synthesized during the evolution of stars. Since the supernovae are regarded as the last stage of stellar evolution, the composition of cosmic rays may indicate the supernova origin of cosmic rays (Hayakawa 56).

An abundance anomaly is also seen in other objects. In some Wolf–Rayet stars carbon is overabundant and the $^{13}C/^{12}C$ ratio is rather large, close to that expected from the C-N cycle. Since the red giants are ejecting matter, they may also be responsible for the acceleration of cosmic rays (Hayakawa 58*a*). Elements that have been synthesized in the inner part of the star may be contained in the ejected matter as a result of the mixing of inner matter through a convection layer.

In any case it is quite natural to assume that the generation of cosmic rays is associated with stellar instabilities, such as the supernova explosion and the helium flash. In either case a part of the unstable region that is eventually ejected is heated up as high as several times 10^8 °K or higher, and rapid nuclear reactions occur. It is expected that in rapid nuclear reactions carbon may become more abundant than oxygen as a result of the 3α reaction, the rapid C-N-O cycle, and the freezing (slow) reactions. The H_2-elements may become much less abundant than the H_1-elements as a result of α-capturing with insufficient supply of helium (Hayakawa 60*a*). It seems impossible, however, to account for the cosmic-ray abundances by a single source because even in rapid reactions the conditions necessary for the overabundance of carbon are different from those necessary for the H_2-valley. There may be complex sources, each of which is responsible for a part of the abundances.

6.3.8 Composition of Solar Cosmic Rays

In attempting to determine whether or not the source composition of cosmic rays that is derived through analysis of fragmentation processes represents the chemical composition of matter at sources, the observation of solar cosmic rays provides an important clue because the source can be indentified unambiguously. Studies of solar cosmic rays associated with solar flares were made with considerable success in the last solar maximum in the years 1957 through 1962, and the results were summarized by Biswas and Fichtel (Biswas 65).

The relative abundances among heavy nuclei are essentially constant in many events, whereas the ratio of protons to heavy nuclei varies from one event to another because of the Z/A effect. It is therefore meaningful to derive the average composition of solar heavy nuclei and to compare it with the composition of galactic cosmic rays as well as with the composition of solar matter, the solar abundances. The abundances normalized to the oxygen abundance are shown in Table 6.17.

Table 6.17 Relative Abundances of Solar and Galactic Cosmic Rays and Matter

		Cosmic Rays		Matter	
Element	Z	Solar[1]	Galactic[2]	Solar[3]	Galactic[4]
Helium	2	107 ± 14	48	?	150
Lithium	1	—	0.31	$<10^{-5}$	4.0×10^{-6}
Beryllium, Boron	4, 5	<0.02	0.69	$<10^{-5}$	1.8×10^{-6}
Carbon	6	0.59 ± 0.07	1.37	0.6	0.37
Nitrogen	7	0.19 ± 0.04	0.70	0.1	0.096
Oxygen	8	1.0	1.0	1.0	1.0
Iron	9	<0.03	0.11	0.001	6.4×10^{-5}
Neon	10	0.13 ± 0.02	0.16	?	0.032
Sodium	11	—	0.19	0.002	1.8×10^{-3}
Magnesium	12	0.043 ± 0.011	0.30	0.027	0.036
Aluminum	13	—	0.07	0.002	3.8×10^{-3}
Silicon	14	0.033 ± 0.11	0.15	0.035	0.040
Phosphorus to scandium	15 to 21	0.057 ± 0.017	0.17	0.032	0.024
Titanium to nickel	22 to 28	$\lesssim 0.02$	0.21	0.006	5.2×10^{-3}

References:

1. S. Biswas and C. E. Fichtel, *Space. Sci. Rev.*, **4**, 709 (1965).
2. M. Koshiba et al., *Phys. Rev.*, **131**, 2692 (1963).
3. L. H. Aller, *The Abundances of the Elements*, Interscience, New York, 1961.
4. A. G. W. Cameron, *Astrophys. J.*, **129**, 672 (1959).

It is remarkable that the relative abundances of solar cosmic rays, which are essentially the same as the solar abundances, are at considerable variance with those of galactic cosmic rays. This demonstrates that the composition of cosmic rays is a good representative of the composition of matter at a source and that the sources of galactic cosmic rays are different in nature from ordinary stars such as the sun. This is not in favor of the preferential acceleration of heavy nuclei.

Further inferences may be drawn from the composition of solar cosmic rays. The close resemblence in composition of solar cosmic rays to solar matter suggests that cosmic-ray data provide information that is as good as the spectroscopic determination of the solar abundances. Some elements of high excitation energies, such as helium and neon, are not suitable for spectroscopic observation, but their abundances can be determined by means of cosmic rays. The ratio of helium to medium nuclei in solar cosmic rays is 60 ± 7, and the ratio of protons to medium nuclei is 650 from spectroscopic data (Aller 61). This gives

$$\frac{p}{\text{He}} = 11^{+7}_{-5}, \tag{6.3.16}$$

not in disagreement with the p/He ratio of 6.6 in galactic abundances.* The ratio of neon to oxygen in the sun is slightly smaller than that in galactic matter, but not by much.

The intensity of the light group in solar cosmic rays is negligible, and only its upper limit is given. This is in sharp contrast to its finite intensity in galactic cosmic rays. This fact indicates that the thickness of matter traversed by solar cosmic rays is very small, smaller than 0.2 g-cm^{-2}.

6.4 Energy Spectrum of Primary Cosmic Rays

Sections 4.8 and 5.8 discuss the energy spectrum of primary cosmic rays and give its general shape. The energy spectrum is represented by a power law, and the power index varies very little over a wide range of energy; the power index of the integral spectrum varies from -1.1 to -2.2 in the energy range from 10^9 to 10^{20} eV. This fact must be taken serious in discussing the origin of cosmic rays.

The energy spectrum is found by measuring the intensities in a number of energy ranges. We shall describe several representative methods suitable to different energy regions and give the results thus obtained.

* According to revised spectroscopic data, the solar p/He ratio is significantly higher than the galactic value. This raises a very important question on cosmology.

Before presenting experimental results we have to remark that the energy spectrum at energies smaller than some tens of GeV is subject to solar modulation, thus changing with solar activity. The modulation effect increases as energy decreases, and the intensity at energy below a few GeV per nucleon is believed to be significantly smaller than that outside the solar system even at the solar minimum. We are therefore unable to obtain direct information on the spectrum of galactic cosmic rays of low energies but merely infer the galactic spectrum after correcting for the modulation effect. In order to minimize the correction we refer here mainly to the spectrum at the solar minimum.

6.4.1 Geomagnetic effect

Particles of rigidities below 60 GV are subject to the geomagnetic effect, and the rigidity spectrum could be, in principle, obtained in the geomagnetically sensitive region. Practically, however, the upper limit of rigidity that is suitable for this purpose is 17 GV, the cutoff rigidity for vertical incidence at the geomagnetic equator; the lower limit lies at a few GeV, since the cutoff is not sharp at high latitudes.

Even in this rigidity region the rigidity spectrum of protons is not always reliable because the contribution of albedo protons is considerable. At the geomagnetic pole the albedo flux is as large as 20 percent of the intensity measured. This consists entirely of the upward or splash albedo, and the re-entrant albedo is added at lower latitudes. Since the albedo effect is difficult to estimate, the measurement of intensity alone does not give accurate values for the intensity of primary cosmic rays. Correction for the albedo effect can be made experimentally by distinguishing upward particles from downward ones and measuring the the energies of downward particles; re-entrant albedo particles have energies smaller than the cutoff energy. The proton spectrum may therefore be obtained by direct measurement rather than by measuring the geomagnetic effect.

For heavy primary particles, however, the albedo effect is believed to be of minor importance, since heavy nuclei produced in the upper atmosphere are fewer than protons, and they are more strongly absorbed by nuclear collisions and energy loss by ionization. Even if the albedo effect is neglected, the spectrum derived from the geomagnetic effect is subject to the time variation of intensity. In order to avoid the time variation, which is appreciable even at the solar minimum, simultaneous measurements at different latitudes are required. Two emulsion flights made in early 1962 at Texas and Hyderabad met this requirement, and the energy spectra of heavy nuclei with $Z \geq 6$ and of protons were obtained (Badhwar 63*b*, Daniel 63). If the energy and the rigidity spectra are represented by

$$j(>E) \propto E^{-\alpha}, \qquad j(>R) \propto R^{-\alpha'}, \tag{6.4.1}$$

where E is the total energy per nucleon and R is the rigidity of a particle. The values of α or α' for protons and heavy nuclei are essentially equal, as shown in Table 6.18. The Z-independence of the spectrum is also

Table 6.18 Values of Power Indices of the Energy Spectra of Protons and Heavy Nuclei

Component	Kinetic Energy (GeV per Nucleon)	Energy Exponent α	Rigidity (GV)	Rigidity Exponent α'
Proton	3.9 to 15.8	1.45 ± 0.14	4.8 to 16.8	1.46 ± 0.14
Heavy nuclei ($Z \geq 6$)	1.6 to 7.5	1.59 ± 0.09	4.8 to 16.8	1.50 ± 0.09

observed with Čerenkov counter borne on a polar orbiting satellite (Ginzburg 63).

6.4.2 Direct Measurement

Although simultaneous measurements of cosmic ray intensities at different latitudes provide a means of avoiding the time-variation effect, only a limited number of points can be obtained on the intensity-energy curve. By means of a detector with a suitable energy resolution, the spectrum over a wide energy range can be obtained free of the effect of time variation since the time variation of the spectral shape is of secondary importance. A counter telescope can be used to measure the energy-loss rate and the total energy dissipated; thus yielding the charge, the mass, and the energy of a particle. The measurement of individual tracks in nuclear emulsion is also suitable for this purpose. These instruments, borne on balloons and space vehicles, have made it possible to measure the energy spectra of protons and heavy nuclei at various phases of solar activity. Here we summarize experimental results obtained in 1955, 1956, and 1963, all near the solar minimum.

The integral rigidity spectra of protons and α-particles measured in 1955 and 1956 by means of a Čerenkov-scintillator are shown in Fig. 6.8. (McDonald 59). In this figure the intensities of α-particles are multiplied by a factor of 6.8, so that the rigidity spectra of protons and α-particles

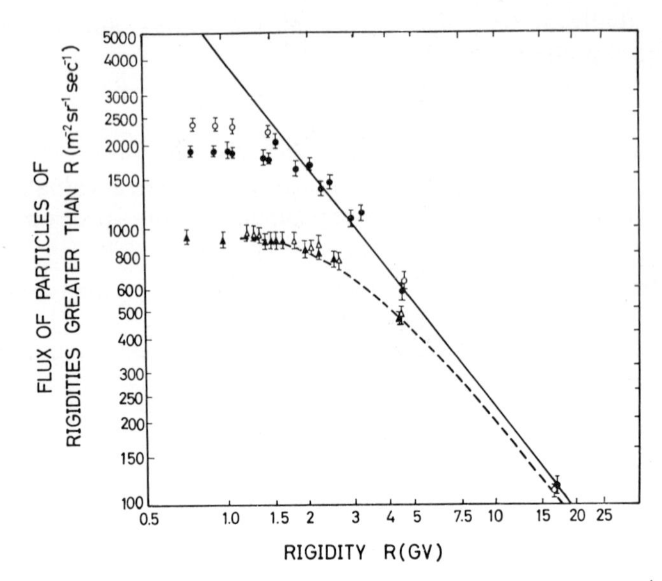

Fig. 6.8 Integral rigidity spectra of protons (●, ▲) and α-particles (○, △). The intensity of α-particles is multiplied by a factor of 6.8. The data at the solar minimum are indicated by ● and ○; those at the solar maximum, by ▲ and △. The dashed curve represents a fit to the solar maximum data, whereas the solid curve is the spectrum expected without solar modulation (McDonald 59).

are represented by a common power law:

$$j(>R) = 0.40\ R^{-1.25}\ \text{cm}^{-2}\ \text{sec}^{-1}\ \text{sr}^{-1} \tag{6.4.2}$$

for $R > 2.5$ GV. The parallelism of the rigidity spectra of various nuclear components has been found for heavy nuclei also in a solar active period (Aizu 60).

The differential rigidity spectrum has been found to deviate from the power law as represented by (6.4.2) at rigidities smaller than 2.5 GV, to reach a maximum at about 1.7 GV, and to decline toward low rigidity. As solar activity increases, the intensity decreases at rigidities up to 5 GV or possibly at higher rigidities; the lower the rigidity, the greater the amount of decrease in intensity. Hence the rigidity at maximum intensity increases with solar activity.

Measurement of the intensity has been extended to rigidities lower than 1 GV by means of counter telescopes borne on a satellite IMP–1 (McDonald 64, Fan 65) as well as by means of balloon-borne emulsion

(Freier 65) in 1963. With the addition of high-rigidity data obtained in the same year (Balasubrahmanyan 64, Fichtel 64, Ormes 64), the rigidity spectra of protons and α-particles are constructed in Fig. 6.9, where the intensity of α-particles is multiplied by a factor of 7.0 (Silberberg 66). The spectra of these two agree with each other above 1 GV, but the helium spectrum is lower at lower rigidities. However, the spectra in the energy-per-nucleon scale are in fair agreement with each other at kinetic energies smaller than 1 GeV per nucleon if the helium intensity is multiplied by a factor of 5.7, as shown in Fig. 6.10 (Balasubrahmanyan 65). This spectrum is analytically represented by a solid curve as

$$\begin{aligned} j(W)\,dW &= 10^4\, W^{1.4}(W+500)^{-4} \quad \text{cm}^{-2}\,\text{sec}^{-1}\,\text{sr}^{-1}\,\text{MeV}^{-1} \\ &\simeq 1.6\times 10^{-7}\, W^{1.5} \quad \text{cm}^{-2}\,\text{sec}^{-1}\,\text{sr}^{-1}\,\text{MeV}^{-1} \end{aligned} \tag{6.4.3}$$

for small W, where W is the kinetic energy per nucleon, and the last expression holds in the limit of low energy.

It should, however, be noted that 1963 was in a declining phase of solar activity. The intensity of α-particles had been increasing in this period; it seems to have reached a maximum in March 1964 for the

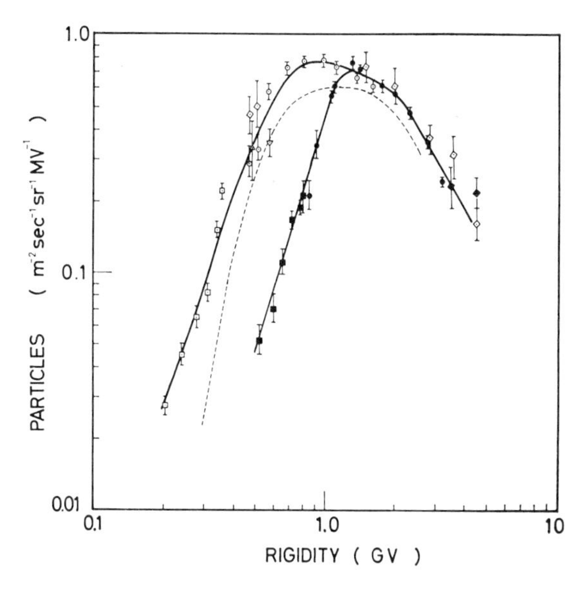

Fig. 6.9 Differential rigidity spectra of protons and α-particles at the solar minimum. The solid curves represent smooth fits to respective spectra, and the dashed curve is a modulated spectrum given by $M(\beta)\, j(E) \propto \exp(-K/\beta)/\beta E^{2.5}$. Dotted symbols: protons. Solid symbols: α-particles. ⊙, ●—Ormes 64; ◇, ◆—Freier 65; —△ Fichtel 64; ▽—Balasubrahmanyan 64; ⊡—McDonald 6.4; ■—Fan 65.

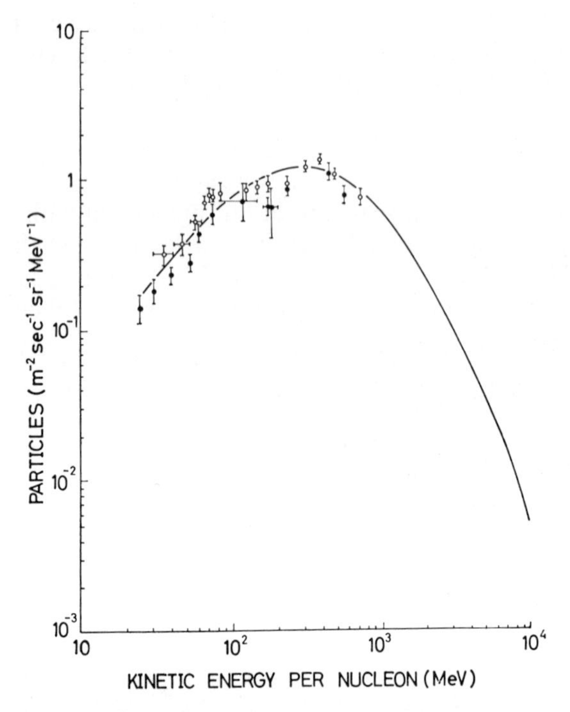

Fig. 6.10 Differential energy spectra of protons (●) and α-particles (○) at the solar minimum. The intensity of α-particles is multiplied by a factor of 5.7. The solid curve represents the spectrum (6.4.3) (Balasubrahmanyan 65).

lowest energy particles, and a little later for higher energy particles (Gloeckler 65). The energy spectra of α-particles differ significantly in two periods before and after February 1964, as shown in Fig. 6.11.

Solar modulation is considered to be effective even at the solar minimum, and the spectrum observed may be modulated by a factor (Parker 63),

$$M(\beta) = \exp(-K/\beta), \tag{6.4.4}$$

where $c\beta$ is the velocity. The values of K at the solar minimum that have been derived by various authors are divergent ($K = 0.8 \sim 2$), depending on the unmodulated spectrum assumed. According to Parker, this holds for particles with Larmor radii smaller than the linear scale of magnetic irregularities in interplanetary space, l. At high rigidities the Larmor radius is larger than l, and the modulation factor takes on the form

$$M(\beta, R) = \exp\left[-\frac{K}{\beta}\left(\frac{lH}{R}\right)^2\right], \tag{6.4.5}$$

where H is the magnetic-field strength. The values of K and l change with solar activity, thus resulting in variations in the intensity and spectral shape. If the distribution of l is taken into account, the modulation factor becomes

$$M(\beta, R) = \exp\left[-(K/\beta)(R_c/R)\right]. \tag{6.4.5'}$$

This is in good agreement with observation.

If these modulation factors are taken into account, the differential energy spectrum of protons outside the solar system may be expressed as (McDonald 64)

$$j(E) \simeq \frac{1}{\beta E^{2.5}} \text{ cm}^{-2} \text{ sec}^{-1} \text{ sr}^{-1} \text{ GeV}, \tag{6.4.6}$$

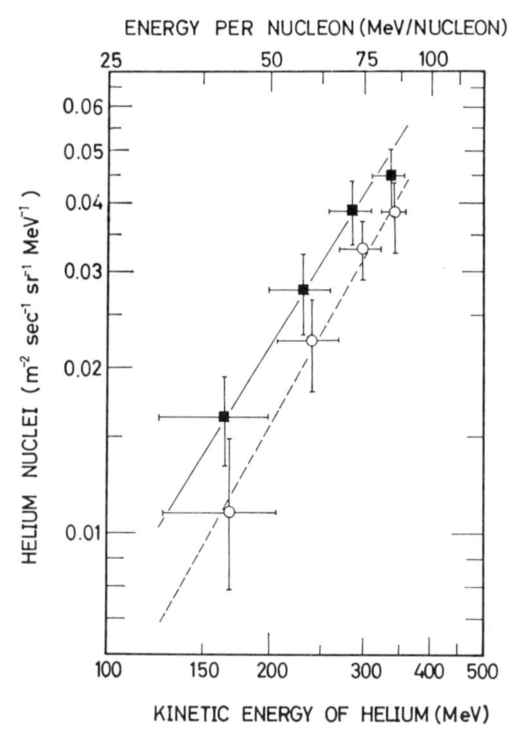

Fig. 6.11 Differential energy spectrum of α-particles at low energies. (open circle) —data obtained during the period Nov. 27, 1963, to February 17, 1964; dashed curve: $j(W) = 0.85 \times 10^{-6} W^{1.85}$; (filled square) —data obtained during the period February 18, 1963, to May 15, 1964; solid curve: $j(W) = 10.5 \times 10^{-6} W^{1.45}$ (Gloeckler 65).

where E is the total energy in GeV. The modulated spectrum $M(\beta)j(E)$ with $K = 0.8$ is shown by a dotted curve in Fig. 6.9. This reproduces a gross feature of the observed spectrum near the maximum.

For heavy nuclei the effect of energy loss by ionization is not negligible at such low energies, and their spectrum is expected to fall off more steeply than the spectrum of lighter nuclei. It is, however, remarkable that the heaviest nuclei, such as iron, are as abundant as in the relativistic region and that the energy spectrum of heavy nuclei is flat down to a level of 20 MeV per nucleon (Comstock 65). This fact seems to be related to the relative abundances different from those in the relativistic region.*

6.4.3 *Spectrum in the Very High-Energy Region*

The energy spectrum above 10 GeV can be obtained by determining the energy through the observation of secondary particles. For energies between 10 GeV and 30 TeV the spectrum is obtained through the analysis of nuclear interactions as in Section 4.8.3. For still higher energies information on extensive air showers is referred to, as discussed in Section 5.8. The results are conventionally expressed by power laws as

$$j(>E) = K\left(\frac{E}{E_0}\right)^{-\alpha}. \tag{6.4.7}$$

The values of K, E_0, and α are given in Table 6.19. The spectrum

Table 6.19 Parameters in the Primary Energy Spectrum (6.4.7)

Energy Region (eV)	E_0 (eV)	K (cm^{-2} sec^{-1} sr^{-1})	α
(1) 10^{10} to 3×10^{13}	10^{12}	$(1.6 \pm 0.8) \times 10^{-5}$	1.60 ± 0.05
(2) 8×10^{14} to 4×10^{17}	10^{17}	$(2.0 \pm 0.4) \times 10^{-14}$	2.20 ± 0.15
(3) 10^{17} to 10^{20}	10^{19}	$(2 \pm 1) \times 10^{-18}$	~ 1.7

covering the whole energy range is drawn in Fig. 6.12.

The spectrum given above holds for primary nuclear particles without discriminating heavy nuclei from protons. This is not immaterial, since the relative intensity of heavy nuclei does not exceed the uncertainty of the values of K.

* The spectra of protons and α-particles are found to rise again toward low energy. These low energy particles may be of different origin from high energy cosmic rays.

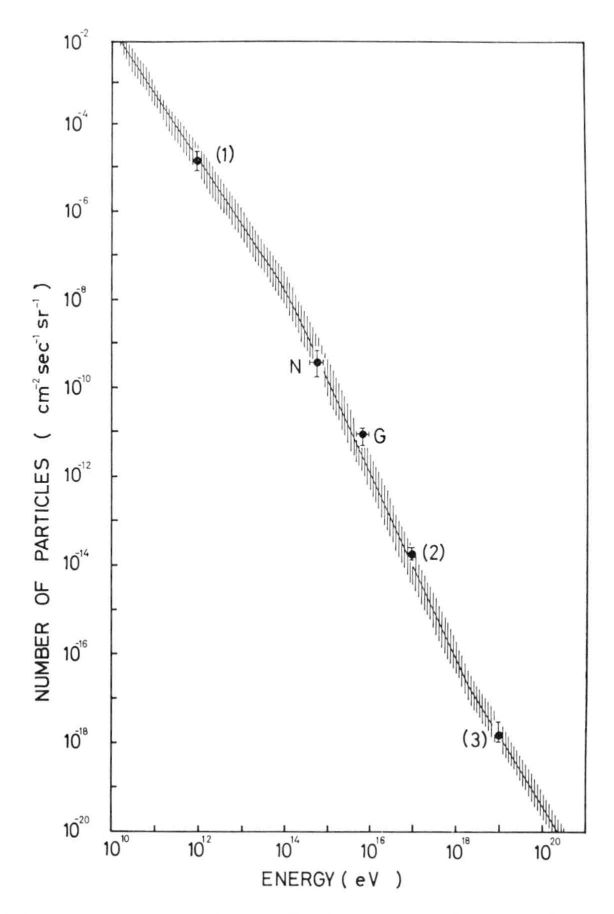

Fig. 6.12 Integral energy spectrum of primary cosmic rays. Curves 1, 2, and 3 represent the spectra given in respective lines in Table 6.19. *N*—EAS data by Nikolsky 62; *G*—EAS data by Greisen 56; the EAS data are given in Table 5.9.

6.4.4 Solar Cosmic Rays

Like the composition of solar cosmic rays, their spectrum also provides rather direct information on the spectrum of particles associated with violent disturbances. The emission of relativistic particles associated with solar flares has been known since 1942, as described in Section 1.9.5. The geomagnetic effect on solar cosmic rays has indicated that they are considerably softer than galactic cosmic rays. It has been also known that the spectrum changes with time, probably because it is affected by the propagation of solar particles through interplanetary space; in the initial phase of a solar event most solar particles are those coming directly along magnetic lines of force, whereas in the later phase particles stored in a magnetic bottle are predominant. During the last solar active period, mainly in 1960, direct measurements of the energy spectra of solar protons and heavy nuclei were performed, and the results are summarized below (Biswas 65).

The differential energy spectra of protons and heavy nuclei are conventionally expressed by a power law as

$$j^{(w)} = K_W(A, t)\left(\frac{W}{AMc^2}\right)^{-\alpha(A,t)-1}, \tag{6.4.8}$$

where W is the total kinetic energy and AM is the mass of the particle concerned. Both K_W and α depend on time t in one and the same event, due both to the time dependences of acceleration and modulation mechanisms. They also change from one event to another. They further depend on the mass number A or the nuclear charge Z; the energy spectrum of heavy nuclei is steeper than that of protons. Moreover, α is energy dependent; only for kinetic energies greater than 35 MeV per nucleon is the power spectrum found to hold, with different slopes for protons and heavy nuclei.

It is therefore more convenient to describe the rigidity spectra by

$$j(R) = K_R(A, t) \exp\left[\frac{-R}{R_0(A, t)}\right]. \tag{6.4.9}$$

Now two parameters, K_R and R_0, may depend on A and t. Actually, however, R_0 is essentially independent of A and is constant over a wide rigidity range covered by observations. This demonstrates an advantage of using the rigidity spectra.

Both the energy and the rigidity spectra obtained at an event on November 16, 1960, are shown in Figs. 6.13 and 6.14, respectively. A number of characteristic features of the solar particles in several events thus far observed are summarized in Table 6.20. From this table we see that the value of R_0 is about 100 MV. The variation of R_0 from event to event is rather small for heavy nuclei, whereas it ranges from 80 to 170 MV for protons. Since the abundance ratio of protons and heavy nuclei is compared in the rigidity range around $6R_0$, a change of R_0 by about 10 percent results in a change in intensity by a factor of about 2. The fluctuations in the proton-heavy nuclei ratio are due mainly to fluctuations in the proton spectrum. This also explains the fact that fluctuations of the proton-heavy nuclei ratio in the same energy-per-nucleon interval are relatively small, as seen in Table 6.20, because the

Table 6.20 Features of Solar Particles

Flare Date and onset Time (UT)	Date and Time (UT) of measurement	Proton/Helium		R_0 (MV)	
		42.5 to 95 MeV per Nucleon	0.57 to 0.87 GV	Proton	He and M
Sept. 3, 1960, 0037	Sept. 3, 1408	43 ± 14	20 ± 7	169.6 ± 6.9	132.6 ± 7.2
	Nov. 12, 1840	32 ± 6	5 ± 1	119.3 ± 7.0	103 ± 9.9
Nov. 12, 1960, 1322	Nov. 13, 1603	36 ± 7	1 ± 0.2	78.0 ± 3.9	109.2 ± 6.7
	Nov. 16, 1951	23 ± 5	1.7 ± 0.5	103.3 ± 10.0	107.2 ± 10.4
Nov. 15, 1960, 0207	Nov. 17, 0600	59 ± 13	2.5 ± 0.6	89.7 ± 5.7	102.9 ± 12.4
	Nov. 18, 0339	51 ± 11	1.7 ± 0.5	81.3 ± 6.4	113.9 ± 15.5

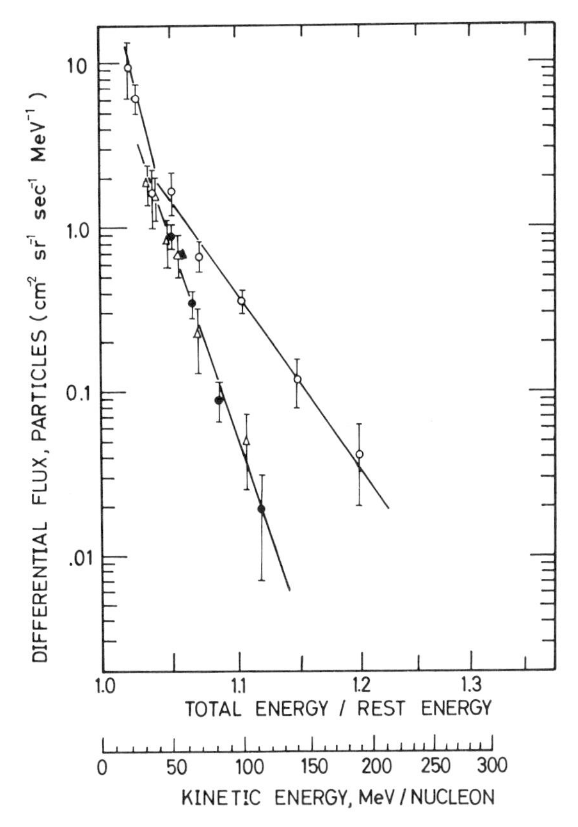

Fig. 6.13 The energy spectrum of solar cosmic rays associated with the event on November 6, 1960 (Biswas 65). ○—protons; △—He nuclei × 10; ●—medium nuclei × 600.

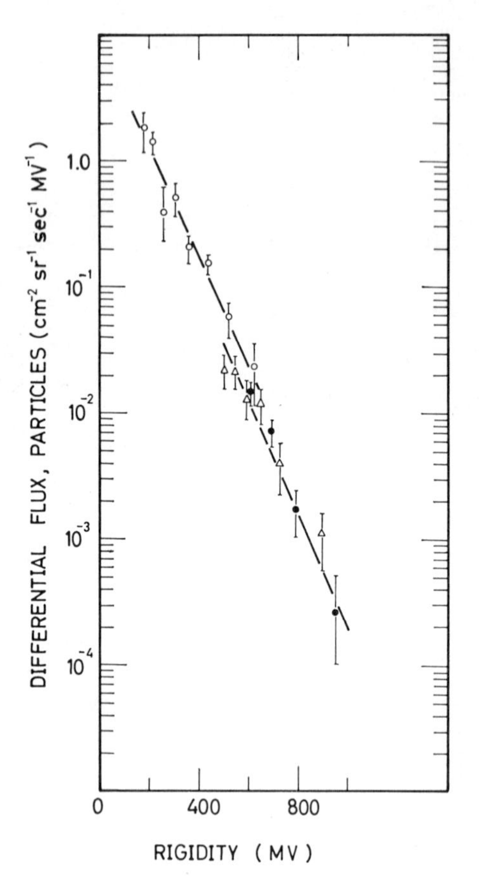

Fig. 6.14 The rigidity spectrum of solar cosmic rays associated with the event on November 16, 1960 (Biswas 65). ○—protons; △—He nuclei; ●—medium nuclei × 60.

same energy-per-nucleon interval corresponds to a low-rigidity range of protons and a high-rigidity range of heavy nuclei.

In addition to solar cosmic rays at solar flare events, protons and α-particles of possibly solar origin have been observed under stationary conditions during the last solar active period (Vogt 62). These particles are mostly of relatively low energies, around 100 MeV per nucleon, and they produce a bump in the energy spectrum. The bump seems to be masked by galactic cosmic rays in the solar minimum period, and this is interpreted as being due both to an increase in the intensity of galactic cosmic rays and to a decrease in the intensity of stationary solar cosmic rays. It is open to question whether these solar cosmic rays are produced from the sun under stationary conditions, or whether they are emitted at flare events and stored in interplanetary space for a considerable period.

6.4.5 *Implications of the Energy Spectrum*

It is a striking fact about cosmic rays that a single elementary particle has macroscopic energy; the highest energy exceeds 1 joule. The way in which particles gain such high energy will be discussed in Section 6.9. Here we remark only that acceleration is most likely due to the interactions of particles with varying magnetic fields while particles are magnetically trapped in a certain spatial region. In a magnetic field of strength H the maximum radius of gyration is given by

$$\rho = \frac{pc}{300H} \text{ cm}, \tag{6.4.10}$$

where p is the momentum of a singly-charged particle in eV/c and H is in gauss. In order for the trapping to be effective the value of ρ has to be smaller than the linear dimension of the trapping region concerned.

In interplanetary space, a typical value of H is a few gammas, and the linear dimension is on the order of 10^{15} cm. Since particles of momenta greater than 10^{13} eV/c are not trapped at all in interplanetary space, the concept of interplanetary cosmic rays is meaningless for particles of momenta larger than 10^{13} eV/c.

In the interstellar space of our Galaxy, a typical value of H is a few μgauss, and the linear dimension is not larger than 10^{23} cm. Hence particles of momenta greater than 10^{20} eV/c cannot be called Galactic cosmic rays.

Actually, however, the upper limits for momenta of trapped particles are smaller than the values given above, since perfect trapping is not generally possible in irregular magnetic fields. In general particles are subject to random walk with a step equal to the average distance between magnetic irregularities. If this is smaller than the radius of gyration, such irregularities are not responsible for the scattering of particles, and consequently a particle diffuses away faster than lower energy particles. It is therefore practical to estimate the maximum momentum for trapping by comparing the radius of gyration with the average distance between magnetic irregularities. This is, according to the discussion in Section 6.2, inferred as being about 10^{12} cm in interplanetary space and as about 3×10^{17} cm or 3×10^{19} cm in interstellar space. Therefore the momenta for cosmic-ray particles to be trapped are

$$p \lesssim 10^{10}\ \text{eV}/c \text{ in the solar system,}$$
$$p \lesssim \begin{cases} 3 \times 10^{14}\ \text{eV}/c \\ 3 \times 10^{16}\ \text{eV}/c \end{cases} \text{in the Galaxy} \tag{6.4.11}$$

respectively if the radius of curvature in the magnetic field is greater than the average distance between magnetic irregularities, l, and the deflection angle is as small as l/ρ. Since the direction of deflection is random, the mean-square angle of deflection while the particle travels over a distance r is about $(r/l)(l/\rho)^2$. The scattering mean free path may therefore be defined as the value of r given by

$$\left(\frac{r}{l}\right)\left(\frac{l}{\rho}\right)^2 \simeq 1 \quad \text{or} \quad r \simeq \frac{\rho^2}{l}. \tag{6.4.12}$$

The rigidity dependence of the modulation factor in (6.4.5) arises from the relation that the scattering mean free time, r/v, is proportional to $\rho^2 \propto R^2$.

According to the argument above, the storage time given in (6.2.4) is modified above the cutoff rigidity to

$$\tau \simeq \frac{L^2}{rv} \simeq \frac{L^2 l}{\rho^2 v} \propto R^{-2}; \tag{6.4.13}$$

namely, the storage time decreases as the rigidity of a particle increases beyond the cutoff rigidity. Since the rigidity spectrum observed is expressed in the stationary state as given in (6.2.6), the rigidity dependence of the storage time results in

$$j(R) \propto q(R)\tau \propto q(R)R^{-2}. \tag{6.4.14}$$

This may explain the steepening of the energy spectrum of Galactic cosmic rays, as observed at energies greater than 10^{15} eV. Actually the change of the slope is not abrupt but gradual, because of the distribution of l and other complicated effects.

Steepening of the spectrum also results if acceleration takes place continuously while particles travel in interstellar space. Assuming that the spectrum is given by the Fermi acceleration as in (6.2.2), the substitution of (6.4.13) into (6.2.2) yields for $R > R_c$ the rigidity spectrum

$$j(>R) \propto \exp\left[-\alpha\left(\frac{R}{R_c}\right)^2 \ln\left(\frac{R}{R_0}\right)\right], \tag{6.4.15}$$

where R_c is the cutoff rigidity. The steepening of the spectrum might take place if cosmic rays were not trapped before a certain time. If the halo were absent, say 10 million years ago (Burbidge 63), few Galactic cosmic rays would have ages that are greater than the age of the halo, provided that they are now trapped in the halo. Hence the age distribution of cosmic rays could not be represented by any exponential form as in (6.2.2), but might have a cutoff at a certain age. This would result in the steepening of the spectrum beyond an energy corresponding to the cutoff age.

Although the detail of the Galactic cutoff will have to be discussed, the steepening of the spectrum beyond 10^{15} eV represents the cutoff of the Galactic component, and the intensity of the *metagalactic component* competes with that of the Galactic component as energy increases. The metagalactic cosmic rays of such high energies may be produced mainly at radio galaxies. In radio galaxies both the magnetic-field strength and the linear dimension of magnetic irregularities may be by an order of magnitude larger than in the Galaxy; since most radio galaxies are spherical and of lower gas density, the interstellar gas therein may be less turbulent than in the Galaxy. Hence the cutoff rigidity may be greater by two or more orders of magnitude, and the metagalactic spectrum in such a high-energy region may have a slope similar to the Galactic spectrum below 10^{15} eV. This seems to be the case, as seen in Fig. 6.12.

The competition between Galactic and metagalactic cosmic rays depends on their relative intensities. If the metagalactic spectrum is extrapolated toward low energy with a constant slope, the metagalactic intensity would be smaller by three orders of magnitude than the Galactic intensity, in rough agreement with the estimate in (6.2.13). Since metagalactic cosmic rays are supposed to be produced mainly in the phase of the radio Galaxy, the intensity of metagalactic cosmic rays may be better estimated by referring to the radio-emission power.

According to Fujimoto et al. (Fujimoto 64), the cosmic-ray yield F is assumed to be proportional to the radio luminosity L. Introducing the radio-luminosity function, $n(>L)$, the density of radio galaxies with luminosities greater than L, we can express the cosmic-ray yield of radio galaxies as

$$F = k \int \frac{-dn}{dL} L \, dL, \tag{6.4.16}$$

where k is a proportionality constant. Hence the cosmic-ray yield per galaxy is given by F/n_g, where n_g is the density of galaxies. This is compared with the cosmic ray yield of our Galaxy, kL_G. The ratio of the densities of metagalactic and Galactic cosmic rays is given by

$$\frac{j_{Mg}}{j_G} = \frac{F/n_g}{kL_G} n_g V_G \frac{\tau_U}{\tau_G}, \tag{6.4.17}$$

where $\tau_U \simeq 10^{10}$ years is the cosmic age.

In evaluating (6.4.17) we have to know the radio-luminosity function. This is expressed, according to Aizu et al. (K. Aizu 64*b*), as

$$n(>L) \simeq \frac{4 \times 10^{30}}{L^{0.85}} \, (\text{Mpc})^{-3} \quad \text{for} \quad 10^{41} < L < 10^{44} \text{ erg-sec}^{-1}. \tag{6.4.18}$$

Because $L_G \simeq 10^{38}$ erg-sec^{-1}, $V_G/\tau_G \simeq 3 \times 10^{-14}(\text{Mpc})^3$/year, we obtain

$$\frac{j_{Mg}}{j_G} \simeq 3 \times 10^{-4}. \tag{6.4.17'}$$

This demonstrates the plausibility that metagalactic cosmic rays of very high energies are mainly produced in radio galaxies.

Now we ask how far the metagalactic cosmic rays produced in a galaxy can reach, or how far away the sources of the metagalactic cosmic rays we are observing are located. Since the magnetic-field strength may be appreciable in intergalactic space, cosmic-ray particles do not travel in a straight line but undergo diffusion, as in interstellar space. If the average distance of magnetic irregularities is as large as $l \simeq 1$ Mpc, nearly equal to the average distance between galaxies, the distance over which particles can travel within the cosmic age τ_U is given by

$$r \simeq \sqrt{c\tau_U l} \simeq 2 \times 10^2 \text{ Mpc}. \tag{6.4.19}$$

Thus only nearby galaxies contribute to the metagalactic cosmic rays that are observed.

If the spectrum of metagalactic cosmic rays has the same shape as that of Galactic cosmic rays, the densities of particles and the energy of cosmic rays in intergalactic space can be obtained by multiplying (6.4.17′) by those in the Galaxy. In the Galaxy they can be evaluated from (6.4.6) as

$$n_G \simeq 10^{-10} \text{ cm}^{-3}, \qquad \langle W_G \rangle \simeq 1 \text{ eV cm}^{-3}, \tag{6.4.20}$$

respectively. As briefly mentioned in Section 6.2, the energy density of cosmic rays in the Galaxy is nearly equal to the energy densities of turbulent motion and magnetic fields. In intergalactic space the equality would hold for a magnetic-field strength of 10^{-7} gauss and a turbulent velocity of 10^7 cm-sec^{-1}. However, no experimental evidence for these values has yet been obtained, although they would give upper limits.

6.5 Anisotropy

It is well known that primary cosmic radiation is essentially isotropic over the celestial sphere. The amplitude of anisotropy, if real at all, is not greater than 1 percent even at extremely high energies, for which observations have become available only recently. It is also known that the anisotropy really exists with respect to the direction of the sun; its amplitude is a fraction of a percent, and the maximum intensity is observed on the day side—both depending on energy and solar activity. Here we are interested only in anisotropy without regard to the solar direction; namely, in the *sidereal-time variation.*

In the discussion of anisotropy it is important to distinguish the sidereal-time variations in the low- and high-energy regions. In the low-energy regions observations are made with neutron monitors, ionization chambers, and counter telescopes, so that the primary energies concerned lie mostly between 10^{10} amd 10^{11} eV. The angular resolution of these detectors is not good enough to determine the sidereal-time variation by means of the directional dependence. In the high-energy region observations are made by detecting extensive air showers, so that the primary energies are higher than 10^{14} eV. Since the air showers have a steep zenith-angle dependence, even a direction-insensitive detector can define the direction of arrival with excellent resolution. In addition, most modern detectors are able to locate the arrival directions of individual showers with an angular resolution as good as 5°. Hence the quality of the directional dependence is better in the high-energy region, but the statistics are better in the low-energy region.

In addition to technical reasons, there is an essential difference between the directional dependences of low- and high-energy cosmic rays. Cosmic-ray particles with low energies are deflected by the geomagnetic field, and the amount of deflection depends rather sensitively on energy and on the latitude at which an observation is made. Since the conventional detectors of low-energy cosmic rays have poor energy resolution, the original directions of particles that arrive from a given direction at a detector are distributed over a wide region outside the geomagnetic field—for example, 60° in longitude and 30° in latitude. Even outside the geomagnetic field these particles are scattered by interplanetary magnetic fields. All these effects smear out the anisotropy of Galactic origin, even if it exists. Therefore we shall put less emphasis on low-energy data than on air-shower data.

In the intermediate-energy region few observations are available for this purpose. Some data have been obtained with counter telescopes underground, but they are statistically poor. A counter telescope with a good angular resolution has supplied some information on anisotropy, mainly concentrated to a few point sources, one of which has been thought to give a significant peak (Sekido 54, 59). A gas Čerenkov telescope with good angular resolution has recently come into operation, and the result so far obtained shows no positive indication of such a peak (Sekido 62).

6.5.1 Anisotropy at Low Energies

The results of the sidereal-time variation, summarized by Conforto (Conforto 62), indicate amplitudes* of 0.06 percent or smaller, except for one measured as early as 1946 with a counter telescope underground. If we refer to data at the solar minimum in order to avoid solar disturbances as much as possible, the maximum intensities are found to appear at about 21 h in sidereal time in the Northern Hemisphere, whereas most of the results in the Southern Hemisphere give maxima at about 7 h, though their existence seems to be barely significant. The direction of the maximum coincides with that at high energies and lies in the Galactic plane. If this is corrected for deflection by the geomagnetic field, the directions of maxima distribute in a plane that includes the spiral arm and the normal to the Galactic plane, These directions are limited to one quadrant, and no maximum is found in the other three quadrants. Observations at stations in the Southern Hemisphere indicate the directions of maxima lie in a plane that is nearly normal to the spiral arm, again limited within one quadrant. It is, however, doubtful how much significance can be given these results.

6.5.2 Anisotropy at High Energies

The sidereal-time variations of extensive air showers analyzed in the first and the second harmonics have been summarized by Greisen (Greisen 56). Although a few observations have found amplitudes that exceed three standard errors, the times of maxima, t_{max}, lie at about 20 h for energies smaller than 5×10^{15} eV and at about 12 h for greater energies. The data with the best statistics obtained for energies between 3×10^{14} and 4×10^{15} eV by Daudin and Daudin (Daudin 53) indicate that the amplitudes of the first and the second harmonics are as large as the standard errors and no greater than 0.4 percent. A recent study carried out very carefully by Delvaille et al. (Delvaille 62) has found for energies of about 10^{15} eV a rather significant peak at about 15 h with an amplitude of about 0.5 percent. More recently Cachon (Cachon 62) has observed a significant peak of 0.15 percent near 21 h at about 2×10^{15} eV, and the sidereal-time variation has been found to

* The amplitude here is defined as

$$A = \frac{\text{maximum intensity} - \text{average intensity}}{\text{average intensity}}.$$

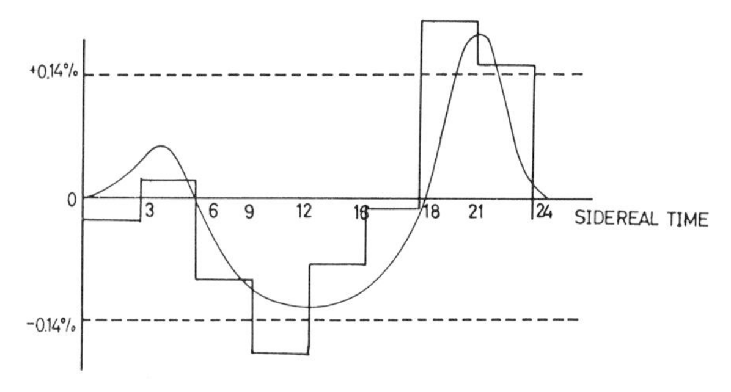

Fig. 6.15 Sidereal-time variations of EAS and galactic radio intensity. ⌐¬ —the fraction of EAS deviating from the average frequency in 3 hours. Solid curve—the intensity of radio noise (arbitrary unit) (Cachon 62).

coincide with that of Galactic radio emission, as shown in Fig. 6.15, thus indicating the maximum to be toward the Galactic center. At a lower energy, about 2×10^{13} eV, however, no variation has been found within the statistical error of 0.1 percent.

If we take all data with amplitudes greater than the standard errors, we see a systematic variation of $t_{\max}$ with energy, as shown in Fig. 6.16*b*; $t_{\max} \simeq 20$ h for $E < 5 \times 10^{15}$ eV and $t_{\max} \simeq 12$ h for 5×10^{17} eV $>$ $E > 5 \times 10^{15}$ eV (Sakakibara 65). The energy at which the rather abrupt change in $t_{\max}$ occurs seems to coincide with the break in the energy spectrum, discussed in Section 6.4.

The amplitude of the sidereal-time variation as given in Fig. 6.16*a* increases from 0.1 to 10 percent; as energy increases from 10^{14} to 10^{17} eV, however, the amplitudes are not much greater than the standard deviations, and most of them lie between one and two standard deviations. Hence these experimental results would not indicate the presence of anisotropy. Summarizing these experimental data, Sakakibara has argued that the distribution of $t_{\max}$ may be regarded as evidence for anisotropy. If the maxima observed were due simply to statistical fluctuations, the times of the maximum amplitudes would have to be distributed uniformly. According to a statistical test the observed distribution of $t_{\max}$ is significantly different from the random one (Sakakibara 65).

At still higher energies EAS of sizes greater than 10^8 that have been observed by the Cornell group appeared to be anisotropic (Delvaille 62). Fourier analysis gives an amplitude of 38 percent, with the maximum at 2 h for the first harmonics; and an amplitude of 20 percent, with the maximum at 4 h for the second harmonics. The probabilities of occurrence by chance are 0.25 and 18 percent, respectively. If these events are

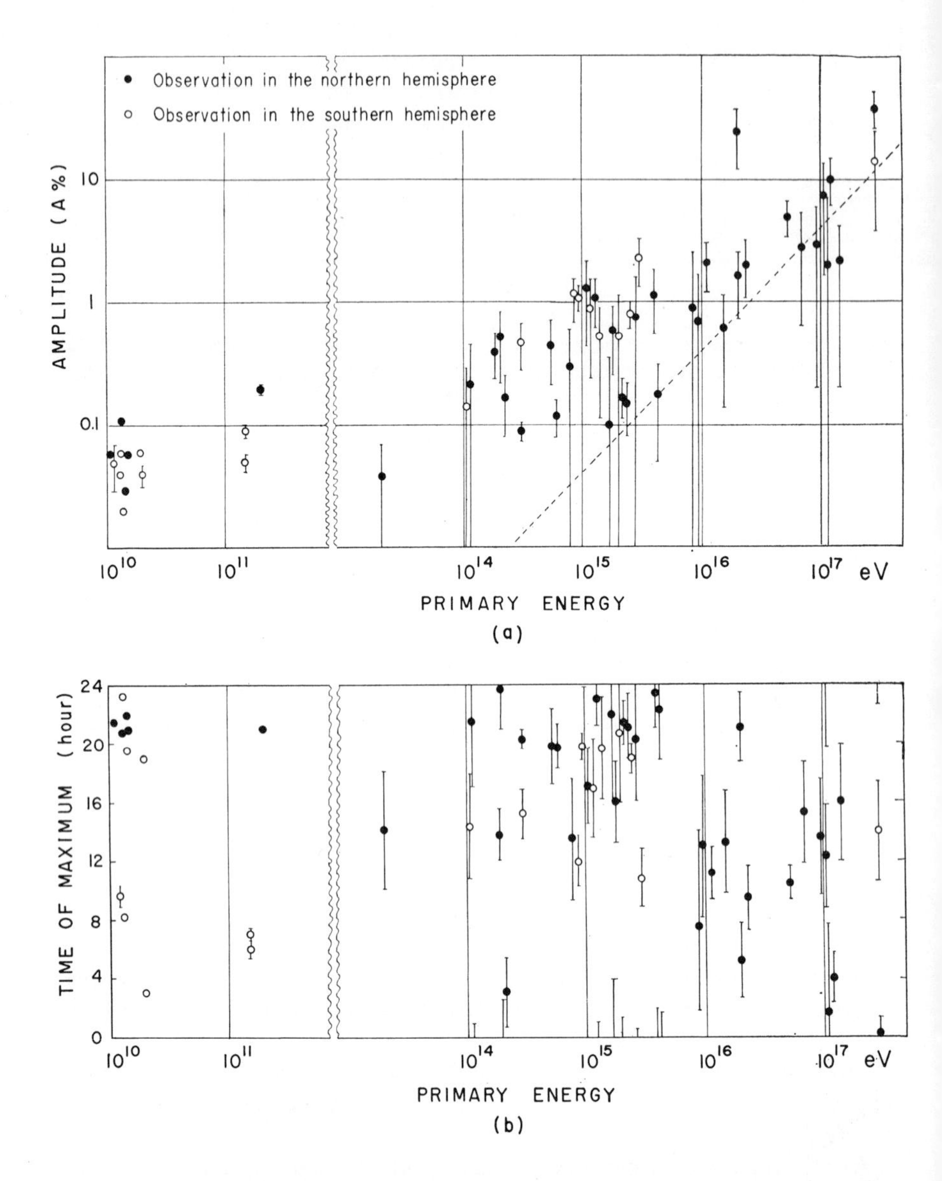

Fig. 6.16 Anisotropy of the cosmic-ray intensity (Sakakibara 65). (*a*) the amplitude $A \equiv (j_{\max} - \bar{j})/\bar{j}$ percent versus energy in eV. Experimental data obtained in the Northern and the Southern Hemispheres are distinguished by full and open circles, respectively; the vertical bars indicate standard errors; the dashed line indicates the amplitude expected from the diffusion normal to the arm field, as discussed in Section 6.5.4. (*b*) the sidereal time of the maximum intensity versus energy in eV.

plotted against the Galactic declination, they are more concentrated in the Galactic plane than they would be if they were isotropic. However, the distribution of EAS with respect to right ascension observed in the same size region by Linsley (Linsley 63) could compensate for the anisotropy obtained by the Cornell group, as seen from Fig. 6.17. The essential isotropy was observed also at an average primary energy of 10^{19} eV. More quantitatively, the upper limit of anisotropy is estimated as 10 percent for an average energy of 10^{18} eV and as 30 percent at 10^{19} eV. These results are neither inconsistent with isotropy* nor with the possible anisotropy found at primary energies between 5×10^{15} and 5×10^{17} eV.

6.5.3 Anisotropy of High-Energy Heavy Nuclei

An interesting feature appears if certain types of air showers are selected. A significant anisotropy seems to have been found for muon-rich EAS (Hasegawa 62*b*). Since EAS containing many muons are

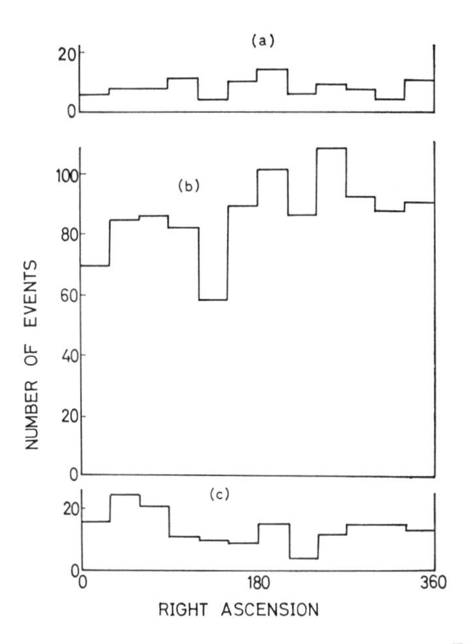

Fig. 6.17 Arrival directions of large EAS (Linsley 63). (*a*) $\bar{E} = 10^{19}$ eV (Massachusetts Institute of Technology data); (*b*) $\bar{E} > 10^{18}$ eV (Massachusetts Institute of Technology data); (*c*) $\bar{E} = 10^{18}$ eV (Cornell University data).

* Recent results obtained by British and Soviet groups give an amplitude smaller than 1 percent at energies around and above 10^{17} eV.

likely to be due to heavy primaries, as discussed in Section 5.7, this may be regarded as indicating an anisotropy of heavy primaries with energies greater than 10^{16} eV per nucleus. A similar anisotropy is observed for showers that are rich in nuclear active particles (Kameda 63).

Although individual nucleon-rich EAS do not coincide with individual muon-rich EAS, their distributions against right ascension look similar. The distributions observed in Tokyo and at Mount Chacaltaya are shown in Fig. 6.18. One sees at a glance a high concentration between 6 and 12 h, whereas few such EAS come from a right-ascension region between 15 and 21 h. This distribution seems to be somewhat analogous

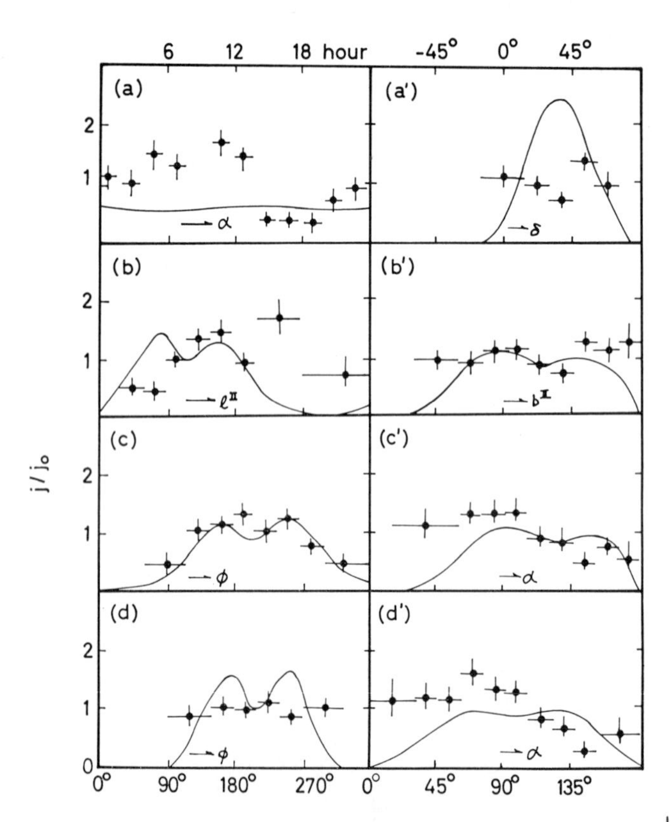

Fig. 6.18 Distribution of the relative intensities of muon-rich EAS. —the observed intensity versus the average intensity; solid curves—expected from isotropy. Equatorial coordinates: (*a*) right ascension (α), (*a′*) declination (δ); Galactic coordinate: (*b*) Galactic longitude (l^{II}), (*b′*) Galactic latitude (b^{II}); Orion arm coordinates: (*c*) azimuth angle (ϕ), (*c′*) pitch angle (α); Sagittarius arm coordinates: (*d*) azimuth angle (ϕ), (*d′*) pitch angle (α) (Sekido 63).

to the sidereal-time variation of ordinary EAS of primary energies between 5×10^{15} and 5×10^{17} eV, but the degrees of anisotropy differ greatly.

The highly anisotropic feature of muon-rich EAS has provided a basis for the belief that the primary particles responsible for the muon-rich EAS are different from those that are responsible for the majority of EAS; namely, the former may be heavy nuclei, and they may behave in a different way because of their short collision mean free path in interstellar matter. The relative intensity of heavy nuclei in high-energy primary cosmic rays has been derived in Section 6.4 on the basis of this interpretation.

6.5.4 *Interpretation of Anisotropy*

As qualitatively discussed in Section 6.2, the degree of anisotropy is of the order of the ratio of the linear dimension of a stored region to the average path length of cosmic rays therein. If cosmic-ray particles undergo scattering by interstellar magnetic fields with a mean free path l, the average path length in a stored region of a linear dimension L is as large as L^2/l. Hence the amplitude of anisotropy is expected to be

$$A \simeq \frac{L}{L^2/l} = \frac{l}{L} \tag{6.5.1}$$

If cosmic rays are stored in the Galactic disk, the linear dimension L should be taken as equal to the thickness of the disk, which is about 10^{21} cm. An anisotropy amplitude smaller than 0.1 percent would have to result in a value of l as small as $1ly$, which is far smaller than the linear size of interstellar clouds. The scattering mean free path can be eliminated if we take account of the density of matter in the stored region, ρ, and the average thickness of matter traversed by cosmic rays, X. Since the path length is expressed by X/ρ, the relation (6.5.1) can be rewritten as

$$A \simeq \frac{L}{X/\rho}. \tag{6.5.2}$$

By introducing $X \simeq 3$ g-cm^{-2} obtained from the fragmentation of heavy nuclei in Section 6.3 and $\rho \simeq 10^{-24}$ g-cm^{-3} given in Section 6.1.1 into (6.5.2), we obtain

$$A \simeq 3 \times 10^{-4}, \tag{6.5.3}$$

which is consistent with the experimental evidence presented in Section 6.5.1. However, the value of l that is required seems to be very small.

If cosmic rays are stored in the halo, which has a linear dimension of 10^{23} cm, the scattering mean free path is

$$l \lesssim 10^{20} \text{ cm}, \tag{6.5.4}$$

consistent with the size of interstellar clouds of about 10 pc. The value of A that can be expected from (6.5.2) is also consistent with experimental results since $\rho \simeq 10^{-26}$ g-cm^{-3} in the stored region, including the halo; thus we have a value of A as small as that given in (6.5.3).

The sense of anisotropy is determined by the density gradient. Since cosmic-ray density outside the Galaxy is supposed to be much smaller than that inside, the density must decrease toward the outer part of the Galaxy. It is therefore expected that the intensity would have a maximum in the direction of the Galactic center if interstellar space were an isotropic scatterer. Actually the Galactic magnetic field has a

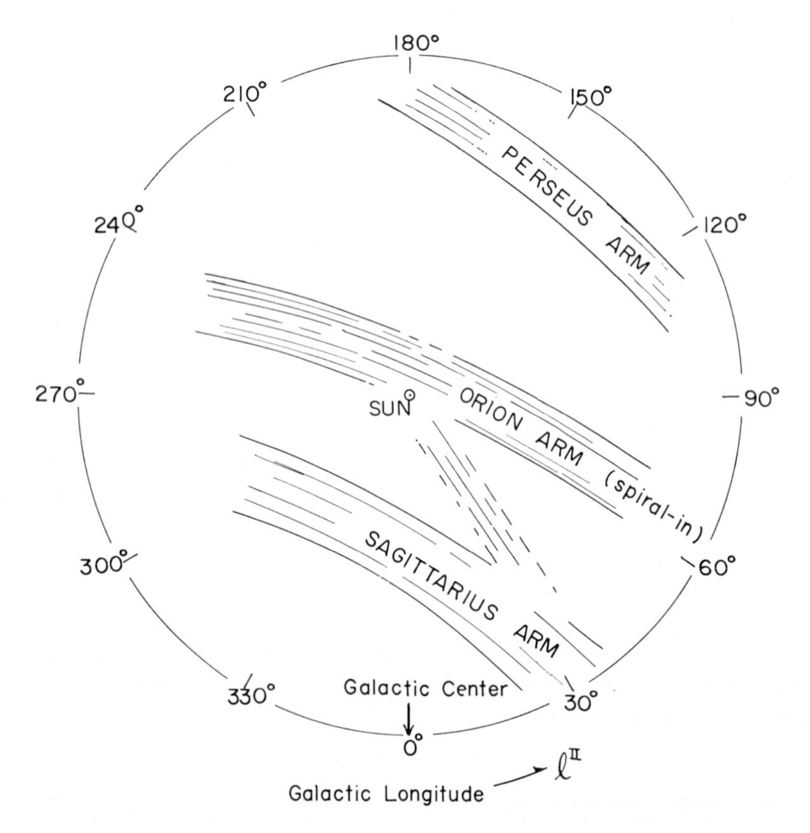

Fig. 6.19 Configuration of spiral arms in the Galactic plane.

spiral structure, and the solar system is located in one of the spiral arms, the Orion arm, as schematically shown in Fig. 6.19. The direction of the intensity maximum at right ascension 20 h may coincide with the spiral-in direction of the Orion arm. This suggests that cosmic-ray sources are more densely populated in the inner part of the Galactic disk and that a considerable part of cosmic rays flow outward along spiral arms (Sakakibara 65).

As energy increases the radius of curvature of a particle in the arm becomes greater than the scattering mean free path, so that particles do not undergo diffusion along the arm. In this case cosmic rays may be regarded as a gas of freely gyrating particles in a uniform magnetic field. The density of the gas is not necessarily uniform but may decrease toward the edge of an arm. The density gradient in the radial direction gives rise to anisotropy, but the intensity maximum in the cross section of the arm is not in the direction of the center. As illustrated in Fig. 6.20, the maximum is in the direction perpendicular to the density gradient,

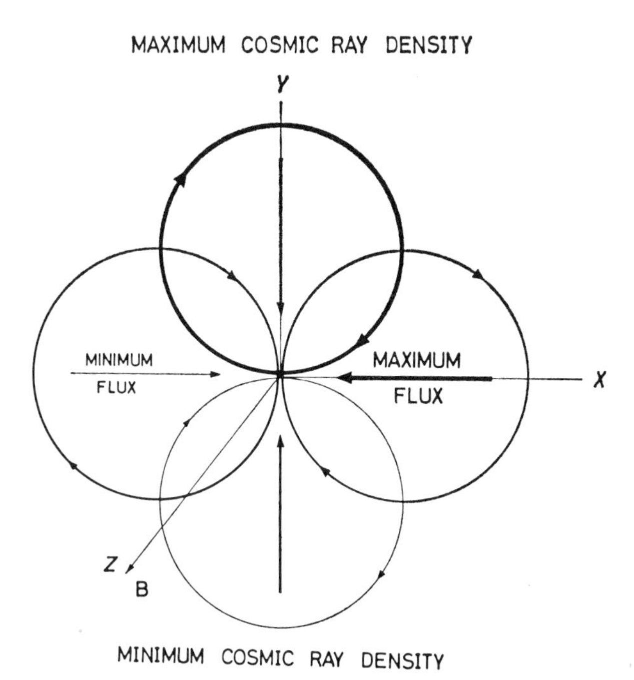

Fig. 6.20 Anisotropy in flux due to inhomogeneity in density. The circles are projections on the xy plane of helices (the plane normal to the magnetic field); their widths indicate the number of particles involved. The straight arrows indicate the fluxes in the different directions (Davis 54).

and the sense of the maximum direction depends on the sense of the magnetic field (Davis 54). The maximum that is observed near right ascension 12 h for energies greater than 5×10^{15} eV corresponds to the Galactic North, and this results from the sense of the arm's magnetic field pointing to the direction of spiral-out, since the sun is located near the inner edge of the Orion arm, as illustrated in Fig. 6.19. The amplitude thus expected is shown by the dashed line in Fig. 6.16*a*.

For electrons the sense of gyration is the opposite; and if we observe synchrotron radiation from electrons outside an arm in which the gyrating electrons emit radio waves, the radio intensity from one edge is stronger than that from another. The radio brightness distribution in the North Polar Spur has this feature, if the spur is identified with the Sagittarius branch, and this is consistent with the magnetic field pointing to us, the same sense as indicated for the Orion arm.

According to Oda and Hasegawa (Oda 62, 63) the electrons responsible for the spur were produced at a strong cosmic-ray source in the Sagittarius branch several thousand years ago. If this is the case, it is quite likely that such a strong source may also be found in the arm in which we are located, and nuclear particles therefrom may show anisotropy. The anisotropy is expected to be pronounced for heavy nuclei, since their mean free path for nuclear interactions is smaller than that of protons (Fukui 62). The quantitative analysis of the anisotropy of muon-rich EAS given in Section 6.5.3 is in favor of this picture (Sekido 63).

Let us consider that particles are produced at time $t=0$ and then propagate along a uniform magnetic field. A particle of pitch angle α travels with a velocity of $c \cos \alpha$ along the magnetic field, so that it reaches us at time

$$t = \frac{r}{c \cos \alpha}, \tag{6.5.5}$$

where r is the distance of the source. Initially therefore the accessible particles are of small pitch angles. As the pitch angle increases the path length traversed is so large that heavy nuclei may be subject to collision with interstellar matter. If the source were located at a large distance, say $r \simeq 100$ kpc, few heavy nuclei would have large pitch angles, irrespective of the time of their generation. However, it is unlikely that the uniformity of the arm fields holds as far as 100 kpc. Nevertheless, anisotropy is expected for heavy nuclei only and not for protons—if most heavy primaries are due to such a source. In accordance with the small relative intensity of heavy primaries at high energies, the protons due to this source comprise but a minor fraction of all protons, and the anisotropy for the former may be masked by the near isotropy of the latter.

Whatever the interpretation may be, the distribution of arrival directions of muon-rich EAS has a characteristic feature, as demonstrated by Sekido and Sakakibara (Sekido 63). Figure 6.18 compares the distributions with respect to four coordinate systems with the isotropic ones. In (*a*) the distribution in the equatorial coordinate system is shown to be evidently different from isotropy. In the Galactic coordinate system, (*b*), a deviation from isotropy is seen in the direction of spiral-out. This is made clear by referring to the arm coordinate systems. Defining the zenith angle with respect to the spiral-out direction as α and the azimith angle as ϕ, where $\phi = 0$ lies in the Galactic plane and near the direction of the Galactic center, the distribution observed is compared with the isotropic distribution expected in the coordinate systems referring to the Orion arm in (*c*), and the Sagittarius branch in (*d*). In both cases the distribution is ϕ are isotropic, whereas significant deviations are seen in the directions of spiral-out, or $\alpha < 45°$. This indicates that there may be a strong cosmic-ray source in the spiral-out direction, although it is not certain whether it belongs to the Orion arm or the Sagittarius branch.

6.6 Secular Variation

The anisotropy of cosmic-ray intensity results in the time variation thereof measured at one point on the earth. In addition to the time variation associated with anisotropy, there is the genuine time variation of the average level of cosmic-ray intensity. The time variation of the latter type is most pronounced when it is associated with solar activity. This is more or less recurrent but not secular. The *secular variation* is concerned with the time variation over many years, so many that the intensity of cosmic rays far before their discovery has to be considered. Since no observation of cosmic rays was made so many years ago, we have to depend on the products left by the effects of cosmic rays. They may be fossils of odd animals that had suffered from the strong radioactivity due to cosmic rays (Krassovskij 58). Practically, however, the products are the nuclides, produced by cosmic-ray interactions, that are radioactive or stable and in general of small abundances. Such nuclides may include those that are hardly produced by thermonuclear reaction in stars, such as deuterium, lithium, beryllium, and boron. They were probably produced in particular local systems by energetic particles less directly connected with general cosmic rays (Hayakawa

54). We are here interested mainly in nuclides produced by general cosmic rays on earth as well as in meteorites.

6.6.1 Possible Causes of Secular Variation

Although no definite evidence for secular variation has yet been found, the following causes may be considered to give rise to the variation of cosmic-ray intensity on and in the neighborhood of the earth:

1. Geomagnetism. Paleomagnetic studies indicate that the geomagnetic field has changed not only its strength but also its sense. Accordingly the intensity of cosmic rays, which is subject to the geomagnetic effect, varies slowly in time.

At the periods when the geomagnetic field changed its sense the magnetic-field strength is expected to have been very low, and consequently cosmic rays falling on the earth were stronger than they are now. However, this period is expected to have been so short that its effect may be quite difficult to detect.

2. Solar activity. Solar activity is recurrent, but the activity in one solar period is appreciably different from that in another period. If the variation in activity is not random but systematic for a considerably long time, it may be regarded as secular and causes the secular variation of cosmic rays.

3. Nonuniformity of Galactic cosmic rays. The intensity of cosmic rays is not necessarily equal everywhere in the Galaxy, as is inferred from the inhomogeneity in radio emission. If, for example, a supernova exploded not far from us, the solar system would be irradiated by strong cosmic rays. Stars of the T-tauri type may be strongly emitting high-energy particles, as is inferred from the overabundance of lithium (Bonsack 60). The sun may once have been at such a stage and thus may have produced strong cosmic rays responsible for the generation of deuterium and other rare nuclides (Hayakawa 60b). Cosmic-ray intensity in the Galactic arms may be greater than that outside, and the solar system may have passed through an arm, as is inferred from the differential rotation of the Galaxy. Then the earth would have been irradiated by strong cosmic rays when it was in the arm (Oda 63).

4. Time variation of general cosmic rays. There are a number of galaxies whose radio-wave emission is far stronger than that of our Galaxy. The strong radio emission may take place at a particular stage in the evolution of a galaxy, and this may have been the case in our Galaxy at some earlier time (Burbidge 62). Whatever it may be, it is possible that the Galaxy was at one time quite active, so that cosmic rays were much stronger. The cosmic rays observed at present could be those that survived. If so, the average level of Galactic cosmic rays would

not be constant in time. More speculatively, cosmic rays in the universe might have been very strong at an earlier epoch in its evolution.

It is not easy to disentangle these causes from observations, but the first cause can be discarded as far as meteorites are concerned. Causes 2, 3, and 4 may be distinguished according to their time scales, which are guessed very roughly to be on the order of 10^5, 10^8, and 10^{10} years, respectively. Such time scales are determined by measuring the abundances of radioactive nuclei with lifetimes of the same orders of magnitude, as well as their decay products. This is a standard method of dating on the assumption that the decay rate of a nucleus is a genuine property that is independent of environmental conditions and constant in time within the age of the universe.

6.6.2 *Methods of Determining the Cosmic-Ray Flux in the Past*

The matter that solidified at the earliest epoch of the solar system, about 4.5×10^9 years ago, is believed to have formed large bodies—much larger than the sizes of meteorites we now see and as large as asteroids. Hence most of the matter was shielded from cosmic rays. The density of such bodies in the asteroid belt was so large that they collided with each other and split into smaller bodies. If the sizes of the fragments were as large as the interaction mean free path of cosmic rays, the entire bodies were exposed to cosmic rays.

Let the exposure have started at time t_e later than t_s, the time of solidification. A part of element A has been converted into element B by irradiation with cosmic rays. The fraction of an element converted to others is given by $\int_{t_e}^{t} \sigma_A J(t)\, dt$, where σ_A is the effective cross section for the nuclear transformation considered and J is the omnidirectional intensity of cosmic rays. Since $\sigma_A \simeq 10^{-24}$ cm^2 for heavy elements and $J \simeq 10$ cm^{-2} sec^{-1} near the maximum of the transition curve in dense matter, the fraction converted would be

$$\int_{t_e}^{t} \sigma_A J(t)\, dt \lesssim 10^{-6}. \tag{6.6.1}$$

This indicates that, if meteorites were exposed to cosmic rays of a nearly constant flux for some giga years, only a negligible fraction would be destroyed by cosmic-ray bombardment.

The effect of cosmic-ray exposure is appreciable if the pre-exposure abundance of element B, $(B)_e$, is much smaller than that of element A. In the case where B is stable we expect

$$\frac{(B)_t - (B)_e}{\sum_A (A)} \int_{t_e}^{t} \sigma_{AB} J(t)\, dt \lesssim 10^{-6}, \tag{6.6.2}$$

where $\sum_A$ represents the summation over parent elements capable of producing element B with the cross section σ_{AB}.

As an example we consider the generation of the light elements—lithium, beryllium, and boron—from heavier ones, mostly the medium-weight elements carbon, nitrogen, and oxygen. Denoting their abundances respectively as (L) and (M) and taking their relative abundances (given in Table 6.2) into account, we are able to set the upper limit on the integrated cosmic-ray flux as

$$\frac{(L)}{(M)} > \sigma_{ML} \int_0^t J(t)\, dt. \tag{6.6.3}$$

Since $(L)/(M) \simeq 3 \times 10^{-6}$ and $\sigma_{ML} \simeq 10^{-25}$ cm^2, we have

$$\int_0^t J(t)\, dt < 3 \times 10^{19} \text{ cm}^{-2}. \tag{6.6.4}$$

Hence the average cosmic-ray flux should not be greater than 10^2 cm^{-2} sec^{-1}.

If we refer only to meteorites, the upper limit would be greater because an appreciable fraction of medium-weight elements are volatile and could have been lost, so that the ratio $(L)/(M)$ in meteorites is larger than 3×10^{-6}.

In the case where B is *radioactive*, the abundance of B changes with time as

$$\frac{d(B)}{dt} = -\lambda(B) + \sigma_{AB}(A)\, J(t)(A), \tag{6.6.5}$$

where $\sum_A$ is dropped by regarding (A) as the total amount of parent elements and σ_{AB} as the average cross section for these parent elements. This yields

$$\frac{(B)_t}{(A)} = e^{-\lambda t} \int_{t_e}^t e^{\lambda t'} \sigma_{AB'}\, J(t')\, dt'$$

$$\simeq \frac{1}{\lambda} \sigma_{AB}\, J[1 - e^{-\lambda(t - t_e)}]. \tag{6.6.6}$$

The last expression holds for the case where the flux intensity does not appreciably change between $t - 1/\lambda$ and t, and J is the intensity during this period. If the lifetime is much longer than the exposure time, this is reduced to the case of a stable nuclide in (6.6.2). If, on the other hand, the lifetime is much shorter than the exposure time, we can neglect the last term in the square brackets in (6.6.6), and the counting rate is equal to the production rate; thus

$$\lambda(B)_t \simeq \sigma_{AB}\, J(A); \tag{6.6.7}$$

namely, element B is in equilibrium with element A.

The relative yields of *cosmogenic nuclides* as given in (6.6.2) and (6.6.6) are not easily measured, since the abundances of elements A and B are greatly different. Moreover, the effective cross section can hardly be obtained with good accuracy. These difficulties can be avoided by comparing the yields of two or more nuclides.

The relative yield of two cosmogenic nuclides, B and B', is given by

$$\frac{(B)_t}{(B')_t} = \frac{\int \sigma_{AB}\, J(t)\, dt}{\int \sigma_{AB'}\, J(t)\, dt} \simeq \frac{\sigma_{AB}}{\sigma_{AB'}} \tag{6.6.8}$$

if both products are stable. The last expression holds again in the case of constant irradiation, and the ratio of the cross sections is known with better accuracy than their absolute values. If both are radioactive, the ratio of their yields is given for constant irradiation by

$$\frac{(B)_t}{(B')_t} \simeq \frac{\lambda'}{\lambda}\frac{\sigma_{AB}}{\sigma_{AB'}}\frac{1-e^{-\lambda(t-t_e)}}{1-e^{-\lambda'(t-t_e)}}. \tag{6.6.9}$$

For $t - t_e \gg 1/\lambda,\ 1/\lambda'$, this is proportional to the ratio of the lifetimes.

If one element, S, is stable and the other, R, is radioactive, the ratio of their yields for constant irradiation is

$$\frac{(S)}{(R)} \simeq \frac{\sigma_{AS}(t-t_e)}{\lambda^{-1}\sigma_{AR}\{1-\exp[-\lambda(t-t_e)]\}} \simeq \frac{\sigma_{AS}}{\sigma_{AR}}\lambda(t-t_e), \tag{6.6.10}$$

where the last expression is obtained for $\lambda^{-1} \ll t - t_e$.

In several cases a pair of cosmogenic nuclides are isobaric, so that the radioactive nuclide decays into the stable one. Then

$$(S) \simeq (\sigma_{AS} + \sigma_{AR})J(t-t_e) \tag{6.6.11}$$

if $\lambda^{-1} \ll t - t_e$, and the ratio is given by

$$\frac{(S)}{(R)} \simeq \frac{\sigma_{AS}+\sigma_{AR}}{\sigma_{AR}}\lambda(t-t_e). \tag{6.6.12}$$

The relations (6.6.8) through (6.6.12) hold if the cosmic-ray flux is constant. In order to illustrate the effect of the intensity variation, we shall consider the following typical cases (Geiss 63).

If a short burst of cosmic rays took place at time $t - t_0$, so that the intensity is expressed as

$$J(t') = J_0 + J_1 t_0\, \delta(t' - t + t_0),$$

the counting rate due to radioactive nuclides produced is proportional to

$$\lambda \frac{(B)_t}{(A)\sigma_{AB}} = [1 - e^{-\lambda(t-t_e)}]J_0 + \lambda t_0 e^{-\lambda t_0} J_1. \tag{6.6.13}$$

Hence the contribution of the burst increases with t_0 for $\lambda t_0 < 1$, reaches a maximum at $t_0 = 1/\lambda$, and decreases exponentially for large t_0. The contribution depends on λ in the same way as t_0 as long as $1/\lambda \ll t - t_e$, and this enables us to find the time of occurrence of such a burst, if any.

If the intensity has changed by J_1 after time $t - t_0$, we obtain

$$\lambda \frac{(B)_t}{(A)\sigma_{AB}} = [1 - e^{-\lambda(t-t_e)}]J_0 + (1 - e^{-\lambda t_0})J_1. \tag{6.6.14}$$

The contribution of the additional flux is $\lambda t_0 J_1$ for $\lambda t_0 \ll 1$ and tends to a saturation value J_1. One can therefore distinguish between the two cases above if nuclides of $1/\lambda < t_0$ are used.

Finally, we consider the effect of a periodic change in cosmic-ray intensity:

$$J(t') = J_0 + J_1 \sin(\omega t' + \varphi).$$

By substituting this into (6.6.6) we obtain

$$\lambda \frac{(B)_t}{(A)\sigma_{AB}} = [1 - e^{-\lambda(t-t_e)}]J_0 + \frac{\lambda}{\sqrt{\lambda^2 + \omega^2}} \times [\sin(\omega t + \varphi - \alpha) - e^{-\lambda(t-t_e)} \sin(\omega t_e + \varphi - \alpha)]J_1, \tag{6.6.15}$$

where $\tan \alpha = \omega/\lambda$. If the phase φ is taken as that at time t, the contribution of the periodic part is $(\lambda/\sqrt{\lambda^2 + \omega^2}) \sin(\varphi - \alpha)$ for $\lambda(t - t_e) \gg 1$. If the irradiation ceases at time $t - t_0$—for example, due to the fall of a meteorite or to the death of a living cell—the radioactivity concerned is reduced to

$$\lambda \frac{(B)_t}{(A)\sigma_{AB}} \simeq J_0 + \frac{\lambda}{\sqrt{\lambda^2 + \omega^2}} e^{-\lambda(t-t_0)} \sin(\omega t_0 + \varphi - \alpha)J_1. \tag{6.6.16}$$

Hence the radioactivities of many samples of different t_0 show a damped periodic change with an amplitude of $\lambda/\sqrt{\lambda^2 + \omega^2}$. The periodicity is observable for $\omega > \lambda$, but in this case the amplitude is very small. Therefore the observation of periodicity is a difficult task.

We have so far considered nuclides produced by cosmic rays. We here mention briefly that past cosmic rays are recorded also by radiation damage in solid matter. Since the probability for radiation damage to take place is proportional to Z^2 of an incident particle, the observation of radiation damage provides a method that is favorable to heavier cosmic-ray nuclei and therefore favorable to a determination of secular variation of the composition of cosmic rays (Walker 65).

6.6.3 Samples of Cosmogenic Elements

Good samples by which cosmogenic products are investigated need to have the following properties. They have been exposed to cosmic rays long enough to accumulate cosmogenic products. If they have not been directly exposed to cosmic rays, they have to be reservoirs of cosmogenic products. The samples have been so stable over a long period that physicochemical fractionation processes can be neglected; the ratio $(B)/(A)$ in the preceding subsection depends only on irradiation by cosmic rays. It is preferable if exposure age is known by the other means—for example, through the ratio of the yields of different radioactive nuclides given in (6.6.9).

In other branches of science, in particular in geology and archeology, cosmogenic products are used for dating on the assumption that the intensity of cosmic rays was constant in the past. Studies of the secular variation in cosmic-ray intensity are also recognized as important with regard to such applications.

Geological Samples. Cosmic rays impinging on the atmosphere produce cosmogenic nuclides mainly by interactions with air nuclei. These nuclides—the products of fragmentation and of neutron and muon capture of nitrogen, oxygen, and argon—are subject to various geophysical effects in the atmosphere, on the ground surface, and in the sea, as has been extensively discussed by Lal and Peters (Lal 62). Among the cosmogenic products radioactive nuclides of relatively short lifetimes, such as ^{7}Be and ^{3}H, are used for geophysical purposes rather than for the study of cosmic rays. A long-lived nuclide, ^{10}Be, may have become concentrated in deep-sea sediments, and has been used to find the secular variation of cosmic-ray intensity over millions of years (Peters 57). A carbon isotope, ^{14}C, which has been used extensively for archeological studies, is concentrated in biological bodies, and its concentration depends on complicated processes of metabolism, molecular exchange, and the burning of fossil fuel.

The production rates of various nuclides can be estimated from the intensity and the energy spectrum of nuclear active particles and from the reaction cross sections concerned. Some of the cross sections are available from laboratory experiments, and others are estimated by the semiempirical method described in Section 2.10. The results are checked for some nuclides at particular altitudes and latitudes. In evaluating the production rates it is important to distinguish those in the stratosphere from those in the troposphere, since nuclides are supposed to stay in the stratosphere as long as a year or longer. The global average of the production rate in the stratosphere is found to be about twice as great as that in the troposphere, and their ratio increases rather rapidly with latitude, as is readily expected from the latitude effect. The global inventories of radioactive isotopes are summarized in Table 6.2.1.

Table 6.21 Cosmic-Ray-Produced Nuclides on the Earth

Nuclide	Half-life	Production Rate (Atoms cm^{-2} sec^{-1})	Global Inventory
^{10}Be	2.7×10^6 years	4.5×10^{-2}	470 tons
^{36}Cl	3.1×10^5 years	1.1×10^{-3}	15 tons
^{14}C	5730 years	2.9	87 tons
^{32}Si	710 years	1.6×10^{-4}	1.4 kg
^{3}H	12.5 years	0.25	3.5 kg
^{22}Na	2.6 years	5.6×10^{-5}	1.2 kg
^{35}S	87 days	1.4×10^{-3}	4.5 g
^{7}Be	53 days	8.1×10^{-2}	3.2 g
^{33}P	25 days	6.8×10^{-4}	0.6 g
^{32}P	14.3 days	8.1×10^{-4}	0.4 g
^{34}Cl	55 minutes	1.4×10^{-3} (nuclear) 2.0×10^{-4} (μ-capture)	2.5 mg

In Table 6.21 we have not included nuclides that are produced by neutrinos. The intensity of general cosmic neutrinos seems to be too weak to produce a measurable amount of nuclides, but solar neutrinos may be strong enough for radiochemical studies. The measurement of radioactive argon, ^{37}A, produced by the $^{37}Cl(\nu_e, e)^{37}A$ reaction is being attempted at an underground station with a target containing a great amount of chlorine (Davis 64, Reines 64).

It should be noted that radioactive isotopes on earth are not exclusively of cosmic-ray origin. It is well known that a considerable amount of radioactive isotopes are attributed to nuclear bombs. A well-known example is ^{90}Sr, and a significant part of ^{35}S originates from nuclear explosions.

Some stable and radioactive nuclides that are rare on earth may have been brought by solar plasma streams and cosmic dust; for example, the neon on earth may have been supplied by the solar wind, as suggested by the helium-neon ratio of 750 as well as the loss rate of neon from earth (Wänke 65). The fall rate of cosmic dust is estimated to be about 10^6 tons per year, and the fraction of ^{10}Be in dust is expected to be about 10^{-11} according to (6.6.6). Hence the contribution of cosmic dust to the total inventory of radioactive nuclides is not always negligible and is probably on the order of 10 percent.

It is also interesting to note that some fundamental particles as yet undiscovered may be produced by cosmic rays and have been accumulated on earth. Examples are the magnetic monopole and the fundamental triplets or quarks. The former is an analogue of the electric charge and would exist if the electric and the magnetic fields were perfectly symmetric, whereas the latter is introduced as a fundamental particle, of which baryons and mesons consist, and has a charge of $\frac{1}{3}e$ or $\frac{2}{3}e$. Since both the monopole and the triplets are not annihilated unless they meet their antiparticles, they should behave like cosmogenic stable nuclides. Both of them should have large masses and consequently high threshold energies, and their production rates should be smaller by many orders of magnitude than those of nuclides produced from nitrogen and oxygen. Moreover, their fractionation processes should be much different from those of nuclear particles because of their extraordinary magnetic and electric properties. The magnetic monopoles could therefore be found in ferromagnetic materials (Goto 63), whereas the triplets should form ions. It would not be appropriate to discuss their inventories here in view of their speculative nature.

Biological Samples. The radioactive carbon isotope ^{14}C has the largest production rate, since the flux of thermal neutrons is high, as described in Section 4.17, and the $^{14}N(n, p)^{14}C$ cross section is large. The ^{14}C thus produced forms carbon dioxide and is accumulated in plants or dissolved in the oceans. The exchange processes of carbon among the atmosphere, the biosphere, and the hydrosphere are complex, and the exchange rates depend on climate and other geophysical conditions, although their average values have been derived (Craig 57). Moreover, the $^{14}C/^{12}C$ ratio is affected by volcanic activity, fossil fuel combustion, nuclear bomb explosion, metabolism, and so forth. Some of the ambiguities can be removed by referring to the content of ^{13}C.

One of the well-dated samples of ^{14}C is the tree-ring, because an appreciable amount of dead cells is left each year (although a deficiency of rings is sometimes found). By measuring the ^{14}C activity in each ring, therefore, it is possible to determine the amount of ^{14}C present in the corresponding year. A number of systematic measurements of ^{14}C have been attempted on old trees such as sequoia, pine, and cedar (Suess 65, Damon 66, Kigoshi 66). With these samples the ^{14}C production rates since 1000 B.C. have been obtained. At still remoter times, archeological samples in Egypt dated by reference to astronomical and historical evidence, and ancient agricultural products are available, though the accuracy of dating decreases for older samples.

Experimental data obtained with various tree samples are in essential agreement with each other. Figure 6.21*a* gives representative data based mainly on an old cedar tree from the Yaku island in southern Japan (Kigoshi 66). Here the relative change in the ^{14}C production rate,

$$\Delta = \left[\frac{Q(t)}{Q(0)}\right] - 1 \tag{6.6.17}$$

is plotted against the years before the present, where $Q(t)$ is the production rate at t deduced by using the half-life of 5730 years. A sharp increase is seen in the seventeenth century, and a general decrease since the beginning of the Christian era. The production rate is recovered and gradually increases until $t \simeq 5000$ years, as indicated by the archeological samples shown in Fig. 6.21*b*.

There are at least two alternative ways of explaining this feature. One is the climate change (Damon 66), and the other is the geomagnetic change (Kigoshi 66). When the climate was warmer plants grew more and trapped a greater quantity of carbon. The decrease of the production rate since 3000 B.C. seems to follow the general deterioration of climate in this period. A sharp maximum, or the oscillation of the production rate after A.D. 1500, seems to correspond to the "little ice age" with a cycle of about 200 years.

Apart from the oscillatory feature in the last 500 years, however, the general trend may also agree with the geomagnetic change based on a paleomagnetic study (Nagata 63). The effect of the geomagnetic change is estimated to lie between the two curves drawn in Fig. 6.21.

These two effects probably coeixst, since both are supported by respective geophysical evidences. This indicates that the secular variation in cosmic-ray intensity can hardly be obtained from the measurement of ^{14}C; we can only give the approximate constancy in the past 5000 years.

Meteorites. In order to avoid such geophysical effects as are discussed above it is necessary to use targets outside the earth. Some space vehicles have been used for this purpose, but they are suitable only for investigating the effects of solar cosmic rays because of their short flight times.

Meteorites are natural targets that have been exposed to cosmic rays over many years. Their life history is, however, complicated. They may have been formed from primordial gas when the solar system was

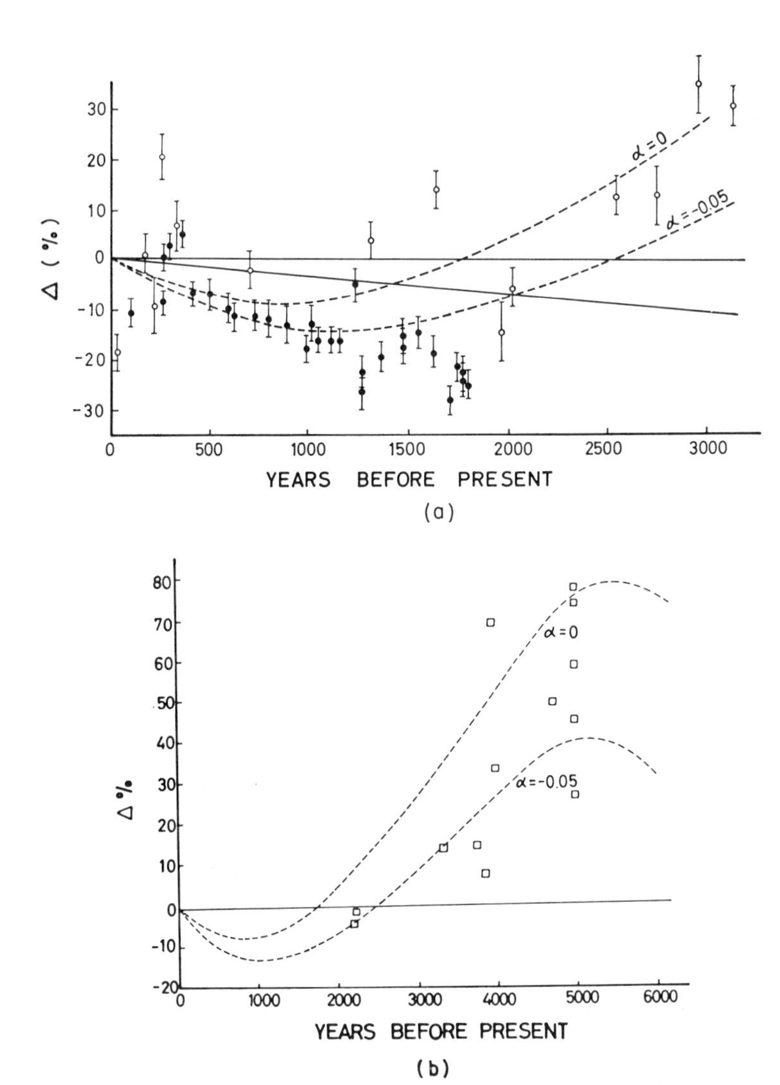

Fig. 6.21 The relative change of the production rate of ^{14}C versus time, $\Delta = [Q(t)/Q(0)] - 1$ in (6.6.17). The line of $\Delta = 0$ represents the constant production rate expected for $T_{1/2} = 5730$ years. The dashed curves represent the time variation of the production rate expected from the geomagnetic-field variation obtained from a paleomagnetic study. Two curves correspond respectively to the cases $\alpha = 0$ and $\alpha = -0.05$, where

$$\alpha = \int_{-\infty}^{0} \Delta(t) e^{\lambda t} \, dt$$

and λ is the decay constant of ^{14}C. (*a*) $\Delta(t)$ up to $t \lesssim 3000$ years before present. ●—Yaku Sugi (Kigoshi 66); ○—sequoia and Pinus ponderosa (Damon 66). (*b*) $\Delta(t)$ up to $t \lesssim 6000$ years before present; □—data from archeological samples.

formed. It is generally believed that in the earliest epoch the solid bodies formed were not as small as the present-day meteorites, and a greater part of their matter may have been shielded from cosmic rays. Small bodies now observed as meteorites are considered to have been split in later periods (Anders 62). Through the whole history of meteorites, their structure has been subjected to complicated physicochemical processes; even at their entry into the atmosphere and at the impact of the ground the friction with air and a shock associated with the impact may cause a considerable change in their properties. Studies of cosmogenic nuclides provide an important means of investigating the history of meteorites.

We first discuss the case of interstellar gas solidifying to form the solar system and the changes that would have occurred at and before that time. Let us consider that the abundances of elements $A, B, \ldots$ were $(A)_f, (B)_f \ldots$ at the time of formation of solid bodies. If A is radioactive and decays into B' with decay constant λ, their abundances at time t is given by

$$(A)_t = (A)_f \, e^{-\lambda(t-t_f)}, \qquad (B')_t = (B')_f + (A)_t(e^{\lambda(t-t_f)} - 1).$$

Since the second equation contains two unknown quantities, $(B')_f$ and $t - t_f$, no solution can be obtained from the single case. Hence the abundance of B_f is compared with that of a stable isotope B with the same atomic number as B', assuming that both B and B' have behaved in the same way in chemical fractionation processes:

$$\frac{(B')_t}{(B)} = \frac{(B')_f}{(B)} + \frac{(A)_t}{(B)} (e^{\lambda(t-t_f)} - 1). \qquad (6.6.18)$$

If the values of $(B')_t/(B)$ for many samples are plotted against those of $(A)_t/(B)$, a straight line is expected to result, so long as $(B')_f/(B)$ and $t - t_f$ are the same for these samples. This method has been applied for the rubidium-strontium and rhenium-osmium series in meteorites and the results thus obtained are summarized in Table 6.22 (Anders 64).

Table 6.22 Solidification Ages of Meteorites

A	B'	B	$T_{1/2}$ (10^9 years)	$t_f - t$ (10^9 years)
^{87}Rb	^{87}Sr	^{86}Sr	46	4.4 ± 0.2
^{187}Re	^{187}Os	^{186}Os	43	4.0 ± 0.8
^{235}U	^{207}Pb	^{204}Pb	0.71	4.6 ± 0.2 (235U and 238U)
^{238}U	^{206}Pb	^{204}Pb	4.51	

The difference in the values of $(A)_t/(B)$ in various samples is considered to be caused by fractionations before and in the course of solidification, and thereafter $(A)_t/(B)$ depends entirely on the radioactive decay. If fractionations continued after solidification, the effect of chemical fractionations can be eliminated by comparing two series of decays. The uranium-lead series is suited to this method, in which the abundances of ^{207}Pb and ^{206}Pb, the decay products of ^{235}U and ^{238}U respectively, are compared. The solidification age thus determined is also shown in Table 6.22.

The results in Table 6.22 show that the solidification of meteorites took place 4.5×10^9 years ago. This is in essential agreement with the terrestrial age determined by similar methods, thus supporting the theory that both the earth and meteorites condensed from interstellar matter at the same time.

If the earth and meteorites were of common origin, they would have the same relative abundances of elements, after corrections for the effects of cosmic-ray irradiation and fractionation. These effects are believed to be of little importance for isotopes of the same chemical properties. However, there are a number of sets of isotopes that have different relative abundances in meteorites.

Such an example has been found in the xenon isotopes. A higher abundance of ^{129}Xe was observed in stone meteorites (Reynolds 60). This is a daughter of ^{129}I with a half-life of 16.4×10^6 years. The excess of ^{129}Xe indicates either that the primordial gases that formed the earth and meteorites had different compositions or that ^{129}I was produced in meteorites or in their parent bodies. The abundances of ^{124}Xe and ^{126}Xe were also found to be anomalous. It is thus suggested that there was an epoch of strong cosmic-ray irradiation after terrestrial and meteorite elements were separated. However, no conclusive evidence for this suggestion has yet been obtained.

Anomalous abundances are also found for chromium isotopes (Shima 65). The abundances of ^{53}Cr and ^{54}Cr are large in a nickel-rich phase of iron meteorites. Since the production cross sections of ^{53}Cr and ^{54}Cr are insensitive to the concentration of nickel, they may have been produced by strong irradiation before the formation of a large body, in which the concentration of chromium occurred along with the formation of the nickel-rich phase.

Evidence for strong irradiation may be provided by the isotopic ratio of lithium in an iron meteorite (Hintenberger 65). The $^7Li/^6Li$ ratio of 1.25 ± 0.2 is significantly smaller than the terrestrial value of 11.8 and is nearly equal to that resulting from the spallation of heavier nuclei. This would require a very high flux, as given in (6.6.4). The irradiation would have taken place when the earth was shielded but meteorites were not.

The origin of deuterium is even more complex. The cosmic-ray flux that could be responsible for the production of lithium would not be strong enough for the production of as much deuterium as 10^{-4} of the hydrogen—if the relative abundances were the same as those of the sun. However, the D/M ratio in the earth and meteorites is smaller than the L/M ratio. Accordingly, deuterium may have been produced by cosmic-ray irradiation after most of the volatile elements such as hydrogen had escaped out of the protoearth and protometeorites (Hayakawa 60b).

All the discussions above would suggest the presence of *early radiation* at the time of the separation of matter in the solar system. This would also tell us that elements affected by the early fractionation would not be suitable for studying the secular variation of general cosmic rays.

If elements were generated after solidification, the results would be free from such complexity. The solidification ages determined by using the decay series of (U, Th)-^{4}He and ^{40}K-^{40}A are much shorter than those given in Table 6.22, about 1.5×10^9 years. This is attributed to the loss of the radiogenic rare gases ^{4}He and ^{40}A. Fortunately, the gas-retention age is longer than the exposure age, and therefore rare gases can be used for studying cosmic-ray exposure.

The exposure ages have been measured for about 20 meteorites. The ages of stone meteorites have been determined mostly by applying the relation (6.6.12) to an isobar pair, ^{3}H-^{3}He. Although their ages are scattered from a few million years to a few hundred of million years, the clustering of ages is found at about 20 million years. The ages of iron meteorites are more widely scattered but appear to cluster in two groups, one about 0.6×10^9 years, and the other at about 0.9×10^9 years. Thus the exposure age is significantly shorter than the solidification age. This would indicate that meteorites are the breakup products of large parent bodies.

After the breakup meteorites have been rather stable, but not entirely so. Further breakup and gradual erosion could have taken place, and they may be considered as a cause of the variation in the exposure ages.

An experimental indication for this is given by comparing the exposure ages of several iron meteorites determined by the ^{36}Cl-^{36}A and ^{40}K-^{41}K methods; the abundance of normal potassium in iron meteorites is so low that ^{40}K and ^{41}K are considered cosmogenic. These pairs are suitable for determination of cosmic-ray intensities over their lifetimes, the half-lives being 3.1×10^5 years for ^{36}Cl and 1.25×10^9

years for ^{40}K. The exposure age determined by ^{36}Cl-^{36}A is systematically shorter than the ^{40}K-^{41}K age (Wänke 63):

$$\frac{^{36}\text{Cl-}^{36}\text{A age}}{^{40}\text{K-}^{41}\text{K age}} = 0.68 \pm 0.05. \tag{6.6.19}$$

This would indicate an increase in cosmic-ray intensity in the last million years.

However, an alternative interpretation is possible and even plausible. If meteorites were subject to erosion, a part near the surface of a meteorite at the time of its fall was located deep below the surface in the remote past. Since cosmic-ray intensity in a deep part is smaller than near the surface due to the attenuation effect, the yield of cosmogenic nuclides in the deep part is smaller. Most of the ^{40}K nuclei could have been produced in the region while it was deep, whereas the ^{36}Cl nuclei were produced while it was close to the surface. Such an erosion effect may be important if we consider cosmogenic elements produced earlier than the exposure age of stone meteorites, which are susceptible to erosion.

Erosion takes place also when a meteorite enters the atmosphere, a surface layer being lost by friction with air. Hence cosmogenic nuclides that are produced mainly in a thin top layer by low-energy particles are hardly seen. Cosmic rays responsible for cosmogenic nuclides in meteorites are therefore considered to have relativistic energies, and the secular variation of low-energy cosmic rays cannot be investigated by means of meteorites. For this purpose cosmic dust may provide a suitable sample, since friction with air is considered to be inappreciable. Studies of cosmic dust have started only recently, and fruitful results are left for future investigation.

6.6.4 Variation of the Cosmic-Ray Intensity

The geophysical and archeological applications of cosmogenic nuclides as well as the determination of the exposure ages of meteorites are based on the assumption that the intensity of cosmic rays is essentially constant in time as well as in the space through which the samples have passed. On the other hand, we have already learnt that the intensity of galactic cosmic rays is anticorrelated with solar activity, and solar cosmic rays are occasionally produced. These two factors also result in the spatial variation in cosmic-ray intensity in interplanetary space. These effects have to be examined in order to see how meteorites have sampled cosmic rays.

The comparison of the radioactivities of ^{37}A and ^{39}A is suitable for this purpose since their half-lives are 35 days and 270 years, respectively

(Fireman 60, Stoenner 60). Most meteorites have elliptic orbits with perihelion distances of 1 A.U. and aphelion distances of 3 to 5 A.U., the latter lying in the asteroid belt. Hence ^{39}A samples the average effect of cosmic rays over many solar cycles along the orbit, whereas ^{37}A samples that near the earth shortly before the fall. If there is a gradient in cosmic-ray intensity, which is supposed to increase with distance from the sun according to the model of solar modulation, the ^{37}A/^{39}A ratio would be smaller than that expected from the zero-gradient case. This effect would be greater as solar activity increases. If, on the other hand, the strong solar cosmic-ray production took place shortly before the fall, the ^{37}A/^{39}A ratio would be larger, and this was the case for a stone meteorite from Hamlet in 1959 (Hayakawa 61a).

If there were no such effects, the ^{37}A/^{39}A ratio would have to be equal to the cross-section ratio, as given in (6.6.9). The cross sections are measured for individual meteorites by using accelerator beams, and their ratios are compared with the ^{37}A/^{39}A ratios in Fig. 6.22 (Schaeffer 63). The results seem to verify the two effects mentioned above, although they may compensate for each other in the solar active period.

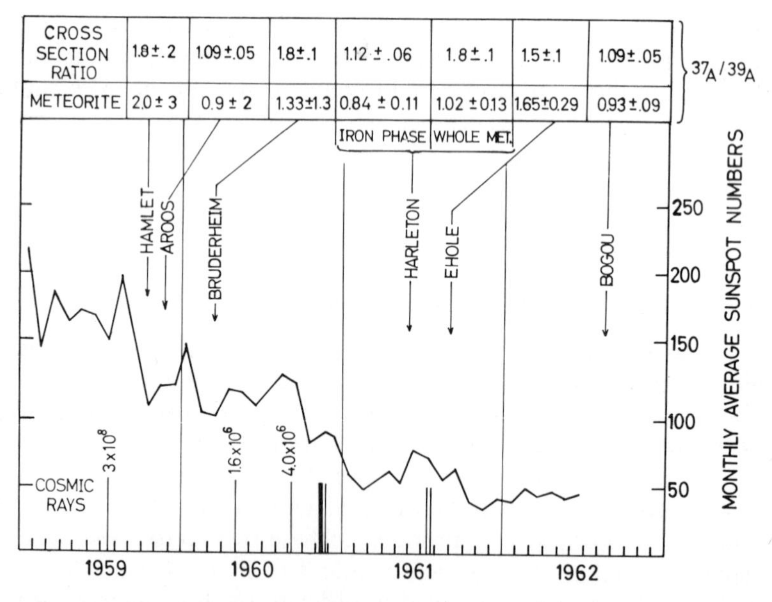

Fig. 6.22 The ^{37}A/^{39}A ratio in recently fallen meteorites. The observed ^{37}A/^{39}A ratios and the cross-section ratios are shown at the tope of the figure; the integral fluxes of solar cosmic rays and their times of occurrence are indicated at the bottom. The monthly average sunspot numbers are shown by the solid line (Schaeffer 63).

The effect of the temporal and spatial variations in cosmic-ray intensity on the yield of cosmogenic nuclides is thus found to be at most 50 percent. A long-term change due possibly to climatic and geomagnetic effects has been measured by means of ^{14}C as being smaller than a factor of 2. Hence the intensity variation of general cosmic rays may be measured within this uncertainty.

As an example, the exposure ages of the stone phase of the Bruderheim meteorite measured by a number of methods by different authors are summarized in Table 6.23. Here the ages are given on the assumption

Table 6.23 Cosmic-Ray-Exposure Age of the Bruderheim Meteorite (Stone Phase)

Method	Half-life (years)	Age (10^6 years)
^{22}Na-^{22}Ne	2.6	27
^{3}H-^{3}He	12.5	24
^{3}H-^{3}He	12.5	35
^{39}A-^{38}A	325	26
^{36}Cr-^{36}A	3.1×10^5	33

of constant irradiation, and the cross sections are obtained by bombardment with accelerator beams as well as by the semiempirical method described in Section 2.10. The agreement of the ages demonstrates the essential constancy of the cosmic-ray intensity over a million years.

A similar result was obtained for an iron meteorite from Aroos (Honda 60). The ratio of the observed activity to the theoretical production rate is plotted against the half-life of each nuclide in Fig. 6.23 (Geiss 63). The result again indicates the constancy of the cosmic-ray intensity over 10^7 years within a factor of 2.

For a longer period, during the last 10^8 to 10^9 years, the comparison of the ^{36}Cl and ^{40}K ages provides information. A slow increase at a rate of about 1/1 Gy is indicated, but this may (though not necessarily in its entirety) be attributed to the erosion effect, as discussed in the preceding subsection. Even if this effect were absent, the variation in intensity would not exceed a factor of 2.

For a still longer period, the meteorites are not usable, since none of them have exposure ages greater than 10^9 years. However, we may be able to know the effects of the early radiation discussed in the preceding subsection.

Summarizing the arguments above, we may say that the intensity of cosmic rays has been essentially constant at least in the last 10^8 years,

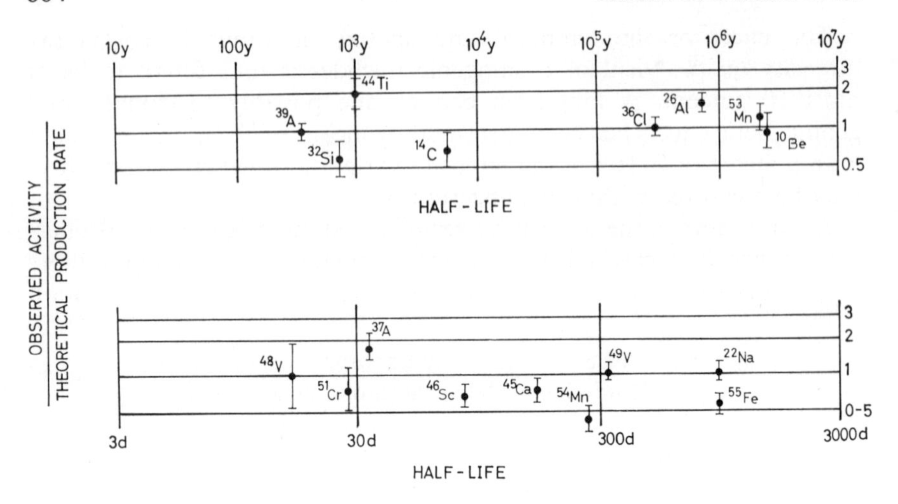

Fig. 6.23 Ratio of observed activities to the theoretical production rates of spallation products in Aroos iron meteorite versus half-life (Geiss 63).

after the Galaxy attained its present structure. The intensity did not radically change when Galactic arms changed their configuration. During the formation of the solar system a strong emission of cosmic rays could have taken place.

6.7 Electrons and Radio Emission

We have thus far discussed the behavior of the nuclear component, of which the cosmic radiation mainly consists. As nuclear particles are accelerated to cosmic-ray energy, one may expect that electrons are accelerated as well. However, the intensity of electrons was found to be negligible in comparison with that of protons (Hulsizer 49, Critchfield 50). This has been attributed either to the acceleration efficiency favoring particles having greater mass or to elimination processes being more effective for electrons while they propagate in interstellar space. Even if no acceleration takes place electrons should be produced through nuclear interactions with interstellar matter (Hayakawa 52).

The energy loss of high-energy electrons is due mainly to the magnetic bremsstrahlung or synchrotron radiation of electrons in interstellar magnetic fields as well as to inverse Compton collisions with low-energy photons (Feenberg 48). The former is responsible for radio emission—and the general Galactic radio emission is interpreted in these terms (Ginzburg 53)—whereas the latter may be responsible for the general

X-ray emission that possibly takes place in intergalactic space. Thus high-energy electrons are found to play an important role in high-energy astrophysics.

The rate of electron energy loss due to elimination processes is proportional to the densities of magnetic and radiation energies, and also to the square of electron energy. Hence the intensity of electrons is expected to decrease as energy increases, and this feature will tell us how long the electrons have traveled in interstellar space. This information is complementary to the thickness of matter traversed, as discussed in Section 6.3 on the basis of fragmentation processes, since the latter gives the matter thickness but not the absolute path length.

6.7.1 Synchrotron Radiation

In designing an electron accelerator, synchrotron, the electron energy loss by radiation provides an important factor. The radiation is caused by the circular motion of an electron in a magnetic field in which the electron is continuously subjected to centrifugal acceleration. In the nonrelativistic region the radiation has a line spectrum at integral multiples of the gyration frequency. This is called the *cyclotron frequency* and is given for a magnetic-field strength H in gauss by

$$\nu_c = \frac{1}{2\pi}\frac{eH}{mc} = 2.81 \times 10^6\, H \quad \text{Hz}. \tag{6.7.1}$$

The power emitted by cyclotron radiation is given by

$$S_n = \frac{8\pi^2 e^2}{c}\, \nu_c^2 \left(\frac{v}{c}\right)^2 \frac{n+1}{(2n+1)!}, \tag{6.7.2}$$

where v is electron velocity and $n = 1, 2, \ldots$ the harmonic order.

In the relativistic region the Doppler effect increases the frequency by a factor of $(E/mc^2)^2$, where E is the total energy of the electron, and higher harmonics form a continuous spectrum. The radiation is emitted within an angle mc^2/E with respect to electron motion, and is linearly polarized preferentially in the direction perpendicular to the magnetic field. This is called *synchrotron radiation.* Its essential features will be given in what follows; for details the reader should consult the book by Ginzburg and Syrovatskii (Ginzburg 64).

In describing synchrotron radiation it is convenient to introduce a frequency

$$\begin{aligned}\nu_s &= \frac{3eH_\perp}{4\pi mc}\left(\frac{E}{mc^2}\right)^2 = \tfrac{3}{2}\nu_c \sin\theta \left(\frac{E}{mc^2}\right)^2 \\ &= 6.26 \times 10^{18}\, H_\perp[E(\text{erg})]^2 = 1.60 \times 10^{13}\, H_\perp[E(\text{GeV})]^2 \quad \text{Hz},\end{aligned} \tag{6.7.3}$$

where $H_\perp = H \sin\theta$, and θ is the angle between the magnetic field and electron motion. The spectrum of radiation emitted by an electron of energy E is expressed as

$$S(E, \nu)\, d\nu = \sqrt{3}\, \frac{e^2}{\hbar c}\, \nu_c \sin\theta\, F\!\left(\frac{\nu}{\nu_s}\right) h\, d\nu, \tag{6.7.4}$$

where $h = 2\pi\hbar$ is the Planck constant and

$$F(x) = x \int_x^\infty K_{5/3}(x')\, dx'. \tag{6.7.5}$$

Here $K_{5/3}(x)$ is the modified Bessel function of the second kind, and the asymptotic expressions of $F(x)$ are

$$F(x) \simeq \sqrt{\pi/2}\, x^{1/2} e^{-x}\left(1 + \frac{55}{72}\frac{1}{x} - \cdots\right) \quad \text{for} \quad x \gg 1, \tag{6.7.5a}$$

$$F(x) \simeq \frac{4\pi}{\sqrt{3}\,\Gamma(\frac{1}{3})}\left(\frac{x}{2}\right)^{1/3}\left[1 - \frac{\Gamma(\frac{1}{3})}{2}\left(\frac{x}{2}\right)^{2/3} + \cdots\right] \quad \text{for} \quad x \ll 1. \tag{6.7.5b}$$

This has a maximum at

$$\nu_m \simeq 0.29\nu_s = 1.2 \times 10^6\, H_\perp\left(\frac{E}{mc^2}\right)^2 \quad \text{Hz}$$

$$= 1.8 \times 10^{18}\, H_\perp[E(\text{erg})]^2 = 4.6 \times 10^{-6}\, H_\perp[E(\text{eV})]^2. \tag{6.7.6}$$

The spectral density at the maximum is

$$S(E, \nu_m) \simeq 1.60\, \frac{e^3 H_\perp}{mc^2} = 2.16 \times 10^{-22}\, H_\perp \quad \text{erg-sec}^{-1}\ \text{Hz}^{-1}. \tag{6.7.7}$$

The degree of polarization is given by

$$\Pi(\nu) = \frac{(\nu/\nu_s) K_{2/3}(\nu/\nu_s)}{F(\nu/\nu_s)} \simeq \begin{cases} \frac{1}{2} & \text{for } \nu \ll \nu_s, \\ 1 - \left(\dfrac{2\nu_s}{3\nu}\right) & \text{for } \nu \gg \nu_s. \end{cases} \tag{6.7.8}$$

The total emission power gives the rate of energy loss,

$$-\frac{dE}{dt} = \int S(E, \nu)\, d\nu = \frac{4}{9}\frac{e^2}{\hbar c} h\nu_s^2 = \frac{16\pi}{3} r_e^2 c\, \frac{H_\perp^2}{8\pi}\left(\frac{E}{mc^2}\right)^2, \tag{6.7.9}$$

where $r_e = e^2/mc^2$ is the classical electron radius. Since this depends on E^2, the energy of an electron decreases as

$$E = \frac{E_0}{1 + t/t_s}, \tag{6.7.10}$$

where E_0 is the initial energy. The characteristic time t_s is the half-life of an electron and is given by

$$t_s = \frac{3mc^2}{2r_e^2 cH_\perp^2}\frac{mc^2}{E} = \frac{5.13 \times 10^8}{H_\perp^2}\frac{mc^2}{E} \quad \text{sec.} \tag{6.7.11}$$

Next we consider the synchrotron radiation emitted by electrons of energies spreading over a wide energy range. Around energy E_0 the energy spectrum of electrons can be approximated by a power law; the unidirectional intensity is expressed as

$$j(E)\,dE = \alpha j_0 \left(\frac{E}{E_0}\right)^{-\alpha} \frac{dE}{E}. \tag{6.7.12}$$

If these electrons are moving toward an observer, the radiation spectrum emitted per unit volume is

$$\begin{aligned} Y_0(\nu)\,d\nu &= \frac{1}{c}\int j(E)\,S(E,\nu)\,dE\,d\nu \\ &= \sqrt{3}\,2^{\alpha/2}\frac{\alpha}{\alpha+2}\Gamma\left(\frac{3\alpha+2}{12}\right)\Gamma\left(\frac{3\alpha+22}{12}\right)\frac{e^2}{\hbar c} \\ &\quad \times h\nu_c \sin\theta\left(\frac{\nu_0 \sin\theta}{\nu}\right)^{\alpha/2} 4\pi\frac{j_0}{c}\,d\nu, \end{aligned} \tag{6.7.13}$$

where $\nu_0 = (\frac{3}{2})\nu_c(E_0/mc^2)^2$. This depends on θ as $\sin^{1+\alpha/2}\theta$. If the directions of magnetic fields are random, the radiation power (6.7.13) is averaged over θ. Thus we obtain for $\alpha > 0$

$$\begin{aligned} Y(\nu) &= y(\alpha)\frac{e^2}{\hbar c}h\nu_c\left(\frac{\nu_0}{\nu}\right)^{\alpha/2}\alpha n_0 \\ &= 1.35 \times 10^{-22} y(\alpha)\left(\frac{1.6 \times 10^{13}}{\nu}\right)^{\alpha/2} H^{1+\alpha/2}\alpha n_0 E_0^{\alpha} \\ &\qquad \text{erg sec}^{-1}\,\text{cm}^{-3}\,(\text{Hz})^{-1}, \end{aligned} \tag{6.7.14}$$

where $n_0 = 4\pi j_0/c$ is the density of relativistic electrons with energies greater than E_0 in GeV, and

$$y(\alpha) = \frac{\sqrt{3}\,2^{\alpha/2}\Gamma\left(\frac{3\alpha+2}{12}\right)\Gamma\left(\frac{3\alpha+22}{12}\right)\Gamma\left(\frac{\alpha+6}{4}\right)}{8\sqrt{\pi}(\alpha+2)\Gamma\left(\frac{\alpha+8}{4}\right)}. \tag{6.7.15}$$

The values of $y(\alpha)$ are given in Table 6.24. It should be noted that for

Table 6.24 Values of y, k_1, and k_2 in (6.7.15) and (6.7.16)

	α						
	0	$\frac{1}{2}$	1	$\frac{3}{2}$	2	3	5
$y(\alpha)$	0.283	0.147	0.103	0.0852	0.0742	0.0725	0.0922
$k_1(\alpha)$	0.80	1.3	1.8	2.2	2.7	3.4	4.0
$k_2(\alpha)$	0.00045	0.011	0.032	0.10	0.18	0.38	0.65

$\alpha \leq 0$, n_0 no longer represents electron density, and in this case we have to regard αn_0, as a whole, as a coefficient of the differential spectrum.

The radiation spectrum emitted by electrons of a power spectrum has a power law with a spectral index $\alpha/2$. The power spectrum of radiation holds in the frequency range $\nu_1 \leq \nu \leq \nu_2$ if the electron spectrum obeys a power law over the energy range

$$\left[\frac{2\nu_1}{3\nu_c k_1(\alpha)}\right]^{1/2} \leq \frac{E}{mc^2} \leq \left[\frac{2\nu_2}{3\nu_c k_2(\alpha)}\right]^{1/2}. \tag{6.7.16}$$

The values of k_1 and k_2 are also given in Table 6.24.

6.7.2 *Inverse Compton Effect*

Let us consider that an electron of energy E collides with a photon of energy ε. In the electron rest system the photon energy is expressed (by reference to Appendix B) as

$$\tilde{\varepsilon} = \varepsilon(1 - \beta \cos \theta) \frac{E}{mc^2}, \tag{6.7.17}$$

where $c\beta$ is the electron velocity and θ is the angle between the momenta of the colliding photon and the electron. The collision is equivalent to the *Compton scattering* of a photon of energy $\tilde{\varepsilon}$. In order to obtain the energy spectrum of scattered photons it is convenient to start with the invariant expression of the Compton cross section (Ginzburg 64).

The cross section is expressed as a function of two invariant quantities $\tilde{\varepsilon}$ and

$$\tilde{\varepsilon}' = \varepsilon'(1 - \beta \cos \theta') \frac{E}{mc^2}, \tag{6.7.18}$$

where $\tilde{\varepsilon}'$ is the energy of the scattered photon and θ' is the angle between the momenta of the scattered photon and the initial electron. The differential cross section for the photon to be scattered in the solid-angle element $d\Omega'$ is expressed as

$$\sigma_c(E, \varepsilon, \varepsilon', \theta, \theta')\, d\Omega'$$

$$= \tfrac{1}{2} r_e^2 \left(\frac{\tilde{\varepsilon}'}{\tilde{\varepsilon}}\right)^2 \left[\left(\frac{mc^2}{\tilde{\varepsilon}} - \frac{mc^2}{\tilde{\varepsilon}'}\right)^2 + 2\left(\frac{mc^2}{\tilde{\varepsilon}} - \frac{mc^2}{\tilde{\varepsilon}'}\right) + \left(\frac{\tilde{\varepsilon}}{\tilde{\varepsilon}'} + \frac{\tilde{\varepsilon}'}{\tilde{\varepsilon}}\right)\right] d\Omega'. \quad (6.7.19)$$

The energy of the scattered photon is readily given by

$$\varepsilon' = \frac{(1 - \beta \cos \theta)\varepsilon}{1 - \beta \cos \theta' + (1 - \cos \Theta)(\varepsilon/E)} = \varepsilon'(E, \varepsilon, \theta, \theta', \Theta), \quad (6.7.20)$$

where Θ is the angle between the momenta of the initial and scattered photons. For $\varepsilon \ll \varepsilon' < E$ this is approximately expressed by

$$\varepsilon' = \frac{(1 - \beta \cos \theta)\varepsilon}{1 - \beta \cos \theta' + (1 - \beta \cos \theta) \cos \theta'(\varepsilon/E)} = \varepsilon'(E, \varepsilon, \theta, \theta'). \quad (6.7.20')$$

In practical cases the initial photons are isotropic, and their spectrum is represented by the Planck distribution of temperature T as

$$n(\varepsilon)\, d\varepsilon = \frac{n_0}{2.4(kT)^3} \frac{\varepsilon^2\, d\varepsilon}{\exp(\varepsilon/kT) - 1}, \quad (6.7.21)$$

where n_0 is the density of photons and the average photon energy is $\bar{\varepsilon} = 2.7\, kT$. Moreover, the energies of the photons of interest lie in the visible and radio-frequency regions, so that

$$4\varepsilon E \ll (mc^2)^2. \quad (6.7.22)$$

After substituting this relation into (6.7.17) we see that the scattering process in the electron rest system is nonrelativistic. Under this assumption the energy of the scattered photon is limited by

$$\varepsilon \leq \varepsilon' \leq 4\varepsilon(E/mc^2)^2. \quad (6.7.23)$$

For the isotropic distribution of initial photons the cross section averaged over the initial directions and integrated over the scattered directions is found to be

$$\begin{aligned}
\bar{\sigma}_c(E, \varepsilon, \varepsilon')\, d\varepsilon' &= \frac{1}{4\pi} \iint (1 - \beta \cos \theta)\sigma_c(E, \varepsilon, \varepsilon', \theta, \theta') \\
&\quad \times \delta[\varepsilon' - \varepsilon'(E, \varepsilon, \theta, \theta', \Theta)]\, d\Omega\, d\Omega'\, d\varepsilon' \\
&\simeq \frac{\pi}{4} r_e^2 \frac{(mc^2)^4}{E^3 \varepsilon^2} \\
&\quad \times \left[2\frac{\varepsilon'}{E} - \frac{(mc^2)^2 \varepsilon'^2}{E^3 \varepsilon} + 4\frac{\varepsilon'}{E} \ln \frac{(mc^2)^2 \varepsilon'}{4E^2 \varepsilon} + \frac{8E\varepsilon}{(mc^2)^2}\right] d\varepsilon'.
\end{aligned} \quad (6.7.24)$$

This gives a nearly flat spectrum in the energy range given in (6.7.23), and the average energy of the scattered photons is

$$\bar{\varepsilon}' = \frac{4}{3}\bar{\varepsilon}\left(\frac{E}{mc^2}\right)^2. \tag{6.7.25}$$

Hence the scattered photons belong, in most practical cases, to the X-ray region. By integrating (6.7.24) over ε' we obtain, of course, the Thomson cross section

$$\sigma_{\text{Th}} = \int \bar{\sigma}_c(E, \varepsilon, \varepsilon')\, d\varepsilon' = \frac{8\pi}{3} r_e^2. \tag{6.7.26}$$

The spectrum of the scattered photons can be obtained by integration over the spectra of initial electrons and photons. The production spectrum is thus given by

$$\begin{aligned} p(\varepsilon') &= \int n(\varepsilon)\, d\varepsilon \int_{E_{\min}}^{\infty} j(E)\, dE\, \bar{\sigma}_c(E, \varepsilon, \varepsilon') \\ &= f(\alpha)\tfrac{1}{2} n_0\, \sigma_{\text{Th}}\, \alpha j_0 \left(\frac{E_0}{mc^2}\right)^{\alpha} \left(\frac{4\bar{\varepsilon}}{3\varepsilon'}\right)^{\alpha/2} \frac{1}{\varepsilon'}, \end{aligned} \tag{6.7.27}$$

where $E_{\min} = mc^2(\varepsilon'/4\varepsilon)^{1/2}$, and the power electron spectrum $j(E) = \alpha j_0(E_0/E)^{\alpha} E^{-1}$ is assumed as before. The numerical coefficient $f(\alpha)$ is calculated as

$$f(\alpha) = 4.51(1.05)^{\alpha} \frac{\alpha^2 + 6\alpha + 16}{(\alpha+2)(\alpha+4)^2(\alpha+6)} \Gamma\left(\frac{\alpha+6}{2}\right) \zeta\left(\frac{\alpha+6}{2}\right), \tag{6.7.28}$$

where $\zeta(x)$ is a Riemann function defined as

$$\zeta(x) \equiv \sum_{n=1}^{\infty} \frac{1}{n^x}.$$

The values of $f(\alpha)$ are given as

$$f(0) = 0.84, \qquad f(1) = 0.86, \qquad f(2) = 0.99, \qquad f(3) = 1.4. \tag{6.7.29}$$

Since $f(\alpha)$ is a slowly varying function near unity, the result above can be reproduced with good accuracy if the spectrum of the scattered photons in (6.7.24) is assumed to be monochromatic:

$$\int \bar{\sigma}_c(E, \varepsilon, \varepsilon') n(\varepsilon)\, d\varepsilon = n_0\, \sigma_{\text{Th}}\, \delta\left[\varepsilon' - \tfrac{4}{3}\bar{\varepsilon}\left(\frac{E}{mc^2}\right)^2\right]. \tag{6.7.30}$$

The production spectrum is readily obtained as

$$p(\varepsilon') = \int_E n_0 \sigma_{\mathrm{Th}} \delta\left[\varepsilon' - \tfrac{4}{3}\bar{\varepsilon}\left(\frac{E}{mc^2}\right)^2\right] j(E)\, dE$$

$$= \frac{\sqrt{3}}{4} \frac{n_0 \sigma_{\mathrm{Th}} mc^2}{(\bar{\varepsilon}\varepsilon')^{1/2}} j\left(mc^2 \sqrt{\frac{3\varepsilon'}{4\bar{\varepsilon}}}\right)$$

$$= \tfrac{1}{2} n_0 \sigma_{\mathrm{Th}} \alpha j_0 \left(\frac{E_0}{mc^2}\right)^{\alpha} \left(\frac{4\bar{\varepsilon}}{3\varepsilon'}\right)^{\alpha/2} \frac{1}{\varepsilon'}. \tag{6.7.31}$$

It should be noted that the slope of the spectrum is the same as in the case of synchrotron radiation.

As a result of the inverse Compton process, the energy of a photon increases from ε to $\varepsilon' \simeq (\frac{4}{3})\bar{\varepsilon}(E/mc^2)^2$, by a factor of nearly $(E/mc^2)^2$. This results in energy loss by an electron at a rate

$$-\frac{dE}{dt} = c \int n(\varepsilon)\, d\varepsilon \int \bar{\sigma}_c(E, \varepsilon, \varepsilon')\varepsilon'\, d\varepsilon'$$

$$\simeq c\sigma_{\mathrm{Th}} \left(\frac{E}{mc^2}\right)^2 \tfrac{4}{3} \int n(\varepsilon)\varepsilon\, d\varepsilon$$

$$= \tfrac{4}{3} c\sigma_{\mathrm{Th}} \left(\frac{E}{mc^2}\right)^2 W_{\mathrm{ph}}, \tag{6.7.32}$$

where $W_{\mathrm{ph}} = n_0 \bar{\varepsilon}$ is the energy density of photons. This is also analogous to the energy-loss rate due to synchrotron radiation—if the magnetic energy density is replaced by the radiation energy density. Since the energy-loss rate is proportional to E^2, we can again define the half-life as

$$t_c = \left(\frac{3}{4c\sigma_{\mathrm{Th}} n_0}\right)\left[\frac{(mc^2)^2}{E\bar{\varepsilon}}\right]. \tag{6.7.33}$$

In the relativistic region $\varepsilon \gg (mc^2)^2/4E$, the cross section decreases from the Thomson cross section as energy increases, and consequently the inverse Compton cross section decreases as well. The production spectrum holds for ε' up to $(mc^2)^2/\bar{\varepsilon}$; above this the spectrum decreases much faster as $\varepsilon'^{-\alpha-2}$. Likewise the energy-loss rate (6.7.32) holds for E up to $(mc^2)^2/4\bar{\varepsilon}$, and above this the loss rate depends only logarithmically on E; that is,

$$-\frac{dE}{dt} = \frac{3}{8} c\sigma_{\mathrm{Th}} W_{\mathrm{ph}} \left(\frac{mc^2}{\bar{\varepsilon}}\right)^2 \left[\ln\left(\frac{2E}{(mc^2)^2}\right) + \frac{1}{2}\right] \tag{6.7.34}$$

Hence the inverse Compton effect in the relativistic region is of minor importance in practical applications.

6.7.3 *Energy Loss of Electrons*

There are other energy loss processes than synchrotron radiation and the inverse Compton effect. In the relativistic region *bremsstrahlung* also contributes to the energy loss. The rate of energy loss is given, in accord with (2.5.15), by

$$-\left.\frac{dE}{dt}\right|_b = 4Z(Z+1)n\frac{e^2}{\hbar c}vr_e^{\,2}\left[\ln\left(\frac{2E}{mc^2}\right)-\frac{1}{3}\right]E$$

for $mc^2 \ll E \ll 137mc^2Z^{-1/3}$, (6.7.35*a*)

$$-\left.\frac{dE}{dt}\right|_b = 4Z(Z+1)n\frac{e^2}{\hbar c}vr_e^{\,2}\left[\ln(191Z^{-1/3})+\frac{1}{18}\right]$$

for $E \gg 137mc^2Z^{-1/3}$, (6.7.36*b*)

in a medium of atomic density n. The bremsstrahlung loss in the case of complete screening is compared, for example, with the energy-loss rate by the inverse Compton effect in (6.7.32) as

$$\frac{-\left.\dfrac{dE}{dt}\right|_b}{-\left.\dfrac{dE}{dt}\right|_c} = \frac{9}{4\pi}\frac{n_H}{n_0}\frac{e^2}{\hbar c}\beta\frac{(mc^2)^2}{\bar{\varepsilon}E}\left[\ln(191)+\frac{1}{18}\right] \tag{6.7.36}$$

in a hydrogen medium of density n_H. Since the energy dependences of these two processes are different, a major contribution at high energies comes from the inverse Compton loss and loss by synchrotron radiation, whereas at low energies loss by bremsstrahlung exceeds the other two. In the latter energy region, however, the energy loss by ionization gives the largest contribution.

The energy loss by ionization in a medium of neutral hydrogen is given in (2.3.3′) as

$$-\left.\frac{dE}{dt}\right|_{i,\,\mathrm{I}} = 2\pi Zncr_e^{\,2}\frac{mc^2}{\beta}\left[\ln\left(\frac{E^3\beta^2}{2I^2mc^2}\right)+\tfrac{9}{8}-\beta^2\right], \tag{6.7.37}$$

where I is the average ionization potential. This is compared for $Z=1$ with the bremsstrahlung loss as

$$\frac{-\left.\dfrac{dE}{dt}\right|_b}{-\left.\dfrac{dE}{dt}\right|_{i,\,\mathrm{I}}} \simeq \frac{4}{3\pi}\frac{e^2}{\hbar c}\frac{\beta^2E}{mc^2}F, \tag{6.7.38}$$

where F is a factor weakly dependent on energy. This is greater than unity for $E > 350$ MeV, as shown in Table 2.7.

In an interstellar medium the degree of ionization is so considerable that the loss of energy loss by ionization in a plasma has to be used. In the energy-loss formulas (2.3.2′) we put

$$\varepsilon(\omega) = 1 - \left(\frac{\omega_p}{\omega}\right)^2, \quad \bar{\omega} = \omega_p = \sqrt{\frac{4\pi e^2 n_e}{m}}, \tag{6.7.39}$$

where $\omega_p/2\pi$ is the plasma frequency in a medium of electron density n_e. Hence the energy-loss rate is expressed as

$$-\frac{dE}{dt}\bigg|_{i,\,\mathrm{II}} = 2\pi n_e c r_e^{\,2}\frac{mc^2}{\beta}\left[\ln\frac{mc^2\beta^2 E}{(\hbar\omega_p)^2} + 0.43\right]. \tag{6.7.40}$$

Since $\hbar\omega_p$ is much smaller than I, the ionization loss in a fully ionized medium is generally greater than that in a neutral medium.

Summarizing the energy-loss rates above, we give their numerical values for relativistic electrons:

$$-\frac{dE}{dt}\bigg|_{i,\,\mathrm{I}} \equiv A_n = 7.62 \times 10^{-9}\, n\left[3\ln\frac{E}{mc^2} + 18.8\right] \quad \text{eV-sec}^{-1}, \tag{6.7.41a}$$

$$-\frac{dE}{dt}\bigg|_{i,\,\mathrm{II}} \equiv A_i = 7.62 \times 10^{-9}\, n_e\left[\ln\frac{E}{mc^2} - \ln n_e + 73.4\right] \quad \text{eV-sec}^{-1}, \tag{6.7.41b}$$

$$-\frac{dE}{dt}\bigg|_{b} \equiv BE = \begin{cases} 1.37 \times 10^{-16}\, n_H\left[\ln\dfrac{E}{mc^2} + 0.36\right]E \quad E\text{-sec}^{-1} & \text{for no screening} \quad (6.7.41c) \\ 7.26 \times 10^{-16}\, n_H E \quad E\text{-sec}^{-1} & \text{for complete screening,} \quad (6.7.41d) \end{cases}$$

$$-\frac{dE}{dt}\bigg|_{s} \equiv C_s E^2 = 0.98 \times 10^{-3} H_\perp^{\,2}\left(\frac{E}{mc^2}\right)^2 \quad \text{eV-sec}^{-1}, \tag{6.7.41e}$$

for synchrotron radiation, and

$$-\frac{dE}{dt}\bigg|_{c} = C_c E^2 = 2.67 \times 10^{-14} W_{\mathrm{ph}}\left(\frac{E}{mc^2}\right)^2 \quad \text{eV-sec}^{-1}. \tag{6.7.41f}$$

for the inverse Compton effect. These energy-loss rates are shown for two typical interstellar conditions in Fig. 6.24.

In comparing these four processes we see that loss by ionization is dominant at low energies—whereas the E^2-dependent processes, the synchrotron and inverse Compton processes, are dominant at high energies. Loss by bremsstrahlung is appreciable only at intermediate energies, near the critical energy. These two E^2-dependent energy-loss rates are equal if

$$W_{\text{ph}} = \frac{1}{8\pi} H^2 = \frac{1}{12\pi} H_\perp^2, \tag{6.7.42}$$

as may be seen by comparison between (6.7.9) and (6.7.32).

In order to obtain the numerical values of the energy-loss rates we have to know the astrophysical quantities, n_H, n_e, H, and W_{ph}. A number of estimates of their values have given divergent results, so that an uncertainty of a factor of 2 is unavoidable. Table 6.25 gives typical values of the astrophysical quantities for the disk and a region of disk plus halo; the energy-loss rates in Fig. 6.24 are drawn by using these

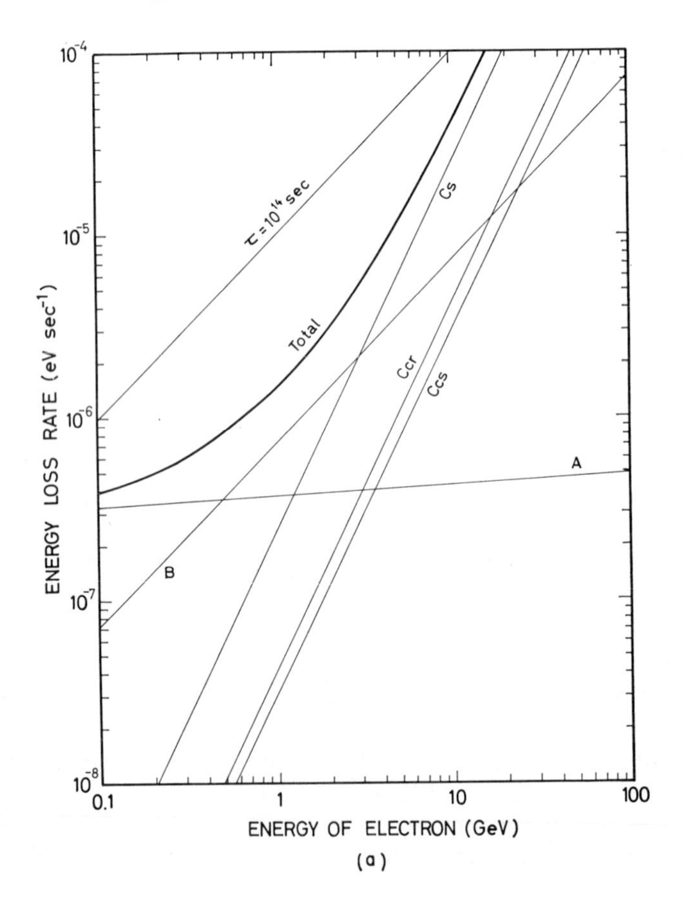

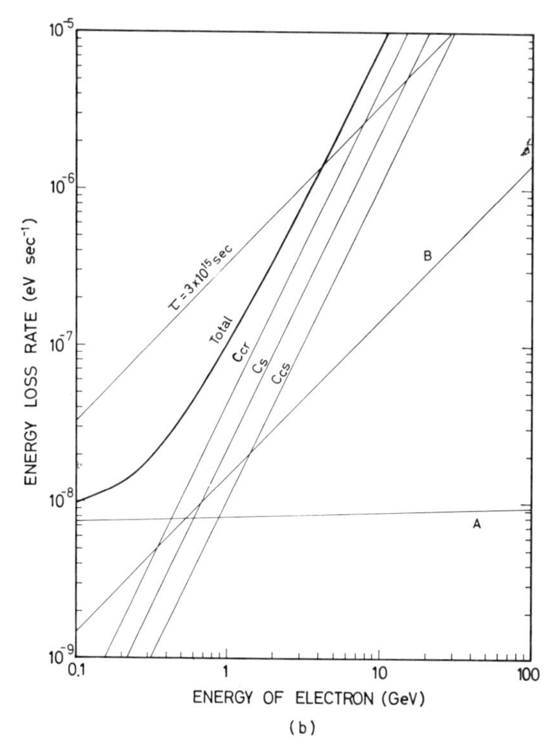

Fig. 6.24 The energy-loss rates of the electron by ionization (A), bremsstrahling (B), synchrotron radiation (C_s), the inverse Compton collisions with starlight photons (C_{cs}) and with microwave photons (C_{cr}). The sum of these effects is also shown. The properties of media are those given in Table 6.25; the energy density of microwaves is assumed as 0.4 eV cm^{-3}. (*a*) The energy-loss rates in the disk. The contribution of escape with the lifetime of $\tau = 10^{14}$ sec is indicated by a line E/τ; (*b*) the energy-loss rates in the ragion of disk + halo. The contribution of escape is indicated by a line E/τ with $\tau = 3 \times 10^{15}$ sec.

Table 6.25 Values of Astrophysical Quantities Relevant to Energy Loss by Electrons

Parameter	Disk	Halo	Disk + Halo
Density of protons (cm^{-3})	1	$\lesssim 10^{-2}$	2×10^{-2}
Density of electrons (cm^{-3})	10^{-1}	$\lesssim 10^{-2}$	1×10^{-2}
Magnetic-field strength (μgauss)	10	3	3
Energy density of starlight (eV-cm^{-3})	0.3	0.1	0.1
Energy density of microwaves (eV-cm^{-3})		0.4	

values. Cosmic-ray studies will provide means of obtaining more reliable values for these quantities, as will be discussed in what follows.

In Figs. 6.24*a* and *b*, which correspond to the *disk* and the *disk plus halo* components, respectively, of Galactic electrons, we draw lines representing

$$-\frac{dE}{dt}=\frac{E}{\tau}, \tag{6.7.43}$$

where τ is the lifetime of cosmic rays in each region, and the values of τ are taken to be 10^{14} and 3×10^{15} sec. If the energy-loss rate is greater than E/τ, the spectrum is essentially determined by energy loss. This may be the case for low and high energies—say, $E < 20$ MeV and $E > 4$ GeV, according to Fig. 2.24*b*. The spectrum is expected to bend as these critical energies are crossed. The bending points depend rather sensitively on the values of the parameters in Table 6.25. The low-energy cutoff is dictated by the loss of energy by ionization and depends on the degree of ionization; whereas the high-energy cutoff is dictated by synchrotron and inverse Compton losses, and depends on the energy densities of magnetic field and radiation as well as on the path length traversed by cosmic rays.

The energy-loss rate given in (6.7.41) is approximately expressed by a polynomial as

$$-\left(\frac{dE}{dt}\right)_{\text{loss}} \equiv L(E) = A + BE + CE^2. \tag{6.7.44}$$

The rate of energy gain may be proportional to energy, as predicted by the Fermi mechanism, but for generality this is also expressed by a polynomial function $G(E)$. Hence the rate of energy change is also expressed by a polynomial as

$$\frac{dE}{dt} = G(E) - L(E) = -(A + B'E + CE^2). \tag{6.7.45}$$

Here A and C are positive, whereas B' is negative or positive, depending on whether acceleration is effective or not. If $B'^2 < 4AC$, the rate of energy change is always negative; that is, energy loss is dominant over energy gain. If $B'^2 > 4AC$, the opposite is the case in an energy range between E_1 and E_2, where

$$E_{1,2} = \frac{1}{2C}\left[-B' \mp \sqrt{B' - 4AC}\right] \qquad (B' < 0). \tag{6.7.46}$$

In this energy range the energy of particles approaches E_2, and consequently the spectrum may have a bump at E_2.

The energy spectrum of electrons at time t after injection, $j(E, t)$, is described by a diffusion equation

$$\frac{\partial j(E, t)}{\partial t} = -\frac{\partial}{\partial E}[(G-L)j(E, t)] - \frac{1}{\tau}j(E, t) + vp(E), \quad (6.7.47)$$

where τ is the mean lifetime of electrons for escape from the spatial region concerned and $p(E)$ is the injection spectrum. The solution is complicated, and we give a reference (Felten 66*a*). In the stationary state we put $\partial j/\partial t = 0$ and have

$$(G-L)j' + \left(G' - L' + \frac{1}{\tau}\right)j = vp. \quad (6.7.48)$$

If the spectrum is assumed to have a power shape, $j \propto E^{-\alpha-1}$, this can be solved as

$$\begin{aligned} j(E) &= \frac{v\tau Ep(E)}{(\alpha+1)(L-G)\tau + E(G'-L')\tau + E} \\ &= \frac{v\tau p(E)}{(\alpha-1)CE\tau + \alpha B'\tau + 1 + (\alpha+1)A\tau/E}, \\ &= \frac{v\tau p(E)}{(\alpha-1)CE\tau - \alpha(E_1+E_2)C\tau + 1 + (\alpha+1)E_1E_2C\tau/E}, \end{aligned} \quad (6.7.49)$$

where the last expression holds for $B'^2 > 4AC$ and $B' < 0$.

From (6.7.49) we can obtain the characteristic features of the spectrum. At low energies the last term in the denominator is so dominant that the spectrum is *flattened* toward low energy as $Ep(E)$. At high energies the first term is so dominant that the spectrum is *steepened* as $p(E)/E$. These critical energies are nearly equal to those at which the straight line crosses the energy-loss curve in Figure 6.24; that is,

$$J(E) \simeq \begin{cases} \dfrac{v\tau Ep(E)}{(\alpha_1+1)A} & \text{for} \quad E \ll A\tau, \\[2ex] \dfrac{vp(E)}{(\alpha_2-1)CE} & \text{for} \quad E \gg 1/C\tau, \end{cases} \quad (6.7.50)$$

where α_1 and α_2 are the differential power exponents in the respective energy regions. If acceleration is effective, the denominator has a minimum or a negative region. In the latter case the electrons injected in the negative region are accelerated to higher energy, and no electrons are left in this region. In the former case the spectrum has a bump at about

$$E \simeq \tfrac{1}{2}(E_1 + E_2) + \frac{1}{2C\tau} = -\frac{B'}{2C} + \frac{1}{2C\tau}. \tag{6.7.51}$$

at which the second term in (6.7.48) vanishes. In either case we would expect a strange feature of the spectrum if acceleration were effective enough. Hence the electron spectrum provides a means of investigating the acceleration rate.

If both protons and electrons are accelerated in the same way in interstellar space, the value of $1/(-B'\tau)$ is determined by the integral exponent to be about $\frac{3}{2}$. The value of C given in Fig. 6.24*b* is 8×10^{-26} $\mathrm{eV^{-1}\ sec^{-1}}$. Thus we have a bump at about

$$E\tau \simeq \frac{1 - B'\tau}{2C} \simeq 1 \times 10^{25} \text{ eV-sec}.$$

The bump would be found at several GeV. However, the injection spectrum may be different from that of protons, so that the argument above would not necessarily hold.

6.7.4 *Galactic Radio Emission*

The energy spectrum of electrons can be derived from the spectrum of general Galactic radio emission if synchrotron radiation is responsible for radio emission. However, thermal radio emission is not negligible at high frequencies. In the radio-frequency region the spectrum of blackbody radiation of temperature T is expressed by the Rayleigh–Jeans law. Referring to the blackbody spectrum, the radio spectrum is customarily expressed by

$$I_\nu = 2\left(\frac{\nu}{c}\right)^2 kT_{\mathrm{eff}}(\nu) = 3.07 \times 10^{-37}\nu^2 T_{\mathrm{eff}}(\nu) \quad \mathrm{erg\ cm^{-2}\ sec^{-1}\ (Hz^{-1})\ sr^{-1}}. \tag{6.7.52}$$

Here T_{eff} is the effective temperature, which may depend on frequency ν except in the case of blackbody radiation. The thermal emission by free-free transition in a transparent plasma has the frequency dependence of $T_{\mathrm{eff}} \propto \nu^{-2}$, and therefore the radio spectrum is expected to be flat. The synchrotron spectrum can be expressed by a power law as

$$I_\nu \propto \nu^{-\beta}, \qquad T_{\mathrm{eff}} \propto \nu^{-2-\beta}, \tag{6.7.53}$$

with $\beta = \alpha/2$ as given in (6.7.13) and (6.7.14). Hence the nonthermal component is masked at high frequencies by the thermal component.

The radio spectrum observed is shown in Fig. 6.25. Up to several hundred MHz the spectrum has a negative slope, and may thus be

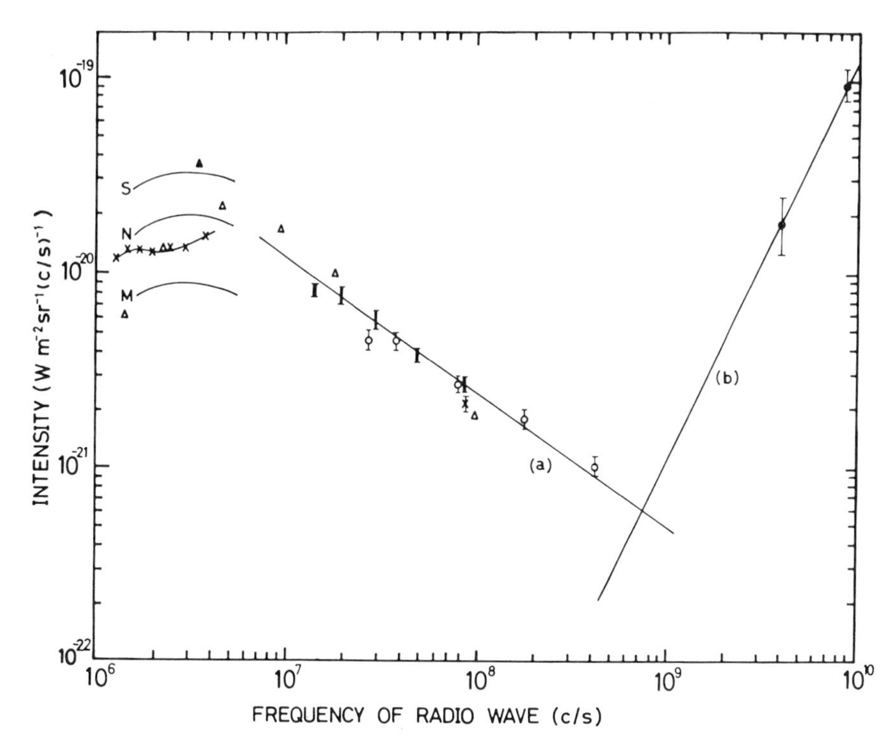

Fig. 6.25 The radio spectrum in the direction of the Galactic pole; line (*a*) represents a nonthermal spectrum of $\nu^{-0.7}$; and line (*b*), a thermal spectrum of ν^2 with $T = 3°$K. ▲—J. H. Chapman, *Space Research, Vol. II*, North Holland, (1961) p. 597 (S. halo); ⌒—T. R. Hartz, *Ann. Astrophys.*, **27**, 823 (1964) (S. halo; N. halo; minimum); △—G. R. A. Ellis, *Nature*, **169**, 1079 (1962) (minimum); ×—J. Hugill and F. G. Smith, *Mon. Not. Roy. Astro. Soc.*, **131**, 137 (1965); I—R. Wielebinski and K. W. Yates, *Nature*, **208**, 64 (1965) (South Pole); ϕ—A. J. Turtle et al., *Mon. Not. Roy. Astro. Soc.*, **124**, 297 (1962); ×—B. Y. Mills, *Publ. Astro. Soc. Pacific*, **71**, 267 (1959); ϕ—A. A. Penzias and R. W. Wilson, *Astrophys. J.*, **142**, 419 (1965); ●—P. G. Roll and D. T. Wilkinson, *Phys. Rev. Letters*, **16**, 405 (1966).

considered as nonthermal in origin, but two points in the microwave region show a blackbody nature (Penzias 65, Roll 66). The latter is regarded as of metagalactic origin and has important bearing on various cosmological problems.

The *nonthermal radiation* is highly anisotropic. It has a latitude dependence strongly peaked in the direction of the Galactic disk. The intensity in the disk direction is 6 to 10 times higher than in other directions, and the thickness of the radio disk is about 500 pc, which is compared with that of the optical disk of about 300 pc or less. The central radio region and the Galactic nucleus are found by thermal emission.

The radio intensity at high latitudes is due almost exclusively to emission outside the disk; namely, from the halo and metagalaxy. If the high-latitude component is attributed to radio emission in the Galactic halo, the difference between the intensities of the disk and the halo components may be due either to the difference in the magnetic-field strengths, or in electron intensities, or both. If the electron intensities in these regions are different, cosmic rays will be stored in the disk to some extent.

As seen from a comparison of Figs. 6.24*a* and *b*, the bending point of the energy spectrum of the disk component is several tens of GeV, whereas that of the halo component is 10 times smaller (Fujimoto 64). The steepening of the electron spectrum results in the steepening of the radio spectrum. The bending points of the radio spectra occur at about 100 GHz and at several hundred MHz. The former is so high that it is masked by thermal radiation, but the latter is in the observable region.

The nonthermal radio waves observed are emitted mainly from electrons with energies of about 1 GeV. The radio intensity of the halo component is about

$$I(100\text{ MHz}) \simeq 2 \times 10^{-21}\text{ W } m^{-2}\text{ sr}^{-1}\text{ Hz}^{-1}, \tag{6.7.54}$$

if the intensity in the pole direction given in Fig. 6.25 is considered as due to the halo component. If the depth of the radio-emitting regions is $L \simeq 2 \times 10^{22}$ cm, the radio-generation rate is

$$\frac{1}{4\pi} Y(\nu) = \frac{I_\nu}{L} \simeq 1 \times 10^{-40}\text{ erg cm}^{-3}\text{ sec}^{-1}\text{ sr}^{-1}\text{ Hz}^{-1}. \tag{6.7.55}$$

The spectral index at about 100 MHz is found to be (Turtle 62)

$$\beta \simeq 0.7 \quad \text{or} \quad \alpha = 1.4. \tag{6.7.56}$$

This corresponds to an electron intensity of

$$j_0 = \frac{1.4 \times 10^{-2}}{H}\text{ cm}^{-2}\text{ sec}^{-1}\text{ sr}^{-1} \quad \text{for} \quad E \geq \frac{2.5}{\sqrt{H}}\text{ GeV}, \tag{6.7.57}$$

where H is in μgauss. The differential spectrum of electrons is given by

$$\alpha j_0 E^{-\alpha-1} = \frac{8.3 \times 10^{-2}}{H^{1.75}} E^{-\alpha-1} \quad \text{cm}^{-2}\text{ sec}^{-1}\text{ sr}^{-1}\text{ GeV}^{-1}, \tag{6.7.58}$$

where E is in GeV. The power spectrum of electrons would hold for

$$1.0\ H^{-1/2}\ \text{GeV} < E < 13\ H^{-1/2}\ \text{GeV}, \tag{6.7.59}$$

according to (6.7.16), if the power radio spectrum holds from 30 to 300 MHz. However, the upper limit could be slightly lowered, because of the contribution of thermal radiation, and the lower limit is subject to disagreement among experimental results. The spectrum observed by Turtle et al. (Turtle 62) begins to bend at about 100 MHz, whereas that observed by Ellis et al. (Ellis 62) gives a much higher intensity at several tens of MHz, and bending takes place at several MHz. At low frequencies the effects of the medium are presumed to play a dominant role.

In a medium of electron density n_e and magnetic-field strength H, the plane of polarization is rotated by an angle

$$\phi = \frac{e^3 n_e H \cos\theta}{2\pi m^2 c^2 \nu^2} L = 2.36 \times 10^4 n_e \frac{LH \cos\theta}{\nu^2}, \tag{6.7.60}$$

where θ is the angle between the directions of propagation and the magnetic field. This is the angle of Faraday rotation and is proportional to the distance of propagation L and the inverse square of frequency. Hence a slight irregularity of fields may depolarize the linear polarization of synchrotron emission.

The spectrum of synchrotron emission is modified in a medium in such a way that the intensity is reduced for

$$\nu \lesssim \frac{ecn_e}{H_\perp} \simeq \frac{15 n_e}{H_\perp}. \tag{6.7.61}$$

This effect begins at about 10 MHz in the disk, whereas it is inappreciable in the halo.

In a medium of temperature T radio waves are absorbed due to electron-ion collisions with a coefficient

$$\begin{aligned} \mu &= \frac{8}{3\sqrt{2\pi}} \frac{e^6}{(kTm)^{3/2}} \frac{n_e^2}{c\nu^2} \ln\left[\frac{2(2kT)^{3/2}}{\gamma^{5/2} e^2 \sqrt{m} 2\pi\nu}\right] \\ &\simeq \frac{0.92 \times 10^{-2} n_e^2}{T^{3/2}\nu^2}\left[17.7 + \ln\left(\frac{T^{3/2}}{\nu}\right)\right]. \end{aligned} \tag{6.7.62}$$

where $\ln\gamma$ is Euler's constant. This gives the optical depth

$$\tau(\nu) = \mu L \simeq \left(\frac{n_e}{10^{-1}}\right)^2 \left(\frac{10^4}{T}\right)^{3/2} \left(\frac{10^7}{\nu}\right)^2 \frac{L}{10^{23}}, \tag{6.7.63}$$

and the effective temperature

$$T_{\text{eff}} = T(1 - e^{-\tau}). \tag{6.7.64}$$

The value of τ is greater than unity for $\nu < 1$ MHz, and the spectrum approaches the blackbody spectrum if radio waves pass through H II regions as thick as 300 pc. For $\tau < 1$ the effective temperature is expressed as $T_{\text{eff}} \simeq T\tau$ and the spectrum is flat according to (6.7.52). The spectrum given in Fig. 6.25 shows such features.

6.7.5 Sources of electrons

In this section and hereafter the electrons produced by the nuclear component of cosmic rays are called *secondary electrons*, whereas those accelerated from thermal or epithermal energy are named *primary electrons*. The secondary electrons are produced, for example, by protons colliding with interstellar hydrogen through the following processes:

$$p + p \to \pi^{\pm} \to \mu^{\pm} \to e^{\pm} \quad \text{(meson decay)}, \tag{6.7.65a}$$

$$\to n \to e^{-} \quad \text{(neutron decay)}, \tag{6.7.65b}$$

$$p + e \to p + e \quad \text{(knock-on)}. \tag{6.7.65c}$$

Other processes are considered to be of minor importance. Among these processes, electrons from meson decays have energies mostly around 0.1 to 1 GeV and are responsible for radio emission, whereas those from neutron decays and knock-ons have relatively low energies and contribute little to radio emission unless they are efficiently accelerated afterwards. These processes are further distinguished by the positive-negative ratio of electrons. A large positive excess is expected for the meson-decay electrons, whereas two other processes give negative electrons only.

The production spectrum of secondary particles through process i is expressed as

$$p_i(E)\,dE = \sum_{Z,A} \int dE_p\, j_Z(E_p)\sigma_{ZA,i}(E_p) f_{ZA,i}(E_p, E) n_A\, dE, \tag{6.7.66}$$

where Z and A represent the charge of cosmic-ray nuclei of the unidirectional intensity $j_Z(E_p)\,dE_p$ and the mass number of target nuclei of density n_A, respectively. Here we define $\sigma_{ZA,i}(E_p)$ as the total cross section for a cosmic-ray nucleus of energy per nucleon E_p, and $f_{ZA,i}(E_p, E)\,dE$ as the effective energy distribution of electrons. In the case of meson decays, therefore,

$$m_{ZA,i}(E_p) = \int f_{ZA,i}(E_p, E)\,dE \tag{6.7.67}$$

gives the multiplicity of electrons produced by a collision.

Nuclear Interactions. Both the processes (6.7.65*a*) and (6.7.65*b*) are caused by nuclear interactions, among which those other than *p-p*, *p-α*, and *α-p* collisions are insignificant. Cross sections suitable for our purpose are partially available, as shown in Sections 2.9, and 3.2 to 3.10. From several experimental results, plausible values of $m\sigma(\pi^+, \pi^-, n)$ are given against the incident proton energy in Fig. 6.26 (Ramaty 66).

Below 10 GeV the energy dependences of yields of secondary particles are steep, and the π^+/π^- ratio for *p-p* collisions is quite large. The ratio

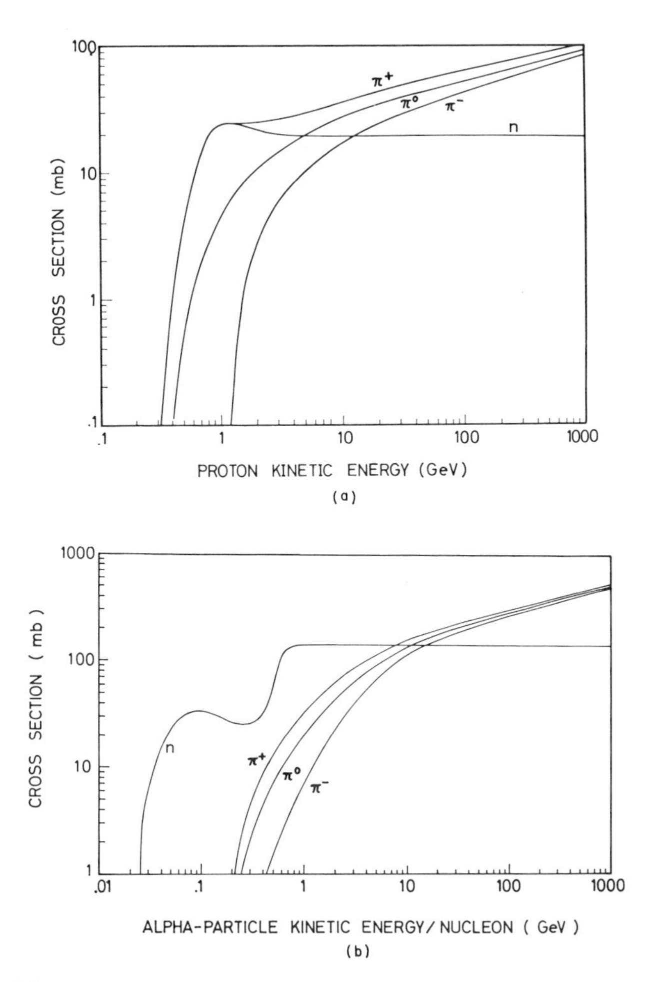

Fig. 6.26 The cross sections (times multiplicities) of production of π^+ π^-, π^0 and *n* by (*a*) *p-p* and (*b*) *p-α* collisions.

of $m\sigma(p\text{-}\alpha)$ to $m\sigma(p\text{-}p)$ is about 4, though it depends on energy (particularly at low energies). The energy dependence of pion multiplicity tends to the $E_p^{1/4}$ law above 20 GeV. Below 20 GeV the energy dependence is steeper and close to $E_p^{1/2}$, as is expected from (2.6.4′). The neutron multiplicity is assumed to be 0.75 for *p-p* collisions.

The energy distribution of secondary particles is complex, and no complete information is yet available. The distribution given in (2.6.4′) is not suitable for our purpose, because this is concerned only with forward pions. The one-fire ball model found to hold at about 100 GeV may be a fair approximation at lower energies, and this predicts an energy distribution concentrated about the average energy. However, the asymmetric emission of fireballs gives an average energy distribution spreading over a wide energy range. In view of such uncertainty it may be a tentative way to approximate the energy spectrum as

$$f(E_p, E) = m(E_p)\,\delta(E - \bar{E}), \tag{6.7.68}$$

where $\bar{E}$ is the average energy of secondary particles.

The average kinetic energies are approximately expressed by (Ramaty 66)

$$\bar{W}_\pi(p\text{-}p) = 0.175\ W_p^{3/4}, \tag{6.7.69a}$$

$$\bar{W}_\pi(p\text{-}\alpha) = 0.134\ W_p^{3/4}, \tag{6.7.69b}$$

$$\bar{W}_\pi(\alpha\text{-}p) = 4\left[0.134\ W_p^{3/4} - \frac{(0.467\ W_p^{3/4} + 0.490)}{(W_p + 4.67)} + 0.105\right], \tag{6.7.69c}$$

$$\bar{W}_n(p\text{-}p) = 0.35\ W_p, \tag{6.7.69d}$$

$$\bar{W}_n(p\text{-}\alpha) = 0.12\ W_p, \tag{6.7.69e}$$

$$\bar{W}_n(\alpha\text{-}p) = 4\left[0.12\ W_p^{3/4} - \frac{(0.42\ W_p^{3/4} + 0.33)}{(W_p + 4.7)} + 0.70\right], \tag{6.7.69f}$$

where all energies are measured in GeV. The above expressions correspond to an inelasticity of 0.3. In deriving the mean secondary energies for α-*p* collisions it is assumed that the α-*p* collision is identical with the *p*-α collision in the center-of-mass system of a proton and an α-particle. This may be questioned, since nucleons in an α-particle may behave independently at high energies. If this is the case, the mean secondary energies in the α-*p* collision would be essentially equal to those in the *p-p* or *p*-α collision, and the final result of the electron intensity would otherwise be overestimated by a factor of about 2. The true value may lie in between; at low energies—say, $W_p < 1$ GeV—the expressions (6.7.69*c*) and (6.7.69*f*) may be a good approximation, whereas at high energies they may have to be replaced by

$$\bar{W}_{\pi,n}(\alpha\text{-}p)=\bar{W}_{\pi,n}(p\text{-}\alpha) \quad \text{for} \quad W_p>1\ \text{GeV}. \tag{6.7.69g}$$

By virtue of (6.7.68), the integration of (6.7.66) is readily performed and gives

$$p_{\pi,n}(E_{\pi,n})=\frac{j(E_p)\,\sigma(E_p)m(E_p)n_A}{(d\bar{E}_{\pi,n}/dE_p)}, \qquad E_p=E_p(\bar{E}_{\pi,n}). \tag{6.7.70}$$

The production spectrum of muons is given as in (4.11.1), but we make the approximation that the muon energy is represented by the average energy; that is, $(m_\mu/m_\pi)E_\pi$. The electron spectrum could also be obtained in the same way—namely, as in Section 4.15. However, this approximation fails for $E_e<m_\pi c^2/4$. Hence we use the method described in Appendix B.

In the rest system of the muon the energy distribution of decay electrons of relativistic energies is given by

$$g_e(E_e^*)=\frac{2E_e^{*2}(3E_0-2E_e^*)}{E_0{}^4}, \tag{6.7.71}$$

where $E_0=105\,m_e c^2$ is the maximum electron energy. Since the angular distribution is nearly isotropic, isotropy is assumed for simplicity. Thus the energy distribution of electrons in the laboratory system is obtained, by utilizing the kinematic relations given in Appendix B, as

$$g_e(E_e)\simeq\frac{1}{2\beta_\pi\gamma_\pi}\int g(E_e^*)\frac{dE_e^*}{E_e^*}\simeq\frac{1}{2\beta_\pi\gamma_\pi\bar{E}_e} \tag{6.7.72}$$

where $\bar{E}_e\simeq 70\,m_e c^2$ and $\gamma_\pi=(1-\beta_\pi{}^2)^{-1/2}$ is the Lorentz factor of the parent pion. Since these electrons are produced by pions having the spectrum (6.7.70), the electron spectrum is given by

$$p_e(E_e)=\int_{E_\pi^-}^{E_\pi^+} g_e(E_e)p_\pi(E_\pi)\,dE_\pi=\frac{m_\pi c^2}{2\bar{E}_e}\int_{E_\pi^-}^{E_\pi^+} p_\pi(E_\pi)\frac{dE_\pi}{p_\pi c}, \tag{6.7.73}$$

where $E_\pi^\pm$ are the maximum and the minimum energies of pions giving an electron of energy E_e;

$$\frac{m_e{}^2c^2E_\pi^\pm}{m_\pi}=\bar{E}_eE_e\pm\sqrt{\bar{E}_e{}^2-m_e{}^2c^4}\,\sqrt{E_e{}^2-m_e{}^2c^4}. \tag{6.7.74}$$

For neutron decay a simpler method is applicable since the maximum energy of electrons is only 0.782 MeV. Thus we can put

$$p_e\left(\frac{E_e}{m_e c^2}\right)=p_n\left(\frac{E_n}{m_n c^2}\right). \tag{6.7.75}$$

The production spectra of electrons (positons and negatons separately) for the above sources are shown in Fig. 6.27. Here the absolute values are given for the choice of

$$n_H = 10^{-2}\ \mathrm{cm}^{-3}, \qquad \frac{n_{\mathrm{He}}}{n_{\mathrm{H}}} = \frac{j_\alpha}{j_p} \simeq 0.15.$$

The production spectra in Fig. 6.27 show that the contribution of neutron decays is important only for electron energies below 10 MeV. From 10 MeV to 1 GeV the production of positons predominates over that of negatons, and the positive excess would be considerable if most of the electrons were due to meson decays. The positive excess decreases as energy increases, and the energy spectrum is approximately expressed

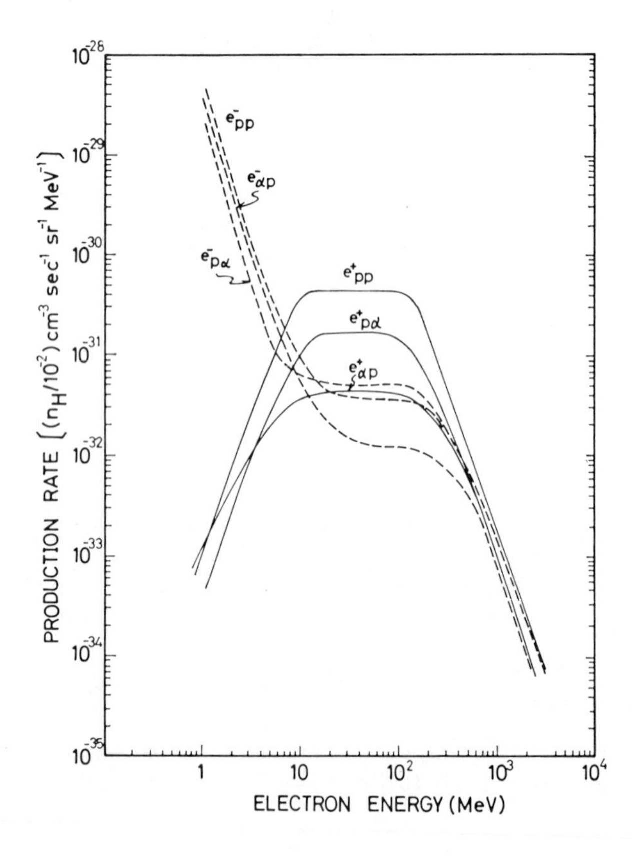

Fig. 6.27 Production spectra of electrons from different nuclear-collision processes.

by use of the proton spectrum as follows:

$$p_e(E_e) \simeq 5 \times 10^{-33} E_e^{-2.67} \left(\frac{n_{\mathrm{H}}}{10^{-2}} \right) \quad \mathrm{cm^{-3}\,sec^{-1}\,sr^{-1}\,GeV^{-1}}, \quad (6.7.76)$$

where E_e is in GeV.

The accuracy of the numerical result above is not better than a factor of 2, mainly because of our poor knowledge about the energy spectrum of pions produced by nuclear collisions, in particular by p-α and α-p collisions. The slope of the spectrum, which is dictated by the multiplicity law and the average secondary energies, is subject to uncertainty as well. If the one-fireball model were valid, the differential exponent would be -3, the spectrum being steeper than (6.7.76). It may be more realistic to consider pion production mainly through excited baryons and mesons. As experimental results are accumulated a better estimate of the production spectrum will be available in the near future.

Knock-on Electrons. The importance of the knock-on process has been pointed out in connection with the presence of electrons of several MeV observed with the IMP-1 satellite (Cline 64), but the observed intensity was found to be larger than that expected from the knock-on process (Hayakawa 64*b*, Brunstein 65).*

The cross section for the knock-on process given in (2.2.21*b*) is proportional to Z^2, and summation over Z gives

$$\sum_z j_z \, \sigma_{ZA} \simeq 1.75 \, j_p \, \sigma_{pA} \, . \quad (6.7.77)$$

We need therefore consider only the proton-electron collision and multiply the final result by factor of 1.75.

The knock-on cross section (2.2.21*b*) is expressed, because of the maximum energy transfer of a proton of total energy E_p and velocity $c\beta_p$, as

$$\sigma(E_p, W)\, dW = 2\pi r_e^2 \frac{mc^2}{\beta_p^2 W^2} \left[1 - \frac{M}{2m} \frac{Mc^2 W}{E_p^2} + \frac{1}{2} \left(\frac{W}{E_p} \right)^2 \right] dW. \quad (6.7.78)$$

Here the last term in the square brackets is negligible compared with the other terms. The range of momenta of protons giving an electron of kinetic energy W is

$$M\sqrt{W/2m} \leq p < \infty. \quad (6.7.79)$$

The cross section $\sigma(E_p, W)$ corresponds to $\sigma(E_p) f(E_p, W)$ in (6.7.66).

* A recent experiment has given an intensity consistent with the knock-on process.

For relativistic protons we have

$$p_k(W) = 1.75\, n_e \int_{Mc^2\sqrt{W/2mc^2}}^{\infty} j_p(E_p)\, \sigma(E_p, W)\, dE_p$$

$$= 1.75\, n_e \pi r_e^2 j_0 \left(\frac{1}{\alpha} - \frac{1}{\alpha+2}\right) \frac{(2mc^2)^{1+\alpha/2}}{W^{2+\alpha/2}}, \qquad (6.7.80)$$

where the proton spectrum given in (6.4.6) is now expressed as

$$j_p(E_p) = j_0 \left(\frac{Mc^2}{E_p}\right)^{\alpha} \frac{1}{E_p}, \qquad j_0 \simeq 1.1\ \text{cm}^{-2}\ \text{sec}^{-1}\ \text{sr}^{-1} \qquad (\alpha = 1.5) \qquad (6.7.81)$$

in the relativistic region.

In the nonrelativistic region the proton spectrum (6.4.6) is expressed as

$$j_p(p)\, dp = j_0 \frac{dp}{Mc}, \qquad (6.7.82)$$

and the electron spectrum for $W < 2mc^2$ is given by

$$p_k(W) = 1.75\, n_e 2\pi r_e^2 j_0 \left[1 + \frac{W}{2mc^2} - \sqrt{\frac{W}{2mc^2}}\right] \frac{mc^2}{W^2}. \qquad (6.7.83)$$

The spectrum of knock-on electrons given in (6.7.80) is numerically represented as

$$p_k(W) = 1.9 \times 10^{-27} \left(\frac{n_e}{10^{-2}}\right) W^{-2.75} \quad \text{cm}^{-3}\ \text{sec}^{-1}\ \text{sr}^{-1}\ \text{MeV}^{-1}, \qquad (6.7.84)$$

where the electron kinetic energy W is measured in MeV. This is also plotted in Fig. 6.28.

Summary of the Production Spectra of Electrons. The production spectra of positons and negatons arising from the processes in (6.7.65) are shown for the hydrogen density of $10^{-2}\ \text{cm}^{-3}$ in Fig. 6.28. Their sum, also shown in Fig. 6.28, may be compared with the radio spectrum and the directly measured electron spectrum, as will be discussed below. The shape of the spectrum is approximated by a power law above 200 MeV, flat between 10 and 100 MeV, and again represented by a power law in the lower energy region.

A characteristic feature of secondary electrons is a large positive excess in the intermediate energy region. The *positive-negative ratio*

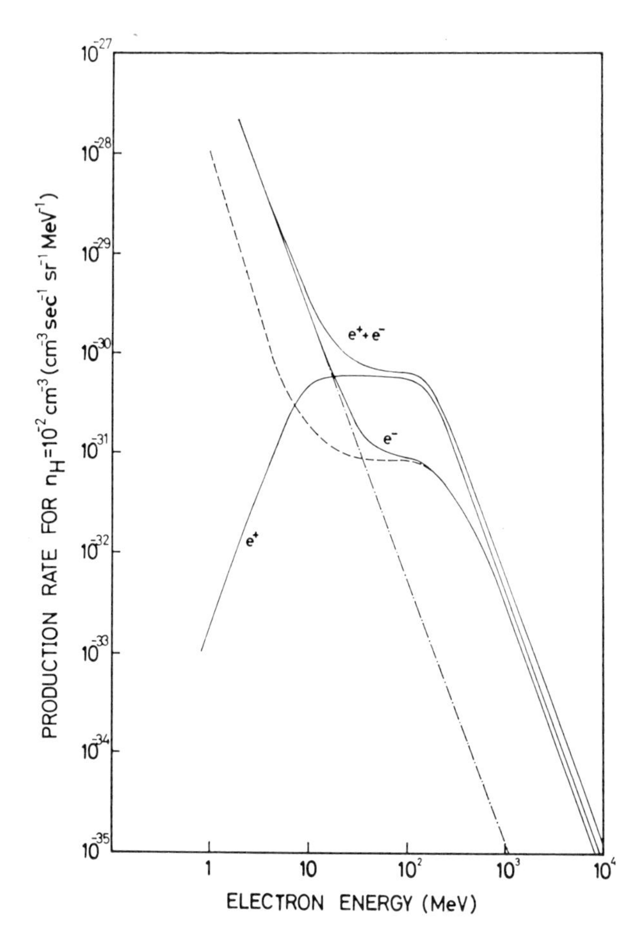

Fig. 6.28 Production spectra of electrons, summed over all nuclear processes and the knock-on process. Dashed line represents e^- from nuclear-collision processes; the dot-dash line, e^- from the knock-on process.

expected is shown in Fig. 6.29. If this were in agreement with the observed data, a substantial part of the electrons would be secondary electrons produced by meson decays. Otherwise, the acceleration of ambient electrons would give a considerable contribution. At low energies we see a negative excess of secondary electrons, and the distinction between primary and secondary electrons has to be made by comparing the calculated secondary electron spectrum with the observed spectrum.

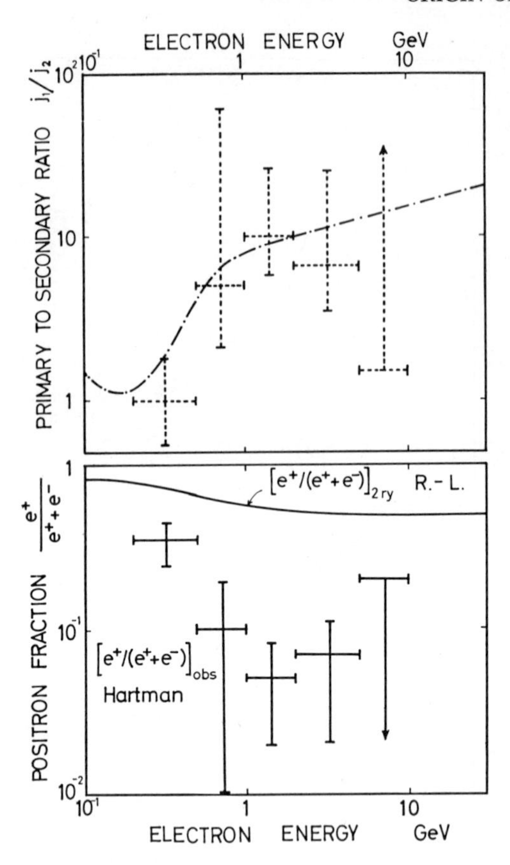

Fig. 6.29 The positive fraction $e^+/(e^+ + e^-)$ and the primary-secondary (j_1/j_2) ratios of electrons. ┼ —the positive fraction observed by R. C. Hartman, *Astrophys. J.* (1967); ┼ —the primary-secondary ratio give by (6.7.89); solid curve—the positive fraction of secondary electrons based on a calculation by Ramaty 66; dot-dash curve—the primary-secondary ratio obtained from the observed electron spectrum and the secondary spectrum assuming $X = 3$ g-cm^{-2}

6.7.6 *Energy Spectrum of Electrons*

Once the production spectrum is given, the energy spectrum of electrons is obtained from (6.7.49). If neither acceleration nor energy loss is appreciable, the energy spectrum is expressed as

$$j(E) = v\tau p(E) = v\tau n_H m_H \left(\frac{p}{n_H m_H}\right). \tag{6.7.85}$$

Since the factor p/n_H is independent of n_H, the electron flux is proportional to the path length traversed

$$X = v\tau n_H m_H, \tag{6.7.86}$$

which may be equal to 2.5 g-cm^{-2} according to (6.3.7). By substituting (6.7.76) into (6.7.85), we obtain the intensity of secondary electrons at 1 GeV:

$$j_2 \simeq 7.5 \times 10^{-4}\ \text{cm}^{-2}\ \text{sec}^{-1}\ \text{sr}^{-1}\ \text{GeV}^{-1}. \tag{6.7.87}$$

This is consistent with the electron intensity expected from Galactic radio emission only if the magnetic-field strength is greater than 10^{-5} gauss, as we can see from the comparison between (6.7.58) and (6.7.87).

The above consideration suggests that not all the electrons responsible for radio emission are secondary. The relative contributions of primary and secondary electrons may be obtained by observing the charge ratio of electrons. The primary electrons are exclusively negative, whereas the secondary electrons have a finite positive-negative ratio. Hence the positive-negative ratio to be observed is

$$\frac{j_+}{j_-} = \frac{j_{2+}}{(j_1 + j_{2-})} = \frac{j_{2+}/j_{2-}}{[1 + (j_{2+}/j_{2-})](j_1/j_2) + 1}. \tag{6.7.88}$$

The primary-secondary ratio, j_1/j_2 is expressed as

$$\frac{j_1}{i_2} = \frac{(j_{2+}/j_{2-}) - (j_+/j_-)}{(j_+/j_-)[1 + (j_{2+}/j_{2-})]}. \tag{6.7.89}$$

The positive-negative ratio for secondary electrons can be obtained from that at production by correcting for energy modulation as

$$\left[\frac{j_{2+}}{j_{2-}}\right]_E = \left[\frac{p_+}{p_-}\right]_{E+(G-L)\tau}. \tag{6.7.90}$$

However, the energy-modulation effect is of little importance if the positive-negative ratio depends only weakly on energy. The values of $j_{2+}/(j_{2+} + j_{2-})$ are compared with the observed positive fractions (Hartman 65). A significant difference between them demonstrates the existence of the primary electrons. The primary-secondary ratio obtained from (6.7.89) is also plotted in Fig. 6.29. The ratio increases with energy. It should also be noted that contamination by atmospheric and albedo electrons modifies the positive-negative ratio; an incomplete subtraction of these electrons results in too large a positive-negative ratio. Hence the observed result gives only an upper limit on the flux of secondary electrons.

The positive-negative ratio is also shown in the *east-west asymmetry* of electron intensity. If electrons were exclusively of secondary origin, the asymmetry would be inappreciable, whereas a large east-west ratio would be expected if primary electrons were predominant (Okuda 63). However, this method is subject to difficulties resulting from the complexity of the geomagnetic effect as well as from uncertainty about the contribution of albedo electrons.

The dominance of primary electrons at energies higher than several GeV is seen also from the energy spectrum. For secondary electrons the differential energy spectrum is represented by (6.7.76), provided that modulation effects are inappreciable. This is steeper than the observed energy spectrum; the differential intensities obtained by various authors are collected in Fig. 6.30; and the spectrum between 3 and 30 GeV is

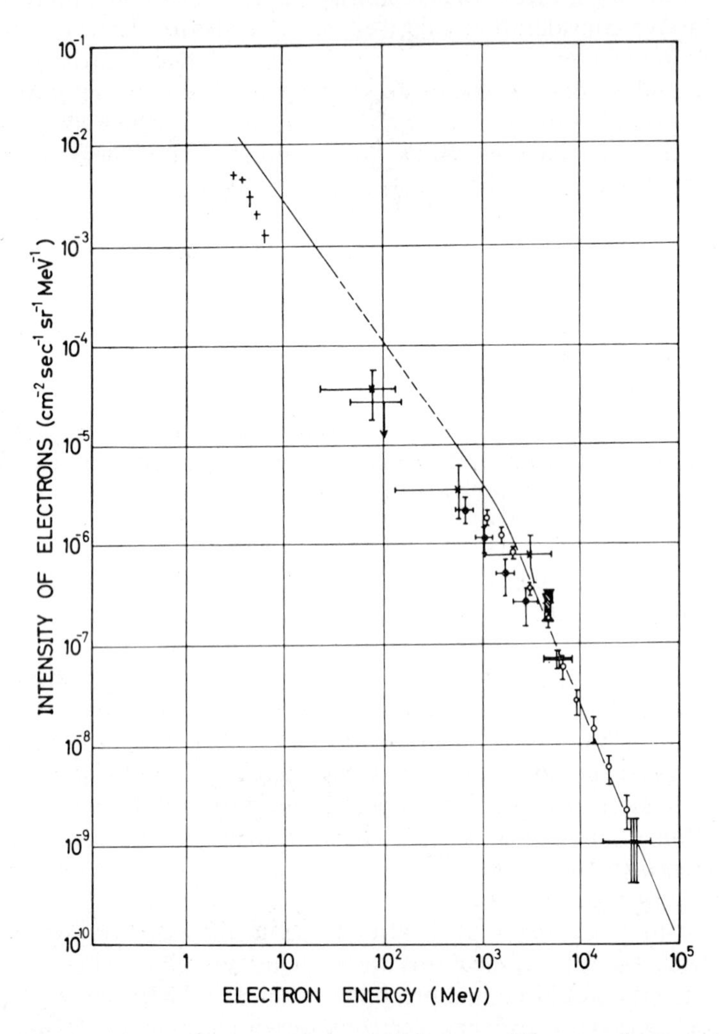

Fig. 6.30 The observed energy spectrum of electrons. The curve drawn represents the electron spectrum corrected for the solar modulation. (Legend is given in table on facing page.)

Symbol	Authors and Reference	Date of Observation	Altitude	Detector
	T. L. Cline, G. H. Ludwig, and F. B. McDonald, *Phys. Rev. Letters*, **13**, 786 (1964)	November 27 to 30, 1963	12,5000 km (IMP—1 satellite)	Scintillator telescope
	P. S. Freier and C. J. Waddington, *J. Geophys. Res.*, **70**, 5753 (1965)	July 28, 1963	2.1 g-cm^{-2} (balloon)	Emulsion, no secondary correction
	J. W. Schmoker and J. A. Earl, *Phys. Rev.*, **138**, *B*300 (1965)	August 1962 to October 1963	5 g-cm^{-2} (balloon)	Magnetic cloud chamber [upper limit]
	J. L'Heureux and P. Meyer, *Phys. Rev. Letters*, **15**, 93 (1965)	July 22, 29, 1964	4.1 g-cm^{-2} (balloon)	Scintillator, gas Čerenkov counter
	J. A. M. Bleeker et al., *Can. J. Phys.* **46**, S 522 (1968)	September 1965 June, August 1966	8 g-cm^{-2} (balloon)	Lead-aluminum, Čerenkov detector (shower)
	B. Agrinier et al., *Phys. Rev. Letters*, **13**, 377 (1964)	November 5, 1963	4.9 to 4.3 g-cm^{-2} 36 to 37 km (balloon)	Spark chamber (shower) [integral]
	B. Agrinier et al., *Proc. Int. Conf. on Cosmic Rays in London*, **1**, 331 (1965)	May 1965	2 to 3 mb (balloon)	Lead plates, spark chamber (shower) [integral]
	V. I. Rubtsov, *Proc. Int. Conf. on Cosmic Rays in London*, **1**, 324 (1965)	October 1964	7.5 g-cm^{-2} (balloon)	Lead plates, scintillators [integral at 2 energies]
	R. R. Daniel and S. Z. Stephens, *Phys. Rev. Letters*, **15**, 769 (1965)	April 6, 1963	10.2 g-cm^{-2} (balloon)	Emulsion [integral *E-W* effect]

represented by a power law as

$$j_e(E) \simeq 5 \times 10^{-3} E^{-2.4} \text{ cm}^{-2} \text{ sec}^{-1} \text{ sr}^{-1} \text{ GeV}^{-1}. \qquad (6.7.91)$$

Since no bump is appreciable except for an apparent bend of the spectrum at a few GeV, we may, as a first approximation, neglect the acceleration effect and take only the energy-loss effect into account. The energy spectrum of secondary electrons thus derived is compared in Fig. 6.31. This is definitely lower than the observed spectrum below 10 MeV and above 1 GeV; in the latter energy range the spectrum of secondary electrons is steeper than the observed one, regardless of whether the electrons are trapped in the disk or fill up the halo.

It should, however, be noted that the solar-modulation effect is considerable below several GeV. Unfortunately the absolute value of the modulation factor is not obtainable, and the deduced unmodulated spectrum depends considerably on the choice of the modulation factor (Hayakawa 65). If the modulation factors given in (6.4.4) and (6.4.5′) are used with $K = 0.8$ for the solar minimum, as derived from the modulation of protons and α-particles (Nagashima 66), the correction for solar modulation can be made as shown in Fig. 6.30. This favors the interpretation that the knee in the spectrum at about 2 GeV is due to solar modulation.* If, on the other hand, the knee at 2 to 3 GeV is due to the energy loss at a rate of CE^2 as represented by (6.7.49), we should have

$$C\tau \simeq 5 \times 10^{-10} \text{ eV}^{-1}. \qquad (6.7.92)$$

This is obtained for a magnetic-field strength of 5 μgauss and $\tau = 3 \times 10^{15}$ sec. This choice of the field strength brings the electron intensity derived from the radio intensity in (6.7.58) to agree with the observed one. However, the radio spectrum with a spectral index of about 0.7 should correspond to the electron spectrum in the electron energy range between 0.4 and 5.4 GeV, as indicated by (6.7.59). In this energy range the electron spectrum may be less steep, and the radio spectrum would be correspondingly flatter.

An alternative interpretation is to assume that the knee takes place above 10 GeV and that the unmodulated spectrum below 2 GeV has the same slope as that at higher energies; for example, we have

$$C\tau < 10^{-10} \text{ eV}^{-1}. \qquad (6.7.93)$$

* It has, however, been found that the 11-year variation in electron intensity is inappreciable. This may lead us to interpret the knee as due to the energy-loss effect in interstellar space The observed electron spectrum seems to be directly related to the radio spectrum. (See Y. Tanaka, *Can. J. Phys.* **46**, S 536 (1968).)

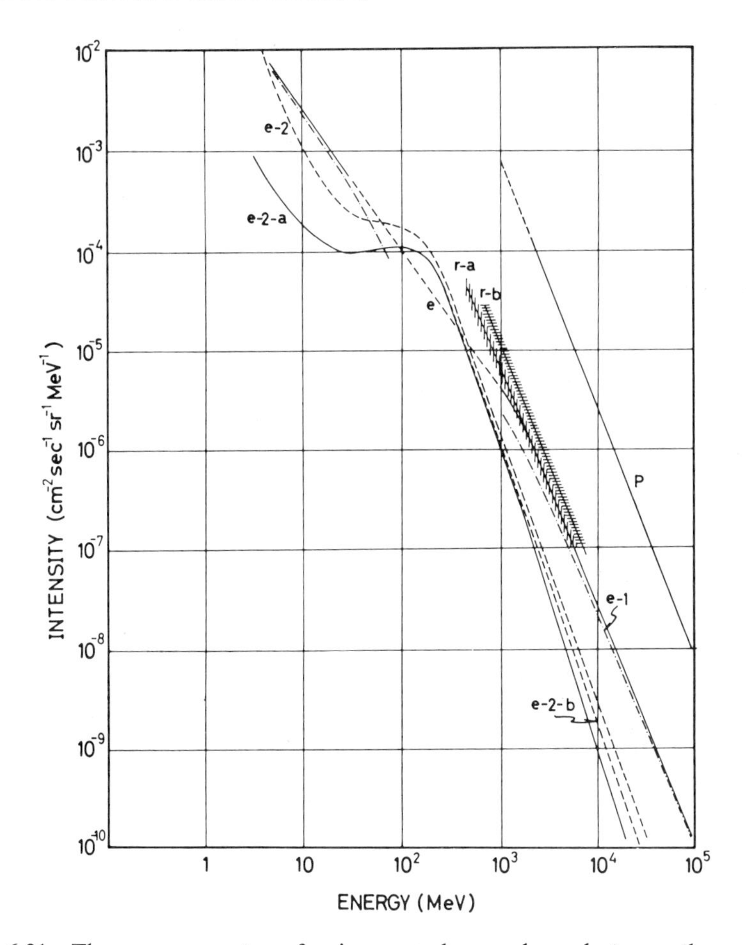

Fig. 6.31 The energy spectra of primary and secondary electrons, the proton spectrum being compared. *e*: the electron spectrum represented by the curve in Fig. 6.30; *e*-2: the secondary electron spectrum without energy-loss correction, taken from Fig. 6.28; *e*-2-*a*: the secondary electron spectrum corrected for energy loss with $C\tau = 5 \times 10^{-10}$ eV^{-1}; *e*-2-*b*: the secondary electron spectrum corrected for the energy loss with $C\tau = 1 \times 10^{-10}$ eV^{-1}; *e*-1: the primary electron spectrum, (*e*) — (*e*-2-*a*). *p*: the energy spectrum of protons; *r*-*a*: the electron spectrum derived from the radio spectrum with $H = 5$ μgauss; *r*-*b*: the electron spectrum derived from the radio spectrum with $H = 3$ μgauss.

This hardly holds for the halo component of electrons, since $C \simeq 2 \times 10^{-25}$ eV^{-1} sec^{-1}, and $\tau \geq 3 \times 10^{15}$ sec, but supports the possibility that electrons are trapped preferentially in the disk. If this is the case, the small amplitude of anisotropy as discussed in Section 6.5 would barely result for a scattering mean free path as small as, or smaller than, 1 light year.

At low energies the situation is more obscure because experimental information is poorer. Below 10 MeV the loss of energy by ionization is so predominant that the energy spectrum should become less steep. The change of slope may take place between 10 and 100 MeV; if this be observed, we shall obtain information on propagation properties. However, there may be another modulation mechanism; that is, the energy gain resulting from a heliocentric electric field (Abraham 66). A potential difference of 1.5 MV would displace the spectrum of knock-on electrons so as to fit the observed spectrum.

A further remark may be relevant concerning low-energy electrons. Since their range is very short they may not represent features of general electrons but may come mainly from nearby sources, as may be the case for low-energy heavy nuclei. Then the spectrum in the low-energy region is not necessarily connected with that in the high-energy region.

In summary, the energy spectrum of electrons may be decomposed into primary and secondary components, as shown in Fig. 6.31. The parallelism between the spectra of primary electrons and protons is demonstrated in the same figure.

6.7.7 Radio Sources in the Galaxy

A considerable part of cosmic electrons are of primary origin. This is supported by the fact that there are many radio sources in our Galaxy and that their nonthermal emission can be interpreted in terms of synchrotron radiation (Hayakawa 58*a*).

The sun emits various types of radio waves, designated types I to V (Wild 62). Among them the type IV radio burst is undoubtedly attributable to synchrotron radiation (Boischot 57), associated with the acceleration of electrons. The majority of these electrons may have energies as low as several MeV, but electrons with energies of about 100 MeV have occasionally been observed at the top of the earth's atmosphere (Meyer 62). As in the case of the nuclear component of cosmic rays, however, electron emission from ordinary stars such as the sun can account for only a minor part of the Galactic electrons.

There are many discrete sources of strong radio emission, most of which are identified with *supernova remnants*. In fact, three such sources can be identified with supernovae recorded in history. They belong to type I, whereas about 10 others whose angular sizes have been measured belong to type II. Their characteristic features are shown in Table 6.26.

Table 6.26 Typical Features of Supernovae

	Type I	Type II
Population	II	I
Mass of presupernova	$\sim M\odot$	$\sim 30\ M\odot$
Hydrogen abundance	Small	Large
Energy release	$\sim 10^{49}$ ergs	10^{51} to 10^{52} ergs
Luminosity at maximum	$-17^m \sim -18^m$	$-16^m \sim -17^m$
Expansion velocity	~ 1000 km-sec^{-1}	~ 5000 km-sec^{-1}
Light curve	Exponential decay	Irregular
Frequency	$1/300 \sim 1/100$ years	$1/100 \sim 1/50$ years

A few tenths of the mass is ejected with a high velocity, so that an expanding shell is characteristic of a supernova remnant. The Cygnus loop consists entirely of such a shell and is regarded as one of the oldest supernova remnants. The age of a remnant can therefore be estimated from its size, assuming the expansion velocity given in Table 6.26. A substantial part of the energy released by a supernova explosion goes into the expansion, and a relatively small fraction of energy is shared with many other modes, such as turbulent motion and relativistic electrons.

The radio flux at frequency ν is given by

$$I_\nu = \frac{1}{4\pi R^2} \int Y(\nu)\, dl\, dS = \frac{1}{4\pi} \int Y(\nu)\, dl\, d\Omega \simeq \frac{1}{4\pi R^2}\, Y(\nu) V, \quad (6.7.94)$$

where R is the distance from the source of volume V, and the integrals over a distance along the line of sight dl and over a source area dS or an angular size $d\Omega$ are extended over the source region. By substituting the expression of $Y(\nu)$ in (6.7.14) into (6.7.94) we obtain

$$I_\nu = y(\alpha) \frac{e^2}{\hbar c} h\nu_c \left(\frac{\nu_0}{\nu}\right)^{\alpha/2} \frac{\alpha N_0}{R^2} = 1.35 \times 10^{-22} y(\alpha)$$

$$\left(\frac{6.26 \times 10^{18}}{\nu}\right)^{\alpha/2} \frac{H^{1+\alpha/2} \alpha N_0 E_0^{\alpha}}{R^2}, \quad \text{erg sec}^{-1}\, \text{cm}^{-2}\, \text{Hz}^{-1}, \quad (6.7.95)$$

where $N_0 = n_0 V$ is the total number of electrons with energies greater than E_0. The number of electrons responsible for radio emission in the frequency range between ν_1 and ν_2 is obtained, with the aid of (6.7.16), as

$$N_e = \int_{E_1}^{E_2} \alpha N_0 \left(\frac{E_0}{E}\right)^{\alpha} \frac{dE}{E} = \frac{I_\nu R^2}{\alpha y(\alpha)(e^2/\hbar c) h\nu_c} \left[\left(\frac{k_1 \nu}{\nu_1}\right)^{\alpha/2} - \left(\frac{k_2 \nu}{\nu_2}\right)^{\alpha/2}\right]$$

$$= \frac{7.4 \times 10^{21} R^2 I_\nu}{\alpha y(\alpha) H} \left(\frac{k_1 \nu}{\nu_1}\right)^{\alpha/2} \left[1 - \left(\frac{k_2\, \nu_1}{k_1\, \nu_2}\right)^{\alpha/2}\right]. \quad (6.7.96)$$

The total energy of electrons contained in this energy range is

$$W_e = \int_{E_1}^{E_2} \alpha N_0 \left(\frac{E_0}{E}\right)^{\alpha} dE = \frac{I_\nu R^2 mc^2}{y(\alpha)(e^2/\hbar c)h\nu_c} \left(\frac{2\nu}{3\nu_c}\right)^{1/2} F(\nu, \alpha)$$

$$= \frac{2.96 \times 10^{12}}{y(\alpha)} \frac{I_\nu R^2 \nu^{1/2}}{H^{3/2}} F(\nu, \alpha) \quad \text{erg}, \tag{6.7.97}$$

where

$$F(\nu, \alpha) = \frac{1}{(\alpha - 1)} \left(\frac{k_1 \nu}{\nu_1}\right)^{\alpha - 1/2} \left[1 - \left(\frac{k_2 \nu_1}{k_1 \nu_2}\right)^{\alpha - 1/2}\right] \qquad (\alpha > 1),$$

$$F(\nu, \alpha) = \tfrac{1}{2} \ln \left(\frac{k_1 \nu_2}{k_2 \nu_2}\right), \qquad (\alpha = 1), \tag{6.7.98}$$

$$F(\nu, \alpha) = \frac{1}{(1 - \alpha)} \left(\frac{k_2 \nu}{\nu_2}\right)^{(\alpha - 1)/2} \left[1 - \left(\frac{k_2 \nu_1}{k_1 \nu_2}\right)^{(1 - \alpha)/2}\right] \qquad (\alpha < 1).$$

The value of W_e is enormous, as large as 10^{47} ergs for most of the radio sources. The total energy of cosmic rays, W_{CR}, may be still greater, since the intensity of electrons observed is smaller by two orders of magnitude than that of protons. Thus we assume

$$W_{\mathrm{CR}} = \kappa W_e \qquad (\kappa = 10^2). \tag{6.7.99}$$

Since cosmic-ray particles are magnetically trapped, their total energy may be equal to the magnetic energy

$$W_H = \left(\frac{1}{8\pi}\right) H^2 V = \kappa W_e = \frac{\kappa K I_\nu R^2}{H^{3/2}}. \tag{6.7.100}$$

This gives the magnetic-field strength

$$H = \left[\frac{48\kappa K I_\nu}{R\phi^3}\right]^{2/7} \qquad (\phi \equiv L/R), \tag{6.7.101}$$

where ϕ is the angular diameter corresponding to the linear dimension of the source, L. Since this depends on $\kappa^{2/7}$, the field strength is rather insensitive to the ambiguous value of κ. This gives the total cosmic-ray energy

$$W_{\mathrm{CR}} = W_H = \kappa W_e = 0.19[\kappa K I_\nu R^2]^{4/7} (R\phi)^{9/7}. \tag{6.7.102}$$

The values of H and W_{CR} thus derived for $\kappa = 100$ are given for representative sources in Table 6.27 (Ginzburg 64).

Table 6.27 Supernova Remnants

Radio Source	Type	Angular Size (min)	Distance[a] R (kpc)	Age (years)	Spectral Index $\alpha/2$	Radio Flux at 10^8 Hz (10^{-23} Wm^{-2} Hz^{-1})	Radio Power in 10^7 to 10^{10} Hz (10^{33} erg-sec^{-1})	Field Strength H (gauss)	Cosmic-Ray Energy W_{CR} (erg)
Cas A	II	4	3.4	250	0.77 ± 0.08	19	250	1.3×10^{-3}	6.6×10^{49}
Kepler's supernova	I	2	1.0	360	0.54 ± 0.10	0.08	0.14	6.9×10^{-4}	5.7×10^{46}
Tycho's supernova	I	5.4	0.36	400	0.53 ± 0.08	0.24	0.055	5.4×10^{-4}	3.2×10^{46}
Tau A	I	5	1.1	910	0.25 ± 0.08	1.7	10	1.0×10^{-3}	2.7×10^{48}
Cygnus Loop	II	170	0.77	$\sim 10^4$	0.10 ± 0.15	0.42	0.58	2.8×10^{-5}	2.6×10^{49}

[a] Distances given here are those adopted by Ginzburg 64. Different values are given by other workers. Hence the numbers in the last three columns should be read with reservation.

In Table 6.27 we can see that a supernova of type II contains more energy than one of type I. We can also notice age dependences of the spectral index and the field strength. If the expansion is adiabatic, we expect $E \propto L^{-1}$, as will be discussed in Section 6.9. The freezing of a magnetic field in a conductive plasma leads to $H \propto L^{-2}$, according to the conservation of magnetic flux. Thus we have

$$I_\nu \propto E_0^{\alpha} H^{1+\alpha/2} \propto L^{-2(\alpha+1)}. \tag{6.7.103}$$

Hence the rate of decrease of the radio flux is

$$\frac{1}{I_\nu}\frac{dI_\nu}{dt} = -2(\alpha+1)\frac{1}{L}\frac{dL}{dt} = -2(\alpha+1)\frac{2v}{L}. \tag{6.7.104}$$

This gives for Cas A a rate of decrease of 1.2 percent per year, since $v \simeq 7000$ km-sec^{-1} and $L \simeq 1.2 \times 10^{19}$ cm, in rough agreement with the observed value of 1.8 percent per year (Högbom 61). However, this rate of decrease is too large to account for the radio power of the Cygnus loop. This suggests that the adiabatic expansion does not hold, at least in later stages. The regeneration of magnetic fields and the acceleration associated therewith have to be taken into account.

The synchrotron radiation in supernova remnants may also be responsible for optical emission. This was in fact verified for the Crab Nebula by the polarization of light in the continuum (Oort 56), and the optical emission is accounted for in terms of synchrotron radiation and emission lines such as H_α. The spectrum in and near the optical region is steeper than in the radio region. If the radio spectrum is extended with the same slope toward a high frequency, it is connected to the infrared spectrum at about 10^{14} Hz, as seen in Fig. 6.32. Hence the power electron spectrum of $dE/E^{1.5}$ holds up to 10^{12} eV if $H = 10^{-3}$ gauss. The radiation power contained and the total energy of electrons in all energy regions up to optical frequencies are as large as 10^{37} erg-sec^{-1} and 10^{47} ergs, respectively.

An electron of such high energy has a half-life as short as about one hundred years. This implies that electrons of such high energies have to be replenished every hundred years or so, and therefore either that high-energy electrons are produced at a large rate or that electrons are continuously accelerated in a supernova remnant.

The electrons responsible for radio emission have a much longer lifetime, probably longer than the active lifetime of the remnant. The relations among the electron energy, the magnetic-field strength, the frequency of radiation, and the lifetime of an electron are illustrated in Fig. 6.33 (Hayakawa 64*a*). In this figure the spectral regions of radio and optical emission for a number of radio objects are indicated.

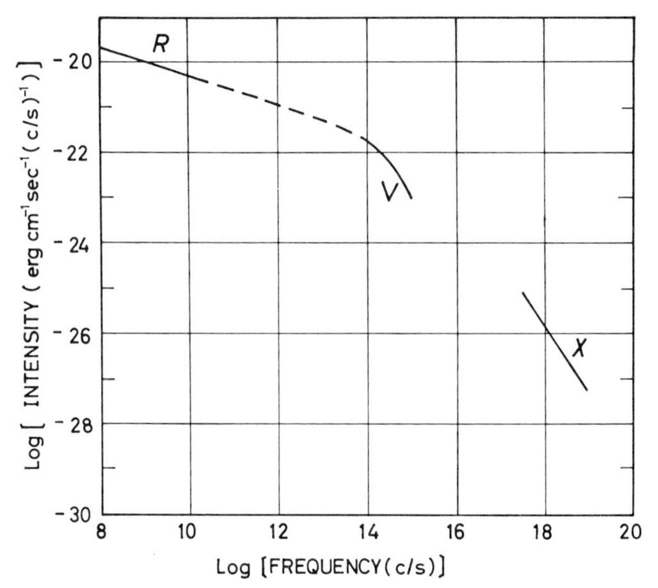

Fig. 6.32 The spectrum of electromagnetic radiation from the Crab Nebula. *R*—radio, *V*—infrared and visible, *X*—X-rays.

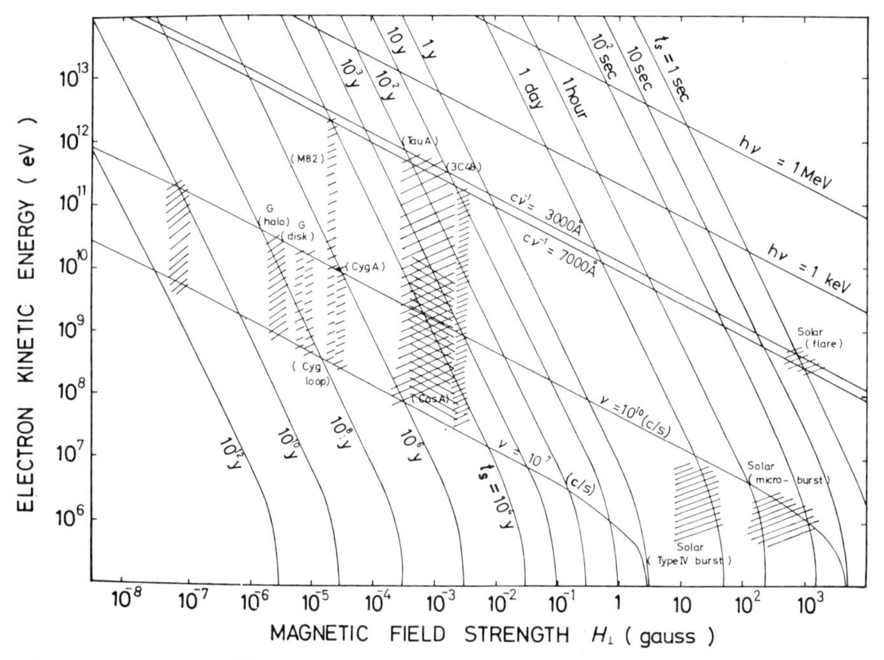

Fig. 6.33 The half-life of an electron against synchrotron radiation. The curves represent the energy bands and the half-lives (t_s). The spectral regions in which synchrotron emissions are observable are hatched for various sources.

The large rate of energy generation needed to maintain the optical emission brings about a difficulty in accounting for the total energy output if the proton-electron ratio κ is assumed to be 100. The total energy of cosmic rays in the Crab Nebula would thus be as large as 10^{49} ergs, nearly equal to the total energy release of a type I supernova, and no replenishment of energy would be permitted. This indicates that the value of κ depends on the spectral index. If the proton intensity is obtained from the electron intensity by shifting the energy scale by a constant factor, the value of $\kappa \simeq 100$ for general Galactic protons and electrons would correspond to $\kappa \simeq 2$ in the Crab Nebula. If this is the case, the total energy of cosmic rays is also on the order of 10^{47} ergs, so that the rate of energy given to relativistic particles cannot exhaust the total energy available over 10^3 years but cannot last more than 10^4 years. This is consistent with the fact that no supernova remnants of type I emitting radio waves have ages greater than 10^3 years. Supernova remnants of type II have greater radio ages, because they have a greater amount of energy available, but their ages do not seem to exceed 10^5 years.

In either case the radio age is smaller than the half-life of radio-emitting electrons, as seen in Fig. 6.33. Hence these electrons will be eventually ejected into outer space. The number of relativistic electrons with energies greater than 10^8 eV in the Crab Nebula is estimated from (6.7.96) to be

$$N_e(\text{Tau A}) \simeq 2 \times 10^{49}. \tag{6.7.105}$$

Because the replenishment of electrons and the difference between type I and II supernovae, the average number of electrons per supernova remnant may be on the order of

$$N_e(\text{SN}) \simeq 10^{50}. \tag{6.7.106}$$

The density of electrons in the Galaxy supplied by supernovae is thus estimated to be

$$n_e(G, \text{SN}) = N_e f_{\text{SN}} \frac{\tau_G}{V_G} \simeq 10^{-12}\ \text{cm}^{-3}, \tag{6.7.107}$$

if we assume the frequency of supernova explosions as $f_{\text{SN}} \simeq 1$ per 50 years, $\tau_G \simeq 10^8$ years, and $V_G \simeq 3 \times 10^{68}$ cm^3. This is just as large as the density of electrons observed and responsible for radio emission. This is consistent with the argument in the preceding subsection that the intensity of primary electrons is greater than that of secondary ones, and that the main source of the former is supernova remnants. It is also worthwhile remarking that the radio spectral indices of supernova remnants lie mostly around 0.7, and consequently the integral power exponents of their electron spectra are about 1.4, also consistent with the spectrum of general Galactic electrons.

For this spectrum we may assume the proton-electron ratio to be about 10^2, and therefore the rate of cosmic-ray generation is estimated as

$$q_G(\text{SN}) = N_p f_{\text{SN}} = \kappa N_e f_{\text{SN}} \simeq 10^{43}\ \text{sec}^{-1}, \tag{6.7.108}$$

in good agreement with the value of q_G in (6.2.11) required for maintaining the intensity of Galactic cosmic rays.

6.7.8 *Radio Galaxies*

There are some radio sources located in our Galaxy, but most of them are identified with distant galaxies. Among them nearly 160 radio sources are optically identified, and about one-third of them may be called *radio galaxies*, Their radio power is large, $\gtrsim 10^{41}$ erg-sec^{-1}, but the distinction between radio and normal galaxies is arbitrary, since the radio-luminosity distribution is continuous.

Our Galaxy is normal and a rather weak radio source. The radio emission is not uniform but consists of the core, the disk and, if any, the halo. The Andromeda Nebula, a galaxy NGC 157 or M 31, is similar to our Galaxy in many respects. Few galaxies have been found to have halos, but some representative galaxies, described in Section 6.1.4, have complicated radio and optical features. Features of these sources are summarized in Table 6.28 (K. Aizu 64*c*).

Table 6.28 lists three distinct classes of galaxies. They may be called normal, weak radio, and strong radio galaxies, which have radio luminosities of about 10^{38}, 10^{41}, and 10^{44} erg-sec^{-1}, respectively. Actually, however, the distribution of luminosity is continuous, as shown in Fig. 6.34. The radio-luminosity function, the density of galaxies with radio luminosities greater than L, shown in Fig. 6.33, can be represented by a simple power law as

$$n(\geq L) \simeq 10^{-1} \frac{10^{38}}{L} (\text{Mpc})^{-3} \tag{6.7.109}$$

for $L > 10^{38}$ erg-sec^{-1}. Hence the distinction between the classes is not clear; they merely represent median features of three groups.

The grouping according to luminosities does not always make sense if we consider a peculiar galaxy of M 82. Although its radio luminosity is small, it has several peculiar features. Its expanding nebulosity has a velocity as high as 10^8 cm-sec^{-1} and may be ascribed to an explosion

Table 6.28 Properties of Galaxies

Galaxy	Type	Optical Size (Kpc)	Distance (Mpc)	Photographic Magnitude (M_{pg})	Radio Flux at 10^8 Hz W-m^2 Hz^{-1}	Spectral Index	Radio power (10^7 to 10^{10}Hz) L (erg-sec^{-1})	Emissivity L/V (10^{-28} erg sec^{-1}cm^{-3})	Magnetic Field (μgauss)	Total Energy (erg)
The Galaxy	Sb	25	—	−20.1	—	0.7	4.4×10^{38}	0.037	6	3×10^{56}
Andromeda (M 31)	Sb	28×7.3	0.60	−19.6	2.7×10^{-24}	0.7	3.3×10^{38}	0.038	3	4×10^{56}
NGC 3034 (M 82)	Irr	6.1×0.9	2.1	−17.4	1.5×10^{-25}	0.3	3.5×10^{38}	2.3×10^3	130	4×10^{54}
NGC 4486 (M 87) (Vir A):										
Extensive	EO	25×20	11	−20.5	6×10^{-24}	1.0	1.2×10^{41}	9.1	27	9×10^{57}
Jet		0.07×0.8			9×10^{-24}	0.4	1.3×10^{41}	1.2×10^4	200	4×10^{56}
3C 405 (Cyg A)		48×32	170	−23.7	1.3×10^{-22}	0.7 1.1	3.4×10^{44}	2.3×10^4	250	8×10^{59}
3C 47 (QSS):										
Extensive			1390	−23	4×10^{-25}	0.89	6.2×10^{43}	0.21	9	2.1×10^{61}
Core							2.1×10^{43}	2.0×10^3	120	1.5×10^{59}

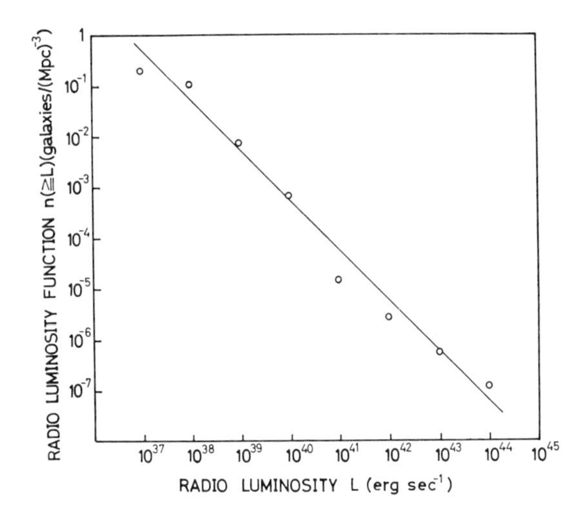

Fig. 6.34 Radio-luminosity function of galaxies. The solid line represents a power law approximation given in (6.7.109).

having taken place in its central region about 10^6 years ago. In fact, the radio lifetime defined by W_e/L is 3×10^6 years, smaller by two orders of magnitude than that of a normal galaxy. This implies that M 82 is more active than normal galaxies, as indicated by its explosive feature, and that a quantity properly representing the activity is not the total luminosity but may be the luminosity per volume or the emissivity L/V; the value of L/V of M 82 is greater by five orders of magnitude than that of a normal galaxy and is close to that of a strong radio galaxy. The importance of the introduction of emissivity may be evident in the comparison between the extensive source and the jet in the central region of M 87. Their luminosities are comparable, and their radio lifetimes are different by a factor of 20. The latter implies that the jet originated long after M 87 was formed. The difference between their emissivities is much greater, thus demonstrating that the emissivity is a good parameter in the sense that it represents a characteristic feature.

Thus it is suggested to parametrize galaxies by emissivity L/V, plotting galaxies on the L versus (L/V) diagram (K. Aizu 64*b*). This is an analogue of the H-R diagram of stars, in which stars are plotted on the L versus T_{eff} diagram. In place of the effective temperature, which represents surface emissivity, the volume emissivity L/V is introduced since a radio galaxy is transparent to radio waves. The L versus (L/V) diagram thus obtained for radio galaxies as well as for supernova remnants is shown in Fig. 6.35. It is interesting to note that strong radio galaxies form a trajectory running from the upper left edge toward lower right. Galaxies in the Virgo cluster form another trajectory that is nearly parallel but shifted downward. Remnants of type I and type II supernovae also form parallel trajectories in the lower part. Each of these trajectories is interpreted as representing an evolutionary

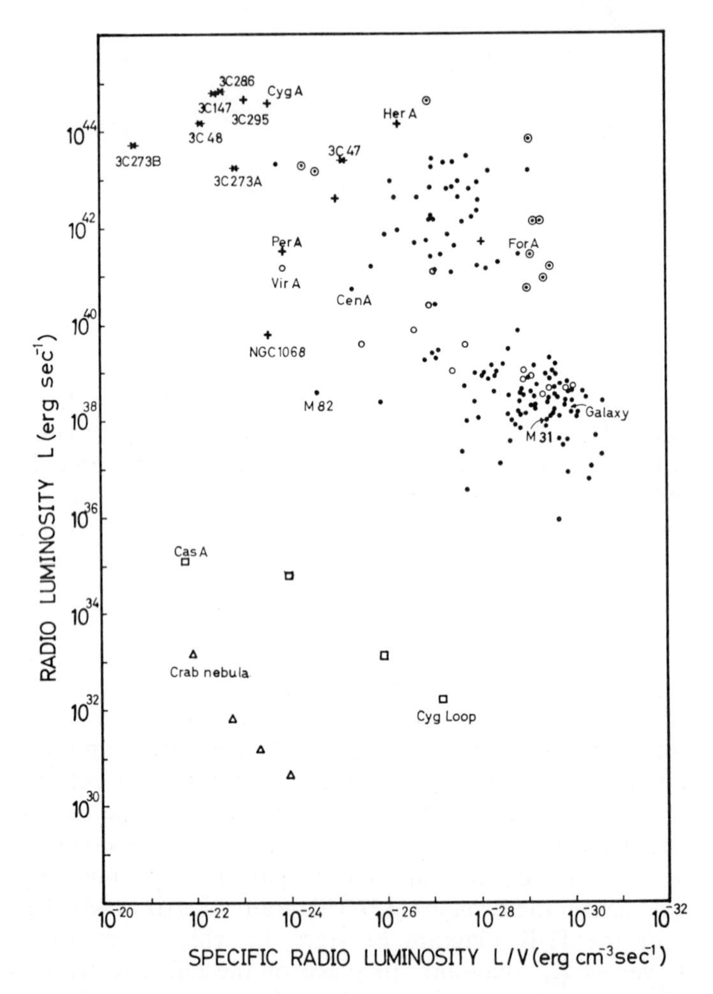

Fig. 6.35 The L versus (L/V) diagram for galaxies and supernova remnants (K. Aizu 64). ✳—QSS; +—strong emission lines; ○—in the Virgo cluster; ●—other galaxies; ⊙—halo; □—type II supernova; △—type I supernova.

sequence of galaxies, starting from upper left and evolving to lower right as the radio-emitting volume expands.

The above theory of evolution implies that the radio galaxy is a stage in the evolution of a galaxy and that QSS represents an early stage of a strong radio galaxy.

The explosion in M 82 and the jet in M 87 indicate that an explosive phenomenon leading to strong radio emission takes place at some stage of evolution and may recur a number of times. This interpretation is consistent with the radio-luminosity functions (6.7.109). The number of galaxies that have luminosities L in dL is proportional to their lifetimes of radio emission with luminosity L, under the assumption that the radio galaxy is a stage in galactic evolution. Since the lifetime is given by

$$\tau_r = \frac{W_e}{L}, \tag{6.7.110}$$

the number of galaxies with lifetimes τ_r in $d\tau_r$ is

$$dn \propto d\tau_r \propto \frac{dL}{L^2}, \tag{6.7.111}$$

provided that W_e is independent of L.

The recurrence of radio emission could also be the case in our Galaxy. Cosmic rays can be generated in association with explosions, and their frequency depends on the scale of explosion. Although such an explosion is suggested to have taken place within 10^8 years in our Galaxy (Burbidge 63), no conclusive evidence has yet been obtained from studies of cosmogenic nuclides, as discussed in Section 6.6.

The parallelism between the trajectories of radio galaxies and supernova remnants in the L versus (L/V) diagram suggests that the properties of these two radio sources are similar in many respects. The lifetime of optical emission is much shorter than the age of the radio galaxy concerned. Hence the supply of electrons has to continue as long as the optical emission lasts, and consequently protons are supplied as well. This provides a basis of the metagalactic origin of cosmic rays discussed in Section 6.4.5.

The same argument may hold for metagalactic electrons. The energy of electrons responsible for radio and optical emission lies between 10^8 and 10^{12} eV, and a power spectrum seems to hold at least in the energy range between 10^9 and 10^{11} eV. Thus we would expect the metagalactic electron intensity relative to the Galactic one to be as given in (6.4.17′). However, energy loss in intergalactic space considerably modifies the electron intensity.

6.7.9 *Electrons in Intergalactic Space*

Magnetic fields are inferred to be much weaker in intergalactic space than in galaxies, and the density of starlight photons may also be smaller. Hence the energy loss of electrons is due almost exclusively to collisions with microwave photons. Hence the cutoff of the electron spectrum is expected to take place for $\tau \simeq 10^{10}$ years at

$$E_c \simeq -\frac{dE}{dt}\frac{\tau}{(\alpha-1)} \simeq 3 \times 10^7 \text{ eV}. \qquad (6.7.112)$$

The energy spectrum of metagalactic electrons may be expressed, if acceleration is negligible, as

$$j_{\mathrm{Mge}}(E) = \left(\frac{j_{\mathrm{Mgp}}}{j_{\mathrm{Gp}}}\right) j_{\mathrm{Ge}}(E) \frac{E_c}{(E+E_c)}$$
$$\simeq 5 \times 10^{-8} E^{-3.4} \quad \mathrm{cm^{-2}\,sec^{-1}\,sr^{-1}\,GeV^{-1}}, \qquad (6.7.113)$$

where the metagalactic-Galactic ratio of proton intensities, $j_{\mathrm{Mgp}}/j_{\mathrm{Gp}} \simeq 3 \times 10^{-4}$, given in (6.4.17′), is taken into account.

These electrons could be responsible for radio emission if the intergalactic magnetic field were strong enough. The radio intensity expected is

$$I_\nu = \frac{1}{4\pi} Y_\nu R_U \simeq 2 \times 10^{-22}\left(\frac{1.6 \times 10^7}{\nu}\right)^{\alpha/2}$$
$$\times \left(\frac{H}{\mu\text{gauss}}\right)^{1+\alpha/2} \quad \mathrm{W\text{-}m^{-2}\,sr^{-1}\,Hz^{-1}} \qquad (6.7.114)$$

with $\alpha \simeq 2.4$. This is weaker than the general Galactic radio intensity, even if the magnetic-field strength is 10^{-6} gauss. Since the intergalactic magnetic field is inferred to be much weaker than 10^{-6} gauss, no appreciable contribution of the intergalactic radio emission is expected. If, conversely, the magnetic-field strength were given, the upper limit of the electron intensity would be obtained from (6.7.114).

The contribution of galaxies to the metagalactic radio intensity is much greater than that of intergalactic radio emission. This is estimated to be

$$I_\nu \text{ (all galaxies at 100 MHz)} \simeq 10^{-22} \mathrm{\ W\text{-}m^{-2}\,sr^{-1}}. \qquad (6.7.115)$$

This is slightly smaller than the general Galactic radio intensity given in (6.7.54). This means that information on intergalactic electrons is hardly obtainable from radio observations.

6.8 Gamma-Rays and X-Rays

Relativistic electrons in cosmic rays give us much information, as has been discussed in the previous sections, but the quality of the information increases in great deal as we know about γ-rays and X-rays. Their origins are shared with the origins of electrons in the following respects.

Nuclear collisions produce mesons as in (6.7.65*a*), and the counterpart of this process gives γ-rays through the process

$$p + p \rightarrow \pi^0 \rightarrow 2\gamma. \tag{6.8.1a}$$

Positons annihilate with negatons, giving γ-rays according to

$$e^+ + e^- \rightarrow 2\gamma, \tag{6.8.1b}$$

which mainly takes place at rest, so that the γ-rays have a line at 0.51 MeV. Electrons colliding with starlight and microwave photons produce X-rays by the inverse Compton effect:

$$e + \gamma \rightarrow e' + \gamma'. \tag{6.8.1c}$$

In addition to these processes, the bremsstrahlung of electrons, the inner bremsstrahlung associated with knock-on processes, the K X-ray excitation by electrons and the high-energy end of synchrotron radiation could be responsible for X-ray emission. Apart from these processes associated with electrons, the K X-ray excitation by nuclear particles and nuclear γ-rays may contribute to the emission of X-rays and γ-rays.

Here γ-rays and X-rays are not clearly distinguished, just roughly according to whether their energy is above or below 0.1 MeV.

6.8.1 Absorption and Scattering

One advantage of using X-rays and γ-rays for the investigation of celestial objects is that they propagate in straight lines, so that their sources can be identified with a detector of suitable angular resolution. This is, however, limited by the absorption of radiation in interstellar and intergalactic media. If the optical depth is greater than unity, one can no longer locate a distant source. Even if the optical depth is smaller than unity, scattering makes a source diffuse.

The absorption of photons with energies below 10 keV is due mainly to the photoelectric effect. Since the photoelectric cross section, as given in (2.4.3), is proportional to Z^5, helium gives the largest contribution to the photoelectric absorption in interstellar space, below the K-absorption edge of oxygen, and the contributions of oxygen and neon are comparable to that of helium above it. The absorption cross section of X-rays in interstellar gas is shown in Fig. 6.36, in which the composition of interstellar elements is assumed to be equal to the solar composition supplemented by the composition of high-temperature stars in Table 6.2 (Felten 66*b*).* If hydrogen were entirely molecular, the absorption coefficient would increase by a considerable factor, since the photoelectric cross section of a hydrogen molecule is estimated as 0.8 times that of a helium atom. The absorption coefficient in this extreme case is shown by a dashed line in Fig. 6.36.

The optical depth for X-rays is obtained if we know the effective absorption cross section $\sigma(E)$ per hydrogen atom, as

$$\tau(E) = \sigma(E) \int n_{\mathrm{H}}(r)\, dr, \tag{6.8.2}$$

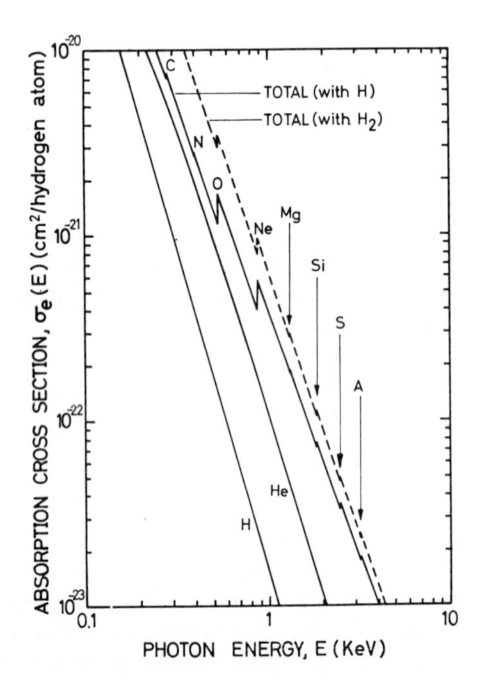

Fig. 6.36 Absorption cross section of X-rays in interstellar space. The cross sections due to hydrogen alone and helium alone are indicated by H and He, respectively. The solid line with the absorption edges due to C. N. O., etc., corresponds to the case where all hydrogen forms atoms, whereas the dashed line corresponds to the case where all hydrogen forms molecules.

* The helium abundance in interstellar matter may be smaller than that assumed in Table 6.2 and Fig. 6.36.

where the integral is taken along the line of sight. For $n_{\rm H} \simeq 1\ {\rm cm}^{-3}$ the absorption length is smaller than 1 kpc for X-rays of energies smaller than 1 keV. Hence such soft X-rays are absorbed considerably in their passage through interstellar space.

In intergalactic space the matter density is so low that the absorption effect might be regarded as negligible. For $n_{\rm H} \simeq 10^{-5}\ {\rm cm}^{-3}$ the absorption length would be about 10^{27} cm for X-rays of 2 keV—if hydrogen were not ionized. Actually, however, hydrogen and possibly helium are completely ionized, but oxygen is probably not ionized. Then the absorption length is as large as 3×10^{27} cm at 2 keV (appreciably smaller than the Hubble radius) if the abundance of oxygen and neon is as large as that in interstellar gas. It is interesting to see from the absorption effect of X-rays whether the intergalactic abundances are the same as the interstellar one.

As energy increases the photoelectric cross section decreases so rapidly that Thomson scattering plays the main role. The scattering mean free path is $1/n_e \sigma_{\rm Th} = 1.5 \times 10^{24}/n_e$ cm, greater than the sizes of most of the objects concerned.

As energy increases further beyond 1 MeV, the Compton cross section decreases and becomes lower than the bremsstrahlung cross section above 100 MeV. Since the cross sections for various processes in the energy range between 10 MeV and 100 GeV are quite small, one need not consider the absorption and scattering effects of γ-rays in this energy region.

At still higher energies collisions with starlight and microwave photons result in strong absorption through the creation of electron pairs with the cross section given in (2.6.10). If the photon spectrum is represented by the Planck distribution for temperature T, the absorption coefficient is obtained by a method analogous to that in Section 6.7.2 as (Nikishov 61)

$$\mu(E) = 2\left(\frac{m^2c^4}{E}\right)^2 \frac{n_0}{4(kT)^3} \int_0^\infty \frac{\psi(E\varepsilon/m^2c^4)}{\exp(\varepsilon/kT) - 1}\, d\varepsilon, \tag{6.8.3}$$

where

$$\psi(x) = \int_1^x s\sigma(s)\, ds,$$

$\sigma(s)$ being the pair-creation cross section as a function of $s = (E\varepsilon/m^2c^4)\sin^2(\theta/2)$. The absorption coefficient thus obtained is shown as a function of γ-ray energy in Figure 6.37 (Gould 66, Jelley 66). The photon-photon scattering also contributes to scattering even below the threshold energy of pair creation, but the cross section decreases rapidly as

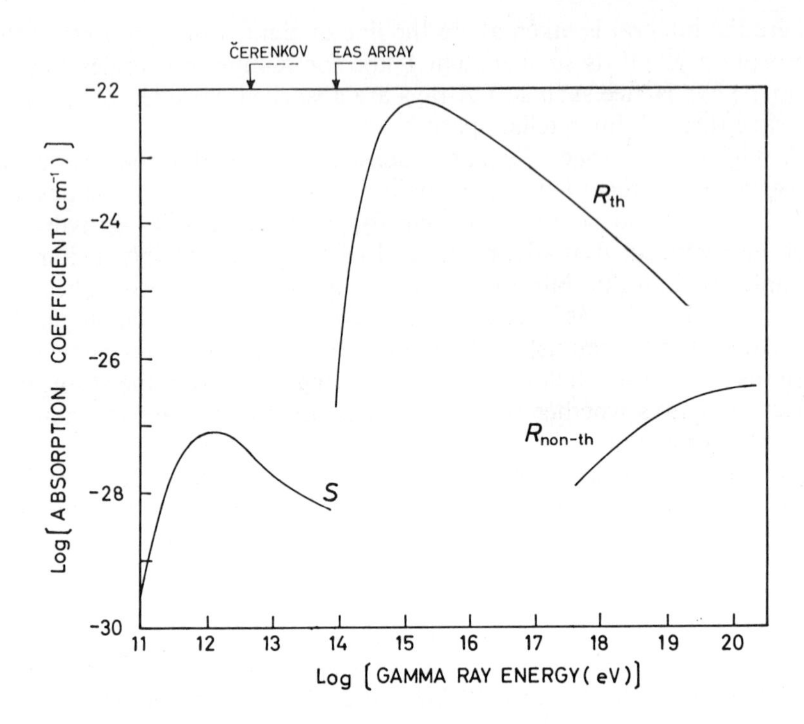

Fig. 6.37 The absorption coefficient of γ-rays in space due to pair creation by the photon-photon collision. S—collisions with starlight; R_{th}—collisions with thermal radio waves of 3°K; R_{nonth}—collisions with nonthermal radio waves. The energy regions in which the measurements of primary γ-rays have been attempted by the Čerenkov light of EAS (Chudakov 63) and an EAS array at Bolivia (Toyoda 65) are indicated on the top of the figure.

E^3, and its absolute value is smaller by a factor of $(1/137)^2$. Hence its contribution is of little practical significance in comparison with the contribution of pair creation.

In calculating the absorption coefficient due to collisions with starlight photons we have to know their density. There are divergent estimates of the photon density, lying between 10^{-1} and 10^{-3} eV-cm^{-3}. We adopt a median value W_{ph}(starlight) $= 10^{-2}$ eV-cm^{-3}.

With this choice of the starlight density we find an appreciable absorption effect for γ-rays with energies between 1 and 10 TeV. Above 100 TeV, the effect of microwave photons is so large that space is highly opaque for γ-rays of energies greater than 10^{15} eV. The absorption length is smaller than the radius of the Galactic halo at energies around 10^{15} to 10^{16} eV.

The absorption coefficient decreases as energy increases, because of a rapid decrease of the pair-creation cross section with increasing energy. The decrease ceases at about 10^{19} eV, since the contribution of non-thermal radio photons to pair creation becomes significant (Goldreich 63), and the creation of meson pairs by collisions with microwave photons can take place. Hence the region beyond 10^{26} cm cannot be seen by means of very-high-energy γ-rays unless the γ-rays are red-shifted to energies below 10^{12} eV.

6.8.2 Gamma-Rays from Neutral Pions Produced by Nuclear Collisions

The rate of neutral-pion production by the collision of cosmic-ray protons and α-particles with interstellar matter can be calculated in the same manner as in Section 6.7.5. The production spectrum of neutral pions is thus given by (6.7.70), in which $m\sigma$ is replaced by the one for neutral pions as shown in Fig. 6.26. The production spectrum of γ-rays is evaluated by

$$p_\gamma(E_\gamma) = \int_{E_\pi^-}^{\infty} 2p_{\pi^0}(E_{\pi^0}) \frac{dE_{\pi^0}}{p_{\pi^0}}, \qquad (6.8.4)$$

where $E_\pi^- = E_\gamma + m_\pi^2 c^4/4E_\gamma$. This is numerically nearly equal to (6.7.73) if we put $E_\gamma = 2E_e$. Therefore the production spectrum of γ-rays can be obtained from that of electrons given in Fig. 6.27 by shifting the energy scale by a factor of 2.

At high energies our information on pion production is not quantitative enough to carry out a calculation like the above. Hence we refer to the γ-ray spectrum observed in the atmosphere. As discussed in Section 4.9, γ-ray intensity is proportional to the thickness of air at small atmospheric depths, and the gradient of the intensity is proportional to the pion-production cross section. Hence the production rate per nucleon is expressed as

$$\frac{p_\gamma(E_\gamma)}{n_\mathrm{H}} = j_\gamma(E, x) \frac{m_\mathrm{H} A^{1/3}}{x}, \qquad (6.8.5)$$

where $j_\gamma(E, x)$ is the differential intensity at the atmospheric depth x and A is the average mass number of air nuclei.

The production spectrum of γ-rays obtained by (6.8.4) and (6.8.5) is shown in Fig. 6.38. This has a broad peak centered at about 70 MeV and decreases toward high energy more steeply than the proton spectrum.

It may be of some practical use to know the total yield of the γ-rays. This is obtained as

$$\frac{q_\gamma}{n_\mathrm{H}} \simeq 1 \times 10^{-26} \ \mathrm{sec}^{-1} \ \mathrm{sr}^{-1}. \qquad (6.8.6)$$

This is nearly as large as the total yield of positive pions. Positive pions decay into positons, most of which annihilate at rest with negatons, whereby two γ-rays of 0.51 MeV are emitted in each annihilation event. Hence the yield of such γ-rays is

$$\frac{q_\gamma(0.51\ \text{MeV})}{n_\text{H}} \simeq 2 \times 10^{-26}\ \text{sec}^{-1}\ \text{sr}^{-1}. \tag{6.8.7}$$

The intensity of γ-rays coming from a given direction can be obtained by integrating the production rate along a line of sight as

$$j_\gamma(E) = \int_0^\infty p_\gamma(E, r, \Omega)\, dr, \tag{6.8.8}$$

where $p_\gamma(r, \Omega)$ is the differential production rate at distance r in the direction Ω. If we are interested in the γ-rays produced in the Galaxy, the expression (6.8.8) is reduced, because of the essential uniformity of the cosmic-ray density in interstellar space, to

$$j_\gamma(E, \Omega) = \left(\frac{\overline{p_\gamma(E)}}{n_\text{H}}\right) \overline{n_\text{H}\, R(\Omega)}, \tag{6.8.9}$$

where $\overline{n_\text{H}\, R(\Omega)}$ is the number of interstellar protons in a unit column of length $R(\Omega)$. In practice one uses a detector of a finite angular resolution. With an effective solid angle $\Delta\Omega$ the intensity expected is

$$\begin{aligned} \int_{\Delta\Omega} j_\gamma(E, \Omega)\, d\Omega &= \left(\frac{\overline{p_\gamma(E)}}{n_\text{H}}\right) \int_{\Delta\Omega} d\Omega \int_0^{R(\Omega)} n_\text{H}(r, \Omega)\, dr \\ &= \left(\frac{\overline{p_\gamma(E)}}{n_\text{H}}\right) \langle \overline{n_\text{H}\, R(\Omega)} \rangle\, \Delta\Omega. \end{aligned} \tag{6.8.9'}$$

The values of $\langle \overline{n_\text{H}\, R(\Omega)} \rangle\, \Delta\Omega$ are given in Table 6.29 for three representative directions and different cone half-angles of detectors assuming that cosmic rays fill up the halo (Pollack 63). One can thus obtain the expected intensity of γ-rays by multiplying the value of $\langle \overline{n_\text{H}\, R(\Omega)} \rangle\, \Delta\Omega$ in Table 6.29 by the production rate in Fig. 6.38.

Table 6.29 Amount of Interstellar Matter Covered by a Counter Telescope ($\overline{n_\text{H}\, R(\Omega)} \times 10^{21}$ cm^{-2})

Cone Half-angle (degrees)	Toward the Center	Toward the Anticenter	Perpendicular to the Disk
5	0.6	0.3	0.02
10	1.5	0.8	0.07
15	2.4	1.3	0.16
20	3.5	1.9	0.29
30	6.0	3.2	0.65

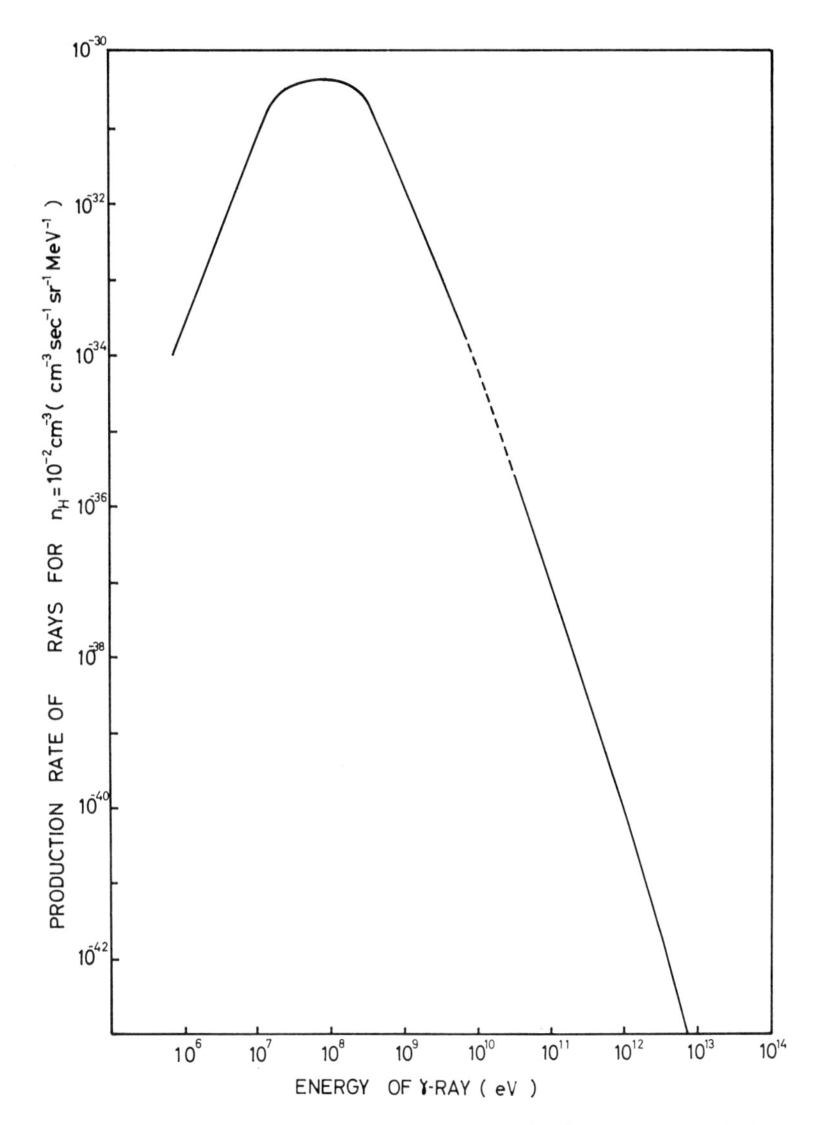

Fig. 6.38 Production spectrum of γ-rays due to the decays of neutral pions.

It may be of some use to give the γ-ray intensity averaged over all directions. Since

$$\langle \overline{n_{\mathrm{H}} R} \rangle_{4\pi} = \frac{1}{4\pi} \int d\Omega \int_0^{R(\Omega)} n_{\mathrm{H}}(r, \Omega)\, dr \simeq 3 \times 10^{21}\ \mathrm{cm}^{-2}, \quad (6.8.10)$$

we have

$$\langle j_\gamma(\pi^0)\rangle \simeq 3\times 10^{-5}, \qquad \langle j_\gamma(0.51\ \mathrm{MeV})\rangle \simeq 6\times 10^{-5}\ \mathrm{cm^{-2}\ sec^{-1}\ sr^{-1}}. \tag{6.8.11}$$

According to the discussion above, observation of γ-ray flux gives the cosmic-ray flux times the matter density along a line of sight. If the former is given, one can determine the distribution of matter in the Galaxy. The numerical values in Table 6.29 are based on the density distribution of hydrogen atoms in the disk as given in Section 6.1 and the assumed density of $n_\mathrm{H} = 10^{-2}\ \mathrm{cm}^{-3}$ in the halo. In the disk it is suggested that the yet invisible matter such as hydrogen molecules is as abundant as hydrogen atoms (Gould 63). The amount of such matter as well as its density distribution will be obtained by observing the γ-rays. Although information on invisible matter will be available in ultraviolet and infrared observations as well, the matter in the halo may be observable only through γ-rays, provided that the cosmic-ray flux is sufficiently high in the halo.

This would also be the case for intergalactic matter if the cosmic-ray intensity were high enough therein. If the intergalactic density is $n_\mathrm{H} \simeq 10^{-5}\ \mathrm{cm}^{-3}$, the value of $n_\mathrm{H} R_U$ is as large as $10^{23}\ \mathrm{cm}^{-2}$, larger than (6.8.10) by a factor of 30. Since, however, the cosmic-ray intensity in intergalactic space is presumably smaller by a factor of 3×10^3 or so than that in the Galaxy, the intensity of γ-rays of intergalactic origin may be smaller than that of Galactic origin in the plane of the Galactic disk, whereas they are comparable in the direction perpendicular to the Galactic plane.

The difficulty of observation of intergalactic γ-rays increases due to the contribution of γ-rays produced in galaxies. The γ-ray yield per galaxy is proportional to the total number of cosmic-ray particles times the density of diffuse matter. The former may be proportional to the number of high-energy electrons and consequently to the radio luminosity. Since the number of protons in a radio source is proportional to that of electrons, $N_p = \kappa N_e$, and the radio yield is $Y \propto N_e H^{1+\alpha/2}$, the γ-ray yield is

$$P_\gamma \propto Y \frac{n_\mathrm{H}}{H^{1+\alpha/2}}. \tag{6.8.12}$$

The γ-ray yield relative to that of the Galaxy can be obtained as

$$P_\gamma = \frac{P_{\gamma,G}\left(Y \dfrac{n_\mathrm{H}}{H^{1+\alpha/2}}\right)}{\left(Y \dfrac{n_\mathrm{H}}{H^{1+\alpha/2}}\right)_G}, \tag{6.8.13}$$

where the subscript G indicates the value for the Galaxy. Referring to the radio-luminosity function $n(>Y)$, we find the γ-ray intensity due to all galaxies to be

$$j_{\gamma,\,\mathrm{Mg}} = P_{\gamma,\,G} \frac{n_{\mathrm{H}}}{n_{\mathrm{H},\,G}} \frac{H_G^{1+\alpha_G/2}}{H^{1+\alpha/2}} \frac{1}{Y_G} \int \left(-\frac{dn}{dY}\right) Y\,dY R_U$$

$$= \frac{p_{\gamma,\,G}}{n_{\mathrm{H},\,G}} \frac{H_G^{1+\alpha_G/2}}{H^{1+\alpha/2}} n_{\mathrm{H}} V_G \frac{1}{Y_G} \int \left(-\frac{dn}{dY}\right) Y\,dY\,R_U, \qquad (6.8.14)$$

where R_U is the Hubble radius and V_G the volume of the Galaxy. This approximately gives

$$j_{\gamma,\,\mathrm{Mg}} \simeq \frac{\left(\dfrac{p_\gamma}{n_{\mathrm{H}}}\right)_G \left(\dfrac{n_{\mathrm{H}}}{H^{1+\alpha/2}}\right)}{\left(\dfrac{n_{\mathrm{H}}}{H^{1+\alpha/2}}\right)_G} \times 10^{21} \quad \mathrm{cm}^{-2}\,\mathrm{sec}^{-1}\,\mathrm{sr}^{-1}\,\mathrm{energy}^{-1}, \qquad (6.8.15)$$

greater than the intergalactic γ-ray intensity and comparable to the Galactic γ-ray intensity in the polar direction, provided that $H \simeq H_G$ and $n_{\mathrm{H}} \simeq n_{\mathrm{H},\,G}$.

Radio sources may be located as γ-ray sources. The γ-ray intensity expected for a source at distance R is given by

$$j_\gamma(E_\gamma) = \frac{P_\gamma(E_\gamma)}{R^2} = \left(\frac{\kappa c}{4\pi}\right) \bar{\sigma}(\bar{E}_p) f(\bar{E}_p, E_\gamma) \frac{n_{\mathrm{H}} N_e}{R^2}$$

$$= \frac{1.8 \times 10^{31}\kappa}{\alpha y(\alpha)} I_\nu \frac{\bar{\sigma}(\bar{E}_p) f(\bar{E}_p, E_\gamma) n_{\mathrm{H}}}{H} \left(\frac{k_1 \nu}{\nu_1}\right)^{\alpha/2} \left[1 - \left(\frac{k_2 \nu_1}{k_1 \nu_2}\right)^{\alpha/2}\right], \qquad (6.8.16)$$

where all quantities are expressed in cgs units, and σf is the differential cross section for γ-ray production. For the strongest radio source the value of the radio flux is $I_\nu \simeq 10^{-19}$ erg-cm^{-2} Hz^{-1} at about 100 MHz, and the magnetic-field strength is $H \simeq 10^{-4}$ gauss. Because $\bar{\sigma} f \simeq 10^{-26}$ cm^2-GeV^{-1}, the intensity of γ-rays expected is

$$j(E_\gamma) \simeq 10^{-8} \kappa n_{\mathrm{H}} \left(\frac{10^{-4}}{H(\mathrm{gauss})}\right) \left[I_\nu \frac{\mathrm{W\text{-}m^{-2}\,sec^{-1}Hz^{-1}}}{10^{-22}}\right] \quad \mathrm{cm}^{-2}\,\mathrm{sec}^{-1}\,\mathrm{GeV}^{-1} \qquad (6.8.17)$$

at $E_\gamma \simeq 0.1$ GeV.

The expression (6.8.17) shows that the γ-ray intensity will give the proton-electron ratio and the matter density in a source. The values of $\kappa \simeq 10^2$ and $n_{\mathrm{H}} \simeq 1$ cm^{-2} could be reasonable estimates for most radio sources. Hence the intensity of the strongest source would be on the order of 10^{-6} cm^{-2} sec^{-1} GeV^{-1} at the peak of the γ-ray spectrum.

A maximum estimate of the γ-ray flux is obtained if all electrons responsible for radio emission are the products of meson decays. The γ-ray flux is proportional to the rate of generation of electrons, dN_e/dt, as

$$j_\gamma(\text{all energies}) = \frac{1}{2}\frac{(dN_e/dt)}{4\pi R^2} = \frac{(N_e/2\tau_e)}{4\pi R^2}. \tag{6.8.18}$$

Comparing this with (6.8.16), the γ-ray flux under this assumption is greater than that given by (6.8.16) by a factor

$$\frac{(68.18)}{\int (68.16)\, dE} \simeq \frac{1}{2\tau_e \kappa c \sigma n_{\text{H}}} \simeq \frac{\tau_c}{2\kappa\tau_e}, \tag{6.8.19}$$

where τ_c is the collision mean free time of protons in the source. In the early stage of the radio source the matter density is so large that one can put $\tau_e = \tau_c$. The flux of γ-rays from QSS may be thus estimated to be 10^{-6} cm^{-2} sec^{-1} or smaller.

In a very early stage the energy loss of electrons may be caused mainly by the inverse Compton effect due to a high radiation density. If this is the case, the γ-ray flux would be higher than that estimated from the radio flux, since only a minor fraction of the energy would be spent for synchrotron radiation.

The inverse Compton effect due to collisions with radio photons in radio sources gives a minimum estimate of the photon flux, as will be discussed in Sections 6.8.4 and 6.8.9.

6.8.3 Gamma Rays from Neutral Pions Produced by Photon-Nucleon Collisions

As the energy of a proton increases, the energy of a starlight photon in the rest system of the proton becomes so large that it can exceed the threshold energy of pion production. The neutral pions that are thus produced decay into γ-rays of extremely high energies. Since the photon density is much higher than the particle density in intergalactic space, the main contribution to intergalactic γ-rays of very high energies is suggested to come from the photopion-production process (Hayakawa 62, 63).

If one-pion production is dominant, the whole process is quite analogous to the inverse Compton effect. If an approximation such as (6.7.30) is used—that is, if the energies of pions produced are represented by their average value as

$$E_\pi = f\bar{\varepsilon}\left(\frac{E_p}{Mc^2}\right)^2, \tag{6.8.20}$$

where $\bar{\varepsilon}$ is the average energy of photons in the laboratory system and f is a numerical factor on the order of unity, the production spectrum of pions is expressed as

$$p_\pi(E_\pi) = \frac{1}{2\sqrt{f}} \frac{Mc^2}{(\bar{\varepsilon}E_\pi)^{1/2}} n_0 \sigma_{\gamma\pi} j_p (\sqrt{(E_\pi/f\bar{\varepsilon})}Mc^2), \qquad (6.8.21)$$

where $j_p(E_p)$ is the differential proton spectrum, which may be represented by a power law as $j_p(E_p) = \alpha j_p(E_0)(E_0/E_p)^\alpha E_p^{-1}$, and $\sigma_{\gamma\pi}$ is the photopion cross section averaged over the energy region concerned. By substituting the power spectrum into (6.8.21) we obtain

$$p_\pi(E_\pi) = \frac{f^{\alpha/2}}{2} n_0 \sigma_{\gamma\pi} \alpha j_p(E_0)\left(\frac{E_0}{Mc^2}\right)^\alpha \left(\frac{\bar{\varepsilon}}{E_\pi}\right)^{\alpha/2} \frac{1}{E_\pi}. \qquad (6.8.21')$$

If E_0 is taken as equal to the average threshold energy

$$E_0 \simeq m_\pi c^2 \frac{Mc^2}{\bar{\varepsilon}} \simeq 10^{17} \text{ eV}, \qquad (6.8.22)$$

the energy of a pion corresponding to E_0 is

$$E_{\pi 0} = f\bar{\varepsilon}\left(\frac{E_0}{Mc^2}\right)^2 = \frac{f(m_\pi c^2)^2}{\bar{\varepsilon}} \simeq f \times 10^{16} \text{ eV}. \qquad (6.8.22')$$

After introducing (6.8.22′) into (6.8.21′) and further into (6.8.4) we have the flux of γ-rays for $f = 1$:

$$j_\gamma(E_\gamma) = \frac{2}{\alpha + 2} n_0 \sigma_{\gamma\pi} R(E_\gamma) j_p(E_0)\left(\frac{E_{\pi 0}}{E_\gamma}\right)^{\alpha/2} E_\gamma^{-1}, \qquad (6.8.23)$$

where $R(E_\gamma)$ is the absorption length of γ-rays with energy E_γ, as given in Fig. 6.37. If the absorption due to microwave photons were absent, we would have a strong γ-ray intensity relative to the proton intensity, since

$$n_0 \sigma_{\gamma\pi} R \simeq 10^{-2} \times 10^{-28} \times 10^{28} \simeq 10^{-2}.$$

Taking correct values of R into account, we draw the γ-ray spectrum in Fig. 6.39.

Photopion production results in the energy loss of protons. The rate of energy loss is approximately given for $E_p > m_\pi c^2(Mc^2/\bar{\varepsilon})$ by

$$-\frac{dE_p}{dt} = cn_0 \sigma_{\gamma\pi} E_\pi \simeq fcn_0 \bar{\varepsilon}\sigma_{\gamma\pi}\left(\frac{E_p}{Mc^2}\right)^2. \qquad (6.8.24)$$

The energy loss for a proton traveling over the age of the universe is

$$\Delta E_p \simeq fn_0 \bar{\varepsilon}\sigma_{\gamma\pi} R_U\left(\frac{E_p}{Mc^2}\right)^2. \qquad (6.8.25)$$

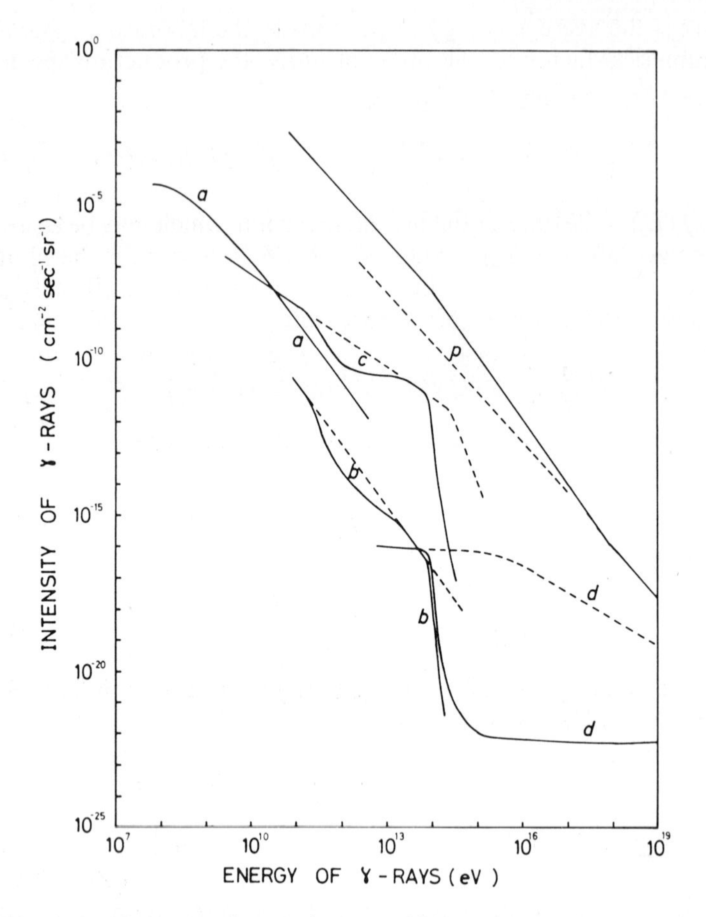

Fig. 6.39 Integral spectra of general γ-rays. (*a*) γ-rays produced by *p-p* collisions in the Galaxy, with $\overline{\langle n_H R \rangle} = 3 \times 10^{21}$ cm^{-2}; (*b*) γ-rays produced by *p-p* collisions in intergalactic space, the density of matter being assumed as $n_H = 10^{-5}$ cm^{-3}; (*c*) γ-rays produced by the inverse Compton effect in intergalactic space; (*d*) γ-rays produced by the collisions of protons with starlight photons in intergalactic space, the density of starlight photons being assumed as 10^{-2} cm^{-3}; (*p*) the integral spectrum of protons; the dashed line represents that expected in intergalactic space. The dashed curves represent the intensities without correction for intergalactic absorption. The solid curves represent the intensities corrected for the intergalactic absorption given in Fig. 6.37.

This becomes comparable to E_p for $E_p \simeq 10^{20}$ eV. Pair creation by collision with starlight photons also contributes to the energy loss, but the energy lost per collision is on the order of $2m_e c^2$, compared with $m_\pi c^2$ in the case of photopion production. Although the pair-creation cross section is greater by one order of magnitude than the photopion cross section, the contribution of the former to the energy-loss rate is smaller than (6.8.24).

Collision with microwave photons causes similar effects at energies greater than 10^{20} eV. Since the value of $n_0 \bar{\varepsilon}$ is greater, the energy loss is quite large above 10^{20} eV. Hence the energy spectrum of protons is expected to show a cutoff at about 10^{20} eV (Greisen 66).

6.8.4 X-Ray Emission by Nonthermal Particles

Various mechanisms of X-ray emission have been discussed elsewhere (Hayakawa 64c, 66). Here results are merely summarized.

Nonthermal electrons with the differential energy spectrum $j_e(E_0/E)^\alpha \times dE/E$ emit photons with energies between ε and $\varepsilon + d\varepsilon$ at a rate of $p(\varepsilon)\, d\varepsilon$. The production spectrum $p(\varepsilon)$ is expressed for *synchrotron radiation* as follows:

$$p_s(\varepsilon)\, d\varepsilon = 4y(\alpha)\sigma_{\mathrm{Th}}\left(\frac{H^2}{8\pi\varepsilon_c}\right) j_e \left(\frac{E_0}{mc^2}\right)^\alpha \frac{\varepsilon_c^{\alpha/2}\, d\varepsilon}{\varepsilon^{1+(\alpha/2)}}$$

$$\simeq 4y(\alpha)\sigma_{\mathrm{Th}}\left(\frac{H^2}{8\pi\varepsilon_c}\right) j_e \left(\frac{1\ \mathrm{keV}}{\varepsilon}\right)^{\alpha/2} \frac{d\varepsilon}{\varepsilon}, \qquad (6.8.26)$$

where $y(\alpha)$ is given in Table 6.24 and

$$\varepsilon_c = \tfrac{3}{2} h\nu_c = 1.75 \times 10^{-8}\, H\,(\mathrm{gauss})\ \mathrm{eV} = 2.8 \times 10^{-20}\, H\,(\mathrm{gauss})\ \mathrm{erg}. \qquad (6.8.27)$$

The photon energy ε is related to the electron energy E through

$$\varepsilon_s = 6.6 \times 10^{-2}\, H_\perp\,(\mathrm{gauss})\, E^2\,(\mathrm{GeV})\ \mathrm{eV}. \qquad (6.8.28)$$

For the *inverse Compton* effect the production spectrum is given in (6.7.31) or

$$p_c(\varepsilon)\, d\varepsilon = \tfrac{1}{2}\sigma_{\mathrm{Th}}\left(\frac{n_0\,\bar{\varepsilon}}{\bar{\varepsilon}}\right) j_e \left(\frac{E_0}{mc^2}\right)^\alpha (\tfrac{4}{3}\bar{\varepsilon})^{\alpha/2} \frac{d\varepsilon}{\varepsilon^{1+(\alpha/2)}}, \qquad (6.8.29)$$

and corresponding to (6.8.28)

$$\varepsilon_c = 1.20 \times 10^3\, T\,(^\circ\mathrm{K})\, E^2\,(\mathrm{GeV})\ \mathrm{eV}. \qquad (6.8.30)$$

Comparison between (6.8.28) and (6.8.30) shows that the energy of electrons responsible for X-ray emission is much lower for the inverse Compton effect than for synchrotron radiation.

The emission of X-rays by *bremsstrahlung* is caused by still lower energy electrons. In the nonrelativistic region the photon-production rate is expressed, for the power spectrum of electrons, $j(w)\,dw = j_e(w_0/w)^\alpha\,dw/w$, as follows:

$$p_b(\varepsilon) \simeq \frac{2}{\pi}\,\sigma_{\mathrm{Th}}\,n_{\mathrm{H}}\,\frac{e^2}{\hbar c}\,mc^2\,\frac{1}{\alpha+1}\,j_e\,\frac{w_0{}^\alpha}{\varepsilon^{\alpha+2}}\,. \tag{6.8.31}$$

The yield of a particular K-line of energy E_K is given by

$$q_{Ke}(E_K) = \frac{3}{4}\,\sigma_{\mathrm{Th}}\,n_{\mathrm{H}}\left(\frac{n_Z}{n_{\mathrm{H}}}\right)\left(\frac{mc^2}{E_K}\right)f_e(\bar{w}, Z)\,\omega_K(Z)\,\frac{1}{\alpha+1}\,j_e\left(\frac{w_0}{E_K}\right)^\alpha, \tag{6.8.32}$$

where $\bar{w}$ is the average energy of the electrons, $f_e(\bar{w}, Z)$ is a slowly varying function on the order of unity and $\omega_K(Z)$ is the K-fluorescence yield. Since the relative abundance of elements responsible for the K-emission in the keV region is small—say, $n_Z/n_{\mathrm{H}} \simeq 10^{-4}$—this is comparable to the emission rate by bremsstrahlung if the bandwidth of a detector is several keV.

The characteristic X-rays are excited also by nuclear particles. Since the cross section is proportional to the square of the nuclear charge of the incident particle, the cross section per proton is 1.75 times the cross section for the proton component, as in (6.7.77). The rate of X-ray emission for the proton spectrum of $j_p(W_0/W)^\alpha\,dW/W$ is given by

$$q_{Kp}(E_K) = 1.75\,\frac{3}{4}\,\sigma_{\mathrm{Th}}\,n_{\mathrm{H}}\left(\frac{n_Z}{n_{\mathrm{H}}}\right)\left(\frac{mc^2}{E_K}\right)^2 f_p(\overline{W}, Z)\,\frac{1}{\alpha+1}\,j_p\left(\frac{W_0}{(M/m)E_K}\right)^\alpha \omega_K(Z), \tag{6.8.33}$$

where $\overline{W}$ is the average energy of the protons.

The analogue of bremsstrahlung in the case of proton impact is *inner bremsstrahlung* associated with the knock-on process. A knock-on electron of velocity $c\beta$ emits photons with the probability given in (2.5.5).

The production rate of X-rays is evaluated by taking account of the Z^2-dependence of the knock-on cross section as

$$p_{\mathrm{ib}}(\varepsilon) = 1.75\,\frac{8}{\pi}\,\sigma_{\mathrm{Th}}\,\frac{e^2}{\hbar c}\,n_e \ln\left(\frac{m\overline{W}}{M\varepsilon}\right)\frac{1}{\alpha+1}\,j_p\left(\frac{W_0}{(M/m)\varepsilon}\right)^\alpha \frac{mc^2}{\varepsilon^2}, \tag{6.8.34}$$

where n_e is the electron density. Here the energy of protons responsible for inner bremsstrahlung is higher than $M\varepsilon/m$. If this is compared with the production rate by electron bremsstrahlung in (6.8.31), the yields are comparable, provided that the velocity spectra of protons and electrons are alike. The latter seems to be the case as long as the loss of energy by ionization is negligible, as will be discussed in Section 6.9.

All the emission processes considered above are responsible for optical emission as well. Since few X-ray sources are optically visible, as will be described in Section 6.8.7, the spectrum of radiation should not be steeper than $d\varepsilon/\varepsilon$. This results in $\alpha \lesssim 0$ in the relativistic region for the synchrotron and the inverse Compton processes, and $\alpha \lesssim -1$ in the nonrelativistic region for bremsstrahlung, inner bremsstrahlung, and the excitation of characteristic X-rays. In other words, the spectrum of nonthermal particles decreases with increasing energy less steeply than

$$j(E)\,dE = j_0 \frac{dE}{E} \qquad \text{for } E \gg mc^2, \tag{6.8.35a}$$

$$= j_0 \frac{dW}{mc^2} \qquad \text{for } W < mc^2, \tag{6.8.35b}$$

where E is the total energy and W is the kinetic energy. For the spectrum like (6.8.35) the maximum energy must exist. Denoting this as E_m or W_m, the energy density of nonthermal particles is estimated as

$$u = \frac{4\pi j_0}{c} E_m \quad \text{or} \quad \frac{\sqrt{2}\pi j_0}{c} \sqrt{W_m/mc^2}\, W_m, \tag{6.8.36}$$

according to whether the maximum energy is in the relativistic or the nonrelativistic region.

Using (6.8.35) and (6.8.36), we can express the X-ray emission rates in (6.8.26) and (6.8.29) respectively as

$$p_s(\varepsilon) = \frac{1}{\pi} y(\alpha) c\sigma_{\mathrm{Th}} \left(\frac{H^2}{8\pi}\right) u_e \frac{3}{4\pi} \frac{E_m}{\varepsilon_m} \frac{1}{(mc^2)^2} \frac{1}{\varepsilon}, \tag{6.8.37}$$

$$p_c(\varepsilon) = \frac{1}{8\pi} c\sigma_{\mathrm{Th}} (n_0 \bar{\varepsilon}) u_e \frac{E_m}{\varepsilon_m} \frac{1}{(mc^2)^2} \frac{1}{\varepsilon}, \tag{6.8.38}$$

where ε_m is the X-ray energy corresponding to E_m. The comparison between the above two demonstrates that they are equally efficient in terms of energy economy. The energy density of radiation due to the synchrotron process is

$$(n\bar{\varepsilon})_s = \frac{4\pi}{c} L \int^{\varepsilon_m} p_s(\varepsilon)\varepsilon\, d\varepsilon = 4y(\alpha) L\sigma_{\mathrm{Th}} \left(\frac{H^2}{8\pi}\right) \frac{u_e}{mc^2} \frac{3}{4\pi} \frac{E_m}{mc^2}, \tag{6.8.39}$$

where L is the radius of the X-ray source concerned. If this is substituted into $n_0 \bar{\varepsilon}$ in (6.8.38), we can obtain the emission rate of γ-rays with energies as high as $\varepsilon_m(E_m/mc^2)^2$. Such γ-rays may be detectable if the X-ray source is large enough.

A similar argument holds if the optical emission rate $p_0(\varepsilon_0)$ is known. The ratio of the X-ray emission to the optical emission rate is given, if they are related through the inverse Compton effect, by

$$\frac{p_c(\varepsilon)}{p_0(\varepsilon_0)}=\tfrac{1}{2}\sigma_{\mathrm{Th}}L\frac{u_e}{mc^2}\frac{E_m}{mc^2}\frac{\varepsilon_0}{\varepsilon_m}\frac{\Delta\varepsilon_0}{\varepsilon}\,. \tag{6.8.40}$$

This implies that the optical luminosity of an X-ray source would have to be strong if X-rays were emitted by the inverse Compton effect in a small source.

The electrons responsible for X-ray emission through the inverse Compton effect can also produce bremsstrahlung. The yield of X-rays for the electron spectrum (6.8.35) is approximately evaluated as

$$p_b(\varepsilon)=\frac{1}{2\pi^2}c\sigma_{\mathrm{Th}}\frac{e^2}{\hbar c}n_{\mathrm{H}}u_e\ln\left(\frac{E_m}{\varepsilon}\right)\frac{1}{E_m\varepsilon}\,. \tag{6.8.41}$$

This can be compared with (6.8.39), taking account of

$$\varepsilon_m=\bar{\varepsilon}\left(\frac{E_m}{mc^2}\right)^2,$$

as

$$\frac{p_b(\varepsilon)}{p_c(\varepsilon)}=\frac{4}{\pi}\frac{e^2}{\hbar c}\frac{n_{\mathrm{H}}}{n_0}\left(\frac{E_m}{\varepsilon}\right). \tag{6.8.42}$$

Due to the smallness of $e^2/\hbar c=\frac{1}{137}$, this is in general smaller than unity in the X-ray region, but the energy of photons by bremsstrahlung is extended to E_m, so that the photon emission at high energies is significant.

The electrons responsible for bremsstrahlung excite K X-rays. If a detector has an energy resolution $\Delta\varepsilon$, the rate of K X-ray emission is compared with that of bremsstrahlung as

$$\frac{p_b(\varepsilon)\,\Delta\varepsilon}{q_{Ke}(E_k)}=\frac{8}{3\pi}\frac{e^2}{\hbar c}\frac{n_{\mathrm{H}}}{n_Z}\frac{1}{f(\overline{W},Z)\omega_K(Z)}\frac{\Delta\varepsilon}{mc^2}\,. \tag{6.8.43}$$

Since elements of $Z\gtrsim 10$ can contribute to K X-rays in the keV region, we have $n_Z/n_{\mathrm{H}}\lesssim 10^{-4}$ and $\omega_K(Z)\simeq 0.1-0.2$. Hence one can resolve K lines only if $\Delta\varepsilon<1$ keV.

Characteristic lines of lighter elements lie in the visible and ultraviolet (UV) regions. The excitation cross section in general increases as the characteristic energy decreases. Hence the emission of K X-rays would be associated with the stronger emission of characteristic lines of oxygen, helium, and hydrogen in the UV and the optical regions, unless the energy spectrum of electrons is cut off at a low energy and/or the electron temperature is so high that the lighter elements are completely ionized. This may, in fact, be the case if the flux of nonthermal particles is very high.

The energy of a nonrelativistic particle of mass M is lost mainly by the ionization process at a rate of

$$-\frac{dW}{dt}\bigg|_I = \tfrac{3}{4}\sqrt{2}\, c\sigma_{\text{Th}}\left(\frac{Mc^2}{W}\right)^{1/2} mc^2 \ln\left(\frac{4W}{I}\right) n_e, \qquad (6.8.44)$$

where I is the average ionization potential in a neutral gas and may be taken as equal to the thermal energy in an ionized medium. For electrons the logarithmic factor is replaced by $\ln[(W/I)\sqrt{e/2}]$, where e is the natural base of logarithms.

The time scale for the particle to lose its energy to the thermal energy is approximately estimated as

$$\begin{aligned}\tau_I &= \int_I^W \frac{1}{(-dW/dt)_I}\, dW \\ &\simeq \frac{8}{9\sqrt{2}\, c\sigma_{\text{Th}} n_e}\left(\frac{W}{Mc^2}\right)^{1/2}\left(\frac{W}{Mc^2}\right)\Big/\ln\left(\frac{4W}{I}\right) \\ &\simeq \frac{3.2\times 10^{13}}{n_e}\left(\frac{W}{Mc^2}\right)^{1/2}\left(\frac{W/mc^2}{\ln 4W/I}\right) \text{ sec.} \qquad (6.8.45)\end{aligned}$$

For $n_e \simeq 10^2$ all nonrelativistic electrons are thermalized within 10^4 years, whereas only protons of energies below 10 MeV are thermalized within this period.

The ineffectiveness of the nonthermal processes for electrons may be revealed by comparing the rates of energy loss by ionization and bremsstrahlung. Only a small fraction of energy is emitted as bremsstrahlung, whereas most of the nonthermal energy is spent for ionization.

The argument above demonstrates that only protons of energies greater than some tens of MeV and relativistic electrons can be considered as the source of the nonthermal processes. Even if nonthermal particles of low energies are present, they are rapidly thermalized and heat up the gas in the source. Thus we are led to consider thermal radiation as a possible mechanism of X-ray emission.

6.8.5 Thermal X-Rays

A high-temperature plasma is able to emit electromagnetic radiation in the X-ray region (Hayakawa 66). If the source is an optically thin object, the *free-free*, the *free-bound* and the *bound-bound* transitions for ions in various stages of ionization take part in thermal radiation.

The free-bound and the bound-bound intensities depend sensitively on the abundances of elements and, in the X-ray region, on the abundances of heavy elements. The chemical composition of X-ray sources may be different from that of the ambient gas, but similar to that of cosmic rays in view of the possible relation of X-ray sources with cosmic-ray sources. Because the hydrogen-helium ratio in the latter case depends on the acceleration process, the cosmic-ray abundances are normalized to the cosmic abundances at the abundance of helium. The cosmic-ray abundances are characterized by an overabundance of heavy elements, as discussed in Section 6.3.

Among the three thermal emission processes mentioned above the free-free process is less sensitive to the chemical composition and determines the general features of emission. A plasma of electron density n_e and ion density n_Z emits radiation by the free-free transition at a rate of

$$p_{\rm ff}(\varepsilon) = \frac{1}{6\pi^3}\frac{e^2}{\hbar c}\sigma_{\rm Th}\, cn_e\left(\frac{mc^2}{kT}\right)^{1/2}\left[\sum_Z Z^2 n_Z \bar{g}_{\rm ff}(Z, T, \varepsilon)\right]\frac{1}{\varepsilon}e^{-\varepsilon/kT},$$

$$= 0.81\times 10^{-12} n_e T^{-1/2}\left(\sum_Z Z^2 n_Z \bar{g}_{\rm ff}\right)\frac{1}{\varepsilon}e^{-\varepsilon/kT}$$

$$\text{cm}^{-3}\ \text{sec}^{-1}\ \text{sr}^{-1}\ \text{keV}^{-1}, \quad (6.8.46)$$

where Ze is the ionic charge and $\bar{g}_{\rm ff}$ is the average Gaunt factor, which is nearly constant (around 0.6 in the temperature range of interest) in the range $1 \lesssim \varepsilon/kT \lesssim 3$. The weighted mean of the Gaunt factor for the Cameron abundance has been found to be about 10-percent larger than the Gaunt factor for a pure hydrogen plasma (Hovenier 65).

Neglecting the above effect of the Gaunt factor, the rates of emission from gases with different compositions are compared, for the same electron densities, by the values of $\sum Z^2 n_z/n_e$. These values are shown in Table 6.30 as $p_{\rm ff}/p_{\rm ff}$(hydrogen), for the cases of the Cameron abund-

Table 6.30 Comparison of the Rates of Emission at Energies above 9.2 keV

	Abundances	Temperature (°K)		
		2×10^7	5×10^7	10^8
$p_{\rm ff}/p_{\rm ff}$ (Hydrogen)	Cameron 59	1.3	1.4	1.4
	Cosmic rays	1.9	1.9	2.0
$p_{\rm fb}/p_{\rm ff}$	Cameron 59	0.4	0.1	0.05
	Cosmic rays	9	3	1

ances and of the cosmic-ray abundances. The small variations with temperature are due to different degrees of ionization. The ratio of the free-bound to the free-free emissions is given approximately by

$$\frac{p_{\mathrm{fb}}(\varepsilon)}{p_{\mathrm{ff}}(\varepsilon)} \simeq \left(\frac{e^2}{\hbar c}\right)\left(\frac{mc^2}{kT}\right)\frac{\sum_{\chi_Z \le k} Z^4 n_{Z,i+1}}{\sum Z^2 n_Z \bar{g}_{\mathrm{ff}}}, \qquad (6.8.47)$$

where χ_Z is the absorption edge for an element of atomic number Z; the numerical values are given in Table 6.31. As the energy of radiation

Table 6.31 Energies of Absorption Edges

Ion	Energy of Absorption Edges (keV)	Ion	Energy of Absorption Edges (keV)
H I	0.0136	Fe XXIV	2.0
He II	0.0544	Si XIII	2.44
C VI	0.490	Si XIV	2.67
N VII	0.667	S XV	3.22
O VIII	0.871	S XVI	3.48
Ne X	1.36	Fe XXV	8.8
Fe XXIII	1.9	Fe XXVI	9.19
Mg XII	1.96		

increases, the absorption edges appear successively, and the above ratio increases discontinuously. In the energy region with no absorption edge the free-bound spectrum has the same energy dependence as the free-free spectrum if the energy dependence of the Gaunt factor is neglected. Thus the ratio levels off to a certain value beyond the edge of Fe XXVI at 9.2 keV; the value of (6.8.47) is shown in Table 6.30 for various chemical compositions and temperatures, with an approximation of $\bar{g}_{\mathrm{ff}} = 0.6$. It is found that the contribution of the free-bound emission is appreciable for the case of the cosmic-ray abundance.

The rate of the line emission relative to the free-free emission rate is given approximately by Elwert (Elwert 54) as

$$\frac{q_{\mathrm{bb}}(\varepsilon)}{p_{\mathrm{ff}}(k)D} \simeq \frac{3\sqrt{3}}{16}\left(\frac{\hbar c}{e^2}\right)\left(\frac{mc^2}{D}\right)\frac{n_{z,i}}{\sum Z^2 n_Z \bar{g}_{\mathrm{ff}}}, \qquad (6.8.48)$$

where D is the average level distance for a mixture of elements. If there are abundant heavy elements in the intermediate stages of ionization, the contributions of their bound-bound transitions are appreciable, since D is on the order of 1 keV.

The emission rates thus calculated by reference to (Kawabata 60) are shown in Table 6.32. For $T = 2 \times 10^7$ °K the contribution of Fe XXIV

Table 6.32 Bound-Bound Emission Rates[a]

Ions	k (keV)	2×10^7 °K	5×10^7 °K	10^8 °K	[b,c] n_Z/n_e
Ne X	1.0	0.2	—	—	6×10^{-4}
Fe XXIII	1.1	0.6	—	—	1×10^{-3} (Fe XXIII–Fe XXIV)
Fe XXIV	1.1	2	0.04	—	
Mg XII	1.5	0.9	0.1	—	1×10^{-3}
Si XIII	1.8	0.2	—	—	7×10^{-4} (Si XIII–Si XIV)
Si XIV	2.0	0.6	0.2	—	
S XV	2.4	0.5	—	—	3×10^{-4} (S XV–Ca XX)
S XVI	2.6	0.3	0.1	—	
Ca XIX	3.8	0.1	0.2	0.03	
Ca XX	4.1	0.01	0.3	0.1	
Fe XXV	6.6	0.04	0.7	0.7	1×10^{-3} (Fe XXV–Fe XXVI)
Fe XXVI	6.9	—	0.04	0.2	

[a] In units of 10^{-16} photons cm^{-3} sec^{-1} sr$^{-1}/(n_e$ cm$^{-3})^2$.
[b] n_Z is the relative abundances of cosmic-ray nuclei at the source.
[c] Elements of $Z = 15$ to 20 are represented either by sulpher or by calcium.

is remarkable because of a large final state density in the L-shell. For $T = 5 \times 10^7$ and 10^8 °K the K X-rays of Fe are appreciable.

For the lines, absorption in the source may be important. When the Doppler broadening is dominant the optical thickness at the line center is given by

$$\tau = \tfrac{3}{8}\sqrt{2\pi}\,\sigma_{\text{Th}}\left(\frac{\hbar c}{e^2}\right)\frac{mc^2}{\varepsilon}\sqrt{(mc^2/kT)}\,gf\,nL$$

$$= 8.6 \times 10^{-23}\left(\frac{mc^2}{\varepsilon}\right)\sqrt{(mc^2/kT)}\,gf\,nL, \qquad (6.8.49)$$

where n is the density of the relevant element in the relevant ionization stage and energy level, M its mass, gf its weighted oscillator strength, and L the linear dimension of the source. If the source has a dimension of 1 ly and an electron density of 300 cm^{-3}, the optical depths for several strong lines are on the order of unity. If this is the case some of the photons cannot escape from the source, and the spectrum has its blackbody value at the line center. However, if the turbulent velocity is comparable to the expansion velocity in a supernova remnant, the optical depth may be smaller than unity, and the photons may escape freely from the source.

The rates of emission in several energy bands due to the free-free, free-bound, and bound-bound transitions are summarized in Table 6.33

Table 6.33 Nuclear Burning from Hydrogen to Iron

Temperature °K)	Transition	ε (keV)			
		$1 \sim 2$	$2 \sim 4$	$4 \sim 8$	$8 \sim 16$
	f-f	0.62	0.14	0.015	0.00045
2×10^7	f-b	0.6	0.8	0.1	0.004
	b-b	6	0.9	0.01	—
	f-f	0.66	0.24	0.062	0.0095
5×10^7	f-b	0.2	0.3	0.1	0.3
	b-b	0.4	0.2	0.2	—
	f-f	0.62	0.25	0.090	0.023
10^8	f-b	0.03	0.1	0.06	0.04
	b-b	—	—	0.2	—

[a] In units of 10^{-16} photons cm^{-3} sec^{-1} keV^{-1} sr$^{-1}/(n_e\ \text{cm}^3)^2$. Cosmic-ray abundances are assumed.

for the cosmic-ray abundances. These spectra are shown in Fig. 6.40, where the spikes of the line emissions are shown by $q_{bb}(\varepsilon)$/keV taking into account the energy resolution of detectors.

For the thermal emission including the above three transitions the spectra up to 4 keV are found to be approximately as $q \propto \varepsilon^{-1}$, rather than exponential. In other words, the spectral index measured with a wide energy resolution may not be used to determine the temperature in this energy region. In the energy range between 4 and 8 keV the contribution of K X-rays of Fe is appreciable for temperatures higher than 5×10^7 °K, though the contribution of the bound-bound transitions depends on the dimension of the source.

The above discussion depends on the abundances of heavy elements. However, the spectra above 9.2 keV hardly depend on the abundances, and the temperature may be determined from the observed spectra.

If the abundances are normal as the Cameron abundance, the spectra are expressed by the free-free spectra to a good approximation.

If the source is optically thick, the radiation is given by the well-known Planck function. In this case the yield per unit area is meaningful, that is,

$$\pi \int q_B(\varepsilon)\, dD = \frac{2\pi\varepsilon^2}{h^3c^2}\,(e^{\varepsilon/kT} - 1)^{-1}, \tag{6.8.50}$$

where the integral is taken over the depth. The differences between the

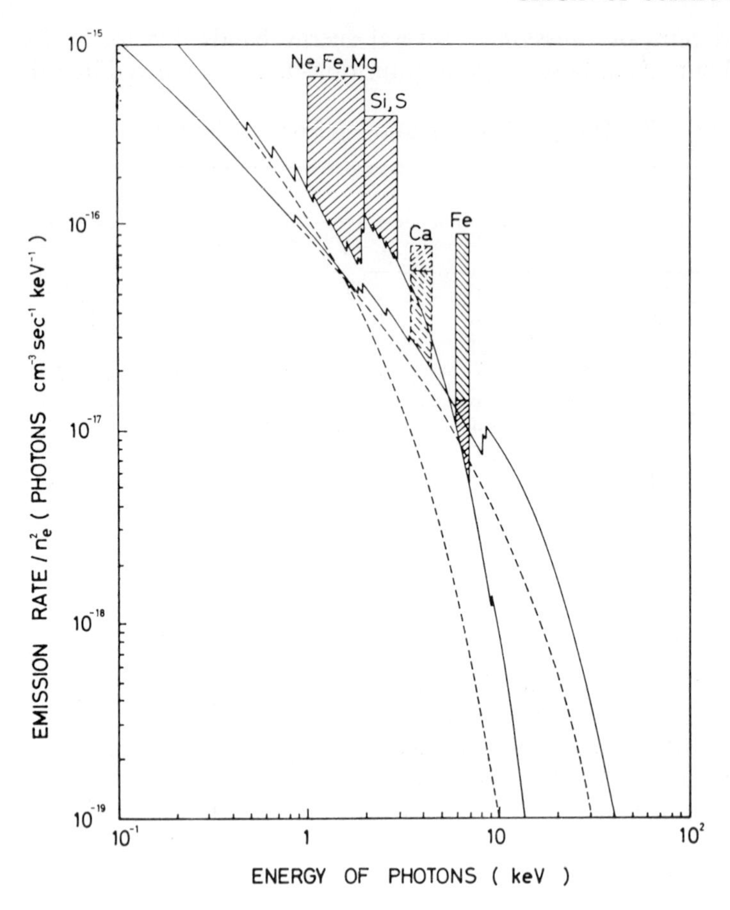

Fig. 6.40 Thermal emission rate of X-rays at 2×10^7 and 1×10^8 °K. Dashed curves—free-free emission; solid curves—free-free + free-bound emission; ▨ —bound-bound emission at 2×10^7 and 1×10^8 °K, respectively.

spectra in this case and emission from an optically thin plasma are characterized by the Rayleigh-Jeans law at low energies, and by an energy dependence of temperature, if the spectrum observed is fitted by a wrong assumption.

Both thermal spectra could be distinguished from the nonthermal spectra discussed in Section 6.8.4, because of different energy dependences. However the distinction between them is not always unambiguous. First, a nonthermal spectrum may fall sharply in the region of the cutoff energy. Second, the tail of the energy distribution of thermal particles may be tapered toward high energy due to the "runaway" phenomenon, since the energy loss by a Coulomb collision decreases as the energy of a particle increases. Since the second effect is a common feature of a tenuous plasma, the thermal spectrum does not necessarily have an exponential tail, but may have a tapered tail, as will be discussed in Section 6.9.

6.8.6 *Nuclear Gamma Rays*

Nuclear particles of nonthermal energies can produce nuclear γ-rays by nuclear interactions. The cross section for 14-MeV neutrons to produce γ-rays of energies between E and $E + dE$ is expressed approximately as

$$\sigma_{A\gamma}(E)\, dE \simeq 1.5 \times 10^{-24}(A^{1/3} - 2)^2 \exp\left(\frac{-E}{T}\right)\frac{dE}{T} \quad \text{cm}^2, \qquad (6.8.51)$$

where A is the mass number of a target nucleus and

$$T \simeq 4A^{-1/4} \text{ MeV}. \qquad (6.8.52)$$

This expression holds for $10 \leq A \leq 150$ and $1 \leq E \leq 8$ MeV, and its energy dependences for light nuclei show broad maxima around 20 MeV.

The excitation of particular lines may also make significant contributions. Among them the 4.4-MeV line of ^{12}C and the 6.1-MeV line of ^{16}O are expected to be the strongest. The excitation cross sections for these lines are

$$\sigma(^{12}\text{C}, 4.4 \text{ MeV}) \simeq 5 \times 10^{-25} \text{ cm}^2 \quad \text{at} \quad W_p \simeq 10 \text{ MeV}, \qquad (6.8.53a)$$

$$\sigma(^{16}\text{O}, 6.1 \text{ MeV}) \simeq 3 \times 10^{-25} \text{ cm}^2 \quad \text{at} \quad W_p \simeq 10 \text{ MeV}. \qquad (6.8.53b)$$

These cross sections also decrease as energy increases beyond 20 MeV. The 1.6-MeV line of ^{20}Ne may also give a comparable contribution, in view of the great abundance of neon and the similarity of its level structure to the 4.4-MeV level of ^{12}C.

The production rate of the continuous γ-rays is thus obtained as

$$p_\gamma(E) = \sum_{ZA} \int j_Z(W)\sigma_{A\gamma}(W, E) n_A \, dW$$

$$\simeq 1 \times 10^{-24} n_\text{H} \left(\sum_A \frac{n_A}{n_\text{H}}\right) \bar{T}^{-1} \exp\left(\frac{-E}{\bar{T}}\right) \quad \text{cm}^{-3}\,\text{sec}^{-1}\,\text{sr}^{-1}\,\text{MeV}^{-1}, \qquad (6.8.54)$$

where $\bar{T} \simeq 2$ MeV and $\sum_A n_A/n_\text{H} \simeq 1 \times 10^{-3}$, provided that the proton spectrum is as given by (6.4.6). The integration of (6.8.54) over the γ-ray energy gives a production rate smaller by one order of magnitude than that of 0.51-MeV γ-rays given in (6.8.7). The production rate of each strong line is also as small as 10^{-27} cm^{-3} sec^{-1} sr^{-1}.

The yield of γ-rays increases if the emission from cosmic-ray nuclei is taken into account. In this case the production rate may be expressed as

$$p_\gamma(E) \simeq 1 \times 10^{-24} n_{\mathrm{H}} \left(\sum_{Z \geq 6} \frac{j_Z}{j_p} \right) \bar{T}^{-1} \exp\left(\frac{-E}{\bar{T}}\right) \quad \mathrm{cm}^{-3}\ \mathrm{sec}^{-1}\ \mathrm{sr}^{-1}\ \mathrm{MeV}^{-1}. \tag{6.8.55}$$

Since $\sum_Z j_Z/j_p \simeq 5 \times 10^{-3}$, the production rate is slightly greater than that in (6.8.54). The yields of the line γ-rays also increase by one order of magnitude and become comparable to the production rate of 0.51-MeV γ-rays.

As a result the intensity of nuclear γ-rays produced in interstellar space is estimated to be about 10^{-4} cm^{-2} sec^{-1} sr^{-1} in the energy band between 1 and 10 MeV, depending, of course, on direction and angular resolution, as indicated in Table 6.29. It should, however, be kept in mind that the intensity of cosmic rays may increase with decreasing energy more rapidly than expected from (6.4.6), as mentioned toward the end of Section 6.4.3. The observation of nuclear γ-rays can therefore shed light on the energy spectrum of low-energy cosmic rays outside the solar system. Since the continuous γ-rays may be mixed with the high-energy tail of X-rays from other origins, the observations of γ-ray lines is suitable for this purpose.

There may be another line at 2.2 MeV due to the ^{1}H$(n, \gamma)^2$H reaction if strong irradiation by nuclear particles takes place on the stellar surface, since a considerable fraction of neutrons are captured before their decay in a medium of density higher than 10^{-9} g-cm^{-3}. However, such γ-rays are hardly observable except from the sun, since the angular size of such a source may be very small.

A greater flux of γ-rays is expected from a supernova remnant if the rapid neutron-capture processes associated with the supernova explosion produce a great amount of radioactive nuclides (Burbidge 57). The flux of γ-rays produced by radioactive decay is estimated for the Crab Nebula to be as large as 10^{-4} cm^{-2} sec^{-1} in the energy range between 60 keV and 1.8 MeV (Haymes 65). The strongest line is expected at 0.39 MeV arising from ^{249}Cf. The intensities of lines vary with time, depending on the lifetimes of radioactive nuclides.

6.8.7 X-Ray Sources

An unexpectedly high intensity of X-rays coming from space other than the sun was discovered by means of rocket-borne Geiger-Müller counters (Giacconi 62).

Space X-rays consist of two components—the general X-rays, which are nearly isotropic, and X-rays from discrete sources. The intensity of the former is observed to be as large as several photons cm^{-2} sec^{-1} sr^{-1} between 2 and 8 Å. Discrete sources with intensities greater than 0.2 photon cm^{-2} sec^{-1} have been resolved, as listed in Table 6.34 (Bowyer

Table 6.34 Positions and Upper Limits on Sizes of X-Ray Sources[a]

Sources	Position (1950) Right Ascension (α)	Position (1950) Declination (δ)	Flux[b] Counts (cm^{-2} sec^{-1})		Remarks
Tau XR–1	$05^h31.5^m$	22.0°	2.7		Size 1 min; coincident with Crab Nebula
Vir A	12^h28^m	12.7°		0.2*	
Sco XR–1	16^h15^m	[c] −15.2°	18.7		Size <20 sec
Sco XR–2	17^h08^m	[d] −36.4°	1.4		Size <30 min
Sco XR–3	17^h23^m	−44.3°	1.1		
Oph XR–1	17^h32^m	[e] −20.7°	1.3		1.1° from Keplers SN 1604[g]
Sgr XR–1	17^h55^m	[e] −29.2°	1.6		2.3° from Galactic center[g]
Sgr XR–2	18^h10^m	[f] −17.1°	1.5		
Ser XR–1	18^h45^m	5.3°	0.7		
Cyg XR–1	19^h53^m	34.6°	3.6	0.9*	
Cyg A	19^h58^m	40.6°		0.4*	Not found by a later observation
Cyg XR–2	21^h43^m	38.8°	0.8	1.0*	
Cas A	23^h21^m	58.5°		0.3*	

[a] Based mainly on results obtained in June 1964, (Bowyer 65) and in April 1965 (Byram 66). The flux values in the latter are indicated by aserisks.
[b] Uncorrected for atmospheric absorption. Measured with a 0.25-mil Mylar window.
[c] Position derived by (Clark 65): $\alpha = 16^h12^m$, $\delta = -15.6°$ or $\alpha = 16^h19^m$, $\delta = -14.0°$. Position derived by (Fisher 66)†: $\alpha = 16^h14^m \pm 1^m$, $\delta = -15°36' \pm 15'$.
[d] Position derived by (Clark 65): $\alpha = 16^h50^m$, $\delta = -39.6°$.
[e] Position derived by (Clark 65): $\alpha = 17^h44^m$, $\delta = -23.2°$.
[f] This source was resolved as two sources by (Fisher 66).†
[g] No identification of Kepler's SN 1604 and of the Galactic center with X-ray sources has been confirmed by (Clark 65) and (Fisher 66).†
† P. C. Fisher et al., *Astrophys. J.*, 143, 203 (1966).

65, Clark 65*b*, Byram 66). They designated in order of strength. For example, Sco XR–1, means the strongest X-ray source in Scorpio.

Few of these X-ray sources have been found to coincide with noticeable optical and radio objects. The coincidence of Tau XR–1 with the Crab Nebula, the remnant of SN 1054, has been verified by repeated observations, especially during lunar occultation (Bowyer 64). Radio sources Cas A, Cyg A, and Vir A are also found to be X-ray sources by one observation (Byram 66), but not always by others. This would suggest that X-ray sources could be identified with active objects such as supernova remnants and radio galaxies, although the identification has not yet been successful for Oph A, Sgr A, and other radio sources. A blue star of magnitude 12.5 seems to have been identified with Sco XR–1, the strongest X-ray source (Sandage 66).

Although the coincidence of X-ray sources with visible objects is rather poor, there seem to be at least three types of X-ray sources; namely, supernova remnants, radio galaxies, and the Sco XR–1 type. Past supernovae recorded in history (Hsi 64) and supernova remnants identified with radio sources in our Galaxy (Harris, 62) are plotted together with X-ray sources on a sky map in Fig. 6.41. Most of them lie in, or close to, the Galactic plane, thus indicating that they belong to our Galaxy. Sco XR–1 lies appreciably away from the Galactic plane, but may be inside the disk if its distance is not greater than several hundred parsecs.

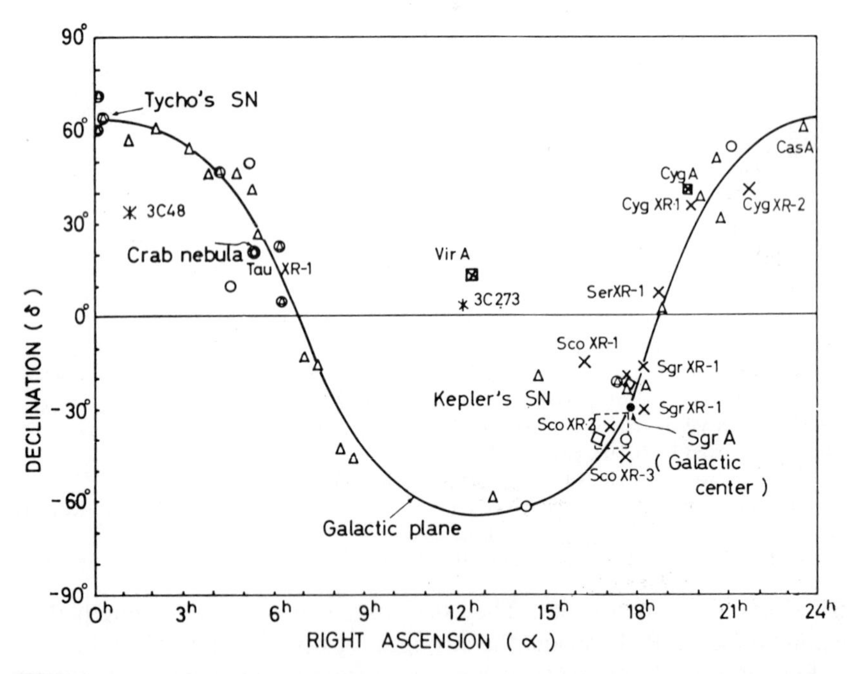

Fig. 6.41 Positions of X-ray sources (×), Galactic radio sources △), past supernovae recorded (○, ⬚), radio galaxies (□), and QSS (✳).

The angular size of an X-ray-emitting region is located in the central part of the Crab Nebula. The size is about 1 minute of arc, corresponding to a linear dimension of 1 light year. This implies that the radius at which the intensity decreases to one-half the maximum value decreases as the energy of photons increases; radii are about 2 and 1 light years for radio and optical emissions, respectively. However, this does not necessarily imply that X-ray emission is due to the same mechanism as the optical and radio emissions—synchrotron radiation—since the intensity contours of the optical and radio emissions are flattened toward the center, thus being in favor of the concentration of synchrotron radiation in the expanding shell.

The energy fluxes from the Crab Nebula in the X-ray, the optical, and the radio regions are 3×10^{-8}, 8×10^{-9}, and 3×10^{-12} erg-cm^{-2} sec^{-1}, respectively. It is rather surprising that the energy flux in the X-ray region is the highest. Using the distance of the Crab Nebula of 1.1 kpc, we estimate the energy-generation rate in the X-ray region to be 4×10^{36} erg-sec^{-1}.

The features of Sco XR–1 are different, particularly in its angular size. Its angular size is found to be less than 16″, possibly as small as or smaller than 8″ (Gursky 66). Its distance is estimated to be about 200 pc, and consequently its size would not exceed 10^{-2} pc. This would indicate a small X-ray-emitting region comparable to the size of the planetary system.

In connection with the fact that most X-ray sources are not detectable in the optical and radio region, we have to answer a question whether or not an X-ray source can emit optical or radio radiation that is strong enough for detection, because any mechanism of X-ray emission is associated with the emission of lower energy radiation.

For the comparison of the radiation intensities in different spectral regions, it is convenient to use the flux in the same relative bandwidth. The flux intensity for our purpose is thus defined as

$$I(\varepsilon)\frac{\Delta\varepsilon}{\varepsilon} = \int_{\Delta\varepsilon} F(\varepsilon)\, d\varepsilon, \tag{6.8.56}$$

where $F(\varepsilon)$ is the energy flux in unit energy range. For $\Delta\varepsilon/\varepsilon = 1$ the values of $I(\varepsilon)$ are compared. Most of the X-ray sources listed in Table 6.34 have flux intensities of

$$I \gtrsim 10^{-8} \text{ erg-cm}^{-2} \text{ sec}^{-1} \tag{6.8.57}$$

in the energy region around 4 keV.

For a source to be undetectable by optical observation it has to be fainter than a certain limit of optical detection. The detection limit depends on the angular diameter of the object and the luminosity of the nearby field. Since X-ray sources lie in the Galactic plane, the field luminosity is high, and the faintest nebula observed in this region is Cyg A of $m_{pg} = 15.1$. If an X-ray source has a larger angular diameter than the optical size of Cyg A, the detectable limit may be as bright as $m_{pg} \simeq 12$. This means that the optical flux intensity from an X-ray source would be

$$I_0 < 10^{-10} \text{ erg-cm}^{-2} \text{ sec}^{-1} \tag{6.8.58}$$

in the energy region around 3eV. Under fortunate conditions the upper limit will be reduced by orders of magnitude.

For radio sources the same considerations apply as in the case of optical sources. Toward the Galactic plane a radio source detectable without too much difficulty may have an intensity higher than 10^{-25} W-m^{-2} Hz^{-1} around 100 MHz. For a spectral index of about 0.7, this gives

$$I_r < 10^{-14} \text{ erg-cm}^{-2} \text{ sec}^{-1} \tag{6.8.59}$$

in the energy region around 1 μeV. The upper limit will be reduced by more careful searching at higher frequencies.

The flux range of X-rays and the upper limits of optical and radio fluxes are shown in Fig. 6.42. The curves drawn in the figure represent the fluxes expected from the free-free transition at a temperature of 10^8 °K and from blackbody radiation at 5×10^7 °K adjusted to fit the X-ray flux. The former is linear in energy in the energy range concerned.

The comparison is also made for the X-ray region of the Crab Nebula. The optical and radio fluxes from this region are estimated as 2×10^{-10} and 3×10^{-14} erg-cm^{-2} sec^{-1}, respectively. These values lie above the upper limits given in (6.58) and (6.59), as plotted in Fig. 6.42.

The energy spectra in the X-ray region have been measured between 1 and 100 keV for Sco XR–1, Tau XR–1, and others.

The Tau XR–1 spectrum shown in Fig. 6.43 is represented by a power law as

$$J(\text{Tau XR–1}, \varepsilon) \simeq 18\varepsilon^{-\beta-1} \quad \text{cm}^{-2} \text{ sec}^{-1} \text{ keV}^{-1}, \tag{6.8.60}$$

with $\beta = 1.3$. The power law is favored by a significant flux above 30 keV observed by means of balloons.

The rate of energy generation is determined by the flux above 1 keV and is about 4×10^{36} erg-sec^{-1} or on the order of 4×10^{-18} erg-cm^{-3} sec^{-1} in the X-ray-emitting region. If the spectrum (6.8.60) is extended to the optical region, its luminosity exceeds the total optical luminosity of the Crab Nebula. If, on the other hand, the spectrum below 1 keV

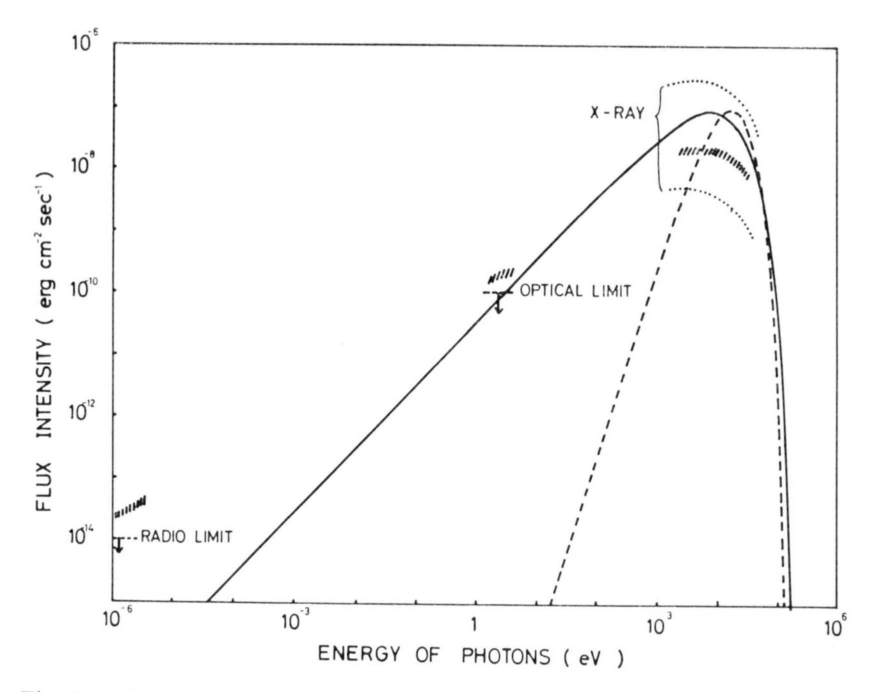

Fig. 6.42 Energy fluxes of radiation in the X-ray, optical, and radio bands. Solid curve—the spectrum expected from the free-free emission at 10^8 °K; dashed curve—the blackbody spectrum of 5×10^7 °K; //////—flux from the Crab Nebula.

is represented as $d\varepsilon/\varepsilon$, the optical luminosity is as faint as that of the central star.

If the X-ray emission is due to synchrotron radiation, the electron spectrum is very steep at energies higher than $10^2/\sqrt{H_\perp}$ GeV, the integral exponent being $\alpha \simeq 2.6$. This requires the electron flux of $(10^{-2}/H)\,\text{cm}^{-2}\,\text{sec}^{-1}\,\text{sr}^{-1}$ for energy higher than $10^2/\sqrt{H}$ GeV. This could be smoothly connected with the spectrum of electrons responsible for radio and optical emission if the spectrum becomes less steep as energy decreases and the magnetic field is stronger in the X-ray-emitting region; for example, $\alpha = 2$ between 10^2 and 10^3 GeV and $H = 10^{-2}$ gauss. Difficulties are caused by the length of electron lifetime and the magnetic energy; the lifetime is shorter than a few days with the above choice of parameters according to (6.7.11), and the total magnetic energy is as large as 10^{49} ergs. Even if the magnetic field is as small as 10^{-3} gauss, the lifetime is so short (about 1 year) that the acceleration of electrons has to be exceedingly efficient.

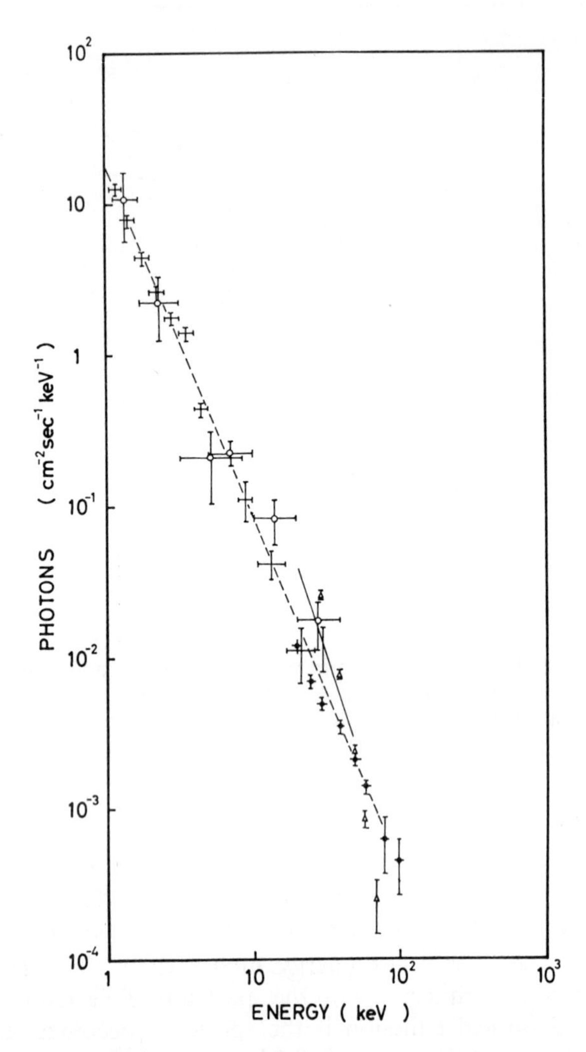

Fig. 6.43 X-ray spectrum of Tau XR–1. The dashed line represents a power law approximate $j(\varepsilon) = 18/\varepsilon^{2.3}$ cm^{-2} sec^{-1} keV^{-1}. ┼ —R. J. Grader et al., *Science*, **152**, 1499 (1966); ○ —S. Hayakawa et al., *Space Research*, VII, 1306 (1966); ● —L. E. Peterson, A. S. Jacobson, and R. M. Pelling, *Phys. Rev. Letters*, **16**, 142 (1966); △—R. C. Haymes and W. L. Craddock, Jr., *J. Geophys. Res.*, **71**, 2361 (1966); ╲—G. W. Clark, *Phys. Rev. Letters*, **14**, 91 (1965).

If synchrotron radiation is mainly responsible for X-ray emission, there may be a close relation between radio luminosity and X-ray intensity (Aizu 65). Assuming their proportionality, we can estimate the X-ray intensities for radio sources, as given in Table 6.35. These are

Table 6.35 Expected X-Ray Fluxes from Radio Sources

Radio Source	Radio Luminosity[a] (erg-sec^{-1})	X-Ray Luminosity[b] (erg-sec^{-1})	X-Ray Intensity[b] Photons cm^{-2} sec^{-1}	
			Estimated	Observed
Crab Nebula	1.0×10^{34}	[c]3.0×10^{36}	[c]2.7	[c]2.7
Cas A	2.5×10^{35}	7.5×10^{37}	7.0	0.3
3C 273B	5.8×10^{43}	1.7×10^{46}	0.086	—
Cyg A	3.4×10^{44}	1.0×10^{47}	2.9	0.4
Vir A (core)	1.3×10^{41}	3.9×10^{42}	0.3	0.2
Cen A (central)	5.5×10^{40}	1.7×10^{43}	0.9	—
Sgr A (central)	4.8×10^{35}	1.4×10^{38}	1.1	—

[a] Integrated over the frequency range between 10^7 and 10^{10} Hz, as given by (K. Aizu 64b).
[b] In the wavelength range between 1.5 and 8 Å.
[c] Normalized.

normalized to the X-ray intensity of the Crab Nebula, and the intensities of others seem to be too high. This may be understood if we consider that the radio spectra are steeper than in the Crab Nebula, so that they may give weaker intensities if extended to the X-ray region.

The inverse Compton effect would give a very small X-ray yield because of the optical photon density of about 1 cm^{-3}. If, however, resonance lines are excited as strongly as the H_α line, the density of resonant photons may be as large as

$$n_{\rm res} = n_0 \, \tau_{\rm res}, \tag{6.8.61}$$

where $\tau_{\rm res}$ is the optical depth for the resonance line given in (6.8.49). For the L_α line the optical depth is as large as $10^5 n_{\rm H}$, where $n_{\rm H}$ is the density of hydrogen atoms in the ground state. If the emission rate of L_α is comparable to that of the optical continuum, the X-ray emission could be economically accounted for in terms of the inverse Compton effect.

If a substantial part of the energy of a supernova—say, 10^{47} ergs—is eventually thermalized, the temperature becomes as high as 10^7°K. Since the lifetime for the free-free transition is considerably large,

$$\tau_{ff} \simeq 1 \times 10^{11} \frac{T^{1/2}}{n_e} \text{ sec,} \tag{6.8.62}$$

the efficiency for X-ray emission is rather poor unless the electron density is as high as 10^4 cm^{-3}. It is also difficult to prevent such a hot plasma from expanding (Hayakawa 66).

The Sco XR–1 spectrum is shown in Fig. 6.44, based on experimental data obtained by several groups. Unlike the spectra of other sources thus far measured, this is approximately expressed by the free-free transition spectrum as

$$\gamma(\text{Sco XR–1}, \varepsilon) \simeq 1.2 \times 10^2 \exp\left(-\frac{\varepsilon}{kT}\right) \varepsilon^{-1} \quad \text{cm}^{-2}\ \text{sec}^{-1}\ \text{keV}^{-1},$$

$$T \simeq 5 \times 10^7\ °\text{K}. \tag{6.8.63}$$

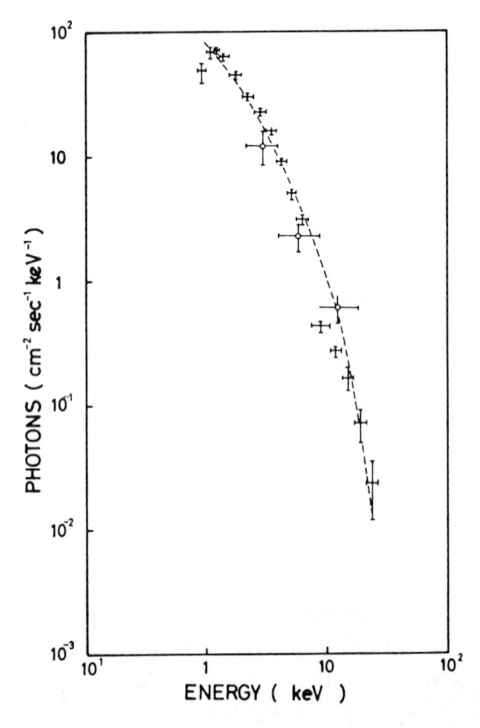

Fig. 6.44 X-ray spectrum of Sco XR–1. The dashed line represents an exponential law approximation $j(\varepsilon) = 1.2 \times 10^2 \exp(-\varepsilon/kT)\varepsilon^{-1}$. cm^{-2} sec^{-1} keV^{-1}, with $T = 5 \times 10^7$ °K. ┼ —R. J. Grader et al., *Science*, **152**, 1499 (1966); ⌖ — S. Hayakawa, M. Matsuoka, and K. Yamashita, *Proc. Int. Conf. on Cosmic Rays in London*, **1**, 119 (1965).

The spectrum appears to fall off toward low energy. This might be regarded as indicating the blackbody spectrum, but the spectrum that fits the low-energy part decreases too steeply toward high energy. Moreover, the slope at high energy could be accounted for only in terms of an energy-dependent temperature. The low-energy behavior may be explained by absorption if the X-rays traverse a gas of 3×10^{21} atoms cm^{-2}, according to Fig. 6.26.

If this is accepted, the absorption may be due to that taking place in the source or to a high abundance of hydrogen molecules in the Galactic disk, since the path length to Sco XR–1 is less than 10^{21} cm.

The rate of energy generation is estimated as several times 10^{36} erg-sec^{-1}, and the rate of energy generation per volume would be enormous, as large as several times 10^{-14} erg-cm^{-3} sec^{-1}. This may be compared with 4×10^{-18} erg-cm^{-3} sec^{-1} for Tau XR–1.

The X-ray emission may be accounted for in terms of thermal emission if $T \simeq 5 \times 10^7$ °K and $n_e \simeq 10^5$ cm^{-3}. The energy density in the X-ray-emitting region is as high as 1 GeV cm^{-3}. This is much higher than that required for synchrotron radiation, but the rate of energy supply is much smaller, since the cooling time is about 10^{11} sec compared with the electron lifetime of 10^6 sec against synchrotron radiation. The lifetime of such a hot plasma is determined by instability rather than by cooling. Unless the instability is suppressed, the lifetime of the plasma would be as short as 10 years. There are many peculiar features in the optical counterpart of Sco XR–1. It has emission lines but no absorption lines. The intensities of both continuum and lines vary in time, a few percent within minutes and a fraction of one magnitude within days, but with little correlation between different colors and lines.

6.8.8 Isotropic X-Rays

In addition to the X-rays from localized sources, X-rays are found to be strong in all directions. Since the angular resolution of detectors thus far used is not good enough, the general X-rays could be attributed to the sum of X-rays from many localized sources. Since the general X-rays are essentially isotropic, in contrast to the distribution of localized X-ray sources concentrated in the Galactic plane, it is unlikely that they are of Galactic origin. It is therefore likely that the general X-rays are of metagalactic origin.

The intensity and the energy spectrum of isotropic X-rays are shown in Fig. 6.45. This may be expressed approximately by

$$j(\varepsilon) \simeq \frac{18}{\varepsilon^2} \quad \text{cm}^{-2}\,\text{sec}^{-1}\,\text{sr}^{-1}\,\text{keV}^{-1} \tag{6.8.64}$$

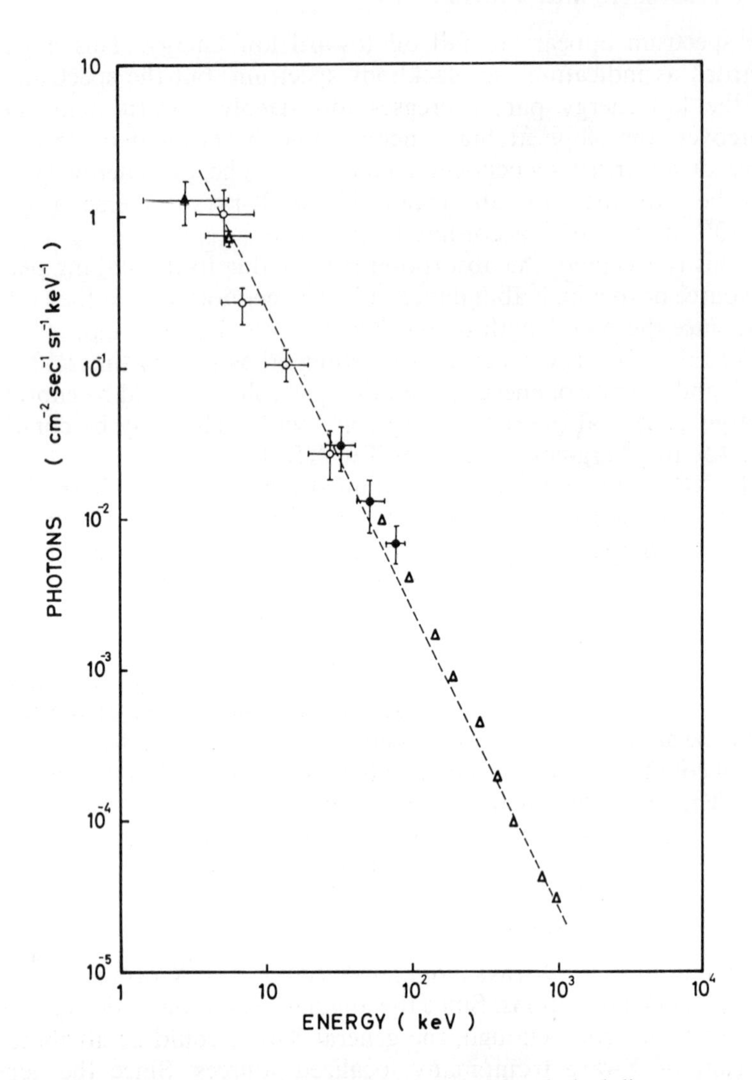

Fig. 6.45 Energy spectrum of isotropic X-rays. The dashed line represents a power law approximation $j(\varepsilon) = 18/\varepsilon^2$ cm^{-2} sec^{-1} sr^{-1} keV^{-1}. ▲ —S. Bowyer et al., *Nature*, **201**, 1307 (1964); △ —P. C. Fisher et al., *Astrophys. J.*, **143**, 203 (1966); ● J. A. M. Bleeker et al., *Phys. Letters*, **21**, 301 (1966); △ A. E. Metzger et al., *Nature*, **204**, 766 (1964); ○ —S. Hayakawa, M. Matsuoka, and K. Yamashita, *Rep. Iones. Space Res. Japan*, **20**, 480 (1966).

for 10 keV $< \varepsilon <$ 100 keV. It tends to be flattened as energy decreases below 10 keV.

It is worthwhile remarking that the flux of γ-rays in the MeV region may possibly lie on the extension of the spectrum (6.8.64).

Although the spectral shape is not finalized yet, the spectrum of isotropic X-rays may be a mixture of the Tau XR–1 type and the Sco XR–1 type. It is therefore natural to ask whether the isotropic X-rays are the sum of X-rays from many localized sources in all galaxies (Hayakawa 66).

If there are N_g X-ray sources per galaxy and their average X-ray yield is Q_x, such X-ray sources in all galaxies in the universe give the isotropic intensity

$$j_{\text{iso}} = \frac{1}{4\pi} N_g Q_x n_g R_U, \tag{6.8.65}$$

where $n_g \simeq 2 \times 10^{-75}$ cm^{-3} is the density of galaxies and $R_U \simeq 10^{28}$ cm is the Hubble radius. If the yield of Tau XR–1 is a standard value, we can put

$$Q_x = 3 \times 10^{44} \text{ photons-sec}^{-1} \tag{6.8.66}$$

using the flux value given in Table 6.34.

The number of X-ray sources in our Galaxy depends on the assumption, on the basis of which N_g is estimated. If the sources are distributed uniformly on the Galactic disk and only those which are located within distances R_x are detectable, we obtain the number of X-ray sources in our Galaxy

$$N_G = N\,(\text{observed}) \left(\frac{R_G}{R_x}\right)^2, \tag{6.8.67}$$

where R_G is the radius of the disk and N (observed) is the number of X-ray sources observed. Since about 20 sources have been found possibly belonging to the Sagittarius and Orion arms, we may put N (observed) $=$ 20 and $R_x = 2.5$ kpc. Thus we obtain

$$2\,N\,(\text{observed}) \left(\frac{R_G}{R_x}\right)^2 \simeq 10^3, \tag{6.8.68}$$

where the factor 2 appears because only half the Galactic plane has been scanned with sufficient care. Assuming that the value of N_G represents average number of X-ray sources per galaxy, we put

$$N_G = N_g. \tag{6.8.69}$$

Although these assumptions ($N_g = N_G = 10^3$ and $Q_x = 3 \times 10^{44}$ sec^{-1}) are subject to future critical examination, a tentative estimate of the intensity of general X-rays is

$$j_{\rm iso} \simeq 0.5 \text{ photons cm}^{-2} \text{ sec}^{-1} \text{ sr}^{-1}. \qquad (6.8.70)$$

This is smaller by one order of magnitude than the observed value, in spite of a possible overestimate of N_g.

However, the X-ray production rate of $N_g Q_x$ per galaxy used above is not always an overestimate and may be an underestimate for radio galaxies like Cyg A and Vir A. If the X-ray emission rate is proportional to the radio power (Aizu 65), the intensity of isotropic X-rays can be estimated by taking the X-ray emission of a radio source as a standard and comparing its radio power with the isotropic radio intensity given in (6.7.115). The upper limits of X-ray emission from radio galaxies indicate that their contribution could hardly give an isotropic intensity exceeding that given in (6.8.70).

Since the summation of all X-ray sources does not seem to give the isotropic intensity comparible to the observed value, we next investigate the X-ray emission in intergalactic space. The thermal bremsstrahlung in intergalactic space would give strong enough X-rays if the electron temperature and density were 1.5×10^8 °K and 3×10^{-6} cm^{-3}, respectively (Hayakawa 64*c*). Although the kinetic pressure of such an intergalactic gas is high enough to balance the pressure of interstellar gas in our Galaxy, the temperature adopted seems too high to be consistent with various physical conditions (Field 64).

The inverse Compton effect in intergalactic space can also produce isotropic X-rays (Felten 63). Collisions with starlight photons are insufficient to account for the X-ray intensity unless the intensity of intergalactic electrons is as high as that in the Galaxy. The collisions with microwave photons give

$$j_{\rm iso}(\text{inverse Compton}) \simeq 0.5\varepsilon^{-2.2} \quad \text{cm}^{-2} \text{ sec}^{-1} \text{ sr}^{-1} \text{ keV}^{-1} \qquad (6.8.71)$$

if we adopt the electron spectrum given in (6.7.113). Although the slope is in good agreement with the observed power spectrum, the absolute intensity is smaller by a large factor than the observed one given in (6.8.64). It should, however, be noted that the absolute flux of electrons estimated in (6.7.113) is subject to considerable uncertainty and the cosmological effect is not fully taken into account.

All the above estimates of the isotropic flux are smaller by one order of magnitude than the observed one. The disagreement should not be taken too seriously, since the values of parameters adopted in our estimates are rather arbitrary. The discussions above suggest that the intensity of isotropic X-rays will serve to obtain several quantities of cosmological importance.

6.8.9 Observation of Gamma Rays

As seen in Fig. 6.45, the flux of X-rays could be smoothly extended to the MeV region with a power spectrum expressed as in (6.8.64) (Metzger 64). The flux value in the MeV region is therefore found to be about 10^{-2} cm^{-2} sec^{-1} sr^{-1}, although the absolute flux of γ-rays in this energy region is subject to large background effects arising from nearby materials, the terrestrial atmosphere, and the detector itself.

If the intensity of isotropic X-rays in the keV region is accounted for in terms of inverse Compton collisions with microwave photons in intergalactic space, multiplying a large factor by a theoretical estimate in (6.8.71), the flux in the MeV region is explicable in the same terms. This would indicate that the flux of intergalactic electrons could be about one-hundredth of that of Galactic electrons.

This would also result in a rather high intensity of intergalactic protons if the proton-electron ratio is universal and the production rate of nuclear γ-rays is about one-hundredth of the Galactic value estimated in (6.8.55), or

$$q_\gamma(\text{intergalactic}) \simeq 10^{-28} n_{\mathrm{H}} \quad \text{cm}^{-3}\,\text{sec}^{-1}\,\text{sr}^{-1}. \tag{6.8.72}$$

This gives a γ-ray flux of about 10^{-5} cm^{-2} sec^{-1} sr^{-1} for $n_{\mathrm{H}} = 10^{-5}$ cm^{-3}. This is comparable to the estimated flux of Galactic γ-rays but much smaller than the observed upper limit.

The same argument would apply also for the γ-rays from the decay of neutral pions. Then the contribution of intergalactic γ-rays would be comparable to that of Galactic γ-rays. However, the high intergalactic-Galactic ratio of cosmic rays is rather questionable, according to the argument given in Section 6.4.5. Thus we take the ratio of 3×10^{-4} and draw the intensity of intergalactic γ-rays in Fig. 6.39.

In the same figure we draw the intensity of Galactic γ-rays derived from (6.8.9) with $\overline{n_{\mathrm{H}} R} = 3 \times 10^{21}$ cm^{-2}. The average intensity of the pion-decay γ-rays given in (6.8.11) may be compared with the observed upper limit (Kraushaar 65)

$$j_\gamma(\gtrsim 100\ \text{MeV}) < 3 \times 10^{-4}\ \text{cm}^{-2}\,\text{sec}^{-1}\,\text{sr}^{-1}, \tag{6.8.73}$$

which is about 10 times higher than the expected one. If the energy spectrum of intergalactic electrons is extended to 5×10^{11} eV with the high absolute value mentioned above, the γ-ray flux expected from the inverse Compton effect is just as large as the upper limit given in (6.8.73).

If the electron spectrum is further extended to 3×10^{14} eV, the inverse Compton effect could give high-energy γ-rays of a very high intensity. If intergalactic space were transparent for such high-energy γ-rays, the flux could be as high as 10^{-10} cm^{-2} sec^{-1} sr^{-1}; that is, only about one-hundredth of the proton flux at about 10^{14} eV. However, strong intergalactic absorption starts at about this energy, as shown in Fig. 6.37, and the γ-ray flux should be reduced by a large factor.

Above 3×10^{14} eV, the inverse Compton effect becomes relativistic, so that the γ-ray spectrum falls off very steeply; the flux of inverse Compton γ-rays with energy greater than 3×10^{15} eV is about 10^{-4} of that with energies 3×10^{14} eV.

Above 10^{15} eV pairs of electrons produced by the collision of high-energy γ-rays with microwave photons have a mean free path comparable to that of γ-rays and radiate high-energy γ-rays by inverse Compton collisions. These processes take place successively until the energy of an electron falls below $E_c \simeq 10^{15}$ eV, below which the range of the electron is smaller than the mean free path for pair creation, and a cascade shower of electrons and γ-rays is formed. This results in an increase of the effective absorption length of photons. However, the increase in the γ-ray flux does not exceed a factor of E/E_c.

In such a high-energy region the observation of γ-rays has been attempted by detecting few muon air showers, as described in Section 5.8.4. If we take the lower value of two experimental results (Toyoda 65), the flux of γ-rays with energies between 10^{15} and 10^{16} eV would be about 3×10^{-14} cm^{-2} sec^{-1} sr^{-1}. This is what we would expect from the inverse Compton effect without intergalactic absorption. However, intergalactic absorption should cut down the γ-ray intensity by a large factor. No indication of γ-rays has been observed above 10^{17} eV.

In comparison with isotropic γ-rays, those from point sources may be observed more reliably. The available results have given only upper limits, as shown in Table 6.36. They may be compared with those estimated in (6.8.16). By using the radio fluxes and the magnetic-field strengths given in Tables 6.27, and 6.28, the expected intensities of γ-rays are compared with the observed upper limits in Table 6.37. Here an undetermined factor κn_{H} enters into the γ-ray flux. The hydrogen density may be considerable—say on the order of 10^2 cm^{-3}—in supernova remnants, whereas it may be lower in radio galaxies. The proton-electron ratio, κ, is uncertain, but may be inferred to be about 10^2 for a radio spectral index greater than 0.7. Even if we make maximum esti-

Table 6.36 Upper Limits of γ-Ray Fluxes

Energy of γ-Rays (eV)	Upper Limit of the Flux: From Sources ($cm^{-2}\ sec^{-1}$))	Upper Limit of the Flux: Isotropic ($cm^{-2}\ sec^{-1}\ sr^{-1}$)	Reference
$\gtrsim 10^8$	$(2 \text{ to } 23) \times 10^{-4}$	3×10^{-4}	1
$\gtrsim 10^8$	$(2 \text{ to } 10) \times 10^{-5}$	—	2
$\gtrsim 10^9$	$(1 \text{ to } 6) \times 10^{-4}$	—	3
$\gtrsim 10^{11}$	$(3 \text{ to } 80) \times 10^{-7}$	—	4
$\gtrsim 5 \times 10^{12}$	$(5 \text{ to } 20) \times 10^{-11}$	—	5
$\gtrsim 1 \times 10^{13}$	$(0.8 \text{ to } 14) \times 10^{-11}$	—	6

References:
1. W. Kraushaar et al., *Astrophys. J.*, **141**, 845 (1965).
2. R. Cobb et al., *Phys. Rev. Letters*, **15**, 507 (1965).
3. H. B. Ögelman, *Phys. Rev. Letters*, **16**, 491 (1966).
4. Y. Sekido et al., *Proc. Int. Conf. on Cosmic Rays at Jaipur*, **3**, 194 (1963).
5. A. E. Chudakov et al., *Proc. Int. Conf. on Cosmic Rays at Jaipur*, **3**, 199 (1963).
6. C. D. Long et al., *Proc. Int. Conf. on Cosmic Rays at London*, **1**, 318 (1965).

Table 6.37 Intensities of γ-Rays from Radio Sources[a]

Source	$E \gtrsim 10^8$ eV: Estimated	$E \gtrsim 10^8$ eV: Upper Limit[b]	$E \gtrsim 5 \times 10^{12}$ eV: Estimated	$E \gtrsim 5 \times 10^{12}$ eV: Upper Limit[b]
Tau A	$2 \times 10^{-10} \kappa n_H$	2×10^{-5} (2)	$1 \times 10^{-12} \kappa n_H$	5×10^{-11} (5)
Cas A	$2 \times 10^{-9} \kappa n_H$	2×10^{-3} (1)	$2 \times 10^{-16} \kappa n_H$	
Vir A	$5 \times 10^{-10} \kappa n_H$	3×10^{-4} (1)	$1 \times 10^{-13} \kappa n_H$	
Cyg A	$5 \times 10^{-9} \kappa n_H$	5×10^{-4} (1)	$3 \times 10^{-18} \kappa n_H$	

[a] In $cm^{-2}\ sec^{-1}$.
[b] Figures in parentheses indicate reference in Table 6.36.

mates, the estimated flux for $E \gtrsim 10^8$ eV is smaller by one or more orders of magnitude than the observed upper limits.*

At higher energies the estimated γ-ray flux depends on the energy spectrum that is assumed. The γ-ray flux for $E \gtrsim 5 \times 10^{12}$ eV is estimated by extrapolating the electron spectrum with constant slope. This method is approximately correct if the production spectrum in Fig. 6.38 is taken

* The upper limits for some sources were, by the summer of 1967, reduced by an order of magnitude in comparison with those in Table 6.37.

into account, as long as the radio spectrum is extrapolated to high frequency. For Tau A a bending of the synchrotron-radiation spectrum is observed. This takes place at an electron energy of about 10^{12} eV. Since the bending of the proton spectrum may take place at a higher energy, the γ-ray flux for $E > 5 \times 10^{12}$ eV is estimated without regard to the bending.

The γ-ray flux of the Crab Nebula estimated for $E > 5 \times 10^{12}$ eV could exceed the observed upper limit if $\kappa n_{\mathrm{H}} \gtrsim 10^2$. This supports a small proton-electron ratio, $\kappa \simeq 1$, as discussed in Section 6.7, and supports one model (Hayakawa 58*b*) over others; namely, that electrons are accelerated with nearly the same efficiency as protons.

It may be worthwhile to mention that the inverse Compton effect may compete with the meson-decay effect at such high energies. The high-energy electrons collide with radio-wave photons and produce γ-rays. The intensity of γ-rays with energies greater than E is related to the radio flux at 100 MHz by

$$j_\gamma(>E) \simeq 1 \times 10^{-14}[\phi(\text{min})]^{-2}[H(10^{-4}\text{ gauss})]^{\alpha/2}$$
$$[I_{100}(10^{-22}\text{ W-}m^{-2}\text{Hz}^{-1})]^2[E(\text{GeV})]^{-\alpha/2}, \quad (6.8.74)$$

where ϕ is the angular diameter of the source concerned. This is negligible for $E \simeq 1$ GeV, but could be comparable to that given in Table 6.37 if the electron spectrum were extended to high energy with a small value of the integral exponent α. However, bending of the electron spectrum takes place for the Crab Nebula and possibly for other sources such as the core of Vir A.

6.9 Acceleration Mechanisms

The discovery of a particle of ultrahigh energy, measurable in macroscopic units, came as a great surprise. It was at first thought that the particle had been accelerated accidentally by a peculiar event. It has subsequently been realized that such highly energetic cosmic rays are not accidental but are part of the very nature of cosmic space. It has also been recognized that nonthermal particles are in general associated with a rarefied plasma.

The reason that cosmic rays are so common in nature may be stated in the following way. The equilibrium of a system is not determined by the energy exchange due to the collisions between particles alone but is greatly affected by that due to the collisions between particles and turbulent units. For low-energy particles the particle collisions are overwhelming, whereas for high-energy particles the mean free path for the energy exchange by particle collisions may be longer than that for collisions with turbulent units. These particles can gain energies higher than those expected from the Maxwell distribution and are considered as nonthermal or suprathermal in the sense that particles obeying the Maxwell distribution are thermal. For the suprathermal particles an equilibrium is reached when the total energy of these particles is equal to whole turbulent energy in a given volume.

Actually, however, the time to reach equilibrium is in general long compared with the mean lifetimes of particles and turbulence in the spatial region under consideration. What we observe is therefore not an equilibrium state but a stationary state. This was first demonstrated by Fermi (Fermi 49), as briefly described in Section 2.1.

Questions have been raised as to whether or not a correct value of $a\tau$ can be obtained under actual astrophysical conditions, and whether or not particles can gain energies high enough for (6.2.1) to hold. In interstellar space, at least in the disk of our Galaxy, the answers have been found negative for several reasons. It is, however, unnecessary to attribute the entire acceleration to turbulence in interstellar space; for example, there are the solar particles accelerated at flares, although the efficiency is not enough to account for the observed cosmic rays. There is good evidence that the Crab Nebula, a supernova remnant, contains a great amount of relativistic electrons, which are believed to be accelerated more efficiently.

Thus we are led to assume the sequence that particles are injected within a short period by a strong magnetic disturbance and are accelerated efficiently in a local, active region. Then they escape to general space, in which acceleration may be moderate. In the Galaxy there are at least three stages in this sequence, exemplified by explosive disturbances at stars, active nebulosities of supernova remnants, and interstellar space, respectively. If we consider the whole universe, they may be replaced by explosions in galactic nuclei, radio galaxies, and general metagalactic space, respectively.

In all stages of the sequence there seem to be common mechanisms of acceleration due to energy exchange between charged particles and magnetic fields. They are classified into *secular* and *statistical* processes. In the former a particle gains energy continuously in time; whereas in the latter the gain and the loss of energy take place in a random manner, but as a result the energy of some particles increases. The secular acceleration results in a rapid increase in energy and may be responsible for the injection process, whereas the statistical acceleration is so slow that it is effective mainly in a large volume. In both the secular and statistical processes the coupling of a particle with a magnetic field takes place in two ways. In one, there is a variation of the momentum component perpendicular to the magnetic field, in the other there is a variation of the component parallel to it. These processes are often called the *betratron* and the *Fermi processes*, respectively.

6.9.1 Statistical Acceleration

Statistical acceleration consists of individual acceleration and deceleration processes that are perturbed by modulation. The inverse of an acceleration process causes a loss of energy equal to the gain of energy, and no net energy change results after their simple average. Owing to a random modulation, however, they do not always compensate for each other, and in general a small amount of energy gain is left. This was illustrated by Fermi (Fermi 49) for the collisions of cosmic rays with magnetic clouds.

Let us consider a cosmic-ray particle of velocity c colliding with a cloud of velocity u. If the collision is head on, the particle gains energy by a fraction $2u/c$, whereas in an overtaking collision the particle loses energy by a fraction $2u/c$. If these two kinds of collisions take place equally, the particle neither gains nor loses energy. Actually, however, the frequency of collisions is proportional to the relative velocity, $(c \pm u)/[1+(u/c)]$, according to whether the collision is head on or overtaking. As a result of their average, therefore, the fractional energy change is

$$2\left(\frac{c+u}{1+u/c}\frac{u}{c}-\frac{c-u}{1+u/c}\frac{u}{c}\right)\Big/\frac{2c}{1+u/c}=2\left(\frac{u}{c}\right)^2.$$

If such collisions take place with a mean free path l, the rate of energy gain is

$$\bar{a}=2\left(\frac{u}{c}\right)^2\left(\frac{c}{l}\right). \tag{6.9.1}$$

This can be expressed in a somewhat different way, since l is the wavelength of a magnetic disturbance. The quantity u/l is the frequency of magnetic-field variation. A particle moving with velocity c encounters a coherent change of magnetic field for the time $\tau_c = l/c$. There are two frequencies, u/l and $1/\tau_c = c/l$; the former is responsible for the energy change—whereas the latter, for modulation. The combination of these effects results in the statistical average of the rate of energy change

$$\bar{a}\simeq\left(\frac{u}{l}\right)^2\tau_c=\left(\frac{u^2}{cl}\right). \tag{6.9.1$'$}$$

The second expression is familiar in the statistical mechanics of irreversible processes.

6.9.2 *Thermodynamic Considerations*

In the discussion we need to introduce a number of parameters. Since the values of such parameters are difficult to determine it may be advisable to consider this problem by referring only to energetics.

Let the energy contained in a volume V be W. The rate of energy change is equal to the energy-input rate Q—presumably of thermonuclear or gravitational origin—minus the radiation rate J as well as the work done by the volume change; that is,

$$\frac{dW}{dt} = Q - J - P\frac{dV}{dt}, \tag{6.9.2}$$

where P is the pressure due mainly to turbulence, magnetic fields, and cosmic rays. We distinguish these modes by the subscript i, and express W and P as

$$W = \sum_i W_i, \qquad P = \sum_i P_i. \tag{6.9.3}$$

In the stationary state (6.9.2) is reduced to $Q = J$. Since we are not interested in thermal radiation, Q and J are regarded as representing nonthermal modes. Hence Q consists mainly of the rate of energy transfer to the hydromagnetic mode, Q_{HM}, and that to cosmic rays. The latter represents the acceleration, so that the total energy of cosmic rays, U, increases at a rate of aU. Thus

$$Q = Q_{\mathrm{HM}} + aU. \tag{9.6.4}$$

The contributors to J are the nonthermal radio emission, the bright-line emission, as well as the escape of cosmic rays. Among them the last one is known to be the largest for most active objects. With the mean lifetime of cosmic rays, τ, therefore, J is expressed as

$$J = \frac{U}{\tau}. \tag{6.9.5}$$

The balance of (6.9.4) and (6.9.5) gives us (Kraushaar 63)

$$Q_{\mathrm{HM}} + aU = \frac{U}{\tau}. \tag{6.9.6}$$

From this we obtain the spectral index given by (6.1.3) as

$$\alpha = \frac{1}{a\tau} = 1 + \frac{Q_{\mathrm{HM}}}{aU}. \tag{6.9.7}$$

Thus the power index is expressed by the relative inputs of energy to the hydromagnetic and the cosmic-ray modes. The fact that α lies between 1 and 2 indicates the approximate equality of the rates of energy transfer to these two.

In the last phase of activity the energy input ceases so that in (6.9.2) only dW/dt and J give significant contributions. In this case nonthermal energy may be *equipartitioned* in the following three modes; that is, turbulent, magnetic, and cosmic-ray modes. Hence we may write

$$W = 3U = 3\bar{E}N, \tag{6.9.8}$$

where $\bar{E}$ is the average energy of cosmic rays. Since

$$\frac{dW}{dt} = 3\left(\frac{d\bar{E}}{dt} N + \bar{E}\frac{dN}{dt}\right), \qquad J = -\bar{E}\frac{dN}{dt},$$

we have

$$\frac{\dfrac{dN}{dt}}{N} = -\frac{3}{2}\frac{\dfrac{d\bar{E}}{dt}}{\bar{E}}.$$

This gives us the spectral index (Syrovatsky 61)

$$\alpha = \tfrac{3}{2}. \tag{6.9.9}$$

It should be remarked that the radio spectra from most radio sources are explained by synchrotron radiation by such electrons as have the spectrum expected from (6.9.9).

It should be also noticed that most strong radio sources exhibit violent hydromagnetic motions induced by explosions. Hence the active region is expanding rapidly at least in its early phase. As an extreme case we assume that the energy supply takes place only during the very initial epoch. Since cosmic rays are accelerated effectively in later stages, Q in (6.9.2) can be neglected. Equation (6.9.2) can be solved if the equation of state is given as

$$W = \beta PV, \qquad PV^{\gamma} = \text{constant}. \tag{6.9.10}$$

This results in

$$P\frac{dV}{dt} = \frac{1}{\beta(1-\gamma)}\frac{dW}{dt}. \tag{6.9.11}$$

By substituting the last relation into (6.9.2) with $Q = 0$ we obtain (Sato 63)

$$\alpha = \frac{3 + 3\beta(1-\gamma)}{3 + 2\beta(1-\gamma)}. \tag{6.9.12}$$

The absence of the pressure term as leading to (6.9.9) corresponds $\gamma = \infty$. Since $\beta = \frac{4}{9}$ and $\gamma = 2$ are plausible for the equipartition of energy into the three modes under consideration, we have $\alpha = \frac{15}{19}$.

One has, however, to be cautious in equating the value of α thus derived in (6.9.9) and (6.9.12) to the spectral index actually observed. Since the energy spectrum may change in time in association with the change in the particle number, the spectrum observed outside the source is the superposition of the spectra at different times. Thus the spectrum inside the source is different from that outside.

6.9.3 Dynamic Description of the Acceleration Mechanism

Since the region where cosmic-ray particles move about is highly conductive, there is hardly any intrinsic electric field, and the acceleration of particles is due mainly to a varying magnetic field. For the motion of a particle in a magnetic field there are *adiabatic invariant quantities* associated with gyration about a magnetic line of force, drift along a line of force, and drift encircling a magnetic flux. Of these, the first has in general the highest frequency and is best conserved. The motion of a particle can be described by introducing a set of action and angle variables for the respective modes (Hayakawa 64*a*).

The action variable for the gyration mode is one-half of the angular momentum about the magnetic line of force of strength H and is defined as

$$J \equiv \frac{cp_{\perp}^{2}}{2ZeH} = \mu\left(\frac{E}{Zec}\right) \tag{6.9.13}$$

for a particle of charge Ze, the total energy E, and the momentum perpendicular to the line of force $p_{\perp}$. This is proportional to the *magnetic moment of gyration*

$$\mu = \frac{Zec\,J}{E} = \frac{c^2 p_{\perp}^{2}}{2H} = \frac{c^2 p^2 \sin^2 \alpha}{2H}, \tag{6.9.14}$$

where α is the pitch angle, the angle between the particle velocity and the line of force. The total energy of a particle of mass m is thus expressed as

$$E = (m^2c^4 + p_{\perp}^{2}c^2 + p_{\parallel}^{2}c^2)^{1/2} = (m^2c^4 + 2\,ZecHJ + p_{\parallel}^{2}c^2)^{1/2}. \tag{6.9.15}$$

Let us consider that the magnetic field moves without changing its configuration. Figure 6.46 shows two typical examples of moving magnetic fields, whose velocities $\mathbf{u}$ are parallel and perpendicular to the field axes, respectively.

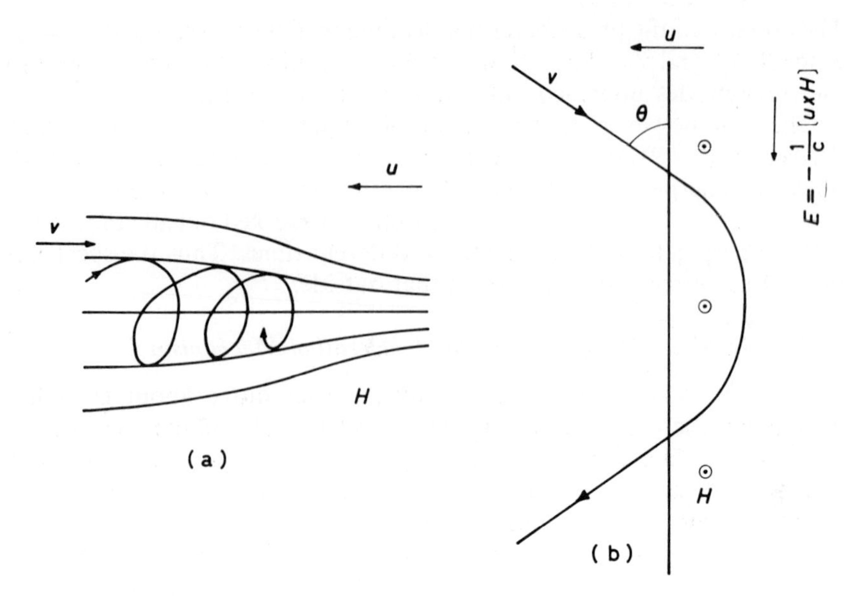

Fig. 6.46 Collisions of charged particles with magnetic irregularities. (*a*) Reflection by a magnetic mirror moving with velocity u perpendicular to the magnetic line of force; (*b*) Reflection by a magnetic wall moving with velocity u parallel to the magnetic line of force.

These are considered as models of magnetic clouds or interstellar plasmas frozen in magnetic fields. When a charged particle collides with a moving cloud there occurs a kinematic energy exchange between them.

Let a particle with energy-momentum $(E, p_{\parallel}, p_{\perp})$ impinge on the magnetic field of Fig. 6.46*a*. If we transform the system into the system moving with the magnetic field, and denote the transformed dynamical variables by bars, the energy-momentum vector becomes

$$(\bar{E}, \bar{p}_{\parallel}, \bar{p}_{\perp}) = \left[\Gamma(E + up_{\parallel}),\ \Gamma\left(p_{\parallel} + \frac{u}{c^2}E\right), p_{\perp}\right],$$

where $\Gamma = [1 - (u/c)^2]^{-1/2}$. In this system the magnetic field is at rest and the electric field is still absent, so that the particle undergoes no energy change, but its momentum components are intermixed in such a way that

$$(\bar{E}, \bar{p}_{\parallel}, \bar{p}_{\perp}) \rightarrow (\bar{E}, \bar{p}_{\parallel} + \Delta\bar{p}_{\parallel}, \bar{p}_{\perp} + \Delta\bar{p}_{\perp}) = (\bar{E}, \bar{p}'_{\parallel}, \bar{p}'_{\parallel}).$$

Returning to the original system, we have

$$(E', p'_{\parallel}, p'_{\perp}) = \left[\Gamma(\bar{E} - u\bar{p}'_{\parallel}),\ \Gamma\left(\bar{p}'_{\parallel} - \frac{u}{c^2}\bar{E}\right), \bar{p}'_{\perp}\right].$$

Thus the variations in energy and momentum components are

$$\begin{aligned} \Delta E &= E' - E = \Gamma u\, \Delta \bar{p}_{\parallel}, \\ \Delta p_{\parallel} &= p'_{\parallel} - p_{\parallel} = \Delta \bar{p}_{\parallel}, \\ \Delta p_{\perp} &= p'_{\perp} - p_{\perp} = \Delta \bar{p}_{\perp}. \end{aligned} \tag{6.9.16}$$

For the case of reflection we have $\Delta \bar{p}_{\parallel} = -2\bar{p}_{\parallel}$, and

$$\Delta E = 2\Gamma^2 u(v_{\parallel} + u)E/c^2, \tag{6.9.17}$$

which is the typical energy change in the Fermi process.

On the other hand, for the moving edge of a plasma frozen in a perpendicular magnetic field, as shown in Fig. 6.46*b*, we can expect an induced electric field, $\mathbf{E} = -(1/c)(\mathbf{u} \times \mathbf{H})$, which makes the particle accelerate (Ginzburg 64). As the total change of energy of the particle is equal to the work done by the induced electric field during the particle passing through the magnetic region, we obtain

$$\begin{aligned} \Delta E &= Ze \int_{-\theta}^{\theta} \left(-\frac{u}{c} H \right) \frac{pc}{ZeH} \cos\theta \, d\theta \\ &= 2up \sin\theta = 2E\left(\frac{uv}{c^2}\right) \sin\theta, \end{aligned} \tag{6.9.18}$$

where θ is the incoming angle of the particle. This result shows that the energy change is also proportional to E.

These examples show that the *Fermi mechanism* is inherent in any collision in which the particle is subject to kinematic constraints. It is important that in each step the correlation with the preceding motion be removed, in order to attain a net change in energy.

Another important process other than the Fermi mechanism is the betratron mechanism (Swann 33), which is expected in a changing magnetic field H and is usually described by means of the adiabatic invariance of J.

As may readily be seen from the adiabatic invariance of J in (6.9.13), we have

$$c^2 \frac{dp_{\perp}{}^2}{dt} = \frac{c^2 p_{\perp}{}^2}{H} \frac{\partial H}{\partial t}. \tag{6.9.19}$$

This is readily integrated to give

$$\frac{p_{\perp}(t)^2}{p_{\perp}(0)^2} = \frac{H(t)}{H(0)}. \tag{6.9.20}$$

There are more complicated acceleration processes arising from nonadiabatic effects, such as the induction effect and the transit-time effect.

They consist of those parts which are nonvanishing after the phase average, and accordingly they may be considered as statistical processes. Their contribution is smaller than the contribution of statistical Fermi and betatron processes, unless the orbit of a particle encircles the symmetry axis of the magnetic field, so that the action variable J is comparable to the total angular momentum of the system.

The statistical average of (6.9.17) and (6.9.18) gives

$$(\Delta E)_F = \zeta\left(\frac{u}{c}\right)^2 E, \tag{6.9.21}$$

as discussed in Section 6.9.1, where ζ is a numerical factor on the order of unity and depends on the geometrical conditions of magnetic scattering.

The betatron acceleration is reversible, and no net change of energy takes place when the magnetic-field strength comes back to the original value. If the gyro-relaxation is superposed on the betatron process, however, one always obtains a net increase of energy (Alfvèn 59). In the initial state the energy of particles is equipartitioned into three modes as

$$p_\perp{}^2 = \tfrac{2}{3}p^2(0), \qquad p_\parallel{}^2 = \tfrac{1}{3}p^2(0). \tag{6.9.22}$$

After a time τ, the magnetic-field strength changes from $H(0)$ to $H(\tau)$, and as a consequence of the betatron acceleration we have

$$p^2(\tau) = \tfrac{1}{3}p^2(0) + \tfrac{2}{3}p^2(0)\frac{H(\tau)}{H(0)} = \tfrac{1}{3}p^2(0)\left[1 + 2\frac{H(\tau)}{H(0)}\right],$$

according to (6.9.20). Then equipartition as in (6.9.22) takes place. In the next time interval of τ, the magnetic-field strength changes to $H(2\tau)$, and we have

$$p^2(2\tau) = \tfrac{1}{3}p^2(\tau)\left[1 + 2\frac{H(2\tau)}{H(\tau)}\right] = \tfrac{1}{9}p^2(0)\left[1 + 2\frac{H(\tau)}{H(0)}\right]\left[1 + 2\frac{H(2\tau)}{H(\tau)}\right].$$

If $H(2\tau) = H(0)$, there is a net gain of energy

$$p^2(2\tau) = p^2(0)\left[1 + \frac{2}{9}\frac{(\Delta H)^2}{H(0)H(\tau)}\right]. \tag{6.9.23}$$

This has the same momentum dependence as the secular betatron acceleration in (6.9.18). Both in the secular and the stochastic cases, therefore the energy gain from *betatron acceleration* is

$$(\Delta E)_B = \bar{b}vp, \tag{6.9.24}$$

where

$$\bar{b} = \left\langle \frac{\Delta H}{2H} \sin^2 \alpha \right\rangle \quad \text{for secular acceleration,} \tag{6.9.25a}$$

$$\bar{b} = \left\langle \frac{\Delta H^2}{9H^2} \right\rangle \quad \text{for stochastic acceleration.} \tag{6.9.25b}$$

The energy gain due to the Fermi mechanism given by (6.9.21) takes place with a mean collision time with magnetic clouds, l/v, where l is the mean free path for the collisions. The energy gain due to the betatron process takes place at a rate proportional to $\partial H/\partial t$ or $\partial\langle\Delta H^2\rangle/\partial t$, according to whether the acceleration is secular or stochastic. In both cases the rates of energy gain are proportional to the velocity of the particle. The velocity dependence can be eliminated in the rate of momentum gain,

$$\frac{dp}{dt} = \left(\frac{a}{c}\right) E + bp, \tag{6.9.26}$$

where

$$a = \zeta \frac{(u/c)^2}{l}, \qquad b = \frac{\partial \bar{b}}{\partial t}. \tag{6.9.27}$$

6.9.4 Mass Dependence of the Acceleration Efficiency

We have shown in Section 6.4 that the energy spectra of heavy nuclei are similar to each other but not to the proton spectrum. This is most clearly demonstrated by the spectra of solar cosmic rays. In Section 6.4 we have argued that electrons are accelerated also, but that their intensity is smaller than the intensity of protons. All such results seem to be related to the mass dependence of the acceleration rate.

The acceleration rate (6.9.26) can be expressed without the mass dependence as

$$\frac{d(\beta\gamma)}{dt} = a\gamma + b\beta\gamma. \tag{6.9.28}$$

Since $\gamma = (1 - \beta^2)^{-1/2}$, this equation depends only on velocity $c\beta$. Hence all kinds of particles should have the same velocity spectrum, provided that the initial velocity is the same for all particles. This is, in fact, found to be the case for galactic protons and α-particles, as shown in Fig. 6.10. The same conclusion has been obtained for the spectra of nuclear particles and electrons, as compared in Fig. 6.47 (Brunstein 66). For

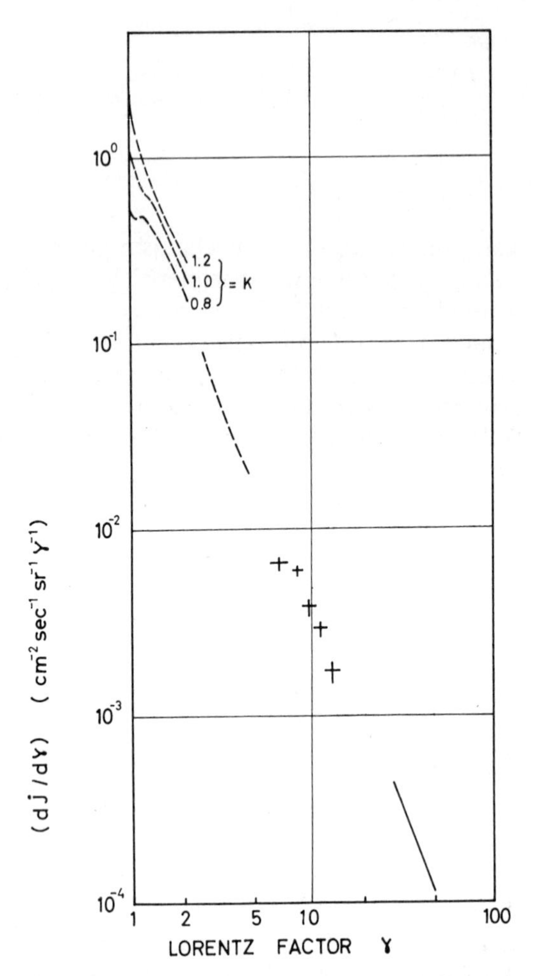

Fig. 6.47 The differential energy-per-mass (γ) spectrum of electrons, protons, and α-particles (Brunstein 66). +—the electron spectrum given in Fig. 6.30. The solid curve represents the proton spectrum given in (6.4.7) with Table 6.19. The dashed curve represents the proton (α-particle) spectrum, correcting (6.4.3) by the modulation factor (6.4.4) with $K = 1.0$. The spectra with different modulation factors with $K = 08.$, 1.0, and 1.2 are also shown. The whole spectrum is approximately represented by $dj/d\gamma = (1.15 \pm 0.15)\gamma^{-2.5 \pm 0.1}$ $\text{cm}^{-2}\ \text{sec}^{-1}\ \text{sr}^{-1}\ \gamma^{-1}$.

solar cosmic rays, however, the velocity dependence is clearly observed, as seen in Fig. 6.13. Solar cosmic rays seem to have the same rigidity spectrum rather than the same velocity spectrum.

The velocity dependence comes from the injection energy, the energy loss, and the modulation process. These effects can be studied more unambiguously if the velocity independence of the spectrum without these effects is assumed.

The rate of energy loss by ionization is expressed as

$$-\frac{d(\beta\gamma)}{dt} = \frac{Z^2 U}{A\beta^2} \simeq \frac{Z^2}{A\beta^2} n_e \, 2 \times 10^{-16} \text{ sec}^{-1}, \tag{6.9.29}$$

thus depending on Z^2/A, where n_e is the density of ambient electrons. If this is greater than the acceleration rate, particles are not accelerated but thermalized. Since this is important mainly in the nonrelativistic region, the acceleration is possible for

$$a\beta^2 + b\beta^3 > \frac{Z^2}{A} U. \tag{6.9.30}$$

If the velocity is small, only the first term can be retained on the left-hand side. Hence the *critical injection velocity* is given by

$$\beta_{ic}^2 = \frac{Z^2}{A} \frac{U}{a}. \tag{6.9.31}$$

For electrons the *critical injection energy* may lie in the relativistic region; in this case we have

$$\gamma_{ic} = \frac{(Z^2/A)U}{a+b}. \tag{6.9.31$'$}$$

The loss of energy by ionization is insignificant for $\beta \gg \beta_{ic}$ or $\gamma \gg \gamma_{ic}$ in the course of propagation. Other energy-loss processes are also Z and A dependent. For nuclear particles the nuclear collision rate depends on $A^{2/3}$ and other loss processes are negligible. For electrons the energy losses due to bremsstrahlung, synchrotron radiation, and the inverse Compton effect are significant, since they depend on A^{-2}, as discussed in Section 6.7. Since they increase rapidly with energy, these effects are insignificant in obtaining the critical injection energy.

The injection energy may have any value greater than its critical value, and depends on the injection mechanism. Above the critical injection energy, the acceleration equation (6.9.28) can be solved for the initial values β_i and γ_i. If the acceleration starts with a nonrelativistic energy, we can put $\gamma_i = 1$. In the nonrelativistic region we have

$$\beta = \left(\frac{a}{b}\right)(e^{bt} - 1) + \beta_i e^{bt}. \tag{6.9.32}$$

Hence the momentum per unit mass of a particle is independent of A and Z if $a \gg b\beta_i$, or the Fermi mechanism is more efficient than the betatron process in the initial stage. Hence the composition of particles accelerated is the same as that of the ambient gas. If, on the other hand, the betatron acceleration is so effective that $b\beta_i > a$, the momentum of a particle accelerated depends on A and Z. If the injection is caused by an electric field such as that due to the charge separation at a shock front, the injection velocity is

$$\beta_i = \left(\frac{Z}{A}\right)^{1/2} \left(\frac{2\,eV}{Mc^2}\right)^{1/2}, \tag{6.9.33}$$

where V is the potential difference. If the sudden heating, such as that observed as the coronal condensation, is responsible for the injection, we have

$$\beta_i = \sqrt{\frac{2kT}{AMc^2}}. \tag{6.9.34}$$

Both (6.9.33) and (6.9.34) show dependences on Z and A.

Whether such dependences can be seen or not depends also on the time during which acceleration continues. For $t \ll 1/b$ the expression (6.9.32) is simplified to

$$\beta \simeq at + \beta_i, \tag{6.9.35a}$$

whereas for $t > 1/b$ we have

$$\beta \simeq \frac{a + \beta_i b}{b}\, e^{bt}. \tag{6.9.35b}$$

If we are concerned with the nonrelativistic region, the time after injection is limited to $at < 1$, and therefore the second alternative in (6.9.35b) does not take place unless $b > a$ and $t < (1/b) \ln [b/(a + \beta_i b)]$. Hence the region in which (6.9.35b) holds is rather narrow.

As time increases to t_r,

$$t_r = \frac{(1 - \beta_i)}{a} \quad \text{or} \quad \left(\frac{1}{b}\right) \ln \left[\frac{b}{(a + \beta_i b)}\right], \tag{6.9.36}$$

particles become relativistic. Thereafter we can put $\beta = 1$ and solve (6.9.28) as

$$\gamma = \exp [(a + b)(t - t_r)] \tag{6.9.37}$$

$$\simeq \exp \left[\frac{-(1 - \beta_i)}{a}\right] \exp [(a + b)t] \quad \text{for} \quad a > b, \tag{6.9.37a}$$

$$\simeq \left(\frac{a + \beta_i b}{b}\right)^{(a+b)/b} \exp [(a + b)t] \quad \text{for} \quad b \gg a. \tag{6.9.37b}$$

In both cases the energy of a particle increases exponentially with time.

The acceleration process is effective only if particles are trapped in the accelerating region. The particles diffuse away due to the scattering by magnetic irregularities. The average number of such scatterings is estimated as

$$n \simeq \left(\frac{L}{l}\right)^2, \tag{6.9.38}$$

where L is the linear dimension of the accelerating region and l is the mean distance between irregularities. The time needed for n scatterings is obtained from

$$n = \frac{c}{l}\left(\frac{at^2}{2} + \beta_i t\right) \simeq \left(\frac{L}{l}\right)^2. \tag{6.9.39}$$

For $\beta \gg \beta_i$ we have the average time for acceleration

$$\tau \simeq L\sqrt{\frac{2}{acl}}. \tag{6.9.40}$$

This relation as well as the derivation above hold independent of the rigidity of a particle.

Since the probability for a particle being subject to acceleration is $\exp(-t/\tau)$, the integral velocity spectrum is obtained, on account of (6. 9.35*a*), as

$$\exp\left(-\frac{t}{\tau}\right) = \exp\left[-\frac{(\beta-\beta_i)}{a\tau}\right]. \tag{6.9.41a}$$

Thus we have an exponential spectrum, as has been seen for the rigidity spectrum of solar cosmic rays in Fig. 6.14. If the betatron acceleration becomes effective as in (6.9.35*b*), the velocity spectrum has a power shape

$$\exp(-t/\tau) \simeq \left(\frac{a + b\beta_i}{b\beta}\right)^{1/b\tau}. \tag{6.9.41b}$$

In this case $b\beta_i$ is not always negligible compared with a, and the spectrum depends on A and Z through the injection velocity β_i. This may be responsible for the Z/A dependence of the energy-per-nucleon spectrum of solar cosmic rays at high energies.

It may be worthwhile remarking that the spectrum (6.9.41*a*) resembles to the Maxwell distribution for small $\beta - \beta_i$, but has a tapered tail as energy increases, as represented by (6.9.41*a*,*b*). Since hydromagnetic disturbances are likely to be associated with a tenuous plasma, the energ. distribution of particles may have a tapered tail, as discussed in Section 6.8.

In the relativistic region the spectrum is obtained from (6.9.37) as

$$\exp\left(-\frac{t}{\tau}\right) = \exp\left(-\frac{t_r}{\tau}\right)\gamma^{-1/(a+b)\tau}; \tag{6.9.42}$$

this is the well-known *power law*. If the integral spectrum is expressed as $\gamma^{-\alpha}$, the flux at equal energy depends on the particle mass as

$$\kappa \propto A^{\alpha}. \tag{6.9.43}$$

This seems to be in fact the case for protons and electrons, as shown in Fig. 6.47. In the GeV region, however, the proton-electron ratio is much smaller, as shown in Fig. 6.31. This could be due to the injection effect represented by the factor $\exp(-t_r/\tau)$. This again implies that the sources of low- and high-energy cosmic rays are different, as has been discussed in Section 6.4 in connection with the spectrum of low-energy heavy nuclei.

The small proton-electron ratio is expected, if the injection velocity is given either by (6.9.33) or by (6.9.34). In either case the injection velocity is so small for protons and heavy nuclei that its effect is inappreciable in the comparison between protons and heavy nuclei. For electrons, on the contrary, the injection velocity is so large that the betatron acceleration is dominant even in the initial stage. This requires (Hayakawa 64*a*)

$$mc^2 < \left(\frac{b}{a}\right)^2 (2kT \quad \text{or} \quad 2\,eV) < Mc^2, \tag{6.9.44}$$

where m is the electron mass. If the betatron acceleration is dominant, the flux ratio has to be multiplied by $(M/m)^{1/2b\tau}$, thus compensating the flux ratio (6.9.43) by a considerable factor. The flux ratio depends on the details of acceleration processes.

The absolute value of the acceleration rate is inferred from the integral exponent of the spectrum to be

$$a + b = \frac{1}{\alpha\tau} \tag{6.9.45}$$

or from the exponential spectrum as

$$a = \frac{\beta_0}{\tau}, \tag{6.9.46}$$

where β_0 is the e-fold value of the velocity spectrum. The quantity $(a+b)\tau$ or $a\tau$ may be defined as the *acceleration efficiency*. This is on the order of unity for galactic cosmic rays and for radio sources, and about 1/10 for solar cosmic rays.

6.9.5 *Properties of Accelerating Regions*

The discussions above demonstrate that an important quantity in acceleration is $a\tau$. In the nonrelativistic region this is given on account of (6.9.27) and (6.9.40) by

$$a\tau \simeq \frac{L}{l}\frac{u}{c}, \tag{6.9.47}$$

whereas in the relativistic region we have

$$a\tau \simeq \left(\frac{L}{l}\right)^2 \left(\frac{u}{c}\right)^2. \tag{6.9.48}$$

In both cases the value of $a\tau$ is close to unity, thus giving a rather flat spectrum.

The other important quantities are the *cutoff rigidity*, which may be as large as Hl, as discussed in Sections 6.4 and 6.5, and the critical injection velocity given approximately by $c\sqrt{(Z^2/A)U\tau}$. The values of these quantities, though they are only order-of-magnitude estimates, are summarized for representative cosmic-ray sources in Table 6.38.

Table 6.38 Characteristic Quantities in Cosmic-Ray Sources

	Solar Flare	Super-nova	Radio Galaxy	Galactic Halo
Density (g-cm^{-3}) ρ	10^{-15}	10^{-23}	10^{-24}	10^{-26}
Magnetic-field strength (gauss) H	$10^{1.5}$	10^{-3}	10^{-4} to 10^{-5}	$10^{-5.5}$
Velocity of irregularities (cm-sec^{-1}) u	10^8	10^8	$10^{6.5}$ to $10^{7.5}$	10^7
Linear dimension (cm) L	$10^{10.5}$	10^{19}	10^{23}	10^{23}
Mean free path (cm) l	10^9	10^{17}	10^{20}	10^{20}
Accelerating time $\tau = L^2/lc$	$10^{1.5}$	$10^{10.5}$	$10^{15.5}$	$10^{15.5}$
$(u/c)^2$	10^{-5}	10^{-5}	10^{-6} to 10^{-8}	10^{-7}
$(L/l)^2$	10^3	10^4	10^6	10^6
$a\tau = (L/l)^2(u/c)^2$	10^{-2}	10^{-1}	1 to 10^{-2}	10^{-1}
Hl (volt)	10^{13}	$10^{16.5}$	$10^{18.5}$ to $10^{17.5}$	10^{17}
$U\tau$	$10^{-5.5}$	$10^{-4.5}$	$10^{-0.5}$	$10^{-2.5}$

These sources are considered to be representatives respective stages in the sequence discussed at the beginning of this section.

6.9.6 Energy Balance Between Cosmic Rays and Other Modes

It is remarkable that the value of $a\tau$ is close to unity in most sources. This implies that hydromagnetic energy is efficiently converted into cosmic-ray energy, thus supporting the equipartition of energy into these modes, as discussed in Section 6.9.2. However, this does not mean an equilibrium among these modes, because the acceleration time τ is not long enough for the relaxation of the cosmic-ray energy. Cosmic rays take away hydromagnetic energy but cannot return energy to the hydromagnetic mode.

Although the relaxation of energy takes place rather slowly, the distributions of velocity and density may be quickly relaxed. Any anisotropy in an isotropic medium produces a magnetic field that suppresses the anisotropy. Likewise, any inhomogeneity produces an electric field that tends to eliminate the inhomogeneity. The latter point is related to the equality of the velocity spectra of protons and electrons in the nonrelativistic region. If they were different, the densities of protons and electrons would have to be different. This should result in a net charge density and give rise to a strong electric field, thus recovering the equality of the velocity spectra.

Although the energy relaxation is slow for high-energy cosmic rays, it is rather fast for low-energy cosmic rays, since they quickly dissipate their energy through the ionization process. The cosmic-ray energy thus lost is spent in heating up interstellar gases. This may be responsible for keeping the temperature of interstellar HI clouds as high as 100°K (Hayakawa 61*b*). This is important in the condensation of the interstellar gas to form stars, since the cooling of the gas is required for the condensation. A high flux of cosmic rays suppresses the formation of stars, and a low rate of star formation results in a low energy output, a part of which can be used for the acceleration of cosmic rays (K. Aizu 64*a*). This energy cycle tends to keep the cosmic-ray flux at a certain level, since any deviation from this level is counterbalanced by the feedback mechanism above

The above discussions demonstrate that cosmic rays play the role of "cushion" in cosmic space. Without this stabilizer the behavior of the universe would be much different. Cosmic rays serve not only as a transmitter of astrophysical information but also as an indispensable mode of transmitting celestial energy.

REFERENCES

Abraham 66	Abraham, P. B., Brunstein, K. A., and Cline, T. L., *Phys. Rev.*, **150**, 1088 (1966).
Agrinier 65	Agrinier, B., et al., *Proc. Int. Conf. on Cosmic Rays at London*, **1**, 331 (1965).
Aizu, H. 60	Aizu, H., et al., *Prog. Theor. Phys. Suppl.* No. 16, 54 (1960).
Aizu, H. 64	Aizu, H., Ito, K., and Koshiba, M., *Prog. Theor. Phys. Suppl.* No. 30, 134 (1964).
Aizu, K. 64a	Aizu, K., et al., *Prog. Theor. Phys. Suppl.* No. 30, 2 (1964).
Aizu, K. 64b	Aizu, K., et al., *Prog. Theor. Phys. Suppl.* No. 31, 35 (1964).
Aizu, K. 64c	Aizu, K., and Tabara, H., *Prog. Theor. Phys. Suppl.* No. 31, 62 (1964).
Aizu 65	Aizu, K., et al., The Sixth COSPAR Symposium at Mar del Plata, (1965).
Alfvèn 59	Alfvèn, H., *Tellus*, **11**, 1 (1959).
Aller 61	Aller, L. H., *The Abundance of the Elements*, Interscience, New York, 1961.
Anders 62	Anders, E., *Rev. Mod. Phys.*, **34**, 287 (1962).
Anders 64	Anders, E., *Space Sci. Rev.*, **3**, 583 (1964).
Appa Rao 63	Appa Rao, M. V. K., and Kaplon, M. F., *Nuovo Cim.*, **27**, 700 (1963).
Badhwar 62	Badhwar, G. D., Daniel, R. R., and Vijayalakshmi, B., *Prog. Theor. Phys.*, **28**, 607 (1962).
Badhwar 63a	Badhwar, G. D., and Daniel R. R., *Prog. Theor. Phys.*, **30**, 615 (1963).
Badhwar 63b	Badhwar, G. D., et al., *Proc. Int. Conf. on Cosmic Rays at Jaipur*, **3**, 38 (1963).
Balasubrahmanyan 63	Balasubrahmanyan, V. K., et al., *Proc. Int. Conf. on Cosmic Rays at Jaipur*, **3**, 110 (1963).
Balasubrahmanyan 64	Balasubrahmanyan, V. K., and McDonald, F. B., *J. Geophys. Res.*, **69**, 3289 (1964).
Balasubrahmanyan 65	Balasubrahmanyan, V. K., Boldt, E., and Palmeira, R. A. R. *Phys. Rev.*, **140**, B1157 (1965).
Balasubrahmanyan 65	Balasubrahmanyan, V. K., et al., *J. Geophys. Res.*, **71**, 1771 (1966).
Biswas 65	Biswas, S., and Fichtel, C. E., *Space Sci. Rev.*, **4**, 709 (1965).
Boischot 57	Boischot, A., and Denisse, J. F., *Compt. Rend.*, **245**, 2194 (1947).
Bonsack 60	Bonsack, W. K., and Greenstein, J. L., *Astrophys. J.*, **131**, 83 (1960).
Bowyer 64	Bowyer, S., et al., *Science*, **146**, 912 (1964).
Bowyer 65	Bowyer, S., et al., *Science*, **147**, 394 (1965).
Bradt 50	Bradt, H. L., and Peters, B., *Phys. Rev.* **77**, 54 (1950).
Brunstein 65	Brunstein, K. A., *Phys. Rev.*, **137**, B757 (1965).
Brunstein 66	Brunstein, K. A., and Cline, T. L., *Nature*, **209**, 1186 (1966).

Burbidge 57	Burbidge, E. M., *et al.*, *Rev. Mod. Phys.*, **29**, 547 (1957).
Burbidge 62	Burbidge, G. R., *Prog. Theor. Phys.*, **27**, 999 (1962).
Burbidge 63	Burbidge, G. R., *Proc. Int. Conf. on Cosmic Rays at Jaipur*, **3**, 229 (1963).
Byram 66	Byram, E. T., Chubb, T. A., and Friedman, F., *Science*, **152**, 66 (1966).
Cachon 62	Cachon, A., *Fifth Interamerican Seminar on Cosmic Rays*, **2**, XXXIX (1962).
Cameron 59	Cameron, A. G. W., *Astrophys. J.*, **129**, 672 (1959).
Cline 64	Cline, T. L., Ludwig, G. H., and McDonald, F. B., *Phys. Rev. Letters*, **13**, 786 (1964).
Clark 65	Clark, G., et al., *Nature*, **207**, 584 (1965).
Comstock 65	Comstock, G. M., Fan, C. Y., and Simpson, J. A., *Proc. Int. Conf. on Cosmic Rays at London*, **1**, 383 (1965).
Conforto 62	Conforto, A. M., *J. Phys. Soc. Japan*, **17**, Suppl. A–III, 144 (1962).
Craig 57	Craig, H., *Tellus*, **9**, 1 (1957).
Critchfield 50	Critchfield, C. L., Ney, E. P., and Oleska, S., *Phys. Rev.*, **79**, 402 (1950).
Dahanayake 64	Dahanayake, C., Kaplon, M. F., and Lavakare, P. J., *J. Geophys. Res.*, **69**, 3681 (1964).
Damon 66	Damon, P. E., Long, A., and Grey, D. C., *J. Geophys. Res.*, **71**, 1055 (1966).
Daniel 62	Daniel, R. R., and Durgaprasad, N., *J. Phys. Soc. Japan*, **17**, Suppl. A–III, 15 (1962).
Daniel 63	Daniel, R. R., and Sreevivasan, N., *Proc. Int. Conf. on Cosmic Rays at Jaipur*, **3**, 60 (1963).
Daudin 53	Daudin, A., and Daudin, J., *J. Atmos. Terr. Phys.*, **3**, 245 (1953); *J. Phys. Radium*, **14**, 169 (1953).
Davis 54	Davis, L., *Phys. Rev.*, **96**, 743 (1954).
Davis 64	Davis, R., *Phys. Rev. Letters*, **12**, 302 (1964).
Delvaille 62	Delvaille, J., Kendziorski, F., and Greisen, K., *J. Phys. Soc. Japan*, **17**, Suppl. A–III, 76 (1962).
Ellis 62	Ellis, G. R., Waterworth, M. D., and Bessel, M., *Nature*, **196**, 1079 (1962).
Elwert 54	Elwert, G., *Z. Naturforschung*, **9a**, 637 (1954).
Fan 65	Fan, C. Y., Gloeckler, G., and Simpson, J. A., *J. Geophys. Res.*, **70**, 3515 (1965).
Fan 66	Fan, C. Y., Gloeckler, G., and Simpson, J. A., *Phys. Rev. Letters*, **17**, 329 (1966).
Feenberg 48	Feenberg, E., and Primakoff, H., *Phys. Rev.*, **73**, 449 (1948).
Felten 63	Felten, J. E., and Morrison, P., *Phys. Rev. Letters*, **10**, 453 (1963).
Felten 66a	Felten, J. E., and Morrison, P., *Astrophys. J.*, **146**, 686 (1966).
Felten 66b	Felten, J. E., and Gould, R. J., *Phys. Rev. Letters*, **17**, 401 (1966).
Fermi 49	Fermi, E., *Phys. Rev.*, **75**, 1729 (1949).
Fichtel 64	Fichtel, C. E., et al., *J. Geophys. Res.*, **69**, 3293 (1964).
Fichtel 66	Fichtel, C. E., and Reams, D., *Phys. Rev.*, **149**, 991, 995 (1966).

Field 64 — Field, G. B., and Henry, R. C., *Astrophys. J.*, **140**, 1002 (1964).

Fireman 60 — Fireman, E. L., and DeFelice, J., *J. Geophys. Res.*, **65**, 3035 (1960).

Freier 65 — Freier, P. S., and Waddington, C. J., *J. Geophys. Res.*, **70**, 5753 (1965).

Fujimoto 64 — Fujimoto, Y., Hasegawa, H., and Taketani, M., *Prog. Theor. Phys. Suppl.* No. 30, 32 (1964).

Fukui 62 — Fukui, S., et al., *J. Phys. Soc. Japan*, **17**, Suppl. A–III, 169 (1962).

Geiss 63 — Geiss, J., *Proc. Int. Conf. on Cosmic Rays at Jaipur*, **3**, 434 (1963).

Gerasimova 60 — Gerasimova, N. M., and Zatsepin, G. T., *Soviet Phys. JETP*, **11**, 899 (1960).

Gerasimova 61 — Gerasimova, N. M., and Rozental', I. L., *Soviet Phys. JETP*, **14**, 350 (1961).

Giacconi 62 — Giacconi, R., et al., *Phys. Rev. Letters*, **9**, 439 (1962).

Ginzburg 53 — Ginzburg, V. L., *Usp. Fiz. Nauk*, **51**, 343 (1953).

Ginzburg 63 — Ginzburg, V. L., et al., *Proc. Int. Conf. on Cosmic Rays at Jaipur*, **3**, 41 (1963).

Ginzburg 64 — Ginzburg, V. L., and Syrovatskii, S. I., *The Origin of Cosmic Rays*, Pergamon, New York, 1964.

Gloeckler 65 — Gloeckler, G., *J. Geophys. Res.*, **70**, 5333 (1965).

Goldreich 63 — Goldreich, P., and Morrison, P., *J. Exp. Theor. Phys. USSR*, **45**, 344 (1963).

Goto 63 — Goto, E., Kolm, H. K., and Ford, K. W., *Phys. Rev.*, **132**, 387 (1963).

Gould 63 — Gould, R. J., Gold, T., and Salpeter, E. E., *Astrophys. J.*, **138**, 408 (1963).

Gould 66 — Gould, R. J., and Schreder, *Phys. Rev. Letters*, **16**, 253 (1966).

Greisen 56 — Greisen, K., *Progress in Cosmic Ray Physics*, III, 1 (1956).

Greisen 66 — Greisen, K., *Phys. Rev. Letters*, **16**, 748 (1966).

Gursky 66 — Gursky, H., et al., *Astrophys. J.*, **144**, 1249 (1966).

Harris 62 — Harris, D. E., *Astrophys. J.*, **135**, 661 (1962).

Hartman 65 — Hartman, R. C., Meyer, P., and Hildebrand, R. H., *J. Geophys. Res.*, **70**, 2713 (1965).

Hasegawa 61 — Hasegawa, H., and Ito, K., *Prog. Theor. Phys.*, **26**, 418 (1961).

Hasegawa 62a — Hasegawa, H., et al., *J. Phys. Soc. Japan*, **17**, Suppl. A–III, 86 (1962).

Hasegawa 62b — Hasegawa, H., et al., *Phys. Rev. Letters*, **8**, 284 (1962).

Hasegawa 63 — Hasegawa, H., Aizu, H., and Ito, K., *Proc. Int. Conf. on Cosmic Rays at Jaipur*, **3**, 83 (1963).

Hayakawa 52 — Hayakawa, S., *Prog. Theor. Phys.*, **8**, 517 (1952).

Hayakawa 54 — Hayakawa, S., *Prog. Theor. Phys.*, **13**, 464 (1954).

Hayakawa 56 — Hayakawa, S., *Prog. Theor. Phys.*, **15**, 111 (1956).

Hayakawa 58a — Hayakawa, S., Ito, K., and Terashima, Y., *Prog. Theor. Phys. Suppl.* No. 6, 1 (1958).

Hayakawa 58b — Hayakawa, S., *Prog. Theor. Phys.*, **19**, 219 (1958).

Hayakawa 60a — Hayakawa, S., Hayashi, C., and Nishida, M., *Prog. Theor. Phys. Suppl.* No. 16, 169 (1960).

Hayakawa 60b — Hayakawa, S., *Publ. Astr. Soc. Japan*, **12**, 115 (1960).

Hayakawa 61a — Hayakawa, S., *J. Geomag. Geoel.*, **8**, 61 (1961).

Hayakawa 61b — Hayakawa, S., Nishimura, S., and Takayanagi, K., *Publ. Astr. Soc. Japan*, **13**, 184 (1961).

Hayakawa 62 — Hayakawa, S., *Phys. Letters*, **1**, 234 (1962).

Hayakawa 63 — Hayakawa, S., and Yamamoto, Y., *Prog. Theor. Phys.*, **30**, 71 (1963).

Hayakawa 64a — Hayakawa, S., et al., *Prog. Theor. Phys. Suppl.*, No. 30, 86, (1964).

Hayakawa 64b — Hayakawa, S., et al., *Prog. Theor. Phys. Suppl.* No. 30, 153 (1964).

Hayakawa 64c — Hayakawa, S., and Matsuoka, M., *Prog. Theor. Phys. Suppl.* No. 30, 204 (1964).

Hayakawa 65 — Hayakawa, S., and Obayashi, H., *Proc. Int. Conf. on Cosmic Rays at London*, **1**, 116 (1965).

Hayakawa 66 — Hayakawa, S., Matsuoka, M., and Sugimoto, D., *Space Sci. Rev.*, **5**, 109 (1966).

Hayashi 62 — Hayashi, C., Hoshi, R., and Sugimoto, D., *Prog. Theor. Phys. Suppl.* No. 22, 1 (1962).

Haymes 65 — Haymes, R. C., Craddock, W. L., and Clayton, D. D., *Space Research VI*, **80** (1966).

Hildebrand 66 — Hildebrand, B., and Silberberg, R., *Phys. Rev.*, **141**, 1248 (1966).

Hintenberger 65 — Hintenberger, H., Voshage, H., and Sarkar, H., *Z. Naturforschung*, **20a**, 965 (1965).

Högbom 61 — Högbom, J. A., and Shakeshaft, J. R., *Nature*, **190**, 705 (1961).

Honda 60 — Honda, M., Shedolovsky, J. P., and Arnold, J. R., *Geochim. Acta*, **22**, 133 (1960).

Hovenier 66 — Hovenier, J. W., *Bull. Astr. Inst. Netherland*, **18**, 185 (1966).

Hsi 64 — Hsi Tse-tsung and Shu-jen Po, *Contributions at the 1964 Peking Symposium*, 1964.

Hulsizer 49 — Hulsizer, R., *Phys. Rev.*, **76**, 164 (1949).

Jelley 66 — Jelley, J. V., *Phys. Rev. Letters*, **16**, 479 (1966).

Kameda 63 — Kameda, T., Toyoda, Y., and Maeda, T., *Proc. Int. Conf. on Cosmic Rays at Jaipur*, **4**, 254 (1963).

Kawabata 60 — Kawabata, K., *Rep. Ionos. Space Res. Japan*, **14**, 405 (1960).

Kigoshi 66 — Kigoshi, K., and Hasegawa, H., *J. Geophys. Res.*, **71**, 1065 (1966).

Korchak 58 — Korchak, A. A., and Syrovatsky, S. I., *Dokl. Akad. Nauk. USSR*, **12**, 792 (1958).

Koshiba 63 — Koshiba, M., et al., *Phys. Rev.*, **131**, 2692 (1963).

Krassovskij 58 — Krassovskij, V. I., and Šklovskij, I. S., *Nuovo Cim.*, Suppl. **8**, 440 (1958).

Kraushaar 63 — Kraushaar, W. L., *Proc. Int. Conf. on Cosmic Rays at Jaipur*, **3**, (1963).

Kraushaar 65 — Kraushaar, W. L., et al., *Astrophys. J.*, **141**, 845 (1965).

Lal 62	Lal, D., and Peters, B., *Progress in Elementary Particle and Cosmic Ray Physics*, **6**, 1 (1962).
Linsley 62	Linsley, J., and Scarsi, L., *Phys. Rev. Letters*, **9**, 123 (1962).
Linsley 63	Linsley, J., *Proc. Int. Conf. on Cosmic Rays at Jaipur*, **4**, 77 (1963).
McDonald 59	McDonald, F. B., and Webber, W. R., *Phys. Rev.*, **115**, 194 (1959).
McDonald 64	McDonald, F. B., and Ludwig, G. H., *Phys. Rev. Letters*, **13**, 783 (1964).
McIlwain 63	McIlwain, C. E., and Pizzela, G., *J. Geophys. Res.*, **68**, 1811 (1963).
McCusker 63	McCusker, C. B. A., *Proc. Int. Conf. on Cosmic Rays at Jaipur*, **4**, 35 (1963).
Metzger 64	Metzger, A. E., et al., Nature, **204**, 766 (1964).
Meyer 62	Meyer, P., and Vogt, R., *Phys. Rev. Letters*, **8**, 387 (1962).
Nagashima 66	Nagashima, K., Duggal, S. P., and Pomerantz, M. A., *Planet. Space Sci.*, **14**, 177 (1966).
Nagata 63	Nagata, T., Arai, Y., and Momose, K., *J. Geophys. Res.*, **68**, 5277 (1963).
Naugle 56	Naugle, J. E., and Freier, P. S., *Phys. Rev.*, **104**, 804 (1956).
Nikishov 61	Nikishov, A. J., *J. Exp. Theor. Phys. USSR*, **41**, 549 (1961).
Oda 62	Oda, M., and Hasegawa, H., *Phys. Letters*, **1**, 239 (1962).
Oda 63	Oda, M., and Hasegawa, H., *Proc. Int. Conf. on Cosmic Rays at Jaipur*, **3**, 370 (1963).
Okuda 63	Okuda, H., *Rep. Ionos. Space Res. Japan*, **17**, 1 (1963).
Oort 56	Oort, J. H., and Walraven, Th., *Bull. Astr. Inst. Netherland*, **12**, 285 (1956).
Ormes 64	Ormes, J., and Webber, W. R., *Phys. Rev. Letters*, **13**, 106 (1964).
Parker 63	Parker, E. N., *Interplanetary Dynamical Processes*, Interscience, New York, 1963.
Penzias 65	Penzias, A. A., and Wilson, R. W., *Astrophys. J.* **142**, 419 (1965).
Peters 57	Peters, B., *Z. Physik*, **148**, 93 (1957).
Pollack 63	Pollack, J. B., and Fazio, G. G., *Phys. Rev.*, **131**, 2684 (1963).
Ramaty 66	Ramaty, R., and Lingenfelter, R. E., *J. Geophys. Res.*, **71**, 3687 (1966).
Reines 64	Reines, F., and Kropp, W. R., *Phys. Rev. Letters*, **12**, 457 (1964).
Reynolds 60	Reynolds, J. H., *Phys. Rev. Letters*, **4**, 8 (1960).
Roll 66	Roll, P. G., and Wilkinson, D. T., *Phys. Rev. Letters*, **16**, 405 (1966).
Rudstam 62	Rudstam, G., Bruninx, E., and Pappas, A. C., *Phys. Rev.*, **126**, 1852 (1962).
Sakakibara 65	Sakakibara, S., *J. Geom. Geoel.*, **17**, 99 (1965).
Sandage 66	Sandage, A. R. et al., Astrophys. J., **146**, 316 (1966).
Sato 63	Sato, H., *Prog. Theor. Phys.*, **30**, 804 (1963).

Schaeffer 63	Schaeffer, O. A., et al., *Proc. Int. Conf. on Cosmic Rays at Jaipur*, **3**, 408 (1963).
Sekido 54	Sekido, Y., Yoshida, S., and Kamiya, Y., *J. Geomag. Geoel.*, **6**, 22 (1954).
Sekido 59	Sekido, Y., Yoshida, S., and Kamiya, Y., *Phys. Rev.*, **113**, 1108 (1959).
Sekido 62	Sekido Y., et al., *J. Phys. Soc. Japan*, **17**, Suppl. A–III, 131, 137, 139 (1962).
Sekido 63	Sekido, Y. and Sakakibara, S., *Proc. Int. Conf. on Cosmic Rays at Jaipur*, **4**, 189 (1963).
Shima 65	Shima, M., and Honda, M., *Technical Report of the Institute for Solid State Physics*, the University of Tokyo, Ser. A., No. 170 (1965).
Silberberg 66	Silberberg, R., *Phys. Rev.*, **148**, 1247 (1966).
Stoenner 60	Stoenner, R. W., Schaeffer, O. A., and Davis, R., *J. Geophys. Res.*, **65**, 3025 (1960).
Suess 56	Suess, H., and Urey, H. C., *Rev. Mod. Phys.*, **28**, 53 (1956).
Suess 65	Suess, H. E., *J. Geophys. Res.*, **70**, 5937 (1965).
Swann 33	Swann, W. F. G., *Phys. Rev.*, **43**, 217 (1933).
Syrovatsky 61	Syrovatsky, S. I., *Soviet Phys. JETP*, **13**, 1257 (1961).
Toyoda 65	Toyoda, Y. et al., *Proc. Int. Conf. on Cosmic Rays at London*, **2**, 708 (1965).
Turtle 62	Turtle, A. J., et al., *Month. Notices Roy. Astron. Soc.*, **124**, 296 (1962).
Vogt 62	Vogt, R., *Phys. Rev.*, **125**, 366 (1962).
Waddington 60	Waddington, C. J., *Progress in Nuclear Physics*, **8**, 1 (1960).
Waddington 62	Waddington, C. J., *J. Phys. Soc. Japan*, **17**, Suppl. A–III, 63 (1962).
Walker 65	Walker, R. M., Fleischer, R. L., and Price, P. B., *Proc. Int. Conf. on Cosmic Rays at London*, **2**, 1086 (1965).
Wänke 63	Wänke, H., *Proc. Int. Conf. on Cosmic Rays at Jaipur*, **3**, 473 (1963).
Wänke 65	Wänke, H., *Z. Naturforschung*, **20a**, 946 (1965).
Wild 62	Wild, J. P., *J. Phys. Soc. Japan*, **17**, Suppl. A–II, 249 (1962).
Zatsepin 51	Zatsepin, G. T., *Dokl. Akad. Nauk. USSR*, **80**, 577 (1951).

APPENDIX A

General Constants and Units

Fundamental Physical Constants

Speed of light in vacuum	$c = 2.997925(1)^* \times 10^{10}$ cm sec^{-1}

* The digits in parentheses represent the standard deviation error in the final digits, as given by E. R. Cohen and J. W. M. DuMond, *Rev. Mod. Phys.*, **37**, 537 (1965).

Elementary charge	$e = 4.80298(6) \times 10^{-10}$ esu
Planck's constant	$h = 6.62559(15) \times 10^{-27}$ erg sec
	$\hbar = h/2\pi = 1.054494(25) \times 10^{-27}$ erg sec
Fine-structure constant	$\alpha = e^2/\hbar c = 1/137.0388(6)$
Avogadro's constant†	$N = 6.02252(9) \times 10^{23}$ mole^{-1}

† The mass of $^{12}C = 12$ amu (atomic mass units).

Boltzmann's constant	$k = 1.38054(6) \times 10^{-16}$ erg deg^{-1}
Gravitational constant	$G = 6.670(5) \times 10^{-8}$ dyn cm^2 g^{-1}

Length

Proton Compton wavelength	$\lambda\!\!\!^{-}_p \equiv \hbar/m_p c = 2.10 \times 10^{-14}$ cm
Pion Compton wavelength	$\lambda\!\!\!^{-}_\pi = \hbar/m_\pi c = 1.41 \times 10^{-13}$ cm
Classical electron radius	$r_e \equiv e^2/m_e c^2 = 2.82 \times 10^{-13}$ cm
Electron Compton wavelength	$\lambda\!\!\!^{-}_e \equiv \hbar/m_e c = 3.86 \times 10^{-11}$ cm
Bohr radius	$a_0 \equiv \hbar^2/me^2 = 0.529 \times 10^{-8}$ cm
Reciprocal Rydberg constant	$1/R_\infty = 0.912 \times 10^{-5}$ cm
Wavelength of 1-eV photon	$\lambda_0 = hc/1$ eV $= 1.24 \times 10^{-4}$ cm $= 1.24 \times 10^4$ Å
Mean radius of the earth	$R_\oplus = 6.37 \times 10^8$ cm
Solar radius	$R_\odot = 6.96 \times 10^{10}$ cm

Astronomical unit	$AU = 1.50 \times 10^{13}$ cm
Light year	$ly = 0.946 \times 10^{18}$ cm
Parsec	$pc = 3.09 \times 10^{18}$ cm $= 3.26$ light years $= 2.06 \times 10^{5}$ AU
Distance of the sun from the Galactic center	$\simeq 10$ kpc
Diameter of the Galaxy	$\simeq 25$ kpc
Average diameter of clusters of galaxies	$\simeq 3$ Mpc
Hubble radius	$R_U = c/H \simeq 0.93 \times 10^{28}$ cm
Hubble constant	$H = 100$ km sec^{-1} Mpc^{-1} $= 3.2 \times 10^{-18}$ cm sec^{-1} cm^{-1}

Area

Pion cross section	$\pi(\hbar/m_\pi c)^2 = 6.28 \times 10^{-26}$ cm^2
Thomson cross section	$\sigma_{\text{Th}} = 8\pi\, r_e^{\,2}/3 = 6.65 \times 10^{-25}$ cm^2
Barn	$1\ \text{b} \equiv 10^{-24}$ cm^2
Area of the Bohr orbit	$\pi a_0^{\,2} = 0.880 \times 10^{-16}$ cm^2
Earth surface area	$4\pi R_\oplus^2 = 5.10 \times 10^{18}$ cm^2
Solar surface area	$4\pi R_\odot^2 = 6.09 \times 10^{22}$ cm^2

Time

Pion proper time	$\hbar/m_\pi c^2 = 4.71 \times 10^{-24}$ sec
Electron proper time	$\hbar/m_e c^2 = 1.29 \times 10^{-21}$ sec
Period of the Bohr orbit/2π	$\hbar^3/me^2 = 2.42 \times 10^{-17}$ sec
Mean solar day*	$8.64 \times 10^4 + 0.0016\,T$ sec
Mean sidereal day*	$8.62 \times 10^4 + 0.0016\,T$ sec
Tropical year*	$3.1557 \times 10^7 - 0.530\,T$ sec
Sidereal year*	$3.1558 \times 10^7 + 0.010\,T$ sec

* T = Julian centuries (in units of 36,525 days from 1900).

Hubble age	$1/H = 3.1 \times 10^{17}$ sec $= 0.98 \times 10^{10}$ years

Mass

Electron mass	$m_e = 0.911 \times 10^{-27}$ g
Mass unit	$m(^{12}\text{C})/12 = 1.66 \times 10^{-24}$ g
Nucleon mass	$m_p = 1.67 \times 10^{-24}$ g

Earth mass	$M_{\oplus} = 5.98 \times 10^{27}$ g
Solar mass	$M_{\odot} = 1.99 \times 10^{33}$ g
Mass of the Galaxy	$M_G \simeq 2 \times 10^{11}\ M_{\odot}$
Mass of the universe	$10^{54}\text{ g} \lesssim M_U \lesssim 10^{56}$ g

Energy

Energy associated with 1 °K	$kT(T = 1\ °\text{K}) = 1.38 \times 10^{-16}$ erg $= 0.862 \times 10^{-4}$ eV
Electron volt	$1\text{ eV} = 1.60 \times 10^{-12}$ erg
Frequency of 1-eV photon	$2.42 \times 10^{14}\ \text{sec}^{-1}$
Rydberg	$13.6\text{ eV} = 2.18 \times 10^{-11}$ erg
Photon energy associated with wavelength λ	$hc/\lambda = 1.99 \times 10^{-8}/\lambda(\text{Å})$ erg
Electron rest energy	$m_e c^2 = 0.511$ MeV
Mass unit	1 amu $= 0.931$ GeV
Proton rest energy	$m_p c^2 = 0.938$ GeV
Neutron rest energy	$m_n c^2 = m_p c^2 + 1.29$ MeV
Radiation density constant	$a = \pi^2 k^4/15c^3\hbar^3 = 7.56 \times 10^{-15}$ erg $\text{cm}^{-3}\ \text{deg}^{-4}$
Stefan-Boltzmann constant	$\sigma = ac/4 = 5.67 \times 10^{-5}$ erg $\text{cm}^{-2}\ \text{sec}^{-1}\ \text{deg}^{-4}$
Gas constant	$R_0 = Nk = 8.31 \times 10^7\ \text{erg mole}^{-1}$ deg^{-1}

Power

Solar radiation	$L_{\odot} = 3.90 \times 10^{33}\ \text{erg-sec}^{-1}$
Star ($M_{\text{bol}} = 0$) radiation	$3.02 \times 10^{35}\ \text{erg-sec}^{-1}$
Star ($m_{\text{bol}} = 0$) radiation	$2.5 \times 10^{-5}\ \text{erg cm}^{-2}\ \text{sec}^{-1}$
Absolute bolometric magnitude	$M_{\text{bol}} = 4.72 - 2.5 \log (L/L_{\odot})$
Apparent magnitude	$m = M + 5 \log$ (distance in pc) $- 5$ + space absorption in magnitudes

APPENDIX B

Kinematics in Collisions and Decays*

* Here we use the units of $c = 1$, so that energy, momentum, and mass are expressed in the same unit.

In this Appendix we give kinematic relations that often appear in the text. For a comprehensive treatment the readers should refer to Baldin (Baldin 59). For a simplified version refer also to Kaplon (Kaplon 60).

B.1 Lorentz Transformation

Let us consider a coordinate system moving with uniform velocity $\boldsymbol{\beta}$. The energy and the momentum in the rest system, E and p, are transformed to the moving system as

$$\begin{aligned} p_\perp^* &= p_\perp, \; p_\parallel^* = \gamma(p_\parallel + \beta E) \\ E^* &= \gamma(E + \beta p), \end{aligned} \tag{B.1}$$

where $p_\perp$ and $p_\parallel$ are the components of p perpendicular and parallel to $\boldsymbol{\beta}$, respectively; γ is called the *Lorentz factor*, which is defined by

$$\gamma = (1 - \beta^2)^{-1/2}. \tag{B.2}$$

Transformation back to the rest system is readily obtained by changing the sign of $\boldsymbol{\beta}$. We sometimes regard the rest system as the *laboratory system* (LS) and the moving one as the *center of mass system* (CMS), or vice versa. We use the convention that quantities in CMS and LS are distinguished by an asterisk on the former.

The electric- and magnetic-field strengths are transformed as

$$\begin{aligned} E_\parallel^* &= E_\parallel, & H_\parallel^* &= H_\parallel, \\ E_\perp^* &= \gamma(E_\perp - \boldsymbol{\beta} \times \mathbf{H}), & H_\perp^* &= \gamma(H_\perp + \boldsymbol{\beta} \times \mathbf{E}). \end{aligned} \tag{B.3}$$

It is almost needless to remark that the transformations (B.1) and (B.3) apply for any vector and tensor quantities, respectively.

If one chooses a coordinate system moving with velocity $c\boldsymbol{\beta}$, which is given by

$$\gamma^2 \boldsymbol{\beta} = \frac{\mathbf{E} \times \mathbf{H}}{E^2 + H^2},$$

the electric and magnetic fields become parallel. Hence one can always find a system in which the electric and magnetic fields are parallel, except in the case where two invariants about electromagnetic field strengths in B.2 are zero.

B.2 Invariant Relations

If we denote the energy-momentum four-vector by $\tilde{p}$ with components (E, p), the four-scalar product of two vectors $\tilde{p}_1$ and $\tilde{p}_2$ is invariant under the Lorentz transformation:

$$\tilde{p}_1 \cdot \tilde{p}_2 = E_1 E_2 - \mathbf{p}_1 \mathbf{p}_2 = \tilde{p}_1^* \cdot \tilde{p}_2^*. \tag{B.4}$$

In particular

$$\tilde{p}\tilde{p} = E^2 - p^2 = M^2, \tag{B.5}$$

where M is the rest mass of a particle.

Another important invariant is $d\mathbf{p}/E$, which may be called the *invariant phase volume*. With use of this invariance property the energy and angular distributions in two different systems are connected as

$$\frac{d\mathbf{p}}{E} = \frac{p^2\, dp\, d\Omega}{E} = p\, dE\, d\Omega = p^*\, dE^*\, d\Omega^*, \tag{B.6}$$

where $d\Omega$ and $d\Omega^*$ are the solid-angle elements in the respective systems.

In collision problems there sometimes appears a collision of two moving particles. Their relative velocity can be expressed as

$$v = \frac{[(\mathbf{p}_1 E_2 - \mathbf{p}_2 E_1)^2 - (\mathbf{p}_1 \times \mathbf{p}_2)^2]^{1/2}}{E_1 E_2 - \mathbf{p}_1 \mathbf{p}_2}, \tag{B.7}$$

where subscripts 1 and 2 refer to the respective particles. This is also an invariant relation, so that the right-hand side can be evaluated in any system.

It is almost needless to mention that quantities perpendicular to $\boldsymbol{\beta}$ are invariant. Among them the transverse momentum and the cross section are well known.

For electromagnetic fields there are two important invariant relations:

$$H^2 - E^2 = H^{*2} - E^{*2},$$

$$\mathbf{E} \cdot \mathbf{H} = \mathbf{E}^* \cdot \mathbf{H}^*.$$

The former implies that $E^* \geq H^* (E^* \leq H^*)$ in one system always gives $E \geq H (E \leq H)$ in any other system, whereas the latter indicates that if $\mathbf{E}^*$ and $\mathbf{H}^*$ are mutually perpendicular in one system, this always holds in any other system.

B.3 Threshold Energy in a Collision and Q-Value

Consider a collision process

$$A + B \rightarrow \sum_{i=1}^{n} (i),$$

in which outgoing particles, i's, are produced by the collision of two incident particles A and B. The energy-momentum conservation, $\tilde{p}_A + \tilde{p}_B = \sum_{i=1}^{n} \tilde{p}_i$, leads us, by an application of (B.4), to an invariant relation

$$(\tilde{p}_A + \tilde{p}_B)(\tilde{p}_A + \tilde{p}_B) = \left(\sum_{i=1}^{n} \tilde{p}_i\right)\left(\sum_{i=1}^{n} \tilde{p}_i\right). \tag{B.8}$$

We evaluate the left-hand side in LS and the right-hand side in CMS; in the latter $\mathbf{p}^* = \sum_{i=1}^{n} \mathbf{p}_i^* = 0$ and $E = \sum_{i=1}^{n} E_i^*$, whereas in the former $\mathbf{p}_B = 0$ and $E_B = M_B$. The Lorentz transformation (B.1) is now expressed as

$$p_A = \beta_c \gamma_c E^*, \qquad E_A + M_B = \gamma_c E^*,$$

where β_c and γ_c indicate the velocity and Lorentz factor connecting CMS to LS, respectively.

From the above we immediately obtain the velocity of CMS and the Lorentz factor as

$$\beta_c = \frac{p_A}{E_A + M_B}, \qquad \gamma_c = \frac{E_A + M_B}{\sqrt{2E_A M_B + M_A^2 + M_B^2}}. \tag{B.9}$$

If two colliding particles are alike, $M_A = M_B = M$, we have

$$\gamma_c = \sqrt{\frac{E_A + M}{2M}}, \qquad \beta_c \gamma_c^2 = \frac{p_A}{2M}. \tag{B.10}$$

In the extremely relativistic (e.r.) case $E_A \gg M_A, M_B, \beta_c$ and γ_c are reduced to

$$\beta_c \simeq 1, \gamma_c \simeq \sqrt{E_A/2M_B}. \tag{B.9, e.r.}$$

Figure B.1 shows β_c and γ_c plotted versus E/M.

The reaction under consideration has a threshold when all secondary particles are at rest in CMS; namely, $\sum_i E_i^* = \sum_i M_i$. By substituting this into (B.8) we obtain the threshold kinetic energy

$$W_A = E_A - M_A = \frac{(\sum_{i=0}^n M_i)^2 - (M_A + M_B)^2}{2M_B}. \tag{B.11}$$

If a reaction is exothermic, energy released by the reaction is defined as the Q-value; that is,

$$Q = (M_A + M_B) - \sum_{i=1}^n M_i. \tag{B.12}$$

This corresponds to (B.8) at zero incident energy; hence in any Lorentz frame

$$Q = \left[\left(\sum_{i=1}^n E_i\right)^2 - \left(\sum_{i=1}^n p_i\right)^2\right]^{1/2} - \sum_{i=1}^n M_i. \tag{B.12'}$$

This is applicable to the decay of a particle if particles A and B are regarded as being amalgamated, so that the mass of a decaying particle is $M_0 = \sum_i M_i + Q$.

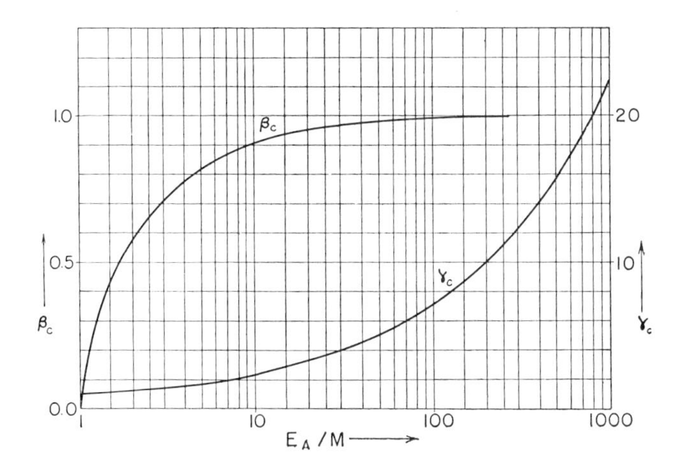

Fig. B.1 The velocity β_c and Lorentz factor γ_c in CMS versus the energy (E_A/M) in LS for the collision of particles of the same mass.

B.4 Energies and Angles of Outgoing Particles

If quantities in CMS are given, those in LS are obtained with the aid of (B.1). Defining the angles of emission in CMS and LS by

$$\mathbf{p}^*\boldsymbol{\beta} = p^*\beta_c \cos\theta^*, \qquad \mathbf{p}\cdot\boldsymbol{\beta}_c = p\beta_c\cos\theta,$$

respectively, we obtain

$$E = \gamma_c E^*(1+\beta_c\beta^*\cos\theta^*), \quad \text{or} \quad \gamma = \gamma_c\gamma^*(1+\beta_c\beta^*\cos\theta^*), \tag{B.13}$$

$$\tan\theta = \frac{\sin\theta^*}{\gamma_c(\cos\theta^* + \beta_c/\beta^*)}, \tag{B.14}$$

where $\beta^* = p^*/E^*$, $\gamma^* = E^*/M$, and $\gamma = E/M$. Quantities in CMS are likewise given in terms of laboratory quantities by changing the sign of β_c in (B.13) and (B.14).

The angle of emission behaves differently according to whether $\beta^* < \beta_c$ or $\beta^* > \beta_c$. For $\beta^* < \beta_c$ the denominator of (B.14) never becomes negative, so that $\theta < \pi/2$. This means that even a particle emitted backward in CMS goes forward in LS, because its velocity in CMS is smaller than the velocity of CMS. Hence the angle of emission and the energy in LS are restricted as

$$\left(\frac{\gamma^2 - \gamma^{*2}}{\gamma_c^2 - 1}\right)^{1/2} \le \cos\theta \le 1 \quad \text{for} \quad \beta^* < \beta_c, \tag{B.15a}$$

$$-1 \le \cos\theta \le 1 \quad \text{for} \quad \beta^* > \beta_c, \tag{B.15b}$$

$$\gamma_c\gamma^*(1-\beta_c\beta^*) \le \gamma \le \gamma_c\gamma^*(1+\beta_c\beta^*). \tag{B.15c}$$

For $\beta^* < \beta_c$, therefore, the energy of an emitted particle takes double values for given θ, whereas it takes a single value for $\beta^* > \beta_c$:

$$\gamma_\pm(\theta) = \frac{\gamma^* \pm \beta_c\cos\theta[\gamma^{*2} - \gamma_c^2(1-\beta_c^2\cos^2\theta)]^{1/2}}{\gamma_c(1-\beta_c^2\cos^2\theta)} \quad \text{for} \quad \beta^* < \beta_c, \tag{B.16a}$$

$$\gamma(\theta) = \frac{\gamma^* + \beta_c\cos\theta[\gamma^{*2} - \gamma_c^2(1-\beta_c^2\cos^2\theta)]^{1/2}}{\gamma_c(1-\beta_c^2\cos^2\theta)} \quad \text{for} \quad \beta^* > \beta_c. \tag{B.16b}$$

Now we consider the energy-angular distribution in LS. Since γ_+ and γ_- are respectively increasing and decreasing functions of γ^* in the case of $\beta^* < \beta_c$, the upper and lower limits of LS energy are given by $\gamma_+(\theta, \gamma_m^*)$ and $\gamma_-(\theta, \gamma_m^*)$, respectively, where $\gamma_m^* M$ is the maximum energy in CMS. In the case of $\beta^* > \beta_c$, on the other hand, the limits are readily obtained as $\gamma(\theta, \gamma_m^*)$ and $\gamma(\theta, \gamma_l^* > \gamma_c)$, where $\gamma_l^* M$ is the minimum energy in CMS.

Taking the above analysis into account, we are now able to derive the energy and angular distributions. The energy-angular distribution in LS is connected with that in CMS, because of (B.6), by

$$f(E, \theta) = f^*(E^*, \theta^*) \frac{dE^*\, d\Omega^*}{dE\, d\Omega} = \frac{p}{p^*} f^*(E^*, \theta^*), \tag{B.17}$$

In the third expression the quantities in CMS are functions of those in LS; that is,

$$E^* = \gamma_c(E - \beta_c p \cos\theta), \qquad \tan\theta^* = \frac{\sin\theta}{\gamma_c(\cos\theta + \beta_c/\beta)}. \tag{B.17$'$}$$

The energy and angular distributions are then obtained by

$$f_1(E) = \iint_{\theta_l}^{\theta_m} f(E, \theta)\, d\theta, \qquad f_2(\theta) = \int_{E_l}^{E_m} f(E, \theta)\, dE. \tag{B.18}$$

The limits on the integrals and the ranges of the arguments are found to be as follows:

1. If $\beta^* \leq \beta_c$,

$$\text{integration from } \cos\theta = \frac{\gamma\gamma_c - \gamma_m^*}{\beta_c \beta \gamma_c \gamma} \text{ to 1 for given } E = \gamma M$$

$$\gamma_c \gamma_m^*(1 - \beta_c \beta_m^*) \leq \gamma \leq \gamma_c \gamma_m^*(1 + \beta_c \beta_m^*) \quad \text{in} \quad f_1(E); \tag{B.19a}$$

$$\text{integration from } \gamma = \gamma_-(\theta) \text{ to } \gamma_+(\theta) \text{ for given } \theta,$$

$$\left(\frac{\gamma_c^2 - \gamma_c^2}{\gamma_c^2 - 1}\right)^{1/2} \leq \cos\theta \leq 1 \quad \text{in} \quad f_2(\theta). \tag{B.19b}$$

2. If $\beta^* > \beta_c$,

$$\text{integration from } \cos\theta = -1 \text{ to 1 for given } E = \gamma M,$$

$$\gamma_c \gamma_m^*(1 - \beta_c \beta_m^*) \leq \gamma \leq \gamma_c \gamma_m^*(1 - \beta_c \beta_m^*) \quad \text{in} \quad f_1(E); \tag{B.20a}$$

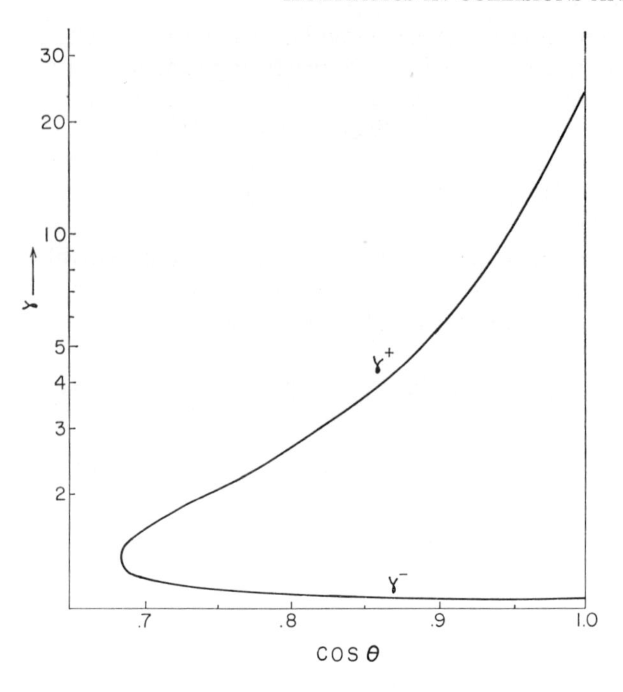

Fig. B.2 The limits of the LS Lorentz factor of a particle produced at LS angle θ for the case where the CMS velocity of the particle is smaller than β_c; from (Kaplon 60).

and

$$\text{integration from } \gamma = \gamma(\theta, \gamma_l^*) \text{ to } \gamma = (\theta, \gamma_m^*) \text{ for given } \theta, \\ -1 \leq \cos\theta \leq 1 \text{ in } f_2(\theta). \tag{B.20b}$$

If the value of β^* is not restricted as above, the lower limit of energy occurs at $\beta^* = \beta_c$, so that both the lower limit of the argument in (B.20*a*) and that of the energy integral are $\gamma = 1$. These allowed regions of energy and angle are shown in Fig. B.2 for case 1.

B.5 Transverse Momentum

Since the *transverse momentum* is invariant under the Lorentz transformation, its distribution is discussed in many practical cases. From the energy-angular distribution $f^*(E^*, \theta^*)$ a variable E^* is replaced by

p^*, and then p^* by the transverse momentum p_T by taking account of

$$f^*(E^*, \theta^*)\, dE^*\, d\Omega^* = f^*[E^*(p^*, \theta^*), \theta^*] \frac{p^*\, dp^*}{E^*}\, d\Omega^*,$$

$$E^* = \frac{(p_T{}^2 + M^2 \sin^2 \theta^*)^{1/2}}{\sin \theta^*},$$

and

$$p_T = p^* \sin \theta^* = p \sin \theta. \tag{B.21}$$

Thus we obtain the transverse-momentum distribution $f(p_T)\, dp_T$ as

$$f(p_T) = 2\pi p_T \int_0^{\pi} \frac{f^*[E^*(p^*, \theta^*), \theta^*]}{(p_T{}^2 + M^2 \sin^2 \theta^*)^{1/2}}\, d\theta^*. \tag{B.22}$$

In LS the transverse-momentum distribution is obtained only by changing the integrand in (B.22) into that in LS, but the limits of the integral have to be considered with the same caution as in Section B.4. Corresponding to (B.16), the allowed range of p_T is given by

$$\sin \theta \sqrt{E_-^2 - M^2} \quad \text{or} \quad 0 \le p_T \le \sin \theta \sqrt{E_+^2 - M^2}. \tag{B.23}$$

Now the upper limit of p_T has a maximum at small θ due to the factor $\sin \theta$ multiplied by $\sqrt{E_+^2 - M^2}$, which increases with decreasing θ. Hence the upper limit takes double values for a given value of p_T. This is illustrated in Fig. B.3, where the two branches of the upper limit are denoted by $p_T^+(1)$ and $p_T^+(2)$; and the lower limit, by p_T^-.

Then the limits of angular integral are given by

$$0 \le \frac{p_T}{M} \le \beta\beta_m^* \left[1 - \left(\frac{\gamma_m^*}{\gamma}\right)^2\right]^{1/2} \quad \text{for} \quad \begin{cases} p_T^- = \sin \theta \sqrt{E_-^2 - M^2} \\ p_T^+ = \sin \theta \sqrt{E_+^2 - M^2} \end{cases} \tag{B.24a}$$

$$\beta\beta_m^* \left[1 - \left(\frac{\gamma_m^*}{\gamma}\right)^2\right]^{1/2} \le \frac{p_T}{M} \le \frac{p_m^*}{M} \quad \text{for} \quad \begin{cases} p_T^- = 0 \\ p_T^+(1, 2) = \sin \theta \sqrt{E_+^2 - M^2}. \end{cases} \tag{B.24b}$$

Explicit expressions for the limits of θ are so complicated that they are not given here.

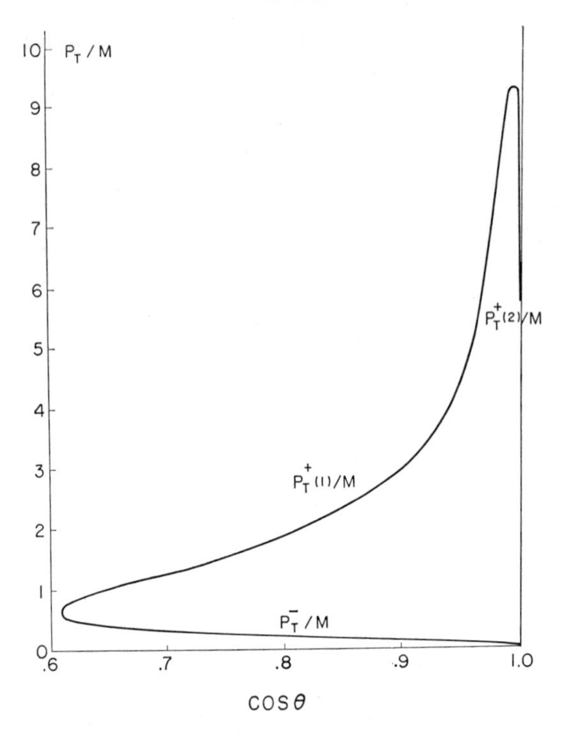

Fig. B.3 The limits of the transverse momentum of a particle produced at LS angle θ; from (Kaplon 60).

B.6 Polarization

In the collision between unpolarized particles outgoing particles may be polarized, depending on the nature of interactions. The polarization is expressed by a unit pseudovector ζ, which is parallel to the spin pseudovector $\mathbf{s}$ in the rest system of the particles concerned:

$$\zeta = \frac{\bar{\mathbf{s}}}{|\bar{\mathbf{s}}|},$$

where the bar indicates quantities in the rest system and hereafter $|\bar{\mathbf{s}}|$ is taken as unity for convenience. In finding the relativistic transformation of the spin we have to know the time component of the spin, s_0. Since $\tilde{s}(s_0, \mathbf{s})$ forms a four-pseudovector, there holds

$$\tilde{s} \cdot \tilde{p} = s_0 E - \mathbf{s} \cdot \mathbf{p} = 0, \tag{B.25}$$

because the scalar product of a vector and a pseudovector is a pseudoscalar quantity that must vanish because of its mirror inversion. From this there results $\bar{s}_0 = 0$ in the rest system.

The value of $\tilde{s}$ in motion can be obtained by the Lorentz transformation of $(0, \mathbf{s})$ as

$$\mathbf{s} = \left[\zeta - \frac{(\zeta \cdot \mathbf{p})\mathbf{p}}{p^2}\right] + \frac{E}{M}\frac{(\zeta \cdot \mathbf{p})\mathbf{p}}{p^2} = \zeta + \frac{(\zeta \cdot \mathbf{p})\mathbf{p}}{M(E+M)},$$
$$s_0 = \frac{(\mathbf{s} \cdot \mathbf{p})}{E} = \frac{(\zeta \cdot \mathbf{p})}{M}. \tag{B.26}$$

The degree of polarization is defined as the projection of ζ onto a certain direction that is represented by a unit vector $\mathbf{n}$. Since we know only two vectors parallel and perpendicular to the direction of flight $\mathbf{p}$, $\mathbf{n}_{\parallel}$ and $\mathbf{n}_{\perp}$, respectively,

$$\zeta \cdot \mathbf{n}_{\perp} = \mathbf{s} \cdot \mathbf{n}_{\perp}, \qquad \zeta \cdot \mathbf{n}_{\parallel} = \frac{M}{E}(\mathbf{s} \cdot \mathbf{n}_{\parallel}) = \frac{M}{p} s_0,$$

are sufficient to express the polarization. The latter vanishes, unless the conservation of parity is violated in the interaction responsible for the production of the particle concerned.

Now let us assume that (B.25) holds in CMS. In the case of strong interactions in which $\mathbf{s}^*$ is perpendicular to $\mathbf{p}^*$ we have $\mathbf{s}^* = \zeta$ and $s_0^* = 0$. Then the spin in LS is given by (B.26) with $\gamma = E/M$ and $\boldsymbol{\beta} = \mathbf{p}/E$. If $\mathbf{s}^* \| \mathbf{p}^*$ as in the case of two-body decays with neutrino emission, we have

$$s^* = \left(\frac{E^*}{M}\right)\zeta, \qquad s_0^* = \frac{p^*}{M}, \tag{B.27}$$

where we have assumed $\mathbf{p}^* = p^* \zeta$ for convenience. Then the spin in LS is given by

$$\mathbf{s} = \frac{E^*}{M}\zeta + (\gamma - 1)\frac{E^*}{M}\frac{(\zeta \cdot \boldsymbol{\beta})\boldsymbol{\beta}}{\beta^2} + \gamma\frac{p^*}{M}\boldsymbol{\beta},$$
$$s_0 = \gamma\left[\frac{p^*}{M} + \frac{E^*}{M}(\zeta \cdot \boldsymbol{\beta})\right]. \tag{B.28}$$

Comparing (B.28) with (B.26) and eliminating $\zeta \cdot \boldsymbol{\beta}$ because of $E = \gamma[E^* + (\boldsymbol{\beta} \cdot \mathbf{p}^*)]$, we have the *longitudinal polarization*

$$\zeta \cdot \mathbf{n}_{\parallel} = s_0\left(\frac{M}{p}\right) = \left(\frac{EE^*}{pp^*}\right) - \left(\frac{\gamma M^2}{pp^*}\right). \tag{B.29}$$

B.7 Spontaneous Decay

In the decay of a parent particle (0)—its energy-momentum vector being denoted as $\tilde{p}_0$,—there holds the energy-momentum conservation, $\tilde{p}_0 = \sum_i \tilde{p}_i$, which leads to an invariant relation like (B.8). Evaluating the left-hand side in the rest system of the parent particle (RS) and the right-hand side in LS, we have

$$\tilde{p}_0^* \tilde{p}_0^* = M_0{}^2 = \sum_i \tilde{p}_i \sum_i \tilde{p}_i . \tag{B.30}$$

In the two-body decay this is simply expressed as

$$2(E_1 E_2 - p_1 p_2 \cos \Theta) = M_0{}^2 - M_1{}^2 - M_2{}^2, \tag{B.31}$$

where Θ is the angle between the directions of two daughter particles. If the daughter particles are massless, a simple relation holds, as follows:

$$\sin^2\left(\frac{\Theta}{2}\right) = \frac{M_0{}^2}{4} E_1 E_2 \leq \frac{M_0{}^2}{E_0{}^2}, \tag{B.32}$$

where $E_0 = E_1 + E_2$. The equality in the last relation gives the maximum opening angle.

In RS the energy of a daughter particle can be calculated by squaring $\tilde{p}_0^* - \tilde{p}_1^* = \tilde{p}_2^*$, as

$$E_1^* = \frac{(M_0{}^2 + M_1{}^2 - M_2{}^2)}{2M_0} \quad \text{and } (1 \rightleftarrows 2). \tag{B.33}$$

From this we obtain

$$p_1^* = p_2^* = \frac{[(M_0{}^2 - M_1{}^2 - M_2{}^2)^2 - 4M_1 M_2]^{1/2}}{2M_0} . \tag{B.34}$$

Since $F^*(E_1^*, \theta_1^*)$ represents a line energy spectrum and the isotropic angular distribution, the energy-angular distribution in LS is given from (B.17) by

$$f(E_1, \theta_1) = \frac{\beta_1\, \delta[\gamma_c(E_1 - \beta_c p_1 \cos \theta_1) - E_1^*]}{4\pi[\gamma_c{}^2(1 - \beta_c \beta_1 \cos \theta_1)^2 - (1/\gamma_1{}^2)]^{1/2}} . \tag{B.35}$$

The integration of (B.35) over Ω results in

$$f(E_1) = \frac{1}{2\beta_c \gamma_c p_1^*}, \quad \gamma_c(E_1^* - \beta_c p_1^*) \leq E_1 \leq \gamma_c(E_1^* + \beta_c p_1^*). \tag{B.36}$$

This is the well-known flat spectrum for a decay product.

If we know the energy and the momentum of one of the daughter particles, the energy of its parent particle is limited as

$$E_0^- \equiv (E_1 E_1^* - p_1 p_1^*) \frac{M_0}{M_1^{\ 2}} \leq E_0 \leq (E_1 E_1^* + p_1 p_1^*) \frac{M_0}{M_1^{\ 2}} \equiv E_0^+ ,$$

$$\gamma_1 \gamma_1^* (\beta_1 - \beta_1^*) \leq \beta_c \gamma_c \leq \gamma_1 \gamma_1^* (\beta_1 + \beta_1^*). \qquad \text{(B.37)}$$

Hence the energy spectrum of the daughter particles is connected with that of the parent particles as

$$f_1(E_1) = \int_{E_0^-}^{E_0^+} f_0(E_0) \frac{M_0}{2p_0 p_1{}^*} dE_0 . \qquad \text{(B.38)}$$

If the daughter particle is massless, we have, instead of (B.37),

$$E_0 \geq \frac{E_1{}^2 + E_1^{*2}}{2E_1 E_1^*} M_0 \equiv E_0^- . \qquad \text{(B.37')}$$

Hence instead of (B.38) we obtain

$$f_1(E_1) = \int_{E_0^-}^{\infty} f_0(E_0) \frac{M_0}{2p_0 p_1^*} dE_0 . \qquad \text{(B.38')}$$

If the angular distribution is taken into account, the situation is rather complicated. The reader should refer to Baldin (Baldin 59) and Kaplon (Kaplon 60).

In three-body decay the third daughter particle often cannot be observed. Then we write the energy-momentum conservation as

$$\tilde{p}_0 - \tilde{p}_3 = \tilde{p}_1 + \tilde{p}_2 .$$

By squaring the left-hand side in RS and the right-hand side in LS, we obtain

$$M_0{}^2 + M_3{}^2 - 2M_0 E_3^* = M_1{}^2 + M_2{}^2 + 2E_1 E_2 - 2\mathbf{p}_1 \mathbf{p}_2 . \qquad \text{(B.39)}$$

Since the value of E_3^* is limited, this formula can be used for checking consistency.

It is sometimes convenient to introduce the CMS of two of the daughter particles. The transformation to this system can be made for velocity

$$\boldsymbol{\beta} = \frac{-(\mathbf{p}_1 + \mathbf{p}_2)}{E_1 + E_2} . \qquad \text{(B.40)}$$

In the case of nuclear β-decay this may be taken as the velocity of a daughter nucleus. In this case $E_3^* \simeq M_3$ and $M_0 - M_3$ is the Q-value for β-decay. If the second particle is a neutrino, $M_2 = 0$ and (B.39) is reduced to $E_2 = Q - E_1$.

B.8 Two-Body Collision

In the two-body collision there are three invariant quantities, defined as follows:

$$s = (\tilde{\mathbf{p}}_A + \tilde{\mathbf{p}}_B)^2 = (\tilde{\mathbf{p}}_1 + \tilde{\mathbf{p}}_2)^2, \tag{B.41s}$$

$$t = (\tilde{\mathbf{p}}_A - \tilde{\mathbf{p}}_1)^2 = (\tilde{\mathbf{p}}_B - \tilde{\mathbf{p}}_2)^2, \tag{B.41t}$$

$$u = (\tilde{\mathbf{p}}_A - \tilde{\mathbf{p}}_2)^2 = (\tilde{\mathbf{p}}_B - \tilde{\mathbf{p}}_1)^2. \tag{B.41u}$$

These three are related to the masses of participating particles as

$$s + t + u = M_A{}^2 + M_B{}^2 + M_1{}^2 + M_1{}^2. \tag{B.42}$$

Here s is the squared total energy in CMS and is also expressed for the particle (B) at rest in LS as

$$s = (E_A^* + E_B^*)^2 = (E_1^* + E_2^*)^2 = M_A{}^2 + M_B{}^2 + 2E_A M_B$$

$$= M_1{}^2 + M_2{}^2 + 2(E_1^* E_2^* - \mathbf{p}_1^* \cdot \mathbf{p}_2^*). \tag{B.43}$$

The CMS momenta and energies of outgoing particles are obtained from (B.41s) on account of $\mathbf{p}_1^* = -\mathbf{p}_2^* \equiv \mathbf{p}^*$ as

$$s = M_1{}^2 + M_1{}^2 + 2(\sqrt{M_1{}^2 + p^{*2}}\sqrt{M_2{}^2 + p^{*2}} + p^{*2}).$$

Hence

$$p^{*2} = \frac{1}{4}\frac{s^2 - 2(M_1{}^2 + M_2{}^2)s + (M_1{}^2 - M_2{}^2)^2}{s},$$

$$E_1^* = \frac{s + M_1{}^2 - M_2{}^2}{2\sqrt{s}}, \tag{B.44}$$

$$E_*^2 = \frac{s + (M_2{}^2 - M_1{}^2)}{2\sqrt{s}}.$$

The squared *four-momentum transfer* t is expressed as a function of s and the scattered angle θ^* in CMS. Because of $\mathbf{p}_A^* \cdot \mathbf{p}_1^* = \mathbf{p}_B^* \cdot \mathbf{p}_2^* = p_A^* p_1^* \cos\theta^*$, (B.41$t$) is reduced to

$$t = \tfrac{1}{2}[(M_A{}^2 + M_B{}^2 + M_1{}^2 + M_2{}^2) - 2(E_A^* E_1^* + E_B^* E_2^*) + 2p_A^* p_1^* \cos\theta^*].$$

Since the second term in the square brackets is expressed by reference to (B.44) and similar relations for the incident state as

$$2(E_A^* E_1^* + E_B^* E_2^*) = s + \frac{(M_A{}^2 - M_B{}^2)(M_1{}^2 - M_2{}^2)}{s},$$

we obtain

$$t = -\tfrac{1}{2}\left[s - (M_A{}^2 + M_B{}^2 + M_1{}^2 + M_2{}^2) + \frac{(M_A{}^2 - M_B{}^2)(M_1{}^2 - M_2{}^2)}{s}\right] + p_A^* p_1^* \cos\theta^*. \quad \text{(B.45)}$$

In this expression we find that $-t$ is an increasing function of θ^*, and its maximum and minimum values are given by

$$[-t]_{\min}^{\max} = \tfrac{1}{2}\left[s - (M_A{}^2 + M_B{}^2 + M_1{}^2 + M_2{}^2) + \frac{(M_A{}^2 - M_B{}^2)(M_1{}^2 - M_2{}^2)}{s}\right] \pm 2p_A^* p_1^*. \quad \text{(B.46)}$$

Hence we have

$$-t = [-t]_{\min} + 2p_A^* p_1^*(1 - \cos\theta^*) = [-t]_{\min} + 4p_A^* p_1^* \sin^2\frac{\theta^*}{2}. \quad \text{(B.45')}$$

For elastic collisions $[-t]_{\min}$ vanishes, and the last term represents the squared momentum transfer.

It can be proved that if $(M_A - M_1)(M_B - M_2) > 0$, then $t < 0$ (Salzman 62). When the collision is symmetric with respect to two participating particles both in the incoming and the outgoing states, t is always negative. In other words, a particle exchanged between two vertices of collision is virtual.

The value of t can be obtained from the scattered angle in CMS or from the recoil energy in LS:

$$t = M_A{}^2 + M_1{}^2 - 2E_A^* E_1^* + 2p_A^* p_1^* \cos\theta^* = M_B{}^2 + M_2{}^2 - 2M_B E_2. \quad \text{(B.46')}$$

In the case of elastic scattering this is reduced to

$$t = -2p_A^{*2}(1 - \cos\theta^*) = -2M_B(E_2 - M_2). \quad \text{(B.46'')}$$

Likewise $u = -2p_A^{*2}(1 + \cos\theta^*)$. Thus the square of the transverse momentum is given by

$$p_T^2 = p_A^{*2} \sin^2\theta^* = \frac{-tu}{(t+u)} = \frac{tu}{(s - 2M_A^2 - 2M_B^2)}. \tag{B.47}$$

If scattering is preferentially in the forward direction, $t \ll u$, and consequently

$$p_T^2 \simeq -t. \tag{B.47$'$}$$

Conversely, the smallness of the transverse momentum results in (B.47′).

If, on the other hand, the differences between M_A and M_1, and between M_B and M_2 are very large, we may evaluate s and t in a system where a lighter particle is at rest. Since the momentum of a heavier particle is negligible in comparison with its rest energy, we can obtain

$$-st \simeq (M_A^2 - M_1^2)(M_B^2 - M_2^2). \tag{B.48}$$

The energy and angular distributions in LS are of simple forms, because that the energy of an outgoing particle in CMS has a unique energy. If the angular distribution in CMS is denoted by $g^*(\cos\theta^*)$, the energy-angular distribution in LS is, according to (B.17), given by

$$F(E, \theta) =$$

$$\frac{\beta\delta[\gamma_c(E - \beta_c p \cos\theta) - E^*]}{[\gamma_c^2(1 - \beta_c\beta\cos\theta)^2 - (1/\gamma^2]^{1/2}} g^*\left[\frac{1}{\beta\gamma}\sqrt{\gamma^2\gamma_c^2(1 - \beta\beta_c\cos\theta)^2 - 1}\right]. \tag{B.49}$$

The integral over the solid angle gives us

$$F_1(E) = \frac{2\pi}{\beta_c\gamma_c p^*} g^*\left(\frac{p^*}{p}\right), \tag{B.50a}$$

$$\gamma_c(E^* - \beta_c p^*) \leq E \leq \gamma_c(E^* + \beta_c p^*).$$

The integral over energy yields

$$F_2(\theta) = \frac{\beta\gamma}{\beta^*\gamma^*\gamma_c[1 - (\beta_c/\beta)\cos\theta]} g^*\left(\frac{\beta^*\gamma^*}{\beta\gamma}\right), \tag{B.50b}$$

in which γ, and consequently β, are given in (B.16) as functions of θ.

The above results are applied for some cases of practical importance.

1. *Elastic scattering* ($M_A = M_1$, $M_B = M_2$).

$$p_A^{*2} = p_B^{*2} = \frac{s^2 - 2(M_A{}^2 + M_B{}^2)s + (M_A{}^2 - M_B{}^2)^2}{4s},$$

$$\cos\theta^* = 1 - \frac{\sqrt{-t}}{p_A^*}. \qquad \text{(B.51)}$$

The kinetic energy of the recoil particle is simply given by

$$W_2(\theta_2) = E_2(\theta_2) - M_B = \frac{2\beta_c{}^2 \cos^2\theta_2}{1 - \beta_c{}^2 \cos^2\theta_2} M_B \leq 2\beta_c{}^2 \gamma_c{}^2 M_B$$

$$= \frac{2p_A{}^2 M_B}{2E_A M_B + M_A{}^2 + M_B{}^2}. \qquad \text{(B.52)}$$

The maximum energy transfer is $2M_B v_A{}^2$ in the nonrelativistic limit and tends to E_A in the high-energy limit. Equation (B.52) also gives the value of $-t/2M_B$. The expressions of γ_2 and β_2 as functions of θ_2 are given by

$$\gamma_2(\theta_2) = \frac{1 + \beta_c{}^2 \cos^2\theta}{1 - \beta_c{}^2 \cos^2\theta}, \qquad \beta_2(\theta_2) = \frac{2\beta_c \cos\theta}{1 + \beta_c{}^2 \cos^2\theta} \qquad \text{(B.53)}$$

These should be introduced into (B.50*b*) to obtain the angular distribution. For the scattered particle we have

$$\gamma_1^* = \sqrt{(M_A/M_1)^2 + \beta_c{}^2\gamma_c{}^2(M_B/M_1)^2}, \qquad \beta_1^* = \frac{\beta_c \gamma_c M_B}{M_1 \gamma^*}, \qquad \text{(B.54)}$$

and $\gamma_1(\theta_1)$ is obtained by introducing these into (B.16).

2. *Reactions induced by a massless particle* ($M_A = 0$). Now we have

$$\gamma_c = \frac{E_A + M_B}{\sqrt{2E_A M_B + M_B{}^2}}, \qquad \beta_c = \frac{E_A}{E_A + M_B}. \qquad \text{(B.55)}$$

The energy of a recoil particle is identical with (B.52), and that of a scattered particle (photon) is given by

$$E_1(\theta_1) = \frac{\beta_c M_B}{1 - \beta_c \cos\theta_1} = \frac{E_A}{1 + (E_A/M_B)(1 - \cos\theta_1)}. \qquad \text{(B.56)}$$

This is simply the energy of a Compton-scattered photon.

B.9 Multiple Production

The kinematics in multiple production is so complicated that few relations of practical use are obtainable without simplifying assumptions or specific models. Many relations given in this section are therefore derived by referring to the *fireball model.*

B.9.1 Angular Distribution in the Center-of-Mass System

The angle of emission of the ith particle in LS, θ_i, is related to that in CMS, θ_i^*, through (B.14). By counting the number of particles with emission angles smaller than θ we can construct the integral angular distribution $F(\theta)$, which is normalized as $F(\pi) = 1$. This is equal to the integral angular distribution in CMS, $F(\theta^*)$, where θ^* is the CMS angle corresponding to θ, if no particle emitted has CMS velocity smaller than β_c. This is because θ is an increasing function of θ^* in the case of $\beta^* \geq \beta_c$. More loosely, this is satisfied if $1 + (\beta_c/\beta^*) \cos \theta^* \geq 0$.

Under this condition the function $F(\theta)$, called the *F-plot* (Duller 54) represents the angular distribution in CMS. Since the solid angle in CMS is given by

$$2\pi(1 - \cos \theta^*) = 4\pi \sin^2\left(\frac{\theta^*}{2}\right),$$

the isotropic distribution in CMS is represented by

$$F(\theta) = \sin^2\left(\frac{\theta^*}{2}\right) \quad \text{or} \quad \frac{F(\theta)}{[1 - F(\theta)]} = \tan^2\left(\frac{\theta^*}{2}\right). \tag{B.57}$$

Under the condition above, β_c/β^* in (B.14) can be put equal to unity, and (B.14) is reduced to

$$\tan \theta = \gamma_c^{-1} \tan\left(\frac{\theta^*}{2}\right). \tag{B.14$'$}$$

Therefore the plot of log $[F/(1 - F)]$ against log tan θ gives a straight line of slope 2, for the isotropic distribution. The angle corresponding to the median angle in CMS is given by the angle at which log $[F/(1 - F)]$ intersects the x-axis, or $F = \frac{1}{2}$. If the angular distribution in CMS is symmetric with respect to the forward and backward directions, the shapes of log $[F/(1 - F)]$ above and below the x-axis are symmetric with respect to the median point.

In the symmetric case $\theta^* = \pi/2$ gives the median angle, which implies that the number of particles emitted with $\theta^* < \pi/2$ and $\theta^* > \pi/2$ are equal. The corresponding angle in LS, called also the *median angle*, $\theta_{1/2}$, is given through (B.14) by

$$\tan \theta_{1/2} = \frac{\beta^*}{\beta_c \gamma_c} \simeq \frac{1}{\gamma_c}. \tag{B.58}$$

This provides a means of obtaining the Lorentz factor γ_c. However, the value of the median angle can be defined only with considerable ambiguity, especially unless the number of secondary particles is very large. Moreover, much of the information on the angular distribution is not fully utilized.

Both the F-plot and the median-angle method are based only on the integral angular distribution. More detailed information can be obtained by referring to the differential angular distribution. This is conventionally represented by the plot of log tan θ_i on a horizontal line, called the log tan θ plot. According to (B.14), the emission angle in CMS is expressed as

$$\log \tan \theta_i = \log \tan\left(\frac{\theta_i^*}{2}\right) - \log \gamma_c - \log\left[\frac{1 + (\beta_c/\beta^*) - 1}{2 \cos^2(\theta^*/2)}\right] \tag{B.59}$$

The last term vanishes for $\beta^* = \beta_c$ and is negligibly small in most cases. If this is the case, the Lorentz factor has little effect in shifting the distribution of log tan θ_i. This means that the shape of the log tan θ plot is Lorentz invariant and represents the angular distribution in CMS; for example, the CMS angular distribution is symmetric, and the values of log tan θ_i distribute symmetrically about $-\log \gamma_c$.

The last property is used for evaluating the value of γ_c in the symmetric case. Since $\sum_i \log \tan (\theta_i^*/2)$ vanishes owing to symmetry, we have (Castagnoli 53)

$$-\log \gamma_c = \frac{1}{n} \sum_{i=1}^{n} \log \tan \theta_i, \tag{B.60}$$

where n is the number of secondary particles. The method of evaluating γ_c by reference to the center of gravity of the log tan θ plot is called the *Castagnoli method.*

Even if the emission angles of n particles are taken into account, the value of γ_c thus determined is subject to statistical fluctuations. The latter may be represented by the second moment of the log tan θ distribution, defined by

$$\sigma^2 \equiv \sum_{i=1}^{n} \frac{(\log \tan \theta_i - \overline{\log \tan \theta})^2}{(n-1)}. \tag{5.61}$$

In the isotropic distribution the value of σ is given by

$$\sigma = \pi \log\left(\frac{e}{2\sqrt{3}}\right) = 0.39. \tag{B.62}$$

The rms error of $\log \gamma_c$ due to statistical fluctuations is $\sigma/\sqrt{n}$. The deviation of σ from (B.62) represents the degree of anisotropy.

The Castagnoli method of evaluating γ_c gives large weight to the particles with small emission angles. This would give rise to serious error if any asymmetry exists. This disadvantage may be removed by the following method (Yazima 65). Writing (B.14′) as

$$\gamma_c \tan \theta_i = \tan\left(\frac{\theta_i^*}{2}\right) = \frac{1 - \cos \theta_i^*}{\sin \theta_i^*} = \operatorname{cosec} \theta_i^* - \cot \theta_i^* \tag{B.63a}$$

and its inverse as

$$\gamma_c^{-1} \cot \theta_i = \cot\left(\frac{\theta_i^*}{2}\right) = \frac{1 + \cos \theta_i^*}{\sin \theta_i^*} = \operatorname{cosec} \theta_i^* + \cot \theta_i^*, \tag{B.63b}$$

and noticing that $\sum_i \cot \theta_i^* = 0$ in the symmetric case, we can obtain from (B.63a) and (B.63b)

$$\sum_{i=1}^{n} \operatorname{cosec} \theta_i^* = \left(\sum_{i=1}^{n} \tan \theta_i \sum_{i=1}^{n} \cot \theta_i\right)^{1/2} \tag{B.64}$$

and

$$\gamma_c = \left(\frac{\sum_{i=1}^{n} \cot \theta_i}{\sum_{i=1}^{n} \tan \theta_i}\right)^{1/2}. \tag{B.65}$$

The expression (B.64) is useful for obtaining the sum of the energy of particles emitted in CMS as

$$\sum_{i=1}^{n} p_i^* = \sum_{i=1}^{n} p_{Ti} \operatorname{cosec} \theta_i^* = \bar{p}_T \sum_{i=1}^{n} \operatorname{cosec} \theta_i^* = \bar{p}_T \left(\sum_{i=1}^{n} \tan \theta_i \sum_{i=1}^{n} \cot \theta_i\right)^{1/2}. \tag{B.66}$$

In the second equality use has been made of the fact that the transverse momentum is independent of θ^*.

In the asymmetric case it is more convenient to use the expressions

$$\gamma_c \tan \theta_i = \frac{p_i^* \sin \theta_i^*}{E_i^* + p_i^* \cos \theta_i^*} = \frac{p_T}{p_T^2 + \mu^2} (E_i^* - p_i^* \cos \theta_i^*), \tag{B.67a}$$

$$\gamma_c^{-1} \cot \theta_i = \frac{1}{p_T} (E_i^* + p_i^* \cos \theta_i^*), \tag{B.67b}$$

rather than (B.63), where E_i^* is the total energy of an emitted particle in CMS and μ is its mass; both p_T and μ are assumed to be the same for all secondary particles. Now (B.66) is modified to

$$\sum_i p_i^* = p_T \frac{\sqrt{1+\mu^2/p_T{}^2}}{\sqrt{1-Z^2}} \left(\sum_i \tan\theta_i \sum_i \cot\theta_i\right)^{1/2}, \tag{B.67}$$

where Z is the *asymmetry factor* defined by

$$Z \equiv \frac{\sum_{i=1}^n p_i^* \cos\theta_i^*}{\sum_i E_i^*}. \tag{B.68}$$

Equation (B.65) is modified to

$$\gamma_c = \frac{1}{\sqrt{1+\mu^2/p_T{}^2}} \sqrt{\frac{1-Z}{1+Z}} \left(\frac{\sum_i \cot\theta_i}{\sum_i \tan\theta_i}\right)^{1/2}. \tag{B.69}$$

If the value of Z is known, this should give a better estimate of γ_c than other methods.

If the angular distribution is expressed by an analytic form, $\cos^m\theta^*\, d(\cos\theta^*)$, the average value of $\operatorname{cosec}\theta^*$, which appears in (B.66), is evaluated as

$$\overline{\operatorname{cosec}\theta^*} = \frac{1}{n}\sum_{i=1}^n \operatorname{cosec}\theta_i^* = \frac{\pi(m+1)!}{2^{m+1}\left(\frac{m}{2}-1\right)!\left(\frac{m}{2}\right)!} \simeq (m+1)\sqrt{\frac{\pi}{2m}} \simeq \tfrac{5}{4}\sqrt{m}, \tag{B.70}$$

where the last two expressions hold for large m. Since this relation holds in the symmetric csse, comparison with (B.64) gives

$$m \simeq \frac{2}{n^2\pi}\left(\sum_{i=1}^n \tan\theta_i \sum_{i=1}^n \cot\theta_i\right). \tag{B.71}$$

B.9.2 Inelasticity

The fraction of energy transferred to secondary particles is defined as *inelasticity*. The energy transferred to secondary particles can be expressed through (B.67*a*) as

$$\sum_i E_i^* = \frac{p_T}{1-Z}\left(1+\frac{\mu^2}{p_T{}^2}\right)\gamma_c \sum_i \tan\theta_i. \tag{B.72}$$

This is compared with the energy available in CMS; from (B.8) we obtain

$$(E_A^* + E_B^*) - (M_A + M_B) = \sqrt{2\gamma_L M_A M_B + M_A{}^2 + M_B{}^2} - (M_A + M_B),$$
$$\simeq 2\gamma_c M_B, \tag{B.73}$$

where $\gamma_L = E_A/M_A$, and the last expression holds in the extremely relativistic case.

Dividing (B.72) by (B.73) we obtain the inelasticity in CMS:

$$K^* \equiv \frac{p_T}{2M_B} \frac{1}{1-Z} \left(1 + \frac{\mu^2}{p_T{}^2}\right) \sum_i \tan \theta_i. \tag{B.74}$$

It is also convenient to introduce the inelasticity in LS. The energy transferred to secondary particles in LS is given by

$$\sum_i E_i = \gamma_c \left(\sum_i E_i^* + \beta_c \sum_i p_i^* \cos \theta_i^* \right) = \gamma_c (1 + \beta_c Z) \sum_i E_i^*. \tag{B.75}$$

Dividing this by the kinetic energy of the incident particle, $(\gamma_L - 1)M_A$, we obtain the inelasticity in the extremely relativistic case:

$$K_L \equiv \frac{\sum_i E_i}{(\gamma_L - 1)M_A} \simeq \frac{1+Z}{2\gamma_c M_B{}^2} \sum_i E_i^* = (1+Z)K^*. \tag{B.76}$$

The difference between K_L and K^* arises from asymmetry.

The asymmetry effect is made clear if we work out kinematics in the mirror system, in which the incident particle is at rest. In the *mirror system*, abbreviated as MS, the energy of a secondary particle is given by

$$E_{im} = \gamma_L (E_i - \beta_L p_i \cos \theta_i). \tag{B.77}$$

Hence the mirror inelasticity is obtained as

$$K_M \equiv \frac{\sum_i E_{iM}}{(\gamma_L - 1)M_B} \simeq \frac{\sum_i (E_i - p_i \cos \theta_i)}{M_B}. \tag{B.78}$$

The last expression, holding in the extremely relativistic case, contains quantities in LS only, and consequently K_M can be obtained rather more directly than K_L and K^*.

The numerator of the last expression may be interpreted as the mass of a fictitious particle participating in the collision; it is sometimes called the target mass (Birger 59). Thus

$$M_t = \sum_i (E_i - p_i \cos \theta_i). \tag{B.79}$$

In order to compare K_M with K_L and K^* we rewrite (B.77) for $\gamma_L \gg 1$ as follows:

$$E_{iM} \simeq \tfrac{1}{2} p_{Ti}\left(1 + \frac{\mu^2}{p_{Ti}^2}\right)\gamma_L \tan\theta_i + \frac{E_i}{2\gamma_L}. \tag{B.77$'$}$$

Neglecting the last term and taking again into account the constancy of p_T, we have

$$K_M \simeq \frac{p_T}{2M_B}\left(1 + \frac{\mu^2}{p_T{}^2}\right)\sum_i \tan\theta_i = (1 - Z)K^*. \tag{B.79$'$}$$

The asymmetry effect is explicitly given.

From (B.74), (B.76), and (B.79′) we obtain the following useful relations:

$$K_L = \frac{1+Z}{1-Z} K_M, \tag{B.80}$$

$$\sqrt{K_L K_M} = \sqrt{1 - Z^2} K^*, \tag{B.81}$$

$$K_L + K_M = 2K^*, \tag{B.82}$$

$$K_L - K_M = 2ZK^*. \tag{B.83}$$

In practice K_M may be obtained first by measuring the energies and emission angles of the secondary particles. Then γ_L from (B.78) and consequently $\gamma_c \simeq \sqrt{\gamma_L M_A/2M_B}$ are evaluated, provided that M_A is known for the latter. If the asymmetry factor Z is known, the rest of the quantities can be obtained.

B.9.3 Characteristic Quantities Concerning the Survival and Recoil Particles

It is of practical meaning to distinguish, among many particles emerging from a collision, both the survival and recoil particles from other secondary particles. They are denoted by A' and B', respectively, and quantities concerning them are designated by corresponding subscripts.

The transverse momentum of the survival particle is given as

$$p_{TA'} = p_{A'} \sin\theta_{A'} \simeq (1 - K_L)\gamma_L \sin\theta_{A'} M_A, \tag{B.84}$$

where we have used the approximation $1 - K_L \equiv (E_{A'} - M_A)/(E_A - M_A) \simeq p_{A'}/E_A$. This contains γ_L, which is not always determined without considerable error.

In MS the same relation as (B.84) holds, but γ_L can be eliminated in the following way. If we measure the angle in MS from a direction opposite to the incident one, the emission angle of the recoil particle in LS is related to that in MS as

$$\gamma_L \tan\theta_{B'} = \frac{\sin\theta_{MB'}}{(\beta_L/\beta_{MB'}) - \cos\theta_{MB'}}. \tag{B.85}$$

Since

$$\frac{\beta_L}{\beta_{MB'}} = (1-K)\left[\frac{\gamma_L{}^2 - 1}{(1-K)^2\gamma_L{}^2 - 1}\right]^{1/2},$$

(B.85) is solved as

$$\sin\theta_{B'} = \frac{\gamma_L \tan\theta_{B'}}{1+\gamma_L{}^2\tan^2\theta_{B'}}\frac{\beta_L}{\beta_{MB'}} \pm \sqrt{1 - \left[\left(\frac{\beta_L}{\beta_{MB'}}\right)^2 - 1\right]\gamma_L{}^2\tan^2\theta_{B'}}. \tag{B.86}$$

Hence the transverse momentum of the recoil particle is expressed by

$$\begin{aligned} p_{TB'} &= p_{B'}\sin\theta_{MB'} \simeq (1-K_M)\gamma_L M_B \sin\theta_{MB'} \\ &\simeq \frac{(1-K_M)M_B}{\tan\theta_{B'}}\left\{1 \pm \left[1 - \frac{2K_M(1-K_M/2)}{(1-K_M)^2}\tan^2\theta_{B'}\right]^{1/2}\right\}. \end{aligned} \tag{B.87}$$

Since this has double values, we are unable to remove the ambiguity. Hence we need to know the momentum of the recoil particle as well as its emission angle.

If the former is known, the four-momentum transfer to the recoil particle can be obtained from (B.46″) as

$$\Delta_B = \sqrt{-t} = \sqrt{2M_B(E_{B'} - M_B)}. \tag{B.88}$$

This can also be expressed in quantities in MS as

$$\begin{aligned} \Delta_B{}^2 &\equiv (\mathbf{p}_{MB} - \mathbf{p}_{MB'})^2 - (E_{MB} - E_{MB'})^2 \\ &= 4(1-K_M)M_B{}^2\gamma_L{}^2 \sin^2\frac{\theta_{MB'}}{2} + \frac{K_M{}^2}{1-K_M} M_B{}^2 \cos\theta_{MB'} \\ &\simeq \frac{1}{1-K_M}(K_M{}^2 M_B{}^2 + p_{TB'}{}^2). \end{aligned} \tag{B.89}$$

From (B.88) and (B.89) the dependence of the recoil energy on the transverse momentum and consequently on the emission angle is obtained as

$$\frac{E_{B'} - M_B}{M_B} = \frac{1}{2(1 - K_M)}\left(K_M{}^2 + \frac{p_{TB'}{}^2}{M_B{}^2}\right)$$

$$= \frac{1 - K_M}{\tan^2 \theta_{B'}}\left\{1 + \frac{K_M}{1 - K_M}\tan^2 \theta_{B'}\right.$$

$$\left.\pm\left[1 - \frac{2K_M(1 - K_M/2)}{(1 - K_M)^2}\tan^2 \theta_{B'}\right]^{1/2}\right\}. \tag{B.90}$$

This shows that the upper sign in the last term gives a greater recoil energy. If we can select recoil particles of energies greater than

$$E_{B'} - M_B > \left[\frac{(1 - K_M)^3(3 - 2K_M)}{K(2 - K_M)^2}\right] M_B,$$

their values of $p_{TB'}$ and Δ_B can be evaluated by choosing the upper sign. However, this introduces a bias, by selecting limited cases.

In LS we have the same relation as (B.89); that is,

$$\Delta_A{}^2 = \frac{1}{1 - K_L}(K_L{}^2 M_A{}^2 + p_{TA'}^2). \tag{B.91}$$

If $p_{TA'}$ is assumed as equal to $p_{TB'}$, the value of Δ_A can be obtained. However, we may be interested in obtaining Δ_A by referring only to quantities in LS. In this case it is convenient to rewrite (B.91) with the aid of (B.84) as*

$$\Delta_A{}^2 = \left[\frac{K_L{}^2}{1 - K_L} + \frac{1 - K_L}{K_L{}^2}\left(\frac{\sum_i E_i}{M_A}\sin \theta_{A'}\right)^2\right] M_A{}^2$$

$$\geq 2\sum_i E_i \sin \theta_{A'} M_A. \tag{B.92}$$

The sum of secondary energies $\sum_i E_i$ can be estimated rather easily, and the uncertainty in measuring a very small angle $\theta_{A'}$ is partly diminished because of the relation $\Delta_A \propto \sin^{1/2} \theta_{A'}$. The minimum estimate here is very important, because the value of Δ_B estimated by the above method is apt to be an overestimate.

* K. Yokoi, private communication.

B.9.4 One-fireball Model

A number of characteristic quantities of physical and kinematical interest are introduced with regard to specific models. One of them is the formation of excited states after a collision. If the colliding particle finds itself in such an excited state, the collision can be regarded as a two-body one, as discussed in Section B.8. Since this is known to be of little reality at very high energies and its essential features can be understood by the excited-baryon model described later, we shall no longer discuss this model in further detail. We thus begin with the next simplest model, the *one-fireball model.*

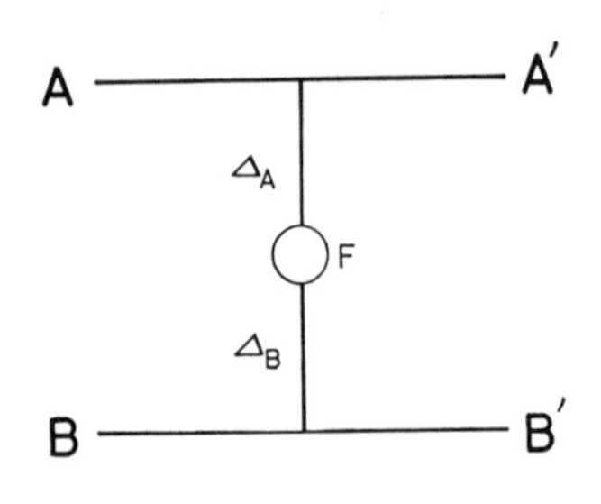

Fig. B.4 Graphical representation of the one-fireball model.

The one-fireball model is represented by the diagram in Fig. B.4, in which a fireball F is formed and two incident particles come out as A' and B', respectively. Hence the quantities with respect to the latter are the same as in Section B.9.3. Now all secondary particles are produced from the fireball F.

The fireball has four-momentum $\tilde{p}_F = (\mathbf{p}_F, E_F)$ and mass M_F, which are related as

$$M_F^2 = \tilde{p}_F \cdot \tilde{p}_F = E_F^2 - p_F^2 = E_F^{*2} - p_F^{*2}. \tag{B.93}$$

Now the asymmetry factor is given by

$$Z = \frac{p_F^* \cos \theta_F^*}{E_F^*}, \tag{B.94}$$

where θ_F^* is the emission angle of the fireball. By virtue of (B.94), the fireball mass is related to the asymmetry factor by

$$M_F^2 = E_F^{*2} - p_F^{*2} \cos^2 \theta_F^* - p_{FT}^2 = E_F^{*2}(1 - Z^2) - p_{FT}^2 \simeq E_F^{*2}(1 - Z^2). \tag{B.95}$$

Since $K^* \simeq E_F^*/2\gamma_c M_B$ and (B.81), the fireball mass is expressed by

$$M_F \simeq E_F^* \sqrt{1 - Z^2} = 2\gamma_c M_B K^* \sqrt{1 - Z^2} = 2\gamma_c M_B \sqrt{K_L K_M}. \tag{B.96}$$

The energy of the fireball in CMS is given by

$$E_F^* \simeq 2\gamma_c M_B K^* = \gamma_c M_B (K_L + K_M).$$

Dividing this by (B.96) we obtain

$$\gamma_F^* = \frac{E_F^*}{M_F} = \frac{(K_L + K_M)}{2\sqrt{K_L K_M}}. \tag{B.97}$$

Thus the mass and velocity of the fireball can be obtained if the inelasticities in LS and MS are known.

B.9.5 Two-fireball Model

The fireball formed as above does not necessarily decay into many mesons but may split into two other fireballs. One of these fireballs flies forward, and the other backward; they are distinguished by the subscripts f and b, respectively. In the rest system of the parent fireball, their energies and momenta are given by (B.33) and (B.34), respectively, which are now denoted as $E_f^{(0)}$, $p_f^{(0)}$, etc. In CMS they are given by the Lorentz transformation with velocity

$$\boldsymbol{\beta}_F^* = \frac{\mathbf{p}_F^*}{E_F^*}$$

as

$$\begin{aligned} E_f^* &= \gamma_F^*(E_f^{(0)} + \beta_F^* p_f^{(0)} \cos\theta_f^{(0)}), \\ p_{fL}^* &= \gamma_F^*(p_f^{(0)} \cos\theta_f^{(0)} + \beta_F^* E_f^{(0)}) \cos\theta_F^* - p_f^{(0)} \sin\theta_f^{(0)} \sin\theta_F^*, \\ p_{fT}^* &= \gamma_F^*(p_f^{(0)} \cos\theta_f^{(0)} + \beta_F^* E_f^{(0)}) \sin\theta_F^* + p_f^{(0)} \sin\theta_f^{(0)} \cos\theta_F^*, \end{aligned} \tag{B.98}$$

where $\theta_f^{(0)}$ is the emission angle of the forward fireball with respect to the parent one. The smallness of the transverse momentum requires that both θ_F^* and $\theta_f^{(0)}$ be small.

Since the parent fireball may be at rest in CMS, the properties of the daughter fireballs reveal themselves in the case of $\beta_F^* = 0$ and $\gamma_F^* = 1$. Then the relations in (B.98) are reduced to

$$E_f^* = E_f^{(0)}, \qquad p_{fL}^* = p_f^{(0)} \cos\theta_f^{(0)}, \qquad p_{fT}^* = p_f^{(0)} \sin\theta_f^{(0)}. \tag{B.98'}$$

The energy independence of transverse momentum together with the last relation should require that $\theta_f^{(0)}$ decrease as energy increases.

The restriction to the mass and the energy of a daughter fireball is obtained by taking account of two invariant quantities; that is,

$$M_F^2 = (\tilde{p}_f + \tilde{p}_b) \cdot (\tilde{p}_f + \tilde{p}_b) = (\gamma_f^* M_f + \gamma_b^* M_b)^2, \tag{B.99}$$

$$\Delta_F^2 = -(\tilde{p}_f - \tilde{p}_b) \cdot (\tilde{p}_f - \tilde{p}_b) = M_F^2 - 2(M_f^2 + M_b^2). \tag{B.100}$$

If two daughter fireballs are symmetric, we have

$$\Delta_F^2 = 4(\gamma_f^{*2} - 1) M_f^2 = 4 p_f^{*2}. \tag{B.100'}$$

The quantity Δ_F defined above is proportional to the angular momentum of the parent fireball. This would depend on the partition of angular momentum between the fireball and the outgoing particles, and also on the effective range of interactions. An important quantity to reveal such features may be the momentum transfer defined by

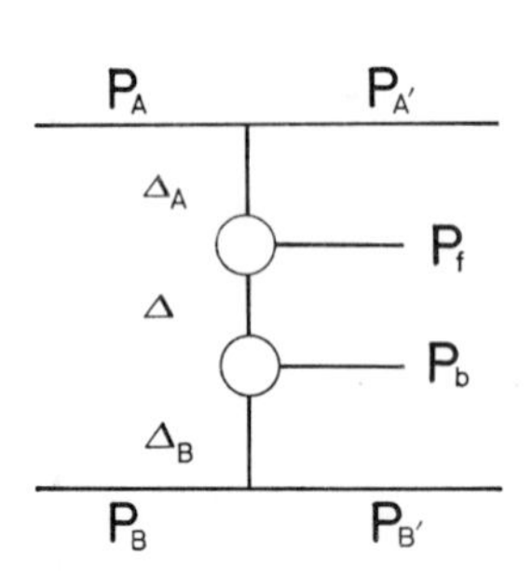

Fig. B.5 Graphical representation of the two-fireball model.

$$\Delta^2 = \tfrac{1}{4}[(\tilde{p}_A - \tilde{p}_B) - (\tilde{p}_{A'} - \tilde{p}_{B'}) - (\tilde{p}_f - \tilde{p}_b)]^2$$
$$= [\tilde{p}_A - \tilde{p}_{A'} - \tilde{p}_f]^2 = -[\tilde{p}_B - \tilde{p}_{B'} - \tilde{p}_b]^2. \tag{B.101}$$

A graphical representation of the *two-fireball model* suitable to the introduction of the momentum transfer Δ is shown in Fig. B.5.

The value of Δ^2 can be evaluated as

$$\Delta^2 = -[(E_A - E_{A'} - E_f) - (\mathbf{p}_A - \mathbf{p}_{A'} - \mathbf{p}_f)]$$
$$\times [(E_A - E_{A'} - E_f) + (\mathbf{p}_A - \mathbf{p}_{A'} - \mathbf{p}_f)].$$

The second factor on the right-hand side is replaced, by using the energy-momentum conservation, by

$$-(E_B - E_{B'} - E_b) - (\mathbf{p}_B - \mathbf{p}_{B'} - \mathbf{p}_b),$$

and the longitudinal and transverse components of the momenta are explicitly separated. Thus we obtain

$$\Delta^2 = -[(E_A - E_{A'} - E_f) - (p_A - p_{A'L} - p_{fL})]$$
$$\times [(E_B - E_{B'} - E_b) - (p_b - p_{B'L} - p_{bL})]$$
$$- [p_{A'T} + p_{fT}][p_{B'T} + p_{bT}].$$

The first term on the right-hand side represents the absolute square of the longitudinal component of momentum transfer, $\Delta_L{}^2$; and the second term is that of the transverse component, $\Delta_T{}^2$, Thus we are able to write Δ^2 as

$$\Delta^2 = \Delta_L{}^2 + \Delta_T{}^2. \tag{B.102}$$

In expressing $\Delta_L{}^2$ in terms of observable quantities we take into account that each fireball decays into secondary particles with energies

E_i or E_j and momentum components p_{iL}, p_{iT} or p_{jL}, p_{jT}. Accordingly we have

$$E_f = \sum_i E_i = \sum_i (M_i^2 + p_{iL}^2 + p_{iT}^2)^{1/2}, \qquad p_{fL} = \sum_i p_{iL},$$
$$E_b = \sum_j E_j = \sum_j (M_j^2 + p_{jL}^2 + p_{jT}^2)^{1/2}, \qquad p_{bL} = \sum_j p_{jL}, \tag{B.103}$$

where M_i and M_j are the masses of these secondary particles. In the extremely relativistic case terms in Δ_L^2 are expressed by

$$E_i - p_{iL} \simeq \frac{(M_i^2 + p_{iT}^2)}{2p_{iL}} p_{iL} = \tfrac{1}{2} p_{iT}\left(1 + \frac{M_i^2}{p_{iT}^2}\right)\tan\theta_i,$$

$$E_j + p_{jL} \simeq 2p_{jL} + \frac{M_j^2 + p_{jT}^2}{2p_{jL}}$$
$$= 2p_{jT}\left[\cot\theta_j + \tfrac{1}{4}\left(1 + \frac{M_j^2}{p_{jT}^2}\right)\tan\theta_j\right],$$

$$(E_{A'} - p_{A'L}) - (E_A - p_A) \simeq \frac{M_{A'}^2 + p_{A'T}^2 - (1 - K_L)M_A^2}{2(1 - K_L)E_A}$$

Thus we obtain

$$\Delta_L^2 = \left[\sum_i p_{iT}\left(1 + \frac{M_i^2}{p_{iT}^2}\right)\tan\theta_i + \frac{M_{A'}^2 + p_{A'T}^2 - (1 - K_L)M_A^2}{(1 - K_L)E_A}\right]$$
$$\times \left\{\sum_j p_{jT}\left[\cot\theta_j + \tfrac{1}{4}\left(1 + \frac{M_j^2}{p_{jT}^2}\right)\tan\theta_j\right] + (E_{B'} + p_{B'L} - M_B)\right\}. \tag{B.104}$$

This can be further reduced because the transverse momenta of secondary particles are equal, and both θ_i and θ_j are small. Thus

$$\Delta_L^2 \simeq \langle p_T^2 \rangle \sum_i \tan\theta_i \sum_j \cot\theta_j. \tag{B.104'}$$

The transverse momentum of a fireball may be estimated by assuming random walk of the transverse momentum of each secondary particle:

$$|p_{fT}| \simeq \sqrt{n_i \langle p_T^2 \rangle}, \qquad |p_{bT}| = \sqrt{n_j \langle p_T^2 \rangle}, \tag{B.105}$$

where n_i and n_j are the numbers of secondary particles emitted from the respective fireballs. If the transverse momenta of scattered and recoil particles are as small as p_T, the value of Δ_T^2 is given approximately by

$$\Delta_T^2 \simeq \sqrt{n_i n_j}\, \langle p_T^2 \rangle. \tag{B.106}$$

If this is not the case, the value of $\Delta_T{}^2$ may be expressed by

$$\Delta_T{}^2 \simeq p_{A'T}^2 + \sqrt{n_i n_j}\langle p_T{}^2\rangle. \tag{B.106'}$$

In both cases the value of Δ_T may not greatly exceed 1 GeV/c.

As shown in Chapter 3, the value of Δ_L is as small as a few GeV/c over a wide range of energy. Consequently Δ^2 may be regarded as constant as is p_T. If the smallness of Δ is assumed, the following useful relation is obtained. Since the time component of $\tilde{\Delta}$ in CMS vanishes in the symmetric case and is small even in the asymmetric case, the mass of the fireball is given by

$$\begin{aligned} M_f{}^2 &= \Delta^2 - \Delta_A{}^2 + 2(|p_{fL}^*|\ \Delta_L + p_{fT}\ \Delta_T) \\ &\simeq 2|p_{fL}^*|\ \Delta_L \simeq 2\gamma_f^* M_f\ \Delta_L . \end{aligned} \tag{B.107}$$

The last two expressions hold at high energies where $|p_{fL}^*| \gg \Delta_L, \Delta_T, \Delta_A, p_{fT}$. This implies that the energy dependences of M_f and γ_f^* are nearly the same. Because of (B.96) and (B.99) we have

$$M_f \propto \gamma_f^* \propto \gamma_c^{1/2}, \tag{B.108}$$

provided that inelasticity K is essentially energy independent.

The relation (B.107) has a further consequence with regard to the relation between M_F and M_f. In the symmetric case (B.99) and (B.107) give

$$M_F \simeq \frac{M_f{}^2}{\Delta_L}. \tag{B.109}$$

If the fireball splits further into two and the four-momentum transfer between these two behaves in an analogous way, the relations (B.108) and (B.109) hold for each pair of fireballs in the rest system of their parent fireball. If the splitting stops at the kth stage and the mass of the kth fireball is M_k, we finally have $N = 2^k$ fireballs. Then the mass M_k is related to M_F as

$$\frac{M_F}{\Delta} \simeq \left(\frac{M_k}{\Delta}\right)^N, \qquad N = 2^k. \tag{B.110}$$

Hence the number of fireballs is given by

$$N \simeq \log\left(\frac{M_F}{\Delta}\right) \Big/ \log\left(\frac{M_k}{\Delta}\right) \propto \log\,(K\gamma_c). \tag{B.111}$$

The last relation arises from (B.96) and the energy independence of the final fireball mass.

REFERENCES

Baldin 59 Baldin, A. M., Gol'danskij, I., and Rosental, I. L., *Kinematics in Nuclear Reactions*, State Publishing House, Moscow, 1959.

Birger 59 Birger, N. G., and Smorodin, Yu. A., *J. Exp. Theor. Phys.*, **37**, 1355 (1959).

Castagnoli 53 Castagnoli, C., et al., *Nuovo Cim.*, **10**, 1539 (1953).

Duller 54 Duller, N. M., and Walker, W. D., *Phys. Rev.*, **53**, 215 (1962).

Kaplon 60 Kaplon, M. F., and Yamanouchi, T., *Nuovo Cim.*, **15**, 519 (1960).

Salzman 62 Salzman, F., and Salzman, G., *Phys. Rev.*, **125**, 1703 (1962).

Yajima 65 Yajima, N., and Hasegawa, S., *Prog. Theor. Phys.*, **33**, 184 (1965).

APPENDIX C

Results of Cascade Theory

Let the number of electrons (or photons) with energies between E and $E+dE$ at thickness t in radiation units be $\pi^{(j)}(E_0, E, t)\,dE$ (or, for photons, $\gamma^{(j)}(E_0, E, t)\,dE$). An initiating particle of energy E_0 is indicated by the prefix $(j), j=\pi$, or γ according to whether it is an electron or a photon. The integral energy spectra are expressed as

$$\Pi^{(j)}(E_0, E, t) = \int_E^{E_0} \pi^{(j)}(E_0, E', t)\,dE',$$
$$\Gamma^{(j)}(E_0, E, t) = \int_E^{E_0} \gamma^{(j)}(E_0, E', t)\,dE' \tag{C.1}$$

The moments of those functions are expressed by

$$p_n^{(j)}(E_0, E) = \int_0^\infty t^n \pi^{(j)}(E_0, E, t)\,dt,$$
$$g_n^{(j)}(E_0, E) = \int_0^\infty t^n \gamma^{(j)}(E_0, E, t)\,dt, \tag{C.2}$$

and $P_n{}^j$ and $G_n{}^j$ correspond to the integral spectra. The zeroth moment is called the track length, and higher moments are related to the center of gravity $\bar{t}$ and the longitudinal spread τ as

$$\bar{t} = \frac{p_1}{p_0} \quad \text{and} \quad \tau^2 = \left(\frac{p_2}{p_0}\right) - \frac{p_1{}^2}{p_0{}^2}, \tag{C.3}$$

respectively.

Under approximation A these quantities are expressed as follows. Using the age parameter s, we have

$$\pi^{(j)}, \gamma^{(j)}(E_0, E, t) = \frac{H_i^{(j)}(s)}{\sqrt{2\pi}\, s^n[\gamma_1''(s)t + n/s^2]^{1/2}} \left(\frac{E_0}{E}\right)^s \frac{1}{E}\, e^{\lambda_1(s)t}, \tag{C.3}$$

where t and E_0/E are related through s as

$$t = -\frac{1}{\lambda_1'(s)}\left[\ln\left(\frac{E_0}{E}\right) - \frac{n}{s}\right]. \tag{C.4}$$

Table C.1 Values of n in (C.3) and (C.4)

Primary	π	γ	Π	Γ
π	0	$\frac{1}{2}$	1	$\frac{3}{2}$
γ	$-\frac{1}{2}$	0	$\frac{1}{2}$	1

The $H_i^j(s)$ depend on a primary particle j and secondary particles i, and they are given by

$$\begin{aligned} H_\pi^{(\pi)} &= \frac{D+\lambda_1(s)}{\lambda_1(s)-\lambda_2(s)}, \qquad H_\gamma^{(\pi)} = \frac{C(s)}{\lambda_1(s)-\lambda_2(s)}, \\ H_\pi^{(\gamma)} &= \frac{B(s)}{\lambda_1(s)-\lambda_2(s)}, \qquad H_\gamma^{(\gamma)} = \frac{D+\lambda_2(s)}{\lambda_1(s)-\lambda_2(s)}, \\ \lambda_1(s)+\lambda_2(s) &= -[A(s)+D], \qquad D = 0.7733. \end{aligned} \tag{C.5}$$

The values of n in (C.3) and (C.4) are given in Table C.1. The values of $A(s)$, $B(s)$, $C(s)$, $\lambda_2(s)$, $\lambda_1(s)$, $\lambda_1'(s)$, $\lambda_2''(s)$, and $H_i^{(j)}(s)$ are given in Table C.2 (Rossi 41, 52).

The integral spectra are expressed by

$$\Pi^{(j)}, \Gamma^{(j)}(E_0, E, t) = \frac{H_i^{(j)}(s)}{\sqrt{2\pi}\, s^n[\lambda_1''(s)t + n/s^2]^{1/2}} \left(\frac{E_0}{E}\right)^s e^{\lambda_1(s)t}, \tag{C.6}$$

in which the values of n are given in Table C.1. The general trend of cascade curves $\Pi^{(\pi)}(E_0, E, t)$ can be seen in Fig. C.1.

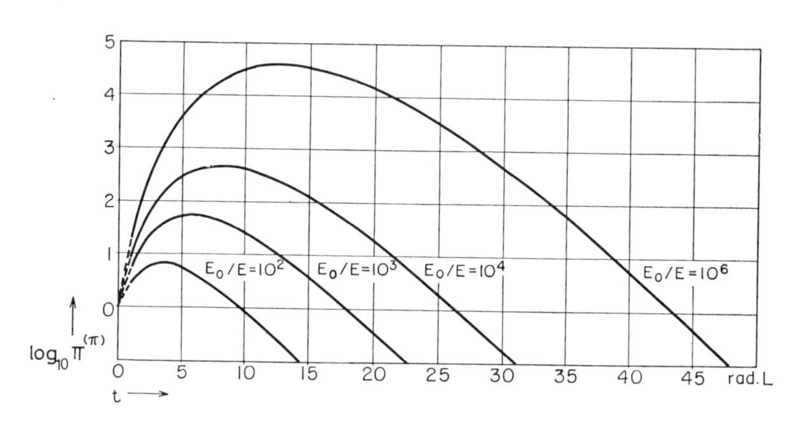

Fig. C.1 The number of electrons of energy greater than E in a shower initiated by an electron of energy E_0, $\Pi^{(\pi)}$ (E_0, E, t) as a function of t. Computed for various values of E_0/E according to approximation A; from (Rossi 41).

Table C.2 Numerical Values of Some Functions That Enter into the Theory of Cascade Showers under Approximation A[a]

s	$A(s)$	$B(s)$	$C(s)$	$\lambda_2(s)$	$\lambda_1(s)$	$(\lambda_1'(s)$	$\lambda_1''(s)$
0.0	0.0000	1.546	∞	$-\infty$	$+\infty$	$-\infty$	$+\infty$
0.1	0.1520	1.400	12.842	−4.715	+3.789	−25.005	—
0.2	0.2863	1.280	6.123	−3.330	2.270	−9.488	+75
0.3	0.4067	1.180	3.923	−2.749	1.569	−5.415	+26
0.4	0.5152	1.095	2.846	−2.415	1.127	−3.654	12.5
0.5	0.6146	1.022	2.214	−2.201	0.813	−2.693	7.6
0.6	0.706	0.959	1,802	−2.055	0.576	−2.093	4.96
0.7	0.791	0.905	1.513	−1.953	0.389	−1.685	3.50
0.8	0.870	0.855	1.3014	−1.878	0.235	−1.389	2.55
0.9	0.943	0.812	1.1400	−1.824	0.108	−1.1660	1.97
1.0	1.0135	0.7733	1.0135	−1.787	0.000	−0.9908	1.563
1.1	1.078	0.7383	0.9112	−1.760	−0.092	−0.8501	1.275
1.2	1.142	0.7065	0.8276	−1.744	−0.171	−0.7333	1.060
1.3	1.200	0.6778	0.7580	−1.734	−0.239	−0.6362	0.893
1.4	1.257	0.6514	0.6988	−1.732	−0.298	−0.5531	0.764
1.5	1.311	0.6272	0.6484	−1.734	−0.350	−0.4825	0.655
1.6	1.363	0.6049	0.6047	−1.741	−0.395	−0.4214	0.565
1.7	1.412	0.5842	0.5666	−1.751	−0.435	−0.3691	0.487
1.8	1.460	0.5650	0.5329	−1.762	−0.470	−0.3238	0.423
1.9	1.506	0.5473	0.5032	−1.780	−0.500	−0.2841	0.370
2.0	1.550	0.5306	0.4767	−1.797	−0.526	−0.2498	0.320
2.1	1.592	0.5148	0.4528	−1.816	−0.550	−0.2202	0.277
2.2	1.634	0.5004	0.4313	−1.837	−0.570	−0.1943	0.241
2.3	1.674	0.4866	0.4117	−1.859	−0.589	−0.1719	0.210
2.4	1.713	0.4736	0.3940	−1.882	−0.605	−0.1523	0.182
2.5	1.750	0.4614	0.3776	−1.904	−0.619	−0.1354	0.159
2.6	1.787	0.4499	0.3627	−1.928	−0.632	−0.1205	0.138
2.7	1.821	0.4389	0.3489	−1.951	−0.643	−0.1077	0.120
2.8	1.857	0.4285	0.3362	−1.977	−0.654	−0.0964	0.107
2.9	1.892	0.4186	0.3243	−2.003	−0.663	−0.0863	0.093
3.0	1.923	0.4093	0.3134	−2.026	−0.671	−0.0777	0.080
4.0	2.211	0.3352	0.2347	−2.264	−0.720	−0.0307	—
5.0	2.448	0.2847	0.1882	−2.480	−0.742	−0.0146	—
6.0	2.648	0.2479	0.1574	−2.669	−0.752	−0.0080	—
7.0	2.822	0.2198	0.1354	−2.837	−0.759	−0.0048	—
8.0	2.977	0.1975	0.1189	−2.988	−0.763	−0.0031	—
9.0	3.115	0.1794	0.1060	−3.123	−0.765	−0.0021	—
10.0	3.239	0.1644	0.0957	−3.246	−0.766	−0.0015	—

[a] From (Rossi 41).

Table C.2—*Concluded.* Numerical Values of Some Functions That Enter into the Theory of Cascade Showers under Approximation A[a]

s	$H_\pi^{(\pi)}(s)$	$H_\gamma^{(\gamma)}(s)$	$H_\gamma^{(\pi)}(s)$	$H_\pi^{(\gamma)}(s)$
0.0	0.500	0.500	0.469	0.533
0.1	0.537	0.463	0.478	0.521
0.2	0.543	0.457	0.489	0.507
0.3	0.542	0.458	0.498	0.499
0.4	0.536	0.464	0.508	0.489
0.5	0.526	0.474	0.520	0.480
0.6	0.513	0.487	0.531	0.471
0.7	0.496	0.504	0.541	0.463
0.8	0.477	0.523	0.551	0.453
0.9	0.456	0.544	0.560	0.443
1.0	0.433	0.567	0.567	0.433
1.1	0.408	0.592	0.573	0.422
1.2	0.383	0.617	0.576	0.410
1.3	0.357	0.643	0.578	0.397
1.4	0.331	0.669	0.577	0.384
1.5	0.306	0.694	0.574	0.370
1.6	0.281	0.719	0.568	0.355
1.7	0.257	0.743	0.561	0.340
1.8	0.235	0.765	0.554	0.325
1.9	0.213	0.787	0.542	0.310
2.0	0.194	0.806	0.530	0.295
2.1	0.176	0.824	0.518	0.280
2.2	0.160	0.840	0.505	0.266
2.3	0.145	0.855	0.492	0.252
2.4	0.132	0.868	0.478	0.240
2.5	0.120	0.880	0.465	0.227
2.6	0.109	0.891	0.451	0.215
2.7	0.099	0.901	0.438	0.204
2.8	0.090	0.910	0.425	0.193
2.9	0.082	0.918	0.412	0.183
3.0	0.075	0.925	0.401	0.173
4.0	0.034	0.966	0.304	0.108
5.0	0.018	0.982	0.242	0.073

[a] From (Rossi 41).

Table C.3 Values of l and m in (C.8)

Function	Primary			
	Electron		Photon	
	l	m	l	m
π	0.137	0	0.137	−0.18
γ	0.180	0.18	0.180	0
Π	0.137	0.37	0.137	0.18
Γ	0.180	0.55	0.180	0.37

The shower curves represented by π, γ, Π, and Γ have maxima at

$$t = t_{\max} = -\frac{1}{\lambda'(1)}\left[\ln\left(\frac{E_0}{E}\right) - m\right], \tag{C.7}$$

and the values at the maxima are

$$\begin{aligned} \pi^{(j)}, \gamma^{(j)}(E_0, E, t_{\max}) &= \frac{l}{[\ln(E_0/E) - m]^{1/2}} \frac{E_0}{E^2} \\ \Pi^{(j)}, \Gamma^{(j)}(E_0, E, t_{\max}) &= \frac{l}{[\ln(E_0/E) - m]^{1/2}} \frac{E_0}{E}\,. \end{aligned} \tag{C.8}$$

The values of m and l are given in Table C.3.

The track lengths are found to be

$$\begin{aligned} p_0^{(\pi)}(E_0, E) &= p_0^{(\gamma)}(E_0, E) = 0.437\left(\frac{E_0}{E^2}\right), \\ g_0^{(\pi)}(E_0, E) &= g_0^{(\gamma)}(E_0, E) = 0.572\left(\frac{E_0}{E^2}\right), \\ P_0^{(\pi)}(E_0, E) &= P_0^{(\gamma)}(E_0, E) = 0.437\left(\frac{E_0}{E}\right), \\ G_0^{(\pi)}(E_0, E) &= G_0^{(\gamma)}(E_0, E) = 0.572\left(\frac{E_0}{E}\right). \end{aligned} \tag{C.9}$$

Other quantities related to moments are

$$\bar{t} = \frac{p_1}{p_0} = -\frac{1}{\lambda_1'(1)}\ln\left(\frac{E_0}{E}\right) + h = 1.01\ln\left(\frac{E_0}{E}\right) + h, \tag{C.10}$$

$$\tau^2 = \frac{p_2}{p_0} - \frac{{p_1}^2}{{p_0}^2} = \frac{\lambda_1''(1)}{[\lambda_1'(1)]^3}\ln\left(\frac{E_0}{E}\right) + k = 1.61\ln\left(\frac{E_0}{E}\right) + k. \tag{C.11}$$

The values of h and k are given in Table C.4.

Table C.4 Values of h and k in (C.10) and (C.11)

Function	Primary			
	Electron		Photon	
	h	k	h	k
π	1.0	−0.1	1.8	1.2
γ	1.2	1.1	2.0	2.3
Π	0.0	−0.7	0.8	0.6
Γ	0.2	0.5	1.0	1.7

The integral spectra under approximation B are expressed by

$$\Pi_i^{(j)}, \Gamma_i^{(j)}(E_0, E, t) = \frac{D_i^{(j)}(s)H_i^{(j)}(s)p(s, \varepsilon)}{\sqrt{2\pi}\, s^n[\lambda_1''(s)t + n/s^2]} \left(\frac{E_0}{\varepsilon_0}\right) e^{\lambda_1(s)t}, \qquad \text{(C.12)}$$

where ε_0 is the critical energy. These are different from those under approximation A by a factor of $D_i^{(j)}(s)p(s, \varepsilon)$. Thus

$$D_i^{(j)}(s) = -\frac{s[d\lambda_1(s)/ds]}{H_i^{(j)}(s)\Gamma(s+1)}, \qquad \text{(C.12)}$$

where $\Gamma(s+1)$ is the gamma function; its numerical values are given in Table C.5 (Belenkij 48). The function $p(s, \varepsilon)$ is

$$p(s, \varepsilon) = e^{\varepsilon} \int_{\varepsilon}^{\infty} e^{-x}\left(1 - \frac{\varepsilon}{x}\right)^s dx, \qquad \varepsilon = f(\lambda)\frac{E}{\varepsilon_0}, \qquad \text{(C.13)}$$

Table C.5 Values of $D(s)$

s	$D(s)$
0	1.000
0.5	1.805
0.7	2.02
0.8	2.11
1.0	2.29
1.1	2.38
1.2	2.46
1.4	2.65
1.6	2.83
1.8	3.06
2.0	3.32

and its numerical values are given in Table C.6. Here

$$f(\lambda) = -\left\{s \frac{d\lambda_1}{ds}[\lambda - \lambda_2(s)]\right\}_{s=s_1} \frac{1}{\lambda + D},$$

and the value of the term inside the braces is taken at $\lambda = \lambda_1(s)$ or $s = s_1(\lambda)$. At $s = 1$ the spectrum is represented by

$$p(1, \varepsilon) = 1 + \varepsilon\, e^{\varepsilon} E_i(-\varepsilon) \simeq \frac{1}{\varepsilon} - \frac{2}{\varepsilon^2}, \tag{C.13'}$$

the last expression holding for $\varepsilon \gg 1$.

For $\varepsilon = 0$ we have $p(s, 0) = 1$ according to (C.13). The total number of electrons reaches a maximum at $s = 1$, at which

$$\Pi^{(\pi)}(E_0, 0, t_{\max}) = \frac{0.314}{[\ln(E_0/\varepsilon_0) - 1]^{1/2}} \frac{E_0}{\varepsilon_0},$$
$$t_{\max} = 1.01\left[\ln\left(\frac{E_0}{\varepsilon_0}\right) - 1\right]. \tag{C.14}$$

The center of gravity and the longitudinal spread are obtained as

$$\bar{t}(E_0, 0) = 1.01 \ln\left(\frac{E_0}{\varepsilon_0}\right) + 0.4,$$

and

$$\tau_\pi^{\,2}(E_0, 0) = 1.61 \ln\left(\frac{E_0}{\varepsilon_0}\right) - 0.2, \tag{C.15}$$

In order to describe the three-dimensional development of cascade showers we have to know the distance of a particle from the shower axis and the angle between its path and the shower axis. These coordinates are represented by two-dimensional vectors $\mathbf{r}$ and $\boldsymbol{\theta}$, respectively. Hence the structure function is written, for example, as $\pi(E_0, E, \mathbf{r}, \boldsymbol{\theta}, t)$. In

Table C.6 Values of $p(s, \varepsilon)$, $\varepsilon = f(\lambda)(E/\varepsilon_0)$

		ε			
s	$f(\lambda)$	0.1	0.2	0.5	2.0
0.7	2.37	0.840	0.768	0.630	0.386
1.0	2.29	0.798	0.685	0.538	0.277
1.3	2.33	0.746	0.628	0.468	0.204

practice either the angular or the lateral structure function is used. They are given by

$$\pi_1(E_0, E, \boldsymbol{\theta}, t) = \int \pi(E_0, E, \mathbf{r}, \boldsymbol{\theta}, t)\, d\mathbf{r},$$

and

$$\pi_2(E_0, E, \mathbf{r}, t) = \int \pi(E_0, E, \mathbf{r}, \boldsymbol{\theta}, t)\, d\boldsymbol{\theta},$$

respectively. Here we give mainly numerical values of the lateral structure functions. Readers who are interested in the details are referred to (Kamata 58) and (Nishimura 67).

Under approximation A the differential lateral structure functions are approximately obtained as

$$\begin{aligned}
&\pi_2 \simeq r^{s-2+(2/3)}, && \lambda_1'(s)t + \ln\left(\frac{E_0}{E}\right) + \ln\left(\frac{Er}{E_s r_M}\right) = 0 && \text{for} \quad \bar{s} \leq 2 - \tfrac{2}{3},\\
&\pi_2 \simeq \text{constant}, && \lambda_1'(s)t + \ln\left(\frac{E_0}{E}\right) = 0 && \text{for} \quad s > 2 - \tfrac{2}{3},\\
&\gamma_2 \simeq r^{s-2}, && \lambda_1'(s)t + \ln\left(\frac{E_0}{E}\right) + \ln\left(\frac{Er}{E_0 r_M}\right) = 0 && \text{for} \quad s \leq 2,\\
&\gamma_2 \simeq \text{constant}, && \lambda_1'(s)t + \ln\left(\frac{E_0}{E}\right) = 0 && \text{for} \quad s > 2,
\end{aligned} \tag{C.16}$$

where $E_s = 21$ MeV and r_M is the Molière unit.

In approximation B the integral lateral structure functions are given by

$$\begin{aligned}
&\Pi_2 \simeq r^{s-2}, && \lambda_1'(s)t + \ln\left(\frac{E_0}{\varepsilon_0}\right) + \ln\left(\frac{\varepsilon_0 r}{E_s r_M}\right) = 0 && \text{for} \quad s \leq 2,\\
&\Pi_2 \simeq \text{constant}, && \lambda_1'(s)t + \ln\left(\frac{E_0}{\varepsilon_0}\right) = 0 && \text{for} \quad s > 2,\\
&\Gamma_2 \simeq r^{s-2} \ln\left(\frac{r_M}{r}\right), && \lambda_1'(s)t + \ln\left(\frac{E_0}{\varepsilon_0}\right) + \ln\left(\frac{\varepsilon_0 r}{E_s r_M}\right) = 0 && \text{for} \quad s \leq 2,\\
&\Gamma_2 \simeq \text{constant}, && \lambda_1'(s)t + \ln\left(\frac{E_0}{\varepsilon_0}\right) = 0 && \text{for} \quad s > 2.
\end{aligned} \tag{C.17}$$

Numerical results in approximations A and B are compared for the integral lateral structure functions for electrons. It is convenient to introduce the normalized function

$$\tilde{\Pi}_2 = \frac{\Pi_2}{\int \Pi_2 \, 2\pi r \, dr} \tag{C.18}$$

and to give the numerical values of

$$\tilde{\Pi}'_2 \equiv \left(\frac{Er}{E_s r_M}\right)^{2-s} \tilde{\Pi}_2 \quad \text{for approximation } A,$$

$$\tilde{\Pi}'_2 \equiv \left(\frac{r}{r_M}\right)^{2-s} \tilde{\Pi}_2 \quad \text{for approximation B.}$$

They are given in Table C.7 and Fig. C.2 for approximation A, and in Table C.8 and Fig. C.3 for approximation B.

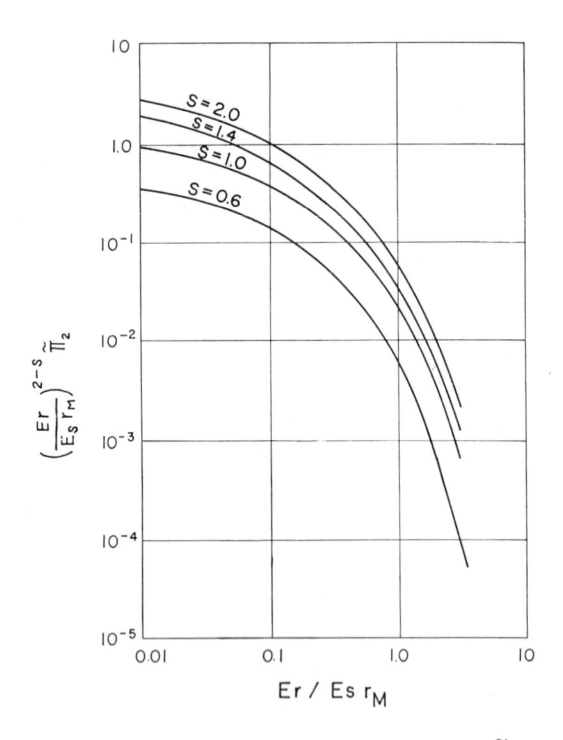

Fig. C.2 Normalized lateral structure function $(Er/E_s r_M)^{2-6}\,\tilde{\Pi}_2$ in approximation A; from (Kamata 58).

Table C.7 Numerical Values of $(Er/E_s r_M)^{2-s}\tilde{\Pi}_2$ in Approximation A

	s			
$Er/E_s r_M$	0.6	1.0	1.4	2.0
0	0.38	1.01	2.5	$\lim_{r/r_m \to 0} [-0.53 \ln (Er/E_s r_M)]$
0.01	0.34	0.91	1.9	2.5
0.03	0.27	0.70	1.35	1.8
0.1	0.17	0.37	0.68	0.99
0.3	0.05	0.15	0.27	0.37
1.0	0.0044	0.019	0.041	0.60
3.0	0.00009	0.00065	0.0019	0.37

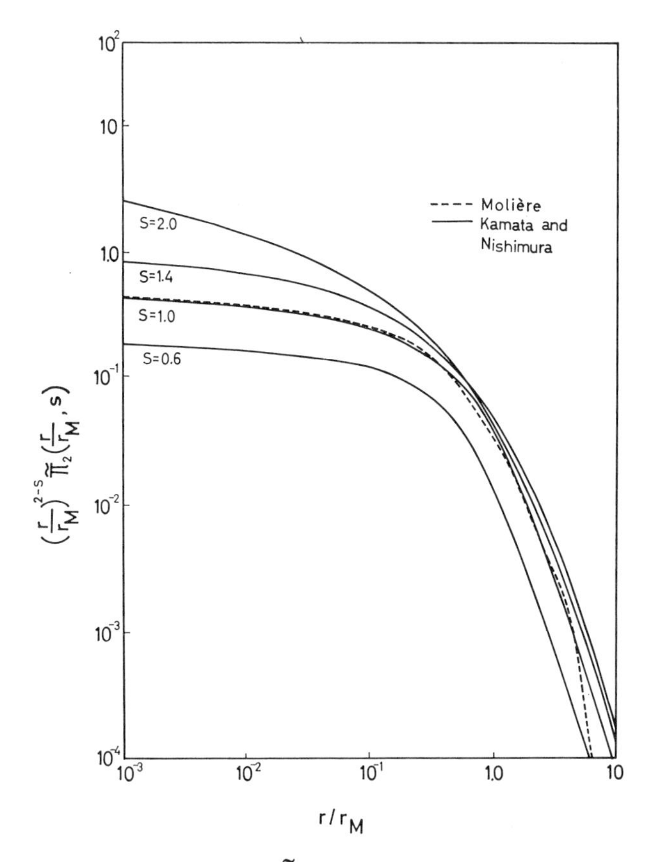

Fig. C.3 Lateral structure function $\tilde{\Pi}_2\,(r/r_M, s)$ in approximation B calculated by Kamata and Nishimura (solid curves for $s = 0.6$, 1.0, 1.4, and 2.0) and by Molière (dashed curve for $s = 1.0$) (Molière 48); from (Kamata 58).

Table C.8 Numerical Values of $(r/r_M)^{2-s}\tilde{\Pi}_2$ in Approximation B

	s			
r/r_M	0.6	1.0	1.4	2.0
0.001	1.95×10^{-1}	4.4×10^{-1}	8.5×10^{-1}	2.4
0.003	1.90×10^{-1}	4.2×10^{-1}	8.0×10^{-1}	1.92
0.01	1.81×10^{-1}	3.9×10^{-1}	7.1×10^{-1}	1.40
0.03	1.58×10^{-1}	3.4×10^{-1}	5.8×10^{-1}	9.3×10^{-1}
0.1	1.34×10^{-1}	2.65×10^{-1}	3.8×10^{-1}	5.3×10^{-1}
0.2	1.08×10^{-1}	2.03×10^{-1}	2.6×10^{-1}	3.2×10^{-1}
0.5	4.9×10^{-2}	1.09×10^{-1}	1.22×10^{-1}	1.21×10^{-1}
1.0	1.35×10^{-2}	4.1×10^{-2}	5.0×10^{-2}	4.4×10^{-2}
2.0	2.64×10^{-3}	8.8×10^{-3}	1.48×10^{-3}	1.18×10^{-2}
5.0	2.16×10^{-4}	7.7×10^{-4}	1.55×10^{-4}	1.32×10^{-3}

Table C.9 Numerical Values of the Mean-Square Lateral Spread at $s = 1$

E/ε_0	$\langle r^2 \rangle$ for π_2	$\langle r^2 \rangle$ for Π_2
∞	$0.723 \left(\frac{E_s}{E}\right)^2$	$0.241 \left(\frac{E_s}{E}\right)^2$
10	0.58	0.20
7	0.53	0.18
5	0.46	0.17
3	0.36	0.14
2	0.28	0.11
1.5	0.22	0.090
1.0	$0.15 \left(\frac{E_s}{\varepsilon_0}\right)^2$	$0.064 \left(\frac{E_s}{\varepsilon_0}\right)^2$
0.75	0.21	0.087
0.5	0.32	0.125
0.4	0.40	0.16
0.3	0.53	0.19
0.2	0.82	0.26
0.15	1.02	0.31
0.10	1.31	0.39
0.05	1.58	0.53

The second moments of π_2 and Π_2 at $s=1$ in approximation B are given in Table C.9. They give us a general idea of lateral spread.

The lateral distributions of energy flow can be evaluated by

$$\Pi_E = \int_0^\infty \pi_2 E\, dE = \varepsilon_0 \left(\frac{r_M}{r}\right) g_e\left(\frac{r}{r_M}\right) \Pi_2,$$

$$\Gamma_E = \int_0^\infty \gamma_2 E\, dE = \varepsilon_0 \left(\frac{r_M}{r}\right) g_r\left(\frac{r}{r_M}\right) \Pi_2. \tag{C.19}$$

Numerical values of $g_e(r/r_M)$ and $g_r(r/r_M)$ are given in Table C10, and those of $(\Pi_E + \Gamma_E)/\varepsilon_0 \Pi_2$, $\Gamma_E/\varepsilon_0 \Pi_2$, and $\Pi_E/\varepsilon_0 \Pi_2$ are given separately in Fig. C.4.

For showers developing in the atmosphere the variation of density with altitude has to be taken into account. If we observe showers at

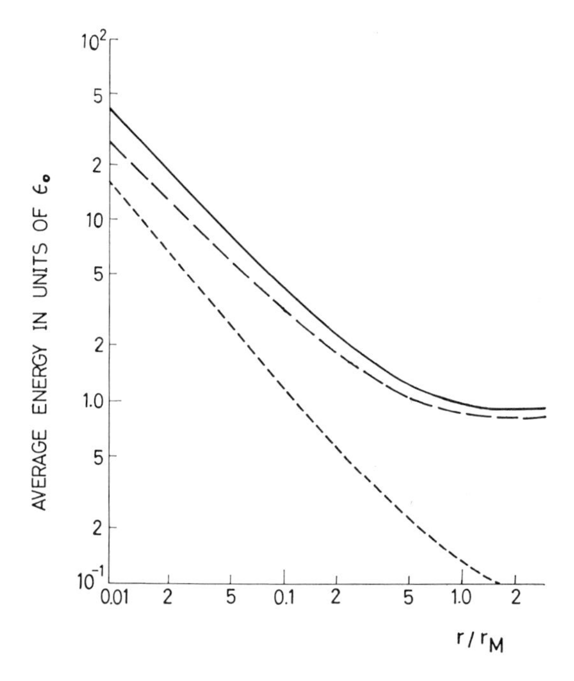

Fig. C.4 Lateral distribuitons of energy flows of electrons and photons relative to those of the numbers thereof; from (Kamata 58)

———— $\dfrac{\Pi_E + \Gamma_E}{\varepsilon_0 \Pi^2}$, ... $\dfrac{\Gamma_E}{\varepsilon_0 \Pi_2}$, ... $\dfrac{\Pi_E}{\varepsilon_0 \Pi_2}$

Table C.10 Values of $g_e(r/r_M)$ and $g_r(r/r_M)$ in (C.19)

	r/r_M						
	0	0.01	0.4	0.1	1.0	2.0	4.0
$g_e(r/r_M)$	0.15	0.15	0.12	0.14	0.14	0.20	0.30
$g_r(r/r_M)$	0.21	0.26	0.45	0.82	0.83	1.8	3.7

depth T, the lateral spread is approximately equal to that in the uniform atmosphere at depth $T - \Delta T$. The value of ΔT depends on the age parameter s, as shown in Table C.11.

Table C.11 Numerical Values of ΔT in Approximation A

s	0.6	1.0	1.4	2.0
ΔT	0.98	1.75	3.3	7.43

The lateral structure functions given above are obtained in the limit of infinite primary energy. For a finite value of E_0 the age parameter is determined by

$$\lambda'(s)t + \ln\left(\frac{E_0}{\varepsilon_0}\right) + \ln\left(\frac{r}{r_M}\right) - \tfrac{1}{2}\Psi\left(1 - \frac{s}{2}\right) = 0, \tag{C.20}$$

where $\Psi(x) \equiv d[\Gamma(x+1)]/dx$. Numerical values of $\tilde{\Pi}_2(E_0, r, s)$ in this case are given for $s = 1.0$ and $s = 1.4$ in Fig. C.5.

Another important approximation implied in the results above is called the Landau approximation, which is valid only if the Williams theory of multiple scattering holds. As discussed in Section 2.1, this is found to be a poor approximation at small and large angles. Without the Landau approximation the characteristic energy for multiple

Fig. C.5 Lateral structure function for finite incident energies. The values of the incident energy divided by ε_0 are shown for the curves (*a*) $s = 1.0$, (*b*) $s = 1.4$; from (Kamata 58).

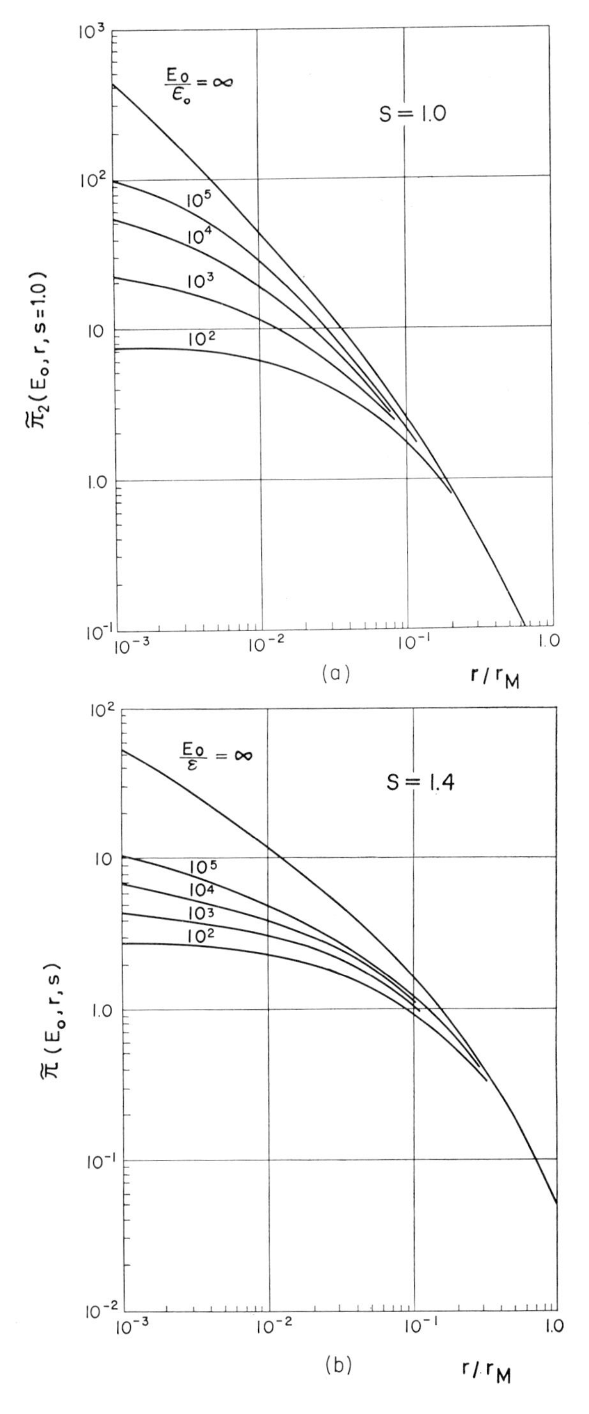

$\frac{E_0}{\varepsilon_0} = \infty$
S = 1.0
10^5
10^4
10^3
10^2
$\tilde{\pi}_2(E_0, r, s=1.0)$
10^3
10^2
10
1.0
10^{-1}
10^{-3}
10^{-2}
10^{-1}
1.0
(a)
r/r_M
$\frac{E_0}{\varepsilon} = \infty$
S = 1.4
10^5
10^4
10^3
10^2
$\tilde{\pi}(E_0, r, s)$
10^2
10
1.0
10^{-1}
10^{-2}
10^{-3}
10^{-2}
10^{-1}
1.0
(b)
r/r_M

scattering, E_s, should be replaced by

$$K = \Omega^{1/2} \frac{E_s}{2} [\ln (191\, Z^{-1/3})]^{1/2}. \tag{C.21}$$

The values of K and Ω are given in Table C.12.

The structure function is now obtained as a power series of $1/\Omega$. Taking the first two terms, $\Pi^{(0)} + \Omega^{-1}\Pi^{1)}$, in approximation A, we com-

Table C.12 Numerical Values of K and Ω

	Material					
	Carbon	Aluminum	Iron	Lead	G–5 Emulsion	Air
Z	6	13	26	82		
K(MeV)	19.2	19.4	19.5	19.1	19.7	19.3
Ω	15.4	14.9	14.3	12.9	14.0	15.2

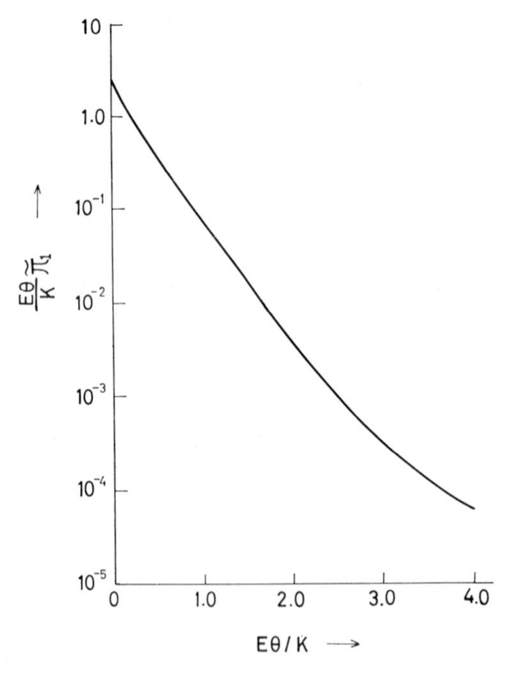

Fig. C.6 Normalized angular structure function without the Landau approximation: $(E\theta/K)\tilde{\Pi}_1(0) + (E\theta/\Omega K)\tilde{\Pi}_1(1)$. $K = 19.3$ MeV, $\Omega = 15.3$; from (Kamata 58).

pare the angular structure functions with and without the second term, as shown in Fig. C.6. The first term alone is equal to the angular structure function with the Landau approximation if K is replaced by E_s. This deviates appreciably from the function with the second term at small and large angles.

REFERENCES

Belenkij 48 Belinkij, S. Z., *Shower Processes in Cosmic Rays*, 1948.

Kamata 58 Kamata, K., and Nishimura, J., *Progr. Theor. Phys. Suppl.*, **6**, 93 (1958) and their papers cited there.

Molière 48 Molière, G., *Z. Naturforschung*, **3a**, 1801 (1948).

Nishimura 67 Nishimura, J., *Handbuch der Physik*, Vol. 46/2, Springer-Verlag, Berlin, 1967, p. 1.

Rossi 41 Nishimura, J., Rossi, B., and Greisen, K., *Rev. Mod. Phys.*, **13**, 240 (1941).

Rossi 52 Rossi, B., *High-Energy Particles*, Prentice-Hall, New York, 1952, Chapter 5.

Author Index

Subject Index